Advances in Intelligent Systems and Computing

Volume 903

Series editor

Janusz Kacprzyk, Systems Research Institute, Polish Academy of Sciences, Warsaw, Poland
e-mail: kacprzyk@ibspan.waw.pl

The series “Advances in Intelligent Systems and Computing” contains publications on theory, applications, and design methods of Intelligent Systems and Intelligent Computing. Virtually all disciplines such as engineering, natural sciences, computer and information science, ICT, economics, business, e-commerce, environment, healthcare, life science are covered. The list of topics spans all the areas of modern intelligent systems and computing such as: computational intelligence, soft computing including neural networks, fuzzy systems, evolutionary computing and the fusion of these paradigms, social intelligence, ambient intelligence, computational neuroscience, artificial life, virtual worlds and society, cognitive science and systems, Perception and Vision, DNA and immune based systems, self-organizing and adaptive systems, e-Learning and teaching, human-centered and human-centric computing, recommender systems, intelligent control, robotics and mechatronics including human-machine teaming, knowledge-based paradigms, learning paradigms, machine ethics, intelligent data analysis, knowledge management, intelligent agents, intelligent decision making and support, intelligent network security, trust management, interactive entertainment, Web intelligence and multimedia.

The publications within “Advances in Intelligent Systems and Computing” are primarily proceedings of important conferences, symposia and congresses. They cover significant recent developments in the field, both of a foundational and applicable character. An important characteristic feature of the series is the short publication time and world-wide distribution. This permits a rapid and broad dissemination of research results.

More information about this series at http://www.springer.com/series/11156

Waldemar Karwowski · Tareq Ahram
Editors

Intelligent Human Systems Integration 2019

Proceedings of the 2nd International Conference on Intelligent Human Systems Integration (IHSI 2019): Integrating People and Intelligent Systems, February 7–10, 2019, San Diego, California, USA

Editors
Waldemar Karwowski
University of Central Florida
Orlando, FL, USA

Tareq Ahram
Institute for Advanced Systems Engineering
University of Central Florida
Orlando, FL, USA

ISSN 2194-5357 ISSN 2194-5365 (electronic)
Advances in Intelligent Systems and Computing
ISBN 978-3-030-11050-5 ISBN 978-3-030-11051-2 (eBook)
https://doi.org/10.1007/978-3-030-11051-2

Library of Congress Control Number: 2018966122

This Springer imprint is published by the registered company Springer Nature Switzerland AG
The registered company address is: Gewerbestrasse 11, 6330 Cham, Switzerland

Preface

This volume, entitled *Intelligent Human Systems Integration 2019*, aims to provide a global forum for introducing and discussing novel approaches, design tools, methodologies, techniques, and solutions for integrating people with intelligent technologies, automation, and artificial cognitive systems in all areas of human endeavor in industry, economy, government, and education. Some of the notable areas of application include, but are not limited to, energy, transportation, urbanization and infrastructure development, digital manufacturing, social development, human health, sustainability, a new generation of service systems, as well as developments in safety, risk assurance, and cybersecurity in both civilian and military contexts. Indeed, rapid progress in developments in the ambient intelligence, including cognitive computing, modeling, and simulation, as well as smart sensor technology, weaves together the human and artificial intelligence and will have a profound effect on the nature of their collaboration at both the individual and societal levels in the near future.

As applications of artificial intelligence and cognitive computing become more prevalent in our daily lives, they also bring new social and economic challenges and opportunities that must be addressed at all levels of contemporary society. Many of the traditional human jobs that require high levels of physical or cognitive abilities, including human motor skills, reasoning, and decision-making abilities, as well as training capacity, are now being automated. While such trends might boost economic efficiency, they can also negatively impact the user experience and bring about many unintended social consequences and ethical concerns.

The intelligent human systems integration is to a large extent affected by the forces shaping the nature of future computing and artificial system development. This book discusses the needs and requirements for the symbiotic collaboration between humans and artificially intelligent systems, with due consideration of the software and hardware characteristics allowing for such cooperation from the societal and human-centered design perspectives, with the focus on the design of intelligent products, systems, and services that will revolutionize future human–technology interactions. This book also presents many innovative studies of ambient artificial technology and its applications, including the consideration of human–machine interfaces with a particular emphasis on infusing intelligence into development of technology throughout the lifecycle development process, with due consideration of user experience and the design of interfaces for virtual, augmented, and mixed reality applications of artificial intelligence.

Reflecting on the above-outlined perspective, the papers contained in this volume are organized into seven main sections, including:

I. Humans and Artificial Cognitive Systems
II. Intelligence, Technology and Analytics
III. Computational Modeling and Simulation
IV. Humans and Artificial Systems Complexity
V. Smart Materials and Inclusive Human Systems
VI. Human-Autonomy Teaming
VII. Applications and Future Trends

We would like to extend our sincere thanks to Axel Schulte, Stefania Campione, and Marinella Ferrara, for leading a part of the technical program that focuses on Human-Autonomy Teaming and Smart Materials and Inclusive Human Systems. Our appreciation also goes to the members of Scientific Program Advisory Board who have reviewed the accepted papers that are presented in this volume, including the following individuals:

J. Platts, UK
U. Schmid, Germany
N. Stanton, UK

Intelligence Technology and Analytics
Chair: A. Ebert, Germany
D. Băilă, Romania
R. Philipsen, Germany

We hope that this book, which presents the current state of the art in Intelligent Human Systems Integration, will be a valuable source of both theoretical and applied knowledge enabling the design and applications of a variety of intelligent products, services, and systems for their safe, effective, and pleasurable collaboration with people.

Orlando, FL, USA
February 2019

Waldemar Karwowski
Tareq Ahram

Contents

Intelligence, Technology and Analytics

Human-Autonomy Teaming

Applications and Future Trends

Humans and Artificial Cognitive Systems

Context Awareness Computing in Smart Spaces Using Stochastic Analysis of Sensor Data

Jae Woong Lee[1(✉)] and Sumi Helal[2(✉)]

[1] Department of Computer Science, SUNY Oswego, Oswego, NY, USA
Jaewoong.lee@oswego.edu

[2] School of Computing and Communications, Lancaster University, Lancaster, UK
s.helal@lancaster.ac.uk

Abstract. In building a smart space, it becomes more critical to develop a recognition system which enables to be aware of contexts, since the appropriate services can be provided under the accurate recognition. As services satisfying for desires of individual human residents are more demanding, the necessity for more sophisticated recognition algorithms is increasing. This paper proposes an approach to discover the current context by stochastically analyzing data obtained from sensors deployed in the smart space. The approach proceeds in two phases, which is to build context models and to find one context model matching the current state space, however we mainly focus on the phase building context models. Experimental validation supports the approach and approved validity

Keywords: Smart spaces · Context awareness computing · Sensors Conditional probability table · K-means clustering · Principal component analysis

1 Introduction

During the last decade, we have been experiencing dramatic transformations in our paradigm of daily living life. Sensors are everywhere to monitor our activities and to enable to recognize contexts, and actuators and smart devices are operated to provide necessary and/or convenient services to us. In the smart space deployed with such devices, many chores are unawarely performed and complicated works are simplified. With outstanding achievements in internet technology, sensors and actuators are integrated into more intelligent devices, which is now known as Internet of Things (IoT) devices. The advent of ubiquitous sensors and IoT is continuously making our living space smarter.

The performance of smart spaces depends on various factors including sensor technology and IoT technology, but this paper addresses recognition of contexts. A smart space provides appropriate services to human residents after recognizing the current context. The recognition process demands accurate analysis of sensor data since the context is realized by the sensors attached in the space. It requires building well-structured profiles for contexts, which can be easily managed and efficiently manipulated.

W. Karwowski and T. Ahram (Eds.): IHSI 2019, AISC 903, pp. 3–9, 2019.
https://doi.org/10.1007/978-3-030-11051-2_1

This paper proposes an approach for context awareness computing utilized by stochastic analysis of sensor data. The key idea comes from two observations: sensors which are triggered in a given context are mostly related, and particular sensors are important and contributive to become the context. Hence, a whole dataset can be divided into a certain number of groups, each of which contains related sensors. Due to the similarity, the structure of the context model developed in our context-driven simulation approach is adapted here with minor changes [1]. Note that the context model requires finding the important sensors.

The main framework of the approach consists of two phases: in the first phase, context models are built and then the context which describes the present state of the space most similarly is discovered among the models in the second phase. In the first phase, the context models are derived from given sensor datasets which were generated under supervised learning. In this learning phase, three methods used in machine learning and statistics are utilized. First, two methods of Conditional Probability Table (CPT) and K-means clustering enables to find *k* numbers of groups, each of which is declared as a context. Last, Principal Components Analysis (PCA) returns important and representative sensors per each context, which finally define the complete context model. Once the context models are built, one context will be discovered by comparing the present state space and all the context models. At this comparison step, two methods of calculating Euclidean distances and cosine similarity are utilized. Due to the page limit, and this paper addresses only the first phase of creating the context models.

This paper is organized as follows. In the Sect. 2, we describe existing work related to context awareness computing and statistical analysis methods. And the principles of the proposed approach are overviewed in the Sect. 3, which is followed by experimental validation. We conclude the paper with a short discussion and future work plan.

2 Related Work

There were many context models proposed in various areas – human computer interaction [2], context awareness computing [3], and activity recognition learning [4]. The proposals attempted to model contexts from the current state of the space which was observed through human senses or electronic sensors. In some research, the context models were defined in the form of rules. In Context-Aware Simulation System for smart home (CASS) [5], for instance, the system defined rules to describe certain conditions, detected the conflicts of rules, and provided the ability to control a character to move it. In Context-driven simulation approach [1], contexts were defined as abstract and representative state spaces by specifying related sensors and their status. Additionally, their context models defined the causality in between other contexts, which enabled to generate the entire daily living scenario.

The research commonly oriented the methods on matching the current state space with manually predefined context models [6]. Ontology facilitated efficient modeling and reasoning for context [7], however it still needed humans' efforts in configuration. To avoid the burdens and increase automaticity, research in activity recognition on deriving meaningful high-level information from low-level information could be used. The goal of the research is to cluster from collected sensor datasets. In CBARS [8], a supervised learning model was built first, and unsupervised learning for new data was applied for new activity recognition. The challenge was that CBARS needed a supervised learning model. AALO [9] addressed that challenge. AALO is an active recognization system that can accurately classify specified activities according to locations and times in which the activities are performed. CBCE [10] proposed a method for combing multiple classifiers including Naïve Bayes (NB) models, hidden Markov models (HMMs), and conditional random fields (CRFs). This ensemble of classifiers can recognize activities in given sensor datasets, however, they do not provide a method to define abstract information of context to represent the other state spaces in a cluster.

3 Principles of Approach

The principal ideas in the defining context models are (1) to cluster sensor datasets into groups which have related sensors, and (2) to discover particular sensors and their values which can represent each group. The values of the sensors for each context are formed in context conditions, since the context begins only if the sensors have the values.

3.1 Context Model

Before we dive further into the details of our approach, it would first be helpful to define context. In the context-driven simulation approach [1], it is an abstracted state space envelope that represents consecutively occurring state spaces. A context is intended to represent an important and meaningful state space in the group of relevant state spaces with respect to activities. It is described by three properties: context conditions, which express conditions to enter the context, context activities, which are activities available in the context (for play back of some of them), and next contexts, which can be transitioned to after activities are performed. In context-awareness computing, context conditions only are needed to define a context.

3.2 Overall Approach

The approach proceeds in three steps, each of which utilizes a statistical method. First, the number of contexts is decided by Conditional Probability Table (CPT), and then meaningful and representative state spaces are defined as contexts by K-means Clustering. Finally, important sensors which contribute to become each context are discovered by using Principal Component Analysis.

Deciding the Number of Contexts by CPT. In order to find the number of contexts, we first capture the probability of consecutive occurrence of each pair of different sensors in the datasets. The idea is that sensors in a context are related and thus the occurrence probability of each other is fair high. In other words, if an occurrence probability of a pair of sensors is low, the sensors are not in a context. This probability can be accurately calculated from the frequency of occurrence of consecutive sensor events. These conditional probabilities are arranged as a ξ x ξ table (ξ being the number of sensors), which is called the Conditional Probability Table (CPT).

CPT is used as the probabilistic fingerprint of the entire dataset. A pair of sensor events with high conditional probability usually contains sensor events that are related and associated together. They could belong to the same context, and therefore are highly likely to occur together in this order. On the other hand, if the pair has low conditional probability, its sensor events are considered unrelated and would rarely occur together. A pair of sensor events with low probability indicates the end of a context and the start of another. Therefore, we divided the dataset between sensor events e_i and e_{i+1} if the conditional probabilities of e_i and the first sensor event e_i satisfy the condition $p(e_i) \leq \theta * p(e_i)$, where θ is a parameter that represents the extent to which ei relates to e_i. We set θ to 0.5 in the experiments. Using this method, we divided the dataset into k groups, which are considered as the number of contexts.

Defining the Contexts by K-means Clustering. By our observation, a meaningful state space is sufficiently distant from other meaningful state spaces, but could be close to other relevant yet non-meaningful state spaces. To find which state spaces are meaningful, all are partitioned into k clusters, in which each state space belongs to the cluster with the nearest mean. Therefore, the universal set of state spaces $S_U = \{S_1, \ldots, S_i, \ldots, S_\omega\}$ is divided into $\{\hat{S}_1, \ldots, \hat{S}_i, \ldots, \hat{S}_\omega\}$, where S_i is a state space and $\hat{S}_i$ is a cluster of state spaces. Each cluster $\hat{S}_i$ minimizes the sum of distances between the within-state space and the mean according to the following formula:

$$\arg\min_{S_U} \sum\nolimits_{i=1}^{k} \sum\nolimits_{S_t \in \hat{S}_j} \|S_t - \mu_i\|^2, \quad (1)$$

where S_i means a state space in cluster $\hat{S}_i$. After S_U is classified into k clusters, cluster centroids are considered meaningful state spaces and are candidates for contexts.

Discovering Context Conditions by PCA. The centroid in context is representative and meaningful, but has insufficient information to define the context. We observed that multiple sensors usually contribute to begin a context. Principal Components Analysis (PCA) enables to discover those important sensors. Once we find the relevant sensors via the stochastic analysis of sensors' high-dimensional data, the original dataset can be projected onto lower-dimensional data. The process is repeated for each cluster and the remaining data is used to build context conditions.

Principal components are sensors that show definite variance patterns that explicitly express the change of states. We want to know in which pattern the dataset is scattered. For this, a matrix of covariances (*cov*) is calculated first. In a ξ-dimensional dataset, covariance *cov* is calculated as

$$cov\left(\hat{s}^i, \hat{s}^j\right) = \frac{\sum_{k=1}^{\xi}\left(\hat{s}_k^i - \mu_i\right)\left(\hat{s}_k^j - \mu_j\right)}{(\xi - 1)}, \quad (2)$$

where $\hat{s}^i$ and $\hat{s}^j$ are the set of sensor values in dimensions i and j, respectively; i and j are the sensor values in each dimension. The total covariances establish a $\xi \times \xi$ covaraince matrix R, shown in the Eq. 3.

From covariance matrix, we calculate the eigenvectors, each of which can conduct linear transformations of sensor data and characterize its variance; the eigenvalues then measure how well the sensor data is scattered. We choose the eigenvectors that show the most variant spread of data as principal components. If data is evenly scattered with an axis transformed by an eigenvector (i.e., the data pattern is recognized explicitly), it is an important eigenvector, which means it's the desired principal component and has a high eigenvalue. The challenge is in determining the threshold for which eigenvalues are high enough to be acceptable. We propose threshold θ_e for total eigenvalues of selected eigenvectors. In our approach, first the eigenvectors are sorted by eigenvalues in descending order; then, eigenvectors with higher values are chosen until the sum of corresponding eigenvalues exceeds θ_e.

$$R(\hat{S}) = \begin{bmatrix} cov(\hat{s}^1, \hat{s}^1) & \dots & cov(\hat{s}^1, \hat{s}^\xi) \\ \vdots & \ddots & \vdots \\ cov(\hat{s}^\xi, \hat{s}^1) & \dots & cov(\hat{s}^\xi, \hat{s}^\xi) \end{bmatrix} \tag{3}$$

Eigenvectors satisfying the condition establish a feature matrix. The original high-dimentional dataset is transformed into a low-dimensional dataset through the feature matrix. Context conditions are created by collecting all sensor and forming them in a range. For instance, the expected condition for s_i, those values are 1, 4, and 2.5, is $1 \leq s_i \leq 4$, which covers all the values.

4 Experimental Validation

To evaluate the performance of the proposed approach, we conducted a few experiments. For this experiment, we first obtained the context models from sensor datasets for 10 days by applying the approach. Then we synthesized sensor datasets by running Persim 3D [11] with the context models. Persim 3D adapted the context-driven simulation approach, and generated sensor data from a scenario which is described by the sequence of contexts. The key of the validation is to compare the actual dataset and its synthetic dataset and to show similarity in between. If the approach is valid and thus the contexts are correctly discovered, Persim 3D should generate the similar dataset as the actual dataset. Therefore, our validation goal was to compare the generated dataset with actual dataset.

For this purpose, we built statistic models of an actual dataset that apply a Bayesian network. The Bayesian network enables to calculate join probability distribution on the entire dataset by using CPT. If the occurrence probability of the simulated dataset is not similar to those of the actual dataset, it says that the dataset was not correctly generated. However, the probability based on sensor events becomes very low as multiplying probability of pairs of sensor events, thus we utilized Activity Playback Model which enables to describe a dataset in the activity level. It prevents the probability from diminishing. Table 1 shows the occurrence probabilities of each dataset. In most experiments, simulated datasets show high similarity and the average similarity is 70.95%.

5 Conclusion

The approach based on stochastic analysis of sensor data reduces humans' efforts in processing recognition of the current context and increases automaticity of the process. Through the experiments, it validly shows the good performance. Our next research will concentrate on developing more efficient statistic methods which can improve the performance. We will also research on unsupervised learning methods, which thus are able to detect contexts without training. It will relate to the real-time recognition.

Table 1. Similarity of occurrence probability of generated dataset based on our approach against actual dataset. Note that different *k* is applied on Nov/05

Dataset	Actual dataset	# of Context (K)	Simulated dataset	Similarity
Nov/03	0.150	5	0.124	82.55%
Nov/04	0.150	5	0.124	82.55%
Nov/05	0.033	4	0.234	70.75%
Nov/07	0.150	5	0.124	82.55%
Nov/10	0.150	5	0.124	82.55%
Nov/11	0.090	5	0.045	50.03%
Nov/13	0.090	5	0.045	50.03%
Nov/14	0.150	5	0.124	82.55%

References

1. Lee, J.W., Helal, A., Sung, Y., Cho, K.: A context-driven approach to scalable human activity simulation. In: ACM SIGSIM Conference on Principles of Advanced Discrete Simulation, pp. 373–378. ACM, New York (2013)
2. Fischer, G.: User modeling in human-computer interaction. J. User Model. User-Adapt. Interact. **11**(1–2), 65–86 (2001)
3. Salber, D., Dey, A., Abowd, G.: The context toolkit: aiding the development of context-enabled applications. In: Conference on Human Factors in Computing Systems, pp. 434–441. ACM New York (1999)
4. Hasan, M., Roy-Chowdhury, A.: Context aware active learning of activity recognition models. In: IEEE International Conference on Computer Vision, pp. 4543–4551. IEEE Express, Washing DC. (2015)
5. Park, J., Moon, M., Hwang, S., Yeom, K.: CASS: a context-aware simulation system for smart home. In: 5th ACIS International Conference on Software Engineering Research, Management & Applications, pp. 461–467. IEEE Express (2007)
6. Lee, J.W., Helal, S.: Inferring context from human activities in smart spaces. In: 29th International FLAIRS Conference, pp. 695–701. AAAI Press, Florida (2016)
7. Wang, X.H., Zhang, D.Q., Gu, T., Pung, H.K.; Ontology based context modeling and reasoning using OWL. In IEEE Annual Conference on Pervasive Computing and Communications Workshops, pp. 18–22. IEEE Washington DC. (2004)

8. Abdallah, Z.S., Gaber, M. M., Srinivasan, B., Krishnaswamy, S.: CBARS: cluster based classification for activity recognition systems. In: International Conference on Advanced Machine Learning Technologies and Applications, pp. 82–91. Springer, Berlin (2012)
9. Hoque, E., Stankovic, J.: AALO: activity recognition in smart homes using active learning in the presence of overlapped activities. In: International Conference on Pervasive Computing Technologies for Healthcare, pp. 139–146. IEEE Express (2012)
10. Jurek, A., Nugent, C., Bi, Y., Wu, S.: Clustering-based ensemble learning for activity recognition in smart homes. Sensors **14**(7), 12285–12304 (2014)
11. Lee, J.W., Cho, S., Liu, S., Cho, K., Helal, S.: Persim 3D: Context-driven simulation and modeling of human activities in smart spaces. IEEE Trans. Autom. Sci. Eng. **12**(4), 1243–1254 (2015)

A Strain Based Model for Adaptive Regulation of Cognitive Assistance Systems—Theoretical Framework and Practical Limitations

Dominic Bläsing(✉) and Manfred Bornewasser

Institute of Psychology, University Greifswald, Franz-Mehring-Str. 47, 17489 Greifswald, Germany
{dominic.blaesing, bornewas}@uni-greifswald.de

Abstract. In order to manage increasing complexity so called cognitive assistance systems are integrated into assembly systems. On the basis of real-time measurement and analysis of physiological signals, these assistance systems help to coordinate efficient behavior and to prevent states of long lasting detrimental workload and strain. With measurement technology getting smaller, more powerful and wearable it's possible to collect and analyze personal physiological data in real-time and detect significant changes at the workplace. It is intended to use these data to control a cognitive assistance systems which as a consequence of a monitored detrimental workload leads to adaptive changes in assembly processes and to a reduction of workload. The underlying principle can be a self-actualizing machine learning algorithm. We want to present a theoretical framework to sketch possibilities of such data-controlled, adaptive systems and to describe some obstacles which have to be overcome before they're ready for use.

Keywords: Cognitive assistance system · Mental workload · Physiological measurement · Acceptance

1 Introduction

A rise of complexity in work tasks is one of the catalysts for a feeling of strain at work. While most work processes in areas like assembly were formerly dominated by physical work, mental or cognitive aspects are taking over. Production on demand, a rising number of variability in products, parts and variants as well as shorter product life cycles are possible drivers for complexity issues in manual assembly [1]. As a consequence workload increases and more and more people feel stressed or strained at work. As stressed workers tend to make more failure and to be less productive [2], it is necessary to find ways to help them avoid such states of detrimental workload. In Situ-Projections, Pick-by-Light Systems, AR-Glasses or simple Monitors can be used as cognitive assistance systems to reduce mental workload when they are well designed, reliable, easy to use and not hindering the work flow.

W. Karwowski and T. Ahram (Eds.): IHSI 2019, AISC 903, pp. 10–16, 2019.
https://doi.org/10.1007/978-3-030-11051-2_2

2 Towards a Framework for Cognitive Assistance Systems

Without a certain amount of effort, work is not possible [3]. Each task requires resources to solve them. This might imply more physical, mental or emotional activation, however, each tasks normally requires a combination of such resources and needs at least a minimal amount of energy. Cognitive assistance systems can help to reduce mental effort by restructuring information, helping to make decisions between alternatives or can support the worker while he or she is attentionally disrupted by side-tasks or unusual demands.

Mental workload describes the relationship between the internal cognitive resources of a person and the cognitive demands of a work task. Work can be seen as a dynamic process of ups and downs in mental workload oscillating between states of overload and underload. In between underload and overload the ideal range of workload to be most productive is located [4]. The optimal range is different for every person depending on a variety of factors. The bonds, borders or red lines could be crossed easily for a shorter or longer duration, leading to a more or less decrease in performance [5].

Strain in general and especially mental workload can be measured via different physiological indicators like heart rate or heart rate variability [6], EEG-signals [7], respiration parameters [8], electrodermal activity/galvanic skin response [9], eye tracking and gaze behavior [10] as well as brain imaging techniques like fNIRS or fMRT [11]. Most reaction patterns are caused by changes in the autonomous nervous system, an increase in sympathetic and/or a decrease in parasympathetic activity, like an increase in heart rate or the pupil size with rising mental workload. Multimodal measurement of mental workload seems to be a more robust and valid process than using only one indicator [12, 13].

3 How to Build an Adaptive Cognitive Assistance System

A cognitive assistance system requires a deep understanding of the processes at the working station. Therefore, a process modulation like Work Domain Analysis [14] or the IMPRINT technique (Improves Performance Research Integration Tool) based on Wicken's Multiple Resource Theory as proposed by Rusnock et al. [15] are an optimal starting point. Process analysis gives an overview of work flows and bottlenecks and leads to a more comprehensive conception of cognitive assistance systems and their chances and risks. It allows a prediction, in which situations mental workload might rise or go down. Thus, process analysis is not only a starting point, it also functions as some kind of baseline model of a distribution of states of high and low mental workload throughout the assembly process [14, 15].

Besides getting this deep understanding of the process it's needed to implement the physiological measurement systems in the working place in an unobtrusive, non-invasive way. The first decision should be about which measurements could be used and how they could be integrated in the working place. When workers are supposed to wear gloves, sensors to measure electrodermal activity can be build into those gloves

[16]. When they are wearing helmets, EEG and Eye Tracking sensors could be integrated in these helmets as well as ECG electrodes in normal work clothing.

Before physiological data can be used to regulate cognitive assistance systems it is necessary to build one common indicator of mental workload and to measure it in an efficient way. So all steps of data cleaning, processing, and computing have to be done in a very short amount of time finally showing a change of state. Personalized mental effort models seem to be more precise than population models [17], so even more important a personal baseline needs to be established. Thus, we have to compare personal baseline and actual states of workload as well as changes in the states of workload during the individual assembly process [6].

Based on the data of process modelling, existing task instructions and a close cooperation with the assembly workers the actual cognitive assistance should be programmed. It's important, that assistance is more than an on/off solution. To be really adaptive it must provide different granularities of information depending on experience and competency of workers. For example, In Situ-Projections can provide help, increase performance and reduce time on task as long as the worker does not feel patronized or gets overassisted [18, 19]. Thus, the level of cognitive assistance must not only be in balance with the actual mental workload level, but also with personal resources.

4 Obstacles and Possible Solutions

In laboratory settings machine learning algorithms achieve astonishing results for discriminating low and high mental workload [6, 10]. They can even be used to adapt task difficulty [10]. In a field situation, however, there are several factors that can't be controlled and discrimination of task difficulty is challenging when subjects show different patterns of competence and experience. Here we try to focus on some important problems for adaptive cognitive assistance systems and present possible solutions.

4.1 Acceptance: Usability and Trust

A cognitive assistance system without acceptance will be used less intensively and can't enfold its full potential. Therefore, it is important to design it in a way that workers don't feel uncomfortable, disrupted or monitored, when they use the system. For this reason it is necessary to integrate end-users of the system into the developmental processes. This starts with a close cooperation and participation of workers in the developmental process. What do they need and expect from such a system and in which ways do they like to use it or to interact with it? For an adaptive system relying on physiological data, it is most important to integrate all sensors needed for the measurement in the work clothing. ECG measurement via Holter-systems or chest straps might deliver good signal quality, but could reduce acceptance. Textile electrodes in compression shirts might find increased acceptance, although signal quality is lower. Using field electrodes to measure heart activity might be an even better solution [20]. Similar examples can be found for other physiological measurement systems.

Most importantly, the system must be a plug and play, workers should just need to put it on like a normal working T-shirt and interact with the assembly system as usual. Especially for measuring physiological data, data protection and the fear of misuse of those data is a topic that should be covered [17]. It can be speculated that cognitive assistance systems are much sought-after sources of information for different organizational ends.

4.2 How to Get a Personal Baseline?

Directly connected to the usability and acceptance question is the question of how to measure a solid baseline for such a system. It should be done in a fast and easy way and nearly doesn't interrupt the daily activity of the worker. There're several ways a baseline measurement can be done. Usually it is a one-time measurement during rest, lasting between 5 and 30 min in a sitting, standing or lying condition. Such kind of baseline measurement helps to validate differences to a resting condition but does not help to understand where the cut-off values or red lines are. Additionally it is always afflicted by daily changes, nervousness of the subject and extraordinary situative influence factors. So for detecting changes in mental workload and to support a machine learning algorithm with data, a baseline should be assessed under rest and under different workload conditions to get minimum and maximum values. In this way it's possible to detect ups and downs during daily work and interpret them. It is possible, too, to miss a baseline measurement and to use another algorithm to check if the status of the subject changed compared to the last segment of data [6]. A combination of both approaches can be useful for a self-updating algorithm which can compare actual data with historical data from the same subject and in combination with a well modeled process even predict when mental workload changes.

4.3 Interpreting Short-Term and Long-Term Workload Changes

But even with non-invasive measurement technology and a good baseline measurement there is still the problem of how to interpret the changes in the data. When is the cut-off value reached for overload, when has it fallen below the optimal range? What does a change of .5 in the "mental workload score" mean and is it the same for two different subjects? There's also a problem in the difference of the physiological response to a stressor or mental strain and the subjectively perceived stress [8]. Physiological arousal might be long gone, but perceived stress is still high, so at which point should the adaptive system readapt to the new state of workload? Additionally, not every change in physiological parameters indicates a change of mental workload or is directly connected to the working place. Changes in posture, emotional arousal as well as changes in surrounding environmental factors, such as sudden noise, light or heat, often lead to strong physiological reactions.

4.4 Latency Problems Using Combined Measurements

A combined classifier has always a better accuracy than a single indicator [8, 13] and for example data from an acceleration sensor can be used to correct the influence of rising heart rate caused by movement to the mental workload score. Additional failures result from the fact, that they're all sampled with different frequencies and from different devices and therefore data alignment can be a problem to solve. Latencies or delays vary between physiological indicators. While changes in pupil-size or brain activity can be measured quickly in seconds or even faster [10], changes in respiration or heart rate might take up to a minute [21] and measurable changes in cortisol can last up to 30 min [22].

5 Conclusion

An adaptive cognitive assistance system can be a useful tool to help workers to reduce mental workload and in consequence to work more efficiently and to make less failures. There is no doubt, that physiological signals offer chances to quantify the state of mental workload, that intelligent self-updating machine learning algorithms might contribute to an adaptive regulation of demand and workload and that stress and strain even in highly complex assembly processes can at least be reduced or even be avoided [23]. However, in order to make such an adaptive assistance system work, we have to overcome a number of obstacles. These might be classified in three categories: Technically, they concern a series of measurement procedures, especially the selection and integration of indicators of mental workload with different frequencies and latencies, the determination of personal baselines, a valid separation of indications of cognitive strain and physical movements, and finally the identification and interpretation of physiological states and change points. Personally, they concern a balance between information assistance and different amounts of resources like knowledge, experience and expertise varying from worker to worker. Also acceptance and usability problems are part of this category. Situationally, there is a strong need for a combination of physiological and behavioral recordings. Physiological signals without context are different to interpret, they get their meaning only in relation to situational events. A general process analysis might be a valid reference frame, however, what is needed are variant specific analyses. Interruptions of normal courses constitute a further challenge.

All these aspects need further research. However, although we can invest a lot of effort in better measurement techniques, better algorithms or personalized assistance, we always have to keep the focus on the user. His adoption of cognitive assistance systems is a crucial factor to ensure return on investments and to conserve cognitive and health resources of the workforce.

Acknowledgements. The authors acknowledge the financial support by the Federal Ministry of Education and Research of Germany in the project Montexas4.0 (FKZ 02L15A261).

References

1. Schuh, G., Gartzen, T., Wagner, J.: Complexity-oriented ramp-up of assembly systems. CIRP J. Manuf. Sci. Technol. **10**, 1–15 (2015)
2. Samy, S.N., ElMaraghy, H.: A model for measuring products assembly complexity. Int. J. Comput. Integr. Manuf. **23**, 1015–1027 (2010)
3. Hacker, W.: Arbeitsgegenstand Mensch: Psychologie dialogisch-interaktiver Erwerbsarbeit: ein Lehrbuch. Pabst Science Publ, Lengerich (2009)
4. Wickens, C.D.: Multiple resources and mental workload. Hum. Factors: J. Hum. Factors Ergon. Soc. **50**, 449–455 (2008)
5. Young, M.S., Brookhuis, K.A., Wickens, C.D., Hancock, P.A.: State of science: mental workload in ergonomics. Ergonomics **58**, 1–17 (2015)
6. Hoover, A., Singh, A., Fishel-Brown, S., Muth, E.: Real-time detection of workload changes using heart rate variability. Biomed. Sign. Process. Control **7**, 333–341 (2012)
7. Zarjam, P., Epps, J., Lovell, N.H.: Beyond subjective self-rating: EEG signal classification of cognitive workload. IEEE Trans. Auton. Ment. Dev. **7**, 301–310 (2015)
8. Plarre K., Raij A.B., Hossain M., et al.: Continuous inference of psychological stress from sensory measurement scollected in the natural environment. In: Proceedings of ACM/IEEE Conference on Information Processing in Sensor Networks, pp. 97–108 (2011)
9. Ma, Q.G., Shang, Q., Fu, H.J., Chen, F.Z.: Mental workload analysis during the production process: EEG and GSR activity. Appl. Mech. Mater. **220–223**, 193–197 (2012)
10. Kosch, T., Hassib, M., Buschek, D., Schmidt, A.: Look into my eyes: using pupil dilation to estimate mental workload for task complexity adaptation. In: Extended Abstracts of the 2018 CHI Conference on Human Factors in Computing Systems - CHI 2018, pp. 1–6. ACM Press, Montreal (2018)
11. Hincks, S.W., Afergan, D., Jacob, R.J.K.: Using fNIRS for real-time cognitive workload assessment. In: Schmorrow, D.D., Fidopiastis, C.M. (eds.) Foundations of Augmented Cognition: Neuroergonomics and Operational Neuroscience, pp. 198–208. Springer International Publishing, Cham (2016). https://doi.org/10.1007/978-3-319-39955-3_19
12. Chen, F., et al.: Robust Multimodal Cognitive Load Measurement. Springer International Publishing, Cham (2016)
13. Seoane, F., et al.: Wearable biomedical measurement systems for assessment of mental stress of combatants in real time. Sensors **14**, 7120–7141 (2014)
14. Li, Y., Burns, C., Hu, R.: Understanding automated financial trading using work domain analysis. In: Proceedings of the Human Factors and Ergonomics Society Annual Meeting, vol. 59, pp. 165–169 (2015)
15. Rusnock, C.F., Borghetti, B.J.: Workload profiles: a continuous measure of mental workload. Int. J. Ind. Ergon. **63**, 49–64 (2018)
16. Valenza, G., Citi, L., Garcia, R.G., Taylor, J.N., Toschi, N., Barbieri, R.: Complexity variability assessment of nonlinear time-varying cardiovascular control. Sci. Rep. **7**, 42779 (2017)
17. Kim, J., Andre, E.: Emotion recognition based on physiological changes in music listening. IEEE Trans. Pattern Anal. Mach. Intell. **30**, 2067–2083 (2008)
18. ElKomy, M., Abdelrahman, Y., Funk, M., Dingler, T., Schmidt, A., Abdennadher, S.: ABBAS: an adaptive bio-sensors based assistive system. In: Proceedings of the 2017 CHI Conference Extended Abstracts on Human Factors in Computing Systems - CHI EA 2017, pp. 2543–2550. ACM Press, Denver (2017)

19. Kosch, T., Abdelrahman, Y., Funk, M., Schmidt, A.: One size does not fit all: challenges of providing interactive worker assistance in industrial settings. In: Proceedings of the 2017 ACM International Joint Conference on Pervasive and Ubiquitous Computing and Proceedings of the 2017 ACM International Symposium on Wearable Computers on - UbiComp 2017, pp. 1006–1011. ACM Press, Maui (2017)
20. Kuronen, E.: EPIC sensors in electrocardiogram measurement. Master thesis, Oulu University of Applied Sciences (2013)
21. Draghici, A.E., Taylor, J.A.: The physiological basis and measurement of heart rate variability in humans. J. Physiol. Anthropol. **35**, 22 (2016)
22. Spencer, R.L., Deak, T.: A users guide to HPA axis research. Physiol. Behav. **178**, 43–65 (2017)
23. Bornewasser, M., Bläsing, D., Hinrichsen, S.: Informatorische Montageassistenzsysteme in der manuellen Montage: Ein Werkzeug zur Reduktion mentaler Beanspruchung. Z. Arb. Wiss. **72**, 264–275 (2018)

Hexagonal Image Generation by Virtual Multi-grid-Camera

Robert Manthey(✉) and Danny Kowerko

Technical University of Chemnitz, 09107 Chemniz, Germany
{robert.manthey, danny.kowerko}@informatik.tu-chemnitz.de

Abstract. The process of capturing of an image is realized by a two-dimensional plane composed of photosensitive elements of almost entirely rectangle shape in technical solutions. However, biological visual systems use almost entirely hexagonal shapes and theoretical research shows the advantages of them but also the lack of usable capturing devices. We address this problem and create a virtual multi-grid-camera to overcome the problem and make further research possible. We create some scenes with common known content to demonstrate the use and show some effects being the result of the different shapes.

Keywords: Hexagonal image processing · Dataset generation Tessellation · Biological visual systems · Human visual system

1 Introduction

Digital cameras and eyes are important optical recognition devices of technical and biological systems to decide for edibility of fruits, the distance of predator and prey, or the quality of a visual inspected component. To realize this, the reflected light being captured by the recognition device and analyzed to get properties like shape, structure and features of the surface of the observed object. The transformation from the continuous domain of the real world scenario with objects and properties to a processible form of information is realized by discretization. This discretization influences the quality of the following steps of feature extraction and analysis. [1].

Several tessellations and their properties were observed and compared by [2] and [3] to find an optimal discretization. They found that the common used rectangle grid has some advantages with simple calculations but disadvantages with information lost during the transformation. Compared to this, the hexagonal grid offers the most efficient information transformation, advantages in pixel-neighborhood and shows some algorithmic improvements [4, 5], but be nearly only used in biological systems like insect compound eyes or human retina, as shown in Fig. 1.

Nearly any previous research indicates the lack of native capture and display devices to perform more studies on the hexagonal grid. This work presents a solution to that problem, provide a tool to create scenes and to capture images of them with self-defined grids.

W. Karwowski and T. Ahram (Eds.): IHSI 2019, AISC 903, pp. 17–22, 2019.
https://doi.org/10.1007/978-3-030-11051-2_3

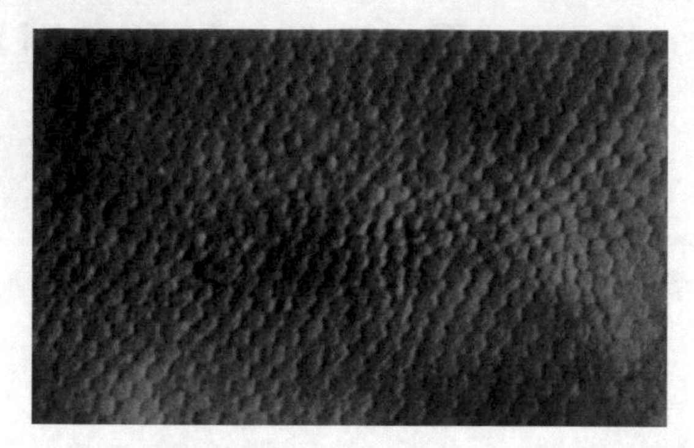

Fig. 1. Image of a human retina showing the mostly hexagonal structure of the visual cells [6]

2 Image Acquisition Principle

The process to acquire an image from a scene, as shown in Fig. 2, depend on radiation emitted from a source like a lamp, the sun or reflected from a object. They are projected to a plane in the imaging system composed of detector elements transforming the properties of the respective radiation to corresponding digital information of the pixel. This transformation encloses an error, which reduce the amount of information depending on the shape of the elements as investigated by [3]. Beside the commonly used approximations of rectangles, also triangles and hexagons may be used, having their specific properties, as shown in Fig. 3.

A common effect is the difference of the resolutions needed to cover the same area at the image plane with different grids as result of the different dimensions of the shapes as shown in Figs. 5, 6 and 6.

3 System Architecture and Technical Realization

To capture images with defined grid structure in a way as closely as possible to real world image acquisition, we use the raytracing-software POV-Ray[1]. They simulate the properties of light distribution, reflection etc. to produce realistic images of a defined scene and allow the definition of a triangle-based mesh representing the grid structure of the virtual capturing device.

The scenes are created by the use of the internal programming-like language allowing the construction of mathematical defined object as well as compositions of them. In a similar way light sources, cameras etc. are defined and positioned inside the scene. This approximates the real world and prevents the introduction of artifacts as resampling might do. [8]

---

[1] http://www.povray.org.

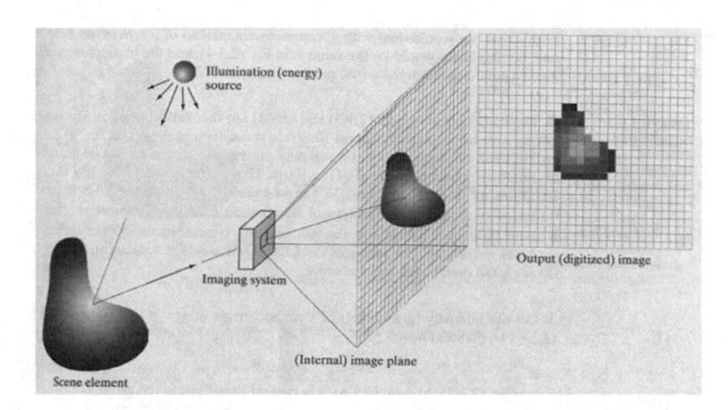

Fig. 2. Principle of image acquisition showing the light source, the object reflecting the light to the imaging system, the projection to the image plane and the rectangle-based discretized, digital representation [7, p.51].

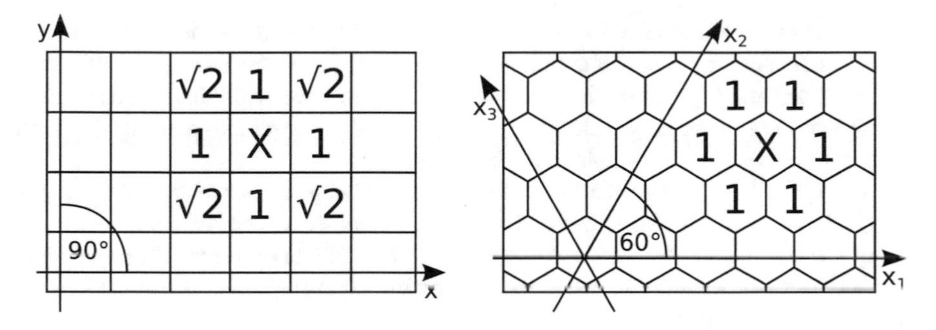

Fig. 3. Main properties of (Left) rectangle and (Right) hexagonal grid. The two orthogonal axis of the first simplifies calculations. However, measuring lengths need corrections because of the non-uniform distances of neighborhood elements. The second contains three axis skewed by 60 degree, which are non-orthogonal but allow some simplifications at angle calculations and result in uniform neighborhood.

During the capturing process, a given scene definition and the designated mesh being used to produce the corresponding triangle-based image as well as a rectangle-based image for comparison at the same moment, as shown in Fig. 4.

With well-defined triangles, other more complex structures like hexagons are possible. We use this to produce the image based on the hexagonal grid in a post-processing step, which combine six equilateral triangles to one respective hexagon. Because of the well-defined size of the triangles, the surfaces of desired grid elements are equal to the elements of the rectangle grid. In addition, the resolution of the images must be well-defined to cover the same amount of space by the image plane as good as possible. This allows the comparison of the results of the different grids.

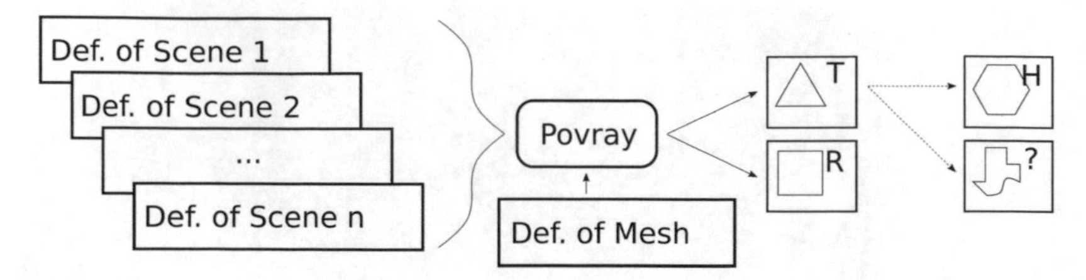

Fig. 4. Workflow of the presented system. Scenes with different content are defined as well as the mesh representing the structure of capture device. Both is processed by POV-Ray to produce images based on grids with rectangle and triangle pixels. The triangle image is post processed to produce the desired grid like hexagons.

4 Comparison and Results

We create various scene definitions of different content with geometric primitives like spheres and cubes, fractals, common known test objects, complex scenes modeling real world samples and high-resolution images. Different meshes with different resolutions were created comparable to selected rectangle-based resolutions covering the same amounts of area at the image plane of the virtual camera.

We produce image of each scene with the designated grids as shown in the samples of Figs. 5, 6 and 7. All hexagon-based images are smaller but higher compared to the rectangle ones, as expected because of the different shape dimensions. They also show fewer artifacts especially at round edges, which clearly present and confirm the advantages of the hexagonal grid.

Fig. 5. Image Lena (http://www.lenna.org) captured with cameras of equal overall dimension of the virtual camera sensors. Left) with 64×64 square pixels. Right) with 59×68 hexagon pixels showing fewer artifacts and better details, despite the fact that 18% fewer pixels be needed.

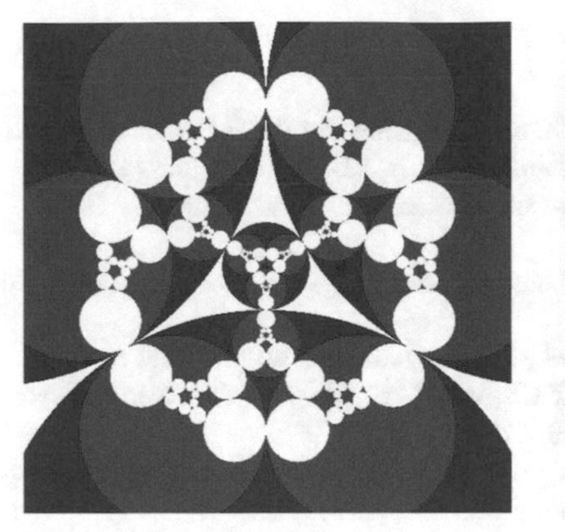
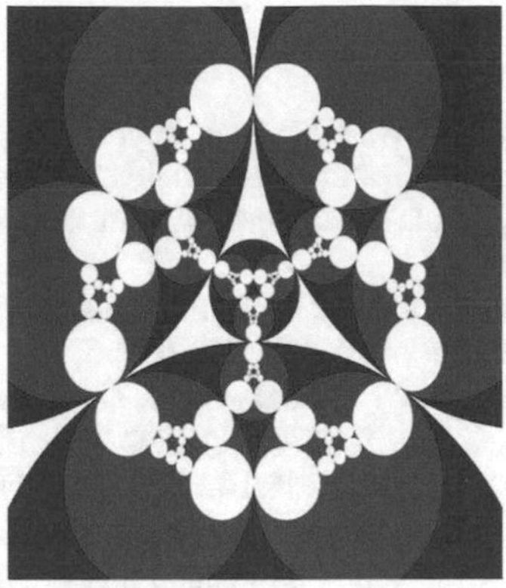

Fig. 6. Fractal captured with Left) 512 × 512 square pixels showing artifacts at the curved edges and Right) with 476 × 549 hexagonal pixels.

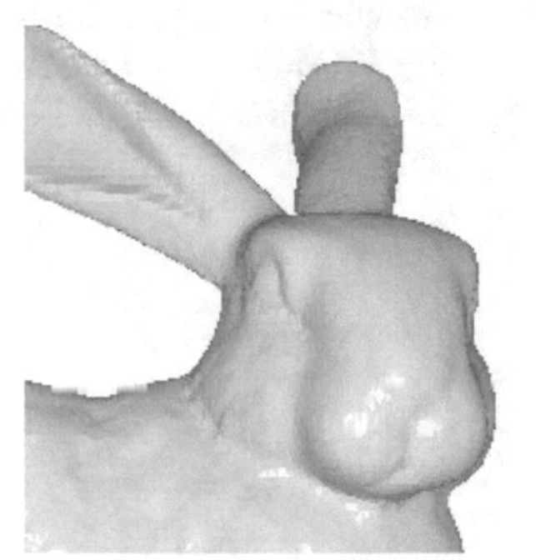
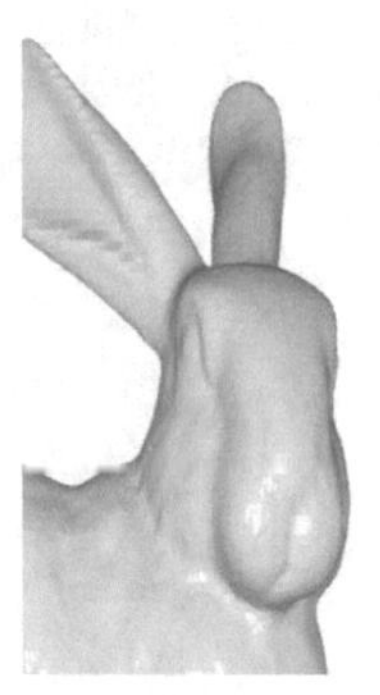
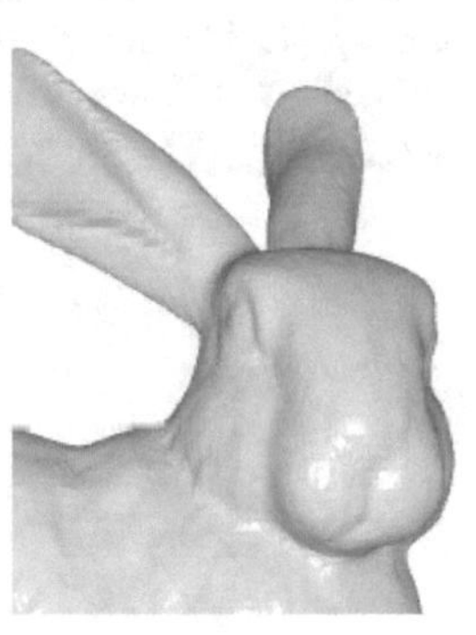

Fig. 7. Zoomed part of a scene with the Stanford Bunny (http://graphics.stanford.edu/data/3Dscanrep/#bunny) captured with cameras of equal overall dimension. Left) Square pixels showing clear aliasing artifacts at the shoulder and the ears. Middle) with triangle pixels. Right) with hexagon pixels showing much fewer artifacts.

5 Summary and Future Work

In the field of image acquisition, grid beside the rectangle one are known but not commonly used and usable devices with these different grids are not easily available. To overcome this problem and to simplify further research on advanced grid structures, we create a system to capture images with hexagonal grids and present some samples showing some advantages over rectangle grid. Further investigations may simplify quantitative comparisons of different grids and focus on the consequences of different grids to high images analysis algorithms.

Acknowledgments. This work was partially accomplished within the project localizeIT (funding code 03IPT608X) funded by the Federal Ministry of Education and Research (BMBF, Germany) in the program of Entrepreneurial Regions InnoProfile-Transfer.

References

1. He, X., Jia, W.: Hexagonal structure for intelligent vision. In: IEEE International Conference on Information and Communication Technologies, pp. 52–64. IEEE Press (2005)
2. Golay, M.J.: Hexagonal parallel pattern transformations. Trans. Comput. **18**(8), 733–740 (1969)
3. Deutsch, E.: Thinning algorithms on rectangular, hexagonal, and triangular arrays. Commun. ACM **15**(9), 827–837 (1972)
4. He, X., Jia, W., Wu, Q., Hintz, T.: Parallel edge detection on a virtual hexagonal structure. In: Cérin, C., Li, K.-C. (eds.) GPC 2007. LNCS, vol. 4459, pp. 751–756. Springer, Heidelberg (2007). https://doi.org/10.1007/978-3-540-72360-8_68
5. Wüthrich, C., Stucki, P.: An algorithmic comparison between square- and hexagonal-based grids. Graph. Models Image Process. **53**(4), 324–339 (1991)
6. Curcio, C.A., Kenneth, J., Sloan, R., Packer, O., Hendrickson, A.E., Kalina, R.E.: Distribution of cones in human and monkey retina: individual variability and radial asymmetry. Science **236**(4801), 579–582 (1987)
7. Gonzalez, R.C., Woods, R.E.: Digital Image Processing, 3rd edn. Pearson, London (1999)
8. Manthey, R., Conrad, S., Ritter, M.: A framework for generation of testsets for recent multimedia workflows. In: Antona, M., Stephanidis, C. (eds.) UAHCI 2016. LNCS, vol. 9739, pp. 460–467. Springer, Cham (2016). https://doi.org/10.1007/978-3-319-40238-3_44

Classification of Different Cognitive and Affective States in Computer Game Players Using Physiology, Performance and Intrinsic Factors

Ali Darzi[1(✉)], Trent Wondra[2], Sean McCrea[2], and Domen Novak[1]

[1] Department of Electrical and Computer Engineering, University of Wyoming, Laramie, WY, USA
{adarzi,dnovak1}@uwyo.edu
[2] Department of Psychology, University of Wyoming, Laramie, WY, USA
{twondra1,smccrea}@uwyo.edu

Abstract. Intelligent systems infer human psychological states using three types of data: physiology, performance, and intrinsic factors. To date, few studies have compared the performance of the data types in classification of psychological states. This study compares the accuracy of three data types in classification of four psychological states and two game difficulty-related parameters. Thirty subjects played nine scenarios (different difficulty levels) of a computer game, during which seven physiological measurements and two performance variables were recorded. Then, a short questionnaire was filled out to assess the perceived difficulty, enjoyment, valence and arousal, and the way the participant would like to change two game parameters. Furthermore, participants' intrinsic factors were assessed using four questionnaires. All combinations of the three datasets were used to classify six aspects of the short questionnaire into either two or three classes using three types of classifiers. The highest accuracies for two-class and three-class classification were 98.4% and 81.5%, respectively.

Keywords: Affective computing · Game difficulty adaptation · Physiological measurements · Task performance · Intrinsic factors

1 Introduction

An intelligent cybernetic system can use affective computing techniques to infer the user's affective and cognitive states, then modify its behavior accordingly to ensure a positive user experience. For example, this approach is frequently used in computer games to intelligently adapt game difficulty to suit the player's mood and ability, thus providing a pleasant gameplay experience [1, 2]. The user's affective and cognitive states can be assessed using three types of data: physiological measurements, task performance, and intrinsic factors [3].

Physiological measures from either the central or peripheral nervous system can be used to quantitatively estimate psychological states in real time (during the task itself)

W. Karwowski and T. Ahram (Eds.): IHSI 2019, AISC 903, pp. 23–29, 2019.
https://doi.org/10.1007/978-3-030-11051-2_4

without the user's active participation. They include the electroencephalogram (EEG) [4], which records the electrical activity of the brain, electrocardiogram (ECG) [5], which monitors electrical activity of the heart (specifically heart rate), galvanic skin response (GSR) [5], which records the activity of the skin's sweat glands, skin temperature, respiration rate [6], and eye movement. All the above physiological signals were also analyzed in this study. *Task performance* is a task-specific concept and is thus not as generalizable as physiology, but is also frequently used to assess psychological states; in the case of games such as Pong, it is often defined simply as the in-game score [7]. In this study, participants' in-game score was recorded as well. Finally, a participant's *intrinsic factors* such as personality can provide significant information, but are generally combined with physiology or task performance to classify cognitive and affective states. In this study, participants were asked to fill out four questionnaires that assessed several intrinsic factors such as extraversion.

Playing a computer game may evoke several complex affective and cognitive states, such as mental workload [2], enjoyment, anger, hate, and love [6]. Alternatively, several studies have used two-dimensional models of emotion such as the valence-arousal system to asses in-game emotion [8]. In contrast, however, when designing systems that adapt game difficulty based on cognitive and affective states, most researchers only focus on a single psychological aspect such as perceived task difficulty. Thus, few studies have compared the ability of affective computing techniques to classify multiple different psychological states within the same game.

This study examines the accuracy of three different input datasets (physiology, performance, intrinsic factors) for classification of four different cognitive and affective variables (perceived difficulty, enjoyment, valence, arousal) and two different desired changes to game difficulty (ball speed, paddle size) in a computer-based game of Pong. Accurate classification of psychological states is perhaps the most critical scientific challenge of any game difficulty adaptation algorithm, and can be done using automated classification algorithms such as support vector machine (SVM) or linear discriminant analysis [4]. Our ultimate goal is to have the computer game react to these states and adapt its difficulty to ensure the optimal game experience or the user; however, as the first step, this paper is limited to offline classification of psychological states. The objective is to find the most informative features for psychological state classification.

2 Materials and Methods

Study Setup: In the study, we evoked different cognitive and affective states in 30 healthy university students (24.2 ± 4.4 years old, 11 females) using different difficulty levels of a computer game that was reused from our previous arm rehabilitation study [7]. It is a Pong game consisting of two paddles and a puck on a board (Fig. 1, left). The bottom paddle is controlled by the participant while the top paddle is controlled by the computer. If the puck passes a player's paddle and reaches the top or bottom of the screen, the other player scores a point and the puck is instantly moved to the middle of the board, where it remains stationary for a second before moving in a random direction. The game difficulty can be adjusted using two parameters: the ball speed and

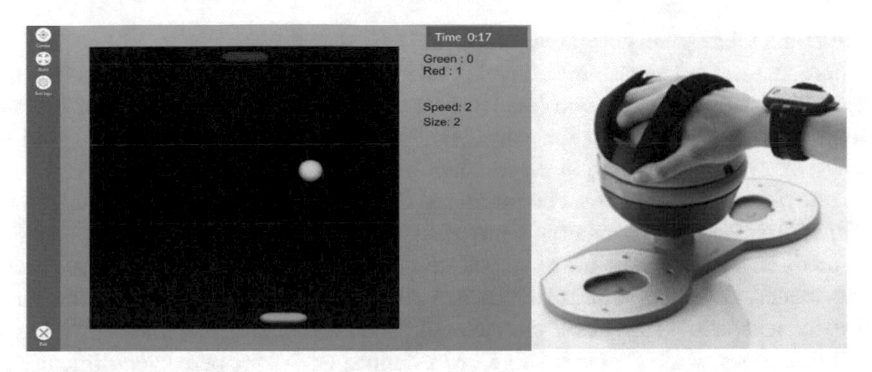

Fig. 1. The Pong game (left) and the BIMEO device (right).

the paddle size (with the paddle size being the same for both paddles at all times). The player moves their paddle left and right by tilting the Bimeo (Kinestica, Slovenia) arm rehabilitation device (Fig. 1, right) left and right.

Measured Data: For classification of the different affective and cognitive states, three types of data were collected: game performance, physiology, and intrinsic factors. The performance dataset includes two game performance measures: in-game score and the amount of arm movement, which is recorded by the Bimeo. To monitor the impact of intrinsic user factors on performance and physiology, four questionnaires were filled out: the learning and performance goal orientation measure [9], behavioral inhibition/activation scales [10], a self-efficacy scale [11], and a Big Five personality measure [12]. Two g.USBamp signal amplifiers and associated sensors (g.tec Medical Engineering GmbH, Austria) were used to record six types of physiological signals: 8-channel EEG, 2-channel electrooculogram (EOG), ECG, respiration [13], GSR, and ST. All physiological signals were sampled at 256 Hz. The EEG channels were recorded from prefrontal, frontal and central areas of brain based on the 10-20 placement system [14]: AF3, AF4, F1, F2, F5, F6, C1, and C2. As EEG signals are severely affected by eye activity, a 2-channel EOG was recorded to not only provide more physiological information but also to use as a reference signal with which to denoise the EEG signals. One EOG channel reflected up-down movement while the other one reflected left-right movement of the eyes. To record the EOG, small ECG electrodes (Kindall) were placed according to suggestions in the literature [4]. Finally, a seventh physiological signal (point of gaze on the screen in two dimensions) was recorded using an eye tracker (Gazepoint, Canada).

Study Protocol: The study protocol started with a 2-minute baseline recording of physiological signals, during which participants did not do anything and were instructed to relax. The main part of experiment then consisted of nine trials (test periods), each two minutes long. The nine trials consisted of all possible combinations of ball speeds (slow, medium, fast) and paddle sizes (small, medium, large), played in random order. After each trial, a short questionnaire was filled out to assess six parameters: perceived difficulty (1-7), enjoyment (1-7), valence (1-9, with 1 being very positive and 9 being very negative), arousal (1-9), desired changes to ball speed (-2 to 2, where -2 means decrease by 2 levels), and desired changes to paddle size (-2 to 2, where 2 means increase by two levels). It should be noted that the order of difficulty settings was preset, and that the participant's desired changes to the ball speed and paddle size were not actually used to adapt difficulty.

Contribution of this study: The perceived difficulty, enjoyment, valence and arousal obtained from the questionnaires were classified into either two (low/high) or three (low/medium/high) classes based on all combinations of the three recorded datasets (performance, physiology, intrinsic factors). All three datasets also included the current ball speed and paddle size. For 2-class classification, the class "low" was defined as 1-3 for all categories while the class "high" was defined as 5-7 for perceived difficulty and enjoyment or 7-9 for valence and arousal. In 3-class classification, the low, medium and high ranges were 1-2, 3-5 and 6-7 for perceived difficulty and enjoyment; they were 1-3, 4-6 and 7-9 for valence and arousal. Similarly, the participants' desired changes to game difficulty settings were mapped into either two (increase/decrease) or three (increase/no change/decrease) classes. For both ball speed and paddle size, 1 and 2 were mapped to the class "increase" while -1 and -2 were mapped to "decrease". For two-class classification, the "no change" class was dropped.

As a basis for classification into two or three classes, we first used the stepwise feature selection algorithm [13] to find the most informative set of features. Then, three different classifiers (SVM with a linear kernel, decision tree, or ensemble decision tree) were used to classify combinations of the different datasets (performance, physiology, intrinsic factors) into two or three classes for each of the six possible outcome variables (perceived difficulty, enjoyment, valence, arousal, desired ball speed and paddle size change) separately. The classifiers were validated using 10-fold crossvalidation method using all available data points, independently of which participant they were measured from.

3 Results

Table 1 presents the 2-class classification accuracies for all combinations of the input datasets. The highest accuracy is obtained for classification of desired changes of ball speed using physiological measurements. For the other five classification cases, the combination of all datasets yields the most accurate classifier, with the lowest classification accuracy (86.9%) obtained for perceived difficulty.

Table 2 presents the 3-class classification accuracies for all combinations of the input datasets. The highest accuracy is obtained for emotional valence using the combination of all datasets. Physiology yielded the most accurate classifier for three of the six classification cases; the other three classification cases, the combination of all three datasets yielded the highest classification accuracy. The lowest classification accuracy was obtained for the desired paddle size change.

Table 1. Two-class classification accuracies for all combinations of datasets. If the classification method is not mentioned, the support vector machine was used. (Ph: Physiology, In: Intrinsic factors, Pe: Performance, *: Ensemble decision tree used)

Classification cases	Ph	In	Pe	Ph & In	Ph & Pe	In & Pe	All
Difficulty level	85.8%	*83.5%	*79.1%	85.7%	86.6%	*83.0%	**86.9%**
Enjoyment	**87.8%**	*80.9%	73.3%	85.7%	85.4%	*81.7%	86.5%
Valence	89.2%	*91.1%	88.8%	92.8%	93.9%	*93.0%	**93.9%**
Arousal	89.0%	*87. %	76.8%	88.8%	87.7%	*86.3%	**89.4%**
Speed change	**98.4%**	*97.2%	92.0%	96.5%	97.9%	*95.3%	96.6%
Paddle size Change	**98.3%**	92.2%	91.4%	97.5%	98.1%	92.4%	97.8%

Table 2. Three-class classification accuracies for all combinations of datasets. If the classification method is not mentioned, the support vector machine was used. (Ph: Physiology, In: Intrinsic factors, Pe: Performance, *: Ensemble decision tree used)

Classification cases	Ph	In	Pe	Ph & In	Ph & Pe	In & Pe	All
Difficulty level	76.6%	*70.0%	65.2%	77.4%	77.9%	*71.1%	**81.1%**
Enjoyment	68.8%	*69.1%	51.8%	70.1%	65.5%	64.8%	**71.4%**
Valence	70.7%	*67.0%	55.6%	75.6%	73.3%	*66.3%	**76.2%**
Arousal	**68.8%**	*67.9%	54.8%	67.8%	66.7%	*63.3%	66.7%
Speed change	80.0%	*77.0%	*70.7%	80.0%	77.4%	*78.5%	**81.5%**
Paddle size change	74.1%	*72.2%	55.9%	75.6%	72.2%	73.0%	**78.1%**

4 Discussion

The obtained results compared the classification accuracy of game players' psychological states using all combinations of physiological signals, performance, and intrinsic factors. The physiological dataset was the most informative of the three individual datasets, and the combination of all three datasets yielded the best accuracy for 8 of the 12 classification cases. As the classifiers are highly accurate, our next step will be to use them in a real-time manner: the participant's psychological state will be classified, and the game will then adapt its difficulty in a way that is expected to increase player motivation. Since the classifiers are not computationally demanding, a real-time version of the classification procedure is feasible.

Prior to real-time implementation, the training dataset should be expanded to include more than three possible discrete values of ball speed and paddle size, thus allowing the psychological state classification to also be useful for very high and very low difficulties. Furthermore, it may be possible to further increase classification accuracy and improve the user experience by including a history of previous difficulty levels and psychological states that they evoked, thus allowing the computer to estimate how participants reacted to certain difficulty levels in the past.

5 Conclusions

In this study, three sets of classifiers are used to classify four affective/cognitive states of Pong game players into either 2 or 3 classes. The proposed classifiers can also determine how participants would like to change the game difficulty to make it more fun. Three data sets (physiological signals, game performance, and intrinsic factors) are used as the input of the classifiers. Among the 2-class classifiers, the highest accuracy was obtained for desired ball speed change (98.4%) while the lowest was obtained for perceived difficulty level (86.9%). Among the 3-class classifiers, the highest accuracy was obtained for desired ball speed change (81.5%) while the lowest was obtained for psychological arousal (68.8%). As the next steps, additional improvements will be made to increase the classifiers' robustness, and the classifiers will then be used to adapt game difficulty in response to players' psychological states, thus improving the gameplay experience.

Acknowledgments. Research supported by the National Science Foundation under grant no. 1717705 as well as by the National Institute of General Medical Sciences of the National Institutes of Health under grant no. P20GM103432.

References

1. Tan, C.H., Tan, K.C., Tay, A.: Dynamic Game Difficulty Scaling Using Adaptive Behavior-Based AI. IEEE Trans. Comput. Intell. AI Games **3**, 289–301 (2011)
2. Zhang, X., Lyu, Y., Hu, X., Hu, Z., Shi, Y., Yin, H.: Evaluating Photoplethysmogram as a Real-Time Cognitive Load Assessment during Game Playing. Int. J. Human-Computer Interact. **34**, 695–706 (2018)
3. Darzi, A., Gaweesh, S.M., Ahmed, M.M., Novak, D.: Identifying the Causes of Drivers' Hazardous States Using Driver Characteristics, Vehicle Kinematics, and Physiological Measurements. Front. Neurosci. 12, (2018)
4. Ma, J., Zhang, Y., Cichocki, A., Matsuno, F.: A Novel EOG/EEG Hybrid Human-Machine Interface Adopting Eye Movements and ERPs: Application to Robot Control. IEEE Trans. Biomed. Eng. **62**, 876–889 (2015)
5. Rodriguez-Guerrero, C., Knaepen, K., Fraile-Marinero, J.C., Perez-Turiel, J., Gonzalez-de-Garibay, V., Lefeber, D.: Improving Challenge/Skill Ratio in a Multimodal Interface by Simultaneously Adapting Game Difficulty and Haptic Assistance through Psychophysiological and Performance Feedback. Front. Neurosci. 11, (2017)
6. Picard, R.W., Vyzas, E., Healey, J.: Toward machine emotional intelligence: analysis of affective physiological state. IEEE Trans. Pattern Anal. Mach. Intell. **23**, 1175–1191 (2001)
7. Goršič, M., Cikajlo, I., Novak, D.: Competitive and cooperative arm rehabilitation games played by a patient and unimpaired person: effects on motivation and exercise intensity. J. Neuroeng. Rehabil. **14**, 23 (2017)
8. Reuderink, B., Mühl, C., Poel, M.: Valence, arousal and dominance in the EEG during game play. Int. J. Auton. Adapt. Commun. Syst. **6**, 45 (2013)
9. Kim, T.T., Lee, G.: Hospitality employee knowledge-sharing behaviors in the relationship between goal orientations and service innovative behavior. Int. J. Hosp. Manag. **34**, 324–337 (2013)

10. Carver, C.S., White, T.L.: Behavioral inhibition, behavioral activation, and affective responses to impending reward and punishment: The BIS/BAS Scales. J. Pers. Soc. Psychol. **67**, 319–333 (1994)
11. Hsia, L.-H., Huang, I., Hwang, G.-J.: Effects of different online peer-feedback approaches on students' performance skills, motivation and self-efficacy in a dance course. Comput. Educ. **96**, 55–71 (2016)
12. Gosling, S.D., Rentfrow, P.J., Swann, W.B.: A very brief measure of the Big-Five personality domains. J. Res. Pers. **37**, 504–528 (2003)
13. Darzi, A., Gorsic, M., Novak, D.: Difficulty adaptation in a competitive arm rehabilitation game using real-time control of arm electromyogram and respiration. In: 2017 International Conference on Rehabilitation Robotics (ICORR). pp. 857–862. IEEE (2017)
14. Klem, G.H., Lüders, H.O., Jasper, H.H., Elger, C.: The ten-twenty electrode system of the International Federation. The International Federation of Clinical Neurophysiology. Electroencephalogr. Clin. Neurophysiol, Suppl (1999)

Deployment of a Mobile Wireless EEG System to Record Brain Activity Associated with Physical Navigation in the Blind: A Proof of Concept

Christopher R. Bennett[1(✉)], Laura Dubreuil Vall[2], Jorge Leite[3], Giulio Ruffini[2], and Lotfi B. Merabet[1]

[1] The Laboratory for Visual Neuroplasticity, Department of Ophthalmology, Massachusetts Eye and Ear Infirmary, Harvard Medical School, Boston, MA, USA
Christopher_bennett@meei.harvard.edu
[2] Neuroelectrics Corporation, Cambridge, USA
[3] Neuromodulation Center, Spaulding Rehabilitation Hospital, Harvard Medical School, Boston, USA

Abstract. Little is known about how the brain processes information while navigating without visual cues. Technical limitations recording brain activity during real-world navigation have impeded research in this field. We have developed a study paradigm that benefits from wireless EEG recording technology. Participants heard a sequence of directional commands instructing them to physically or mentally navigate a 3 × 3 m grid. Data from a sighted control and an individual with profound blindness highlight the viability of the technology. A power spectral density analysis on the alpha frequency band during the physical navigation task revealed diffuse signal fluctuations for the blind participant, while a more robust signal within occipital-parietal regions was seen for the sighted control. Both participants displayed highly similar signal fluctuations during mental navigation. This work demonstrates the feasibility of brain activity recording during navigation-related tasks using a wireless EEG system for identifying brain processing patterns associated with visual experience.

Keywords: Mobile EEG · Visual impairment · Navigation

1 Introduction

Individuals living with profound blindness face numerous challenges in order to remain functionally independent in a world that relies heavily on sight. For example, an individual with congenital blindness must be able to navigate without using visual spatial cues or rely on prior visual experience [1]. Current evidence suggests that blind individuals rely on non-visual sensory information (e.g. from hearing and touch) to generate a cognitive spatial map of their surroundings [2]. However, the neural correlates associated with carrying out this task remain largely unknown.

W. Karwowski and T. Ahram (Eds.): IHSI 2019, AISC 903, pp. 30–36, 2019.
https://doi.org/10.1007/978-3-030-11051-2_5

Previous efforts have attempted to identify brain areas associated with navigation in the blind using functional magnetic resonance imaging (fMRI) [3, 4]. These studies revealed that tasks requiring participants to mentally navigate through a virtual maze (using tactile and/or auditory cues) recruit a large network of brain regions including the frontal cortex, hippocampus, and temporal parietal junction [3, 4]. Intriguingly, mental navigation in the blind also appears to implicate the occipital cortex, that is, the region of the brain normally associated with the processing of visual information. These aforementioned studies have been helpful in identifying key brain regions involved with spatial processing associated with mental navigation. However, the constraints and confines of the scanner environment limit the generalizability of these findings as they relate to real-world physical navigation.

There has been an interest to develop methods to record brain activity directly and capture the neural correlates associated with free physical movement and in an unrestricted manner [5]. Wireless mobile electroencephalography (EEG) recording systems may be ideally suited for this purpose. Unlike fMRI, EEG records the brain's underlying electrical activity directly using electrodes placed on the surface of the scalp and with high temporal resolution [6, 7].

To our knowledge, only one study has used EEG to characterize the electrophysiological correlates associated with mental navigation in blind individuals [8]. Similar to previous fMRI studies, Kober et al. [8] reported that parietal-occipital brain regions were associated with performing a mental navigation task, and individuals with profound blindness showed desynchronization over these areas compared to normally sighted controls. While these results are helpful in identifying the neural networks implicated with navigation in the case of blindness, a number of important questions still remain. First, while all the blind participants in the study were completely blind, only two were blind due to congenital causes while the remaining six became blind much later in life (i.e. between the ages of 10 to 39). Therefore, the effect of prior visual experience cannot be adequately disentangled. Second, navigation performance was based on tasks requiring only mental imagery. Thus, the neural correlates associated with active physical navigation still remain unknown.

In this study, we developed a protocol incorporating a mobile wireless recording EEG system designed to record data during a real-world physical navigation task. As a proof of concept, we identified and compared the neural correlates associated with performance on a physical and mental navigation task in an individual with congenital profound blindness and a normally sighted control.

2 Methods

2.1 Participants

The blind participant was a 36 year old male born with Leber's congenital amaurosis. Residual functional vision was clinically assessed as light perception in both eyes, and denies having any past visual memories. He is a highly experienced traveler, using a cane as his primary assistive device. The sighted control was a 34 year old male with normal visual acuity (20/20 Snellen) and with no prior history of neurological or cognitive impairment. Written informed consent was obtained prior to participation and the study was approved by the Institutional Review Board at the Massachusetts Eye and Ear, Boston, MA, USA.

2.2 Recording Montage and Data Analysis

EEG data was collected using a wireless 20-channel Enobio system (Neuroelectrics, Barcelona, Spain) with a sampling rate of 500 Hz. Placement of the EEG electrodes corresponded to the standard 10–20 international system. Once the cap was placed on the participant, conductive gel was added to each electrode site to increase conductivity between the electrode and scalp surface. The electrodes were made of a solid gel material and each channel fed to a wireless Bluetooth transmitter. Signals were captured on-line and recorded by software running on a standard laptop computer positioned within 10 meters from the test participant.

The wireless transmitter also houses a motion capture accelerometer unit (using three scaling constants; one for each axis of motion). Information from this unit was used as part of signal post-processing analysis to filter (i.e. mathematically regress) the recorded signals and remove artifacts related to the motion of the head caused during walking. The purpose for signal cleaning was to minimize the signal energy by subtracting scaled versions of the accelerometer data (using a Kalman filter).

EEGLAB toolbox for Matlab was used for further post-processing and analysis [9]. A band pass filter between 0.1 and 35 Hz was applied to the data. Independent component analysis (ICA) was performed on each subject using a *runica* decomposition [10] in order to identify and remove eye blink/movement components from the data. The data was then segmented into 3 s epochs (ranging from 1 s before stimulus onset to 2 s after). Each epoch was baseline corrected using the 1 s immediately prior to cue onset. Epochs that exceeded 100 μV following an automatized artifact rejection tool were signaled for visual inspection before removal.

2.3 Tasks

The study participants were required to carry out two behavioral tasks while wearing the mobile EEG unit. These tasks were experimenter controlled and auditory commands were presented using Presentation software running on a second laptop computer used to trigger the data acquisition laptop device. The presentation and data acquisition computers were synced through the Lab Streaming Layer (LSL) protocol, with a time resolution of 2 ms to provide sufficient resolution for ERP extraction. Both participants completed the tasks while wearing a blindfold.

Navigation Task: Physical Walking. Participants were instructed to stand in a 3×3 m grid (area of 9 m^2) of nine separate positions numbered 1 through 9 (see Fig. 1). The task was comprised of an initial planning (i.e. following auditory commands) stage, followed by an active walking (physical execution) stage requiring a turn decision to be made. On each trial, the participant would start at the central position of the grid (position “5” and facing position “2”). Following a series of 4 auditory commands, there was a another verbal cue (“walk”) with a 2 s duration, and the participant then had 3 s to walk towards the target grid point. There were six possible auditory commands: right, left, straight, turn (i.e., 180°) and then right, turn and then left, or turn and then move straight forward. There were 64 trials in total. Each trial consisted of four turns, in which participants could finish either at the initial starting point (i.e. closed loop route) or at any other point of the grid (i.e. open loop route). At the end of the sequence, participants were asked to verbally report their position on the grid. Behavioral performance was recorded in both participants to ensure that they understood the task.

Fig. 1. Top-down view of the 3 × 3 grid participants mentally and physically navigated

Navigation Task: Mental Imagery. The mental imagery task was identical to the physical walking task, except for that participants were instructed to mentally navigate in the grid after the turn commands.

3 Results

Each participant completed both tasks with near perfect performance (average of 98% and 97% for the blind and sighted participant respectively). Following signal post-processing, data for the mental and physical walking navigation tasks were analyzed using power spectral density. 100% of the processed data was included and head plots were generated at 2 Hz intervals between 6 Hz and 14 Hz to encompass and isolate the alpha frequency band (8–13 Hz). The resulting spectral plots for the physical navigation task are shown in Fig. 2 for the individual with profound blindness (top) and sighted control (bottom). The spectral analysis revealed diffuse signal fluctuations in the alpha frequency band (8–13 Hz) for the blind participant localized within occipital-parietal brain regions during the physical navigation task. In the sighted control however, activation was more robust and focused within the occipital pole (corresponding to early visual processing areas).

Power spectral density analysis for the mental imagery task revealed that both participants exhibit highly similar signal fluctuations that appeared strongest in occipital-parietal cortical areas. The resulting spectral plots for the individual with profound blindness (top) and sighted control (bottom) are shown in Fig. 3.

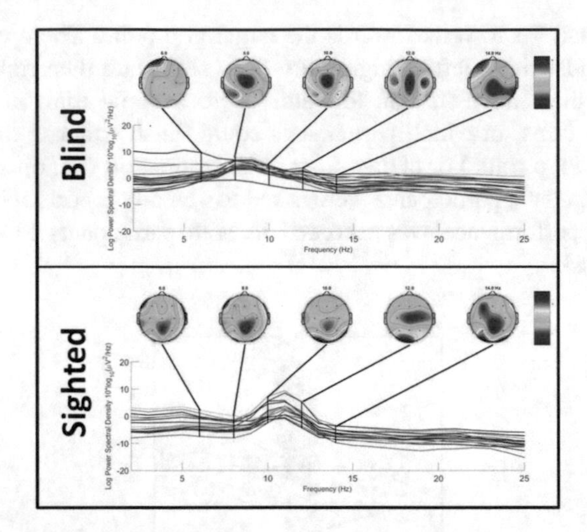

Fig. 2. Power spectral density plot for the blind (top) and sighted control (bottom) participants during the physical walking navigation task

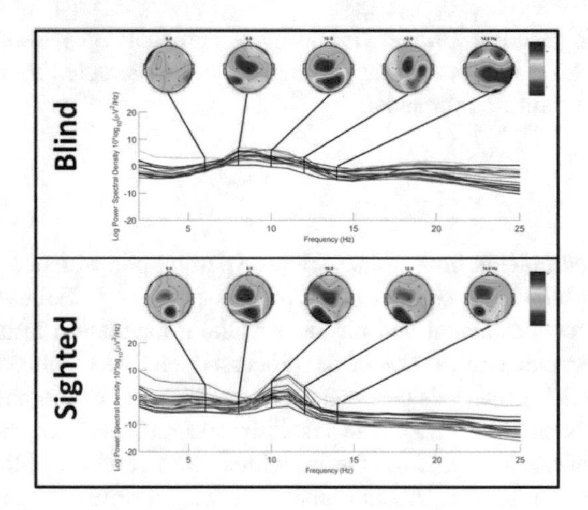

Fig. 3. Power spectral density plot for the blind (top) and sighted control (bottom) participants during the mental "imagined" walking navigation task

4 Discussion

The results of this study demonstrate that the use of a wireless recording apparatus and signal post processing pipeline allow for the characterization of brain activity associated with navigation and comparing performance in blind and sighted individuals. Therefore, there is the potential to characterize the neural correlates associated with navigation in an unrestricted fashion; a feat that previously was not possible with other neuroimaging modalities such as fMRI.

Similar to previous neuroimaging studies [3, 4, 8], our results suggest that navigation is associated with parietal-occipital regions of the brain. In our study, this was evident during both the mental and physical navigation tasks. Interestingly, while signal fluctuation during the physical navigation task was found predominantly within early visual areas in the sighted control, it was seen within higher order visual areas in the blind participant. This effect was not observed during the mental navigation task where activation within higher order visual areas was observed in both participants. Of note, the peak alpha frequency was somewhat higher and sharper in the sighted control than in the blind participant. This trend could be linked to potential memory performance [11] and/or physical fatigue [12]. Replication of this study in a larger sample of participants will be needed in order to disentangle this issue.

Studying how the brain adapts within the setting of profound vision loss reveals important insights regarding the brain's ability to adapt to sensory deprivation [13]. That is, by understanding how the brain develops and adapts in the setting of visual deprivation, we can gain insight into how to develop novel and more effective rehabilitative approaches and assistive technology for the blind. Refinement of this technology will help further contribute to our understanding of how the brain changes and adapts to the loss of sight and in relation to the development of real word compensatory skills. Furthermore, this same platform technology can serve to investigate brain activity in other behavioral tasks of interest, and in other conditions of visual impairment (such as hemianopia) or causes (e.g. comparing ocular versus cerebral causes of blindness).

Acknowledgements. This work was supported by the Knights Templar Eye Foundation and the National Institutes of Health (R01 EY019924-08).

References

1. Loomis, J.M., Golledge, R.G., Klatzky, R.L.: Navigation system for the blind: auditory display modes and guidance. Presence **7**(2), 193–203 (1998)
2. Lahav, O., Mioduser, D.: Construction of cognitive maps of unknown spaces using a multisensory virtual environment for people who are blind. Comput. Hum. Behav. **24**(3), 1139–1155 (2008)
3. Kupers, R., Chebat, D.R., Madsen, K.H., Paulson, O.B., Ptito, M.: Neural correlates of virtual route recognition in congenital blindness. Proc. Natl. Acad. Sci. **107**(28), 12716–12721 (2010)
4. Halko, M.A., Connors, E.C., Sánchez, J., Merabet, L.B.: Real world navigation independence in the early blind correlates with differential brain activity associated with virtual navigation. Hum. Brain Mapp. **35**(6), 2768–2778 (2014)
5. Cruz-Garza, J.G., et al.: Deployment of mobile EEG technology in an art museum setting: evaluation of signal quality and usability. Front. Hum. Neurosci. **11**, 527 (2017)
6. Luck, S.J.: An Introduction to the Event-Related Potential Technique. MIT press, Cambridge, pp. 45–64 (2005)
7. Lopez-Calderon, J., Luck, S.J.: ERPLAB: an open-source toolbox for the analysis of event-related potentials. Front. Hum. Neurosci. **8**, 213 (2014)

8. Kober, S.E., Wood, G., Kampl, C., Neuper, C., Ischebeck, A.: Electrophysiological correlates of mental navigation in blind and sighted people. Behav. Brain Res. **273**, 106–115 (2014)
9. Delorme, A., Makeig, S.: EEGLAB: an open source toolbox for analysis of single-trial EEG dynamics including independent component analysis. J. Neurosci. Methods **134**(1), 9–21 (2004)
10. Makeig, S., Jung, T.-P., Bell, A.J., Ghahremani, D., Sejnowski, T.J.: Blind separation of auditory event-related brain responses into independent components. Proc. Natl. Acad. Sci. **94**(20), 10979–10984 (1997)
11. Angelakis, E., Lubar, J.F., Stathopoulou, S.: Electroencephalographic peak alpha frequency correlates of cognitive traits. Neurosci. Lett. **371**(1), 60–63 (2004)
12. Ng, S.C., Raveendran, P.: EEG peak alpha frequency as an indicator for physical fatigue. In: 11th Mediterranean Conference on Medical and Biomedical Engineering and Computing 2007. Springer (2007)
13. Merabet, L.B., Pascual-Leone, A.: Neural reorganization following sensory loss: the opportunity of change. Nat. Rev. Neurosci. **11**(1), 44 (2010)

The Effects of Culture on Authentication Cognitive Dimensions

Mona A. Mohamed(✉)

e-Business and Technology Management, Towson University, 8000 York Road, Towson, MD, USA21252-000
mmohamed@towson.edu

Abstract. The purpose of this research is to answer the question: does selecting images from one's culture improve the memorability of Recognition-Based Graphical Password (RBG-P)? The results show that the failure rate of authentication increases through the progression of consecutive three phases. The findings also suggest that cultural groups with higher number of images selected from own culture, inclined to experience lower failure rates and vice versa. In fact, the ISGs that selected from their own culture showed approximately half the means of failure of the groups that selected from other cultures. Resultantly, culture has significant effects on the password memorization, therefore, the designer must be motivated to provide cross-cultural interfaces that reduce the risk and improve usability of RBG-P which minimize the frustration associated with login failures. The quick depreciation of the success rate with the progress of the phases suggests the possibility of memory decay due to temporal effects.

Keywords: Culture · Deflection point · Mental model · Memorability Confounded effects · Memory decay · Trade-off • image sequence Culturally-oriented memorization

1 Introduction and Background

There are inherent vulnerabilities that exist in the alphanumeric password authentication due to the lack of entropy [1], to minimize these vulnerabilities, graphic passwords have been developed. Cognitive password such as Recognition-Based Graphical Password (RBG-P) as a "knowledge-based" authentication has been introduced as alternative to the alphanumerical password due to the fact that human remember images far better than text ([2–7]). On the other hand, [8] and [9] attributed this gained security posture to users finding recognition easier to do than recall. Nevertheless, it is always a dilemma that faces users and security professionals on how to create a password that is difficult to be cracked or guessed by attackers, but at the same time easy for the user to remember. This is a challenging usability and security tradeoff. Memorability is the main characteristics of usability that leads the user to create weak password as has been concluded by [10] that human memory limitation is the main reason for selecting weak password. It has been corroborated by many investigators that graphic interfaces lead to higher usability levels. The authors attributed this to the

W. Karwowski and T. Ahram (Eds.): IHSI 2019, AISC 903, pp. 37–42, 2019.
https://doi.org/10.1007/978-3-030-11051-2_6

fact that graphical images increase password memorability through users' visual memory. Recognition-based techniques such as RBG-P is visual recognition memory that is employed to recognize previously selected during registration phase. However, the sequencing of these images is determined by recalling and not by recognition.

Drawing on a review of literature concerning mental models, [11] found that memorability rate of RBG-P that belongs to the users' culture was much higher than the rate of memorability of RBG-P images that do not belong to the user's culture. The author affirmed that it is well known that users from dissimilar cultures have different mental models, therefore, the significance of cross-cultural design should never be underrated. [12] defined culture as "*the collective programming of the mind distinguishing the members of one group or category of people from others*".

The mental model of both the user and the designer are of great significance to their decisions and their interactions with any User Interface (UI). More often, the mental model is a result of tacit knowledge internalization and culture that affect the user thinking about how the system works. Based on the Mental Model Theory (MMT), [13] attributes human reasoning to the construction and manipulation of mental models. Nevertheless, the validity of these claims has been called into enquiry as graphical password came with its own ambiguities and challenges, especially those related to cultural contexts.

This research is designed to test the following hypothesis:

H_1: μFLAF (First Login Attempt Failure) = μSLAF (Second Attempt Login Failure) = μTLAF (Third Attempt Login Failure)

2 Methods

A between group design is used for 50 participants; 25 from China and 25 from Kingdom of Saudi Arabia to measure the effects on RBG-P authentication. The experiment was carried out over a period of 5 weeks. During the enrollment phase the participants were asked to select 5 images in sequence from a set of 90 images in the webtool develop by the author. 30 of these images from Chinese culture, 30 images from Saudi culture and 30 images are neutral with no reference to any specific culture. Besides the image selection this study added more complexity to the RBG-P by enforcing the sequence of the images. This sequence is expected to make the guessing more difficult to the attacker. On the other hand, the memorability of the password is measured across three phases (times) as images selected and memorized by Image Selection Groups (ISGs). The authentication process conducted three times and pre- and post-assessment questionnaire were conducted using Likert Scale (1-5), where 1 = Strongly Disagree, 2 = Agree, 3 = Neither Disagree nor Agree, 4 = Agree and 5 = Strongly Agree.

3 Results and Discussion

The exploratory analysis of the success rate during the three phases of the study show that Phase I has more participants with less failure compare to phase II and Phase III. Although there do not appear to be any differences in number of participants who failed during the first phase, there is an obvious trend that participants made more failed attempts to log in during phase II and much higher during phase III compared to phase I and phase II (i.e., 1.04, 1.32 and 1.84 respectively). These differences can be attributed to the time effects on the memory of the users during the three phases. For better evaluation of cultural effects, the image selection has been categorized into six cultural Image Selection Groups (ISGs) as show in Table 1 below regardless of the national culture. These six groups run from A to F depending on how many images chosen from own culture. Table 1 clearly indicates that groups that selected images from their own culture have higher numbers.

Table 1. Image selection groups (ISGs) categorization based on the type of cultural image from which the respondents selected their RBG-P

Group	Description	# Participants
Group A	Participants select 5 images from own culture	21
Group B	Participants select 4 images from own culture	6
Group C	Participants select 3 images from own culture	9
Group D	Participants select 2 images from own culture	5
Group E	Participants select 1 image from own culture	4
Group F	Participants select 5 images from other images	5

The multivariate analysis constructs the model by carrying out between-group design and their behavior towards login attempt failure. Table 2 demonstrates that there are significance differences between images selection cultural groups). Hence, we reject Null hypothesis that μFPF = μSPF = μTPF. We conclude that there are significant differences of the login failures due to ISGs behavior and the time of selection (phases). Hence, culture has a significant effect on the RBG-P authentication usage.

The behavior of the ISGs shows a very interesting pattern as depicted in Fig. 1. In general, it was found that participants with higher number of images selected from own culture, resulted in lower failure rates and vice versa. In fact, the groups that selected from their own culture showed approximately half the means of failure of the groups that selected from other cultures. However, the failures do not show large differences across groups during the first phase, while the last phases showed the highest variability. These results are motivation for the designers to provide cross-cultural interfaces that reduce the security risks and the frustration associated with login failures. The quick depreciation of the success rate with the progress of the phases suggests the possibility of *memory decay* due to time effects.

This suggests that culture has significant effects on the memorization of the password which includes both recognition and recalling of the images and their sequence.

Table 2. The *between groups* and the model *intercept* and the ISGs effects in different *phases*

Between group					
BTest	Value	Exact F	DF	DenDF	Prob > F
F Test	0.4579	4.0295	5	44	0.0043*
Intercept					
Test	Value	Exact F	DF	DenDF	Prob > F
F Test	1.4622	64.3399	1	44	<0.0001*
Groups					
Test	Value	Exact F	DF	DenDF	Prob > F
F Test	0.4579	4.0295	5	44	0.0043*
Phases					
Test	Value	Exact F	DF	DenDF	Prob > F
F Test	0.4483	9.6385	2	43	0.0003*

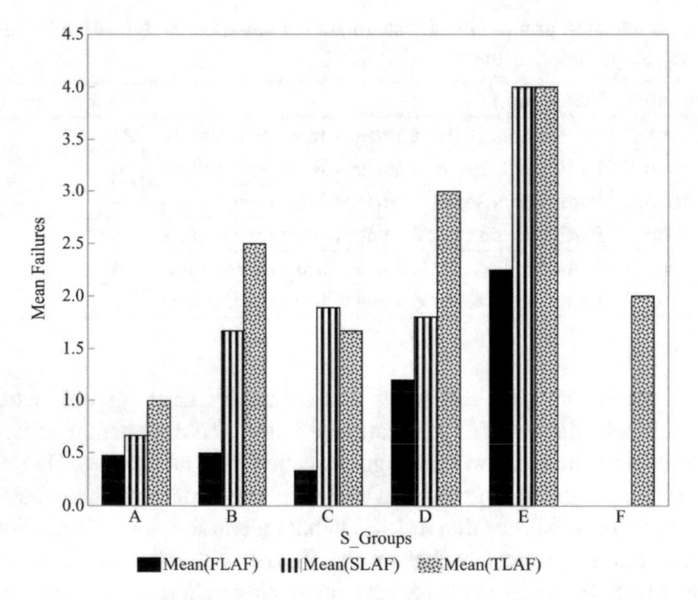

Fig. 1. Mean failures for image selection groups (see Table 1 for the groups assignments)

To visually represent the cultural effects, the participants have been divided into two culturally-categorized groups viz. those who selected from their own culture of 3 images and more; and those who selected less than three images from their own culture. Figure 2 shows the first phase had almost the same performance of success across groups. The differences also started to appear in the second phase and clearer in the third phase, this greatly reflects the effects of time on memorization of the selected images. There is a gradual successive decrease of the mean success in the selection per each group during the three phases of login. It is noteworthy, that the first phase images selection of the both groups shows no mean differences which is indicative to absence of time effects as it is conducted in the same time of the password creation. However, the differences started to appear in the second phase and clearer in the third phase due to time effects on memorization. In addition, the differences between phases within each group are also vastly different. It is obvious that the means of selection from other cultures shows higher differences between phases compared to the means of selections from the same culture. Again, culture has a significant effect on RBG-P memorization. In summary, apparently there are two effects: (1) the effect of the time period; and (2) the effects of the culture. This is another motivation for the designers to provide cross-cultural interfaces that reduce the frustration and the risk associated with login failures.

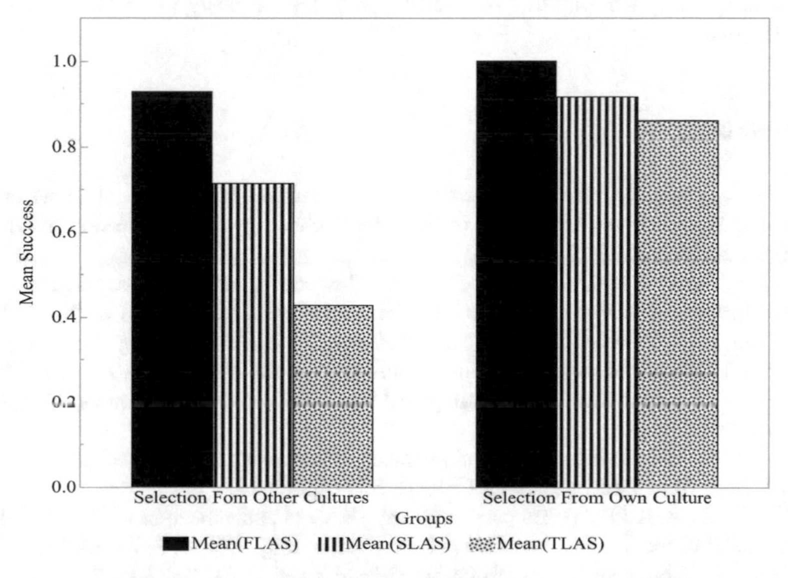

Fig. 2. The *mean success rate* of selection of images at the three login attempts for the two groups

It can be concluded from the above results that culture has clear effects on the memorization of the password, which includes both recognition and recalling of the images and their sequence, respectively. Additionally, there is a temporal effect on the user memorization of the RBG-P across the three phases. Accordingly, designing RBG-P within the context of these considerations is important to align with the cognition and perception of the user in different cultures. Moreover, values for these two variables can be used in determining the policy for the renewal period of the password. Furthermore, it was found in this research that one of RBG-P memorization strategies is to relate the password to previously encountered occurrences. Hence, it is important to narrow the delta between the designer and the user's mental model. Also, the designer must be able to estimate the maximal password complexity or the entropy, to avoid enforcing the user to write the password in a paper. Writing the password down is known to security professional 'something you have' which is weak compared to memorizing it or 'something you know' which stronger. The point at which this transformation occurs is described by the author as the '*deflection point*'. It is noteworthy that, the *deflection point* is purely a user tacit knowledge, the designer must elicit it to determine the maximum entropy. This occurs only if the delta of mental models of the user and the designer is at its minimal.

The sequence, on the other hand, is a test for memorability that may enforce the user to choose an easy sequenced RBG-P. However, the sequence will be truly effectives in reflecting the cognitive dimension as a result of user experience i.e. "*Culturally-Oriented Memorization*". For instance, the user may select a very easy sequence, but very personal that fits the daily activities such as time for prayers, sport, political events, meals etc. which makes it expensive for attackers to guess. The cross-cultural design may use this line of thinking into RBG-P design to make the password more effective.

References

1. Jingbo, Y., Pingping, S.: A secure strong password authentication protocol. Paper Presented at the 2nd International Conference on Software Technology and Engineering (ICSTE), vol. 2, pp. V2-355–V352-357 (2010)
2. Lashkari, A., Farmand, S.: A survey on usability and security features in graphical user authentication algorithms. Int. J. Comput. Sci. Netw. Secur. (IJCSNS) **9**(9), 196–205 (2009)
3. Lashkari, A.H., Zakaria, O.B., Farmand, S., Saleh, R.: Shoulder surfing attack in graphical password authentication. (IJCSIS) Int. J. Comput. Sci. Inf. Secur. **6**(2), 145–154 (2009)
4. Suo, X.: A design and analysis of graphical password. (Master of Science), Georgia State University (2006)
5. Barate, A.K., Shinde, S.S.: Graphical password system using different techniques–a review. Int. J. Eng. Trends Technol. (IJETT) **9**(11), 536–539 (2014)
6. Sahu, S., Singh, A.: Enhanced user graphical password authentication with an usability and memorability. Int. J. Adv. Res. Comput. Sci. Softw. Eng. **5**(6), 477–484 (2015)
7. Bhusari, V.: Graphical authentication based techniques. Int. J. Sci. Res. Publ. **3**(7), 31–38 (2013)
8. Rasekgala, M.; Ewert, S.; Sanders, I., Fogwill, T.: Requirements for secure graphical password schemes. Paper Presented at the IST-Africa Le Meridien Ile Maurice, pp. 1–10 (2014)
9. Bhusari, V.: Graphical authentication based techniques. Int. J. Sci. Res. Publ. **3**(7), 31–38 (2013)
10. Forget, A., Chiasson, S., van Oorschot, P.C., Biddle, R.: Persuasion for stronger passwords: motivation and pilot study. In: Persuasive Technology. Lecture Notes in Computer Science, vol. 5033, pp. 140–150 (2008)
11. Aljahdali, H., Poet, R.: The affect of familiarity on the usability of recognition-based graphical password. Paper Presented at the 12th IEEE International Conference on Trust, Security and Privacy in Computing and Communications, pp. 1528–1534 (2013)
12. Hofstede, G., Bond, M.H.: Hofstede's cultural dimensions: an independent validation using Rokeach's value survey. J. Cross-Cult. Psychol. Mark. **15**(4), 417–433 (1984)
13. Van der Henst, J.-B.: Mental model theory versus the inference rule approach in relational reasoning. Think. Reason. **8**(3), 193–203 (2002)

An Approach on Simplifying the Commissioning of Collaborative Assembly Workstations Based on Product-Lifecycle-Management and Intuitive Robot Programming

Werner Herfs, Simon Storms, and Oliver Petrovic(✉)

Laboratory for Machine Tools and Production Engineering (WZL), Chair of Machine Tools, RWTH Aachen University, Steinbachstr. 19, 52074 Aachen, Germany
{w.herfs, s.storms, o.petrovic}@wzl.rwth-aachen.de

Abstract. Today's trends in the manufacturing industry lead to shorter product-lifecycles and smaller batch sizes with an increasing number of variants in the product range. These trends make it increasingly difficult to implement fully automated production processes economically. One approach that nevertheless makes the advantages of process automation accessible is partial automation through the application of human-robot collaboration (HRC). Small and medium-sized companies, in particular, lack the necessary expertise to successfully implement this technology. Standardized planning systems can bridge these competence gaps. This paper presents a system of this kind. The combination of product-lifecycle-management with collaboration-specific process planning significantly simplifies the commissioning of HRC-processes in dynamic process environments. In addition, a graphical user-interface, makes robot programming more intuitive in order to avoid the tedious training of code-based robot programming.

Keywords: Human-robot-collaboration · Product-lifecycle-management
Intuitive robot programming · Assembly planning

1 Introduction

Contrary to the arguments for the automatization of assembly processes, such as rising quality requirements and labor costs, fully manual assembly is still frequently used in practice [1]. This is mainly due to the usually high number of variants, short product-lifecycles and small batch sizes. These boundary conditions have a negative impact on the economic implementation of fully automated processes, as there is no sufficient payback period available to compensate for the high investment costs. In addition, many companies have a lack of competence in the field of automation technology. One solution is to transform manual assembly into a collaborative assembly process. The skilled worker is supported by a cobot or an automated process unit in a partially automated process [2]. In such a socio-technical system, the assembly partners can

W. Karwowski and T. Ahram (Eds.): IHSI 2019, AISC 903, pp. 43–49, 2019.
https://doi.org/10.1007/978-3-030-11051-2_7

utilize their complementary strengths to compensate for each other's weaknesses. The scalable degree of automation achieved in this way, leads to the fact that partial automation can be justified even with small batches and that the risk of bad investments is reduced. In order to utilize the mentioned potentials, planning-, safety-, acceptance- and competence-related obstacles must first be overcome.

This paper presents a systematic approach to address these challenges. The approach is based on the preliminary work done in [1] and [3] and thus combines two existing planning concepts. On one hand, product-lifecycle-management strategies are used to simplify the implementation of automation processes. On the other hand, an optimized distribution of tasks between man and machine in a collaborative environment is used, which is based on a capability analysis. On this basis a module is built, which first maps the extracted data in standardized data models. The data models serve as the basis for further work preparation, in order to derive a robot program for collaborative applications. The planner or executing worker communicates with the developed module via an intuitive, graphical user interface. The robot program generated in this manner can then be executed and monitored via the developed software.

In summary, the achieved continuous flow of information simplifies the commissioning of collaboration systems in a dynamically changing environment. This enables a wider spectrum of users to benefit from the technology of human-robot-collaboration.

2 State of the Art

Due to the high dynamics in today's production processes, there are numerous individual approaches to make the commissioning of automation solutions more economical, flexible and adaptive. For example, the approach [3] uses product data in a framework to adapt automation processes to dynamic requirements in, thus avoiding dedicated ramp-up phases. Another example is the use of machine-learning to avoid unnecessary iterations during the commissioning of collaboration systems [4]. In addition, the fast and above all simple programming of the robotic system is a basic requirement for a broad and agile use of the technology. There are some solutions directly from robot manufacturers that bypass the previous, mostly code-based generation of the robot program. An example is the graphical user interfaces of the Panda of Franka Emika GmbH [5]. Such systems reduce the competence hurdle for the programmer enormously. As an example for a hardware-based solution, [6] describes a teach-in tool that allows intuitive teach-in of robot TCP-positions. Current safety standards such as DIN ISO/TS 15066 [7] and employee acceptance of collaboration with the robot are further factors influencing the efficiency of the process. According to [8], the worker's acceptance is primarily achieved by a process that is transparent to the human. The actions of the robot must therefore always be comprehensible for the worker.

3 Systematic Approach

Product-lifecycle-management, competence-based task assignment and intuitive robot programming are the key components of the developed planning system. In this way, various already existing strategies are combined in a comprehensive, integrated system, which is visualized in Fig. 1.

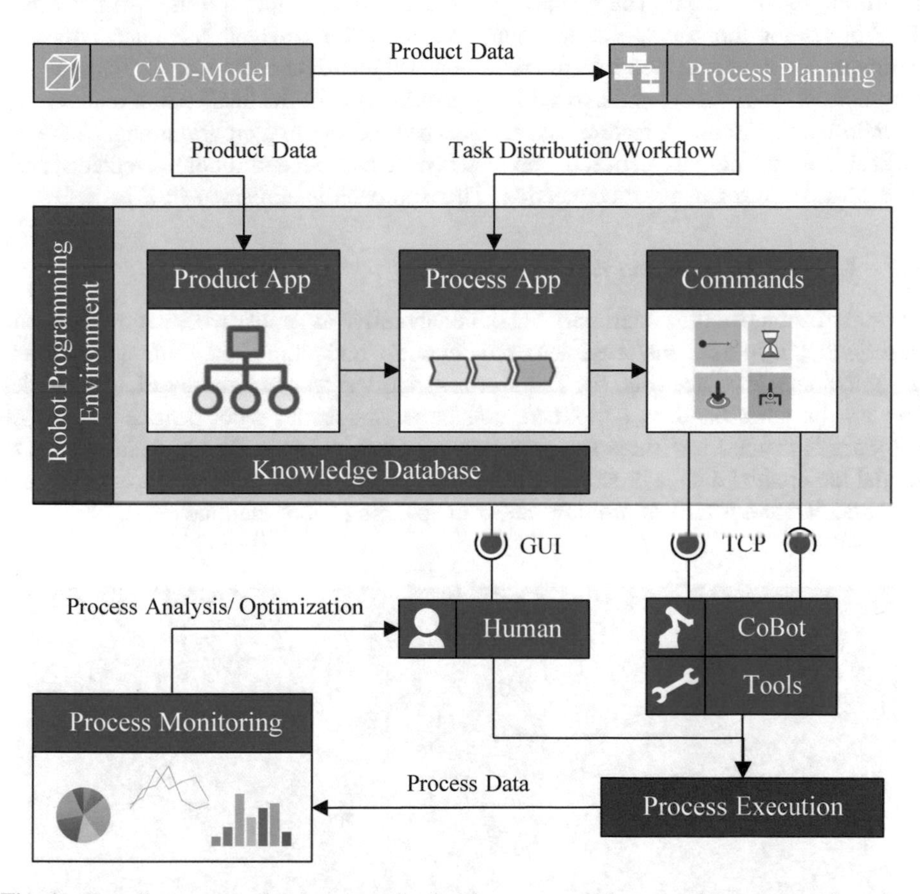

Fig. 1. Overall structure of the systematic planning approach for assembly processes based on human-robot-collaboration

After being enriched, the product data from the CAD-models of the components provide the foundation for the planning system. First, the product structure can be derived which can be further used for the planning of the respective assembly process. Other data types that can be exported from the CAD-model for process planning are physical product data and feature data. This information can be used to deduce the component masses, required forces and torques as well as other process-relevant characteristics. On the basis of this data, a competence-based distribution of tasks between humans and robots can be carried out and optimized in the course of process planning. As described in [1], the capabilities of the process resources are compared with the requirements of the process steps and potential allocations are determined on the basis of this comparison. The optimal distribution is then calculated in a downstream optimization algorithm. The process plan can be transferred to the robot programming software via an interface that has to be implemented. In addition to the process planning unit, the product data is also transferred directly to the robot programming environment. The geometric data can, for example, be used to determine TCP-poses for the robot. For this purpose, component-related reference points or geometries are transformed into the robot's world coordinate system. The coordinates can then be used, for example, as gripping positions or for the localization of assembly operations. Furthermore, feature data can also be used in robot programming. These are utilizable for the derivation of tool parameters or feature-related robot movements, such as the feed rate for a screw connection. This approach is explained in [3].

3.1 Robot Programming Environment

In order to simplify the commissioning of collaborative assembly workstations, the data described above is transferred into data models and combined with an intuitive, graphical user interface (Fig. 2). The user is supported on one hand by PLM strategies and on the other hand by simplified, collaboration-specific robot programming. The software is divided into three sub-applications: a product-related part, which serves to model the product data, a process-related part, in which the process and its environment are modeled and a tool to program the corresponding robot commands.

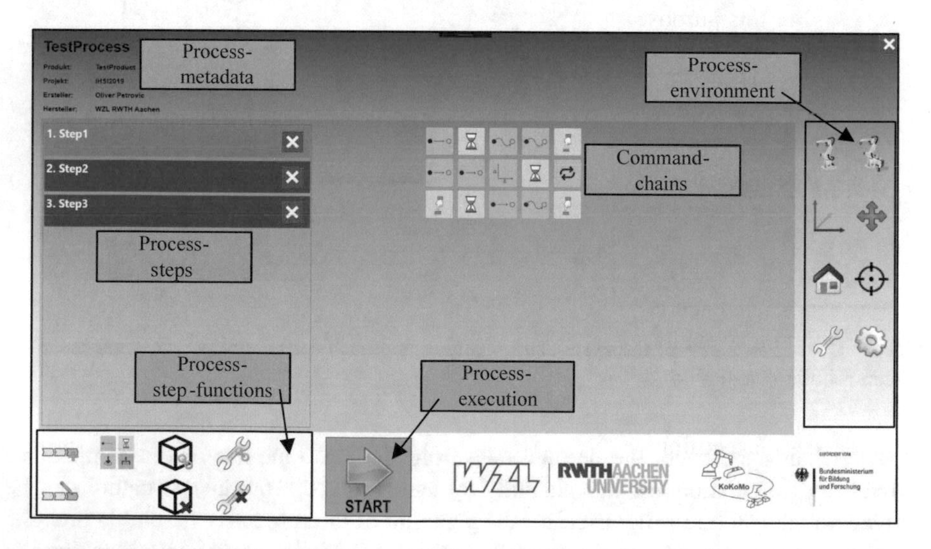

Fig. 2. Implemented graphical user-interface of the robot programming environment

Product Application. As already mentioned, in this part the product data is modeled and thus made usable. This enables a transfer of knowledge from product design to work preparation. Initially, the metadata of the product can be defined. The metadata is primarily used to precisely identify and categorize the product. Once this foundation of the model has been established, product components with their individual properties can be added successively. These components are then linked to each other via interfaces. This procedure creates a complete product structure that can be used as a data model. In particular, the geometry of the components and the position and nature of the interface can be used to subsequently define the robot commands.

Process Application. Once the product model has been completely defined, the correlated assembly process can be modelled on this basis in order to enable a knowledge transfer on the process side as well. First, the individual process steps are created and the components used in this step are linked. The link provides the relevant component and interface information to define the robot commands for each process step. In addition to generating the process sequence, the relevant aspects of the process environment for robot programming can also be created. The data models of the robot, tools and reference coordinate systems can be generated and then filled with information. For example, the kinematic limits of the robot defined in this way serve as the adjustable minima and maxima in the creation and parameterization of robot commands. The created tool functions, however, can be used as additional commands to make these functionalities programmable and integratable into the command-chains.

Robot Command-Chain. Once both the product and the relevant process components have been modeled in sufficient depth, executable robot command-chains can be generated. For this purpose, the user accesses a set of basic commands. The associated user-interface is graphically intuitive, which results in a low competence barrier and a steep learning curve. The most elementary form of commands are the movement commands, which enable the common robot movement types. Buffer times can be inserted by using waiting commands. The robot waits either for a defined time frame or awaits a specific feedback from the operator. The touch of the robot can also be selected as the input, which further increases the degree of interaction and thus the acceptance of the worker. In addition, the previously created process tools can be controlled via tool commands. In addition to the kinematic parameterization of the commands, a reference coordinate system can also be specified. By an implemented coordinate transformation this can for example considerably reduce the recurring programming effort after a displacement of components.

3.2 Process Execution and Monitoring

Once the process resources have been modeled and parameterized, the collaborative assembly can be performed. Three executable modes are available for this purpose. In the automatic mode, all created robot commands are executed under the defined boundary conditions. This mode is intended for the operation of the assembly cell with a validated robot program. In the test mode and step-by-step mode, on the other hand, the process can be validated with reduced kinematics or step-by-step in order to avoid risks and increase process understanding. These modes can also be used to train the operators in order to further increase their acceptance. Transparency can also be promoted through continuous process monitoring. For this reason, the operator is shown the command currently executed by the robot at all times during the process. Furthermore, dashboarding strategies are used to process and analyze process data. This further increases process understanding and can be used for process optimization. For example, frequently recorded collisions, long waiting times or unwanted peaks in process forces can be indicators of an useable potential for optimization.

3.3 Data Management and Interfaces

The data transfer between the modules and applications as well as the data backup is realized by a serialization into the XML-format. In addition to the product and process files, there is a global knowledge database in which process resources can be stored in order to make them applicable across processes. Even if the approach is universally applicable, an interface to a Kuka iiwa robot was implemented to validate the robot programming environment. Via the offered Java interface a TCP interface was set up which is used to communicate with the robot via JSON-strings.

4 Conclusion

This paper presents a comprehensive approach that simplifies the commissioning of collaborative assembly processes. By combining existing approaches to form a linked overall system, one benefits from novel synergy effects. Product-lifecycle-management strategies in the form of knowledge transfer are used to make both process planning and robot programming more efficient and simple. The latter is further reinforced by a graphically intuitive user-interface, which is composed of fundamental basic functions. After only a short training phase, robot programs can be created that are complex enough for numerous applications. These programs are fully executable and can be monitored. The system and user-interface, where validated on a demonstrator, which consists of an assembly station with a collaborating Kuka iiwa.

References

1. Storms, S., Roggendorf, S., Stamer, F., Obdenbusch, M., Brecher, C.: PLM supported automated process planning and partitioning for collaborative assembly processes based on a capability analysis. In: 7th WGP-Jahreskongress Aachen, pp. 241–249 (2017)
2. Finkemeyer, B.: Towards safe human-robot collaboration. In: 22nd International Conference on Methods and Models in Automation and Robotics, pp. 883–888. IEEE Press, New York (2017)
3. Brecher, C., Storms, S., Ecker, C.: An approach to reduce commissioning and ramp-up time for multi-variant production in automated production facilities. In: 3rd International Conference on Ramp-up Management (ICRM). Procedia CIRP **51**, 128–133 (2016)
4. Doltsinis, S., Ferreira, P., Lohse, N.: A symbiotic human–machine learning approach for production ramp-up. IEEE Trans. Hum.-Mach. Syst. **48**(3), 229–240 (2018)
5. Franka Emika GmbH: User Manual: Panda Research. Munich (2018)

6. Do, H.M., Kim, H.-S., Park, D.I., Choi, T.Y., Park, C.: User-friendly teaching tool for a robot manipulator in human robot collaboration. In: 14th International Conference on Ubiquitous Robots and Ambient Intelligence, pp. 751–752. IEEE Press, New York (2017)
7. DIN ISO/TS 15066:2016: Robots and robotic devices - Collaborative robots
8. Zhu, H., Gabler, V., Wollherr, D.: Legible action selection in human-robot collaboration. In: 26th IEEE International Symposium on Robot and Human Interactive Communication, pp. 354–359. IEEE Press, New York (2017)

Discover, Imagine, Change: Community Place-Based Activities Using Unique Mobile Apps

Dalit Levy[1(✉)], Yuval Shafriri[2], and Yael Aleph[3]

[1] Community Information Systems, Zefat Academic College, Zefat, Israel
dalitl@zefat.ac.il
[2] Tel Aviv University, Tel Aviv, Israel
[3] Bar Ilan University, Ramat Gan, Israel

Abstract. Following the results of a study focusing on the unique affordances of mobile technologies that support their informed integration in learning environments, a novel pedagogy has been developed and tried out within the context of cultural, geographical, and archaeological heritage in three different communities of learners in Israel. The first part of this paper briefly presents the main findings of the study and sketches the emergent uniqueness profile of mobile apps for learning. The second part outlines the DICE model (Discover, Imagine, ChangE) for designing educational place-based mobile activities.

Keywords: Blended spaces · Environmental knowledge · Mobile learning

1 Introduction

The consolidation of the web, the mobile and locative media, and the internet of 'smart' things (IOT), opened new possibilities of interaction between the individual, her or his community, and their places of living, working, and visiting. As a result, the processes of developing cultural heritage concepts have changed. These processes not only document, map and teach about the place in which the community operates, but also act within the place. This duality might change both the relations with the place and the meaning given to it, in addition to actual changes (physical and digital) in the local environment. The paper discusses these new possibilities through the pedagogical idea of "Discover, Imagine, Change", abbreviated as DICE. This educational process was formulated as an implication of a recent study on the uniqueness profile of educational mobile applications [1].

The educational use of mobile applications is thought to have significant learning potential. Smartphones are already massively embedded in daily life but integrating mobile technologies within learning environments is a complex and challenging mission which requires innovative pedagogical thinking and strategic changes, beyond merely implement e-learning methods with the aid of mobile devices. In recent years, much research has been conducted on the integration of mobile apps into educational settings [2, 3]. However, the task of identifying the unique features and affordances of mobile technologies has been a complex one [4]. The research from which DICE has

W. Karwowski and T. Ahram (Eds.): IHSI 2019, AISC 903, pp. 50–55, 2019.
https://doi.org/10.1007/978-3-030-11051-2_8

emerged sought to focus solely on learning processes and learning outcomes that are made possible only when using mobile apps, and to identify their unique and exclusive affordances. Within this framework, a unique educational mobile application has been defined as an application that has learning affordances attributed exclusively to mobile devices and apps, with benefits that are unattainable in outdoor learning environments, when no digital technologies are involved. As Fig. 1 suggests, the term MUC has been used to label such a unique mobile app.

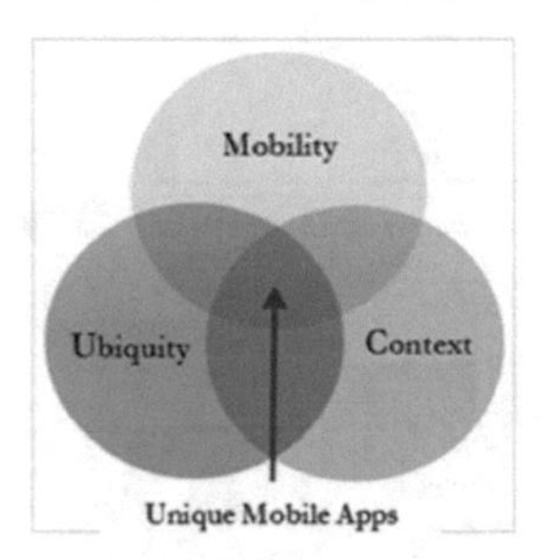

Fig. 1. MUCs - unique mobile apps selected for the study

While both the "M" (mobility) and the "U" (ubiquity) imply independent learning anytime and anywhere, the "C" (context) suggests some dependency on the decisions and actions of the instructional designer and/or the learner. The distinctiveness of MUCs thus lies in their ability to be deployed at any time and in any location, while nevertheless being sensitive to the context - the environment, the user, and the learning activity. This combination is what makes mobile apps unique for learning purposes.

This paper has two parts. The first part sketches the analytic process conducted as part of a study of more than two hundred mobile applications for learning, focusing on selected MUCs. The second part outlines the DICE idea - a novel pedagogy based on collaborative mobile and place-based learning using MUCs, developed and tried out within the context of cultural, geographical, and archaeological heritage in three different communities of learners in Israel. Through participation in such mobile-enhanced activities, learners not only discover the place they live in but might also contribute to actually changing it.

2 Constructing the Uniqueness Profile of MUCs

When discussing and studying mobile learning [5], three types of affordances are often mixed: (1) affordances attributed to non-mobile desktop applications, including complex design systems such as AutoCAD; (2) universal applications operating both on non-mobile and mobile computing devices, thus available anywhere and anytime; and (3) affordances attributed exclusively to mobile apps. While many studies focus on learning with mobile applications as part of using a broader technology-enhanced learning toolbox, our study sought to focus solely on those learning processes and learning outcomes that are made possible only when using mobile apps, and to identify those unique and exclusive affordances.

The analytic process was conducted as part of a study of more than two hundred mobile applications for learning [1]. Most of these apps have been available for free use by any teacher or learner, on any regular mobile device, and did not require special hardware or external gadgets. Two research questions directed the study: first, what makes mobile apps unique for educational purposes? And, second, what are the unique affordances of mobile technologies that support their informed integration in learning environments? The gradual qualitative analytic process began in 2015 and resulted in five emergent themes of uniqueness of mobile apps, organized into three levels, as is shown in Fig. 2.

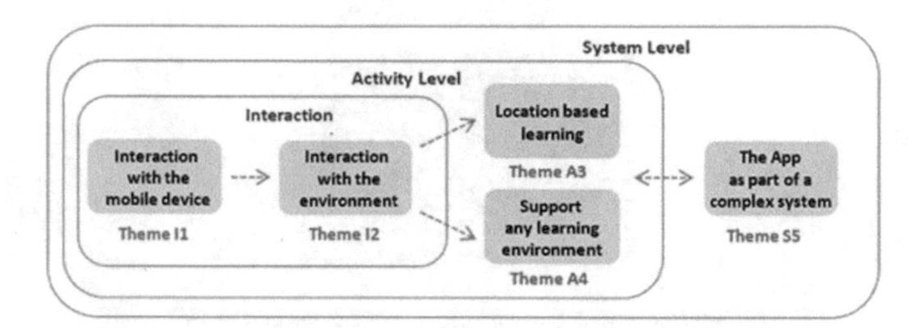

Fig. 2. The emergent categorical system

Common to these emergent categories is the experience of learning in blended spaces [6, 7]. This primary pedagogical principle led to additional principles such as embodied cognition, the device as a discovery machine, and open playful design. Taken together, these principles draw a uniqueness profile for MUCs that supports deep understanding of the environment in which the unique mobile app operates. The emergent profile of uniqueness encircling these principles is drawn in Fig. 3 below.

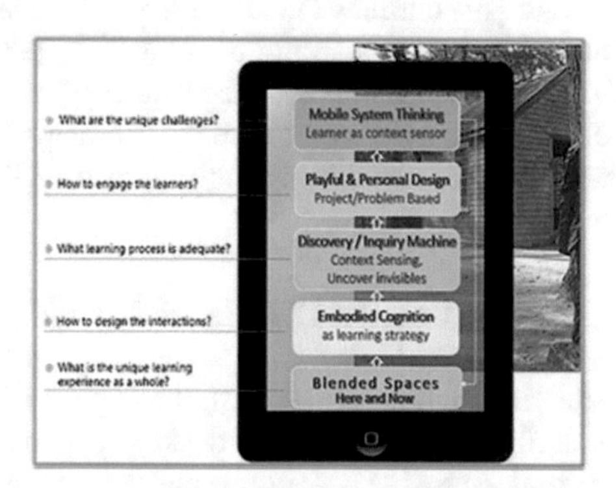

Fig. 3. The uniqueness profile of the learning experience with mobile apps

According to the results of the study, the overall uniqueness profile stems from the fundamental principle *Blended Space – Here and Now*, while each additional principle has some relationship with this major unique learning principle as well as with the others. Therefore, the fundamental principle is drawn at the bottom of the uniqueness profile. In other words, the study suggests that these principles should be treated not just as a list, but as a structure in which the components are placed layer upon layer so that each layer serves as a base for the next, and all are made possible by applying the founding principle.

3 DICE: Smart Learning with Local Places

Along with the structure and the flow of the learning principles in the uniqueness profile, Fig. 3 presents the main question that needs to be considered in applying each of these principles. The uniqueness profile as a whole facilitates utilization of the pedagogical principles for deep understanding of the environment and for promoting new literacy of *Mobile System Thinking*. To enable such utilization, mobile or place-based learning activities should integrate innovative technologies that allow the learner to perform as a “context sensor” of his or her environment [8, 9]. This section outlines DICE (Discover, Imagine, ChangE) as a novel pedagogy aiming to deal with such integration.

The DICE idea has been developed and tried out within the context of cultural, geographical, and archaeological heritage in three different communities of learners in Israel. The pedagogical process is based on collaborative mobile learning, including elements of inquiry and discovery, mapping and documentation using locative media, gamification, and knowledge sharing. These elements are based on the results of the research presented in the first part, namely the uniqueness profile of mobile applications for learning. DICE processes can be used in diverse communities with regard to their unique cultural heritage or specific local theme. Through participation in such mobile-enhanced activities, learners not only discover the place they live in but might also contribute to actual change in it.

As is apparent from its title, the suggested DICE pedagogical model has three legs:

1. Discover: Students conduct an inquiry, document, map, and create knowledge from their findings. This includes mobile field survey, and other inquiry methods such as an archeological or historical survey.
2. Imagine: A design process based on the knowledge students created in the first phase, where groups of students create a model of imaginative change or solution to a problem in their place, using design and collaborative apps.
3. Change: Actual or virtual “place making” phase based on principles of the models suggested in the ‘Imagine’ phase, followed by a community mobile activity created and led by the students.

Based on the uniqueness profile of mobile applications for learning [1], the model can be regarded as an implementation of place-based learning with the assistance of mobile and other ‘smart’ technologies [10]. The DICE model aims to use this uniqueness in such a way that the student becomes more aware of the surrounding while the place becomes a powerful or ‘smart’ object to learn with.

In terms of design considerations, in creating blended space learning experiences in which the merging of multiple sources is crucial to knowledge construction, the environment and the context should be regarded as inherent dimensions to consider. Building upon the TPACK framework [11] we suggest adding the knowledge of the environment to the original framework when designing unique educational mobile apps. As is illustrated in the right side of Fig. 4, the upgraded framework is labeled TEPACK to denote Technological, Environmental, Pedagogical, and Content Knowledge.

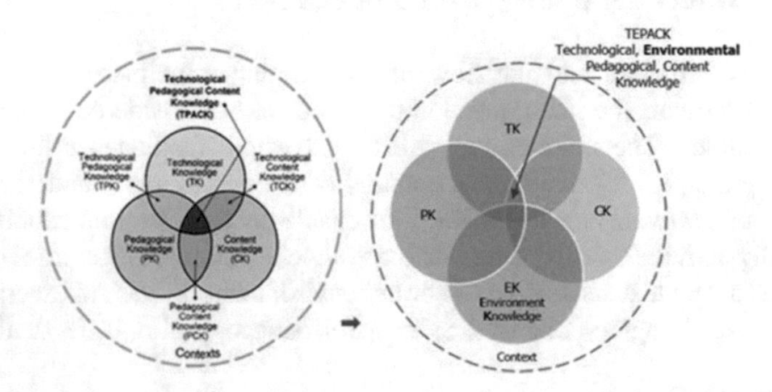

Fig. 4. Adding Environmental Knowledge to the TPACK model

There are known challenges related to the status of technology in human-environmental relations, that arises within the DICE pedagogical model. The interdependence between the environment and the mobile application has been apparent also throughout the study referred to in Sect. 2. Indeed, cognitive and spatial dissonance often occurs when using mobile apps [12]. The interdependence between the external environment and the unique mobile app for learning should be therefore carefully considered by designers, educators, and learners alike.

4 Summary

As is suggested by the Uniqueness Profile (Fig. 3), the mobile device might serve as the context sensor of a *Discovery Machine* in diverse environments, enabling *Embodied Cognition* interactions as a learning strategy using smart learning through sensory affordances. MUC applications and learning activities also enable building *Blended Space* artifacts in context, documenting sensations, and sharing data and messages between near and distant actors in an *Open and Playful approach.* All these lead ultimately to *Mobile System Thinking*, which is necessary in multi-device learning situations and within *Blended Space Experiences*.

A major challenge is to apply MUC affordances not only as 'context sensor' extensions of the human body and mind, but also as a way for learners themselves to become 'context sensors', curious and aware of their surroundings. This is important especially at a time when mobile apps already serve as a digital interface to the world; when the world itself is increasingly turning digitized; and when objects in our world are becoming more connected and 'smarter' [13]. This fascinating combination generates new objects to think with, new blended spaces to live in, and may also generate new ways of being, experiencing and learning. Through participation in mobile-enhanced DICE activities within the context of various cultural, geographical, and archaeological heritage, learners not only discover the place they live in but might also contribute to actual change in it. The learners in such activities are a small step ahead of being just mobile consumers and even producers of digital information. The DICE model suggests using unique mobile apps in such a way that the student becomes more aware of the surrounding, while the place becomes a powerful object to learn with.

References

1. Shafriri, Y., Levy, D.: What are the unique characteristics of integrating mobile applications in learning? J. Interactive Learn. Res. **29**(3), 271–299 (2018)
2. Hirsh-Pasek, K., Zosh, J.M., Golinkoff, R.M., Gray, J.H., Robb, M.B., Kaufman, J.: Putting education in "educational" apps: lessons from the science of learning. Psychol. Sci. Public Interes. **16**(1), 3–34 (2015)
3. Notari, M.P., Hielscher, M., King, M.: Educational apps ontology. In: Churchill, D., Lu, J., Chiu, T.K.F., Fox, B. (eds.) Mobile Learning Design: Theories and Applications, pp. 83–96. Springer, Singapore (2016)
4. Woodill, G.: Unique affordances of mobile learning. In: Udell, C., Woodill, G. (eds.) Mastering Mobile Learning. Wiley, Hoboken (2014). https://doi.org/10.1002/9781119036883.ch15
5. Pegrum, M.: Future directions in mobile learning. In: Churchill, D., Lu, J., Chiu, T.K.F., Fox, B. (eds.) Mobile Learning Design: Theories and Applications, pp. 413–431. Springer, Singapore (2016)
6. Benyon, D.: Presence in blended spaces. Interact. Comput. **24**(4), 219–226 (2012)
7. O'Keefe, B., Benyon, D.: Using the blended spaces framework to design heritage stories with schoolchildren. Int. J Child-Comp. Interact. **6**, 7–16 (2015)
8. FitzGerald, E., Ferguson, R., Adams, A., Gaved, M., Mor, Y., Thomas, R.: Augmented reality and mobile learning: the state of the art. Int. J. Mob. Blended Learn. **5**(4), 43–58 (2013)
9. Kamarainen, A., Metcalf, S., Grotzer, T., Dede, C.: EcoMOBILE: designing for contextualized STEM learning using mobile technologies and augmented reality. In: Crompton, H., Traxler, J. (eds.) Mobile Learning and STEM: Case Studies in Practice, pp. 98–124. Routledge, New York (2015)
10. Zimmerman, H.T., Land, S.M.: Facilitating place-based learning in outdoor informal environments with mobile computers. TechTrends **58**(1), 77–83 (2014)
11. Rosenberg, J.M., Koehler, M.J.: Context and technological pedagogical content knowledge (TPACK): a systematic review. J. Res. Technol. Educ. **47**(3), 186–210 (2015)
12. Beland, L.P., Murphy, R.: Ill communication: technology, distraction & student performance. Labor Econ. **41**, 61–76 (2016)
13. Ally, M., Prieto-Blázquez, J.: What is the future of mobile learning in education? Int. J. Educ. Technol. High. Educ. **11**(1), 142–151 (2014)

Engineering Better Ethics into Human and Artificial Cognitive Systems

John Celona(✉)

Decision Analysis Associates, LLC, San Carlos, CA, USA
JCelona@DecisionAA.com

Abstract. Evolutionary ethics presents a hypothesis for understanding ethics as an evolved social behavior which develops according to the advantages or disadvantages it creates for an individual in its environment. Using this framework, designers of human or artificial cognitive systems can identify the ethical standards they wish to follow and their system to promote, and design into the system the legal, social, or prudential incentives or disincentives to promote the desired choice of ethical standards.

Keywords: Ethics · Evolutionary ethics · Human system design
Artificial cognitive system design

1 Introduction

As I discuss in a chapter written for a forthcoming book [1], designers and engineers of human and artificial systems grapple Frankenstein-like with the ethical implications of their systems and their use. Adding a cognitive dimension (where the system evaluates and possibly learns) furthers the complexity and difficulty of considering consequences and their ethical implications, both anticipated and unanticipated. These consequences may be realized in the operations of the technology itself, or in the system's effect on the behavior of people interacting with the system.

Designers and engineers striving to meet these challenges face daunting challenges in identifying, abstracting and applying the relevant ethical rules. Many lifetimes can be spent studying the various theories and approaches to ethics.

Even then, it is not obvious how to apply those learnings to a human system (like the law or a government benefit) or an artificial learning system. Examples of the latter in a broad sense include Facebook and Google algorithms, facial recognition, cyber-security, and deep learning systems to dispatch human safety inspectors. What are the ethical implications of a boiler in a school exploding after a deep learning system indicated that a safety inspection was not required?

To meet these challenges, we propose a straightforward approach: a hypothesis useful both to define the desired ethical standards and to predict whether the system results (whether from action of the human or technology components) will meet those standards. This approach can be used to engineer the desired ethical standards into the system. The hypothesis is called *Evolutionary Ethics*.

W. Karwowski and T. Ahram (Eds.): IHSI 2019, AISC 903, pp. 56–62, 2019.
https://doi.org/10.1007/978-3-030-11051-2_9

2 Evolutionary Ethics

The hypothesis is straightforward: what if ethics were an evolved social behavior? Like all hypotheses, the question of whether this is true or false does not apply: a hypothesis can only be disproven, never proven. Rather, the relevant question is whether the hypothesis makes testable predictions for observable phenomena of interest. If so, we use it until a better one comes along.

If ethics were an evolved social behavior then, like all evolved characteristics (whether physical or behavioral), behaviors which created advantages for a species in its environment would be passed along either genetically or through teaching. Those which disadvantaged the species would be weeded out.

There is considerable evidence that ethics exists in other species [2] and, indeed, Darwin himself thought that ethics followed the same evolutionary process as other physical characteristics and behaviors. "The following proposition seems to me in a high degree probable- namely, that any animal whatever, endowed with well-marked social instincts, [citation omitted] the parental and filial affections being here included, would inevitably acquire a moral sense or conscience as soon as its intellectual powers had become as well, or nearly as well developed, as in man." [3]

Subsequent work reviewed and conducted by Bekoff and Pierce [2] found an ethical and moral sense in a variety of social species, including elephants, wolves, dogs, rats, cats of all sizes, bats, monkeys, etc. They organize their work on the moral lives of animals into "the *cooperation* cluster (including altruism, reciprocity, honesty, and trust), the *empathy* cluster (including sympathy, compassion, grief, and consolation), and the *justice* cluster (including sharing, equity, fair play, and forgiveness)." [4]

Some of the advantages conferred by cooperation, empathy, or justice are straightforward, others less so. Sharing food may confer an immediate disadvantage (you get less to eat), but a long-term advantage in prompting others to share with you in the future when you may be of need. Many ethically-implicated behaviors share this pattern: they confer an immediate disadvantage to the individual at the promise of a future benefit when others reciprocate. Indeed, this is the essence of the "Golden Rule:" "So whatever you wish that men do to you, do so to them…" [5].

Other advantages are less obvious. Where is the advantage in foregoing an immediate opportunity or sacrificing ones' own interests when there is no foreseeable prospect of future reciprocation by others?

One possible advantage is avoiding sanction for unethical behavior and, indeed, many social species in addition to man have sanctions for unethical behavior (e.g., expulsion from the pack.) In human society, possible sanctions may be by the legal system or social (shunning, disapproval, etc.).

However, actions may raise ethical implications even though they are perfectly legal and socially approved. You have the legal right to kill in self-defense (subject to limitations) and it is socially acceptable to do so, but you may feel it is wrong to do so nonetheless. The sixth commandment is "[t]hou shalt not kill," without any reference to self-defense. Many soldiers experience trauma from society's requirement that they fight and kill in war, even those who do so by controlling remote drones thousands of miles away from the fight. [6] Is there a "moral injury" unethical behavior inflicts on the perpetrator apart from and in addition to the immediate and possible future tangible consequences? Could "moral injury" be part of the evolved response to promote more ethical behavior?

If ethics evolve, we need a framework for specifying ethical standards that (1) captures how the ethical implications of actions may advantage or disadvantage individuals and systems and drive future behavior; (2) clarifies the ethical implications of the behaviors that may result; and (3) enablers designers and engineers to think hard about the ethics they are designing into the system and make informed choices.

3 A Framework for Specifying Ethical Standards

To describe this framework, we need to consider both the objectives of ethics (what do they accomplish?) and for whom (the individual or others). The objectives dimension needs to show how actions may advantage or disadvantage an individual. The "who" dimension needs to capture the self- or other-regarding effects.

For the first dimension (what goal?), consider Maslow's hierarchy of needs. [7] The needs he identified (in order of priority) are: Physiological Needs (food, water, temperature), Safety and Security, Love and Belonging, Self-esteem, and Self-actualization. For the second dimension (for whom?), consider a scope beginning the individual and progressing to larger scope: family, tribe or nation, species, possibly then warm-blooded life and then all life. These two dimensions describe a continuum of possible ethical standards, as shown in Fig. 1.

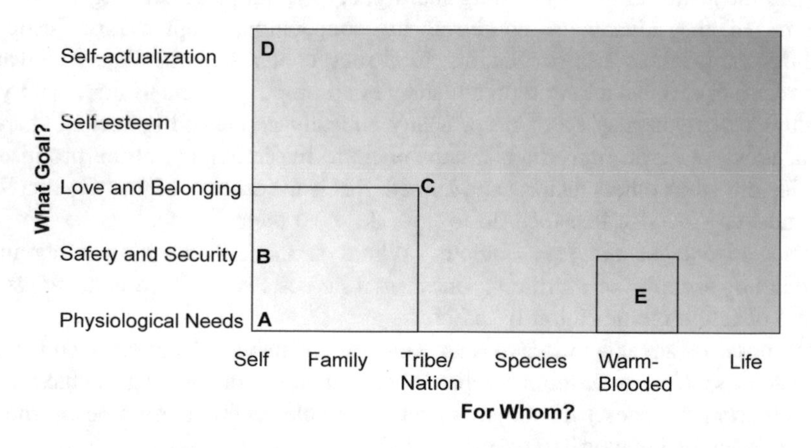

Fig. 1. A continuum for describing possible ethical standards

Given those two dimensions, an individual's ethical standards can be specified by picking a point on the graph. The area between that point and the axes are the interests that person wishes his or her ethical standards to respect. Actions which violate those interests would be unethical.

For example, an individual at point *A* would only be concerned with obtaining food, water, and maintaining temperature. An individual at point *B* would be developed enough to recognize danger and react to it. An individual at point *C* adds other-regarding social behavior to consider the needs of its entire group (family or not). A human sociopath (by definition incapable of feeling for others) would fall at point *D*.

One could also specify unusual ethics. An individual who felt it was unethical to kill animals but fine to kill people would have standards described by the area around point *E*. For the purpose of discussing how system design impacts groups of individuals, we'll focus on the progressive addition of objectives and scope followed by most individuals and groups rather than outliers.

4 Methods of Motivating Observance of Ethical Standards

What drives individuals to follow one ethical standard over another? For action to have an ethical implication, there has to be a choice. You cannot accuse a bacteria which kills its host or a vine which kills the tree it grows on of being unethical because they have no choice; they are only following rules laid down in their genes. Howard and Abbas follow this approach: "**Ethics** are your *personal* standards of right and wrong. Your code of *proper behavior*." [8] An action is ethical if it is right given your choice of ethics. The question, then, is how to drive the choices of ethical standards individuals follow.

One may think of three progressive levels of incentives. Legal systems enforce minimum standards at the risk of formal sanction. Social mores enforce standards at the risk of disapproval or shunning. The most expansive incentive is what an individual perceives to be in his or her best interests, which can be termed prudential behavior. These three methods are illustrated in Fig. 2.

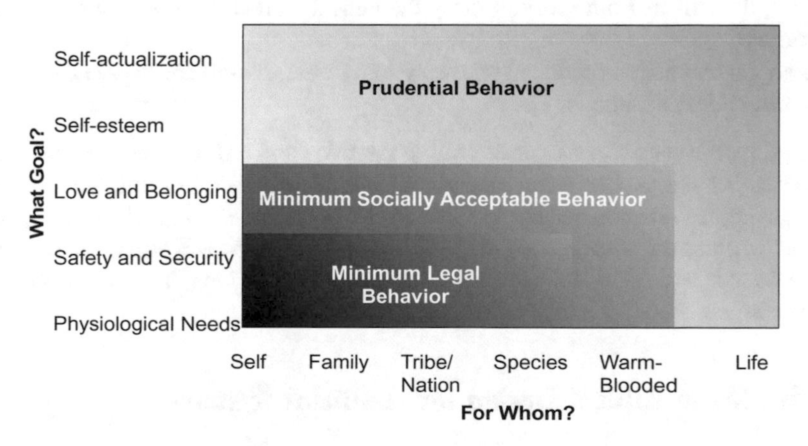

Fig. 2. Progressive methods of incenting choice of ethics

Adhering to one's ethical standards makes one feel good about oneself, helping self-esteem and self-actualization. One the other hand, not living up to them damages both of these, and may inflict the sort of "moral injury" described above. Thus, from this perspective, following one's ethics furthers one's prudential interests even when the behavior deviates from societal or legal standards.

Many difficult ethical dilemmas arise from conflict between acting according to one's lower-level or higher-level goals. Should you lie to get the deal or job? Should you report possibly illegal activity by your best customer, though it may cost you their business? Plus risking future social or legal sanction for your action or inaction.

More serious ethical dilemmas arise with choices sacrificing your own interests to help another, or hurting another's interests to help yourself. Should you donate a kidney to a friend in dire need of a transplant? Should you split your inheritance with disinherited siblings for whom the sum would be life-changing (though it is not for you), even though the deceased specifically excluded them from his will?

At a system level, the question is whether incentives in the system promote and reward higher or lower-level choices of ethics. For example, the rules of professional responsibility require attorneys to zealously pursue any legal objectives their clients may have, not to seek the truth or a just outcome. [1] Likewise, some of mishaps in testing self-driving cars seem to indicate that their operation can lead to greater driver inattention and collisions that most drivers would avoid when driving unassisted. How are system designers to consider the ethical implications in their designs?

5 One-Stage Ethical Design for Human Systems

Considering the ethical framework and incentive structure described above, ethics can be designed into a human system as follows:

1. Using the ethical framework, specify the ethical standards you wish the system to promote.
2. Align the incentives produced by the system (legal, social, and prudential) with the chosen ethical standards.

Legal incentives are a matter of writing the rules and enforcement mechanisms for the desired behaviors. For social incentives, clear identification of undesired behaviors and signaling to others is needed to invoke social sanctions. For prudential incentives, consider whether the system discourages or rewards higher-level ethics (an example of the latter is tax deductions for charitable donations), or unethical behavior that damages higher-level needs (e.g., lying in insurance claims).

6 Two-Stage Ethical Design for Artificial Systems

In designing artificial cognitive systems, a two-stage approach is needed. First, consider the behavior of the artificial portion of the system. Then, consider the impacts of the system on the creatures (including people) which interact with it.

Regarding the system itself, it has no ethics. It simply follows the rules programmed into it. Those rules may prioritize certain feedbacks over others in determining next steps but, barring artificial consciousness (of which there is currently no prospect, a multitude of science fiction writers notwithstanding), the system won't consider whether its actions are right or wrong—only whether they optimize according to the value functions defined for it.

What actions are the system optimized to take? And what impacts do those actions have on the creatures interacting with it? What incentives do those action create for the creatures interacting with it?

Given those incentives, evaluating the impact on people interacting with the system is the same process as described above, just with some of the inputs provided by the actions of an artificial cognitive system in addition to the inputs provided by the relevant human systems and by nature. If this second step identifies incentives for ethically undesirable behaviors, the artificial or human portions of the system may need to be revised to produce different incentives.

7 Summary

Evolutionary ethics presents a framework for understanding ethics as an evolved social behavior which, like all products of evolution, develops according the advantages or disadvantages it creates for individuals in their environment. This framework describes a continuum of possible choices of ethical standards. Ethical standards may be promoted and enforced through legal, social, or prudential means.

To design ethics into human or artificial cognitive systems:

1. If applicable, consider the actions optimized for in the artificial system and their impacts on people interacting with the system.
2. Include these impacts in the total array of effects, incentives, and disincentives, including those from the relevant human systems and from nature.
3. Identify the desired choice of ethical standards.
4. Evaluate the ethical standards promoted by the entire array of incentives and, if necessary redesign the artificial or human portions of the system to incent the desired choice of ethics.

Acknowledgments. The author would like to Dr. Ali E. Abbas of the Neely Center for Ethical Leadership and Decision Making at the University of Southern California for prompting and supporting my thinking and writing on this topic, and Dr. Ronald A. Howard of Stanford University for his insights which spurred my ethical inquiries beginning forty years ago.

References

1. Celona, J.: Evolutionary ethics: a potentially helpful framework in engineering a better society. In: Abbas, A., Gee, S. (eds.) Next Generation Ethics: Engineering a Better Society. Cambridge University Press, Cambridge (2019)

2. Bekoff, M., Pierce, J.: Wild Justice: The Moral Lives of Animals. The University of Chicago Press, Chicago (2009)
3. Darwin, C.: The Descent of Man, and Selection in Relation to Sex, 2nd edn, p. 57. J. Murray, London (1874)
4. Bekoff, M., Pierce, J.: Wild Justice: The Moral Lives of Animals, p. xiv. The University of Chicago Press, Chicago (2009)
5. Throckmorton, B.H.: Gospel Parallels: A Synopsis of the First Three Gospels, 4th edn, p. 29. Thomas Nelson, Nashville (1979). (Citing Matt. 7:12 and Luke 6:31)
6. https://www.nytimes.com/2018/06/13/magazine/veterans-ptsd-drone-warrior-wounds.html
7. Maslow, A.H.: A theory of human motivation. Psychol. Rev. **50**(4), 370–396 (1943)
8. Howard, R.A., Abbas, A.E.: Foundations of Decision Analysis, p. 781. Pearson Education Inc., New York (2016)

A New Method for Classification of Hazardous Driver States Based on Vehicle Kinematics and Physiological Signals

Mickael Aghajarian(✉), Ali Darzi, John E. McInroy, and Domen Novak

Department of Electrical and Computer Engineering, University of Wyoming, Laramie, Wyoming, USA
{maghajar,adarzi,mcinroy,dnovakl}@uwyo.edu

Abstract. Hazardous driver states are the cause of many traffic accidents, and there is therefore a great need for accurate detection of such states. This study proposes a new classification method that is evaluated on a previously collected driving dataset that includes combinations of four causes of hazardous driver states: drowsiness, high traffic density, adverse weather, and cell phone usage. The previous study consisted of four sessions and eight scenarios within each session. Four physiological signals (e.g. electrocardiogram) and eight vehicle kinematics signals (e.g. throttle, road offset) were recorded during each scenario. In both previous and present studies, the presence or absence of the different causes of hazardous driver states was classified. In this study, a new classifier based on principal component analysis and artificial neural networks is proposed. The obtained results show improvement across all classification accuracies, especially when only vehicle kinematics data are used (mean of 12.7%).

Keywords: Hazardous driving state · Artificial intelligence · Driving performance · Physiological measurements · Affective computing

1 Introduction

Hazardous mental or physical states are the cause of many traffic accidents. For instance, fatigued driving resulted in an estimated 800 deaths and 41,000 injuries in the United States in 2015. Distracted driving, on the other hand, caused 1.25 million deaths worldwide in 2015, with an estimated 3,477 deaths in the United States alone [1].

A system with the ability to detect hazardous driving states (HDS) and intervene in dangerous situations can mitigate the number of deaths and injuries, thus increasing driving safety. Although several such systems have been proposed (e.g. [2, 3]), there is room for improvement. One important issue with such automated intervention systems is that they usually monitor a specific gesture or posture of the driver (e.g., hands-on/off the wheels) and do not identify the cause of HDS (e.g., stress). This study therefore aims to develop an automated system that not only detects HDS, but can also identify specific causes of HDS based on a combination of vehicle kinematics and driver

W. Karwowski and T. Ahram (Eds.): IHSI 2019, AISC 903, pp. 63–68, 2019.
https://doi.org/10.1007/978-3-030-11051-2_10

physiology. The obtained results can be used to tailor the response of an intervention system to each cause of HDS.

Many physical/physiological (e.g., fatigue) or cognitive/affective (e.g., anger) conditions may cause HDS, and most causes of HDS therefore have both physical and mental components. In this paper, we discuss four causes of HDS: distractions, fatigue, demanding driving conditions, and the driver's characteristics. Distracted driving is perhaps the most infamous HDS, and can lead to catastrophic situations. Performing secondary tasks in addition to driving (e.g., using a cell phone) results in both visual distractions [4] and cognitive distractions [5]. Drivers need 7–12 seconds to regain situational awareness after each distraction [6], increasing the likelihood of accidents. Fatigue and drowsiness make driving even harder and affect both driver physiology [7] and vehicle kinematics [8]. Even if a driver is alert, difficult driving conditions like blizzards introduce a high level of mental demand. During such conditions, drivers may devote all their mental resources and still not be able to drive effectively [4].

Regardless of their causes, HDS can be assessed using three methods: physiological measurements, vehicle kinematics, and self-report questionnaires. Physiological signals can be unobtrusively recorded during driving and are correlated with many psychological states (e.g. workload). A few examples of such signals in driving studies are the electrocardiogram (ECG) [9], which records heart rate [10], galvanic skin response (GSR), which records the activity of the skin's sweat glands [11], respiration rate (RR) [12], and skin temperature (ST) [13]. Vehicle kinematics, on the other hand, include signals such as longitudinal speed [14], rotation of the steering wheel [15], and the lateral lane position (distance from the lane center) [16].

In our previous study [11], we exposed drivers to four different causes of HDS (mild sleep deprivation, adverse weather, cell phone use, and high traffic density), and collected three different types of information: vehicle kinematics, physiological measurements (RR, ST, GSR, and ECG), and driver characteristics (personality, mood, and stress level). We then created different classifiers to automatically identify the presence or absence of each of the four causes of HDS. The contribution of this study is to propose a new classification method that does not need to extract features from raw data and increases the classification accuracies. Of the three types of data from the previous study, vehicle kinematics and physiological measurements are used in this study.

2 Study Setup and Protocol

Study Setup: The dataset from the previous study includes data from 21 people (25.1 ± 8.7 years old, six females) who participated in four simulated driving sessions in the University of Wyoming driving simulator lab (WYOSIM). Of the four sessions, two were meant to mimic drowsy (mildly sleep-deprived) driving and were held in the early morning while it was still dark outside. In these two sessions, only night scenarios were used in WYOSIM, and the participants were told to have less than 6 hours of sleep the preceding night. The other two sessions were meant to mimic alert driving, and participants were instructed to have more than 7 hours of sleep the preceding night. These two sessions were held between 10 am and 5 pm, and only day scenarios were used in WYOSIM. The order of drowsy and alert driving sessions was random. Each session

Fig. 1. Town scenario (Left) and highway scenario (Right).

consisted of 8 scenarios (4 min/scenario) that represented all possible combinations of traffic density (high/low), weather (sunny/snowy), and cell phone use (phone/no phone), in random order. For low traffic density, participants drove on a highway with few cars (density factor 0.3 in WYOSIM); for high traffic density, they drove in a town with dense traffic (density factor 1.5) (Fig. 1). In snowy weather, visibility was lower than in sunny weather and the friction between the tires and the road was reduced to 60% of the sunny-weather value [17]. Furthermore, in the "cell phone" scenarios, participants used their cell phone to browse the Internet or send text messages [11].

Measured signals: In each scenario, the g.USBamp signal amplifier (g.tec Medical Engineering GmbH, Austria) was used to record 4 physiological signals: electrocardiogram, respiration, skin temperature, and galvanic skin response. Furthermore, 8 vehicle kinematics signals were recorded: throttle force, lañe number, lateral lane position, road offset, longitudinal velocity, vertical velocity, and slip level of front and rear tires. In the previous study, three or more features were calculated from each raw signal (either physiology or vehicle kinematics), and the stepwise algorithm was used to select the best set of features. Then, three types of classifiers (support vector machine, decision tree, and logistic regression) were used to classify the presence or absence of each of the four causes of HDS (traffic density, weather, cell phone, drowsiness) [11]. In this study, the raw signals were directly used as inputs to the classifiers.

The contribution of this study: A classification method based on principal component analysis (PCA) and artificial neural networks (ANN) was implemented. Several binary ANN classifiers were used to classify the presence or absence of each cause of HDS. The computational advantage of the proposed method is that raw physiological and vehicle kinematics signals were used; therefore, there was no need for any preprocessing or feature extraction methods.

Let $\chi \in \mathbb{R}^{n_s \times n_e \times n_t \times n_p \times n_m}$ denotes the previously collected data, where n_s signifies the number of subjects, n_e is the number of sessions per subject, n_t is the number of scenarios within each session, n_p is the number of data samples, and n_m denotes the number of raw signals. In our study, $n_s = 21$, $n_e = 4$, $n_t = 8$, and five physiological signals with a sampling frequency of 512 Hz and eight vehicle kinematics signals with a sampling frequency of 60 Hz were recorded. Therefore, if only physiological signals are used, $n_m = 4$ signals and $n_p = 122,880$ samples; if only vehicle kinematics signals are used, $n_m = 8$ signals, and $n_p = 14,400$ samples. The number of samples is calculated based on the sampling frequency and the length of each scenario (4 min).

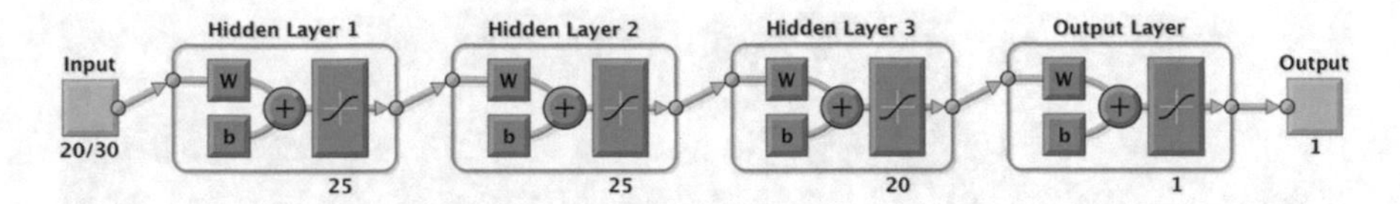

Fig. 2. The proposed artificial neural network classifier. It uses 20 inputs for physiological or vehicle kinematics only, and 30 inputs for the combination of them.

In the PCA-ANN method, we first need to generate a data matrix, D, by stacking the raw signals. The matrix, D, is of size $n_{pm} \times n_{set}$, where $n_{pm} = n_p \times n_m$ and $n_{set} = n_s \times n_e \times n_t$. Since n_{pm} is much larger than n_{set}, the original covariance matrix ($C' = D\,D^T$) is a large-scale square matrix that requires prohibitively extensive computations to calculate its eigenvectors. Instead, the covariance matrix with reduced dimensionality ($C = D^T D$) is used to allow easy calculation of eigenvectors. After calculating the eigenvectors of the covariance matrix with reduced dimensionality, the k best eigenvectors are selected (eigenvectors corresponding to the largest eigenvalues). The value of k is a hyperparameter that is selected by trial and error. In this study, $k = 20$ if only physiology or only vehicle kinematics are used while $k = 30$ if both physiology and vehicle kinematics are used. Let V_k denotes the matrix of k best eigenvectors. In the next step, the input data for the ANN classifiers was then generated using $W = D^T D\ V_k$, where W is the input data for the ANN [18]. Figure 2 shows the structure of three-layer ANN classifiers with either 20 or 30 inputs and one output. For the first, second and third hidden layers, 25, 25 and 20 neurons are used, respectively. The hyperbolic tangent sigmoid function is chosen as the transfer function of all hidden layers as well as the output layer. For each cause of HDS, one ANN is developed using Levenberg-Marquardt backpropagation algorithm as the training method. To train and test the classifiers, 75% and 25% of the data are used, respectively. The 4-fold cross-validation method is used to validate the ANN classifiers, and the mean values of classification accuracies are reported.

3 Result and Discussion

In this section, the classification accuracies obtained from the new PCA-ANN method are compared to the results of the previous study. Table 1 shows the accuracies for three input types: physiology only, vehicle kinematics only, and both physiology and vehicle kinematics. The PCA-ANN method exhibits higher accuracy for classification of drowsiness using any input type, demonstrating a strong potential advantage over the previous methods. Likewise, the proposed method outperforms the previous methods when using vehicle kinematics to classify all four HDS causes, especially high/low traffic density (nearly 100% accuracy). In contrast, when using physiological signals, the accuracy of the new PCA-ANN method varies significantly depending on the cause of HDS – from 16% worse to 23% better than the classification methods from the previous study. Overall, the obtained results show an improvement across all classification accuracies compared to the previous study: vehicle kinematics (mean improvement of 12.7%), physiological responses (mean improvement of 1.2%) and the

Table 1. Classification accuracies obtained with different input data types using the classifiers from the previous study and using the proposed method (Physio: physiology, Vk: vehicle kinematics, Ps: previous study, PCA-ANN: current study).

	Cell phone	Alert vs. drowsy	Highway vs. town	Snowy vs. clear
Physio (Ps)	81.8%	55.2%	86.8%	56.8%
Physio (PCA-ANN)	69.9%	**78.9%**	70.5%	**66.1%**
Vk (Ps)	64.3%	53.1%	83.3%	71.2%
Vk (PCA-ANN)	**74.1%**	**69.6%**	**99.9%**	**79.5%**
Both (Ps)	82.3%	55.2%	91.4%	71.5%
Both (PCA-ANN)	75.9%	**82.7%**	81.5%	71.1%

combination of both (mean improvement of 2.7%). The high variation in the classification accuracies between the PCA-ANN and the previous study methods could be due to the nature of the inputs or difference in the classifiers. The exact reasons for the differences between the new method and the previous methods could be further investigated in future studies.

The new PCA-ANN method does have a few negative aspects as well. For instance, since we do not know what properties of the raw data are being used for classification, it is more difficult to identify the specific effect of each cause of HDS on physiology and vehicle kinematics. Another drawback of the proposed method is the trial-and-error process of choosing the ANN topology.

4 Conclusion

This study uses a previously collected driving dataset to test the performance of a PCA-ANN classification method in categorizing the presence or absence of four causes of HDS. Two types of data (physiological and vehicle kinematics) and their combination are used, and the obtained accuracies are compared with the results of the previous study. The highest classification accuracies of the proposed method were 75.9% for cell phone use, 82.7% for alert vs. drowsy driving, 99.9% for low vs. high traffic density, and 79.5% for snowy vs. clear weather. Generally, the proposed method performed better than the method of the previous study when only vehicle kinematics data was used. In the case of physiological measurements only, however, the results vary significantly – the accuracy of the PCA-ANN method ranges from 16% worse to 23% better than the results of the previous study. This high variation in results indicates that different causes of HDS require different approaches to be classified accurately.

As the next step, the developed HDS detection systems should be combined with intervention systems that will take actions to increase driver safety based on the detected HDS. These intervention systems can then be tested in simulated and real driving to determine their effect on driver safety and satisfaction.

Acknowledgments. Research supported by the National Science Foundation under grant no. 1717705 as well as by the National Institute of General Medical Sciences of the National Institutes of Health under grant no. P20GM103432.

References

1. National Highway Traffic Safety Administration (NHTSA): 2015 Motor Vehicle Crashes: Overview (2016)
2. Zhang, L., et al.: Cognitive load measurement in a virtual reality-based driving system for autism intervention. IEEE Trans. Affect. Comput. **8**, 176–189 (2017)
3. Healey, J.A., Picard, R.W.: Detecting stress during real-world driving tasks using physiological sensors. IEEE Trans. Intell. Transp. Syst. **6**, 156–166 (2005)
4. Recarte, M.A., Nunes, L.M.: Mental workload while driving: effects on visual search, discrimination, and decision making. J. Exp. Psychol. Appl. **9**, 119–137 (2003)
5. Hwang, Y., Yoon, D., Kim, H.S., Kim, K.H.: A validation study on a subjective driving workload prediction tool. IEEE Trans. Intell. Transp. Syst. **15**, 1835–1843 (2014)
6. Lu, Z., Coster, X., de Winter, J.: How much time do drivers need to obtain situation awareness? A laboratory-based study of automated driving. Appl. Ergon. **60**, 293–304 (2017)
7. Chuang, C.-H., Cao, Z., King, J.-T., Wu, B.-S., Wang, Y.-K., Lin, C.-T.: Brain electrodynamic and hemodynamic signatures against fatigue during driving. Front. Neurosci. **12** (2018)
8. Guo, M., Li, S., Wang, L., Chai, M., Chen, F., Wei, Y.: Research on the relationship between reaction ability and mental state for online assessment of driving fatigue. Int. J. Environ. Res. Public Health. **13**, 1174 (2016)
9. Jung, S.-J., Shin, H.-S., Chung, W.-Y.: Driver fatigue and drowsiness monitoring system with embedded electrocardiogram sensor on steering wheel. IET Intell. Transp. Syst. **8**, 43–50 (2014)
10. Collet, C., Clarion, A., Morel, M., Chapon, A., Petit, C.: Physiological and behavioural changes associated to the management of secondary tasks while driving. Appl. Ergon. **40**, 1041–1046 (2009)
11. Darzi, A., Gaweesh, S.M., Ahmed, M.M., Novak, D.: Identifying the causes of drivers' hazardous states using driver characteristics, vehicle kinematics, and physiological measurements. Front. Neurosci. **12** (2018)
12. Darzi, A., Gorsic, M., Novak, D.: Difficulty adaptation in a competitive arm rehabilitation game using real-time control of arm electromyogram and respiration. In: 2017 International Conference on Rehabilitation Robotics (ICORR), pp. 857–862. IEEE (2017)
13. Kajiwara, S.: Evaluation of driver's mental workload by facial temperature and electrodermal activity under simulated driving conditions. Int. J. Automot. Technol. **15**, 65–70 (2014)
14. Jun, J., Guensler, R., Ogle, J.: Differences in observed speed patterns between crash-involved and crash-not-involved drivers: application of in-vehicle monitoring technology. Transp. Res. Part C Emerg. Technol. **19**, 569–578 (2011)
15. Zheng, Y., Hansen, J.H.L.: Lane-change detection from steering signal using spectral segmentation and learning-based classification. IEEE Trans. Intell. Veh. **8858**, 1 (2017)
16. Sun, R., Ochieng, W.Y., Feng, S.: An integrated solution for lane level irregular driving detection on highways. Transp. Res. Part C Emerg. Technol. **56**, 61–79 (2015)
17. Kordani, A.A., Molan, A.M., Monajjem, S.: New formulas of side friction factor based on three-dimensional model in horizontal curves for various vehicles. In: T&DI Congress 2014, pp. 592–601. American Society of Civil Engineers, Reston (2014)
18. Turk, M., Pentland, A.: Eigenfaces for recognition. J. Cogn. Neurosci. **3**, 71–86 (1991)

Lane Change Prediction Using an Echo State Network

Karoline Griesbach, Karl Heinz Hoffmann(✉), and Matthias Beggiato(✉)

Chemnitz University of Technology, Straße der Nationen 62, 09111 Chemnitz, Germany
{karoline.griesbach,hoffmann}@physik.tu-chemnitz.de, matthias.beggiato@psychologie.tu-chemnitz.de

Abstract. Lane change prediction can reduce accidents and increase the traffic flow. An Echo State Network is implemented for the prediction of left lane changes in an urban area. The Echo State Network has three input variables: turn signal, head rotation in y-direction and steering angle. The input variables were generated from a Naturalistic Driving study in the urban area of Chemnitz, Germany. A successful prediction for left mandatory and discretionary lane changes was realized.

Keywords: Echo state network · Lane change · Predictions · Naturalistic driving study

1 Introduction

Lane change maneuvers lead to many accidents in road traffic. The statistics in Germany [1] show a slight increase of accidents due to drivers' fault. The majority of these accidents occur in urban areas (about 65%), so it is important to not only reduce accidents out of town like on highways, but in cities as well. One way to reduce accidents are advance driver assistance systems which are capable to predict lane changes. A lot of accidents in the urban area arise because of drivers' fault during lane change execution, among others due to insufficient safe distances and during overtaking [1]. Overtaking is part of the process during a lane change. The distance to other vehicle is important to survey the possibility of the execution of a lane change, e.g. the gap in the target lane has to be big enough.

Advanced driver assistance systems which are capable to predict lane changes can reduce the risk of traffic accidents. The system could warn the driver of critical maneuvers, switching on helpful driver assistance systems or switching off interfering driver assistance systems to avoid conflicting situations.

Moreover, the traffic density is increasing which leads to an increase in accidents and decrease in traffic flow. One of the reasons for traffic jams are lane changes [2]. It is important to reduce traffic jams and accidents. The increase of traffic congestions can be avoided with an advanced driver assistance system and the upcoming vehicle-to-vehicle communication. The combination of both systems can warn the vehicles in the

W. Karwowski and T. Ahram (Eds.): IHSI 2019, AISC 903, pp. 69–75, 2019.
https://doi.org/10.1007/978-3-030-11051-2_11

surrounding of the maneuver planning of an ego vehicle, e.g. a lane change. Quick and strong breaking which leads to traffic congestions can be avoided, because of an early warning to other vehicles in the environment. They have enough time to adapt to the situation without effecting the traffic flow.

The present work addresses the needs to improve lane change assistance systems. An Echo State Network (ESN) was developed to predict the lane change intention of drivers in an urban area. The data set was generated from a Naturalistic Driving study in Germany [3]. Below, an overview of previous research is given. Afterwards the ESN for lane change prediction is introduced and finally the results are represented and discussed.

2 Previous Research

A driver can execute a lane change due to different reasons. Lane changes are subdivided into mandatory and discretionary lane changes. The mandatory lane change is necessary to maintain a certain route, e.g. if a driver wants to use an exit lane. The discretionary lane change is executed to improve or maintain the driving situation, e.g. overtaking a slow lead vehicle to keep the own speed [4]. Previous research analyzed various variables which are important for lane change prediction. They can be divided in three parts: vehicle attributes (e.g. velocity, gas pedal pressure), driver attributes (e.g gaze behavior, head rotation) and the vehicle environment (e.g slow lead vehicle, gaps between vehicles). A few indicators are discussed in the following subsection.

2.1 Indicators for Lane Change Prediction

Drivers show characteristic gaze behavior 3 s prior to a lane change which were executed in the urban area of Chemnitz, Germany. The gaze behavior varies for left and right lane changes. For left lane changes the majority of the gazes were to the left side, during right lane changes the majority of the gazes were to the right side and into the rear mirror [5].

The head rotation is another relevant indicator for a lane change. If the head rotation was added to a lane change classifier the performance accuracy for lane change prediction increases [6]. Furthermore, [6] found out that the standard deviation of the head rotation in the horizontal direction for left lane changes is significantly higher than for no lane changes.

An obvious indicator would be the turn signal, but several studies showed different results regarding its activation. The turn signal was used in 89% of the lane changes [5], it was switched on in average 2.28 s prior to a lane change [7].

The steering angle could be a promising indicator for a lane change which is similar to a sinus curve [8]. Other important indicators are variables of the surrounding as, for example, the time to collision [6].

To sum up, different variables can be considered for the prediction of lane changes. Previous studies used different combinations of input variables to predict and model lane changes. A selection of these studies is described in the following subsection.

2.2 Previous Algorithm for Lane Change Prediction

Previous studies applied different learning algorithm for lane change prediction. Successful lane change prediction was done with neural networks [4, 6, 9, 14], Bayesian networks [10] and with a fuzzy system [11]. Hidden Markov Models predicted successful lane changes [12] and driver behavior at intersections [13]. One study correctly predicted 94.2% of left lane changes with a neural network using the inputs head position and orientation, possibility of a lane change and distance to other vehicles [9]. In another study a Multilayer Perceptron and a Convolutional Neural Networks was developed to predict lane changes [14]. As input variables speed, positive and negative acceleration, starting movement jerk, cruising track jerk, starting brake jerk and ending brake jerk were used. The Multilayer Perceptron reaches a precision of 82.91% and the Convolutional Neural Network of 89.64%. Both studies used data from a Naturalistic Driving study.

3 Method

In this paper, an ESN was trained with data of a Naturalistic Driving study in Chemnitz, Germany [3] to predict left mandatory and discretionary lane changes.

3.1 Dataset

The data set was provided by the project "Interdisziplinäres Zentrum für Fahrerassisenzsysteme" [3]. It is the same data set as in [9]. The raw data was generated by a Naturalistic Driving. In total 60 test persons, between 20 to 65 years old, drove a specified 40 km long route in the urban area of Chemnitz, Germany. A VW Touran with automatic transmission and equipped with a data logging system was used as test car. In total eight types of lane changes were extracted through the video cameras [5]. A lane change due to: a slow lead vehicle (driver overtakes a vehicle driving with lower speed), turn-off lanes, added lanes, an entering (e.g. acceleration lane), a merging vehicle (a vehicle wants to merge to the left, so the ego vehicle makes space and merge to the left as well), an obstacle (e.g. construction zone),a return (return to right lane after overtaking) and other unknown reasons.

A lane change was annotated, if the center of the vehicle crossed the line between to lanes. The study collected driver attributes (for instance use of the turn signal, gaze direction, head rotation), vehicle attributes (for instance velocity, steering angle) and the vehicle environment (for instance type of the lane change, left or right lane change, vehicles in the surrounding).

3.2 Echo State Network

The prediction is realized with an ESN. In the following we use the definition of an ESN from [15, 16]. The ESN differs from the neural network in learning the weights. For the ESN only the output weights (red arrows in Fig. 1) instead of all weights are trained. The ESN is using a supervised learning algorithm, so the desired output (target) of an input of the ESN is known. The goal of the ESN network is to compute an output which match up with the target and is able to generalize to unknown data [16].

The input weight matrix $\boldsymbol{W}^{\text{in}} \in \mathbb{R}^{N_x \times (1+N_u)}$ (N_x = number of reservoir neurons, N_u = number of input neurons), the recurrent weight matrix $\boldsymbol{W} \in \mathbb{R}^{N_x \times N_x}$ and the leaking rate α specify the reservoir. At first, the matrices $\boldsymbol{W}^{\text{in}}$ and $\boldsymbol{W}$ are initialized, so that the echo state property is fulfilled [15, 16]. The input weight matrix, the recurrent weight matrix, the input vector $\boldsymbol{u}(n) \in \mathbb{R}^{N_u}$ of $n = 1, \ldots, T$ data points and the previous reservoir state vector $\boldsymbol{x}(n) \in \mathbb{R}^{N_x}$ influence the update of the reservoir $\tilde{\boldsymbol{x}}(n)$:

$$\tilde{\boldsymbol{x}}(n) = \tanh\left(\boldsymbol{W}^{\text{in}}[1; \boldsymbol{u}(n)] + \boldsymbol{W}\boldsymbol{x}(n-1)\right). \tag{1}$$

The expression $[\cdot;\cdot]$ defines a vector (or matrix) concatenation. The function tanh is a typical sigmoid wrapper function. The reservoir activation depends on the leaking rate, the update of the reservoir and the previous reservoir state:

$$\boldsymbol{x}(n) = (1-\alpha)\boldsymbol{x}(n-1) + \alpha\tilde{\boldsymbol{x}}(n). \tag{2}$$

The leaking rate α specifies the rate of the reservoir update. When all reservoir neurons are updated, the output weights $\boldsymbol{W}^{\text{out}} \in \mathbb{R}^{N_y \times (1+N_u+N_x)}$ are determined with:

$$\boldsymbol{W}^{\text{out}} = \boldsymbol{Y}^{\text{target}}\boldsymbol{X}^{\text{T}}\left(\boldsymbol{X}\boldsymbol{X}^{\text{T}} + \beta\boldsymbol{I}\right)^{-1} \tag{3}$$

The target output of the training data is given by $\boldsymbol{Y}^{\text{target}} \in \mathbb{R}^{N_y \times T}$ is the concatenation of $[1; \boldsymbol{u}(n); \boldsymbol{x}(n)]$, the matrix $\boldsymbol{I}$ defines the identity matrix and β is the regularization coefficient to avoid overfitting. For further details see [15, 16].

3.3 Implementation of the Prediction

The input variables head rotation in y-direction, the steering angle and the use of the turn signal were extracted from the data set of the Naturalistic Driving study. The input data was preprocessed with min-max normalization.

In total the data set (training and testing) contains around 2.5 h of driving data. The training data set contained 26 left lane changes (9 × turn-off, 6 × entering and 11 × slow lead vehicle). The test data set contained 6 left lane changes (2 × turn-off, entering and slow lead vehicle). The target value of the lane changes were marked with 1. The marking starts 5 s (151 frames) prior to the lane change occurrence. All other target values were set to 0. The ESN was generated with 1000 reservoir neurons, $\alpha = 0.3$, $\beta = 10^{-8}$ and 100 inputs for each input variable, so $N_u = 300$. For each input variable the input was given for the current time and for the prior 100 frames. The input and recurrent weights were initialized with a random uniform distribution $[-0.3, 0.3)$. Each 0.033 s the inputs were given to the ESN and the activation of the reservoir was calculated with (1) and (2). Then the output weights were trained with (3) and afterwards the reservoir was updated for the test data and the output was calculated with

$$Y = W^{out}X \tag{4}$$

The output of lane changes was coded with 1 and of no lane changes with 0. The marking of a lane change depends on a threshold, how often the output exceeds the threshold and a condition which checks if new occurrences still appear. First of all the output of a time point is compared with a threshold. If its greater than a threshold, a counter starts. If it is bigger than a threshold and if the counter is greater 30, the predicted output ypred is set to 1, otherwise it is 0. A further variable checks if the counter is still changing. If the counter is not changing for 151 frames, the counter is set to 0. A lane change is recognized, if the sum of the y_{pred} is bigger than 30.

4 Results

The introduced ESN predicted all left lane changes due to an entering, a slow lead vehicle and one of the turn-off lane changes for testing. Figure 2 shows the target (known output) at the top and the calculated output of the ESN at the bottom. The calculated output matches well with the target output, except for the first lane change. It is a lane change due to turning. The steering angle is much higher compared to the other lane changes and the head rotation is lower. The activation was too untypical for a lane change. The lane change occurred after a sharp right curve and a slight left curve and then the driver immediately switched to the turn-off lane. So, the steering angle was more activated than usual and the driving behavior did not match with the behavior prior to a usual lane change. The ESN generates no false alarms and the true positive rate with 83.33% and the false positive rate with 0.00% are good. In average a lane change could be predicted in advance 3.53 s (SD = 0.58 s).

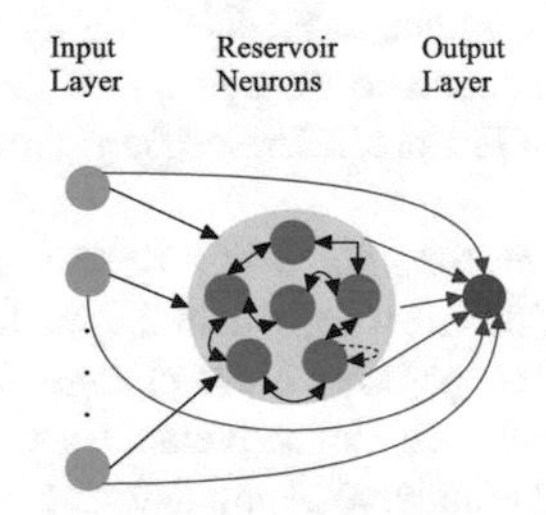

Fig. 1. Scheme of an ESN

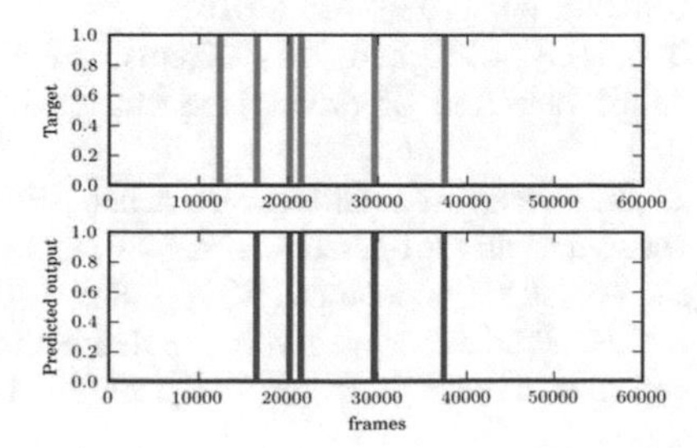

Fig. 2. Output of the ESN vs. target output

5 Conclusion

An ESN was implemented to predict left mandatory and discretionary lane changes. It predicted successfully left lane changes during testing, except for one lane change which showed unusual driving behavior prior to the lane change. The training set with 26 lane changes is smaller than in previous studies [9, 14], nevertheless the results are promising. Especially considering the easy computation of the ESN compared to neural networks, Bayes networks or Hidden Markov Models. Future research will refine the prediction for left lane changes, compare it to other classifiers and add the prediction of right lane changes. Furthermore, a prediction time of 3.53 s is achieved. With this prediction time the drivers have enough time to react to dangerous situations if a warning for an unsafe lane change occurs and an implementation for vehicle-to-vehicle communication is possible as well.

References

1. Statistisches Bundesamt, Verkehr - Verkehrsunfälle. Wiesbaden (2016)
2. Laval, J.A., Daganzo, C.F.: Lane-changing in traffic streams. Transp. Res. Part B: Methodol. **40**(3), 251–264 (2006)
3. I-FAS, Carai-Studie [raw data files]. Technische Universität Chemnitz, Chemnitz (2013)
4. Zheng, J., Suzuki, K., Fujita, M.: Predicting driver's lane-changing decisions using a neural network model. Simul. Model. Pract. Theory **42**, 73–83 (2014)
5. Beggiato, M., Krems, J.F.: Sequence analysis of glance patterns to predict lane changes on urban arterial roads. 6. Tagung Fahrerassistenzsysteme (2013)
6. Peng, J., Guo, Y., Fu, R., Yuan, W., Wang, C.: Multi-parameter prediction of drivers' lane-changing behaviour with neural network model. Appl. Ergon. **50**, 207–217 (2015)
7. Henning, M.J., Georgeon, O., Krems, J.F.: The quality of behavioral and environmental indicators used to infer the intention to change lanes. In: Proceedings of the International Driving Symposium on Human Factors in Driver Assessment, Training, and Vehicle Design, pp. 231–237 (2007)
8. Schmidt, K., Beggiato, M., Hoffmann, K.H., Krems, J.F.: A mathematical model for predicting lane changes using the steering wheel angle. J. Saf. Res. 85–e1 (2014)
9. Leonhardt, V., Wanielik, G.: Recognition of lane change intentions fusing features of driving situation, driver behavior, and vehicle movement by means of neural networks. In: Advanced Microsystems for Automotive Applications 2017, pp. 59–69. Springer (2018)
10. Leonhardt, V., Pech, T., Wanielik, G.: Data fusion and assessment for maneuver prediction including driving situation and driver behavior. In: 19th International Conference on Information Fusion (FUSION), pp. 1702–1708 (2016)
11. Bocklisch, F., Bocklisch, S.F., Beggiato, M., Krems, J.F.: Adaptive fuzzy pattern classification for the online detection of driver lane change intention. Neurocomputing **262**, 148–158 (2017)
12. Li, G., Li, S.E., Liao, Y., Wang, W., Cheng, B., Chen, F.: Lane change maneuver recognition via vehicle state and driver operation signals - results from naturalistic driving data. In: IEEE Intelligent Vehicles Symposium (IV), pp. 865–870 (2015)
13. Streubel, T., Hoffmann, K.H.: Prediction of driver intended path at intersections. In: IEEE Intelligent Vehicles Symposium Proceedings, pp. 134–139 (2014)

14. Díaz-Álvarez, A., Clavijo, M., Jiménez, F., Talavera, E., Serradilla, F.: Modelling the human lane-change execution behaviour through multilayer perceptrons and convolutional neural networks. Transp. Res. Part F: Traffic Psychol. Behav. **56**, 134–148 (2018)
15. Jaeger, H.: The "echo state" approach to analysing and training recurrent neural networks-with an erratum note. Technical Report, vol. 148, no. 34. German National Research Center for Information Technology (GMD), Bonn, Germany (2001)
16. Lukoševičius, M.: A practical guide to applying echo state networks. Springer (2012)

Out of Position Driver Monitoring from Seat Pressure in Dynamic Maneuvers

Alberto Vergnano(✉) and Francesco Leali

Department of Engineering Enzo Ferrari, University of Modena and Reggio Emilia, Via P. Vivarelli, 10, 41125 Modena, Italy
{alberto.vergnano, francesco.leali}@unimore.it

Abstract. An airbag system is designed to reduce the accident outcome on the car occupants. The airbags deployment against manikins is severely tested according to international regulations. The accident scenarios with Out of Position (OP) occupants are critical since they can be hardly expected during design. The airbag deployment in these scenarios can be improved by developing adaptive strategies, provided that the Airbag Control Unit must be aware of the actual occupant position. The present research investigates a sensor system to monitor the occupants in an interactive Human-Car system. The driver position is monitored by pressure sensors, while an accelerometer enables to compensate for acceleration and noise. Real driving experiments in dynamic conditions are reported. The results prove that three OP conditions are effectively identified.

Keywords: Out of position · Driver monitoring · Intelligent vehicle Safety system · Driving experiment

1 Introduction

The airbags are classified as car passive safety, just for helping to reduce the damage in case of accident. However, the airbag deployment in unexpected accident scenarios against Out of Position (OP) occupants can be even additionally harmful [1, 2]. The effectiveness of the airbag system can be increased by making it adaptive to the crash conditions [3]. An airbag with adaptive capabilities can be designed with several solutions [4, 5]. Research works investigate adaptive Airbag Control Units (ACU) taking into account the car speed, impact severity, belt use and occupant body size [3, 6, 7]. Then, the ACU can be improved by considering also the occupants position on the car seats.

A driver position can be monitored through eye and face tracking with artificial vision systems [8]. This equipment is quite complex for actual car applications and prone to heavy noises during real driving. Occupant position can be also identified by monitoring the pressure on the seat surface for ergonomics [9] or safety [10] studies. The occupant OPs can be effectively identified if the pressure field is compared with the normal position one, that is in sequent experiments for the same maneuver. However, a question arises on how to identify the position without the possibility of comparison with many experiments. So, the present research improves the equipment [10] with an

W. Karwowski and T. Ahram (Eds.): IHSI 2019, AISC 903, pp. 76–81, 2019.
https://doi.org/10.1007/978-3-030-11051-2_12

additional sensor mounted on the car, considered as reference moving platform. Real driving experiments investigate the system capabilities in dynamic maneuvers.

The paper is organized as follows. Section 2 presents the sensor equipment and methods. The driving experiments are reported and discussed in Sect. 3, while the concluding remarks are drawn in Sect. 4.

2 Sensor Equipment and Method

The sensor equipment consists of two modules. The relative module monitors the driver pressure on the seat, while the absolute module monitors the acceleration of the car, assumed as the reference moving platform. The driver pressure is clearly affected by inertial effects, that can be evaluated by absolute acceleration monitoring.

In the first module, the pressure is monitored by thin Force Sensing Resistors (FSR). FSRs with different areas were tested. FSRs' with smaller area better detect local pressure changes. However, larger ones are more robust to the interface irregularities due to driver clothes and seat cover seams. The relative module is then equipped with 16 FSR 402 (Interlink Electronics®, Camarillo, California, USA), as 0.45 mm thick and 13 mm diameter active surface. The sensors, shown in Fig. 1a, are arranged in the layout shown in Fig. 1b, named as in the controller software according to Seat, Back, Left and Right initials. Figure 1c shows the equipment as installed on a Quattroporte car (Maserati®, Modena, Italy).

The absolute module uses a BNO 055 System in Package (Bosh®, Reutlingen, Germany). The BNO055 integrates an accelerometer, a gyroscope, a geomagnetic sensor and a microcontroller running Bosch Sensortec sensor fusion software. The BNO 055 is a plug-and-play system that is integrated into the module through the bidirectional I^2C communication protocol. The available fused sensor data are the vectors of angular rate, quaternion, Euler angles, rotation vector, linear acceleration, gravity and magnetic field strength. In order to compensate the relative module for acceleration and noise, the BNO055 is mounted on the car as close to the seat as possible and aligned with the car reference system. In the following experiments, only the 3-axes linear acceleration is monitored, leaving for future research the other data.

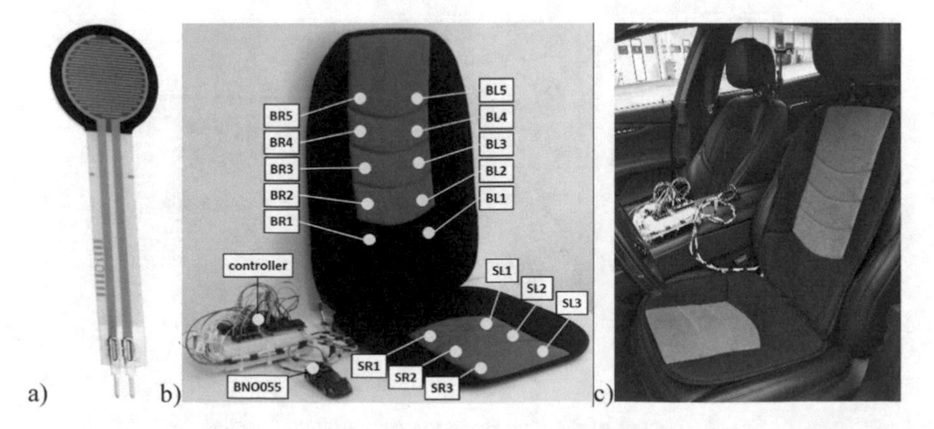

Fig. 1. Sensor equipment: (a) FSR 402, (b) device layout and (c) installed system.

The two modules are controlled through an Arduino Mega controller (Arduino®, Turin, Italy). The complete system includes also LCD screen, LED and pushbutton serving as HMI, SD card slot for data logging and 9 V battery to enable a stand alone device. The prototype is cost effective as 140Eur. Figure 2 shows the signals and energy connections in the system. The complete wiring is omitted for clarity of presentation. The cycle time is 26 ms.

The pressure signals must be processed in order to get useful information. So the, pressure centers in longitudinal and transversal directions are defined as:

$$FSR_{LONG} = \sum_{i=1}^{16} c_{L,i} FSR_i \tag{1}$$

$$FSR_{TRANS} = \sum_{i=1}^{16} c_{T,i} FSR_i \tag{2}$$

where FSR_i are the signal values as read by the analog inputs, $c_{L,i}$ and $c_{T,i}$ are the weighing factors for these signals in longitudinal and translational directions. When the system is switched on, all the FSRs and BNO055 sensors are calibrated. Then, during first driving maneuvers, $c_{L,i}$ and $c_{T,i}$ are determined with a best fitting algorithm with the longitudinal acc_L and translational acc_T accelerations respectively. Then the pressure center coordinates are calculated in the experiments. For results evaluation, note that the pressure centers would move according to the driver body inertia, that is opposite to the acc_L and acc_T accelerations. However, once defined the fitting rule, the center and acceleration values can be compared.

3 Driving Experiments

The driving experiments were performed in the Autodromo di Modena, Italy, within the Modena Automotive Smart Area (MASA) program for connected and autonomous cars [11]. The set up in a racing circuit without any traffic enables to safely test OP

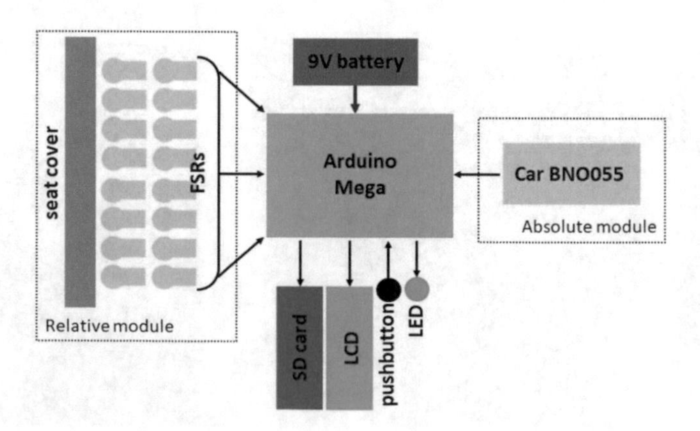

Fig. 2. Layout of the devices in the two modules equipment.

driving. The experiment track consists of a sequence of car maneuvers: straight acceleration (automatic transmission) till about 30 km/h speed, clockwise 360° turn as about 10 m radius at constant speed, counterclockwise 360° turn as about 10 m radius, straight brake. For each driver, the experiments considered 4 postures: normal driving position, forward reclined against steering wheel, left reclined against the side door, right reclined toward the side door distracted by other devices or objects. The experiment results are reported in Fig. 3 considering the longitudinal direction and in Fig. 4 considering the transversal one. The pressure center values can be considered dimensionless.

After calculating the coefficients in (1) and (2) with the best fitting subfunction, the pressure center agrees quite close with the acceleration in case of normal driving position, as in Figs. 3a and 4a. The OPs are clearly detected in the longitudinal direction since the driver back is separated from the seat. However, the OP

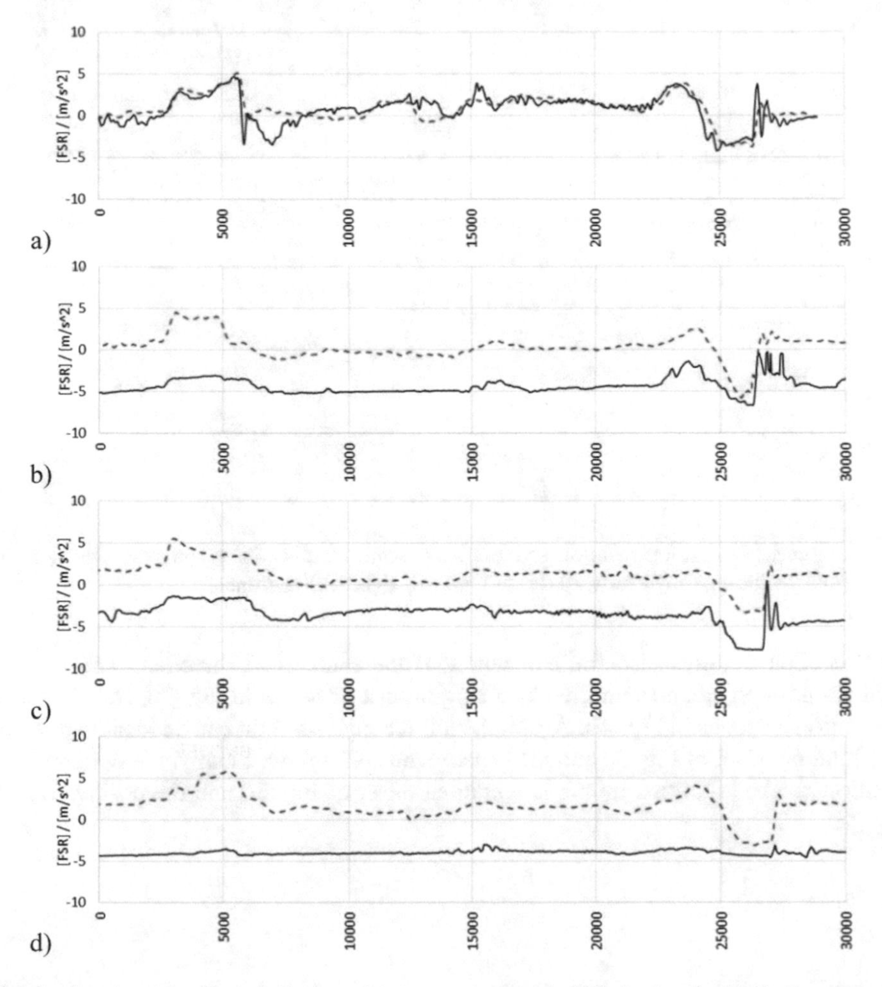

Fig. 3. Acceleration (dotted line) and pressure center (solid) in longitudinal direction for (a) normal position and (b) forward, (c) left and (d) right reclined ones.

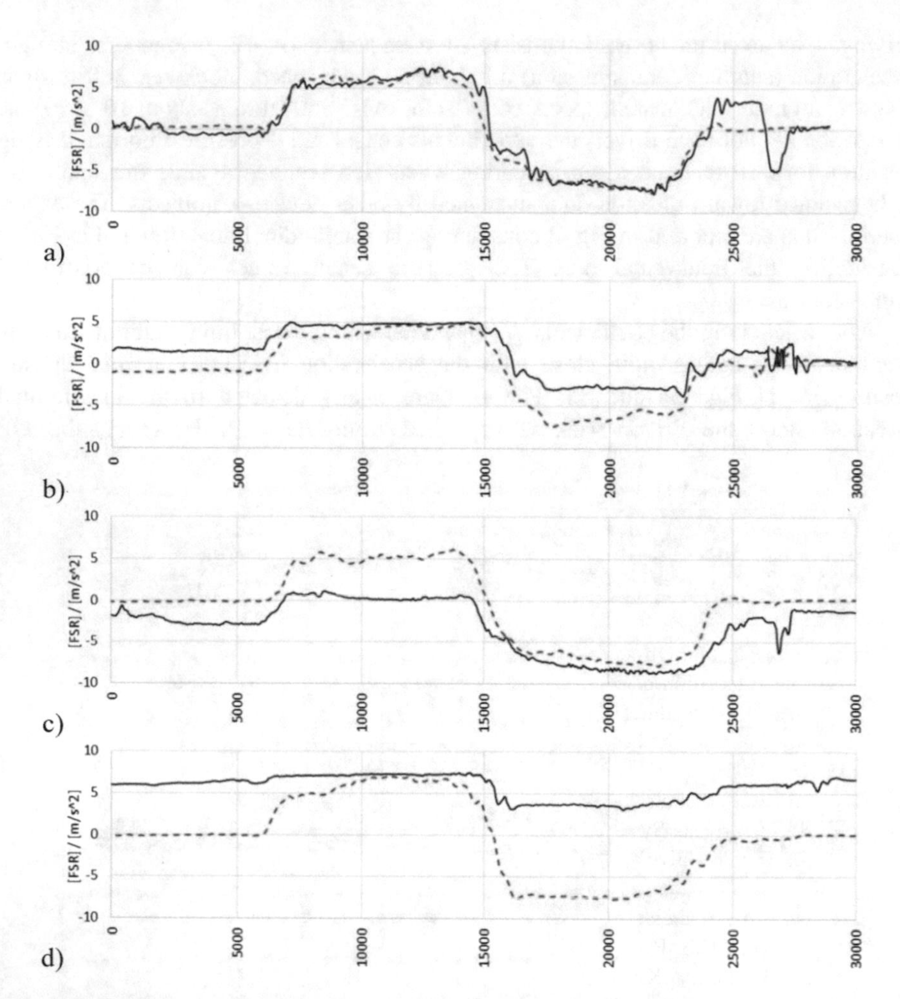

Fig. 4. Acceleration (dotted line) and pressure center (solid) in transversal direction for (a) normal position and (b) forward, (c) left and (d) right reclined ones.

identification requires also the evaluation of the transversal direction. The forward reclined position can be identified by a longitudinal offset, as in Fig. 3b, and almost no translational one, as in Fig. 4b. Again, the left reclined position can be identified by the longitudinal offset of Fig. 3c and the negative one of Fig. 4c. Finally, the right reclined position can be identified by the longitudinal offset of Fig. 3d and the positive one of Fig. 4c.

4 Conclusions

The driver or passenger position is monitored with 16 pressure sensors, mounted both on seat and seatback. An accelerometer enables to compensate the pressure center shift from the inertia effects. Experiments in driving dynamic conditions, with car accelerations and engine noise, prove the reliability of the system. The pressure center shifts in the longitudinal direction clearly identify an OP condition. The center shifts in the transversal direction enable to identify which OP occurs: forward reclined, left reclined or right reclined one.

The occupant OPs enables to develop an adaptive strategy in the ACU for the airbag deployment. So, future works will investigate the introduction of a smart seat into the car system. A pressure pad would enable the driver profiling in order to enhance the driving experience. Machine learning techniques will be introduced to better identify the OPs.

References

1. Potula, S.R., Solanki, K.N., Oglesby, D.L., Tschopp, M.A., Bhatia, M.A.: Investigating occupant safety through simulating the interaction between side curtain airbag deployment and an out-of-position occupant. Accid. Anal. Prev. **49**, 392–403 (2012)
2. Recommended Procedures for Evaluating Occupant Injury Risk from Deploying Airbags. The Side Airbag Out-of-Position Injury Technical Working Group. Alliance, AIAM, AORC, and IIHS (2003)
3. Mon, Y.J.: Airbag controller designed by Adaptive-Network-Based Fuzzy Inference System (ANFIS). Fuzzy Sets Syst. **158**(24), 2706–2714 (2007)
4. Ryan, S.: An innovative approach to adaptive airbag modules. SAE Technical Paper 980646 (1998)
5. Depottey, T.A., Schneider, D.W.: Airbag cushion with adaptive venting for reduced out-of-position effects. U.S. Patent 7,261,319, (2007)
6. Farmer, M.E., Jain, A.K.: Occupant classification system for automotive airbag suppression. In: IEEE Computer Society Conference on Computer Vision and Pattern Recognition 1, pp. I-I. Washington (2003)
7. Yang, J., Håland, Y.: Modeling of adaptive passenger airbag systems in car frontal crashes. In: International Technical Conference on the Enhanced Safety of Vehicles 1996, pp. 486–501. Melbourne (1996)
8. Borghi, G., Venturelli, M., Vezzani, R., Cucchiara, R.: Poseidon: Face-from-depth for driver pose estimation. In: IEEE Conference on Computer Vision and Pattern Recognition, pp. 5494–5503, Honolulu (2017)
9. Andreoni, G., Santambrogio, G.C., Rabuffetti, M., Pedotti, A.: Method for the analysis of posture and interface pressure of car drivers. Appl. Ergon. **33**(6), 511–522 (2002)
10. Vergnano, A., Leali, F.: Monitoring driver posture through sensorized seat. In: 1st International Conference on Human Systems Engineering and Design: Future Trends and Applications, Reims (2018)
11. Modena Automotive Smart Area, http://www.automotivesmartarea.it

Weighing the Importance of Drivers' Workload Measurement Standardization

Eduarda Pereira[1], Susana Costa[2], Nélson Costa[2], and Pedro Arezes[2](✉)

[1] DPS, School of Engineering, University of Minho, Guimarães, Portugal
pereira.eduarda@gmail.com

[2] ALGORITMI Centre, School of Engineering, University of Minho, Guimarães, Portugal
{susana.costa,ncosta,parezes}@dps.uminho.pt

Abstract. Workload is an inescapable topic within the context of Human-Machine Interaction (HMI). Evolution dictates that HMI systems will be all the more attractive to users the more intuitive they are. While attempting to create an optimized workload management system, the authors have encountered a difficulty in gathering a homogeneous definition of workload and a standardized manner of measuring it. In fact, some researchers even call into question the fact that maybe different things are being discussed. Could it be the underlying cause of many of the HMI failures so far recorded? Either way, the weight this concept carries is too heavy to be dealt lightly. It is very important that standardized strategies for measurement of workload are developed so that, for one, different results can be compared and contribute to a more robust understanding of the concept and that, for two, the measurement of workload is globalized and able to be adapted to all users.

Keywords: Measurement · Cognitive workload · ADAS · Driving Human factors · HMI · Autonomous vehicles

1 Introduction

The task of driving a car is extremely demanding. There are many similarities in conduction within the variability of drivers. But the differences are also significant, depending on: the individual characteristics of each driver; the so-called short-term driver states (the temporary impairments such as distraction, high mental workload, or sleepiness [1, 2]; and the "road design (i.e. motorways vs. rural roads vs. city roads), road layout (straight vs. with curves, leveled vs. inclined, junction vs. no junction) and traffic flow (high density vs. low density)" [3, 4]. The task demand and the capacity to execute it are also relevant, as problems arise when efforts induced by surrounding stimuli exceed the processing data capacity of drivers. They may quickly become unable to manage all the difficulties posed by the road context, with potentially harmful consequences [5, 6]. Thus, assessing the driver's state has become imperative for the

W. Karwowski and T. Ahram (Eds.): IHSI 2019, AISC 903, pp. 82–90, 2019.
https://doi.org/10.1007/978-3-030-11051-2_13

development of Advanced Driver Assistance Systems (ADAS), which will ultimately reduce the occurrence of fatal incidents on the road. Exposition of control, guidance and navigation issues, as well as driver's individual capabilities, determine the total amount of workload [5, 7]. The sooner the driver detects a problematic situation, the easier it is for him/her to "manage it". Response time is commonly associated with driver workload. Recent advances in workload applicability in HMI systems have made the concept too broad. It is typically seen as the amount of resources needed to handle the demands of a task, and it is closely related to the characteristics of the task, the situation and the driver. However, there is no universal definition of workload [8–10]. Moreover, several authors do not define workload and their evaluations are based solely on the results of the measurements performed. As it is very important to measure mental workload, in order to understand the problematic regarding the interaction between drivers and cars, and implement HMI systems with algorithms that are able to manage it accordingly, it is important that the measurement methods are validated for harmonization [11] which may, then, render them comparable and provide the bases for the establishment of algorithms for workload management systems within the vehicles' HMIs.

As noted by Hancock and Desmond [12], while the terms stress and workload arise out of somewhat different traditions, there is a great deal of conceptual overlap in describing demands on the individual arising from both internal and external factors. Indeed, the concept of workload is not only closely related to stress, but also to fatigue, burnout and inattention [13]. Now, these are concepts, which, in turn, are related to other major ergonomic indicators such as accidents.

Driving support systems are increasingly involved in autonomous driving systems. It is not certain when cars with level 5 autonomous driving will be available to move legally on the roads. For now, the available autonomy levels require and/or allow driver intervention through the human-machine-interfaces HMI.

The design attributes of new generation driving assistance systems should be influenced by human factors in driving. According to a recent news article, development of 'human-like' self-driving technologies is attracting investor capital [13]. Also, because the design of the driving assistance, if guided by human factors, is likely to enhance driver acceptance, safety benefits will be achieved. When they interact with technology, drivers experience mental workload (MWL), a mental state that has been the focus of neuroeurgonomics research, mainly associated with the autonomous driving paradigm. The looked-but-failed-to-see phenomenon is quite often for drivers, caused by a cognitive workload that is either too high or too low [1, 2]. However, the coupling with accident causation has not been established via a direct link yet [14]. Research on the impact of driving workload is largely dependent on effective experimental methods and evaluation methods [15] and it is important to reference methods that can be used in research but also with respect to real-life applications [1, 2].

Workload measures can be classified by nature as physiological, subjective or performance. Physiological measures include cardiac measures (heart rate (ECG), heart rate variability and blood pressure), respiratory measures (respiratory rate, volume and concentration of carbon-dioxide in air flow), eye activity measures (eye blink rate and interval of closure, horizontal eye activity, pupil diameter, eye fixations), speech measures (pitch, rate, loudness), brain activity (measured through electroencephalogram – EEG - and electrooculogram - EOG) [16]. Subjective measures include questionnaires like the NASA Task Load Index (NASA-TLX).

If a standardization of methods to evaluate workload is envisioned, it is also extremely important that the approach to the workload concept be unanimous so as to enable the response to the characterization and comparison of MWL measurement methodologies that promote the development of efficient driver workload management systems and MWL-adaptive HMIs.

2 Methodology

This work is based on a Systematic Literature Review (SLR), which was previously conducted by the authors in the course of their study on autonomous vehicles (AV). This SLR followed a 3-step approach and tried to answer the research question: "How can the available literature on driver workload measurement be characterized and compared, to allow for a better understanding of the current scenario, and to promote the use of data in future studies and development of efficient driver workload management systems?".

The authors resorted to 3 bibliographic databases were selected (ISI Web of Science, Elsevier and PubMed) and, using the keywords "driver workload measurement" for the time span from 2008 to 2018. Papers from ISI Web of Science were limited to open access papers, for the other two databases, only B-on Consortium of Portugal granted access were considered. Review articles were excluded, because it was important that the authors, themselves, performed the tests. After duplicates were removed, a total of 36 articles were retrieved from the databases. For the present study only a selection of these articles has been considered. The criteria exclusion are:

- The data collected was regarding other types of workload (e.g., physical workload);
- The data collected was regarding workload that was not directly related to driving.

This triage resulted in a final total of 23 articles of interest for this study.

3 Results and Discussion

Results from the SLR are presented in Table 1.

The bulk of information gathered in the most relevant articles devoted to this theme the last decade show some convergence between some authors, but also many dissimilarities.

Table 1. SLR results.

Author	Workload concept	Workload measurements
Karthaus et al. [17]	n.d.	Physiological (EEG)
Zhang and Kumada [3]	"the ratio of demand to allocated resources"	Subjective (NASA-TLX)
Lyu et al. [15]	"measure of the effort expended by a human operator while performing a task, independently of the performance of the task itself" [18] "the demand tasks exerted on a pool of undifferentiated mental resources"	Subjective (NASA-TLX)
Foy et al. [8]	n.d.	Physiological (FNIRS - to measure the PFC activity)
Aricò et al. [19]	"mental workload is a complex construct that is assumed to be reflective of an individual's level of attentional engagement and mental effort" [20] "measurement of mental workload essentially represents the quantification of mental activity resulting from performance of a task or set of tasks."	Physiological (EEG) Subjective workload perception (instantaneous self-assessment)
Platten et al. [21]	"the interaction between the current demands of a situation (task load) and the resources, skills and characteristics of a driver"	Physiological (pupillometry)
Shakouri et al. [22]	"represents the cost of accomplishing a task" [23] and can be defined as the amount of information-processing resources used per unit time to meet the level of performance required for the task [24]	Subjective (NASA-TLX) Physiological (heart rate variability) Performance (variability in steering angle, braking and speed)
Horrey et al. [25]	n.d.	Subjective (NASA-TLX) Physiological (near-infrared spectroscopy, heart monitoring, pupil diameter and eye movement) Performance (lane keeping and headway maintenance, speed variation, response times to critical braking events, and turn signal task response time)
Xing et al. [26]	n.d. "the proportion of an operator's limited capacity that is needed to conduct a specific task." [27]	ERRC-SVM (novel hybrid method for measuring driver workload comprising 15 measured physiological and vehicle-state variables)

(*continued*)

Table 1. (*continued*)

Author	Workload concept	Workload measurements
Ruscio et al. [28]	n.d	Subjective (NASA-TLX) Physiological (Mean Heart Rate, High Frequency Power of heart rate variability, Blood Volume Pulse amplitude and Cardiac Autonomic Balance)
Heine et al. [29]	"the amount of resources needed for the processing of a certain task" [30]	Physiological Performance
Teh et al. [31]	"the amount of information-processing resources used per time unit, to meet the level of performance required" [24]	Subjective (NASA-Task Load Index, NASA-RTLX; Rating Scale Mental Effort and Continuous Subjective Rating) Performance: Mean and standard deviation of speed, Mean time headway, High frequency component of steering angle and Standard Deviation of Lateral Position
Ronen et al. [32]	n.d.	Subjective Physiological Performance
Rose et al. [33]	n.d.	Subjective (NASA-TLX)
Di Stasi et al. [34]	n.d.	Mental fatigue and mental workload Subjectives Physiological: saccadic eye movement parameters (including the peak velocity)
Hajek et al. [35]	"the sum of the costs of cognitive processing and is reflected in physiological measurements" [14]	Physiological (heart rate, galvanic skin response and respiration)
Ko and Ji [36]	"the amount of attentional resources required to complete a task" [24]	Subjective (NASA-TLX and a flow short scale to measure the perceived demand level of the activity)
Ahlström et al. [1, 2]	n.d.	Performance (detection response task) Physiological (EEG and EOG - the most reliable results are obtained when these parameters are analyzed in parallel)
Pecchini et al. [5]	"driver's personal reaction to task demand"	Performance (analysis of steering behaviour of the driver, in particular time evolution of steering wheel angles)
Galy et al. [37]	n.d.	Subjective (NASA-TLX)

(*continued*)

Table 1. (*continued*)

Author	Workload concept	Workload measurements
Brookhuis and Waard [14]	"a consequence of the driving task's demands, among other things"	ECG (physical efforts)
Stuiver et al. [38]	"how much of someone's information-processing capacity is needed during task performance and how this is influenced by task demands" [39]	Physiological (heart rate, blood pressure)
Baldauf et al. [40]	"the overall cognitive effort a person invests in his performance while carrying out a task" [41]	Performance: peripheral detection task Physiological: electrodermal activity Subjective: subjective workload assessment technique Time perception

n.d. (not defined)

It can be seen, for instance, that there is a tendency to use physiological and subjective assessment methodologies in the latter years, being that the NASA-TLX method appears as the favorite within the methodology for subjective evaluation. This may be due to the focus that this method has deserved for several years, having even been revised and improved in subsequent versions, which may grant authors some sense of confidence in its use. But the major similarity that can be found is that almost all authors attempt to establish a methodology that embraces several methods. It seems reasonable to assume, given the SLR carried, that studying MWL often requires to deal with a bulk of knowledge in different fields [19], in hopes of rendering more reliable results, by comparing and inferring one through the other. This seems to be the case for some studies which comprise the use of a set of performance and/or subjective and/or physiological methodologies, but also sets of, for instance, different physiological methodologies [14].

Also noteworthy is the stark lack of consensus and even the high variability in the definition of the workload concept [19, 37], which may explain the difficulty in applying similar methodologies to obtain answers. More problematic, however, is the lack of definition of workload in a significant number of studies, which may even attest to the thesis that, in fact, when talking about workload, researchers may not all be talking about the same concept (at the very least, one does not have evidence of doing so). Indeed, it seems logical to infer that, for the above mentioned, there is a need to converge in one concept of workload, in order to establish a standardized way of measuring it, so as to yield scientific results that enable the development of efficient workload management and workload-adaptive HMI systems for all vehicles, including AV.

4 Conclusions

Long gone is the time when technological state-of-the-art meant complex, intricate systems. The goal is to offer simplicity despite all the necessary technological complexity that lies within. More than a hedonic question, simplicity in HMI is also paramount for addressing more pressing issues such as safety and, therefore, trust in such systems. Referring to the specific case of autonomous driving, this issue takes on a considerable dimension, due to the impact that a fault in this system can have, both in terms of severity and the number of affected. If, on the one hand, recent studies report the lack of confidence of hypothetical purchasers in autonomous technology, the history of accidents and their consequences has not favorably advocated such technology. In high level of autonomy vehicles, where the participation of the driver in the dynamic driving task may be requested, the cooperation between the driver and the HMI must be flawless or, at least, allow the associated error to be so modulated that it does not become a relevant hazard. So, this cooperation has to flow in a natural manner. Design principles state that the system′s interface must be such that the driver does not need training to interact with it, and easily interprets the message that is being transmitted. That is called being intuitive. Moreover, design principles also take into account not only the variability of users, but also variability within the different states of the user, the road conditions, the weather, and so on. That is called being adaptive. For several years, now, a panoply of ADAS have been developed for tackling this variability (e.g., automate lighting, lane departure warning systems) by resorting to a bulk of technology (e.g., LIDAR, computer vision). It would seem logical that the creation of an optimized HMI would be a straightforward process, as it would only imply compliance with the design criteria already studied and established by experts. It so happens, though, that juggling all the criteria is not an easy task. Since it is difficult to establish a comparison between the different methods of measurement, it is important to consider the standardization of the methods and methodologies that have been used so far.

In short, the initial concept of workload, the different measurements and the different scenarios for the realization of the experience, can somehow justify all differences in results and who knows, even the current incidents that occur with the use of ADAS in vehicles, disqualifying their adhesion and use.

Ideally, a standardized methodology for evaluating MWL has to be established, whereby a bulk of reliable, comparable data will emerge, that will allow for identifying decisive parameters in several driving scenarios, and from which processing will arise at least one algorithm that transcribes a MWL-adaptive HMI, capable of guaranteeing that the MWL of the driver remains within optimal limits, above or below which the vehicle's HMI will trigger an alert to the driver.

Acknowledgments. This work has been supported by European Structural and Investment Funds in the FEDER component, through the Operational Competitiveness and Internationalization Programme (COMPETE 2020) [Project nº 039334; Funding Reference: POCI-01-0247-FEDER-039334].

References

1. Ahlström, C., Kircher, K., Fors, C., Dukic, T., Patten, C., Anund, A.: Measuring driver impairments: sleepiness, distraction, and workload. EEE PULSE **3**, 1–9 (2012)
2. Young, M.S., Birrell, S.A., Stanton, N.A.: Safe driving in a green world: a review of driver performance benchmarks and technologies to support 'smart' driving. Appl. Ergon. **42**(4), 533–539 (2011)
3. Zhang, Y., Kumada, T.: Relationship between workload and mind-wandering in simulated driving. PLoS One **12**(5), e0176962 (2017). https://doi.org/10.1371/journal.pone.0176962

4. Costa, S., Simões, P., Costa, N., Arezes, P.: A Cooperative Human-Machine Interaction Warning Strategy for the Semi-Autonomous Driving Context, 1–7 (2017)
5. Pecchini, D., Roncella, R., Forlani, G., Giuliani, F.: Measuring driving workload of heavy vehicles at roundabouts. Transp. Res. Part F **45**, 27–42 (2017)
6. Fitzpatrick, K., Chrysler, S., Park, E.S., Nelson, A., Robertson, J., Iragavarapu, V.: Driver Workload at Higher Speeds. FHWA/TX-10/0-5911-1, FHWA (2010)
7. Ba, Y., Zhang, W.: A review of driver mental workload in driver-vehicle-environment system. Internationalization, Design and Global Development (2011)
8. Foy, H.J., Runham, P., Chapman, P.: Prefrontal cortex activation and young driver behaviour: a fNIRS study. PLoS One **11**(5), e0156512 (2016). https://doi.org/10.1371/journal.pone.0156512Colquhoun
9. Eggemeier, F.T., Wilson, G.F., Kramer, A.F., Damos, D.L.: General considerations concerning workload assessment in multi-task environments. In: Damos, D.L. (ed.) Multiple Task Performance, pp. 207–216. T&F, London (1991)
10. Young, M.S., Brookhuis, K.A., Wickens, C.D., Hancock, P.A.: State of science: mental workload in ergonomics. Ergonomics **58**, 1–17 (2015)
11. Heine, T., Lenis, G., Reichensperger, P., Beran, T., Doessel, O., Deml, B.: Electrocardiographic features for the measurement of drivers' mental workload. Appl. Ergonomics **61**, 31–43 (2017). https://doi.org/10.1016/j.apergo.2016.12.015
12. Hancock, P.A., Desmond, P.A.: Preface. In: Hancock, P.A., Desmond, P.A. (eds.) Stress, Workload, and Fatigue, pp. 13–15. Lawrence Erlbaum Associates, Mahwah (2001)
13. Coughlin, J.F., Reimer, B., Mehler, B.: Driver wellness, safety & the development of an awarecar. Mass Inst. Technol., 1–15 (2009)
14. Brookhuis, K.A., de Waard, D.: Monitoring drivers' mental workload in driving simulators using physiological measures. Accid. Anal. Prev. **42**(3), 898–903 (2010). H Associates, Mahwah, NJ, pp. 13-15
15. Lyu, N., Xie, L., Wu, C., Fu, Q., Deng, C.: Ariver's cognitive workload and driving performance under traffic sign information exposure in complex environments: a case study of the highways in China. Int. J. Environ. Res. Public Health **14**(2), 203 (2017)
16. De Waard, D.: The Measurement of Drivers' Mental Workload. Groningen University, Traffic Research Center, Groningen (1996)
17. Karthaus, M., Wascher, E., Getzmann, S.: Proactive vs. reactive car driving: EEG evidence for different driving strategies of older drivers. PLoS One **13**(1), 19–1500 (2018). https://doi.org/10.1371/journal.pone.0191500. Omation (pp. 37–46). Springer International Publishing
18. Pfeffer, S., Decker, P., Maier, T., Stricker, E.: Estimation of operator input and output workload in complex human-machine-systems for usability issues with iFlow. In: Harris D. (eds.) Engineering Psychology and Cognitive Ergonomics. Understanding Human Cognition. EPCE 2013. Lecture Notes in Computer Science, vol. 8019. Springer, Berlin, Heidelberg (2013)
19. Aricò, P., Borghini, G., Di Flumeri, G., Colosimo, A., Pozzi, S., Babiloni, F.: A passive brain–computer interface application for the mental workload assessment on professional air traffic controllers during realistic air traffic control tasks. Prog. Brain Res. **228**, 295–328 (2016). https://doi.org/10.1177/0018720814542651
20. Wickens, C.D.: Processing resources and attention. Multiple-task performance, 3–34 (1991)
21. Platten, F., Schwalm, M., Hülsmann, J., Krems, J.: Analysis of compensative behavior in demanding driving situations. Transp. Res. Part F: Traffic Psychol. Behav. **26**, 38–48 (2014)
22. Shakouri, M., Ikuma, L.H., Aghazadeh, F., Nahmens, I.: Analysis of the sensitivity of heart rate variability and subjective workload measures in a driving simulator: the case of highway work zones. Int. J. Ind. Ergon. **66**, 136–145 (2018)

23. Hart, S.G.: Nasa-task Load Index (NASA-TLX); 20 Years Later. Moffett Field, Santa Clara County (2006)
24. Wickens, C.D., Hollands, J.G.: Engineering Psychology and Human Performance (2000)
25. Horrey, W.J., Lesch, M.F., Garabet, A., Simmons, L., Maikala, R.: Distraction and task engagement: how interesting and boring information impact driving performance and subjective and physiological responses. Appl. Ergon. **58**, 342–348 (2017). https://doi.org/10.1016/j.apergo.2016.07.011
26. Xing, Y., Lv, C., Cao, D., Wang, H., Zhao, Y.: Driver workload estimation using a novel hybrid method of error reduction ratio causality and support vector machine. Measurement **114**, 390–397 (2017)
27. O'Donnell, R.D., Eggemeie, F.T.: Workload assessment methodology. In: Boff, K.R., Kaufman, L., Thomas, J.P. (eds.) Cognitive Processes and Performance. Wiley, Hoboken (1986)
28. Ruscio, D., Caruso, G., Mussone, L., Bordegoni, M.: Eco-driving for the first time: the implications of advanced assisting technologies in supporting pro-environmental changes. Int. J. Ind. Ergon. **64**, 134–142 (2018)
29. Heine, T., Lenis, G., Reichensperger, P., Beran, T., Doessel, O., Deml, B.: Electrocardiographic features for the measurement of drivers' mental workload. Appl. Ergon. **61**, 31–43 (2017)
30. Eggemeier, F.T., Wilson, G.F., Kramer, A.F., Damos, D.L.: Workload assessment in multi-task environments. In: Damos, D.L. (ed.) Multiple Task Performance, pp. 207–216. Taylor & Francis, London (1991)
31. Teh, E., Jamson, S., Carsten, O., Jamson, H.: Temporal fluctuations in driving demand: The effect of traffic complexity on subjective measures of workload and driving performance. Transp. Res. Part F: Traffic Psychol. Behav. **22**, 207–217 (2013)
32. Ronen, A., Yair, N.: The adaptation period to a driving simulator. Transp. Res. Part F: Traffic Psychol. Behav. **18**, 94–106 (2013)
33. Rose, J., Bearman, C., Dorrian, J.: The low-event task subjective situation awareness (LETSSA) technique: development and evaluation of a new subjective measure of situation awareness. Appl. Ergon. **68**, 273–282 (2018)
34. Di Stasi, L.L., Renner, R., Catena, A., Cañas, J.J., Velichkovsky, B.M., & Pannasch, S.: Towards a driver fatigue test based on the saccadic main sequence: a partial validation by subjective report data. Trans. Res. Part C: Emerg. Technol. **21**(1), 122–133 (2012)
35. Hajek, W., Gaponova, I., Fleischer, K.H., & Krems, J.: Workload-adaptive cruise control? A new generation of advanced driver assistance systems. Transp. Res. Part F: Traffic Psychol. Behav. **20**, 108–120 (2013)
36. Ko, S.M., Ji, Y.G.: How we can measure the non-driving-task engagement in automated driving: comparing flow experience and workload. Appl. Ergon. **67**, 237–245 (2018)
37. Galy, E., Paxion, J., Berthelon, C.: Measuring mental workload with the NASA-TLX needs to examine each dimension rather than relying on the global score: an example with driving. Ergonomics **61**(4), 517–527 (2018). https://doi.org/10.1080/00140139.2017.1369583
38. Stuiver, A., Brookhuis, K.A., de Waard, D., Mulder, B.: Short-term cardiovascular measures for driver support: increasing sensitivity for detecting changes in mental workload. Int. J. Psychophysiol. **92**, 35–41 (2014)
39. de Rivecourt, M., Kuperus, M.N., Post, W.J., Mulder, L.J.M.: Cardiovascular and eye activity measures as indices for momentary changes in mental effort during simulated flight. Ergonomics **51**, 1295–1319 (2008)
40. Baldauf, D., Burgard, E., Wittmann, M.: Time perception as a workload measure in simulated car driving. Appl. Ergon. **40**, 929–935 (2009)
41. Hart, S.G., Wickens, C.D.: Workload assessment and prediction. In: Booher, H.R. (ed.) An Approach to Systems Integration, pp. 257–296. Van Nostrand Reinhold, New York (1990)

Developing Intelligent Multimodal IVI Systems to Reduce Driver Distraction

Ahmed Farooq(✉), Grigori Evreinov, Roope Raisamo, and Arto Hippula

Tampere Unit for Computer-Human Interaction (TAUCHI), Faculty of Communication Sciences, University of Tampere, Tampere, Finland
{Ahmed.Farooq,Grigori.Evreinov,Roope.Raisamo,Arto.Hippula}@uta.fi

Abstract. As research into autonomous vehicles gets mainstream, automobile manufacturers are trying to reinvent the ways we operate and interact with our vehicles. This is now more evident in the central instrument clusters than ever before. While some manufacturers are focusing on adding touchscreens to replace most in-vehicular infotainment (IVI) controls, others are trying to make the IVIS smarter and multimodal. Unfortunately, the lack of reliable tactile feedback for touchscreen interaction in IVI systems can be a major issue. Although this may not be a critical flaw in mobile device interaction, it can be a dangerous limitation for in-car systems where visual distraction can be fatal. For this reason, our research* is focused on exploring and developing new methods of providing tactile actuation using both vibrotactile and pneumatic feedback techniques.

Keywords: Human-systems integration · Multimodal interaction

1 Introduction

Currently, most manufacturers utilize custom yet rudimentary systems for driver-vehicle interaction (DVI) and information exchange. These systems (Fig. 1) utilize either an adapted mobile platform (such as Apple Carplay, Android Auto etc.) or more traditional in-vehicle infotainment platforms (i.e. Genivi). The adapted mobile platforms such as Android Auto and Apple Carplay utilize the visual and auditory I/O mechanism from their mobile platform core (Android, iOS) to facilitate interaction with the driver or a passenger. The most useful attribute of these systems is the seamless integration of Natural Language Processing (NPL). In fact, natural voice interaction through onboard smart assistants is one of the key building blocks for these platforms. However, apart from NLP, these systems provide limited usability advantages over any common IVI system. As these systems create an interaction layer on top of the driver's mobile device, channeling data and services from the phone itself, this can sometimes make the system more difficult to operate [1]. This is because these platforms, in most cases, limit the functionalities and services already available on the mobile device, creating a workspace for the user that may be familiar yet more restricted than the user expects. Furthermore, as these systems are designed by companies with limited

W. Karwowski and T. Ahram (Eds.): IHSI 2019, AISC 903, pp. 91–97, 2019.
https://doi.org/10.1007/978-3-030-11051-2_14

experience of in-vehicle user-system interaction, their usability as well as usefulness may still need to be proven on the road. Lastly, most automobile manufacturers support these mobile platforms on top of their own custom IVI systems, which means that the driver/user must navigate through multiple systems that may have completely counter-intuitive user interfaces just to carry out simple tasks.

Fig. 1. (Left to right and top to bottom) Byton's Shared Experience Display, BMWs Gesture INTERFACE, Audi A8 full touch display, GM's Cadillac User Experience 2018, Mercedes Command Online system and Tesla Model 3 IVI system.

2 Multimodal Driver-Vehicle Interaction

The most traditional way of providing information to the driver is visualization. Graphical user interfaces (GUIs) are still the most popular method of information mediation in GPS systems, radios, and mobile phone as well as other IVI systems. This is problematic, as GUIs reduce driving safety alarmingly. When a complicated GUI captures driver's gaze and attention, it demands more than 20 s for the driver to gain awareness of surroundings and take control over driving [1]. In urgent situations, 10 s is too short a time to prevent collisions or other serious accidents. HUD displays, continuous information mediation in automated driving, and glasses have been proposed to solve the problem, as they seem to be superior to manual driving assisted with dashboard-mounted displays. However, the major problem of visual distraction still remains.

Audio-based interaction reduces visual load as the driver can use speech for input and get a response through audio. In addition to traditional synthetic speech samples and warning signals (beeps), directional and spatial audio has been implemented in IVI systems as feedback methods. Several studies indicated that such specialized audio cues work better than unlocalized sounds. Furthermore, studies show [2] that abstract audio signals worked only when presented from the center of gaze and attention (in this case, tablet) instead of, for instance, behind steering wheel. However, audio might easily be neglected due to environmental noise inside the vehicle, and therefore, it is not sufficient for information mediation.

The use of haptics and touch-actuated interfaces has often been considered as an option to overcome shortcomings of both visual and auditory information mediation. Our research in the field [3], shows that complementary haptic cues may reduce the visual workload and decrease reaction time for driving-related tasks. Haptic feedback or clicks have also shown [4] to significantly improve user satisfaction and task completion times when operating with touchscreens. Even though there has been plenty of progress in haptic technologies recently [5, 6], most available haptic interaction methods still produce vibration making them prone to left unnoticed due to environmental trembling. Another significant problem of the technology is the driver's need to touch the device operated. This means that at least one hand needs to be taken off the steering wheel to be able to operate the UI, which can be seen as somewhat problematic. Therefore, IVI systems should utilize different modalities suitable for the task and particular situation. Our research [3, 4 and 6] has already started to solve this problem showing, that it is possible to provide driving information and improve driving performance by augmenting current environmental and telemetric information through haptic information channel as well as visual and auditory feedback.

3 Multimodal Interaction: System Design

To ensure an intelligent and useful interaction system can be developed for future IVI systems, we focused our efforts in two key areas. Redesigning how users interact and get feedback from the center stack (touchscreen) while reducing the amount and type of information that needs visual confirmation; and proposing novel ways on of interacting with IVIS (via the driver's seat). In this section, we go through each area of focus by discussing the prototype devices developed to enhance user interaction.

3.1 Touchscreen Interaction

To improve haptic feedback in a moving car we developed an advanced layer of tactile feedback that could reduce the need for continuous visual validation. This device uses a transparent overlay that generated skin micro-displacements by moving side to side on the touchscreen. This type of actuation was designed to create more reliable feedback for touchscreen interaction in a haptically noisy environment. The tangential actuation of the screen overlay consisted of three Tectonic actuators (TEAX14C02-8) attached to an L-shaped bracket and fixed to the touchscreen overlay. On actuation of the PET film ($\sim$100 mm thick), the overlay displaced the users skin, in contact with it, creating tactile actuation. The overlay covered the entire screen of the Intel ExoPC slate, used in the prototype, and sat almost flush with the screen.

The second novel device in our research utilized pneumatic feedback to avoid (vibration) signal attenuations. As this method bypasses vibrotactile actuation altogether, therefore, noisy road vibrations do not affect it. The device, Pneumatic Subwoofer (PSW), (Fig. 2) created pressurized air pulses via two hermetically sealed subwoofers in a closed chamber and funneled the air pulses onto the surface of ExoPC tablet's touchscreen. The prototype provided variable magnitude of pneumatic pulses via a modulated digital sine wave generator regulated to translate signal amplitude and frequency into pneumatic haptic signals. The PSW device used two standard (Raptor-6) car-woofers of 140 W each with (2 * 4) 8 Ω load impedance and uses a maximum of

(6.5 A * 2) 13 A of current. The setup was tested in various environments to ensure its reliability and usability in actual moving car was not affected due to environmental noise.

3.2 Novel Methods of Interaction

Improving how drivers navigate on the road is a key component of reducing driver distraction. Our research focused on shifting this visual-attention-heavy task and introducing haptic feedback to reduce driver distraction. The Haptic Seat prototype provided three distinct types of actuation cues; simple event-based cues; spatial navigation (directional); and cues/alerts for warnings. The prototype used Tectonic TEAX25C10-8/HS actuators embedded on either side of the driver's seat, to provide vertical actuation while two Fukoku USR60-E3T ultrasonic motors with asymmetrical arrangement of spherical touch points were used to generate directional actuation (Fig. 2). The motors were used to compliment the Techtonic actuators and provide the directional feedback needed to perform the navigational task. Essentially, the directional rotation of the motors along with the degree of their rotation provided the necessary (complex haptic) information needed to complete the navigational task. Using rotations of 120 and 270 in two distinct bursts, the prototype provided instruction of right and left turnings as well as hard turns or U-turns.

Fig. 2. (Left to right) LSE TS, PSW pneumatic and Haptic Seat prototype devices.

4 User Study and Results

In total, three studies were conducted in the research to evaluate the various techniques and prototypes. The first study was carried out in a Volvo XC60 being driven in a straight line by professional drivers on the Nokian Tyres Testing Track (NTTT). The purpose of the study was to evaluate the effectiveness of providing pneumatic actuation on the central touchscreen for text entry and menu-based selection tasks as compared to simple vibration-based feedback. The second study was conducted to evaluate the tangential and lateral actuation approach on a touchscreen device and it was carried out on a patch of highway within the participants' own vehicle. The third study was conducted in a laboratory setup using the driving simulators (Lane change test software) to compare navigation feedback using conventional audio cues as well as haptic and audio cues using the Haptic Seat prototype. This section details the design of these studies as well as the testing parameters utilized in the design.

4.1 Road Studies 1 & 2

Studies 1 & 2 had identical tasks, which involved completing three usage scenarios (text entry, selecting menu option & performing a gesture). The participants (N = 14, M = 6, F = 8) needed to complete the task as soon as possible using the relevant type of device feedback (PSW in study 1 and LSE in study). To replicate the generic vibrotactile feedback method, we developed a custom vibration-based actuation system (CVAP) and used it as an experiment control in both studies. The participants completed these tasks in two scenarios, (1) while driving on a straight road at 45 km/h in a no traffic area, (2) while the car was stationary in the parking lot.

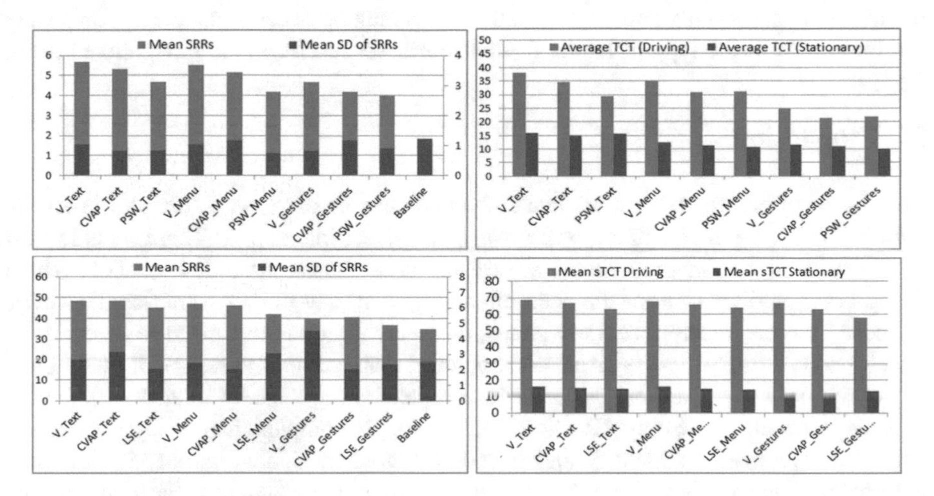

Fig. 3. Primary task (left), and secondary task performance (right) using PSW and LSE device prototypes in studies 1 & 2.

We measured the primary task performance by looking at the Steering Wheel Reversals (as suggested by SAE J2944_201506). To measure the secondary task performance we looked at task completion times and task errors. The results (Fig. 3) show that the participants performed worst in driving tasks where there was no multimodal/haptic feedback. However, amongst the three haptic feedback techniques (CVAP, PSW & LSE), both LSE and PSW improved performance of primary and secondary tasks.

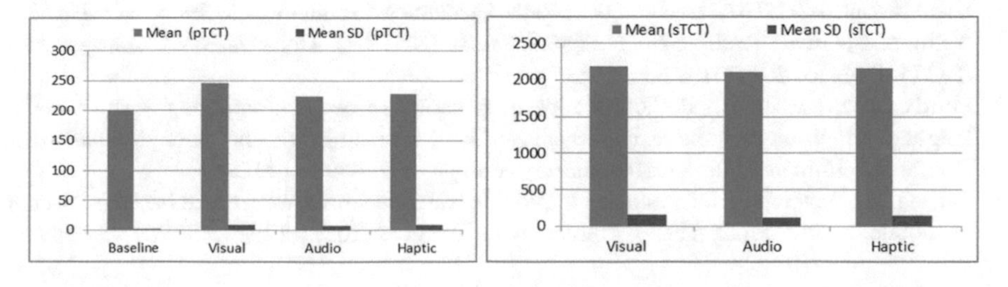

Fig. 4. Primary LCT task (left), and secondary navigation task performance (right)

4.2 Lab Study 3

Study III was conducted in the lab where the primary task was to follow the LCT simulator and the secondary task was to identify the navigational cues by pressing the correct button on the steering wheel. All the participants (N = 24, M = 16, F = 8) utilized independent interaction modalities (Visual-only, Audio-only & Haptic-only) to complete the secondary task, comparing devices as well as modalities in the study. Looking at primary and secondary task performances (Fig. 4), audio-only and haptics-only modalities, yielded fewer errors compared to visual-only modality (VoM). Furthermore, VoM also increased the task completion time. Although haptic interaction was fast and yielded the smallest number of primary and secondary task errors, it was not as fast as audio-only feedback. We think this may be because the participants were not familiar with the type of haptic feedback utilized and its use in navigational tasks.

5 Conclusion

The results of this research show that some tasks can be extremely difficult to perform within the driving environment when using just visual interaction. These include 'text entry' as well as 'layered menu selection' on a touchscreen device. IVI system designers should try to limit the need of such tasks in day-to-day scenarios. However, as these tasks are inherent to the systems, the interaction methods need to be carefully developed to reduce driver's visual and cognitive distraction. The results of Study I and II clearly show that haptics is suitable for supplementing visually intensive IVI tasks. Moreover, results of Study III show that although complex haptic signals require cognitive overheads to identify and decode the applied signals, application of more natural information signals can be as fast to decode as audio cues. Furthermore, the results point towards very limited, if any, performance-degradation of the participants in cognitively demanding primary task (LCT). Although more research is needed to identify the particular usefulness of the tested prototypes, especially outside lab conditions, it is possible to relay complex haptic information to users without the need for extensive training.

Acknowledgements. This research is funded by the Henry Ford Foundation grant.

References

1. Visual-Manual NHTSA Driver Distraction Guidelines for In-Vehicle Electronic Devices, National Highway Traffic Safety Administration (NHTSA). Department of Transportation (DOT), Doc No. 2014-21991
2. Politis, L., Brewster, B., Pollick, F.: To beep or not to beep? Comparing abstract versus language-based multimodal driver displays. In: Proceedings of the 33rd Annual ACM Conference Human Factors Computing System, pp. 3971–3980 (2015)
3. Farooq, A.: Developing technologies to provide haptic feedback for surface based interaction in mobile devices. Ph.D. Thesis, University of Tampere (2017). ISBN 978-952-03-0589-5

4. Lylykangas, J., et al.: Responses to visual tactile and visual–tactile forward collision warnings while gaze on and off the road. J. Trans. Res. Part F: Traffic Psychol. Behav. **40**, 68–77 (2017)
5. Lederman, S.J., Klatzky, R.L.: Haptic perception: a tutorial. Atten. Percept. Psychophys. **71** (7), 1439–1459 (2009)
6. Salminen, K., et al.: Cold or hot? How thermal stimuli are related to human emotional system? In: Oakley, I., Brewster, S. (eds.) Haptic and Audio Interaction Design, HAID 2013. Lecture Notes in Computer Science, vol. 7989. Springer, Heidelberg (2013)

Design Methodology for Flight Deck Layout of Civil Transport Aircraft

Zhefeng Jin(✉), Yinbo Zhang, Haiyan Liu, and Dayong Dong

COMAC ShangHai Aircraft Design and Research Institute, No. 5188 Jinke Road, Pudong New Area, Shanghai, China
jinzhefeng@comac.cc

Abstract. Human-Centered design philosophy has been accepted by Civil Transport Aircraft scopes. Flight deck layout can carry out the human-centered design philosophy. Now we should probe into one method of flight deck layout which can incarnate the human-centered design philosophy. This article provide one method of flight deck layout by the way that according to "design eye point" as benchmark and the design should accommodate and take into account the fundamental characteristics of the pilot population.

Keywords: Civil transport aircraft · Flight deck · Layout

1 Introduction

In the process of the civil aircraft cockpit design, the concept of generally accepted for the cockpit of the "human-centered" design philosophy, the cockpit layout is the specific implementation "human-centered" design philosophy of the cockpit. At the beginning of the domestic civil aircraft design, it is necessary to explore a kind of can reflect the cockpit "human-centered" design philosophy of the cockpit layout design method. By the "design eye point" as a benchmark and base on pilot physiological limits, cognitive features and individual differences as accessibility and visibility design parameters, this paper will explore a kind method of cockpit layout design that can be used to guide the for determining main characteristic parameters. This method can provide a reference for the main civil aviation manufacturers. The cockpit layout design method is in the following of relevant standards and airworthiness regulations and is to be put forward in the process of developing the cockpit layout design and solving specific problems to determine adjustment and optimization. This paper will elaborate the cockpit layout design process, the overall parameters and important control device (in the instrument panel, sidestick and windshield as a typical example) as parameters selection and determine methods. After the plane model validation, in accordance with the method, aircraft manufacturers can effectively avoid the major problems such as key design parameters unreasonable cockpit design lead to major changes in early development process and can efficiently complete the cockpit layout design.

W. Karwowski and T. Ahram (Eds.): IHSI 2019, AISC 903, pp. 98–104, 2019.
https://doi.org/10.1007/978-3-030-11051-2_15

2 Cockpit Layout Design Flow

The cockpit layout design flow chart shown in Fig. 1.

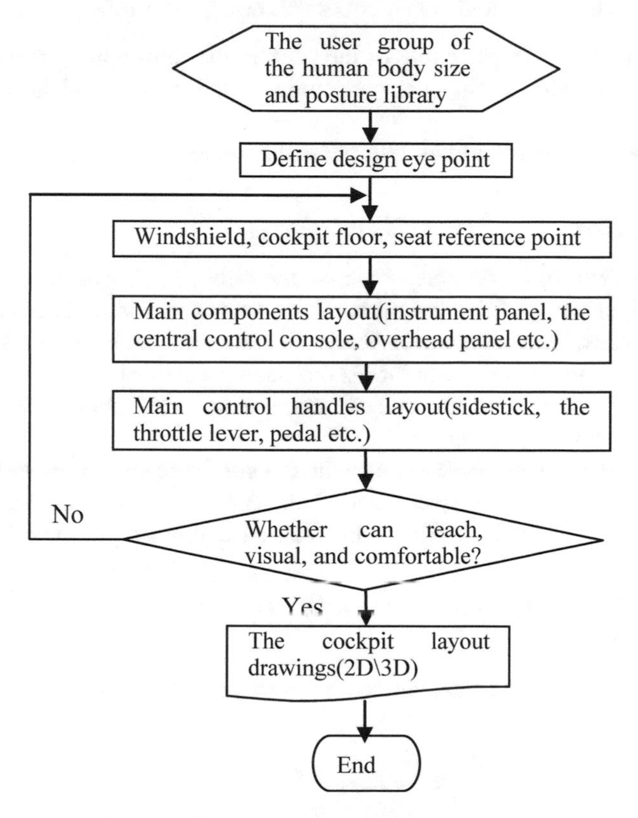

Fig. 1. The cockpit layout design flow chart

3 The Methods to Determining Specific Parameters of Cockpit Layout

By the "design eye point" as a benchmark and base on pilot physiological limits, cognitive features and individual differences as accessibility and visibility design parameters, this paper will explore a kind method of cockpit layout design that can be used to guide the for determining main characteristic parameters. This method can provide a reference for the main civil aviation manufacturers.

This paper will discuss the DEP specific parameters definition method and other main characteristics of the object (e.g., windshield, floor, main components, main control handle, etc.) relative to DEP in the flow chart.

3.1 The User Group of the Human Body Data

Anthropometric data [1] usually use percentile to represent the human body level. The most commonly used data is 5th, 50th and 95th percentile.

The cockpit layout should consider the following three aspects:

1. should make the 50th percentile of the pilot in the most appropriate location;
2. at least 5 percentile to the 95th percentile of the pilots is advantageous for the operation;
3. ensure that 1 percentile to the 99th percentile of the pilots can safe operation.

3.2 Define Design Eye Point (DEP)

Cockpit layout will to be designed base on the pilot design eye point as a reference point. At the beginning of the design, it can assume a point as a design eye point and give it coordinates. This point will be the basis of the basic reference data to design the windshield, floor, seat reference point. At the same time, this point will also be hold all dimensions as reference as the cockpit components and main control devices to determine the basis of location.

The transverse distance between captain and copilot is connected with the arm can reach and a preliminary assessment size of the central control console. This distance should be guaranteed for both pilots can reach all equipment on the central control.

3.3 Seat Reference Point (SRP) Location Relative to the Design Eye Design

SRP will be determined according to the 50th percentile pilot comfortable sitting posture.

$$H_s = H_{e50} - H_{c50} + D_c \tag{1}$$

H_s— Vertical distance between seat SRP and DEP;
H_{e50}— The 50th percentile pilot sitting eye height;
H_{c50}— Vertical distance between the 50th percentile pilot comfortable sitting posture eye site relative to the standard sitting position eye site;
D_c— Thickness of the clothes.

$$L_s = L_{e50} - L_{c50} + D_c \tag{2}$$

L_s— Horizontal distance between SRP and eye design point;
L_{e50}— The 50th percentile pilot ectocanthus to back of head;
L_{c50}—The horizontal distance between the 50th percentile pilot comfortable sitting posture eye site relative to the standard sitting position eye site;
D_c— Thickness of the clothes.

3.4 The Floor Location Relative to the Design Eye Point

The floor location relative to the design eye point is determined according to the 50th percentile pilot comfortable sitting posture that you can think legs are basic parallel with upper body.

$$H_f = H_s + H_{x50} \times cos\alpha_1 + D_s \tag{3}$$

H_f —Vertical distance between the cockpit floor and DEP;
H_s —Vertical distance between seat SRP and DEP
H_{x50} —50th the lower leg length and medial malleolus height of standard sitting posture
α_1 —Seat backrest angle;
$H_{x50} \times \cos\alpha_1$ —50th the lower leg length and medial malleolus height;
D_s —The thickness of the heel.

3.5 The Instrument Panel Location Relative to the Design Eye Point

The instrument panel installation position need to consider the pilot visual characteristics, accessibility of control device on the instrument panel.

$$L_p = L_{ua5} + D_{sh5} - L_{e5} \tag{4}$$

L_p —Horizontal distance between the instrument panel and eye design point;
L_{ua5} —The 5th percentile pilot upper limb length;
D_{sh5} —The 5th percentile pilot back shoulder peak breadth;
L_{e5} —The 50th percentile pilot ectocanthus to back of head.

Note: Consider the human body shoulder activity, we can increase the appropriate correction.

The instrument panel vertical position from design eye point is determined by external visual requirements. the top edge each position should be lower than The lowest line of sight angle of external visual requirements, Under the edge of the instrument panel is determined by knee activity space constraints. The width of the instrument panel should be coordinated with the nose shape layout.

Display interface and the polit line of sight must be at least angle 45°, as shown in Fig. 2.

When human body is sitting comfortable posture to observe display information, experience show that the cone axis is lower than the level of the line of sight about 20°–25° and the angle with the ground about 65°–70° [2].

3.6 Sidestick Location Relative to the Design Eye Point

Side stick location is determined according to the 5th percentile pilot upper extremity length and operating posture.

Manipulation of the sidestick diagram as shown in Fig. 3.

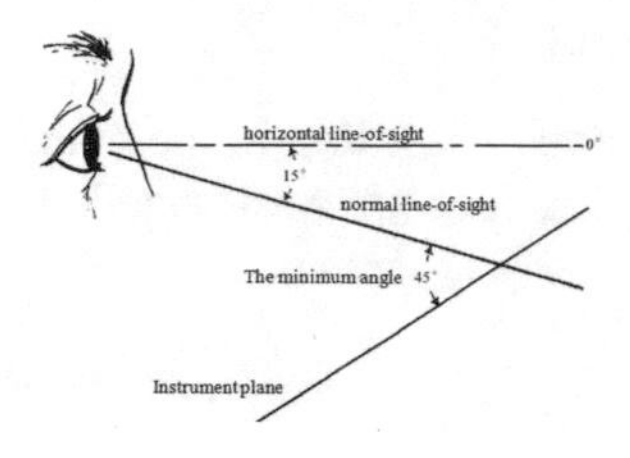

Fig. 2. .

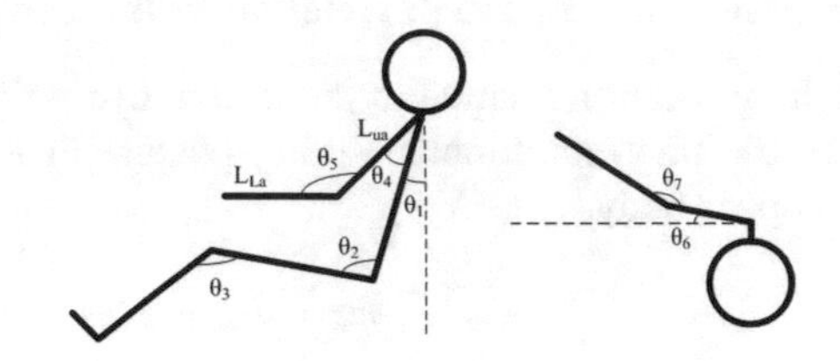

Fig. 3. .

Z direction distance from sidestick to design eye point shown below:

$$W_{sp} = 0.5 \times W_{hw5} + L_{ua5} \times cos\theta_6 + L_{La5} \times cos(\theta_6 + (180 - \theta_7)) \tag{5}$$

W_{sp} —Z direction distance from sidestick to design eye point;
W_{hw5} —The 5th percentile pilot shoulder breadth;
L_{ua5} —The 5th percentile pilot upper arm length;
L_{La5} —The 5th percentile pilot forearm and hand length.

θ6, θ7 definition shown in Fig. 6.
X direction distance from sidestick to design eye point shown below:

$$L_{sp} = L_{ua5} \times sin\theta_6 + L_{La5} \times sin(\theta_6 + (180 - \theta_7)) - L_{e5} \tag{6}$$

L_{sp} —X direction distance from sidestick to design eye point shown below.
L_{e5} —The 5th percentile pilot ectocanthus to back of head.

Y direction distance from sidestick to design eye point shown below:

$$H_{sp} = H_{e5} - H_{sh5} + L_{ua5} \times cos(\theta_1 + \theta_4) - L_{La5} \times cos(\theta_5 - \theta_1 - \theta_4) \tag{7}$$

H_{sp} —Y direction distance from sidestick to design eye point;
H_{e5} —The 5th percentile pilot sitting eye height;
H_{sh5} —The 5th percentile pilot sitting shoulder high.

3.7 The Front Windshield Location Relative to the Design Eye Point

The distance from intersection of windshield and level sight line through design eye pilot to design eye pilot will decide windshield area size. The value of plane windshield generally is about 500 mm to 600 mm, the hyperboloid windshield slightly larger than the number.

3.8 Internal View

Considering the instrument panel is not affected by the outside world light interference, it has a glareshield, the glareshield top of the above is a former direction outside view, from 3.10, the top of the glareshield should be at least 17° below the level of sight.

Important and use of high frequency instrument shall not exceed the head turn biggest view area, located at the optimal visual areas as far as possible.

Instrument (e.g., sidedisplay) which is to flight operation but use frequency is not too high shall not exceed the head eyes turn biggest visual areas.

The human eye optimal field of view and the best field of view refer to MIL-STD-1472G "Fig. 25. Vertical and horizontal visual fields" [3].

3.9 External Vision

Outside the cockpit vision must ensure that the plane in various states has adequate vision scope. The theoretical minimum visual field [3] is shown in Table 1 and Fig. 4 provisions.

Table 1. The theoretical minimum visual field

	Azimuth	−120	−90	−80	−40	−30	0	10	20
AC 25.773-1	Up view	15	30	35	35	32.5	25	22.5	20
	Down view	−15	−27	−27	−20	−17	−17	−17	−10

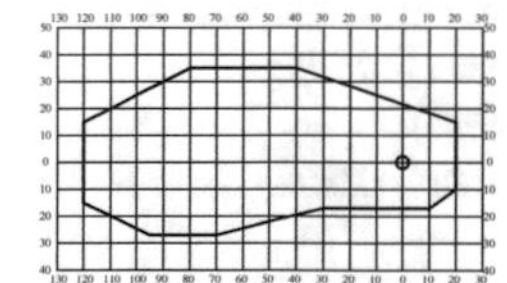

Fig. 4. .

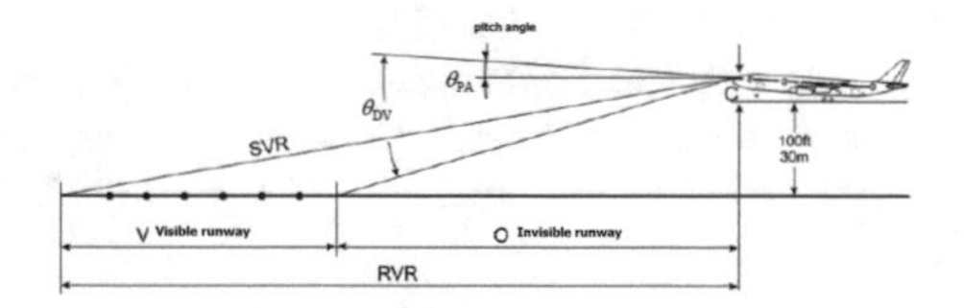

Fig. 5. .

In actual design work, for each specific aircraft configurations, the actual need of vision scope is proper adjustment and correction on the basis of the standard horizon, mainly consider the approach speed and airplane attitude angle [4].

As shown in Fig. 5, the pilot down vision of zero bearing horizon shows below:

$$\theta_{DV} = \theta_{PA} + \arctan\frac{30 + C}{RVR - V} \tag{8}$$

θ_{DV} —The pilot down vision of zero bearing horizon(Down Vision);

θ_{PA} —The pitch angle of the aircraft;

C —The vertical distance between design eye pilot and the lowest point of the main landing gear tires when the aircraft pitch angle is θ_{PA}. This parameter is the aircraft inherent characteristics;

RVR —Runway Visual Range, this value is not less than 366 mm.
V — for the visible part of runway, this value is a length of approach would be covered in three seconds at landing approach speed.

4 Conclusion

This article mainly use the "DEP" as a design benchmark and adopt pilot physiological limits, cognitive features and individual differences as the basis of accessibility and visibility design parameters. This article explore a kind of method that can be used to determine main design parameters in the process of initial cockpit layout design. This method can be used to guide initial design work of the civil aircraft cockpit layout design.

The proposed design method has been used in large aircraft and wide-body aircraft and has been proved that this method is feasible. The result of application is mainly manifested in the design of model(such as C919 and CR929), effectively promote the model development work.

References

1. Ahlstrom, V., Longo, K.: Human Factors Design Standard (HFDS) for Acquisition of Commercial off-the-Shelf Subsystems, Non-developmental Items, and Developmental Systems. DOT/FAA/CT-03/05 (2003)
2. Society of Automotive Engineers. Pilot Visibility from the Flight Deck (1989)
3. U.S. Army Aviation and Missile Command. Human Engineering (1999)
4. Federal Aviation Administration. Pilot Compartment View Design Considerations. ANM-110 (1993)

Motion Capture Automated Customized Presets

William Machado de Andrade(✉), Jonathan Ken Nishida, Milton Luiz Horn Vieira, Gabriel Souza Prim, and Gustavo Eggert Boehs

DesignLab, Departamento de Expressão Gráfica – Centro de Comunicação e Expressão, Universidade Federal de Santa Catarina, Campus Universitário, CCE, bloco A, sala 101, Florianópolis, SC, Brazil
{w.andrade,milton.vieira,gustavo.boehs}@ufsc.br,
{jonathan.nishida,gabriel.prim}@posgrad.ufsc.br

Abstract. Motion Capture technologies transfer coordinate data from the human body movement and locomotion to a digital structure in order to move the avatar according to the actions performed by a person, creating digital animation function curves, marking as a keyframe each frame captured in a timeline. Those keyframes create excess of short movements as they try to correct the coordinates to a distinct virtual character from the real person who originated them, making the avatar quiver each time it performs any action. For animation purposes, in order to produce visually harmonic movements, it is necessary to remove manually the exceeded keyframes. The present study proposes an automated scripted method to reduce the amount of keyframes, keeping the shapes of the function curves, in order to customize the aesthetic gestural properties of characters animated by MoCap. It is presented a graphic comparison from before and after applying the automated customization proposed.

Keywords: Motion capture · Customizable FCurve · MoCap keyframing

1 Introduction

Researchers have approached Motion Capture (MoCap) techniques and technologies for distinct applications with varying characteristics. From health studies to fictional character animation, MoCap offers results that may require adjustments to provide the output needed by its users in their diverse fields.

Some of the treatments related to MoCap data includes calibration strategies [1], correcting occlusion and denoising from markers or missing points [2], lossless compressions [3], and storing the sessions to organize databases of movement.

The current study presents a different approach, concerning the refinement of the animation, originated from MoCap sessions. Such subject derives from the authors' empirical experience while producing an experimental animation series. Some of the problems observed during production met viable solutions such as the ones listed above. One did not: the high frequency of the data makes it difficult for animators to edit captured movements.

W. Karwowski and T. Ahram (Eds.): IHSI 2019, AISC 903, pp. 105–110, 2019.
https://doi.org/10.1007/978-3-030-11051-2_16

MoCap equipments of different sorts register the three-dimensional position and orientation of most joints in the human body, at frequencies of at least 30 Hz. This data can be plotted in function curves (FCurves) that display captured properties in time and space along the X, Y, and Z-axes.

Such digital display is not uncommon to 3d animators, who are accustomed to editing such FCurves with keyframes, and their interpolations, determining the numerical values of the coordinates for each limb in the body. Each keyframe has customizable 'in' and 'out' interpolation schemes so that the animator can determine the aesthetic characteristics of the movement.

However, the high density of MoCap data does not allow animators such freedom. Each frame in the MoCap animation is marked as a keyframe. Without any time in-between keyframes, it is impossible to choose the proper interpolation that will determine the aesthetic properties of the movement. Figure 1 demonstrates a fragment of an FCurve resulted from a Motion Capture session at DesignLab, the host research laboratory in the Federal University of Santa Catarina (Brazil).

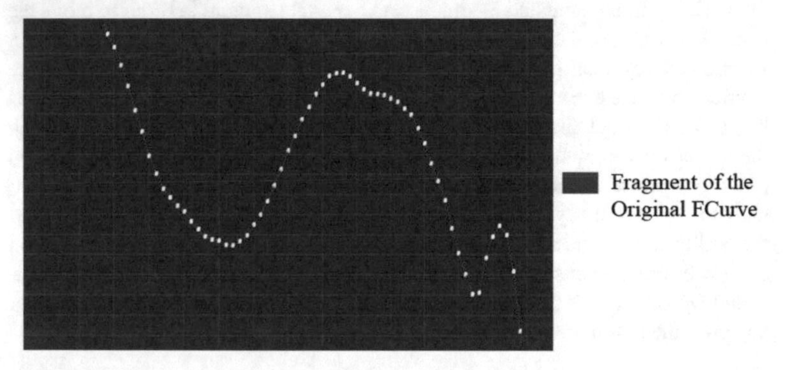

Fig. 1. The horizontal axis represents time. In this fragment, the red spline is the FCurve of x (extracted from Autodesk MotionBuilder); each white dot is a keyframe. There are sixty-one frames displayed in the image; therefore, there are sixty-one keyframes as each frame is also a keyframe

In trying to develop smooth movements, during the production of the experimental animation, four animators were employed in the refinement of FCurves. The goal of this refinement was to diminish the quivering and jerking movements, as well as self-collisions that are a result of differences in the body structures, volume and sizes of the original captured human body and the mapped 3d character.

These animators have manually deleted some of the keyframes and intuitively estimated the position of the remaining ones to make it possible to smooth out, customize, and repurpose animations. In addition to deleting and adjusting existing keyframes, adjustment layers were necessary to determine new biases and speeds enabling the intended results, and adding even more data, keyframes and FCurves, to alter the movement's original properties.

Such strategy did not prove to be cost-effective, besides being time-consuming, which motivated new ideas able to automate the deletion of the excess of keyframes without reconfiguring the shape of the FCurve, keeping the animation according to the originally captured motion, and prone to customization by manipulating the interpolation.

2 Related Work

Motion Capture technologies record and transfer the coordinates (X, Y, and Z) from human and animal bodies, or from inanimate objects, to a digital structure in order to reproduce (using a 3d software) in a virtual character the actions performed by the source [9].

There are different MoCap techniques, grouped by marker-based and markerless approaches. The first distributes markers to pre-determined parts of the body, captured by optical cameras. The second records the performance on video [4]. DesignLab uses marker-based MoCap, considered the most popular one, which computes the different positions of the markers placed on the human body as trajectories followed by the digital character [5]. The laboratory currently has fourteen Vicon T40S (NIR 18) optical cameras covering an area of 10 m × 6 m.

Transferring the coordinates to the trajectories is referred to as mapping [6]. Such process, due to diverse problems during a session (e.g., occlusion of markers or capturing another reflexive object at the same environment), may present missing data or noise, creating a *"mocap data refinement problem"* [7]. When the digital character breaks or quivers because of such problem, it may lose its "realistic motion parameters" [8] or believability.

The research conducted for the current paper did not find any texts including the excess of keyframes and the impossibility to use interpolation 'in' and 'out' properties as a *"mocap data refinement problem"*, although there is one mention enacting that MoCap data may "be adjusted using frame-by-frame animation from the re-positioning of limbs…" [9]. That only seems possible by adding an adjustment layer using the tools from the 3d software used to process the data or with numerous animators manipulating all the keyframes and their properties.

Such "low flexibility" was already noted, in 2004, as the "weakness" [10] of motion capture when it is necessary to edit the captured movement for any other purpose besides the one that guided the session.

The current context provided the problems and stimulus to solutions that could allow editing the number and properties of the keyframes bring back the in-betweens that are created when the animation is done manually and offer possibilities to reshape the FCurves and aesthetically manipulated the captured data.

3 Potential Solution

The studied compared methods from both the literature and tools available in the Autodesk MotionBuilder and 3ds Max software packages. Different tests were conducted on a virtual skeleton that is part of DesignLab's motion capture database [11]. A human walking sequence was chosen due to the action's familiarity with observers

[12], displaying information that may be classified as desired or not by simple observation.

Although samples of keyframe elimination were not found in the literature, it came to our attention that MotionBuilder has built-in tools that may serve this purpose: Smooth Filter and Key Reducing (Fig. 2).

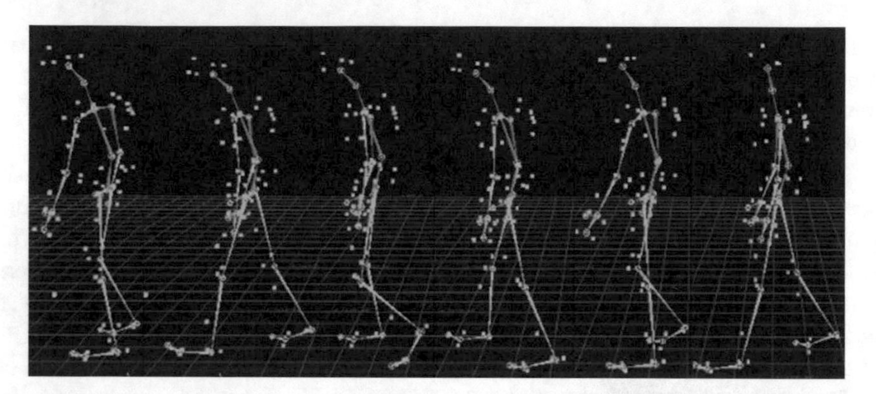

Fig. 2. A human walking captured at the DesignLab's facilities. The fragment in Fig. 1 is from the right foot of the virtual skeleton above

The former removes sudden peaks in the FCurve, but it does not reduce the number of keyframes. It helps to eliminate movements that are dissonant with other parts of the motion, but when applied to the legs and feet, it often results in foot skating. After some testing, it was considered an inadequate solution for this work.

The second, Key Reducing, proved to be closer to this study's goal. It matched the purpose of eliminating excess keyframes, but with a side effect: when applied with maximum intensity (1.0) it significantly changes the shape of the FCurve, visually altering the nature of the movement (Fig. 3).

With lower intensities ($\sim$0.2), the motion was considered of good quality, but the keyframe reduction was not sufficient to allow for proper editing using Bézier Spline (B-Spline) handles, a type of keyframe interpolation.

When applied only to the feet, with lower intensity, the Key Reducing tool kept the trajectories similar to the original capture. While with higher intensities, the feet's trajectories became unstable.

Although displaying satisfactory outputs, Key Reducing still results in FCurves that don't allow further editing. From an animator's point of view this does not meet the concept of an "ideal character animation system (…) able to synthesize believable motion, and, at the same time, provide sufficient control to the generated motion" [13].

A third option was then tested: creating a custom scripted solution in MotionBuilder's Python Editor. On that matter, Forsythe [14] provided invaluable contributions by offering script lines that were the base for the customization possibilities this study aimed to achieve. By adapting Mr. Forsythe's original algorithm, the code below was originated.

```
a = 0
while a < 3:
fcurve = lModel.Rotation.GetAnimationNode().Nodes[a].FCurve
for i in reversed(range(0, len(fcurve.Keys), 2)):
fcurve.KeyDeleteByIndexRange(i, i)
a += 1
```

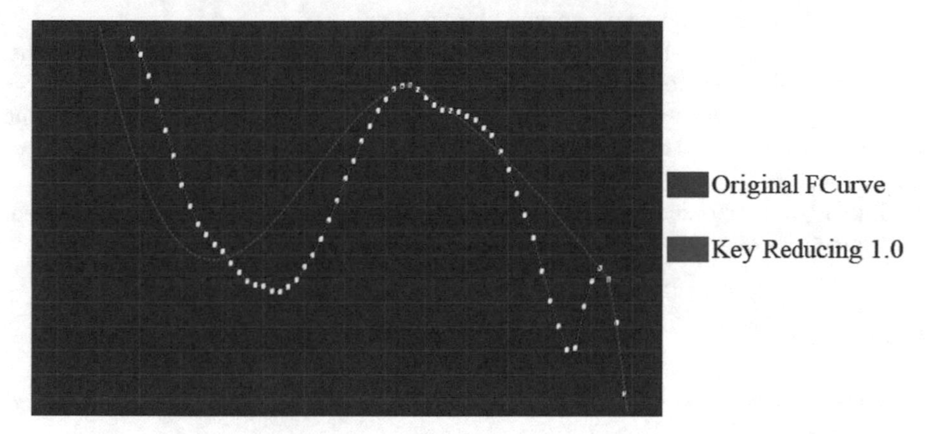

Fig. 3. Despite reducing sixty-one keyframes to only two, Key Reducing changes the shape of the curve and, therefore, the animation

Applying the code will delete the exceeding keyframes at regular intervals (the interval may be controlled by the number marked in bold in the script or by running the script repeated times), but with lesser FCurve deformation. Thus, the result allows for change in the 'in' and 'out' interpolations that was desired by the animators (Fig. 4).

The gait that resulted from this method presented less visual artifacts than the results obtained with Key Reducing (at intensity 1.0), suggesting the script is an adequate solution to provide animators with the proper conditions to customize the FCurves.

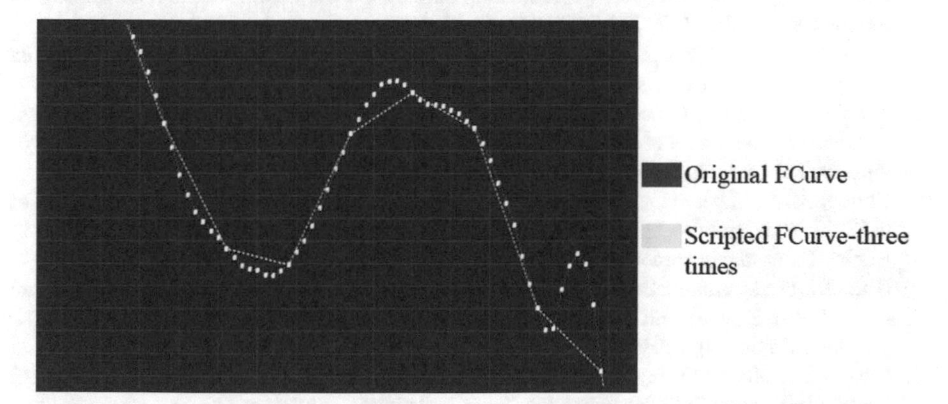

Fig. 4. The sixty-one original keyframes were reduced thirty-one the first time the script was activated, with slight changes on the curve; the second time there was a reduction to sixteen and the third time to eight, with the shape above

4 Conclusions

The current research is the beginning of a deeper study concerning the customization and reuse of MoCap databases. To validate the proposed method, further investigations are required. The preliminary tests indicate a potential track to be followed, providing control to animators and making the FCurves available to aesthetically manipulation until the desired results may be reached by the user.

The advantage of the proposed approach is its potential automation means, according to the selection (of limbs) by the animator.

Proving it is viable and reliable, the next step for the script is incorporating it to the software's interface and combining it with smoothing strategies to the recreated FCurve with enough control for its interpolation.

Finally, it is expected that the current work may join other efforts in the treatment and refinement of MoCap data for its varied applications.

References

1. Estévez-García, R., et al.: Grid open data motion capture: MOCAP-ULL database. Proc. Comput. Sci. **75**, 316–326 (2015)
2. Holden, D.: Robust solving of optical motion capture data by denoising. ACM Trans. Graph **38**, 1 (2018). Article 165
3. Wang, P., et al.: The alpha parallelogram predictor: a lossless compression method for motion capture data. Inf. Sci. **232**, 1–10 (2013)
4. Canton-Ferrer, C., Casas, J.R., Pardàs, M.: Human motion capture using scalable body models. Comput. Vis. Image Underst. **115**, 1363–1374 (2011)
5. Ayusawa, K., Ikegami, Y., Nakamura, Y.: Simultaneous global inverse kinematics and geometric parameter identification of human skeletal model from motion capture data. Mech. Mach. Theor. **74**, 274–284 (2014)
6. Geng, W., Yu, G.: Reuse of motion capture data in animation: a review. In: Kumar, V., et al. (eds.) ICCSA. LNCS, vol. 2669, pp. 620–629. Springer, Heidelberg (2003)
7. Feng, Y., et al.: Exploiting temporal stability and low-rank structure for motion capture data refinement. Inf. Sci. **277**, 777–793 (2014)
8. Güdükbay, U., Demir, I., Dedeoğlu, Y.: Motion capture and human pose reconstruction from a single-view video sequence. Digit. Sig. Process. **23**, 1441–1450 (2013)
9. Nikolai, J., Bennett, G.: Stillness, breath and the spine - dance performance enhancement catalysed by the interplay between 3D motion capture technology in a collaborative improvisational choreographic process. Perform. Enhanc. Health **4**, 58–66 (2016)
10. Chung, H.-S., Lee, Y.: MCML: motion capture markup language for integration of heterogeneous motion capture data. Comput. Stand. Interfaces **26**, 113–130 (2004)
11. Boehs, G., et al.: Locomotion dataset. http://tecmidia.ufsc.br/en/locomotion-dataset/
12. Guo, X., et al.: Automatic motion generation based on path editing from motion capture data. In: Pan, Z., et al. (eds.) Transactions on Edutainment IV. LNCS, vol. 6250, pp. 91–104. Springer, Heidelberg (2010)
13. Gao, Y., et al.: From keyframing to motion capture. In: Hommel, G., Huanye, S. (eds.) Human Interaction with Machines, pp. 35–42. Springer, Dordrecht (2006)
14. Forsythe, A.: Python scripting in motion buider. 07-Dealing with keyframe data: FBFCurve. http://www.awforsythe.com/tutorials/pyfbsdk-7

Assessing Social Driving Behavior

Giorgio Grasso, Pietro Perconti(✉), and Alessio Plebe

Department of Cognitive Science, University of Messina, v. Concezione 8, 98121 Messina, Italy
{gmgrasso, perconti, aplebe}@unime.it

Abstract. Recent advances in Artificial Intelligence are making automated vehicles an ever closer reality. However, we should expect a period when full or partial autonomous vehicles and ordinary cars coexist, during which it would be essential to fully understand the cognitive processes used by ordinary people when driving. Our work attempt to progress in this direction, by designing a system for assessing when and why subjects resort to costly social processes, rather than using quick and automated reactions. In particular, it will be crucial to assess when drivers use mentalizing abilities, in addition to paying attention to other people by means of simpler automated sensorimotor control processes. In our experimental design we investigate the main precursors of mindreading, that is, eye contact and shared attention.

Keywords: Autonomous vehicle · Social cognition · Mentalization

1 Introduction

Research on autonomous vehicles has a long tradition, and for a long time has been dominated by control engineering combined with A.I. components such as automatic decision making and planning [1]. In the last few years, a sudden acceleration has been boosted by the novel family of methods known as deep neural networks [2, 3], quickly established as the solution of choice for a wide range of computational problems [4]. Adoption of deep neural models in the automotive domain is now attainable by exploiting graphics processing units (GPUs), thanks to the CUDA software interface [5], and real-time GPU-based computers like NVIDIA Drive PX. Therefore, there is an increasing enthusiasm in deep neural models solutions for autonomous vehicles [6–9]. Together with technologies enabling dedicated communication (V2x) between vehicles and other vehicles and infrastructures [10], a future of monopoly of algorithms over traffic will be real, with great advantages in terms of traffic efficiency and safety.

Notwithstanding, the shift from human to automatic driving will pose serious challenges. Currently, it is even impossible to demonstrate that automated vehicles are safer than humans. Car accidents account for more than 2% of total mortality, but this is due to the vast usage of cars, in fact humans are very reliable at driving: in the United States there are only about 1,09 deaths and 77 injuries per 100,000,000 miles. To date there are less than 0.5 million miles per year run by autonomous driving vehicles [11], a distance far too short for an empirical evaluation [12].

W. Karwowski and T. Ahram (Eds.): IHSI 2019, AISC 903, pp. 111–115, 2019.
https://doi.org/10.1007/978-3-030-11051-2_17

Bearing this in mind, an important question concerns the cognitive processes used by ordinary people when driving, in particular during critical interactions with other active agents, such as cars, trucks, cyclists, and pedestrians. Several research researches are addressing this question [13, 14], which still remains largely unanswered. Our work attempt to progress in this direction, by designing a system for assessing when and why subjects resort to costly social processes, rather than using quick and automated reactions. In particular, it will be argued for the idea that we have to take into account mentalizing abilities [15, 16] i.e., social cognition processes aiming at inferring intentions of others and not simply at paying attention to other people behavior by means of automated sensorimotor control processes. Our experimental design is intended to test the typical perceptual conditions which elicit the shift from an automated driving style to a mentalizing assessment of the driving scene. We test the main precursors of mindreading, that is, eye contact and shared attention [17]. We argued for the thesis that measurable eye contact and joint attention perceptual patterns are typical of scenarios which elicit social cognition and mentalizing drive. Our attempt ends up in suggesting that these empirical findings could be used to improve advanced driver assistance systems in the cases in which a mentalizing driving style is needed.

2 Mentalization During Driving

Generally speaking, human behavior can be characterized by low or high levels of mentalization. Above all, it depends on how much they are automatic or voluntary. It is, however, also matter of how much the behavior is social. Swimming alone in a swimming pool is a typical low level mentalization behavior. It is, in fact, a non-social and automatic action. On the contrary, when someone is asked for: "And do you see the person, here in this courtroom, who committed this crime?", and she points at the accused, we are facing with an high level mentalization behavior. It is an intentional action and it is directed to another individual and to a social scenario. It is not only matter of attention and will. It depends, in fact, on interpreting (or not) a given behavior by means of the intentional vocabulary and the folk psychology framework, consisting of believes, desires, and propositional attitudes.

Mentalizing can be more or less useful when you are driving a vehicle. Driving a car in an isolated motorway is a completely different experience than trying, to say, to cross a road in downtown Hanoi. Guessing other people intentions is a crucial issue in one case, but not in the other. For this, mental processes which are involved during driving are so various. Sometimes social cognition processing is highly demanding, while in other circumstances the brain works, so to speak, in a solipsistic and automatic way. Traffic rules play a key role in leading people to adopt a more or less mentalizing driving style.

The issue of "shared spaces" in traffic regulation is particularly interesting in this perspective. A shared space is characterized by the removal of observable features such as traffic signs, kerbs, and road surface markings. Instead of controlled crossings and regulated squares, shared spaces are available for multiple uses, according to the contingent needs. It is controversial whether this kind of public spaces is a smart approach to traffic problems or if shared spaces increase the feeling of fear, anxiety, and unsafety [18]. Anyway, what really matters here is again the convenience to adopt a negotiation driving style, or to base our behavior merely on conventional rules and habits. When we will be able to fully model how that "convenience" works, we will be endowed with a fruitful theoretical resource to better deal with the social autonomous vehicle challenges. One can simply provide ordinary cars with "Intention indicators" and "Message displays", as proposed by the Nissan Center for Automotive Research, but color symbolization in different cultures and cross-linguistic varieties suggest to follow another road.

Riaz and Niazi [19] present a social autonomous vehicle (AV), "which interacts with other AVs using social manners similar to human behavior. The presented AVs also have the capability of predicting intentions, i.e., mentalizing and copying the actions of each other, i.e., mirroring". This cognitive architecture includes two modules. The Mentalizing Module "helps the AV to find the intention of neighbouring Avs"; the Mirroring Module "helps the AV to change its trajectory according to the changed trajectory of the nearest AV". The Mentalizing Module is based on Lewis Fry Richardson's studies, in particular on the "Richardson's arms race model". Although Riaz and Niazi maintain that their model will improve the collision avoidance capabilities of AVs, it does not try to simulate the cognitive faculties actually involved in performing the given task. A more biologically inspired account would be theoretically desiderable.

Trying to discriminate the neural basis of spontaneous *vs.* voluntary mentalizing during everyday experiences, Spiers and Maguire [20] used as a case study for their investigation the taxi driver ordinary experience of driving in central London. They found an "increased activity in a number of regions, namely the right pSTS, the mPFC and the right temporal pole", largerly overlapping with many neuroimaging studies examining the neural basis of mentalizing. It seems that when the driver shifts from the "coasting driving style", "where subjects were actively driving and moving through the city, but did not have any directed thoughts", to a negotiating driving style, the brain starts to work in a highly mentalizing mode.

Our experimental design is intended to test the typical perceptual conditions which elicit the shift from an automated driving style to a mentalizing assessment of the driving scene. We test the main precursors of mindreading, that is, eye contact and shared attention [21]. We argued for the thesis that measurable eye contact and joint attention perceptual patterns are typical of scenarios which elicit social cognition and mentalizing drive. Our attempt ends up in suggesting that these empirical findings could be used to improve advanced driver assistance systems in the cases in which a mentalizing driving style is needed.

3 Methods

In order to assess a quantitative measure of the level and the role of mentalization adopted by drivers during car rides, an experimental setup has been developed, employing a set of virtual reality facilities and technologies. The goal has been to recreate a realistic environment to simulate driving in a city context, populated by cars, pedestrians and other key features, typical of an urban scenario.

A driving simulator has been developed from the ground up, exploiting a VR software platform typically employed for videogame implementation, namely Unity. Unity is a cross-platform game engine developed by Unity Technologies and first released 2005. As of 2018, the engine has been extended to support 27 platforms, making it one of the leading game engine both for desktop and mobile applications. This software platform can be used to create realistic three-dimensional videogames as well as simulations.

The simulator leverages on the power and ease of development of Unity, together with the extremely large body of 3D models available for the game engine. Thus detailed models for cars, buildings, road signs, pedestrians and such can be easily obtained and readily deployed in virtual/gaming setups.

The engine supports several methods for animating objects within the virtual environment and can be programmed to produce engaging life-like environments.

Together with the software platform a virtual reality headset, namely Oculus Rift made by Oculus VR, has been employed to project the simulator users into an immersive and sensor rich experience. Oculus Rift, initially released in 2016, is equipped with two Pentile OLED displays, 1080 × 1200 resolution per eye, a 90 Hz refresh rate, and 110° field of view. The device also features rotational and positional tracking, and integrated headphones that provide a 3D audio effect.

The experimental setup, constructed for the purpose of the present research work, consists of several software and hardware components. A graphical workstation, equipped with an Intel Core i7 processor, 16 GB of RAM, a 250 GB SSD and an NVIDIA Titan X 12 GB graphics card is employed to run the simulation, ensuring a smooth high resolution rendering of the virtual environment, being projected onto the VR headset. A Logitech force feedback steering wheel, with pedals and gear shift is also used to provide a complete driving experience during simulation.

The computing hardware, wheels, pedals and gear shift, the graphical workstation with the VR headset are connected to a physical car driving setup, consisting of an actual car seat positioned on a rigid frame, where all the equipment is attached.

The simulator, based upon the Unity engine, uses a purchased model of a typical small US city and is populated by several dozens of car models, which are animated with a simple self-driving algorithm, as well as 50 pedestrians, that are animated as well.

All objects in the simulation are controlled by the simulation engine and exhibit specific behaviors, tailored to reproduce relevant situations where it is required by the user to adopt a specific mentalization task.

A typical view of the virtual environment is visible in Fig. 1, where a boy is crossing the street, not paying attention to the car arriving from the left of scene. In this case the driver is supposed to anticipate this behavior through visual engagement with the pedestrian as it approaches the curb.

Fig. 1. A critical scene, where a boy suddenly crosses the street in spite of the arriving car.

References

1. Dickmanns, E.D.: Vehicles capable of dynamic vision. IJCAI **97**, 1577–1592 (1997)
2. Schmidhuber, J.: Deep learning in neural networks: an overview. Neural Netw. **61**, 85–117 (2015)
3. Goodfellow, I., Bengio, Y., Courville, A.: Deep Learning. MIT Press, Cambridge (2016)
4. Deng, L., Yu, D.: Deep Learning: Methods and Applications. now publishers, Boston (2014)
5. Sanders, J., Kandrot, E.: CUDA by Example: an Introduction to General-Purpose GPU Programming. Addison Wesley, Reading (2014)
6. Gurghian, A., Koduri, T., Bailur, S.V., Carey, K.J., Murali, V.N.: DeepLanes: end-to-end lane position estimation using deep neural networks. In: Proceedings of IEEE International Conference on Computer Vision and Pattern Recognition, pp. 38–45 (2016)
7. Wu, B., Iandola, F.N., Jin, P.H., Keutzer, K.: SqueezeDet: unified, small, low power fully convolutional neural networks for real-time object detection for autonomous driving. CoRR abs/1612.01051 (2016)
8. Rausch, V., Hansen, A., Solowjow, E., Liu, C., Kreuzer, E., Hedrick, J.K.: Learning a deep neural net policy for end-to-end control of autonomous vehicles. In: Proceedings of American Control Conference, pp. 4914–4919 (2017)
9. Bojarski, M., et al.: Explaining how a deep neural network trained with end-to-end learning steers a car. CoRR abs/1704.07911 (2017)
10. Zhao, Y., Yao, S., Shao, H., Abdelzaher, T.: Codrive: cooperative driving scheme for vehicles in urban signalized intersections. In: ACM/IEEE International Conference on Cyber-Physical Systems, pp. 308–319 (2018)
11. Dixit, V.V., Chand, S., Nair, D.J.: Autonomous vehicles: disengagements, accidents and reaction times. PLoS ONE **11**, e0168054 (2016)
12. Kalra, N., Paddock, S.M.: Driving to safety: how many miles of driving would it take to demonstrate autonomous vehicle reliability? Transp. Res. Part A: Policy Pract. **94**, 182–193 (2016)
13. Jorge, C.C., Rossetti, R.J.F.: On social interactions and the emergence of autonomous vehicles. In: International Conference on Vehicle Technology and Intelligent Transport Systems, pp. 423–430 (2018)
14. Bengtsson, P.: Attuning the pedestrian-vehicle and driver-vehicle – why attributing a mind to a vehicle matters. In: Karwowski, W., Ahram, T. (eds.) International Conference on Intelligent Human Systems Integration, pp. 308–319 (2018)
15. Samson, D.: Theory of mind. In: Reisberg, D. (ed.) The Oxford Handbook of Cognitive Psychology. Oxford University Press, Oxford (2013)
16. Vilarroya, O., i Argimon, F.F. (eds.): Social Brain Matters – Stances on the Neurobiology of Social Cognition. Rodopi, Amsterdam (2007)
17. Tomasello, M.: Origins of Human Communication. MIT Press, Cambridge (2009)
18. Karndacharuk, A., Wilson, D.J., Dunn, R.: A review of the evolution of shared (street) space concepts in urban environments. Transp. Rev. **34**, 190–220 (2014)
19. Spiers, H., Maguire, E.A.: Spontaneous mentalizing during an interactive real world task: an fMRI study. Neuropsychologia **44**, 1674–1682 (2006)
20. Riaz, F., Niazi, M.: Towards social autonomous vehicles: efficient collision avoidance scheme using Richardson's arms race model. PLoS ONE **12**, e0186103 (2017)
21. Perconti, P.: Filosofia della mente. Il Mulino, Bologna (2017)

Interaction Patterns for Arbitration of Movement in Cooperative Human-Machine Systems: One-Dimensional Arbitration and Beyond

Daniel López Hernández[1(✉)], Marcel C. A. Baltzer[1], Konrad Bielecki[1], and Frank Flemisch[1,2]

[1] Fraunhofer FKIE, Fraunhoferstraße 20, 53343 Wachtberg, Germany
{daniel.lopez.hernandez,marcel.baltzer, konrad.bielecki,frank.flemisch}@fkie.fraunhofer.de
[2] Institute for Industrial Engineering and Ergonomics (IAW), RWTH Aachen, Bergdriesch 27, 52062 Aachen, Germany

Abstract. Arbitration becomes necessary when two interacting actors have different goals. Movement is one of the most prominent examples for arbitration. Many situations when arbitration is necessary are repetitive problems. One way to design arbitration processes in a structured manner is to structure them as interaction patterns that considers the user's perception of a situation. These internal target states of the user can be influenced by information exchanged between the automated system and the user through different modalities. As a one-dimensional example for this, a device for unplugging a USB stick from a computer was constructed. This device allows the human to interact with the computer via various modalities. A user interface ("pattern designer") allows real time design of messages and parameters. Interaction patterns designed through this method can be used in other domains such as automated driving or drone flying.

Keywords: Human machine cooperation · Arbitration of movement
Interaction patterns · Interaction design

1 Introduction

If two independently thinking entities start planning and acting together, the emergence of conflicts is inevitable. Conflicts are not bad per se, but can be helpful to consider previously overlooked aspects of actions. Likewise, conflicts can be obstructive or even result in deadlocks, e.g. in a movement situation when an automation wants to turn right and the human left: Taking the middle could be bad for both and fatal for the combined human-machine-system. Human – machine arbitration includes a structured negotiation between the human and the automation to solve these conflicts. Arbitration is the method to reach a common unambiguous decision on how to act in due course of time [1, 2].

W. Karwowski and T. Ahram (Eds.): IHSI 2019, AISC 903, pp. 116–122, 2019.
https://doi.org/10.1007/978-3-030-11051-2_18

A way to describe interaction, e.g. negotiating between two actors on the interaction level is to use patterns. Patterns are approved solution strategies to repeating known problems. A pattern describes a problem or a system of different forces or tension poles, e.g. [3] that occur repeatedly in our environment and describes the core of the solution in a way that it can be used again and again [4]. A system of patterns can form a pattern language. Creating new patterns and connecting them with other patterns enlarges the pattern language.

Patterns address certain internal target states of the involved actors. An example for an internal target state is an emotion. Internal states might not be directly quantified, but give an indication to the actors that something is or has changed therefore giving a new interpretation of the current state. A scream from another person can transport an internal target state of danger. A parent can change the interpretation of a child of the internal target state "authority" by a firmer grasp of the hand to indicate that the parent now has more authority over the decision e.g. to safely cross a street. The change of an internal target state is not fixed to a certain modality and can be quite complex. To reduce complexity, it can be helpful to make use of metaphors. Metaphors are built from experience and can be transported using so called Image Schemas. Image schemas originate from our experiences with humans, objects and events in our environment. They are not specific to a certain sensory modality [5, 6]. The so created ambiguity of being abstract (they are schematic) and not abstract (they are embodied) can be solved by understanding image schemas as abstract repeating representations of dynamic patterns of embodied interaction with our environment, that structure our understanding of how the world works [7].

2 Arbitration Patterns

There are certain characteristics that make the use of image schemas and interaction patterns attractive for the arbitration of movement. Interaction patterns that follow the structure presented by [7], incorporate the user's perception into its solutions. This makes the pattern not only more intuitive by using image schemas [8], but the outcome can be more predictable. As such, the result of the arbitration process has a higher probability of being accepted by both interacting actors.

Some situations of arbitration can be very complex, e.g. flying an aircraft (multiple degrees of freedom). Trying to define patterns from such use cases can be complex due to the quantity of variables and the nature of the use cases themselves. To manage complexity, it makes sense to start with 1-dimensional movement, where the number of variables is greatly reduced.

Following the concept of pattern languages, a complex interaction process can be divided into simpler blocks, each one forming a single interaction pattern. These patterns can then be reused in other processes if needed with the advantage having been proved. This means that interaction patterns defined for 1-dimensional movement can also be adapted and extended for more complex 2- and 3D arbitration applications.

The pattern in this paper was developed using participatory design through the pattern designer tool. The tool allows for a dynamic exploration of different cues and patterns with a multimodal approach. The use case of unplugging a USB device from a computer was used. This use case was selected due to its simplicity and commonality so most users would be familiar with the process and the consequences of doing it incorrectly.

3 USB Arbitrator

For the purpose of exploration and definition of patterns that can be used for arbitration, a special device, "USB Arbitrator", was designed and built. In order to properly represent a cooperative system, the Arbitrator needed to be able to communicate with the user through different modalities. Therefore, the device uses a mix of light, sound and haptic cues that can be custom defined according to the wish of the user (Fig. 1 left). The visual channel uses both text and colored LEDs. The purpose of the text field is to inform the user about the current scenario, e.g. writing data, reading data, etc. A piezo buzzer is used for the acoustic channel; adjustments for sound include the pitch and tone duration. To enable haptic feedback and movement, the socket where the USB stick is plugged, is attached to a small platform that can slide in an out (Fig. 1 right). Subsequently, this platform is attached to a high precision motor that controls the movement, the forces and the point when the USB stick is allowed to be unplugged. It is also possible to induce vibrations through the motor to further enhance the haptic feedback. A special USB socket was used to keep the USB stick locked in place until a solenoid was activated releasing it.

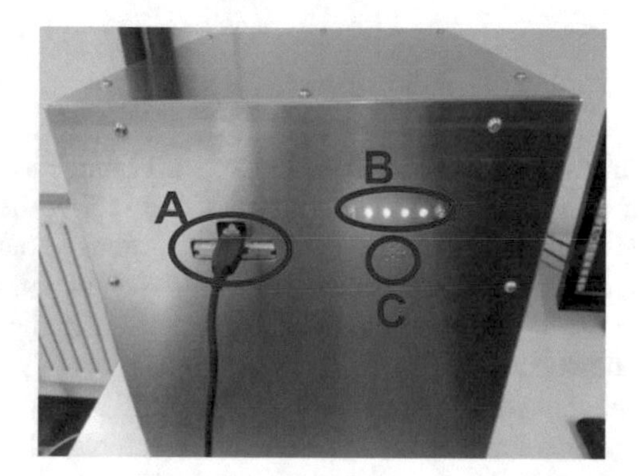

Fig. 1. (Left) Front panel of the Arbitrator. (A) USB socket with haptic feedback; (B) LEDs can be set in different colors and blinking patterns; (C) Piezo buzzer for acoustic cues. (Right) Movement of the USB socket was realized by attaching a platform to a high precision motor.

4 Creating Patterns

The signals or cues that are sent to the Arbitrator are designed in the Pattern Designer. This is a software that allows creating specific cues for each modality. The cues for each modality are defined similarly; they can be a single continuous signal or a series of pulses organized in groups. In addition to this, exclusive parameters for each mode type, e.g. tone for audio or color for visual, can be adjusted as well. Figure 2 shows the basic principle for creating visual cues. Patterns where movement is involved usually consist of a series of steps performed in certain order. Once the cues have been defined, they can be assigned to the desired step of our pattern (Fig. 3).

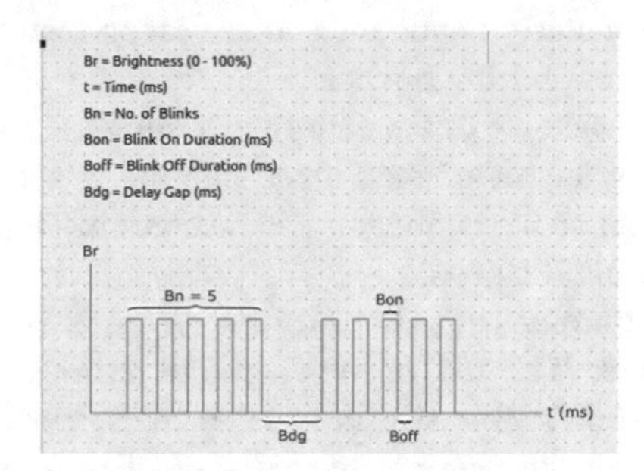

Fig. 2. Visuals tab of the pattern designer. Here the frequency of the signal pulses can be defined as well as other parameters exclusive for this mode type.

Step:	Acoustic	Visual	Haptic	Coupling	Triggers
1:	Pos4-auc	Pos4-off	Pos4-off	Pos1-Loc	Force: N; Position: mm
2.	Pos1-auc	Pos1-soli	Pos4-off	Pos1-Loc	Force: N; Position: 4 mm
3:	Pos3-mic	Pos2-mic	Pos4-off	Pos2-Loc	Force: N; Position: 8 mm
4:	highPitcl	Pos3-fas	Pos2-mic	Pos3-Loc	Force: N; Position: 12 mm
5:	Pos5-off	Pos3-fas	Pos4-off	Pos4-unl	Force: 15 N; Position: 17 mm
6:	Pos1-auc	Pos1-soli	Pos1-shc	Pos1-Loc	Force: N; Position: mm

Fig. 3. The pattern is formed by assigning the modal cues to each escalation step. Position and force triggers define when each step will be activated.

Position and force triggers are used to determine when each escalation step is activated. It is important to mention that the cues as well as the patterns can be previewed individually. This way it is possible to have a more fluid design of cues and patterns while also being able to involve the user in the process.

4.1 A Pattern for Increasing Situation Awareness

The following is an example of how a general interaction pattern for arbitration can be derived with the presented method and tool. The pattern follows the structure used by [7].

Pattern. Inform-Warn.

Problem. Reduced situational awareness can have serious consequences depending the scenario. Sudden notifications by the technical system can startle the user provoking an undesired reaction. The user needs to be made aware of the situation in a controlled manner.

Solution. Using an escalation scheme with increasing levels of interaction can bring the current status of the situation to the attention of the user. Each escalation step incorporates additional cues while making more obvious current ones. The internal target states that are particularly important and are recommended to be stimulated are Urgency, Authority, Power and Importance.

Consequences. The objective of the pattern is to increase situational awareness due to change in perception of the before mentioned internal target states. Overreaction is a possibility. Such cases can create confusion and hesitation. This is especially important to consider in time critical situations.

Implementation Example. The pattern Inform-Warn is exemplified in the use case of unplugging a USB stick. The goal of the user is to retrieve stick from the computer while the automated system's objective is to keep the integrity of the date. The pattern's objective is to make the user aware that loss of data can potentially occur by continuing with the action without preventing him to do so. Acoustic, visual and haptic cues are arranged in four escalation steps. Pulling distance of the USB stick determines the point when the next escalation is reached (Fig. 4).

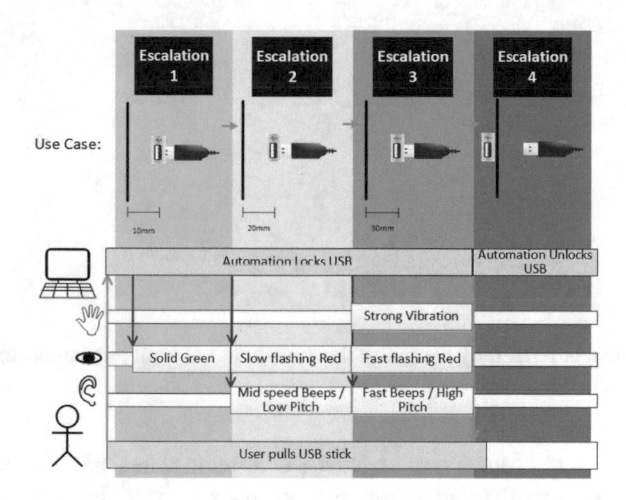

Fig. 4. Inform-Warn pattern is divided in four escalation steps. Each step incorporates additional cues to increase the user's situation awareness. (Color figure online)

- Escalation 1: From 0 to 10 mm of pulling distance, a solid green color is displayed by the front LEDs. USB stick remains locked.
- Escalation 2: From 10 to 20 mm of pulling distance, slow flashing red lights and slow/low pitch beeps are activated. USB stick remains locked.
- Escalation 3: From 20 to 30 mm of pulling distance. Final step before decoupling. Fast flashing red lights, fast/high pitch beeps and strong vibration are activated. USB remains locked.

- Escalation 4: More than 30 mm of pulling distance. Lock is released. The user can now remove the USB stick.

As mentioned before, the objective of the example pattern is to make the user aware of the danger of data loss should he decides to continue with his action. Other patterns can easily be created by simply modifying the escalation steps and cues, for example, a pattern to prevent data loss by keeping the USB lock active all the time.

5 Outlook

The discussed method for defining patterns for use in arbitration of movement is not exclusive to the dimension in which they were initially defined. The patterns have a general approach that can be easily tailored to any other application or combined with others patterns within the same pattern language to solve complex problems of interaction. The Pattern Designer tool allows for dynamic exploration of multimodal interactions and their effects in the internal target states of the user (perception of a situation). By using this approach together with an exploratory device, e.g. USB Arbitrator, a more complete pattern language can be created offering interaction designers a powerful approach to create new interfaces for human-machine cooperation across different domains.

Although the example pattern discussed in this paper has been adapted and tested in 2D environments, e.g. in a driving simulator [9], using the pattern to solve other arbitration problems in 3D movement applications is still pending. More testing and validation to ensure that the pattern is accepted equally by various users need to be done as well.

References

1. Baltzer, M., Altendorf, E., Meier, S., Flemisch, F.: Mediating the interaction between human and automation during the arbitration processes in cooperative guidance and control of highly automated vehicles: basic concept and first study. In: Stanton, N., Landry, S., Bucchianico, G. D., Vallicelli, A. (Eds.) Advances in Human Aspects of Transportation Part I, pp. 439–450. AHFE Conference, Krakow (2014)
2. Kelsch, J., Flemisch, F.O., Löper, C., Schieben, A., Schindler, J.: Links oder rechts, schneller oder langsamer? Grundlegende Fragestellungen beim Cognitive Systems Engineering von hochautomatisierter Fahrzeugführung. In: 48. FAS Anthropotechnik: Cognitive Systems Engineering in der Fahrzeug- und Prozessführung, Braunschweig (2006)
3. Flemisch, F.O.: Pointillistische Analyse der visuellen und nicht- visuellen Interaktionsressourcen am Beispiel Pilot-Assistenzsystem. Dissertation, Universität der Bundeswehr München, Neubiberg (2000)
4. Alexander, C., Ishikawa, S., Silverstein, M., Jacobson, M., Fiksdahl-King, I., Angel, S.: A Pattern Language: Towns, Buildings, Construction. Oxford University Press, Oxford (1977)
5. Lakoff, G.: Women, Fire and Dangerous Things. University of Chicago Press, Chicago (1987)
6. Johnson, M.: The Body in the Mind: The Bodily Basis of Meaning, Imagination, and Reason. University of Chicago Press, Chicago (1987)

7. Baltzer, M., López, D., Flemisch, F.: Towards an interaction pattern language for human machine cooperation and cooperative movement. Cogn. Technol. Work (2018, accepted)
8. Hurtienne, J., Klockner, K., Diefenbach, S., Nass, C., Maier, A.: Designing with image schemas: resolving the tension between innovation, inclusion and intuitive use. Interact. Comput. **27**, 235–255 (2015)
9. Baltzer, M.C.A., Lassen, C., López, D., Flemisch, F.: Behaviour adaptation using interaction patterns with augmented reality elements. In: Schmorrow, D., Fidopiastis, C. (eds.) Augmented Cognition: Intelligent Technologies. LNCS. Springer, Cham (2018)

Pupil Size as Input Data to Distinguish Comprehension State in Auditory Word Association Task Using Machine Learning

Kosei Minami(✉), Keiichi Watanuki, Kazunori Kaede, and Keiichi Muramatsu

Graduate School of Science and Engineering, Saitama University, 255 Shimo-okubo, Sakura-ku, Saitama-shi, Saitama 338-8570, Japan
k.minami.117@ms.saitama-u.ac.jp

Abstract. In communication, it is very important for a speaker to understand the comprehension state of the speaking partner. In this study, the "comprehension state" is defined as whether or not the speaker's message is clearly understood, which is difficult to accurately evaluate. This study aims to evaluate the comprehension state from the pupil size using machine learning. We conduct a word association task using elements that are similar to those used in conversations and measure the pupil size; this pupil size data is used as input data for machine learning. The results show that high accuracy is achieved by learning the low frequency components of the pupil size.

Keywords: Pupil size · Comprehension state · Word association task

1 Introduction

Easy-to-understand explanations are very important in communication. Speaking partners comprehend a speaker's explanation by sharing knowledge with the speaker. Therefore, modifying the explanation according to the comprehension state of a speaking partner is a simple method to ensure understanding [1]. However, it is difficult to accurately estimate the comprehension state of a speaking partner. Therefore, an objective evaluation indicator is necessary to determine this comprehension state. Pupillary responses are controlled by the autonomic nervous system. The activity of the autonomic nervous system can be affected by the mental state. Based on the relationships among pupil size, autonomic nervous system activity, and mental condition, an analysis of pupil size during conversations can be useful to estimate the mental state of an individual [2, 3]. In our previous study [4], we measured the pupil size in auditory word association tasks and suggested that the pupil size varies based on the comprehension state. The purpose of this study is to evaluate the comprehension state during a word association task from the pupil size using machine learning. We use the pupil size data as input data for machine learning because we have already shown the relationship between pupil size and the comprehension state in our previous study. The pupil size data is divided into arbitrary widths and stride intervals to estimate the comprehension state at a random time. It can be assumed that the division width and stride interval

W. Karwowski and T. Ahram (Eds.): IHSI 2019, AISC 903, pp. 123–129, 2019.
https://doi.org/10.1007/978-3-030-11051-2_19

affect the classification accuracy because these change the characteristics of the pupil size. Therefore, we compare the accuracy by changing the width and movement interval and evaluate the validity of this method for estimating the comprehension state.

2 Pupil Size Measurement

2.1 Word Association Task

Word association task is a word game involving multiple associated auditory words with a common theme and participant answers. For example, if the theme is "apple", auditory words such as "red," "fruit," or "sphere" would be presented. Then, participants imagine "red spherical fruits" from these auditory words and associate "apple". This task uses elements that are similar to those used in a conversation, such as listening, word association, and understanding of the meaning.

2.2 Experiment Environment

Figure 1 shows the experiment environment. The pupil size was recorded at 1000 Hz by an eye tracker (SR Research EyeLink 1000 Plus). The participant's head position was stabilized by a chin rest. The participants wore a headphone for listening to auditory words. An 18.5-inch monitor (1366 × 768 pixels) was placed 750 mm away from the participants. A fixation point was displayed at the center of the monitor and the background color was gray. The time in which the participants made the association was measured by a mouse click.

2.3 Experiment Process

Figure 2 shows the experiment process. Sixteen males, who were over 20 years, participated in the experiment. In the task, we presented five auditory words in 3-s intervals using headphones. This number of words and presentation interval time were decided according to a preliminary test such that participants could easily memorize the words. Participants clicked the mouse button when they could associate the words with the theme. This trial was repeated 10 times for each participant. Prior approval for the experiment was obtained from the Saitama University Ethics Review Committee.

2.4 Relationship Between Comprehension State and Pupil Size

In this experiment, we showed trends by statistically analyzing the pupil size [4]. The pupil size data was interpolated and standardized using other data such as those related to blinks and saccade, and eye movement speed. Figure 3 shows the average of all trials (160 trials) extracted from the pupil size data for 10 s with reference to the mouse click time. Figure 3 shows that the pupil size tended to increase before an association was made and decrease after the association was made.

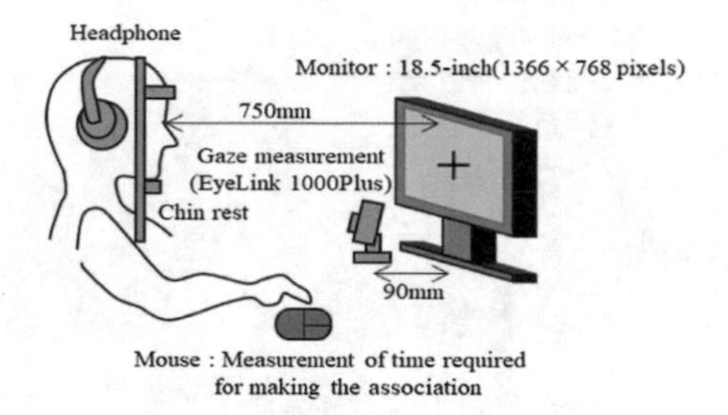

Fig. 1. Experiment environment

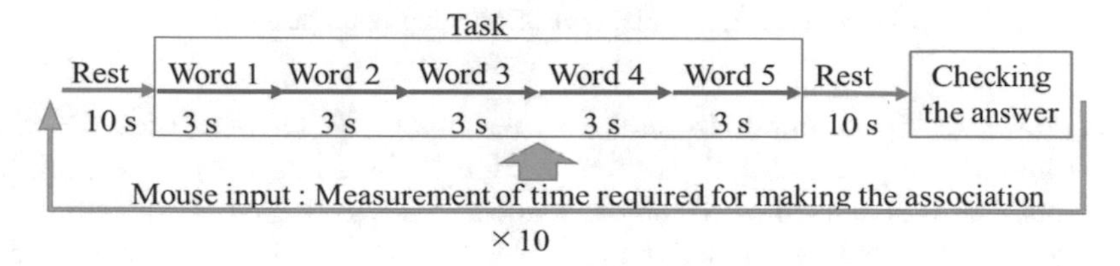

Fig. 2. Experiment process

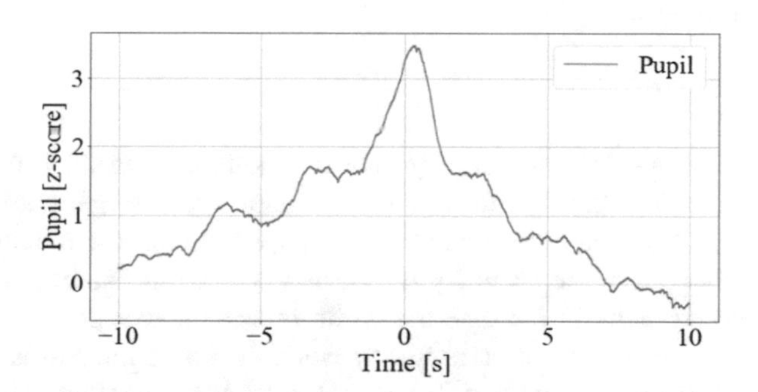

Fig. 3. Average pupil size based on the mouse click time

3 Distinguishing Comprehension State Using Machine Learning

3.1 Method

According to the trend in the change of the pupil size obtained in our previous study, it can be presumed that the comprehension state can be evaluated using the pupil size. Therefore, this experiment attempted to identify the comprehension state in the auditory word association task using a machine learning method with pupil size as input. A support vector machine (SVM) [5], which has been widely used for classification, was used in this experiment. An SVM performs classification by finding an optimal hyperplane that maximizes the margin between the hyperplane and the data points near the hyperplane. The SVM was implemented using the scikit-learn Python library [6].

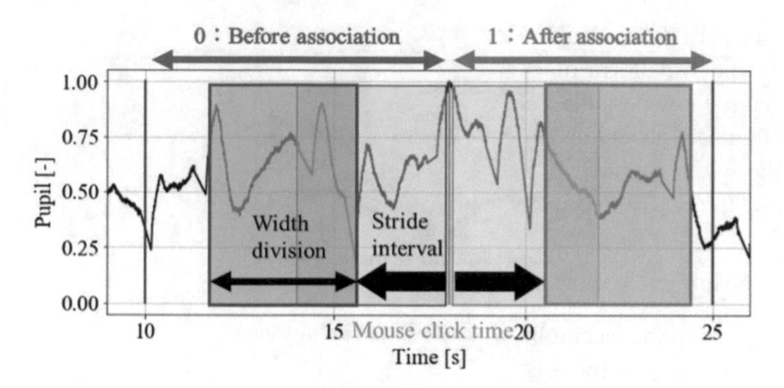

Fig. 4. Conceptual diagram of dividing pupil size data

As shown in Fig. 4, pupil size data was the input data for the SVM, which was divided into arbitrary widths and movement intervals based on the mouse click time. Each input data was labeled as "0 (before association)" or "1 (after association)." The accuracy was calculated by comparing these labels and SVM solutions. There were two ways of dividing the pupil size: the condition in which the width changes from 1–7 s with constant stride interval and the condition in which the stride interval changes from 0.5–2 s with constant width.

3.2 Result

Table 1 shows the classification accuracy for each width with constant stride interval of 0.5 s. These values of accuracy are by cross-validation where one participant data is considered to be test data and the remaining participant data is the training data. It can be seen from Table 1 that the accuracy increases as the width increases. Furthermore, the accuracy exceeds 80% in the case the width is more than 4 s.

Table 2 shows the classification accuracy when the stride interval is changed. The dataset when the stride interval was 0.5 s was used for validation. The accuracy is constant even though stride interval is changed. In particular, high accuracy can be maintained in spite of having few samples when the width is 7 s and stride interval is 2 s.

Table 1. Accuracy for each width (stride interval = 0.5 s)

Width [s]	Number of samples	Training accuracy [%]	Validation accuracy [%]
1	4324	63.2	62.2
2	3694	67.0	65.6
3	3077	75.5	73.4
4	2464	82.0	79.5
5	1878	85.6	83.0
6	1352	92.8	90.6
7	847	97.5	95.2

The cause of low accuracy as the learned width reduced was due to the short changes in the waveform of the pupil size data. Therefore, pupil size was relearned after applying a low pass filter (LPF) using fast Fourier transform (FFT). The results are shown Table 3. The cutoff frequencies of the LPF were 1, 0.33, 0.2, and 0.14 Hz because the width changed from 1–7 s. It can be seen from Table 3 that the accuracy increases as the cutoff frequency decreases.

3.3 Discussion

We can see that the accuracy increases as the width increases. Furthermore, the accuracy also increases by smoothing the pupil size data using an LPF even when the width is short. This is due to the frequency components of the pupil size. The frequency component of the pupil size in the task is shown in Fig. 5. It can be seen from Fig. 5 that the peak frequencies are under 0.1 Hz and approximately 0.3–0.35 Hz. First, the peak under 0.1 Hz is considered. We can see that the frequency components of pupil size data under 0.1 Hz include comprehension state information because the accuracy is high when the cutoff frequency is 0.14 Hz. It can be assumed that there was less fluctuation in the pupil size because the comprehension state changes only once in the word association task. Thus, further studies should be conducted to evaluate the pupil size in case of several changes in the comprehension state. Next, the peak of around 0.3–0.35 Hz is considered. This peak is because the pupil size data fluctuate in the 3-s cycle. It is known that the pupil size increases reflexively by sound stimulation [7]. Based on this knowledge, the reason for this fluctuation may be because auditory words were presented at 3-s intervals in the word association task. The cause of low accuracy in short width is learning the fluctuation by sound stimulation. Last, the result when the accuracy is constant regardless of the number of samples in Table 2 is considered. The SVM generates an optimal hyperplane that separates the data into categories. If the sample distribution is clearly separated, the SVM achieves high accuracy even for a small number of samples [8]. Accordingly, it is inferred that pupil size data has sufficiently different characteristics before and after association.

Table 2. Accuracy for each stride interval (width = 7 s)

Stride [s]	Number of samples	Training accuracy [%]	Validation accuracy [%]
0.5	847	97.5	95.2
1	483	98.0	94.2
1.5	380	97.9	95.3
2	314	98.2	94.7

Table 3. Accuracy for each cutoff frequency (width = 1 s, stride interval = 0.5 s)

Cutoff frequency [Hz]	Number of samples	Training accuracy [%]	Validation accuracy [%]
1	4324	63.4	62.9
0.33	4324	65.7	65.0
0.2	4324	69.9	69.2
0.14	4324	81.1	80.4

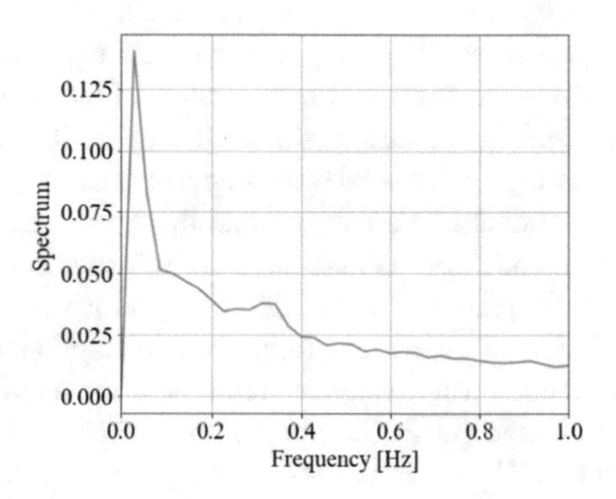

Fig. 5. Spectrum of pupil size data

4 Conclusion

In this study, a method to distinguish the comprehension state in an auditory word association task using machine learning was proposed. The comprehension state of participants was distinguished by an SVM using the pupil size. Input data for learning were the pupil size data divided into several seconds; high accuracy was achieved for longer widths. From this result, it can be assumed that the comprehension state information appears in the low frequency component of the pupil size. Thus, this study suggested the validity of using pupil size to evaluate the comprehension state. However, the pupil size changes depending on external factors such as sound stimulation and light stimulation; therefore, learning considering external factors is important. In the future, we will elucidate the effects of external factors on pupil size and construct a more accurate machine learning method.

References

1. Clark, H.H., Schaefer, E.F.: Contributing to discourse. Cogn. Sci. **13**, 259–294 (1989)
2. Bradlay, M.M., Miccoli, L., Escrig, M.A., Lang., P.J.: The pupil as a measure of emotional arousal and autonomic activation. Psychophysiology **45**, 602–607 (2008)
3. Hess, E.H.: Attitude and pupil size. Sci. Am. **212**, 46–54 (1965)

4. Minami, K., Watanuki, K., Kaede, K., Muramatsu, K.: The effects of listener understanding on pupil and blinks. In: Proceedings of the 28th Design and System Division Conference, The Japan Society of Mechanical Engineers, (JSME D&S 2018) (2018). (in Japanese). (in Press)
5. Vapnik, V.N.: Statistical Learning Theory. Wiley, New York (1998)
6. Pedregosa, F., Varoquaux, G., Gramfort, A., Michel, V., Thirion, B., Grisel, O., Blondel, M., Prettenhofer, P., Weiss, R., Dubourg, V., Vanderplas, J., Passos, A., Cournapeau, D., Brucher, M., Perrot, M., Duchesnay, E.: Scikit-learrn: machine learning in Python. J. Mach. Learn. Res. **12**, 2825–2830 (2011)
7. Shiga, N., Ohkubo, Y.: Pupillary reflex dilatation to the auditory stimuli - the effects of parasympathetic activity on the pattern of the pupillary reflex dilation. Tohoku Psychol. Folia **39**, 31–39 (1981)
8. Yao, J.T., Zhao, S.L., Saxton, L.V.: A study on fuzzy intrusion detection. In: Data Mining, Intrusion Detection, Information Assurance, and Data Networks Security, vol. 5812, pp. 23–30. The International Society for Optics and Photonics (2005)

Research on Product Color Design Under the Cognition of Brand Image

Xin-xin Zhang(✉), Ming-gang Yang, and Xin-ying Wu

School of Art, Design and Media, East China University of Science and Technology, M. Box 286 NO. 130, Meilong Road, Xuhui District, Shanghai 200237, China
823615861@qq.com

Abstract. As one of the paramount visual features in a product system, color conveys not only a visual sense of beauty but also the physical and psychological needs of the human beings. In the design process, the in-depth study of color elements can effectively use the emotional power to increase the competitiveness of products in the market. Focusing on this, the product color brand image is excavated and the related color design elements are extracted by applying the technical methods of Kansei Engineering. At last, taking the color image study of a car as an example to verify the research idea of this examine.

Keywords: Color brand image · Kansei engineering · Product color image cognitive · Color characteristics

1 Introduction

When people choose the commodity, they will determine the color of products according to their own emotional weight for each color [1]. As one of the most important features in the product vision system [2], the color not only represents the visual beauty, but also the physiological and psychological needs of one person [3]. While, the color image cognition process is not a simple "formula" between color and psychological cognitive semantics, but a non-linear "chaos system" that contains various factors which influence the color perception, with the deepening of perceptual information research in the information age. And it is generally difficult to code and measure, just like other cognitive processes to recognize the kansei information. In order to effectively use the emotional power contained in which may increase the market competitiveness of products, the color image knowledge needs to be quantitatively studied in the process of product color design. Simultaneously, the direction of product image design is guided by the accurate extraction of color brand image. And the brand image is obtained from perceived information in the process of brand image cognition which relates to the intrinsic characteristics of "object" and combines the psychological activities and cognitive abilities of "people". LU Zhang-ping et al. explored the intrinsic relevance of color design and image cognition from the perspective of rationality through the relationship drawing between color image and semantic vocabulary [4]. However, due to the complex ambiguity of the user's emotional demand information, the color design process of the product is mostly based on

W. Karwowski and T. Ahram (Eds.): IHSI 2019, AISC 903, pp. 130–136, 2019.
https://doi.org/10.1007/978-3-030-11051-2_20

the user's color sensibility demand and its subjective judgment and industry experience, which is different from the user's real emotional needs. And the design knowledge is cumbersome and not convenient for quick extraction. In order to reduce the color perception gap between designers and users, the technical method of Kansei Engineering is used to gained the product color image design knowledge base in this research.

2 Background

2.1 Brand Image Cognition

People's consumption concept has also changed greatly with the improvement of material and cultural life level. The emotional demand has become diversified and individualized, which reflects people's pursuit of emotional quality and satisfaction of self-realization. The brand image of product color is the brand impression in the minds of users which exists in the concept of consumers', and all the features and beliefs that are imposed upon the color, being formed in the process of users' cognitive experience after interacting with the product. Kansei engineering is a kind of theory and method of using engineering technology to explore the relationship between kansei needs of cognition subject and design elements. This study can accurately grasp the brand image to enhance the brand value of color effectively.

2.2 The Product Color Image

The product color image is generated by the user's product color cognition process, which is the result of acquiring and recreating the color sensibility demand through their own visual sensory channel. It can fully express the user's color sensitivity needs to be an important medium for designers to understand the user's color emotions. From the visual attention mechanism, the color information accounts for about 80%, and the product shape is about 20% in the visual communication between the user and the product [5]. And the color has a great power to appeal directly to emotions in satisfying users' emotional needs, increasing the added value and competitiveness of products. Takahashi et al. analyzed the correlation between color and emotion, through emotional vocabulary and facial expression stimulation [6]. By mining the correlation between product color image and color features, the product color design to meet market demand is effectively guided and shortens the development cycle.

3 Methods

The detailed research process is shown in Fig. 1.

3.1 Quantitative Extraction of the Product Color Image

The product color image set can be described as $A = \{a_1, a_2, \ldots, a_d\}$ and the research sample collection $B = \{b_1, b_2, \ldots, b_h\}$ are obtained by the KJ method, the hybrid

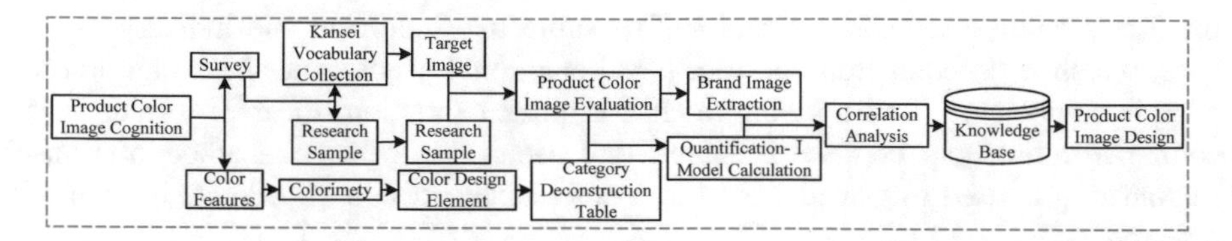

Fig. 1. The process of product color image design under the brand image cognition

clustering algorithm and others. Then, applying the semantic difference method to quantify the user's color sensitivity needs, the product color image evaluation matrix $\boldsymbol{X}$ is got, as shown in:

$$\boldsymbol{X} = \begin{bmatrix} x_{11} & x_{12} & \cdots & x_{1d} \\ x_{21} & x_{22} & \cdots & x_{2d} \\ \vdots & \vdots & x_{ij} & \vdots \\ x_{h1} & x_{h2} & \cdots & x_{hd} \end{bmatrix}. \tag{1}$$

where x_{ij} means to the jth color image evaluation value of the ith sample.

3.2 Evaluate the Brand Image Entropy of the Color

Image Entropy [7] is used to calculate the information amount of the color image for characterizing the degree of uncertainty in people's perception of color image. Then the brand image of the product color is obtained from the calculation results. The calculation formula as follows. In order to reduce the error in the evaluation process, the data in Eq. (1) is normalized to obtain the decision matrix $\boldsymbol{Y}$. Then the probability of the target image p_{ij} is got. Then, the image entropy is acquired by:

$$I_j = -z \sum_{i=1}^{h} P_{ij}\, ln\, P_{ij}. \tag{2}$$

where $\boldsymbol{I}_j$ stands for the image entropy value which its unit for "Nat/symbol", and the constant z = 1/ln m. Then the images are weighted by:

$$w_j = 1 - I_j / \sum_{j=1}^{d} (1 - I_j). \tag{3}$$

Thus, the weight of each image of the product color can be calculated, and the larger one is the brand image.

3.3 Correlation Calculation Between the Brand Image and Design Elements of the Product Color

According to the three elements of color in the principle of colorimetry, the design element project *r* and its category *t* of color are extracted and summarized based on the deconstruction of the color characteristics of the samples. Then the Quantification-I theory is used to calculate the association between the brand image and design elements of the product color combined with the Eq. (1) and the calculation result of Image entropy. Assuming the main color feature of the study sample has a total of *m* color design elements, the category of the *r*th color design element is L_r. So as to all research samples, $\delta_\mathrm{i}(\mathrm{r, t})(\mathrm{r} = 1, 2, \cdots, m; \mathrm{t} = 1, 2, \cdots, Lr)$ is indicated that the *t*th category of the *r*th color design element which appears in the *i*th sample. Thus, the mapping relationship between the color brand image and the color design elements of the product is:

$$\delta_\mathrm{i}(\mathrm{r,t}) = \begin{cases} 1 & \begin{array}{l}(\text{Qualitative description of the rth color design element} \\ \qquad\qquad\qquad \text{is the tthcategory in the ith sample})\end{array} \\ 0 & (\text{else}) \end{cases} \tag{4}$$

and the mathematical model is:

$$\mu_i = \sum_{r=1}^{m} \sum_{t=1}^{L_r} \delta_\mathrm{i}(\mathrm{r, t}) O_{rt} + \varepsilon_i \tag{5}$$

where O_{rt} only depends on the objective coefficient of the type *t* from the color design element *r*. And ε_i refers to the random error in the *i*th sample. Calculating the Eq. (5), the closer the complex correlation coefficient is to 1, the accuracy of the model is higher. And the weights of the *r*th color design element and the *t*th category on the product color brand image are respectively represented by the partial correlation coefficient and the standard coefficient. Then, the design factors and categories with the large partial correlation coefficient and standard coefficient should be more considered in product color development.

4 Case Study

4.1 Identify the Target Images and Research Samples

The sedan was studied as a case in this paper. According to the mainstream car brand and satisfaction rankings in the year of 2015 to 2017 [8], the research samples were selected among seven car brands. Firstly, 36 car pictures and 27 pairs of adjectives that describing the car's color image were initially collected. Secondly, 24 research samples were discussed and selected by the experts according to similarity of product color, as shown in Fig. 2.

Thirdly, six groups vocabularies that describing the car color image were determined by using the KJ method and the hybrid clustering algorithm. Specifically

Fig. 2. The research samples.

described as: Conservative-Stylish, Conservative-Atmospheric, Rough-Delicate, Steady-Dynamic, Earthy-Luxurious, Sharp-Soft. Finally, the questionnaire was developed by using the Semantic difference method based on the 24 research samples and the six sets of color images. Then the image evaluation value of car color was gain from the 22 valid questionnaires. Among these, 13 people from women and 9 were men.

4.2 Extract the Car Color Brand Image

Equations (2) and (3) were calculated based on the car color image evaluation value, and the results were shown as follows: Conservative-Stylish ($w = 0.2044$), Conservative-Atmospheric ($w = 0.1402$), Rough-Delicate ($w = 0.2149$), Steady-Dynamic ($w = 0.2$), Earthy-Luxurious ($w = 0.1271$), Sharp-Soft ($w = 0.1136$). It can be determined that the brand image of car color is Rough-Exquisite, because the weight value of it is the largest.

4.3 Deconstruct the Car Color Features

There are three common color modes named RGB, CMYK and HSB. Among them, because the RGB color mode cannot directly give the component of each color value, the HSB mode that is more accurate in sensing color than the RGB mode is selected. It is a color pattern based on the human eye and still keeps the calculation simple. Where H means hue (0°–360°), S represents saturation (0%–100%), B(Brightness) represents brightness (0%–100%). In this study, due to the complexity of the car color, the design elements and categories of the car's main color were summarized based on the rules of Table 1. Then the HSB values of the 24 study samples were extracted by the image processing software.

4.4 Correlation Calculation Between the Brand Image and Design Elements of Car Color

According to Eq. (4), the data applicable to the Quantification-I theory were obtained from Table 1. And based on this, combined with the evaluation value of automobile color image, the correlation between the brand image (Rough-Exquisite) and each design element and category was calculated by using Eq. (5). The calculation results are shown in Table 2. From Table 2, this model is proved to be appropriate and accurate through the complex correlation coefficient. Among the color features, the design factor that has a greater influence on the car color brand image is the brightness.

Table 1. The deconstruction table of car's main color feature

Three elements	Category	Quantitative indicators
Hue	Warm (H_1)	[red purple, red, red yellow, yellow]
	Neutral (H_2)	[black, grey, white]
	Cool (H_3)	[yellow green, green, blue green, blue, blue purple, purple]
Saturation	Low (S_1)	[0, 33]%
	Medium (S_2)	[34, 67]%
	High (S_3)	[68, 100]%
Brightness	Low (B_1)	[0, 33]%
	Medium (B_2)	[34, 67]%
	High (B_3)	[68, 100]%

And the weights of neutral color, medium saturation and medium brightness in various types of objects are relatively large. These mean that the factors which have a higher impact on the car's color refinement image are neutral colors (including black, gray, white), medium saturation (34%–67%) and medium brightness (34%–67%).

Table 2. The relationship between color brand image and color features

Three elements	Category	Coefficient		
		Standard	Partial	Complex
Hue	H_1	-0.012	0.1150	0.772
	H_2	**0.045**		
	H_3	-0.054		
Saturation	S_1	-0.046	**0.2308**	
	S_2	**0.142**		
	S_3	0.004		
Brightness	B_1	-0.033	0.0677	
	B_2	**0.010**		
	B_3	0.005		

5 Conclusion

The technical methods of perceptual engineering have been applied to acquire the knowledge of product color image design. And the association between color brand image and color features is used to guide the design direction of car's color image. Main conclusions are as follows: (1) Completing the semantic description of the product's color image to explore the brand image. (2) Constructing a knowledge base of automobile color brand image design that contains the color characteristics and the correlation with the color brand image, according to the principle of colorimetry. Thus, in order to better grasp the user's color requirements, other influencing factors such as car color matching and embellishment color will be considered comprehensively based on these research results in the follow-up study.

References

1. Palmer, S.E., Schloss, K.B.: An ecological valence theory of human color preference. J. Proc. Natl. Acad. Sci. U.S.A. **107**(19), 8877–8882 (2010)
2. Liu, X., Wang, K., Wang, Q.: Application of metaphorical characteristics of color in mobile phone UI design. J. Packag. Eng. **39**(8), 200–204 (2018)
3. Hanada, M.: Correspondence analysis of color-emotion associations. Color Res. Appl. **43**, 224–237 (2017)
4. Lu, Z., Gu, Q., Li, M., et al.: Image cognition of automobile color. J. Packag. Eng. **20**, 20–24 (2014)
5. Jian, Z., Xu, B., Jin, L.: The localization algorithm of human body based on omnidirectional vision. In: IEEE Information Technology and Artificial Intelligence Conference, pp. 172–176. IEEE Press, New York (2011)
6. Takahashi, F., Kawabata, Y.: The association between colors and emotions for emotional words and facial expressions. Color Res. Appl. **43**, 247–257 (2018)
7. Zhang, X., Yang, M., Zhou, Y.: Research on the design of smart pension product modeling based on brand image. In: Zhou, J., Salvendy, G. (eds.) ITAP 2017. LNCS, vol. 10297, pp. 304–315. Springer, Cham (2017)
8. Mainstream car brand rankings of the China Brand Power Index. http://www.chn-brand.org/c-bpi/zhuliuch1.html

Design Components of Clinical Work Environments with Computerized Decision Support Systems

Uta Wilkens[1(✉)] and Florian M. Artinger[2,3]

[1] Ruhr-University Bochum (RUB), Institute of Work Science, Chair for Work, Human Resources and Leadership, Universitätsstr. 150, 44780 Bochum, Germany
uta.wilkens@ruhr-uni-bochum.de
[2] Max Planck Institute for Human Development, Lentzteallee 94, 14195 Berlin, Germany
artinger@mpib-berlin.mpg.de
[3] Simply Rational GmbH, Eberhard-Roters-Platz 7a, 10965 Berlin, Germany

Abstract. Computerized Decision Support Systems (CDSSs) can be a vital component in a medical setting to foster the use of evidence based medicine and minimize malpractice. Surprisingly, the adoption rate of CDSSs has remained far below expectations and there has been little impact of CDSSs on measurable health outcomes. We outline the components of clinical work environments in order to elaborate on the driving forces for technology acceptance. The components address issues such as high involvement work systems and distributed intelligence. The reflection of these characteristics leads us to the conclusion that the perceived usefulness of a technology and its ease of use is a necessary but not a sufficient condition. Technological acceptance primarily depends on the perceived mindfulness of individual intelligence in workplace design.

Keywords: Computerized decision support systems · Artificial intelligence
Distributed intelligence · Workplace design

1 Introduction

Computerized decision-support systems (CDSSs) have a large potential to improve decision making processes. In medicine, CDSSs can be central in empowering professionals to practice evidence based medicine and reducing malpractices [1–4]. However, the adoption rate of CDSSs is slow and improvement of measurable health outcomes remain below expectations [1, 5–8]. In order to tackle these problems, research has focused on the proficiency of the tools. Little attention has been given to the working context in which practitioners are embedded in. There is a need not just to search for single influencing factors in the working context but to better understand the design principles of highly professional work environments such as clinics in order to develop integrative approaches for the use of CDSSs. The assumption is that technology acceptance of CDSSs depends on the design components of the work systems. It is therefore necessary to gain a deeper understanding of the characteristics of these

W. Karwowski and T. Ahram (Eds.): IHSI 2019, AISC 903, pp. 137–141, 2019.
https://doi.org/10.1007/978-3-030-11051-2_21

components. This paper develops a theoretical-conceptual framework of design components of professional work settings in hospitals with CDSSs. We relate these components to issues of technology acceptance. For this purpose, we combine research from artificial intelligence, distributed intelligence, and human-centered job design with research on technology acceptance.

2 Outline of Design Components

2.1 Clinical Work Environments with Computerized Decision Support Systems as Socio-Technical System

The use of CDSSs implies a considerable change within the medical system as the functions of CDSSs can complement and possibly even replace those of human actors. As CDSSs become part of the workplace and human interaction, they do not only constitute a technical system but a socio-technical system [9, 10]. Whether CDSSs are used in a manner that improves health outcomes depends on the technological acceptance of CDSSs.

Current CDSSs are often black box tools that make use of machine learning algorithms where it is impossible for the decision maker to assess how a recommendation by a CDSS is derived. Especially professional decision makers are skeptical and exhibit what has been labeled "algorithm aversion" towards such intransparent tools [11–13]. However, it is not only practitioners who are uncomfortable when using intransparent algorithms as part of their decision making process. The European Charter of Patients' Rights guarantees every patient the right to informed consent. This requires transparency as to why a certain treatment is favorable. Moreover, research has identified twelve types of error in human-computer interaction with the general conclusion that machine rules "do not correspond to work organization or usual behaviors" [14, p. 1200].

The technology acceptance model emphasizes the perceived usefulness of a technology and its ease of use [15, 16]. A crucial way to further the ease of use of CDSSs is to construct sufficiently simple, transparent, and intuitive CDSSs that can often perform on-par in comparison to complex algorithms [2, 17, 18]. Beyond such modification of the tools, it is necessary to consider the design component of clinical work environments as a further issue of technology acceptance. Most important characteristics are the high risk/high involvement work systems and the distributed intelligence among individual and artificial actors.

2.2 Semi-standardized High Risk/High Involvement Work Environment

As whether success or failure ensues often has severe consequences in clinical work settings, such an environment can be classified as high risk/high involvement work environment [19, 20]. The decision making process leading to success or failure frequently involves standard procedures which are concurrently repeated. In such a setting, the quality of work can considerably profit from CDSSs that assist standard operations. At the same time, due to variations between patients and the occurrence of non-standard cases such a setting also requires careful attention and mindfulness of the individual professional [21]. That is, technology acceptance will depend on whether and how the socio-technical design allows and furthers mindful action by the individual.

2.3 Distributed Intelligence

A medical work system with CDSSs is based on *distributed intelligence* where "intelligence is distributed 'across minds, persons, and the symbolic and physical environments, both natural and artificial'" [22, p. 192] with reference to [23, p. 47]. This definition can be transferred to socio-technical work environments [24]. As both, the human being and the machine possess intelligence (either human or artificial) and intelligence is distributed among these unequal agents this raises the question, how to best integrate the tool or machine in the work system instead of solely focusing on the tool development without considering individual behavior [22].

Given the increasing attention on intelligent machines and tools in the workplace, it causes a re-distribution of intelligence and can be considered as a matter of perceived justice [25]. Even though the social psychology of justice in resource allocation is primarily related to the distribution of income and material resources [26] there is high plausibility that it also impacts the (re-)distribution of intelligence and the underlying distribution rules in the workplace (see [27] for procedural justice in knowledge sharing). Individual intelligence in a work context does not mean the score of an intelligence test but is the synonym for expert status and knowledge, problem-solving capabilities, role definition, decision making and similar important immaterial resources of job identity [28, 29]. When it potentially comes to a re-distribution of intelligence especially procedural justice is supposed to be important as meta-analyses have shown that individual job performance or counterproductive work behavior are related to perceived procedural justice [26], see also [30]. It can be assumed that the acceptance of CDSSs depend on perceived procedural justice in the implementation process how the use of CDSSs relates to the individual expert role.

2.4 Human-Centered Instead of Tayloristic Job Design

A tayloristic job design results from collecting and systemizing individual data for job descriptions. This method is well known as the first approach in scientific management [31]. The approach started promising due to productivity effects but ended exhibiting rather poor performance due to individual alienation from work, missing individual learning and development perspectives and increased segregation of the working class [32]. Human-centered job design with emphasis on individual autonomy and responsibility in order to keep motivation high [33] could mitigate negative tayloristic side-effects while sustaining or even exceed tayloristic productivity effects [34]. As artificial intelligence is based on methodological components of collecting and systemizing data, which are comparable to Taylorism it at least bears the risk of unintended tayloristic side-effects. As this would destroy the opportunity to profit from distributed intelligence CDSSs have to be adapted to the insights gained from research on human-centered job design.

3 Summary

The central argument that we make is that the acceptance of CDSSs is not a pure function of perceived usefulness of a technology and its ease of use. This is a necessary but not a sufficient condition. Instead, technological acceptance also depends on the perceived mindfulness of individual intelligence in workplace design. The further development of intelligent tools thus needs to reflect their integration in the workplace interaction and implementation process. The higher the treatment of the individual expert role with respect and enhancement of this role with the help of CDSSs the higher the technology acceptance.

As already suggested by Herbert Simon [35], successful decision making is akin to the blades of two scissors. On the one hand, there is the decision making environment, on the other hand the tool that decision makers use. In order to increase the uptake of CDSSs that can be vital to further evidence based medicine where the decision maker does indeed rely on up to date medical knowledge it is necessary to provide an environment that fosters such an uptake but also to use tools that respond to the needs of the practitioner.

References

1. Greenes, R.A.: Clinical Decision Support: The Road to Broad Adoption. Elsevier/Academic Press, Oxford (2014)
2. Djulbegovic, B., Hozo, I., Dale, W.: Transforming clinical practice guidelines and clinical pathways into fast-and-frugal decision trees to improve clinical care strategies. J. Eval. Clin. Pract. **24**, 1–8 (1981)
3. Beck, A.H., et al.: Systematic analysis of breast cancer morphology uncovers stromal features associated with survival. Sci. Transl. Med. **3**, 108ra113–108ra113 (2011)
4. Dawes, R.M., Faust, D., Meehl, P.E.: Clin. vs Actuar. Assess. Sci. **243**, 1668–1674 (1989)
5. Black, A.D., Car, J., Pagliari, C., Anandan, C., Cresswell, K., Bokun, T., McKinstry, B., Procter, R., Majeed, A., Sheikh, A.: The impact of ehealth on the quality and safety of health care: A systematic overview. PLoS Med. **8**(1), e1000387 (2011)
6. Greenes, R.A., Bates, D.W., Kawamoto, K., Middleton, B., Osheroff, J., Shahar, Y.: Clinical decision support models and frameworks: Seeking to address research issues underlying implementation successes and failures. J. Biomed. Inform. **78**, 134–143 (2018)
7. Vrieze, S.I., Grove, W.M.: Survey on the use of clinical and mechanical prediction methods in clinical psychology. Prof. Psychol. Res. Pract. **40**, 525–531 (2009)
8. Jaspers, M.W.M., Smeulers, M., Vermeulen, H., Peute, L.W.: Effects of clinical decision-support systems on practitioner performance and patient outcomes: a synthesis of high-quality systematic review findings. J. Am. Med. Informatics Assoc. **18**, 327–334 (2011)
9. Jones, A.J.I., Artikis, A., Pitt, J.: The design of intelligent socio-technical systems. Artif. Intell. **39**, 5–20 (2013)
10. Wilkens, U., Herrmann, T.: Gibt es eine Arbeitswissenschaft der Digitalisierung? Ein Diskursbeitrag. In: Schlick, C. (ed.) Megatrend Digitalisierung. Potenziale der Arbeits- und Betriebsorganisation, pp. 215–230. GITO Verlag, Berlin (2016)
11. Shteingart, H., Neiman, T., Loewenstein, Y.: The role of first impression in operant learning. J. Exp. Psychol. Gen. **142**, 476–488 (2013)
12. Dietvorst, B.J., Simmons, J.P., Massey, C.: Overcoming algorithm aversion: people will use imperfect algorithms if they can (even slightly) modify them. Manag. Sci. **64**, 1155–1170 (2018)

13. Artinger, F.M., Petersen, M., Gigerenzer, G., Weibler, J.: Heuristics as adaptive decision strategies in management. J. Organ. Behav. **36**, S33–S52 (2015)
14. Koppel, R., et al.: Role of computerized physician order entry systems in facilitating medication errors. JAMA **293**, 1197–1203 (2005)
15. Davis, F.D.: User acceptance of information technology: system characteristics, user perceptions and behavioral impacts. Int. J. Man Mach. Stud. **38**, 475–487 (1993)
16. Venkatesh, V., Davis, F.D.: A theoretical extension of the technology acceptance model: four longitudinal field studies. Manag. Sci. **46**, 186–204 (2000)
17. Jenny, M.A., Pachur, T., Lloyd Williams, S., Becker, E., Margraf, J.: Simple rules for detecting depression. J. Appl. Res. Mem. Cogn. **2**, 149–157 (2013)
18. Jenny, M.A., et al.: Are Mortality and Acute Morbidity in Patients Presenting With Nonspecific Complaints Predictable Using Routine Variables? Acad. Emerg. Med. **22**, 1155–1163 (2015)
19. Bonias, D., Bartram, T., Leggat, S.G., Stanton, P.: Does psychological empowerment mediate the relationship between high performance work systems and patient care quality in hospitals? Asia Pacific J. Hum. Resour. **48**, 319–337 (2010)
20. Scotti, D.J., Driscoll, A.E., Harmon, J., Behson, S.J.: Links among high-performance work environment, service quality, and customer satisfaction: an extension to the healthcare sector. J. Healthc. Manag. **52**, 109–124 (2007)
21. Weick, K.E., Sutcliffe, K.M.: Mindfulness and the quality of organizational attention. Organ. Sci. **17**, 514–524 (2006)
22. Cobb, P.: Learning from distributed theories of intelligence. Mind Cult. Act. **5**, 187–204 (1998)
23. Pea, R.D.: Learning scientific concepts through material and social activities: conversational analysis meets conceptual change. Educ. Psychol. **28**, 265–277 (1993)
24. Fischer, G.: Communities of interest: learning through the interaction of multiple knowledge systems. In: 24th Annual Information Systems Research Seminar in Scan-dinavia (IRIS'24) (Ulvik, Norway), Department of Information Science, Bergen, Norway, pp. 1–14 (2001)
25. Karsh, B.T.: Beyond usability: designing effective technology implementation systems to promote patient safety. Qual. Saf. Health. Care. **13**, 388–394 (2004)
26. Cohen-Charash, Y., Spector, P.E.: The role of justice in organizations: a meta-analysis. Organ. Behav. Hum. Decis. Process. **86**, 278–321 (2001)
27. Kim, W.C., Mauborgne, R.: Procedural justice, strategic decision making, and the knowledge economy. Strat. Mgmt. J. **19**, 323–338 (1998)
28. Crocetti, E., Avanzi, L., Hawk, S.T., Fraccaroli, F., Meeus, W.: Personal and social facets of job identity: a person-centered approach. J. Bus. Psychol. **29**, 281–300 (2014)
29. Skorikov, V., Vondracek, F.W.: Occupational identity. In: Schwartz, S.J., Luyckx, K., Vignoles, V.L. (eds.) Handbook of Identity Theory and Research, pp. 693–714. Springer, New York (2011)
30. Ötting, S.K., Maier, G.W.: The importance of procedural justice in Human-Machine Interactions: Intelligent systems as new decision agents in organizations. Comput. Hum. Behav. **89**, 27–39 (2018)
31. Taylor, F.W.: The Principles of Scientific Management. Harper, New York (1911)
32. Kern, H., Schumann, M.: Industriearbeit und Arbeiterbewußtsein. Frankfurt a. M, Suhrkamp (1985)
33. Hackman, J., Oldham, G.R.: Development of the Job Diagnostic Survey. J. Appl. Psychol. **60**, 159–170 (1975)
34. Adler, P.S., Cole, R.: Designed for learning: a tale of two auto plants. Sloan Manag. Rev. **34**, 85–94 (1993)
35. Simon, H.M.: Rational choice and the structure of the environment. Psychol. Rev. **63**, 12–38 (1956)

Human Resource Development Opportunities in Latvian Health Care Organization

Henrijs Kalkis[1,2(✉)], Ansis Ventins[2], Sandis Babris[3], Zenija Roja[2], and Kristine Bokse[2]

[1] Riga Stradiņš University, Dzirciema Street 16, Riga 1007, Latvia
henrijs.kalkis@gmail.com
[2] University of Latvia, Aspazijas Blvd. 5, Riga 1050, Latvia
zenija.roja@lu.lv
[3] BA School of Business and Finance, K. Valdemara 161, Riga, Latvia
sandis.babris@icloud.com

Abstract. The aim of the research was to investigate the existing human resource management in a health care organization and develop guidelines for improvement, based on the analysis of the theory and empirical research results. In the research a leading health care organization in Latvia was selected. In total, the organization employs over 1300 people, including more than 600 doctors. The study analyzed in two departments: medical center and health center. Several research methods were applied, incl. SWOT analysis, survey and statistical data analysis. The results of the research indicate that employees lack knowledge and motivation that affects their development and career opportunities in the organization. The organization pays a lot of attention to the client rather than the staff because of the existing quality policy and customer service standards, but the organization lacks human resources management and development strategy. The authors offer human resources development guidelines that can be implemented by the investigated healthcare organization.

Keywords: Employee · Human resources · Development · Health care Organization

1 Introduction

Nowadays the employee is the core value of the company, the first prerequisite for the company to operate successfully and create lasting values [1] and [2]. Employers, on the other hand, can provide return on investment in the short and long term by investments in human resources [3]. Human resource management involves several functions, including, staff planning and selection, motivation, mentoring, developing and also strongly paying attention to working conditions, interpersonal relationships, social environment [4]. Already in 2006, the World Health Organization (WHO) put forward a task to reach sufficient number of human resources (staff) in health care [5]. The justification was based on data from a study published in the same WHO document proving that the chance for a mother and a newborn to survive is directly proportional to the ratio of medical staff to population in the country. Also, the latest WHO

W. Karwowski and T. Ahram (Eds.): IHSI 2019, AISC 903, pp. 142–147, 2019.
https://doi.org/10.1007/978-3-030-11051-2_22

recommendations require every country to make an effort to provide a healthcare system with local human resources [6].

Continuing care for human well-being at work, their involvement in processes changes the behavior of the working people in a positive direction, and vice versa - this awareness and trust help to keep the company's financial growth expediting its prosperity as a whole [4]. The consolidation of such awareness and a climate of mutual trust leads to a sharp increase in the financial well-being of the company [7]. The performance of the organization can be assessed with a function that includes the interaction between labor productivity and the health of employees [8]. Labor productivity expresses an obligation on which the worker can achieve the goals of the organization. In turn, the health of the employees refers to the psycho-bio-social nature of the worker, including satisfaction with work, satisfaction with performance, personal well-being, and moral behavior [9].

In the research a leading health care organization in Latvia was selected. Its organizational activities are connected with private health institutions and work to identify patients' needs, plan healthcare offers, purchase new technologies and improve services. In total, the organization employs over 1300 people, including more than 600 doctors. The study involved two departments, medical center and health center.

The aim of the research was to investigate the existing human resource management in a health care organization and develop guidelines for improvement, based on the analysis of the theory and empirical research results.

2 Methods

Several research methods were applied, including SWOT analysis, survey and statistical data analysis.

The study used SWOT (Strength-Weaknesses-Opportunities-Threats) analysis. An analytical tool for assessing strategic plans and corporate planning was created at Stanford University. This analytical tool was called the SOFT analysis, which we know today as a SWOT analysis [10].

The survey was used to find out the opinion of employees about existing human resources management in the health care organization.

The calculated required sample size in each department corresponds to 67 respondents. Data were analyzed and interpreted using statistical data processing: correlation, Hi-square test [11].

3 Results and Discussion

SWOT analysis was carried out to analyze the factors that directly affect the health care organization's performance (see Table 1).

One of the strengths is to work with existing profits, which enables the organization to develop and improve service capabilities. The purchase of new infrastructure and medical equipment enables the organization to improve the range and quality of its services. An important strength lies in the organization's affiliates' location, as it

Table 1. SWOT analysis results

Internal factors	
Strengths	Weaknesses
• Profits every year • Money investments in renovation, equipment, infrastructure and mediation. • Strategic location for affiliates • Competitive price • A customer service and quality policy • Implementation of the "Doctor's Office" • A recognizable brand • Extensive advertising opportunities	• No existing human resource strategy • Unfinished certification process • Lack of medical staff • Underestimated staff • Technical condition of the building • Knowledge and skills of employees
External factors	
Opportunities	Threats
• Introduction of new technologies • New service opportunities • Possibilities to combine other outpatient clinics, health/medical centers under one brand • Find new partners in Latvia and abroad • Invest in new employees in education so they can work in the health care organization later • Opportunity to find out the strengths and weaknesses of competitors	• Not the only one private health care provider in the country • Decreasing number of patients • Insufficient budget for health care and allocation of allowances • Unfavorable legislative changes • Lower bids for equivalent services from competing companies • Patients lose solvency

enables patients to easily access the service they need without having to go a long way. In addition, the organization has great opportunities to advertise on TV, radio, and on the Internet, also on social sites.

The organization also has weaknesses, for example, there is lack of human resources policy/strategy that does not allow monitoring employee performance and perform job analysis. Not only in public hospitals, but also in private health facilities, there is a large shortage of mid-level medical staff, which results in employees having to work in different departments. For this reason, they feel underestimated by the leadership. For some branches, the technical condition of the building is not in perfect order, which causes discomfort to the patients themselves and also to doctors.

The organization has great potential to grow under the network of other outpatient clinics, health/medical centers. To attract new perspective doctors, because of which new services and new, advanced technologies and treatment methods. The organization can work with specific institutions to help them reach new customers and build new partnerships. To sponsor young doctors and when they finish their studies, then offer job opportunities at a health center.

The big threat to the health care organization is that it is not the only private health care provider in Latvia. Political factors are also closely linked to economic factors, because they determine how much funding will be allocated to health and education. The inadequate budget for health care in Latvia and the number of allowances allocated also affect the private structure and its ability to provide services to patients. A decrease in the number of patients affects the company's profits. Various legislative changes also affect the organization and the possibilities for its development, and for the same reason, patients are affected by the financial side, which then reflects the receipt of the service and they are not able to pay for it.

Next step was to carry out the survey and find out the opinion of employees about existing human resources management in the health care organization. The correlation analysis was performed to compare 7 variables:

- Do you think there exist a human resource policy in the health care organization? (0,307); Do you clearly understand your job goals? (0.372); Are your duties consistent with the objectives described in your job description? (0.231); Management takes care of my well-being at work and consider me as value (0.262); I am encouraged to improve my skills and develop my professionalism (0,209).
- How long have you been working for health care organization? Do you think there exists a human resources policy in the health care organization? (−0.212); Do you clearly understand your job objectives? (−0.230); I am encouraged to improve my work skills and develop my professionalism (−0,194).
- Do you clearly understand your job objectives? Does your duties consistent with the purposes specified in your job description? (0.405); Management take care of my well-being at work and consider me as a value (0.229); I am encouraged to improve my ability to work and develop my professionalism (0,258).
- Does your job responsibilities are consistent with the objectives described in your job description? Management provide for my well-being at work and consider me a worthwhile job (0.242); I am encouraged to improve my skills and develop my professionalism (0,207).
- Management takes care of my well-being at work and consider me a as a value. I am encouraged to improve my skills and develop my professionalism (0.615).

The Hi-square test analysis considered socio-demographic characteristics of respondents about age, education, occupation and seniority. The grouping was done considering the training, career development, and an assessment of the outlook for the future. Comparing each variable with a constant, there is a statistically significant difference between the issues:

- What is your position in the health care organization and does learning help drive your career? (sig = 0.049). It can be observed that there is a statistically significant difference between the position of the employee and whether the training contributes to further career development. Larger number of respondents were expected to come from the medical staff, indicating that the training does not contribute to the development than the staff members in the group of other healthcare professionals.
- What is your position in the health care organization and how do you assess your future prospects? (sig = 0.042). One can observe that there is a statistically significant difference between an employee's position and employee's future prospects. One can conclude that more doctors and residents think that they will be able to develop and build a career than expecting, while nurses are half less than expected. Even a small number of employees feel that their work depends on changing management. Doctors and residents do not think about this at all.

- What is your position in the health care organization and with whom do you associate your future career? (sig = 0.000). Doctors acknowledge that they will link their careers with the current health care organization, compared to nurses, where a slightly higher number is expected to recognize that they will continue to work there. A relatively small number of respondents suggests working in another health care organization or already going to another health care institution.

The results lead to understanding the weaknesses in human resource management in the organization and develop potential improvements. In order to eliminate disadvantages in the development of human resources, it is necessary to develop a new approach that is supplemented by the development of existing human resource methods. Authors have developed guidelines to improve the management of existing human resources in the health care organization (see Fig. 1).

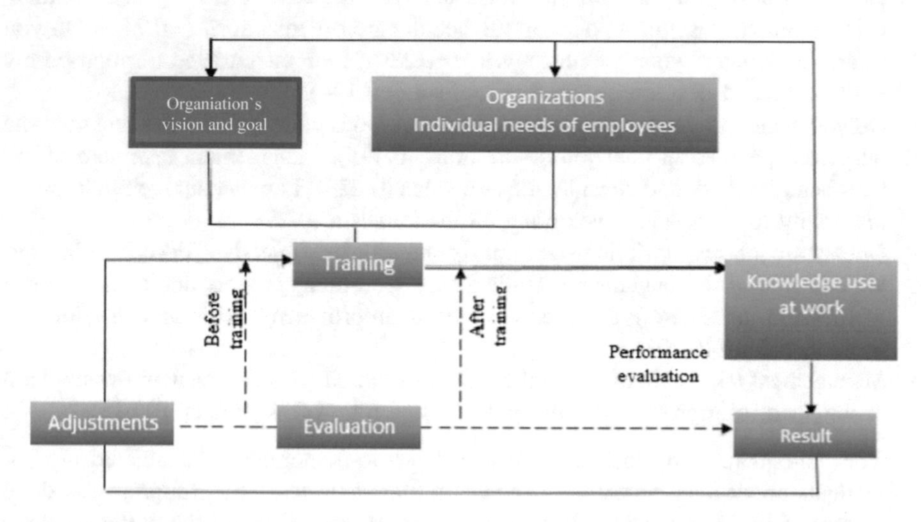

Fig. 1. Guidelines to improve human resource management at the health care organization

The guidelines represented in Fig. 1 show that learning is initiated based on the organization's vision and purpose, including individual needs of employee's. It is followed by the pre-and post-training assessments of employees' knowledge. The employees' existing knowledge is used at their workplace to promote the results of their work and carry out the performance assessment. After the evaluation of the results, the necessary adjustments can be made in the process, the new employees' needs are reconsidered, and the training is determined. The organization must take care of the well-being of employees and treat them as value if their position is not significant enough. According to the survey data, it can be observed that the organization cares more about the welfare of doctors and sees them as more valuable than nurses. Management should be able to appreciate the employees with great potential and knowledge to give them the opportunity to grow and build their own career within the organization and to benefit both parties.

4 Conclusions

The organization pays huge attention to the client rather than the staff because of the development of quality policies and customer service standards, but, accordingly to a SWOT analysis, the organization lacks human resources management strategy and policy. The management of human resources of the organization is significantly influenced by the political factors. An analysis of the survey shows that employees lack knowledge and motivation that affects their development and career opportunities in the organization. There were significant differences between age, education, position and work experience if compared the answers on human resource policy, job objectives, job responsibilities and management attitude towards wellbeing and valuation. The developed guidelines will significantly improve the existing human resource management at the health care organization and the guidelines can be applied in other health care organizations as well.

References

1. Dessler, G.: Human Resource Management, 13th edn, p. 720 p. Pearson, London (2012)
2. Vintisa, K.: Employers' Confederation of Latvia: Human Resources Planning and Assessment Methods for a Small and Medium Business, 31 p. Employers' Confederation of Latvia, Riga (2010)
3. Phillips, J.: Investing in Your Company's Human Capital: Strategies to Avoid Spending Too Little or Too Much. AMACOM, 304 p. (2005)
4. Armstrong, M.: Handbook of Human Resource Management Practice, p. 880 p. Kogan Page, New York (2009)
5. World Health Organization: The World Health Report (2006). http://www.who.int/whr/2006/whr06_en.pdf?ua=1
6. World Health Organization: WHO Global Code of Practice (2010). http://www.who.int/hrh/migration/code/code_en.pdf
7. Roja, Z.: Ergonomic Basics, p. 245 p. Printing house, Riga (2008)
8. Sperry, L.: Effective Leadership: Strategies for Maximizing Executive Productivity and Health, p. 237 p. Brunner-Routledge, New York (2002)
9. Kalkis, H.: Business Ergonomics Management, p. 120 p. Gutenbergs printing, Riga (2014)
10. Thakur, S.: History of the SWOT Analysis (2010). http://www.brighthubpm.com/methods-strategies/99629-history-of-the-swot-analysis/
11. Navidi, W.: Statistics for Engineers and Scientists, 4th edn, p. 928 p. McGraw-Hill Education, New York City (2014)

WeChat Wisdom Medical Treatment Process Based on the Hall Three-Dimensional Structure

Xueman Pan, Chikun Chen, and Fenghong Wang(✉)

School of Design, South China University of Technology, 510006 Guangzhou, China
554994202@qq.com, {egckchen, fhwang}@scut.edu.cn

Abstract. In China, Internet Plus is developing rapidly and the traditional service industry is using the Internet as a platform for integration and innovation. The WeChat wisdom medical treatment process can be the classic case for the deep integration of the Internet and medical health. Based on the theoretical knowledge of Hall's three-dimensional structure, this paper combines three dimensions of time dimension, logic dimension and knowledge dimension to analyze WeChat wisdom medical treatment process systematically and integrally. This research can not only construct an overall research framework of WeChat wisdom medical treatment process, but also study separately the medical treatment process from each dimension. It can be of the guiding significance to combine the overall description with partial analysis for the integration of the traditional medical services and the Internet and the reference on the existing Internet Plus medical health.

Keywords: The hall three-dimensional structure · Internet plus WeChat wisdom medical treatment process · Time dimension · Logic dimension · Knowledge dimension

1 Introduction

In April 2018, Chinese government adopted "The Opinions on Promoting the Development of Internet plus Medical Health" [1]. Internet Plus medical health, which realizes the digital upgrading of the traditional medical industry, combines with mobile Internet, cloud computing and big data, to optimize the allocation of existing medical resources and provide more effective medical health services.

WeChat wisdom medical treatment process, is referred as the classic case for the deep integration of the Internet and medical health, which relieves the "three long and one short problems" of the medical treatment process [2]. It is a solution provided by WeChat for the medical industry about the integration of hospital medical treatment processes and the Internet. Combining with Internet, medical treatment services transform to the combination of online and offline services. With the help of official accounts, city services, and mini program of WeChat, the hospital realizes the online processes of triage and registration, waiting for medical treatment, inspection and the payment service of its fee, electronic report and taking medicine [3, 4].

W. Karwowski and T. Ahram (Eds.): IHSI 2019, AISC 903, pp. 148–154, 2019.
https://doi.org/10.1007/978-3-030-11051-2_23

2 The Hall Three-Dimensional Structure

In 1969, Hall, the system-engineering scholar in American, proposed the structure and form of systems engineering called widely as the Hall three-dimensional structure. Using the technique of morphological analysis, it decomposes engineering system into three fundamental dimensions: time dimension, logic dimension and knowledge dimension [5].

The time dimension depicts a time sequence of engineering activities from inception to retirement. The logic dimension models a problem thinking and solving procedure replied in each continuous activity. The knowledge dimension refers to the discipline, profession or technology related to engineering activities. The Hall three-dimensional structure combines these three dimensions to provide more solving options for the large-scale and complicated systems. This structure can not only achieve the in-depth understanding of the engineering activity, but also find more possibilities with the application of knowledge in various fields [6, 7].

3 Necessity and Feasibility

The essence of WeChat wisdom medical treatment process is the combination of Internet and medical health services. It is the expansion of service design in the context of Internet Plus. Its ontological attribute is the system design of the relationship between people, things, behavior, environment and society [8]. The Hall three-dimensional structure is a systematic engineering methodology, with the integrity of system research (three dimensions), the sequence of engineering progress (time dimension), the problem-oriented method of system engineering activities (logical dimension), and the comprehensiveness of technical expertise (Knowledge dimension) [9]. It is of great significance to the overall description and partial analysis of WeChat wisdom medical treatment process.

In China, there is a phenomenon where takes a long time to register, wait for the medical treatment and take the medicine and a short time to get the medical treatment in hospital. It is called as the "three long and one short problems" and becomes more and more serious. The integration of medical health and the Internet can effectively alleviate this phenomenon. As a classic case, the WeChat wisdom medical treatment process provides a reference for hospitals. When it comes to reference, we need a systematic and holistic analysis of the entire process of WeChat wisdom medical treatment, so that it can really apply in the existing medical services and play an important role.

4 The Construction and Analysis of the Three-Dimensional Model of WeChat Wisdom Treatment Process

Based on the Hall three-dimensional structure and combined with the service features of Internet Plus medical health, the three-dimensional model of WeChat wisdom treatment process is constructed. On this basis, combining with three dimensions of time dimension, logic dimension and knowledge dimension, the medical industries can acquire the overall cognition and systematic understanding of WeChat wisdom medical process. It is of great signification to upgrade the medical service and draw on the experience of the classic case under the background of Internet Plus (Fig. 1).

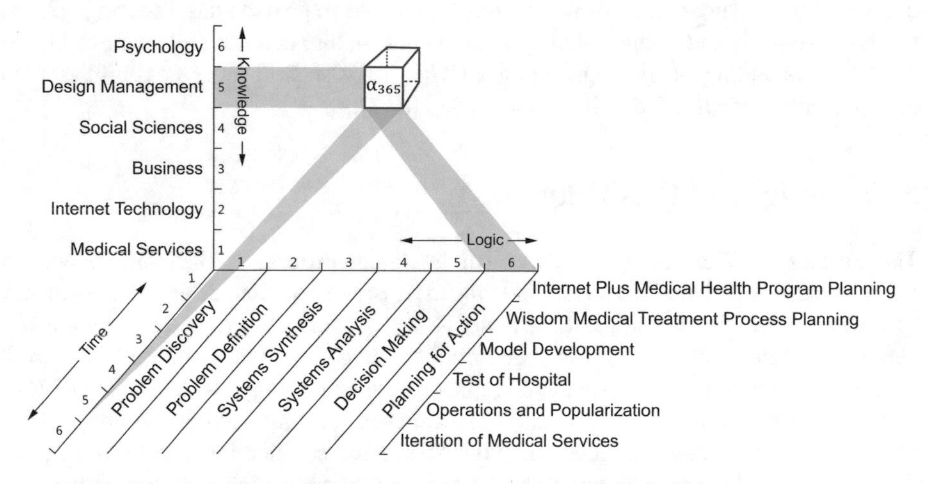

Fig. 1. The three-dimensional model of WeChat wisdom treatment process

4.1 Basic Analysis of the Three-Dimensional Model

WeChat wisdom medical treatment process, including the whole process from triage and registration to the medical treatment and taking medicine, is the combination of online and offline process of medical treatment. As a project of the program of Internet Plus medical health, it portrays the engineering activities of medical treatment service with time and sequence. It divides into six time phases: Internet Plus medical health program planning, medical treatment process planning, model development, test of hospital, operations and popularization, and iteration of medical services.

The logic dimension models a procedure of problem thinking and solving, which applies repeatedly at each stage of the medical service process. Design Council UK describes the creative process as confirming the problem definition and creating the solution called as the Double Diamond [10]. The design of the WeChat wisdom medical process is also a creative process. Its logic dimension is decomposed into six procedures: problem discovery, problem definition, systems synthesis, systems analysis, decision making and planning for action.

The knowledge dimension is the technical expertise and knowledge that applies in the medical service processes. The WeChat wisdom medical treatment process is a fusion of the Internet and the medical service, mainly related to medical services and Internet technology, followed by business, social sciences, design management and psychology.

4.2 The Logic Dimension and the Time Dimension

In the three-dimensional model of WeChat wisdom medical treatment process, the time dimension portrays the time-ordered but irreversible engineering activities. The logic dimension is describe as a procedure about thinking and solving problems, which is reversible and used repeatedly in each engineering activity. Combining the logic dimension with the time dimension, the activity matrix of the WeChat wisdom medical process forms. With the activity matrix, a clear understanding of all hyperfine activities in the project obtains when referring to this case (Table 1).

Table 1. The activity matrix of the WeChat wisdom medical process

Time		Logic					
		1	2	3	4	5	6
		Problem discovery	Problem definition	Systems synthesis	Systems analysis	Decision making	Planning for action
1	Internet Plus medical health program planning	α_{11} The medical health and the Internet Plus	α_{12} Internet Plus medical health	α_{13} Multiple projects proposed of the Internet Plus medical health program	α_{14} Determine the systems evaluation and analyze all projects	α_{15} The wisdom medical treatment process was first implemented	α_{16} Planning for the operations of the wisdom medical treatment processes
2	Wisdom medical treatment process planning	α_{21} The "three long and one short problems" of the hospital	α_{22} Wisdom medical treatment process (The combination of online and offline process)	α_{23} Propose different solutions for the problem definition of wisdom medical treatment process	α_{24} Determine the systems evaluation of wisdom medical treatment process and analyze all projects	α_{25} Determine the final solution of wisdom medical treatment process	α_{26} Planning for the operations of the solution of wisdom medical treatment process
3	Model development	α_{31} Develop the online official accounts and make the offline materials	α_{32} Achieve the online processes of wisdom medical treatment	α_{33} Propose solutions based on the use needs and the medical treatment	α_{34} Determine the systems evaluation based on the technology limitations and analyze	α_{35} The online processes of triage and registration, waiting for medical treatment,	α_{36} Planning for the test of hospital

(continued)

Table 1. (*continued*)

Time		Logic					
		1	2	3	4	5	6
		Problem discovery	Problem definition	Systems synthesis	Systems analysis	Decision making	Planning for action
		(QR code) of hospital		process of hospital	all online processes	inspection and its payment service, and electronic report	
4	Test of hospital	α_{41} Select the offline entrances in the hospital (the position of QR code)	α_{42} Connect the offline entrances with the medical treatment process	α_{43} Find the offline entrances based on the users journey	α_{44} Determine the systems evaluation and analyze the online conversion rate of users	α_{45} Place the QR code on the registration office, payment office, inspection list, etc.	α_{46} Planning for the operations and popularization
5	Operations and popularization	Provide solutions of medical industry, WeChat do not enter the medical industry					
6	Iteration of medical services	α_{61} The application of medical treatment is still fragmented and divisive	α_{62} Follow-up management, support for clinical decision making, etc.	α_{63} Propose the appropriate solutions	α_{64} Determine the systems evaluation and analyze all the solutions	α_{65} Form the closed-loop management of health care services based on the whole life cycle of human	α_{66} Planning for the next phase

4.3 The Time Dimension, Logic Dimension and Knowledge Dimension

The three dimensions of the three-dimensional structure of the WeChat wisdom treatment process are independent and interrelated with each other. Combining the three dimensions, it can form a comprehensive understanding of WeChat wisdom medical treatment process, including the orderly progress of the entire project, the analysis of specific activities at each time phase, and the knowledge and technology used in the whole project (Fig. 2).

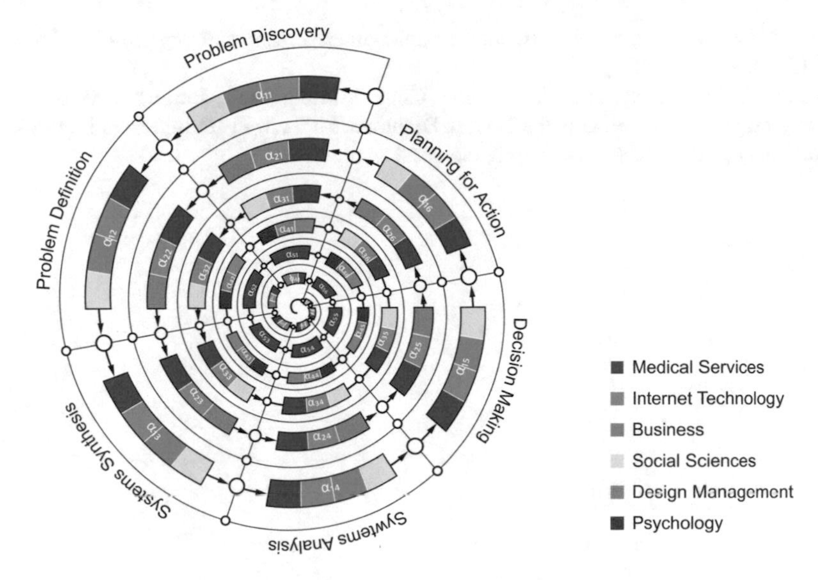

Fig. 2. WeChat wisdom treatment process

5 Conclusion

As the development of Internet Plus, a new mode of thinking and solving the problem is provided for the problems presented in the process of medical health service. The deep integration of medical health and the Internet is in full swing, and the upgrade of the entire medical industry is constantly deepening in the process of exploration and reference. Through comprehensive and systematic induction and analysis of WeChat wisdom medical treatment process, traditional medical services can draw on the advantages of this case and combine their own service characteristics to form a deign service system which is truly suitable for their own development. Finally it can achieve the popularity of Internet Plus medical health.

References

1. "The Opinions on Promoting the Development of Internet plus Medical Health". http://www.gov.cn/zhengce/2018-04/28/content_5286786.htm
2. Innovative Medical Services to Solve the "Three Long and One Short Problems" of the Medical Treatment for the Patients. http://m.xinhuanet.com
3. The Trends Report of the Medical Services improved by WeChat in 2018. http://www.hit180.com/wp-content/uploads/2018/07/baogaoxiazai.pdf
4. Wisdom Hospital. http://act.weixin.qq.com/static/merchant/project.html
5. Duan, H.: The evolution from traditional system engineering to modern system engineering based on the development of the hall model technology system. Intell. Manuf., 88–95 (2016)
6. Hall, A.D.: Three-dimensional morphology of systems engineering. In: IEEE Transactions on Systems Science and Cybernetics, pp. 156–160. IEEE Press, New York (1969)
7. Lv, Y.: Systems Engineering. Tsinghua University Press, Beijing (2006)

8. Xin, X.: Service design drives public affairs management and organization innovation. SHEJI, 124–128 (2014)
9. Wang, Y., Cao, J.: Systems Engineering. China Machine Press, Beijing (2008)
10. The Design Process: What is the Double Diamond. https://www.designcouncil.org.uk/news-opinion/design-process-what-double-diamond

Study on Gait Discrimination Method by Deep Learning for Biofeedback Training Optimized for Individuals

Yusuke Osawa(✉), Keiichi Watanuki, Kazunori Kaede, and Keiichi Muramatsu

Graduate School of Science and Engineering, Saitama University, Shimo-Okubo 255, Sakura-Ku, Saitama-Shi, Saitama 338-8570, Japan
y.osawa.949@ms.saitama-u.ac.jp

Abstract. In this research, to develop a biofeedback training system where trainees can efficiently train inadequacies that do not satisfy ideal walking using a deep learning, we examine a method that discriminates between ideal walking and nonideal walking. In the experiment, to examine the walking components used for the input data, the ground reaction force and joint angle were measured when young people walked normally and when they walked with a brace, to simulate elderly motions. Further, these data were discriminated between conditions as input data using a Convolution Neural Network (CNN). The average accuracy was 79.5% when all walking components were used as input data. In addition, it is thought that it is most suitable to discriminate walking by using all walking components, in consideration of implementation in the system.

Keywords: Walking assistance · Biofeedback training · Motion capture Ground reaction force · Convolution neural network

1 Introduction

In recent years, biofeedback training has focused on rehabilitation and training. In this method, the trainee's biometric information is measured and fed back to the trainee in real time through visual or auditory information. This method reportedly enables us to consciously control biological information, which is normally uncontrollable to an untrained person [1]. Moreover, applying biofeedback training to exercises has been studied, and Stanton et al. reported that running motion was improved by feeding back the motion analysis to the trainee using the motion capture system, as visual information [2]. In general biofeedback training, the actual measured values are approximated to the target values, which are set based on ideal motions. However, the target values set by the trainer are not always the most optimal solution for the trainee in cases where individual differences in physique exist between the trainee and the model that enacts the theoretically ideal movement. Therefore, it is necessary to set a target value considering individual differences between them. For this reason, we have been developing a biofeedback training system where trainees can efficiently train inadequacies that do not satisfy ideal walking through deep learning. The purpose of this study was to examine the walking components used for input data in order to develop a method that discriminate

W. Karwowski and T. Ahram (Eds.): IHSI 2019, AISC 903, pp. 155–161, 2019.
https://doi.org/10.1007/978-3-030-11051-2_24

between ideal walking and nonideal walking. Previous studies reported that influences on floor reaction forces, lower limb joint angles, and toe floor distance are observed due to aging and physical paralysis [3–5]. In addition, considering that the posture of the upper limb during walking has an influence on the walking rate, and that shaking the arm is an operation to antagonize the rotation of the trunk, the upper limb joint angles are also an important factor for evaluating walking [6, 7]. From the above, it can be inferred that features appearing in these walking components can be learned by using ground reaction force, toe floor distance, and joint angles as input data and that the relevancy between them can be learned. In the experiment, the normal walking of young people was defined as ideal walking, and simulated elderly walking was defined as nonideal walking. Further, walking components were measured in these walking states. These data were discriminated between conditions as input data using a CNN.

2 Outline of Biofeedback Training System Optimized for Individuals

The biofeedback training system is shown in Fig. 1. Walking components of ideal walking and nonideal walking are measured, and a model that learned features of ideal walking through input data pertaining to walking components is constructed using a CNN that is suitable for extracting features for matrix data. Moreover, the walking components of the trainee are measured in real time at training, and discriminated based on the model. In addition, the trainee is visually presented with the inadequacies that do not satisfy ideal walking by the Grad-CAM method. This is a method that visualizes the gradient of the error to be back-propagated when learning, and that makes it possible to confirm the part where the feature in the input data appears [8]. By using this system, it is considered that a trainee can perform biofeedback training with a target value considering individual differences.

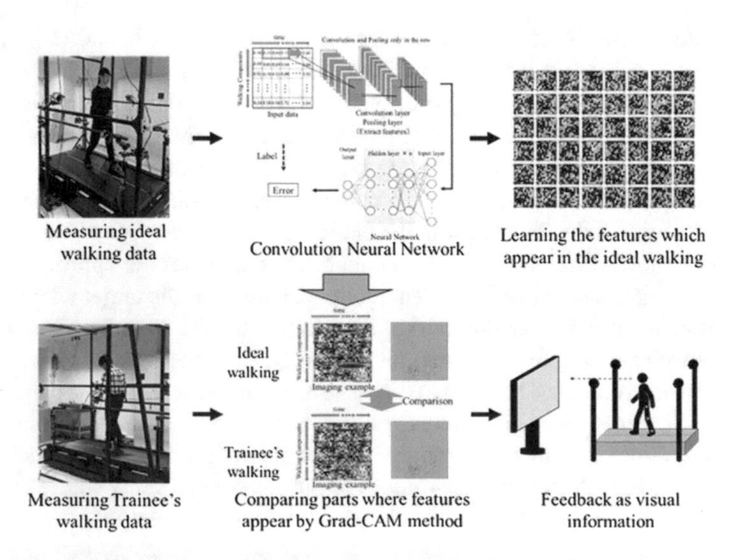

Fig. 1. The biofeedback training system which we have been developed

3 Examination of the Gait Discrimination Method Using a Brace to Simulate Elderly Motions

3.1 Experiment Purpose

The purpose of this experiment was to verify whether ideal walking and nonideal walking can be discriminated from walking components (upper limb joint angles, lower limb joint angles, toe floor distance, and ground reaction force) during a walking cycle. Specifically, each joint angle, toe floor distance, and ground reaction force in normal walking and in simulated elderly walking were measured. Moreover, input data was prepared from the measured data and discriminated using a CNN.

4 Experimental Procedure

The participants were 10 Japanese men (average age 22.8 ± 1.1 years). The experiments were performed under three conditions: (1) normal walking, (2) simulated elderly walking with a straight back, and (3) simulated elderly walking with a bent back. In this study, elderly simulated walking was defined as walking with a brace, to simulate elderly motions (Sanwa Manufacturing Co.). The brace to simulate elderly motions consists of weights (four at the front of the upper limb, two at the wrists, and two at the ankles) for experiencing muscle weakness, supporters for restricting the movement of the elbow joints and knee joints, and a belt for experiencing kyphosis of the lumbar and thoracic regions. In condition (2), the participant wore weights and supporters, and in condition (3), they wore weights, supporters, and a belt, as shown in Fig. 2. Further, the markers of the full body motion capture system were placed on the front, top, and back of head, acromion (R/L), lower corner of shoulder blade (R/L), elbow (R/L), wrist (R/L), first sacral, ileum (R/L), greater trochanter (R/L), knee (R/L), lateral malleolus (R/L), heel (R/L), and fifth metatarsal head (R/L). Under each condition, participants were instructed to walk at the speed that is most easy to walk at, and they walked on a treadmill with a built-in force plate for 120s. Further, the coordinates of each marker during walking were measured using an optical full-body motion capture system, and the ground reaction forces (vertical, longitudinal, and lateral) during walking were measured using the built-in force plate. The speed of the treadmill was calculated from the position of participant relative to the treadmill and automatically adjusted. From the measured marker coordinates, the thoracic vertebra angle, lumbar vertebra angle, shoulder, and elbow joint angles as upper limb data were calculated, and knee and ankle joint angles as lower limb data were calculated. Further, the vertical coordinate of the marker placed on the fifth metatarsal head was defined as the toe floor distance. In condition (1), the floor reaction forces data were normalized by the weight of the participants, and in condition (2) and (3), the floor reaction forces were normalized by the sum value of the weight of the participants and the weight of the brace to simulate elderly motions, respectively. Prior approval for the experiment was obtained from the Saitama University Ethics Review Committee.

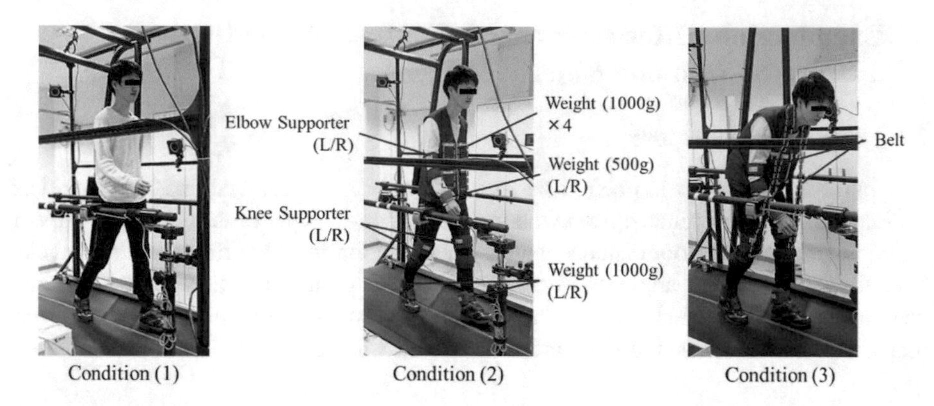

Fig. 2. Conditions of the experiment

5 Construction and Verification of Simulated Elderly Walking

One sample of the input data was the walking component data from the ground of the right foot to the next ground of it. The number of samples obtained from each participant's data is shown in Table 1. Data of one participant, as the verification data, and data of other participants, as training data, were learned and discriminated by using a CNN, and cross validation was performed. Generally, in a CNN, convolution and max pooling are performed in the row and column of the input data. However, in this experiment, convolution and max pooling were performed only in the column, because the row, which is the walking component, has independent values. In addition, discrimination was similarly made using only upper body data, only lower body data, and only ground reaction force data as input data.

Table 1. The number of samples obtained from each participant's data

	Number of samples			
	Condition 1	Condition 2	Condition3	Sum of each participant
Participant 1	71	77	76	224
Participant 2	82	77	81	240
Participant 3	76	70	64	210
Participant 4	80	67	71	218
Participant 5	81	82	76	239
Participant 6	78	78	81	237
Participant 7	78	73	73	224
Participant 8	78	70	66	214
Participant 9	78	82	82	242
Participant 10	84	90	92	266
Sum of each condition	786	766	762	2314

6 Discrimination Results

Table 2 shows the accuracy for each participant when only upper limb data, only lower limb data, only ground reaction force data, and all of the above were used as input data. The average accuracy of all participants was 79.47% when all walking components were used as input data. Moreover, it was 77.30% when only upper limb data were used, it was 77.55% when only lower limb data were used, and it was 66.46% when only ground reaction forces were used.

7 Discussion

As the result of discrimination, the average accuracy of the case where all walking components were used as the input data, where only the upper limb data were used, and where only lower limb data were used were almost the same. However, from the results of each participant, it is understood that the input data where the accuracy was high differs for each participant. In this experiment, muscle weakness and restriction of joint mobility was reproduced by participants who wore weights and braces, but participants could walk as easily because they did not specify the gait. Consequently, participants were able to walk with a small joint motion so as not to get tired, or walk with the same joint motion as normal walking by forcing the muscles to work. Therefore, it can be considered that walking was different under the same conditions for each participant. Further, it is necessary to verify the gait of each participant by clustering gait.

When learning using all walking components as input data, the accuracy was 79.47%. As described above, since participants were able to walk in various gaits, this seems to be because the feature of normal walking was also taken in the condition where participants walked with the brace. Generally, in the case of discrimination by machine learning, learning is performed using more patterns of input data, since it is desirable that the accuracy be high for middle data between labels. However, considering application to the system and visualization of the features by the Grad-CAM method, it is desirable that middle walking, between ideal walking and nonideal walking, is determined to satisfy both of these features. Therefore, it is necessary to learn by using typically ideal and nonideal walking as input data instead of learning including middle walking.

In participant 5, over fitting occurred when all walking components were used. By increasing the batch size, it was possible to prevent over fitting for other participants who had slight over fitting. On the other hand, over fitting could not be prevented even if the batch size was increased for participant 5. From this fact, it can be considered that participant 5 was performing unique walking, so it is necessary to verify this by reproducing a unique walk, for example, walking with one leg injured.

Table 2. The accuracy for each participant when only upper limb data, only lower limb data, only ground reaction force data, and all of them were used as input data

Validation data	Input data			
	Upper limb data	Lower limb data	Ground reaction force data	All of walking components
Participant 1	0.9196	0.6652	0.3438	0.7411
Participant 2	0.5750	0.7375	0.6750	0.7458
Participant 3	0.7286	0.7048	0.6619	0.7333
Participant 4	0.9579	0.9439	0.6495	0.8785
Participant 5	0.4725	0.6606	0.7110	0.4862
Participant 6	0.6541	0.5113	0.6391	0.6504
Participant 7	0.8523	0.7511	0.6667	0.8186
Participant 8	0.9375	0.9777	0.5223	0.8839
Participant 9	0.6987	0.8033	0.9540	0.8410
Participant 10	0.9339	1.0000	0.8223	1.0000
Average	0.7730	0.7755	0.6646	0.7947

8 Conclusion

In this paper, the purpose was to examine the method that discriminates between ideal walking and nonideal walking to develop a biofeedback training system where trainees can efficiently train inadequacies that do not satisfy ideal walking by a deep learning. In the experiment, ground reaction force, toe floor distance, and joint angle were measured when young people walked normally and when they walked with a brace to simulate elderly walking motions. Further, these data were discriminated between conditions as input data using a CNN. As the result, the average accuracy was 79.47% when all walking components were used as input data. In addition, as a result of comparison with the accuracy when only the respective data were used, it is thought that it is most suitable to discriminate walking by using all walking components in consideration of implementation in the system. In the future, we will classify walking and reproduce unique walking to examine whether it can be discriminated. Moreover, we will define ideal walking and construct a model to discriminate between ideal walking and nonideal walking.

References

1. Brown, B.B.: Recognition of aspects of consciousness through association with EEG alpha activity represented by a light signal. Psychophysiology **6**(4), 442–452 (1970)
2. Stanton, R., Ada, L., Dean, C.M., Preston, E.: Biofeedback improves activities of the lower limb after stroke: a systematic review. J. Physiother. **57**(3), 145–155 (2011)
3. Turns, L.J., Neptune, R.R., Kautz, S.A.: Relationships between muscle activity and anteroposterior ground reaction forces in hemiparetic walking. Arch. Phys. Med. Rehabil. **88**(9), 1127–1135 (2007)

4. Murray, M.P., Kory, R.C., Clarkson, B.H.: Walking patterns in healthy old men. J. Gerontol. **24**(2), 169–178 (1969)
5. Watanabe, K., Miyakawa, T.: Gait analysis during stepping over the different height of obstacle in aged persons. Elsevier Sci., 195–198 (1994)
6. Gushiken, S., Oi, N., Tobimatsu, Y., Iwaya, T.: The relevance of standing posture and walking function in elderly people. J. Phys. Med. **14**(3), 241–248 (2003). in Japanese
7. Ballesteros, M.L.F., Buchthal, F., Rosenfalck, P.: The pattern of muscular activity during the arm of natural walking. Acta Physiol. **63**(3), 296–310 (1965)
8. Selvaraju, R.R., Cogswell, M., Das, A., Vedantam, R., Parikh, D., Dhruv, D.: Grad-CAM: visual explanations from deep networks via gradient-based localization. In: ICCV 2017, pp. 618–626 (2017)

How Might Voice Assistants Raise Our Children?

Cezary Biele[1(✉)], Anna Jaskulska[3], Wieslaw Kopec[2], Jaroslaw Kowalski[1], Kinga Skorupska[2], and Aldona Zdrodowska[1]

[1] National Information Processing Institute, al. Niepodległości 188 b, 00-608 Warsaw, Poland
{cezary.biele, jaroslaw.kowalski, aldona.zdrodowska}@opi.org.pl

[2] Polish-Japanese Academy of Information Technology, Koszykowa 86, 02-008 Warsaw, Poland
{kopec, kinga.skorupska}@pja.edu.pl

[3] Kobo Association CAM, Nowolipie 25b, 01-002 Warsaw, Poland
a.jaskulska@kobo.org.pl

Abstract. When mobile devices such as tablets and smartphones were becoming more popular, an important question that psychologists and pediatricians asked was how interactions facilitated by these devices with screens may affect the functioning of children. Nowadays, when technology used by children, such as intelligent voice assistants, does not require a screen at all, these issues seem to fall into the background. Today, concerns are growing about the effect of interacting with voice-driven AI services, as it may potentially have a greater impact on children's cognitive development than engaging with television or smartphones. The purpose of this paper is to outline potentially interesting directions of research in the field of voice assistant technology concerning how this solution may affect the functioning of children, and in particular if, and to what extent, it may redefine the dynamics of social contacts within and outside the family.

Keywords: Intelligent voice assistants · Human factors · Digital parenting

1 Introduction

For several years, artificial intelligence has been the subject of debate far beyond the field of technology itself, as awareness of a wide range of its social consequences has been increasing. AI technology has progressed so much that it is now possible to use it in many processes that required human intervention until recently (e.g. voice recognition or translation). AI-enhanced software is used, for example, in commerce where intelligent systems can talk to customers in call centers without human intervention and external control. Speech recognition systems are becoming widespread, along with voice assistant systems, such as Google Assistant, Siri, Alexa and Cortana. According to recent research [1], 41% of children and young people (aged 9–17) in the UK have a voice-activated personal assistant on their phone, while 23% have such a device at

W. Karwowski and T. Ahram (Eds.): IHSI 2019, AISC 903, pp. 162–167, 2019.
https://doi.org/10.1007/978-3-030-11051-2_25

home. Another 13% of children declare they "would like to have one to talk to". As a consequence, concerns are growing about the potential impact of artificial intelligence and voice assistants on children, as it is thought to be greater than that of television or smartphones, and some groups, such as the Campaign for a Commercial Free Childhood [2], are lobbying against them. These concerns about children's welfare arise with each new mass medium or technology being introduced to a broader public: from a 19th-century novel, through radio and TV, to computers, the Internet and, lately, mobile devices [3, 4].

In light of the above, it is vital to explore the current state of the art, highlighting both the benefits and challenges, and to present opportunities for further research relating to the use of intelligent voice assistants (IVAs) by children. This is even more relevant since the transition from interaction with screens (PCs, smartphones, tablets) to interactions with voice assistants does not merely entail a change in the mode of communication (touch/vision-based versus voice-based). This is in part because voice communication is the most natural way of communicating with other people. What follows in such human-computer interactions (HCI) is a transformation of the perceived status of the machine, which acquires human characteristics [5, 6]. This altered status may have far-reaching consequences for the cognitive aspects of such interaction, and it may necessitate a paradigm shift in HCI studies towards a more interdisciplinary approach to related research.

The purpose of this article is to pose a few open questions and to outline potentially [5, 6] interesting areas of research in the field of voice assistant technology. In particular, it asks questions on how this technology will affect children's lives, e.g., how it may redefine the dynamics of social contacts within and outside the family. Many young children, who have not yet learnt to read and write, may have their psychological world shaped, inter alia, by interactions with voice assistants. These are important issues, especially considering that children are among the largest groups of users of such systems.

2 Considerations and Limitations

As with any other media/technology effects, it is important to acknowledge—and search for—individual and societal factors potentially moderating the impact of the use of IVA on children. Below we present an overview of some possible factors influencing the nature of child-IVA interaction and its potential impact on children's cognition, based on a literature review of related topics [7, 8] and our research and experience in this area. Child-specific factors include: age and gender of the child, neurotype of the brain (if a child is neurotypical or autistic), social context of the use (e.g. in a family setting, supervised or on their own parental/social mediation), the mode of use (mostly as a Smart Home solution, or an Assistant). IVA-specific factors comprise: functionality of the IVA, gender of the IVA's voice, level of possible interactivity, target group of the solution (developed for adults or specifically for children) and the physical casing of the device (a home speaker, a phone or a toy).

3 Promising Areas of Research

3.1 The Effect of Voice Assistants on Linguistic Habits

The activation commands for Google and Apple Home Pod voice assistants are "Hey, Google" or "Hey, Siri", whereas interactions with Amazon's Alexa begin simply with "Alexa…" (also "Echo" or even "Computer"). These forms, stripped of strategies introducing politeness, were chosen for brevity, but it makes them sound like commands. For instance, expressing gratitude with a 'Thank you' was not required until recently and the option to encourage polite exchanges was only added in the 9 May 2018 update of Alexa. Recent changes in this area came as a way to address some parents' concerns connected to the VA impact on children's politeness in interpersonal dealings in later life. Mattel also planned to introduce a voice assistant for children that will require them to speak politely, but they cancelled their plans due to concerns raised by child advocacy groups, lawmakers and parents [9–11]. Related research has already shown that such a requirement changes adults' manner of speech to a more polite one [12]. This impact may be even greater because some children do not consider voice assistants to be devices but, instead, anthropomorphize them to some extent by attributing characteristics such as 'friendly' or 'reliable' [13]. It is, however, worth asking whether voice assistants might reinforce behaviors that parents would like to avoid in their children, i.e. ones that may make children's lives more difficult in adulthood. The effectiveness of voice assistants provides positive reinforcement, reassuring children that the only condition for getting what they want is proper enunciation, whereas politeness and good manners are on the sidelines. The positive reactions of peers observing how they interact with IVAs may also strengthen this effect. All of this might lead to a situation where the child transfers such behaviors to different social contexts, such a classroom or the family environment.

3.2 Voice Assistants' Effects on Social Relationships

During the Google I/O 2018 conference, Google presented the functionalities of its voice assistant which allow it to make phone calls in order to perform tasks assigned by the user. One of the presented examples was booking an appointment with a hairdresser. During the conversation, the person answering the call at the hairdresser's salon was not even aware that she was speaking to an artificial being. Such presentations, which demonstrate the state-of-the-art in voice assistant technology, make us wonder to what extent intelligent voice assistants might supplement or even substitute interpersonal contacts further down the line. A few years ago, when smartphones were becoming more widespread among children and young adults, similar questions were asked as there were fears that contact with screens might negatively impact children's cognitive, emotional, social and even physical development, e.g. suppress contact with peers, lead to physical inactivity, obesity and inadequate sleep quality [14]. Nowadays, we can see that phones or tablets with instant messaging services have merely become just another communication channel.

However, the issue of voice assistants being anthropomorphized emerges here again because communication occurs through voice, a channel associated with interpersonal contacts. In this context, a question arises whether it is possible for voice assistants to become a new sophisticated kind of 'imaginary friend'. It seems that this technology is exceptionally well suited for this role [15], as children themselves prefer the voice interfaces to be personified [16]. Importantly, voice assistants are deprived of all the negative features of a human interlocutor. They will patiently listen to everything, without ridiculing or revealing the secrets 'entrusted' to them (depending on the privacy policy and settings). The fact that a voice assistant may become a kind of a friend or a trusted person for children raises further questions, also explored in the controversial book "To Siri with Love: A Mother, Her Autistic Son, and the Kindness of Machines" [17]. Although the patience of an IVA is a major benefit for some children, the inflexible interaction may sometimes be a source of frustration for others. Overall, when communication with a VA breaks down, children employ repair strategies borrowed from their interpersonal experience, proving that they also have the patience to stay engaged despite difficulties. Here, it is important to note that the transfer of attitudes happens both ways.

Research on the use of voice interfaces in a family setting indicates that assistants can supplement family interactions, as is often the case when novel yet but accessible technology is used in intergenerational contexts of edutainment. This is true even to a point where IVA queries become "embedded in the life of a home" as Porcheron et al. [18] describe. Thus, this use may encourage socializing, with ready access to games, quizzes or other forms of entertainment.

3.3 Voice Assistants' Effect on Children's Everyday Habits

The results of the famous 'marshmallow test' and its subsequent replications and variations [19] indicate that the ability to postpone gratification is correlated with effective functioning in adolescence and adulthood. Children who were able to resist consuming the sweets immediately during those experiments demonstrated better cognitive abilities in adolescence. This can be interpreted as a signal to parents that they should strengthen their children's ability to postpone gratification. This raises the question related to interactions with voice assistants, which produce the opposite effect and, as a result, could teach children to expect immediate responses to their requests [10], for instance, if children want to listen to their favorite songs. Obviously, we should remember that voice-led interactions are still quite complicated even for adults, and sometimes almost impossible for toddlers.

On the other hand, voice assistants may also help children stay on track with digital and shared to-do lists and chores, and it is up to the parents to learn to use these features to their advantage. Moreover, during the interaction with IVA, children are learning how to ask more specific questions, which may teach them that patience and persistence is eventually rewarded. Also, toddlers may be encouraged to improve their pronunciation and learn to speak more slowly [13].

3.4 Ethical and Moral Aspects of Voice Assistants

The questions of decision-making, control and liability always surface in the context of AI-powered technology. Apart from the underlying programming, the actual user settings concerning voice priority, gender, the manner of speaking and censorship, there are also ethical dilemmas connected with the range of interactions, and whether it should include advice or even reporting on negative behaviors. With IVAs, the topic of privacy is vital as all voice interactions are recorded and analyzed. It is unclear whether and how a voice assistant should react when the child provides information indicating a violation of the law, e.g. on taking drugs, being harassed at school or being sexually abused. Should the system use a neural network to process and analyze the content of such utterances or should such quotes be anonymized and flagged for review by ML-supported trained professionals [20] or volunteers [21]? Perhaps neither of these things ought to happen.

Issues related to AI ethics are also widely discussed in the context of autonomous cars, where a controversy arose around the first victims of car accidents. Currently, moral discussions in this area are limited primarily to dilemmas regarding accidents that cannot be avoided (the so-called trolley problem [22]). This shows that insufficient emphasis is placed on broader questions that should have been asked at a very early design stage, e.g., "what design decisions have led to this moral dilemma," "what kind of values should guide the project and how they should be weighed." Once these considerations become more transparent in the specification of available commercial solutions, users would have the possibility to make informed decisions based on their moral compass and privacy preferences.

3.5 Conclusions

As the use of IVAs is becoming more common in many areas of life, it is important to consider the long-term effects it may have on social relationships including aspects such as encouraged attitudes and the use of language. These considerations are especially important when it comes to children who will be growing up surrounded by AI-enhanced technology. The possible cognitive effects of IVAs on children, especially the youngest ones, are hard to predict clearly, but the extent of their positive or negative impact will depend on many child- and solution- specific factors such as the mode of IVA use, and the moderation of caretakers. This is why it is essential to conduct further research on ways to improve IVAs, especially to support children's development or assist them in learning and on potential functions that may help enrich family lives. Overall, when used responsibly, this technology has great potential since it enables natural and hands-free communication which can be seamlessly introduced into many environments, creating user-friendly and accessible experiences both for the older and the younger generation, encouraging them to engage with it together.

References

1. Dataset from research conducted by Engineering UK: For Tomorrow's Engineers 2017 (2018). www.engineeringuk.com
2. Don't let Mattel's new "digital nanny" trade children's privacy for profit. https://www.commercialfreechildhood.org/action/dont-let-mattels-new-digital-nanny-trade-childrens-privacy-profit

3. Critcher, C.: Making waves: historical aspects of public debates about children and mass media. In: The International Handbook of Children, Media and Culture, pp. 91–104 (2008)
4. Gabriel, N.: The Sociology of Early Childhood: Critical Perspectives (2017)
5. Nass, C., Brave, S.: Wired for Speech: How Voice Activates and Advances the Human-Computer Relationship. MIT Press, MA (2007)
6. Schroeder, J., Epley, N.: Mistaking minds and machines: how speech affects dehumanization and anthropomorphism. J. Exp. Psychol. Gen. **145**, 1427–1437 (2016)
7. Piotrowski, J.T., Valkenburg, P.M.: Finding orchids in a field of dandelions: understanding children's differential susceptibility to media effects. Am. Behav. Sci. **59**, 1776–1789 (2015)
8. Valkenburg, P.M., Peter, J., Walther, J.B.: Media effects: theory and research. Ann. Rev. Psychol. **67**, 315–338 (2016)
9. Mattel Pulls Aristotle Children's Device After Privacy Concerns. https://www.nytimes.com/2017/10/05/well/family/mattel-aristotle-privacy.html
10. Wiederhold, B.K.: "Alexa, are you my mom?" The role of artificial intelligence in child development. Cyberpsychol. Behav. Soc. Netw. **21**, 471–472 (2018)
11. On the Heels of Congressional Inquiry, Advocates Ask Mattel to Scrap "Aristotle," AI Device Which Spies on Babies & Kids. http://www.commercialfreechildhood.org/heels-congressional-inquiry-advocates-ask-mattel-scrap-%E2%80%9Caristotle%E2%80%9D-ai-device-which-spies-babies-kids
12. Bonfert, M., Spliethöver, M., Arzaroli, R., Lange, M., Hanci, M., Porzel, R.: If you ask nicely: a digital assistant rebuking impolite voice commands. In: Proceedings of the 20th ACM International Conference on Multimodal Interaction, pp. 95–102. ACM (2018)
13. Druga, S., Williams, R., Breazeal, C., Resnick, M.: "Hey Google is it OK if I eat you?" In: Proceedings of the 2017 Conference on Interaction Design and Children – IDC 2017 (2017)
14. Reid Chassiakos, Y. (linda), Chassiakos, Y. (linda) R., Radesky, J., Christakis, D., Moreno, M.A., Cross, C.: Council on communications and media: children and adolescents and digital media. Pediatrics. **138**, e20162593 (2016)
15. De La Bastide, D.: Research Says Kids Will Be BFFs With Robots In the Future. https://interestingengineering.com/research-says-kids-will-be-bffs-with-robots-in-the-future
16. Yarosh, S., et al.: Children asking questions. In: Proceedings of the 17th ACM Conference on Interaction Design and Children – IDC 2018 (2018)
17. Newman, J.: To Siri with Love: A Mother, Her Autistic Son, and the Kindness of Machines. HarperCollins (2017)
18. Porcheron, M., Fischer, J.E., Reeves, S., Sharples, S.: Voice interfaces in everyday life. In: Proceedings of the 2018 CHI Conference on Human Factors in Computing Systems – CHI 2018 (2018)
19. Flessert, M., Beran, M.J.: Delayed gratification. In: Encyclopedia of Animal Cognition and Behavior, pp. 1–7 (2018)
20. Skorupska, K., Nunez, M., Kopec, W., Nielek, R.: Older adults and crowdsourcing: Android TV App for Evaluating TEDx Subtitle Quality. In: Proceedings of the ACM on Human-Computer Interaction (CSCW 2018), vol. 2 (2018)
21. Kopec, W., Skibiński, M., Biele, C., Skorupska, K., Jaskulska, A., Marasek, K.: Hybrid approach to automation, RPA and machine learning: a method for the human-centered design of software robots. In: Workshop on Industrial Internet of Things at CSCW 2018, NY, USA (2018)
22. Greene, J.D.: Solving the trolley problem. In: A Companion to Experimental Philosophy, pp. 173–189 (2016)

An Estimation Method of Intellectual Concentration State by Machine Learning of Physiological Indices

Kaku Kimura(✉), Shutaro Kunimasa, You Kusakabe, Hirotake Ishii, and Hiroshi Shimoda

Graduate School of Energy Science, Kyoto University, Kyoto, Japan
{kaku, kunimasa, kusakabe, hirotake, shimoda}@ei.energy.kyoto-u.ac.jp

Abstract. Although recent information society has improved the value of intellectual work productivity, its objective and quantitative evaluation has not been established. It is suggested that intellectual productivity can be indirectly evaluated by estimating intellectual concentration states when giving cognitive load. In this study, therefore, the authors have focused on physiological indices such as pupil diameter and heart rate which are supposed to be closely related to cognitive load in office work, and an estimation method of intellectual concentration states from the measured indices has been proposed. Multiple patterns of classification learning methods such as Decision Tree, Linear Discrimination, SVM, and KNN were employed as the estimation method. Based on the estimation method, an evaluation experiment was conducted where 31 male university students participated and the measured psychological indices were given to the classification learning estimators.

Keywords: Intellectual concentration state · Machine learning
Physiological indices

1 Introduction

Recent information society has improved the value of intellectual work and companies have been tackling improvement of intellectual work productivity by improving office environment. The improvement of intellectual productivity can introduce not only their own profit but also social benefit in total. Although the intellectual productivity is very important in this modern society, its objective and quantitative evaluation method has not been established. Considering the objective and quantitative evaluation method of intellectual productivity, most of the actual office work is simple intellectual work and its work efficiency is closely related to intellectual concentration which is also related to cognitive load while working. This means that intellectual productivity can be indirectly evaluated by estimating intellectual concentration states in office work.

In this study, therefore, the authors have focused on physiological indices which are supposed to be closely related to cognitive load in office work [1, 2], and an estimation method of intellectual concentration states from physiological indices has been

W. Karwowski and T. Ahram (Eds.): IHSI 2019, AISC 903, pp. 168–174, 2019.
https://doi.org/10.1007/978-3-030-11051-2_26

proposed. Pupil diameters and heart rate variability were employed as the indices which are supposed to be affected by their cognitive load. In addition, 4 types and 11 patterns of classification machine learning methods such as Decision Tree, Linear Discrimination, SVM (Support Vector Machine) and KNN (K-Nearest Neighbors) were employed as the estimation methods and the concentration states were estimated by the estimator with the highest classification performance among them.

The concentration states to be estimated were one of three states in this study when giving three kinds of cognitive loads which were high, medium and low to artificially generate high, medium and low concentration states, respectively. If it is confirmed that this estimation method is effective, real-time estimation of the intellectual concentration state in the actual office environment can be realized, and an application can be expected as one of the objective and quantitative evaluation method of intellectual productivity.

2 Estimation Method of Intellectual Concentration State

2.1 Physiological Indices During Cognitive Task

Pupil diameter and heart rate are employed in this study as measurement indices for extraction of feature value. Both the pupil diameter and the heart rate are known as indicators which easily vary when giving cognitive load [3, 4]. Regarding the heart rate, the electrocardiogram waveform is measured by a polygraph, Polymate AP216, and the four values, HR (Heart Rate), LF (Spectrum of Low Frequency), HF (Spectrum of High Frequency), LF/HF are extracted. 60 s where the power spectrum of the LF band contains for at least 3 cycles is defined as one section, and this section is shifted every 30 s, and the average value of each section excluding the start of 30 s is extracted as the feature value of the section. The pupil diameter is detected by a measurement device with an infrared camera, FaceLab 5, and the left and right pupil diameters are measured. Also, the extraction of the feature value is the average diameter of each time section. Thus, Totally six values which are HR, LF, HF, LF/HF, left and right pupil diameters are used as explanatory variables for estimation. On the other hand, the objective variable is one of three intellectual concentration states set by changing their cognitive load. The control of the cognitive load is realized by changing answering method of a cognitive task, Receipt-Classification Task. The method of answering tasks is as follows; Task A: High pace, Task B: Slow pace, Task C: Click (do not solve the tasks). Then, the concentration state corresponding to each task's answering method is defined as high, middle, and low concentration state respectively. In this study, the answer time of each task is set to 5 min which is considered to keep the same concentration state. Therefore, the total number of data for each variable in one task is 24.

2.2 Machine Learning

It is supposed that physiological response is greatly depending on the individuals, even though it shows common tendency. In this study, estimators are individually generated by machine learning of measured physiological indices data for three kinds of intellectual concentration states. For the machine learning, Four classification learning methods which are decision tree, linear discrimination, SVM, KNN were employed and realized with MATLAB [5] application. Table 1 shows the classification learning method of all four types and 11 patterns. Regarding SVM, in order to realize multi-class classification with binary classifier, ECOC [6] is used.

Table 1. List of classification methods

Classification method	Remarks
1. Decision tree	
2. Linear discrimination	
3. Quadratic discrimination	
4. Linear SVM	
5. Quadratic SVM	
6. Cubic SVM	
7. Fine Gaussian SVM	$\sigma = 0.6$
8. Middle Gaussian SVM	$\sigma = 2.4$
9. Row Gaussian SVM	$\sigma = 9.8$
10. Fine KNN	k = 1
11. Row KNN	k = 10

2.3 Evaluation Method of Estimation Accuracy

As described in Sect. 2.1, a total of 24 training data are extracted from measuring physiological indices while the cognitive task is performed. 11 patterns of estimators as shown in Table 1 are generated by applying classification learning. The generated estimators are evaluated its classification performance by cross validation of 24 division. Finally, one with the highest classification performance among the 11 patterns is selected as the best estimator and apply it to the estimation of the unknown data. The unknown data means test data measured at different time from the training data. In this way, the correct estimation rate of the intellectual concentration state is evaluated and the average value of all the correct estimation rates is taken as the estimation accuracy.

3 Evaluation Experiment

3.1 Method of Experiment

An evaluation experiment was conducted to show the estimation accuracy of the intellectual concentration state with the estimation method described in Chap. 2. The participants in the experiment were 31 male university students (age: 21.4 ± 1.9). One set of repetition of 1-min resting time and 5-min task time was set as one set, and two sets of data for training and test were obtained in total. In each set, each task was conducted for 5 min with three kinds of answering methods as described in Sect. 2.1. In order to cancel ordering effect, the order of task answer methods was random for each participant.

3.2 Result and Discussion of the Experiment

In the experiment, The data of 6 participants were excluded in the later analysis because of to data loss or so on. The physiological indices data were measured for 2 sets in total, so that estimation accuracy could be calculated when set 1 is as training data, set 2 as test data and vice versa, and the average value of both accuracies was taken as the estimation accuracy.

Table 2. The classification method applied to each participant and the classification performance of the estimator

Participant	Set 1		Set 2	
	Classification method	Correct rate (number)	Classification method	Correct rate (number)
p2	Fine KNN	0.88 (21)	Linear discrimination	0.75 (18)
p3	Linear discrimination	0.96 (23)	Cubic SVM	1.00 (24)
p4	Linear discrimination	1.00 (24)	Linear discrimination	1.00 (24)
p5	Linear discrimination	1.00 (24)	Decision tree	0.88 (21)
p6	Middle Gaussian SVM	0.88 (21)	Decision tree	0.83 (20)
p7	Linear discrimination	0.75 (18)	Middle Gaussian SVM	0.79 (19)
p8	Quadratic SVM	1.00 (24)	Decision tree	0.96 (23)
p11	Linear discrimination	0.83 (20)	Middle Gaussian SVM	0.88 (21)
p12	Decision tree	0.96 (23)	Linear discrimination	1.00 (24)
p13	Cubic SVM	0.71 (17)	Cubic SVM	0.83 (20)
p15	Middle Gaussian SVM	1.00 (24)	Linear discrimination	1.00 (24)
p16	Quadratic SVM	0.96 (23)	Linear SVM	0.92 (22)
p18	Decision tree	0.79 (19)	Roe Gaussian SVM	0.92 (22)
p19	Linear discrimination	1.00 (24)	Middle Gaussian SVM	0.92 (22)
p20	Middle Gaussian SVM	1.00 (24)	Linear discrimination	0.92 (22)
p21	Linear SVM	1.00 (24)	Decision tree	1.00 (24)
p22	Linear discrimination	0.96 (23)	Quadratic SVM	0.96 (22)
p23	Linear discrimination	0.83 (20)	Row Gaussian SVM	0.63 (15)
p24	Quadratic SVM	0.83 (20)	Middle Gaussian SVM	0.71 (17)

(continued)

Table 2. (*continued*)

Participant	Set 1		Set 2	
	Classification method	Correct rate (number)	Classification method	Correct rate (number)
p25	Linear discrimination	1.00 (24)	Linear discrimination	1.00 (24)
p26	Linear discrimination	1.00 (24)	Linear discrimination	1.00 (24)
p27	Middle Gaussian SVM	0.96 (23)	Middle Gaussian SVM	1.00 (24)
p28	Linear discrimination	0.96 (23)	Linear discrimination	0.96 (23)
p29	Linear discrimination	1.00 (24)	Linear discrimination	0.96 (23)
p30	Linear discrimination	0.79 (19)	Linear discrimination	0.71 (17)

Performance Evaluation of Training Data by Classification Learning

In this section, results and discussions of performance evaluation of estimators are described. Table 2 shows the classification learning methods applied to each participant and the generalization performance of the estimators by the cross validation method. The average of the classification performance of all participants' estimators was 91.1%, and it was confirmed that the concentration states were almost correctly classified by choosing the optimum estimator. Among the 2 sets of measurement data for all 25 participants, the highest classification performance was Linear Discrimination (18 data), and the second was Middle Gaussian SVM (9 data). In the physiological indices field, SVM is often applied as an effective method in classification performance. However, in this study, since the number of training data for the test data used for cross validation was as many as 23, it can be supposed that some estimation error occurred due to excessive learning when predicting the remaining one piece of test data. On the other hand, with respect to Linear Discrimination, the number of variables necessary to determine the categorical plane is smaller than that of SVM, and the possibility of excessive learning when predicting test data was low. As a result, estimation errors were less likely to occur and it seems that it became a method with high classification performance. It was also confirmed that the performance of Decision Tree and KNN was low compared with the Linear Discrimination unless they are relatively simple methods and their training speed is fast.

Table 3. Estimation accuracy of test data for each participant

Participant	Correct rate (correct number)		
	Set1 training, Set2 test	Set2 training, Set1 test	Average
p2	0.50 (12)	0.63 (15)	0.56
p3	0.88 (21)	0.79 (19)	0.83

(*continued*)

Table 3. (*continued*)

Participant	Correct rate (correct number)		
	Set1 training, Set2 test	Set2 training, Set1 test	Average
p4	0.63 (15)	0.58 (14)	0.60
p5	0.46 (11)	0.46 (11)	0.46
p6	0.25 (6)	0.33 (8)	0.29
p7	0.46 (11)	0.38 (9)	0.42
p8	0.50 (12)	0.33 (8)	0.42
p11	0.71 (17)	0.67 (16)	0.69
p12	0.96 (23)	0.79 (19)	0.88
p13	0.21 (5)	0.25 (6)	0.23
p15	0.71 (17)	0.71 (17)	0.71
p16	0.33 (8)	0.50 (12)	0.42
p18	0.50 (12)	0.33 (8)	0.42
p19	0.63 (15)	0.66 (16)	0.65
p20	0.63 (15)	0.42 (10)	0.52
p21	0.83 (20)	0.92 (22)	0.88
p22	0.42 (10)	0.63 (15)	0.52
p23	0.29 (7)	0.29 (7)	0.29
p24	0.58 (14)	0.46 (11)	0.52
p25	0.96 (23)	0.67 (16)	0.81
p26	0.92 (22)	0.83 (20)	0.88
p27	0.54 (13)	0.50 (12)	0.52
p28	0.63 (15)	0.54 (13)	0.58
p29	0.58 (14)	0.79 (19)	0.69
p30	0.58 (14)	0.54 (13)	0.56
Average	0587	0.560	0.573

Evaluation of Estimation Accuracy with Test Data in Different Time

Table 3 shows the correct answer rate of intellectual concentration state of each participant. The average accuracy of all valid data was 57.3% which was significantly higher than random estimation ($p < 0.001$). However, there were variations depending on the participants from a correct answer rate of 90% to below 30%. As for the participants who had a high estimation accuracy such as p21, the differences in intellectual concentration state due to the differences in tasks clearly appeared in the difference in physiological responses, and they showed similar responses in both set 1 and set 2. On the other hand, as for the participants with a low estimation accuracy, there were two types those who tended to have a low performance of the estimator like p7, and those who tended not to estimate correctly even though their performance of the estimator were high like p8. As for the former, due to differences in intellectual concentration state due to differences in tasks did not tend to appear as physiological responses, it was difficult to estimate by classification learning and estimation of test data could not be conducted correctly. On the other hand, with regard to the latter, it was supposed that different physiological responses appeared when measured at different times, so that some drift occurred in the physiological indices, and that an incorrect concentration state was made. Psychological burden such as familiarity with cognitive task, stress or fatigue may have caused the drift when measured at different times.

4 Conclusions

In this study, the authors have developed an estimation method of intellectual concentration state by machine learning of psychological indices. An evaluation experiment was conducted where 31 male university students participated and the measured psychological indices were given to the machine learning models. As the result of the evaluation by cross validation, the model which showed the highest classification performance was Linear Discrimination (18 data) and the second was Middle Gaussian SVM (8 data). On the other hand, the estimation accuracy of the test data which were not used as training data of machine learning was only 57.3% in average. There was considerable difference from the participant who had near 90% correct estimation rate to those who was less than 30%. In the future, it is necessary to explore additional indices which well-reflect their concentration states and are robust to other factors such as difference of time or fatigue in order to improve the estimation accuracy.

Acknowledgements. This work was supported by JSPS KAKENHI Grant Number JP17H01777.

References

1. Tryon, W.W.: Pupillometry: a survey of sources of variation. Psychophysiology **12**(1), 90–93 (1975)
2. Jorna, P.G.A.M.: Spectral analysis of heart rate and psychological state: a review of its validity as a workload index. Biol. Psychol. **34**(2), 237–257 (1992)
3. Hess, E.H., Polt, J.M.: Pupil size in relation to mental activity during simple problem-solving. Science **143**(3611), 1190–1192 (1964)
4. Mulder, G., Mulder, L.J.M.: Information processing and cardiovascular control, psychophysiology. Psychophysiology **18**(4), 392–402 (1981)
5. Inc MathWorks: MATLAB. http://www.mathworks.co.jp/products/matlab.html. Accessed May 2018
6. Dietterich, T.G., Bakiri, G.: Solving multiclass learning problems via error-correcting output codes. J. Artif. Intell. Res. **2**, 263–286 (1995)

Smart Home Technology as a Creator of a Super-Empowered User

Jaroslaw Kowalski[1(✉)], Cezary Biele[1], and Kazimierz Krzysztofek[2]

[1] National Information Processing Institute, al. Niepodleglosci 188B, 00-608 Warsaw, Poland
{jaroslaw.kowalski, cezary.biele}@opi.org.pl
[2] SWPS University of Social Sciences and Humanities, Chodakowska 19/31, 03-815 Warsaw, Poland
kkrzyszl@swps.edu.pl

Abstract. Over the last few years Smart Home Technology area has been rapidly growing. The home space saturated with remotely monitored and controlled sensors is different from the same space without these capabilities. The article presents the results of a qualitative study (7 in-depth interviews) conducted in 2018 on users of Smart Home Technology. The research shows that the administrative functions of the system usually become the domain of one household member with the highest technical skills. Such person becomes a "super-empowered user", for whom it is possible to control the house space (e.g. change temperature, light etc.) and watch other householders in real time, even being hundreds of miles away. Such situation changes the relations of power and control in the family. The study shows that the smart building and home space becomes, through the technology, an extension of the senses of the supervisor.

Keywords: Smart home technology · IoT · Social factors

1 Introduction

The dynamic development of ICT makes the human environment increasingly saturated with sensors. Smartphone applications enable the monitoring of various aspects of human life (such as diet, exercise or sleep) and processing the collected data. Smart clothing, smart cars, smart electric grids, and smart cities are already present in public discourse. One of the trends we have seen in recent years is Smart Home Technology (SHT).

Smart homes are defined as residence(s) equipped with a high-tech network, linking sensors and domestic devices, appliances, and features that can be remotely monitored, accessed or controlled, and provide services that respond to the needs of [their] inhabitants [1]. This technology is completely new and redefines the concept of home space, which is no longer a collection of separate items and devices. It becomes an assemblage [2], a network of features and functions that have further emergent features. This phenomenon and trend is so new that there is little research focusing on its importance for users [3]. Previous social research on this subject focused mainly on the

W. Karwowski and T. Ahram (Eds.): IHSI 2019, AISC 903, pp. 175–180, 2019.
https://doi.org/10.1007/978-3-030-11051-2_27

issues of usability or impact of technology on energy efficiency [4]. There are hardly any studies that focus on the sociological aspects of SHT [5].

Functioning in a smart home creates a brand new situation, both for each individual and for the whole family in general. The home space saturated with sensors and smart devices and actuators is qualitatively different from the same space without such equipment.

2 The Study

Previous researches on SHT technology focused mainly on user experience and there is no available research on interaction aspects or meanings given to individual functions, and the impact of these changes on the microcosm of family roles. To fill this gap, we conducted seven individual in-depth interviews (IDI) with smart home users in March and April 2018. The interviews were conducted in respondents' homes so that the ethnographic element was taken into account. The users were able to present the functions of their smart homes and describe them in their own terms. It enabled to reconstruct the meanings attributed to their relationship with the SHT technology. The main goal of the project was to acquire an in-depth understanding of relationships and interactions between smart homes and their users. We define the term "user" in two ways: it is a specific person, but also a micro-group, e.g. a family. In this article, we focus on findings which show how SHT influences the relations of power and control.

3 Smart Home as an Extension of Human Body

In his 1964 book 'Understanding Media: The Extentions of Man,' Marshall McLuhan [6] developed the concept that technology and media are an extension of the human body and mind. In his opinion, the wheel is an extension of the human leg, a book – an extension of the eye, clothing is an extension of the skin while electrical circuits work as extensions of the central nervous system. The essence of McLuhan's insight was that technology extends and complements our senses and bodies. It enhances the existing possibilities, creates new ones, and modifies the essence of humanity (as part of technological progress). What is more, each new invention, which is also an extension of the human being, changes the delicate balance of existing extensions. As he writes: "Any invention or technology is an extension or self-amputation of our physical bodies, and such extension also demands new ratios or new equilibriums among the other organs and extensions of the body" [6]. This quote can be understood as an observation that new inventions produced by people constantly modify and change the scope of their abilities, and the essence of humanity. With every new tool and technological innovation, our set of activities is modified, and so are ourselves. As he claims: "The transformations of technology gave the character of organic evolution because all technologies are extensions of our physical being" [6].

Similar ideas were developed by Andy Clark, and David J. Chalmers the authors of the concept of the so-called "Extended mind" [7]. According to them, man is a "natural born cyborg" adapted in the course of evolution to use tools and extend himself with their help. In their opinion, what distinguishes humans from animals is plasticity, which allows humans to define (or redefine) themselves with new inventions. According to that authors, the prefrontal cortex as well as the extended development and learning period (due to the fact that biological factors and the structure of a woman's pelvis make a longer pregnancy impossible) means that humans have the unique opportunity to incorporate innumerable resources which are non-biologically determined into possible activities. One can say, in summary, that their thought is that every human being is born with an open architecture, "complemented" by the specific historical time, technology and culture. That is why we think that SHT may be perceived as a further extension of the human body.

Of the seven SHT installations in which our research was carried out, four were equipped with cameras with the ability to view what is happening at home in real-time, on a mobile phone. The users used this option several times a day. The desire to monitor the home space was explained by the need for security, and the desire to observe what the pets are doing. This is a very good example of a situation where once a technical opportunity has been generated, it "generates" a need. Thus, Melvin Kranzberg's thesis is confirmed whereby this "invention creates a need" [8].

A similar pattern was observed in the case of photovoltaic panels. In one of the households, such equipment was installed on the roof. The electricity produced was injected into the grid and the operator subtracted its value from the owner's account. Sunny days meant big savings and even net profit to the owner (on very sunny weather). The production of electricity could be tracked on a mobile phone in real time. A special application also visualized the production of electricity in hourly, daily, weekly and monthly summaries. As it turned out, this technology, despite the three years that had passed from its installation, has not become a "background technology" [9] whose existence is forgotten in everyday life. The user said that since he has photovoltaic panels he pays attention to the weather every day ("I wake up and the first thing I do is look out the window to see what kind of weather we have. Sunshine? That is good! We produce it!"). He has the custom of checking electricity production reports several times a day.

In the households visited for the study, cameras often function in tandem with a loudspeaker that allows the voice from a smartphone to be transferred directly to a home-based device. In one of the households, the owner used it to punish a dog which, in the absence of his master, was trying to destroy the furniture. In another one, the owner used it in an emergency. One day, the owner's girlfriend could not be contacted. She was at home, but her smartphone was discharged. Her partner saw that she was at home (thanks to motion sensor reports transmitted to his smartphone) and asked her, via the loudspeaker, to contact her mother immediately.

The householder from another household used the loudspeaker to play a prank on his wife and frightened her (he unexpectedly obtained a much stronger effect than intended because the poorly calibrated loudspeaker distorted his voice and turned it into a chilling growl). In our research, the theme of pranks played by one household member on another one appeared several times. For example, one of the participants, who was returning to his SHT-equipped home from work, deliberately stopped the car in the street and turned off the power at home (using his smartphone), wondering "when my wife calls me to say there has been a power failure and there is no electricity in the entire home".

4 Super-Empowered User

These forms of jokes are possible because several conditions are met. First of all, a SHT installation at home is usually associated with the fact that one person at home (usually a man) understands and controls this technology. Even if the smartphone application which manages various features of the house has several users, there is usually one person who acts as an administrator, configures the system, understands the logic of its operation, and assumes responsibility for repairs and maintenance. Secondly, the network of home functionalities is usually built in such a way that all controlling functions are embedded in a smartphone. Some of the interviewed households also have control panels placed on the wall, but the mobile phone is the default control device. In this situation, a smartphone becomes a device that gives users the power and control over the whole home space. Depending on the installed features, it can enable easy and remote control, e.g. switching appliances on and off; monitoring who is staying at home; speaking to someone at home with a loudspeaker; listening to what is happening at home with the help of microphones; controlling the heating (including a total heating shut-off); raising and lowering the window blinds; turning music on and off etc. A smartphone becomes necessary to control the home space, which is illustrated by an example from one of the respondents. When he came home one night, the smart home system opened the outer gate for him, lighted the house, opened the garage door and turned on the garage light. The owner left the car and went out into the yard. The house automatically closed the garage door and the outer gate and then the owner realized that he had left his smartphone in the car. Because the whole house, including the door locks, was controlled via a smartphone (the owner did not use ordinary keys anymore), he was unable to get inside the house. The only thing he could do was to wait a few hours for his wife's arrival.

5 Discussion

The Smart Home Technology changes the properties of the home space and redefines the relationship of power over it in a qualitatively new way. This has some important consequences. The power over the functions of the house is distributed unevenly. An SHT installation is not a single plug-and-use device. Rather, it is a network of interrelated functions, each of which has a component of hardware and supporting software. This means that the equipment breaks down from time to time and needs to be replaced, the software needs an update, the batteries need replacement and the equipment needs a reset when it freezes. Above all, the system needs to be intellectually controlled, in order for its logic to be understood. This entails some responsibilities that a member of the household must take on. In six of the studied households these tasks were assumed by a man, with a woman performing them in one household. This person functions as the home "admin" who understands how the system works. Such an administrator becomes a "super-empowered user," gaining a new type of power over the home space. One of the respondents said: “my wife cannot surprise me anymore because I have a preview on the cameras and I can see what she does at home when I’m not there.” In such a home, many things become impossible, for example a teenage child playing truant and staying at home. Another respondent gave an example of his children’s wake-up routine. In that house, music starts playing from loudspeakers in the children’s bedrooms at 7 a.m. The system is set up in such a way that when a child tries to mute the music using a control panel on the wall, the music will play even louder.

Thus, we see clearly that the SHT technology and the new ownership of home space have major consequences. The abilities and capabilities of one person in the household are intensified, and the home becomes an extension of the human body and senses. In an extreme case, one may imagine that it will be possible to torment the household members by changing lighting, music, heating, etc.

A smartphone equipped with access codes to subsequent features in the home becomes a tool of power and control. We can think about new challenges which are created with this new technology. The upbringing of children could be one example. How to teach young people to take responsibility when the home space is constantly supervised and every activity is recorded on cameras and through sensors? How to teach agency, when every form of rebellion (truancy, ignoring the alarm clock) is futile and doomed to failure? How to teach children to respect privacy when they are aware that the home space is set to the “default” transparency mode by those who have the highest-level access codes and perform the admin role? How to learn equality and interdependence when the power depends on the person who has access codes to various features at home.

New properties of the home space saturated with smart devices present new challenges that the upcoming era will need to address.

References

1. Balta-Ozkan, N., Boteler, B., Amerighi, O.: European smart home market development: Public views on technical and economic aspects across the United Kingdom, Germany and Italy. Energy Res. Soc. Sci. **3**, 65–77 (2014)
2. Bennett, J.: Vibrant Matter: A Political Ecology of Things. Duke University Press Books, Durham (2010)
3. Hargreaves, T., Wilson, C.: Who uses smart home technologies? Representations of users by the smart home industry. European Council for an Energy Efficient Economy (ECEEE), Summer Study 2013. Toulon/Hye`res, France, 3rd–8th June 2013
4. Darby, S.J., McKenna, E.: Social implications of residential demand response in cool temperate climates. Energy Policy **49**, 759–769 (2012)
5. Wilson, C., Hargreaves, T., Hauxwell-Baldwin, R.: Smart homes and their users: a systematic analysis and key challenges. Pers. Ubiquitous Comput. **19**(2), 463–476 (2015)
6. McLuhan, M.: Understanding Media. The Extensions of Man. The MIT Press, Cambridge (1994)
7. Clark, A., Chalmers, D.J.: The extended mind. Analysis **58**, 10–23 (1998)

8. Kranzberg, M.: Technology and history: ‘Kranzberg’s laws’. Technol. Cult. **27**(3), 544–560 (1986)
9. Ihde, D.: Technology and the Lifeworld. From Garden to Earth. Indiana University Press, Bloomington (1990)

Research on Design Strategy of Sorting Garbage Bins Based on the CREATE Action Funnel

Shaoping Guan(✉), Lu Shen, and Rui Cao

College of Design, South China University of Technology, Guangzhou 510641, China
shpguan@scut.edu.cn, {201721050606, cao.r}@mail.scut.edu.cn

Abstract. This paper proposes an innovative design method to encourage people to classify rubbish, which will provide a systematic theoretical guidance for the design of public garbage bins. First, this paper uses observational method and user interview to obtain the existing design problems of sorting garbage bins in China. Then, based on a persuasive model-the CREATE action funnel and the Hook Model, from the cue stage, reaction stage, evaluation stage, ability stage and timing stage, respectively, it discusses what specific design strategies can be used to facilitate users to classify rubbish and eventually make classification a habit. In conclusion, this paper demonstrates the availability of the method that applies the CREATE action funnel to the design of garbage bins. The paper provides clearer design strategies for sorting garbage bins, which improves the efficiency of sorting and recycling municipal waste.

Keywords: Persuasive design · The CREATE action funnel · User habits Design of sorting garbage bins · Design strategies

1 Introduction

Science and technology not only bring people wealth, but also cause environmental and ecological problems. In order to solve the existing environmental problems and prevent the environment from further damage, green design and sustainable design have become the hot topic in the field of design research. The traditional green design is mainly designed from the perspective of efficient energy, recycled materials, and detachable parts [1]. In fact, designers need to study how to influence users' attitudes and behaviors so sustainable choices can be made to help users live a healthier life.

Garbage classification is a complex behavior related to user's psychology and previous experience. The Create action funnel divides the process of user behavior into five stages: cue, reaction, evaluation, ability, and timing [2]. It explains how to help users to filter through the funnel and finally execute the target action. The Hook Model help designers find strategies for developing user habits [3]. Based on the above situation, we believes that it is possible to provide a clearer design strategy for sorting

W. Karwowski and T. Ahram (Eds.): IHSI 2019, AISC 903, pp. 181–187, 2019.
https://doi.org/10.1007/978-3-030-11051-2_28

garbage bins based on the CREATE action funnel and the Hook model, which improves the efficiency of sorting and recycling municipal waste.

2 Research Status of the CREATE Action Funnel and the Hook Model

The Create action funnel divides the process of user behavior into five stages: cue, reaction, evaluation, ability and timing. The cue is a trigger that attracts users' attention and prompts users to have the intention of executing the target action. The reaction is the brain's intuitive response to the target action. Evaluation means that users weigh the pros and cons of the action. The ability means considering whether you can execute it from the aspects of planning, resources, skills and beliefs. Time is when the user decides when to execute the target behavior. Being ignored, generating negative reactions, spending more than benefits and taking actions that are not urgent can result in users not performing the targeted action (Fig. 1).

Fig. 1. The CREATE action funnel

Nir Eyal proposed the Hook Model that contains four stages: trigger, action, variable reward, and investment in *Hooked: How to Build habit-forming Products*. In his opinion, habit-forming need four conditions: something remind users to take actions; users give positive feedback; variable rewards encourage users to expect the next action; users become "repeat customers" because of their investment. Once the user enters the model, again and again, the user engagement shifts from low to high, and may directly trigger the target behavior based on the clue.

3 Study on Behavior of Garbage Classification

Research on the design of classified trash cans has been progressing at home and abroad. Qi has explained how to design the trash can based on the theory of emotional design proposed by Donald Arthur Norman [4]. Yang has analyzed the psychology of throwing garbage, proposed a more suitable mode for efficient waste sorting and discussed how to design the hardware facilities scientifically of garbage bins [5]. Another successful design case is a rabbit rubbish bin design by Paul Smith [6]. When garbage is thrown in the bag, the big bunny ears of it light up in celebration, which can be a deeper impression and encourage users not to litter. Studies above have produced positive influence. However, we cannot constantly replicate individual successes. We need to find a systematic theoretical guidance, which can be used to facilitate users to classify rubbish and eventually make classification a habit.

We chose the Beijing Road, Yuexiu District, Guangzhou as a pilot for user observation, selecting four different 30 min of a day as a time sample. The results showed that a total of 83 of 131 users were sorted their trash into wrong bins, and only five users stayed for judging. Then, fifteen of them were interviewed. A total of 15 users were street-cleaners, students and white-collar workers of the ratio of 3:5:7. Thirteen of the 15 interviewed users believe that they can classify rubbish correctly, but only three of them are actually completed.

Based on 2036 respondents' field survey data from 6 districts of Ningbo city, Chen investigated two questions: Are you willing to classify rubbish and Is garbage really classified by you in your daily life, in his paper *Paradox between Willingness and Behavior: Classification Mechanism of Urban Residents on Household Waste* [7]. It shows that the proportion of sorting willingness (82.5%) is significantly higher than the actual behavior's (13%).

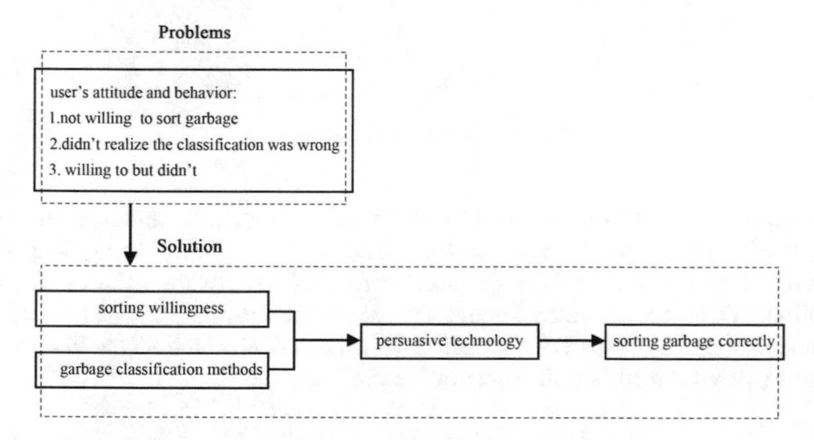

Fig. 2. Structural framework

Studies above revealed that user's behavior can be divided into 3 types, as shown in Fig. 2. It indicated that people do not correctly recognize their classification and waste sorting willingness doesn't necessarily result in behaviors. Therefore, the design of public garbage bins needs to solve two problems: how to transform the user's willingness into behavior; and how to make the garbage classification methods be understood by users.

The systematic research based on the persuasive technology has been successfully introduced into many fields: Zhang introduced persuasive technology into health behavior-oriented product design [8]; Li has put forward the industrial research on the persuasive interactive products for children education [9].

The design of public garbage bins should consider not only the execution of the target action but also the state of mind. The five Stages of CREATE action funnel can be used to shape people's attitudes and behavior. The Hook Model helps designers find strategies for developing user habits. It is necessary to use them comprehensively in the design of garbage bins.

4 Design Strategies of Sorting Garbage Bins

4.1 The CREATE Action Loop of Garbage Classification

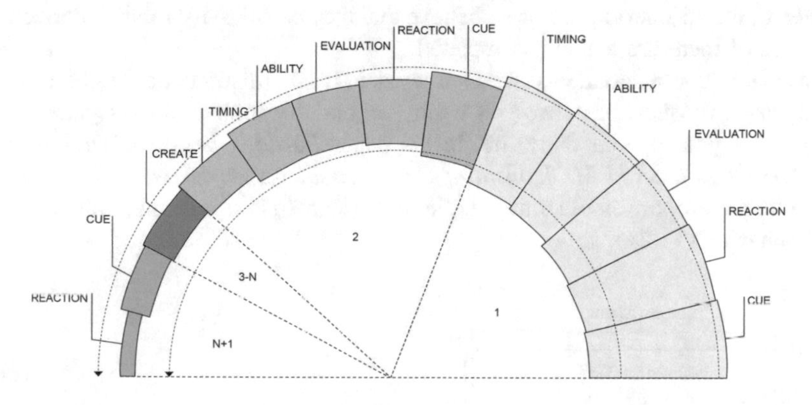

Fig. 3. The CREATE action loop

Based on the CREATE action funnel and the Hook Model, the author preliminarily constructed Create action loop, as shown in Fig. 3. The CREATE action loop shows the structure of the process from the user's first execution of the action to the habit formation. When it is executed through five stages (the part of "1") and be repeated over and over again (the part of "2" and "3-N"), the action, which be familiar to users, will be triggered directly on the basis of the clue.

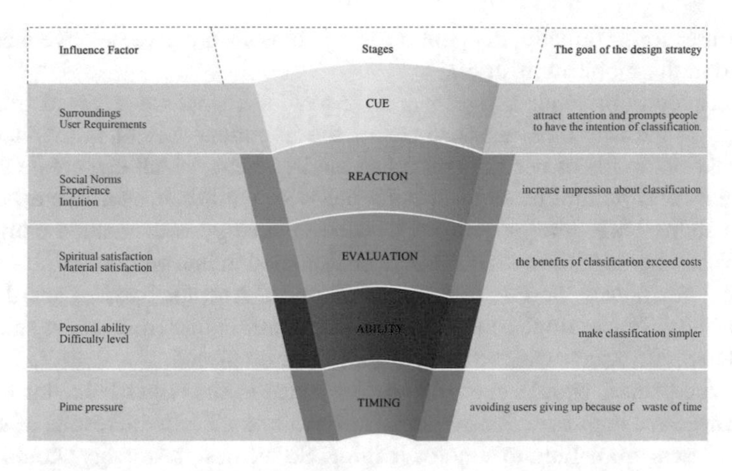

Fig. 4. The CREATE action funnel of garbage classification

Figure 4 shows the "1" part of the CREATE action loop of garbage classification. This model consists of 3 parts From left to right: the influence factor, stages, and the purpose of product design strategies.

4.2 Design Strategy of Sorting Garbage Bins

The goal of the design strategy in the cue stage is to motivate users to classify rubbish The Aesthetic-Usability Effect refers to users' tendency to perceive attractive products as more usable [10]. In other words, beautiful-looking products are more attractive. The exterior design of public garbage bins can be combined with the local particular forms of arts and culture. The different classification boxes can also be distinguished by different colors and materials, which attract users' attention to the classified information and increases the possibility of being persuaded.

Fifteen students, who stood on the edge of a bridge across a highway, held a poster with the letters "why not carpool" to give drivers below advice. The driver will reconsider his or her commute strategy when he or she trapped in the rush hour crawl [11]. The advice given at the proper time will be more acceptable. Therefore, designers can remind users of the harm caused by failing to sort the rubbish with simple but powerful slogans or appearance of the product, which encourages people to classification.

The goal of the design strategy in the reaction stage is to increase user's impression about classification. The liking principle describes that people are more willing to accept the advice of people like them [12]. The garbage bins can be designed to give anthropomorphic feedback such as a smiling or crying face based on the user's behavior. The principle of social identity in economics tells that we use others as a reference point for much of behavior. If the information that most people have begun to classify garbage can be got from garbage bins, users will make the same choice due to public pressure.

People always have a positive response to things that benefits exceed costs. Rewards can inspire use's curiosity about the product to bring about behavior changes.

When the user sorts correctly, the public garbage bins should give positive feedback, as mentioned in the rabbit rubbish bin.

The reciprocity principle suggests that we pay back what we received from others. The design of garbage bins needs to reflect the important role of individual classification behaviors in his or her life and urban environment, which can create a sense of responsibility and prompt users to link the behavior with their own interests.

In the ability stage, the purpose of the design strategy is to make a complex task simpler. Public garbage bins which should be designed in line with principles of human engineering guarantees the comfort of classification. And the symbol should also be easy to identify. For example, the barrel of the mouth is designed as the shape of the garbage to remind users where the garbage should be placed.

In the final stage, people may suspend or abandon the target behavior due to the increased negative impression or other more urgent tasks. From the results of interview, only 4% of users are willing to stay for judging. So the design strategy should focus on avoiding users giving up because of inefficiency. If the information as symbol and instructions can be easy to get by users in the process of walking toward the garbage bin from any direction, The time to stay for judging will be transformed to the time of walking.

A good design strategy may increase a user's self-identification and belief in his or her ability to perform a target behavior. Once the user believes that the classification is a kind of "responsibility", he may have regarded himself as a member of the urban environmentalist, which can help the person to develop a more positive attitude about classification and perform it more frequently.

5 Conclusion

It is important to devote scientific studies to address the issue of environmental conservation. A better design method for sorting garbage bins can improves the efficiency of sorting and recycling municipal waste. Through research above, it is necessary and feasible to provide design strategies for sorting garbage bins based on the CREATE action funnel and the Hook model, which can produce higher quality ideas in a different way and provide a systematic and theoretical guidance in the process of design.

References

1. Cheng, N.L.: General Introduction of Industrial Design. China Machine Press, Beijing (2011)
2. Wendel, S.: Design for Behavior Change. O'Reilly Media, Sebastopol (2013)
3. Eyal, N., Hoover, R.: Hooked: How to Build Habit-Forming Products. Portfolio Penguin, London (2014)
4. Qi, H.: Emotional Design of Categories Trash. Beijing Forestry University, Beijing (2011)
5. Yang, L.: The Design of Household of Waste Classified System in Universities. Southwest Jiaotong University, Chengdu (2017)
6. Rabbit Rubbish Bins Designed by Paul Smith. http://myhouserabbit.com/2009/06/rabbit-rubbish-bins-designed-by-paul-smith/

7. Chen, S.C., Li, R.C., Ma, Y.B.: Paradox between willingness and behavior: classification mechanism of urban residents on household waste. China Popul. Resour. Environ. **25**, 168–176 (2015)
8. Zhang, G.: The Research on Persuasive Design and Its Application in Health Behavior Oriented Product Design. Jiangnan University, Wuxi (2014)
9. Li, Q.: Research on the Persuasive Interactive Products for Children Education. Beijing Institute of Technology, Beijing (2015)
10. Lidwell, W., Holden, K., Butler, J.: Universal Principles of Design. Rockport Publishers, London (2010)
11. Fogg, B.J.: Persuasive Technology: Using Computers to Change What We Think and Do. Morgan Kaufmann, Burlington (2002)
12. Guy, S., Cialdini, R.B.: Influence: The Psychology of Persuasion. Createspace Independent Pub, Charleston (2017)

Culture-Centered Design Enabled by Machine Learning and Digital Transformation

Kevin Clark[1(✉)] and Kazuhiko Yamazaki[2]

[1] Content Evolution, Chapel Hill, NC, USA
ce@contentevolution.net
[2] Chiba Institute of Technology, Chiba Prefecture Narashino, Japan
designkaz@gmail.com

Abstract. Culture-Centered Design (CCD) is a focused version of Human-Centered Design enabled by digital transformation. CCD embraces product, service and experience variations for cultures being served by organizations – and purposefully connecting the culture of the organization with the client and customer cultures it serves.

Keywords: Client · Culture · Culture-Centered · Custom · Customer Data-Decay · Design · Digital · Digital transformation · Experience Global · Human-Centered · Listening · Machine learning · Mass-Customization Multi-national · Simulation learning

1 Introduction

Human-Centered Design (HCD) is poised for growth and new uses and needs enabled by ongoing digital transformation.

Culture-Centered Design (CCD) is a focused version of HCD and an evolutionary step forward in understanding people's needs and addressing them in very specific ways.

CCD makes it possible to design for cultures – multi-nationally vs. globally. Versions of products, services and experiences designed to be culturally appropriate and desirable by people in the diverse places they live and work.

Strategically connecting the culture of the company with the cultures the company serves creates lasting mutual benefit. Shared interests; shared values. Shared interests in experiences outcomes and shared values in how these outcomes are made real. The fusion of interests and values by these core elements: Relevance, Context and Mutual Benefit [1].

This is the promise of culture-customization and user-strategic experience (USE), a new high-relevance and actionable version of mass-customization [2]. It's coming alive in an age where digital transformation allows us to sense and respond to the voices of customers and the choices they make – and the ability to deliver what customers want in truly custom ways informed by the culture they live in: by tribes of common interests and values – all the way to serving the desires of individuals (Fig. 1).

W. Karwowski and T. Ahram (Eds.): IHSI 2019, AISC 903, pp. 188–193, 2019.
https://doi.org/10.1007/978-3-030-11051-2_29

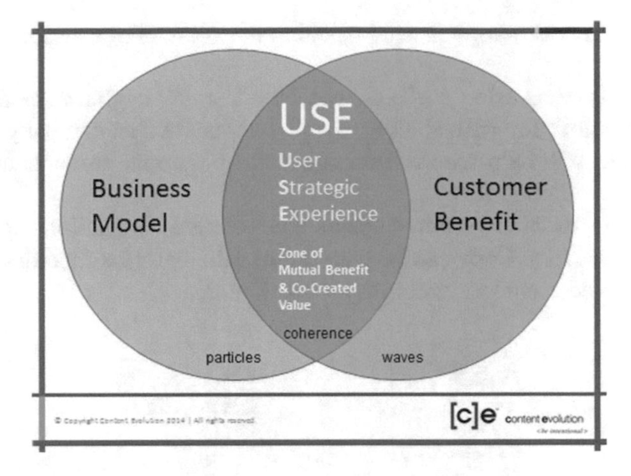

Fig. 1. The USE model: user strategic experience – visually describes the intersection of mutual benefit including the coherence of shared interests and values. Image courtesy of content evolution [3].

2 Listening and Leading [4] with Voice-of-People Using Digital Transformation

Machine learning agents are being used to represent the wants and needs of people in surveys and research. To listen more intently. To lead with more insight and integrity.

Actual data abstracted into clusters as personas is being used to understand what people want faster and in time for critical business decisions. Tanjo.ai[5] machine leaning and its Tanjo Animated Personas (TAPs) technology is one example of this ability to listen with speed.

Amplifying intelligence" is the mission of Tanjo [6]. As a machine learning company, Tanjo is working to help people distinguish what people do well in understanding and acting in the world, and what machine intelligence does well. Tanjo works to amplify and enhance human intelligence, not replace it. Tanjo is language intelligence engine that reads as a person would with prescribed interests. The machine learning platform learned categories of interest by reading in the Dewey Decimal Classification system found in libraries (000–999) [7]. The Tanjo platform can correlate and organize large data sets based on previous reading pattern recognition (Tanjo has been reading the semantic web daily for five years) – and can read and comment on unique text with TAPs.

As co-founder and CEO of Tanjo, Richard Boyd points out, "our technology asks too much of us. It makes us work too hard. In the last two decades we watched schools, hospitals, architecture offices, and companies large and small, implement technology and completely change their habits, alter their focus, re-design living and working spaces to meet technology on its terms. We attended training course for days and weeks. All too often humans would forget the real aim of their effort: educating children, healing people, manufacturing and moving goods and delivering services that make people's lives better and more productive. They would end up in a Faustian bargain serving the technology instead of doing the noble things they originally sought to do" [8].

PersonaPanels [9] is using the Tanjo machine learning engine to model these agent populations in partnership with Nielsen, the global market data company. TAPs offered by PersonaPanels will help with developing voice-of-people products and service for specific cultures and countries.

Further, TAP machine learning agents can be used as stand-ins for respondents evaluating written ideas. Copy can be placed into a dialogue box and one or more TAP agents can read and comment within minutes (Fig. 2).

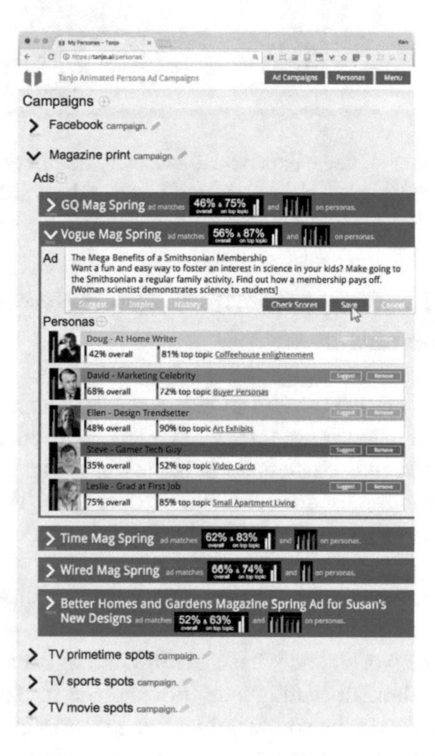

Fig. 2. Screen capture of a Tanjo.ai TAPs, courtesy of Tanjo Inc.

In an age where businesses are increasingly using Agile methods for development and LEAN methods for management, the pace of business is speeding up. TAP machine learning methods intercept decisions in time to inform them.

For example, TAPs are in use in customer service, education and finance. TAPs can understand all digital inputs from customer interaction with an organization and classify them with or without a preexisting taxonomy. In each case a customer segment or world region can be thought of as an addressable culture, one to be uniquely served by insights of machine learning technology.

A machine learning initiative by Tanjo in the U.S. is making possible for all digital knowledge in a community college system to be shared – and make recommendations for similar materials in the collection to teachers, administrators and students. Tanjo is organizing all digital information and resources for the North Carolina Community College System (NCCCS) in the State of North Carolina in the United States [10].

Another example of a platform-based method of evaluation is called Vennli and it uses a Venn Diagram display to visualize results [11]. It also allows for rapid updates of traditional people-based response surveys and showing the relationship of results to the needs of the customer, the offerings of a company, and the offerings of competitors.

3 Experience Vision

"Experience Vision" [12] is a comprehensive design method to innovative services, systems and products which reflect upon potential stakeholders' experiences and company mission and vision. Experience Vision methods are evolving to support Culture-Centered Design (CCD).

"Experience Vision: Vision Centered Design Method" makes it possible to propose new and innovative products, systems and services that are currently unavailable, as well as proposing advances for those that currently exist. It encompasses the entire HCD (Human Centered Design) process and presents a new vision with experiential value for both user and business from an HCD – and now CCD viewpoint.

Experience Vision uses a Structured Scenario-Based Design Method (SSBDM) approach. SSBDM employs personas and scenarios as human-centered representations for the innovative services, systems, and products. We envision these personas and scenarios coming to life every day using the machine learning methods discussed just moments ago in this paper without experiencing what we call "data-decay."

Experience Vision contains three layers of scenarios: value scenario, activity scenario, and interaction scenario. The Experience Vision design method reflects potential stakeholders' experiences and the company mission and vision. Here it is possible to explore the connection in the company culture and the variety of cultures in which customers can be found and embraced.

4 Kaizen Simulation Learning

A game-based learning simulator is developed for Toyota dealerships. Designed by Ultisim [13] simulation learning, it starts out for use in Toyota dealerships in Singapore to make the customer service experience better – and improve service efficiency and profitability better for the dealership.

The initial system design is culture-specific to Singapore and the unique requirements of space and customer timing expectations for this region of the world.

The system is modular and easily updatable, so it can be adapted to other markets quickly and easily. It runs on Virtual World Framework (VWF) that allows the simulation to work on any smartphone, tablet, or internet connected PC in the world without need for a special preload.

The basis for these simulations is ongoing advancement of game technology and the venerable pilot-training flight simulator – a mission-critical application of the technology.

If the task is physical, technical and the training is difficult – and the cost of failure is high in human safety or financial outcomes – simulation learning is the proven way to achieve better learning outcomes. Ultisim is working on similar game-based simulations for use in a variety of industries and professions, including healthcare, clean water, and renewable energy – each of which is its own unique professional culture – benefitting from the pioneering work done earlier for training pilots.

Choiceflows is using Ultisim technology to "test the future" of new products, services and experiences that don't exist yet [14]. The new idea in an R&D stage is prototyped as a game people can play to accelerate familiarity and then respond to questions about the proposed offering, its desirability by culture, and what people might be willing to pay for it in different regions [15].

5 Client Wellness and Fitness

Research was performed to understand what would hold the attention and allow for idea co-creation for executives and professionals visiting IBM in full-day briefings as an outgrowth of design thinking methods. This then led to a study of what experiences companies were having with long-term outsourcing performed by IBM, such as information technology (IT) operations for a company [16].

The findings led to design innovations in customer journey mapping, and insights into what creates customer satisfaction in different markets and client cultures around the world. This is an early spark for CCD and these cultural variations (Fig. 3).

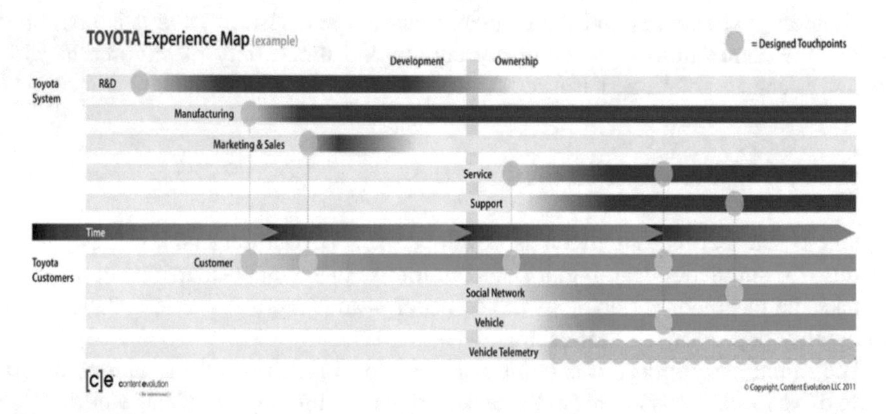

Fig. 3. Example of multi-track journey mapping for Toyota, Courtesy of content evolution [17].

To remain in an outsourcing relationship, the client needs to feel things are getting better (wellness direction). If they entered feeling well and competent, what they want is the ability to get even better and stronger (fitness direction).[18] Early indicators are discovered in this research that show when these long-term relationships are design properly and will last, and when they are fragile and susceptible to disruption.

6 Summary

Culture-Centered Design (CCD) enabled and informed by digital transformation makes it possible to design multi-nationally with increasing confidence. Products, services and experiences with diverse and culturally appropriate designs will be found desirable by people in the places where they live and work. This is the promise of mass-customization and now cultural-customization. CCD and culture-customization are coming alive in an age where digital transformation allows us to both hear the voices of customers with more precision – and the ability to make what customers want truly custom – informed by individual preferences and aggregating into patterns as culture.

References

1. Clark, K.A.: Brandscendence: Three Essential Elements of Enduring Brands. Dearborn/Kaplan, Detroit (2004)
2. Pine, B.J.: Mass Customization: The New Frontier in Business Competition. Harvard Business Review Press, Watertown (1992)
3. Clark, K.A.: USE: User Strategic Experience. Content Evolution (2014) www contentevolution.net
4. © "Listening and Leading" is a copyrighted expression of Content Evolution. www. contentevolution.net
5. www.tanjo.ai
6. www.tanjo.net
7. https://www.britannica.com/science/Dewey-Decimal-Classification
8. Boyd, R.: Superhuman Manifesto. https://wwwtanjo.net/s/SuperhumanAgeManifesto-1.pdf
9. www.personapanels.com
10. North Carolina Community College Brain whitepaper by the North Carolina Community College System (NCCCS) and Tanjo.ai (2018). www.tanjo.net
11. www.vennli.com
12. Yamazaki, K., Ueda, Y.: Experience Vision. Maruzen Publishing, Tokyo (2012)
13. www.ultisim.com
14. www.choiceflows.com
15. Clark, K., Louviere, J., Carson, R.: Familiarity Acceleration white paper, Choiceflows (2017)
16. Clark, K., Smith, R.: Discovering Your WOW. IBM, New York (2006)
17. Multi-track Experience Mapping invented by Kevin Clark (2009)
18. Clark, K., Smith, R.: Client Wellness. IBM, New York (2008)

How Perceived Real-World Danger Affects Virtual Reality Experiences

Shengjie Yao and Gyoung Kim(✉)

S.I. Newhouse School of Public Communications, Syracuse University, 215 University Place, Syracuse, NY 13244-2100, USA
{syao06, gmkim}@syr.edu

Abstract. Since current VR system blocks user's view (vision) of the real world as well as all the wires and physical objects around him/her, major VR manufactures such as HTC suggest securing a large space before experiencing the immersive virtual environment. There are possibilities that these "potential danger" elements could induce a negative effect on their virtual reality experiences. Exploring a relationship between user's percieved danger of the real world and immersion in the virtual reality is the main topic of this paper. In particular, we wanted to see the level of perceived danger when a user in the immersive virtual environment encountered a dangerous situation from the objects in the "real world".

Keywords: Virtual reality · Risk perception · Presence · Risk management

1 Introduction

On December 21st, 2017, a 22-year-old man was reported dead while wearing a virtual reality headset. He fell onto a glass table and passed away because of blood loss [1]. This incident was the first one in which a person died in relation to VR headsets. Human injuries were commonly reported while playing virtual reality games such as selfie tennis [2] alone side with damage to controllers or even to players' bodies in VR playing experiences. It is very likely that the users step onto the wires without noticing their existence and hurt themselves.

The idea of "being there" is one initial concept of a virtual reality system [3] but Current virtual reality systems are not capable of delivering a fully immersive experience. With a real-life potential risk being considered, how people's VR experience is affected by this perceived real-world risk is the main focus of this paper.

2 Virtual Reality Systems

With the market for virtual reality growing as much as $33.9 billion in the next several years [4], various high-fidelity immersive virtual reality platforms are available to consumers, such as Oculus Rift and HTC VIVE. An increasing number of people are using virtual reality headsets to teach, learn and have fun. Virtual reality system creates

W. Karwowski and T. Ahram (Eds.): IHSI 2019, AISC 903, pp. 194–201, 2019.
https://doi.org/10.1007/978-3-030-11051-2_30

a virtual environment where people can play games, simulate working environments or explore where people might not be able to reach.

One primary goal of virtual reality systems is to increase realism. Virtual reality generates a virtual environment which generates the idea of "being there" [3]. The sense of being elsewhere can generate a higher level of immersion which contributes to the simulation process of virtual reality [5]. Virtual reality systems create a higher level of immersion by using higher refresh rate displays (matching human eye) or better tracking devices (to simulate your body movement within the virtual world) [6].

Current virtual reality systems often involve three sensations: visual (Head Mounted Display), auditory (headphones), and tactile (controllers) [7]. These systems aim to enhance the experience of these three sensations and "fooling" people to think that they are elsewhere. With current technology restrictions, it is impossible to provide real-world experiences completely. Therefore, content providers (individuals or organizations) tend to use more engaging content to minimize technology shortcomings.

2.1 Possible Danger Regarding VR HMD Display Systems

"While wearing the VR headset you are blind to the world around you," according to HTC information safety information [9]. Virtual reality headsets block people's view of the outside world to increase realism. Manufacturers use different approaches. One standard solution is to clear a large field. Oculus rift recommends clearing out a play environment and map it within the system. Advanced solutions such as HTC VIVE include front-facing cameras to help identify the objects in real life and generate obstacle signs within the system.

These systems are not without their flaws. The tracking of Oculus Rift is not adapting to real-time changes such as an immediate interrupt by another person or something accidentally enter the playing area. HTC VIVE has only front-facing cameras, and they are not turned on by default which can increase the safety risk during a virtual reality experience.

While the concern of virtual reality generating physical damage to human beings is not being discussed directly, a trend in research of bridging the physical-digital gap between virtual environments and the real world showed the awareness of this topic. Researchers focused on redesigning VR environments to suit the physical world such as redirected walking [10]. Other solutions like pairing real-world objects with virtual counterparts [11] or generating a "Reality skin" [12]. These solutions showed the awareness of potential danger using virtual reality systems. These solutions are excellent for reutilizing objects, but significant brands and companies have not yet adopted them. The danger of physical damage is still present when using virtual reality systems.

3 Literature Review

The idea of virtual reality has been around for 40 years. With the development of computer technology, current immersive virtual reality differs from what a traditional virtual reality or virtual environment. In this chapter, a definition of immersive virtual reality is formed based on previous researches and critical concepts of immersion and presence is explained.

3.1 Definition of Immersive Virtual Reality

Modern virtual reality is mostly treated as a collection of hardware, including computers, head-mounted displays, headphones, and motion-sensing trackers. Steuer [13] argued that this hardware-oriented definition did not provide a conceptual unit of analysis. Based on a hardware collection approach, the virtual reality systems in this paper refer to systems with a high-quality wide field-of-view stereo head-mounted display as well as six degrees of freedom head tracking.

As Steuer [13] described in his article, virtual reality as a concept is referred to every single project of virtual reality experience. Therefore, a virtual reality system should also include content as well. By combining content and hardware, researchers can form a solid definition of virtual reality.

A recent definition of virtual reality came from Pan and Hamilton [14] which is as simple as a "computer-generated world." It covers a critical aspect of this concept which is a computer program that simulates a world that can be presented to people. However, according to this definition, a desktop viewed VR would also be VR [15].

Also, a VR is different in definition to an immersive VR. Immersive virtual reality would be closer to the definition of an immersive virtual environment with a current virtual reality hardware system. An immersive virtual environment is a computer-generated environment which surrounds the user and increase of being within it or sense of presence in particular [16].

A working definition of immersive virtual reality for this paper will be defined as follows:

"An experience generated by computers to surround users and increase their sense of being in the virtual environment using a collection of virtual reality system hardware including a high-quality wide field-of-view stereo head-mounted display and six degrees of freedom head tracking".

3.2 Immersion and Presence

Studies have argued that immersion is a multifaceted concept involving media (medium), users and contexts [17–19]. Early studies treat immersion as a "quantifiable description of a technology" which includes "the extent to which the computer displays are extensive, surrounding, inclusive, vivid and matching" [17]. User "matching" or context relativeness is also important to determine the level of immersion. As "quality of experience" [20], users feel immersed within the VR content based on themselves and social contexts. Based on the definition of an immersive virtual reality experience, feeling immersed is not contributed only by the hardware of the virtual reality, but also the content of this virtual reality. An immersive experience can be judged by its level of immersion as an ongoing procedure [20].

Presence and immersion are strictly mentioned in these studies. Presence is commonly defined as a sense of being in the virtual environment instead of where the people's real body exists [21]. Some scholars treat presence and immersion as a synonymous concept [22] which indicates that adding presence to the concept of immersion is only confusion. Immersion can also be treated as synchronicity of media, user, and contexts where presence is only a human consciousness of being there. Based on Slater and Wilbur's study [17], presence is a function of user psychology of recognizing being inside a virtual setting while immersion is the quality of this experience.

3.3 Attention

Attention is one of the key concepts in communication theories. As suggested in the LC4MP model [23], people have a limited capacity for information processing. That is to say: Even though people can process several tasks simultaneously, they can only process a certain amount of information at the same time.

The engagement level of the activity also affects this allocation of attention. A higher level of engagement leads to higher level of attentional demand of the task. With high engagement required in VR-based activities, an individual needs more attention allocated in one activity than regular flat screen-based activities [24]. As people pay more attention to the VR-based activities, virtual reality systems have long been used for pain relief in the medical field as a distraction.

Previous studies found that VR is a more useful tool in relieving burn pain, wound care and chronic pain [25–28]. The patients feel less pain during a medical treatment as more of their attention were drawn by the virtual reality experience. Explained by the "gate theory" in the medical field, VR reduces the perceptions of the pain and diverts patients' attention away from the pain by providing visual and audio cues which lower patients' actual feeling of the real world [29]. These studies all treat VR as a distraction source to the real-life experiences. Individuals pay attention to the virtual reality story so that they feel less of the real world.

3.4 Risk Perception

People's judgment about the likelihood of negative things such as injury or illness is called risk perception. This judgment will determine people's actual activity towards that risk or hazards. There are two main dimensions: how much people know about the risk and how they feel about them [30]. In the virtual reality experience, people are not capable of knowing the ongoing risk that is happening in the real world. Therefore, they might rely on their memory of what they have already observed before putting up VR headset. From a psychological point of view, people rely on heuristic cues to assess risk level [31]. Lacking sufficient "data" may lower people's risk perception of the situation [32]. That is to say, if people are not able to obtain such heuristic cue, their risk perception could be affected. However, the people have already formed a risk perception before putting up VR headset. Therefore, how pre-formed risk perception will affect virtual reality experience when people are not able to obtain heuristic cue remain to be studied.

4 Method

This study investigated how being aware of the potential danger in the real-world would affect virtual reality experiences. Thus, this study adopted a between-subject design with two conditions (no real-world objects/with real-world objects) to see whether higher level of perceived danger of falling will affect VR experiences.

The study used a VR HMD along with two level of perceived danger of falling (no danger situation vs. danger situation) in the experiment area. The perceived danger of falling in this study were objects (e.g., paper boxes) lying in the VR setting environment.

Before they put up VR HMD, the subjects were shown the experimental field and they were asked to finish a questionnaire asking their perceived danger of falling. The subjects will then be randomly assigned to two conditions and asked to experience the "room escape" game. While they were playing the content in VR, their behaviors were observed by the researcher especially on interactions with real-world objects such as paper boxes or wires.

After the virtual reality experience, the subjects will be asked to finish a questionnaire measuring a level of presence, attention, and enjoyment while playing the game.

4.1 Measurements

Perceived risk was measured using a seven-point Likert scale developed for this study. Items were used to measure the subject's perceived risk of falling using a VR headset (e.g., How safe do you think of using VR headset in the current environment?).

Presence was measured using a fifteen-item seven-point-Likert scale modified from a presence questionnaire [33]. Items were divided into measuring immersion to the virtual world (e.g., How much were you able to control events in the VR headset?), awareness of the virtual world (e.g., How aware were you of events occurring in the virtual world around you?) and involvement of the virtual world (e.g., How involved were you in the virtual environment experience?).

Attention was measured using five items on a seven-point-Likert scale self-reported attention questionnaire revised from the Situational Self-Awareness Scale [34]. Items were adopted from the original surrounding items and revised to measure virtual environment attention (e.g., "I am keenly aware of everything in the virtual environment," "I am conscious of what was going on in the virtual world").

Enjoyment was measured using the Physical Activity Enjoyment Scale (PACES), adopted from Kendzierski and De Carlo's study [35] with a total item of seventeen.

Subject behavior was observed by the researcher. The amount of times when subjects were concerned about getting trapped by the objects in the real-world was coded. Follow-up questions were asked by the researcher such as "Were you aware of the real-world objects when you do ….".

5 Discussion

The results showed that the subjects in the experimental group reported a higher level of perceived level of danger than the control group. We also see tendencies of subjects reporting a lower level of enjoyment, presence, and attention when their perceived level of danger is high. Based on the strong theory backup, though not sufficient subjects were tested at the current stage of this study, we highly expect a stronger correlation will be found when more subjects were tested. Also, we observed subjects feel less hesitant in the virtual reality environment. Therefore, the memory of the environment could also play a role in this relationship between perceived danger and virtual reality experiences. Further studies will be done to measure the subject's memory of the environment as well as using new biological measurements to test attention to the virtual reality experience.

References

1. Wilde, T.: Man dies in VR accident, reports Russian news agency. Pcgamer (2017). https://www.pcgamer.com/man-dies-in-vr-accident-according-to-russian-news-agency/
2. Kuchera, B.: This is the VR game that's hurting players, and they love it, polygon (2016). https://www.polygon.com/2016/4/11/11364904/htc-vive-selfie-tennis-injury-hands-ceilings
3. Biocca, F., Delaney, B.: Immersive virtual reality technology. Lawrence Erlbaum Associates, Inc., Hillsdale (1995)
4. Marketsandmarkets.com. Virtual reality market by component (hardware and software), technology (non-immersive, semi- & fully immersive), device type (head-mounted display, gesture control device), application and geography - global forecast to 2022 (Report Code: SE 3528) (2016)
5. Amasya, A., Solak, E., Erdem, G.: A content analysis of virtual reality studies in foreign language education. participatory educational research. spi15(2), 21–26. https://doi.org/10.17275/per.15.spi.2.3 (2015a)
6. Slater, M.: Place illusion and plausibility can lead to realistic behavior in immersive virtual environments. Philos. Trans. Royal Soc. B: Biol. Sci. **364**(1535), 3549–3557 (2009). https://doi.org/10.1098/rstb.2009.0138(2009
7. Ghosh, S., et al.: NotifiVR: exploring interruptions and notifications in virtual reality. IEEE Trans. Vis. Comput. Graph. **24**(4), 1447–1456 (2018). https://doi.org/10.1109/TVCG.2018.2793698
8. Gonçalves, R., Pedrozo, A.L., Coutinho, E.S.F., Figueira, I., Ventura, P.: Efficacy of virtual reality exposure therapy in the treatment of PTSD: a systematic review. PLoS ONE **7**(12), e48469 (2012). https://doi.org/10.1371/journal.pone.0048469
9. LaMotte, S.: The very real health dangers of virtual reality CNN (2017). https://www.cnn.com/2017/12/13/health/virtual-reality-vr-dangers-safety/index.html
10. Suma, E.A., Krum, D.M.: Impossible spaces: maximizing natural walking in virtual environments with self-overlapping architecture. IEEE Trans. Vis. Comput. Graph. **18**(4), 10 (2012)
11. Adalberto, L., Simeone, E.V., Hans G.: Substitutional reality: using the physical environment to design virtual reality experiences. In: Proceedings of the 33rd Annual ACM Conference on Human Factors in Computing Systems (CHI 2015), pp. 3307–3316. ACM, New York (2015). https://doi.org/10.1145/2702123.2702389

12. Shapira, L., Freedman, D.: Reality skins: creating immersive and tactile virtual environments. In: 2016 IEEE International Symposium on Mixed and Augmented Reality (ISMAR), pp. 115–124, Merida (2016)
13. Steuer, J.: Defining virtual reality: dimensions determining telepresence. J. Commun. **42**(4), 73–93 (1992). https://doi.org/10.1111/j.1460-2466.1992.tb00812.x
14. Pan, X., Hamilton, A.F.C.: Understanding dual realities and more in VR. Br. J. Psychol. (2018). https://doi.org/10.1111/bjop.12315
15. Slater, M.: Immersion and the illusion of presence in virtual reality. Br. J. Psychol. **109**, 431–433 (2018). https://doi.org/10.1111/bjop.12305
16. Oh, C.S., Bailenson, J.N., Welch, G.F.: A systematic review of social presence: definition, antecedents, and implications. Front. Robot. AI **5**, 114 (2018). https://doi.org/10.3389/frobt.2018.00114
17. Slater, M., Wilbur, S.: A framework for immersive virtual environments (FIVE): speculations on the role of presence in virtual environments. Presence – Teleoper. Virtual Environ. **6**(6), 603–616 (1997)
18. Hou, J., Nam, Y., Peng, W., et al.: Effects of screen size, viewing angle, and players' immersion tendencies on game experience. Comput. Hum. Behav. **28**, 617–623 (2012)
19. Shin, D.: Do users experience real sociability through Social TV? J. Broadcast. Electron. Media **60**(1), 140–159 (2016)
20. Shin, D., Biocca, F.: Exploring immersive experience in journalism. New Media Soc. **20**(8), 2800–2823 (2018). https://doi.org/10.1177/1461444817733133
21. Sanchez-Vives, M.V., Slater, M.: From presence to consciousness through virtual reality. Nat. Rev. Neurosci. **6**, 332 (2005)
22. McMahan, A.: Immersion, Engagement, and Presence, **20** (2003)
23. Lang, A.: The limited capacity model of mediated message processing. J. Commun. **50**, 46–70 (2000). https://doi.org/10.1111/j.1460-2466.2000.tb02833.x
24. Singh, A., Uijtdewilligen, L., Twisk, J.W.R., van Mechelen, W., Chinapaw, M.J.M.: Physical activity and performance at schoola systematic review of the literature including a methodological quality assessment. Arch. Pediatr. Adolesc. Med. **166**(1), 49–55 (2012). https://doi.org/10.1001/archpediatrics.2011.716
25. Hoffman, H.G., Patterson, D.R., Seibel, E., Soltani, M., Jewett-Leahy, L., Sharar, S.R.: Virtual reality pain control during burn wound debridement in the hydro tank. Clin. J. Pain **24**(4), 299–304 (2008)
26. Chan, E.A., Chung, J.W., Wong, T.K., Lien, A.S., Yang, J.Y.: Application of a virtual reality prototype for pain relief of pediatric burn in Taiwan. J. Clin. Nurs. **16**(4), 786–793 (2007)
27. Maani, C.V., Hoffman, H.G., Morrow, M., Maiers, A., Gaylord, K., McGhee, L.L., et al.: Virtual reality pain control during burn wound debridement of combat-related burn injuries using robot-like arm mounted VR goggles. J. Trauma **71**(1 Suppl), S125–130 (2011)
28. Van Twillert, B., Bremer, M., Faber, A.W.: Computer-generated virtual reality to control pain and anxiety in pediatric and adult burn patients during wound dressing changes. J. Burn Care Res. **28**(5), 694–702 (2007)
29. Gold, J.I., Belmont, K.A., Thomas, D.A.: The neurobiology of virtual reality pain attenuation. CyberPsychol. Behav. **10**(4), 536–544 (2007). https://doi.org/10.1089/cpb.2007.993
30. Paek, H., Hove, T.: Risk perceptions and risk characteristics. Oxford Research Encyclopedia of Communication (2017). http://communication.oxfordre.com/view/10.1093/acrefore/9780190228613.001.0001/acrefore-9780190228613-e-283. Accessed 18 Nov 2018

31. Slovic, P., Fischhoff, B., Lichtenstein, S.: Perceived risk: psychological factors and social implications. Proc. R. Soc. Lond. A **376**, 17–34 (1981). https://doi.org/10.1098/rspa.1981.0073
32. Fischhoff, B., Slovic, P., Lichtenstein, S., Read, S., Combs, B.: How safe is safe enough? A psychometric study of attitudes toward technological risks and benefits. Policy Sci. **9**, 127–152 (1978). https://doi.org/10.1007/bf00143739
33. Witmer, B.G., Singer, M.J.: Measuring presence in virtual environments: a presence questionnaire. Presence: Teleoper. Virtual Environ. **7**(3), 225–240 (1998)
34. Govern, J.M., Marsch, L.A.: Development and validation of the situational self-awareness scale. Conscious. Cogn. **10**(3), 366–378 (2001). https://doi.org/10.1006/ccog.2001.0506
35. Kendzierski, D., DeCarlo, K.: Physical activity enjoyment scale two validation studies. J. Sport Exerc. Psychol. **13**, 50–64 (1991)

Intelligence, Technology and Analytics

Breaking Down Barriers to Collaboration in Military Satellite Systems

Garrett Wampole(✉) and David Campbell

The MITRE Corporation, Bedford, MA, USA
{gwampole,dcampbell}@mitre.org

Abstract. The current state of the military satellite command-and-control enterprise consists primarily of so-called "stove-piped" acquisitions that provide a complete end-to-end capability to operate a single satellite system, and which typically must be recreated for each new satellite system. In this paper we present several techniques and technologies that we believe can help address challenges of acquiring and operating these types of systems. We will focus on two phases of satellite command and control and show how the adoption of existing commercial technologies can be used to increase the pace of collaboration.

Keywords: Acquisition · Command and control · Collaboration
Systems engineering · Satellite systems · Process management

1 Introduction

The military satellite enterprise is composed of systems that provide various types of capabilities ranging from positioning, navigation, communications, to surveillance, among others. We contend that both the capabilities exhibited by these systems and the nature of the threats to their effective deployment and use are evolving in a way that necessitates a more collaborative command and control approach to their employment [1]. Unfortunately, the primary method of development and acquisition of capabilities in the military satellite enterprise tends to result in institutional, political, and technical impediments to enabling enterprise-wide collaboration. However, we further contend that these issues can be mitigated through close attention by the acquisition authority and that the adoption of readily available commercial technologies will reduce the cost and increase the pace of collaboration.

In this paper we first enumerate and describe the phases of a generalized command and control mission thread as related to the military satellite enterprise. Then we examine two mission phases and identify areas where barriers to collaboration exist. Finally, two examples are given to demonstrate mitigation of collaboration inhibitors with the application of commercial technologies and acquisition procedures.

W. Karwowski and T. Ahram (Eds.): IHSI 2019, AISC 903, pp. 205–210, 2019.
https://doi.org/10.1007/978-3-030-11051-2_31

2 Phases of Command and Control

Even though the capabilities provided by the systems that comprise the military satellite enterprise are various and distinct, it is possible at a high level to generalize the process by which they are employed. We refer to this process as command and control [2] and decompose it into four phases, shown in Fig. 1, that are typically performed regardless of the type of satellite system capability being employed.

1. Ground Resource Management (GRM)
2. Commutation/Decommutation of Digital Data
3. Telemetry, Tracking, and Control (TT&C)
4. Data Analysis and Exploitation

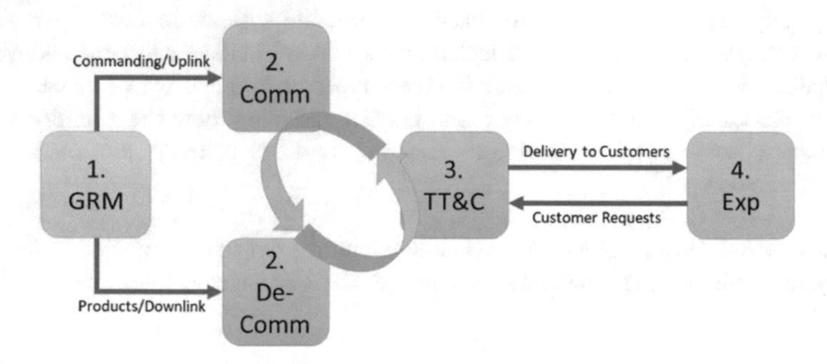

Fig. 1. Satellite command and control phases

2.1 Ground Resource Management

This phase involves identifying, assigning, configuring, and monitoring the operation of the various systems that enable wireless communication with the on-orbit asset. Typically, the equipment includes encryption devices, radio-frequency devices, antenna apertures, antenna control systems, and data recorders. Additionally, the ground equipment is generally a shared resource that must service many assets. As such, scheduling and coordinating the equipment use to maximize its duty cycle is also an aspect of this phase.

2.2 Commutation/Decommutation of Digital Data

For mission-specific digital data to be transmitted to the on-orbit asset via the ground equipment, it must first be transformed from its native format (in which it is most convenient to produce) to an encoded format suitable for transmission over a long-haul radio-frequency communications link, this is termed commutation. Conversely, status information or products from satellite systems are received via the ground equipment in this encoded format and must be translated to a format most suitable for consumption and further processing, this is the decommutation phase.

2.3 Telemetry, Tracking and Control

The TT&C phase encompasses the monitoring of on-orbit asset health and status, resolution of exceptional conditions, management of configuration settings, and performance of necessary periodic maintenance operations. Additionally, the tasking of capabilities hosted by the on-orbit asset is sometimes part of this phase. These operations often result in circular transitions between this phase and the commutation/decommutation phases.

2.4 Data Analysis and Exploitation

Exercising the on-orbit asset capabilities or those of hosted payloads often results in a data product or report that must be shipped to a customer for analysis and exploitation. The transmission format and the networks used are typically mission-specific. The customer's utilization of the product could conceivable result in new requests for capability and therefore result in a new command and control cycle.

3 Barriers to Collaboration

This section focuses on aspects of two phases of the command and control mission that often present barriers to collaboration. First, we identify reasons why collaboration inhibitors tend to appear in military satellite systems, which arise from both technical and socio-political factors; but primarily the latter. In the socio-political category are inhibitors stemming from acquisition processes and business motivations, whereas the technical category includes inhibitors originating in the choice of middleware techniques and a preference for developing new rather than adopting existing technologies. Lastly, we relate these barriers to the commutation/decommutation and analysis/exploitation phases of command and control.

3.1 Socio-Political Factors

The nature of the processes used to specify and acquire military satellite systems almost naturally inclines itself to inhibit collaboration [3]. One of the primary measures of success used to evaluate an acquisition is its ability to execute within a given cost and schedule, which are typically defined at a time in the acquisition lifecycle when the underlying technical requirements and scope of effort may be poorly understood or rapidly evolving. An initially ambitious plan, as schedule and budget are consumed, often shortly gives way to conservative project execution plans that attempt to avoid risk. Any reliance on an external collaboration with a third party whose own project execution plans are subject to change is therefore viewed as risky. This may, however unintentionally, erect barriers to collaboration as the program attempts to move as much responsibility as possible under its own control to reduce uncertainty. Operational efficiencies that could have resulted from collaboration must now wait until later in the system lifecycle, until perhaps much of the system is already built and difficult to modify. By limiting or delaying in this manner interactions with third parties who may benefit from inclusion in the commutation/decommutation phase, potentially unanticipated value resulting from collaboration based on their usage of satellite system products may be altogether overlooked.

Perhaps one of the most valuable of these overlooked enablers of collaboration is the system's own user base. The end users are primarily responsible for the execution of the mission and therefore have the most directly applicable domain knowledge. Collaboration among groups of users performing different missions within the military satellite enterprise may be beneficial. However, the siloed and risk-averse nature of the acquisition processes described above often result in disparate choices of socio-political workflow management and presentation techniques for each separate capability acquisition. This is again due to the perceived risk of reliance on third-parties and presents a barrier to collaboration as the produced capabilities have a vastly different look and feel and limited automated information sharing features, even as they perform very similar mission functions. Enforcement of standardized methods of authoring and performing workflows could create analysis/exploitation value from parts of the enterprise that had previously been unaware of the operations methods in use in each other's mission areas.

3.2 Technical Factors

Ironically, the siloed nature of acquisition processes negatively impacts efforts designed to increase collaboration. The military satellite enterprise has seen several competing efforts to introduce common messaging standards, middleware technologies, and infrastructure stacks that are mostly incompatible at a technical level. In an environment absent of natural market forces to identify a de-facto standard implicitly agreed to by participants in the marketplace, adoption and enforcement of these technical standards will continue to occur in a mostly ad-hoc manner. As a result, message translators or gateways are created to adapt information exchanges or requests between these islands of interoperability. These facilities are often overlooked in terms of both their technical importance and as enablers of collaboration. If the business rules by which the different parts of the enterprise are connected are defined and managed only by groups existing within a given acquisition (as opposed to an enterprise-wide group), then the technical mechanics by which collaboration could take place will not be widely disseminated. This would be especially useful during the commutation/decommutation phase, in which a larger set of participants in the enterprise could benefit from a consistent set of business rules for sharing command or downlinked digital data from a military satellite system.

What is more, these technology standards often partly or almost wholly overlap with commercially developed and maintained standards and technology that have been vetted via competition in the marketplace. While developing these overlapping technologies expands the scope of control of acquisition programs and may reduce perceived risk in accordance with the motivations described above; adoption of existing technologies may make heretofore unknown ecosystems of capabilities (and opportunities for collaboration) available to the enterprise [4]. We believe this is due primarily to the fact that capability development in the commercial space tends to coalesce around the de-facto standards arrived at by competition. Adoption (rather than redefinition) of infrastructure technology is particularly relevant to the analysis/exploitation phase in that larger networks of available and developing technologies (especially in the areas of machine learning and predictive analytics) could be more readily applied to data products created by military satellite system capabilities [5].

4 Mitigation Strategies

Having identified areas where barriers to collaboration exist in the acquisition and utilization of military satellite systems, we detail two examples where the adoption of commercial technologies may increase opportunities for collaboration in the commutation/decommutation and analysis/exploitation phases: Extensible Markup Language (XML) Telemetric and Command Exchange (XTCE) and Business Process Model and Notation (BPMN). Experience with XTCE and BPMN demonstrates the existence of real commercial solutions to improve collaboration in the commutation/decommutation and analysis/exploitation phases.

4.1 XTCE and Commutation/Decommutation Phase

XTCE is a standard method of representing how a digital data stream is sampled to retrieve measurands and data products that comprise the downlink of a satellite system, and the method by which commands or directives to be executed by the satellite system or hosted payload are encoded into a bitstream for transmission via allocated ground equipment and antenna network [6]. The XTCE specification is presented as an XML schema which describes the acceptable form of an instance XML document. This format has the dual advantages of being both machine consumable and somewhat self-describing. This open aspect of the standard's definition and management enables independently-developed capabilities to perform the commutation/decommutation phase of command and control for a given satellite system without having a separate relationship with the original capability developer, as depicted in Fig. 2.

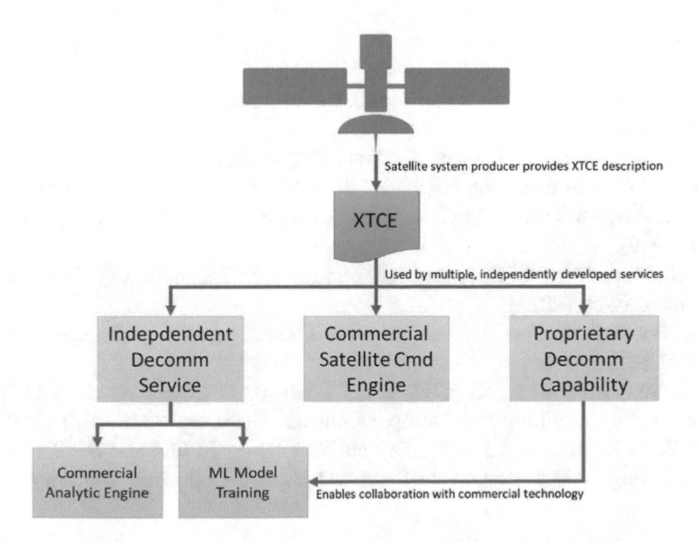

Fig. 2. Commutation/Decommutation phase collaboration

Therefore, we recommend that military satellite acquisition authorities add contractual requirements for delivery of freely distributable XTCE descriptions for satellite systems and hosted payloads. This acquisition process change could result in an increase of opportunities for collaboration, especially in the field of data analytics. The ready availability of decommutated digital data for military satellite systems across the enterprise could enable powerful commercial machine learning algorithms to be trained to uncover important trends and cross-correlations that would not be easily identifiable without this type of collaboration.

4.2 BPMN and Analysis/Exploitation Phase

BPMN is an enabling technology for encoding, automating, and operating business processes. As noted above, the analysis/exploitation phase of command and control often involves what is essentially a business process cycle of providing third parties with data products and fielding their additional requests. There exists a rich set of commercially available BPMN authoring, execution, and optimization tools that have demonstrated competitive value in a variety of similar commercial business processes [7]. Therefore, we recommend that the military satellite enterprise acquire and adopt consistent tooling for running the analysis/exploitation command and control phase business process.

The ability to subject the automated business process to optimization by commercially available tools is particularly relevant to the military satellite enterprise as changing mission requirements and operating environment conditions require shorter time spans and increased pace of workflow between customer analysis and the planning and execution of additional capability requests. Additionally, the ecosystem of commercial BPMN capabilities often provide standardized methods of presenting user interfaces and interacting with human operators which could enable collaboration among operators of different military satellite systems and capabilities, providing benefits in reduced training and cross-pollination of ideas and techniques.

References

1. The Battle Above. https://www.cbsnews.com/news/rare-look-at-space-command-satellite-defense-60-minutes/
2. Chethik, F., Reichner, A., Larson, T.: Emulating a space relay-based system for satellite command and control using the NASA ACTS satellite and VSATs, pp. 169–175 (1994)
3. Johnson, D.: Acquisition of information technology trends within the department of defense, pp. 5–10 (2009)
4. Takai, T.M.: Department of Defense (DoD) Information Technology (IT) Enterprise Strategy and Roadmap, iv–vi (2011)
5. Boja, C., Pocovnicu, A., Batagan, L.: Distributed parallel architecture for big data, pp. 116–127 (2012)
6. Simon, G., Shaya, E., Rice, K., Cooper, S., Dunham, J., Champion, J.: XTCE: a standard XML-schema for describing mission operations databases, pp. 3313–3325 (1994)
7. Juric, M., Pant, K.: Business Process Driven SOA Using BPMN and BPEL: From Business Process Modeling to Orchestration and Service Oriented Architecture, vol. 5, pp. 33 – 38 (2008)

Improving Cyber Situation Awareness by Building Trust in Analytics

Margaret Cunningham(✉) and Dalwinderjeet Kular

Forcepoint, Austin, TX, USA
{Margaret.Cunningham,DKular}@Forcepoint.com

Abstract. Analysts depend on technology to access and understand information, information that ultimately impacts their level of Cyber Situation Awareness (CyberSA). Adoption of advanced analytics, particularly those that generate risk scores or that depend on machine learning, can be impacted by a lack of trust in what the scores represent. Lack of trust in analytics can negatively impact CyberSA and efficient decision making, as analysts who do not trust outcomes from analytic models continue to search for information that confirms the analytic outcome, or continue to seek supplementary environmental information prior to making critical decisions. While human-driven investigative work is, and will remain, critical for security operations, delays in decision making, and increased efforts in information gathering, can negatively impact the efficiency of threat detection. Semi-structured interviews with analysts revealed five avenues for improving trust in analytics, including Context-Based, Case-Based, Model-Based, Ethics-Based, and Human-Centric AI Improvements.

Keywords: UEBA · CyberSA · Analytics · Risk scores · HCI

1 Background

As the cyber landscape expands through the development of more advanced malware and evasion techniques, gaps in security associated with the rapid adoption of cloud and Internet of Things (IoT) services [1], and sophisticated attacks targeting human weaknesses [2], building analytic models and technologies that can capture, consolidate, and communicate information is critical. Cybersecurity analysts depend on technology to understand the environment that they protect, which includes understanding issues such as immediate threats, malignant anomalies, vulnerabilities, and ongoing system status. This also means that how well an analyst's technology represents information directly impacts the analyst's performance in terms of their ability to assess the status of the environment and make decisions based on those assessments.

How analysts gather and process information about system status and emerging threats has changed over time, with industries transitioning away from event-based models that provide analysts with a chronological list of network events, and towards user and entity behavior analytics (UEBA) that assign risk scores to entities using analytic models. However, transitioning from event-based models to the more abstracted UEBA models of cybersecurity is not always a smooth process. One reason for this is that analysts do not always trust the analytics behind UEBA risk scores. This

W. Karwowski and T. Ahram (Eds.): IHSI 2019, AISC 903, pp. 211–216, 2019.
https://doi.org/10.1007/978-3-030-11051-2_32

lack of trust in analytics and risk scores impacts how quickly analysts can process information from their environment, as they may delay decision making to search for supplementary or confirmatory data prior to feeling comfortable enough to make decisions. This delay, and lack of trust in UEBA risk scores, can interfere with the efficiency and accuracy of threat analysis.

To identify strategies to improve trust in risk scores and UEBA, the present study provides results from a series of semi-structured interviews focused on the use of a new analytic model that relies heavily on behavioral analytic risk scores.

1.1 Event-Based and Entity-Based Analytics

For the purpose of this study, it is important to more clearly differentiate between event-based and entity-based analytics, as entity-based analytics require analysts to trust algorithms that they may be unable to fully access or understand. This is especially true for analytics that combine rule-based analytics with more complex statistical approaches.

The present study focuses on analysts' relationships with entity-based analytics. Broadly speaking, event-driven analytic workflows present users with sets of events that are most critical to review, whereas entity-driven analytic workflows present users with a set of entities that have displayed unusual behaviors and/or behaviors consistent with known risky scenarios in the form of a risk score. In the case of entity-based analytics, users are also provided with some relevant context for risk scores.

The word entity can represent more than one thing. For instance, an entity could be a user or a machine, an entity could be monitored or unmonitored, and each entity has its own meta data. The word event can also represent more than one thing, such as an email or a trade, and events can include features that make them more useful (e.g., To, Bcc).

1.2 CyberSA

An analyst's ability to respond quickly to a threat depends on CyberSA, and as attackers can do an enormous amount of damage in a short period of time, improving the speed with which analysts respond is a high priority goal. Use of the term CyberSA is based on Endsley's situation awareness (SA) framework [3] and its application to cyber security [4]. CyberSA is unique in that the first level of SA, perception of environmental information, is intrinsically linked to how the information is provided to the analyst. For analysts, this means that there is always a level of abstraction between their access to information and the ground truth cyber environment. This also means that the first challenge in developing high levels of CyberSA is ensuring that analysts trust the information they are provided.

The stages of CyberSA, illustrated in Fig. 1, includes analytics as its own block to emphasize the impact of analytics on situation awareness. The first stage of CyberSA is perception, the second is comprehension, and the third is projection. Perception includes what people gather from the environment using their senses, and can be impacted by past experiences, expectations, and attentional allocation. It is also impacted by how certain objects or information is presented. Comprehension is a process where people build a better understanding of how information fits together through making classifications, identifying patterns, and fitting information into goal-oriented contexts. Projection is the most advanced stage of SA, and encompasses the ability to dynamically and systemically make projections about what the future state of the environment might be.

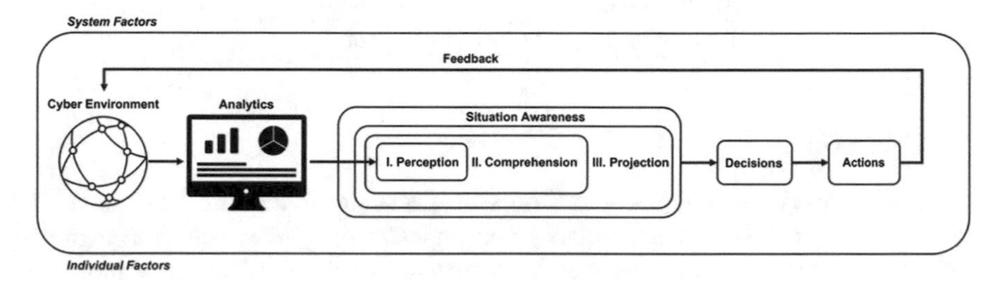

Fig. 1. CyberSA model including the layer of analytic abstraction between analysts and data from the cyber environment (figure adapted from Endsley, 1995)

2 Method

The present interview study was based on the results from a comprehensive literature review of how users and analysts interact with analytics (literature review unpublished). Findings from the literature review were used to develop a semi-structured interview template. Designing content for semi-structured interview questions by integrating findings from existing research has been shown to improve the results of qualitative interview studies [5, 6].

A total of five participants were interviewed using the semi-structured interview guidelines developed by the authors of this paper. The interviews were conducted by both authors, with each interview requiring approximately one hour. After completing the interviews, the authors independently compiled their notes, and then reviewed all findings to ensure that there were no discrepancies. Few discrepancies emerged, and those that emerged were resolved during review without issue. The consolidated results were further explored to establish thematic groupings outlined in the results section of this paper.

3 Results

Interview responses indicated a high level of agreement across five strategies that could improve trust in entity-based analytics. The strategy categories include: Context-Based, Case-Based, Model-Based, Ethics-Based, and Human-Centric AI Improvements. Table 1 provides a summary of which of the five categories each participant discussed during their interview. Details about each strategy category, including some recommendations for future improvements are also summarized below.

Table 1. Summary of strategies for improving trust in analytics by participant

	Participants					
Strategy	1	2	3	4	5	Agreement
Context-based	•	•	•	•	•	100%
Human-centric AI	•	•	•	•	•	100%
Use case-based	•	•	•			60%
Model-based	•		•		•	60%
Ethics-based	•	•		•		60%

3.1 Context-Based

Users need access to what types of data impacts risk scores in the form of context. This means that the analysts expected clarity and understanding of information regarding which policies were violated, as well as access to details regarding data movement violations. To supplement risk scores, analysts who participated in this study currently continue to reference log files to identify why a score may have increased for specific entities. Future design and development of easy-to-use interfaces that provide more context to analysts could improve efficiency.

3.2 Human-Centric AI

The analysts desired the ability to manually override any automatic actions (e.g., blocking of user behaviors, account locking) based on analytic outcomes. The analysts also wanted the freedom to create their own entity-based use cases. Providing control to the users, especially control that impacts the outcome of analytics, was highly desired and also an essential factor for building and maintaining trust with the analytic platform.

3.3 Use Case-Based

Use case-based strategies refer to providing users with information about the analytics within the context of existing use cases relevant to their area of expertise. This means that this strategy requires understanding specific business area needs. Participants noted that use cases are particularly helpful for understanding how to use UEBA, and for interpreting various features (such as timelines) presented in summary dashboards. Participants noted that case-based strategies are primarily critical during training periods, and these strategies should be integrated into all documentation and training materials.

3.4 Model-Based

Certain users or analysts desire a deeper understanding of the models that calculate risk levels. This is also associated with a desire to understand what a "high risk" level means, and how the calculations might be impacted by various data sources. For the analysts interested in this information, providing the rationale behind and details of the model can have an immediate impact on improving trust. This information should be provided in product documentation.

3.5 Ethics

Analysts want to use data in an ethical way while simultaneously retaining the ability to do their jobs well. In light of new regulations such as the (General Data Protection Regulation GDPR), pseudonymization and anonymization of data is critical for all companies. However, some analysts share concern that pseudonymization strategies strip information out of data sources in a way that may shift the results of analytic models. As privacy concerns continue to emerge, and regulations shift, analysts as well as analytic model developers will need to build a collaborative relationship to design systems that meet both privacy and analytic requirements.

4 Conclusion

As the cyber threat landscape grows, analysts' need to rely on abstracted data processed using advanced analytics and machine learning will also grow. Analysts will also be challenged to perceive and respond to threats as quickly as possible, and begin to demand more from their tools and technologies to improve CyberSA. However, reliance on analytics will be tenuous until analysts achieve a higher degree of trust in analytic models and outcomes. While a strong relationship is possible, and likely inevitable, the community may trust more quickly when there is better collaboration between data scientists, software engineers, designers, and cyber security analysts. Meeting analytic needs through sharing expertise across disciplines will make it easier for analysts to "perceive, comprehend, and project" which will result in more efficient and potentially accurate decision-making and faster responses to threats.

The present study has limitations, most notably the small sample size, but our findings emphasize several core avenues for improving trust in advanced analytics. It is notable that one of the categories with full agreement is working towards a more Human-Centric AI strategy, which emphasizes the need for considering the human across all stages of the software development lifecycle, and the need for creating opportunities for empowering users to take control of certain aspects of analytic platforms.

As use of advanced analytics, machine learning, and artificial intelligence progresses, future research should continue to address issues such as trust, and continue to target strategies for making analysts' end-goal decisions more accurate and efficient. Analytic models that are currently under development or currently deployed may benefit from systematic research that reveals what features and capabilities analysts are really using, and what they are ignoring, and why.

References

1. Cisco Systems, Inc.: Annual cybersecurity report. Technical report (2018)
2. Forcepoint: The 2017 state of cybersecurity. Technical report (2017)
3. Endsley, M.R.: Toward a theory of situation awareness in dynamic systems. Hum. Fact. **37**, 32–64 (1995)
4. Barford, P., et al.: Cyber SA: situational awareness for cyber defense. In: Jajodia, S., Liu, P., Swarup, V., Wang, C. (eds.) Advances in Information Security, vol. 46. Springer, Boston (2010)
5. Fylan, F.: Semi-structured interviewing. In: Miles, J., Gilbert, P. (eds.) A Handbook of Research Methods for Clinical & Health Psychology, pp. 65–77. Oxford University, Oxford (2005)
6. Kallio, H., Pietila, A., Johnson, M., Kangasniemi, M.: Systematic methodological review: developing a framework for a qualitative semi-structured interview guide. J. Adv. Nurse. **72**, 2954–2965 (2016)

BIFROST: A Smart City Planning and Simulation Tool

Ralf Mosshammer[1(✉)], Konrad Diwold[1], Alfred Einfalt[1], Julian Schwarz[2], and Benjamin Zehrfeldt[2]

[1] Siemens Austria AG, Siemensstr. 90, 1210 Vienna, Austria
{ralf.mosshammer,konrad.diwold,alfred.einfalt}@siemens.com

[2] pixelart GmbH, Handelszentrum 16, 5101 Bergheim bei Salzburg, Austria
{j.schwarz,b.zehrfeldt}@pixelart.at

Abstract. BIFROST is a persistent, shared design tool and simulation environment for Smart Cities, with a strong focus on powergrid infrastructure. Backed by a reactive server backend and powerful simulation engine, a browser-based, 2.5D user interface empowers researchers, network operators and planning experts to construct and analyze situations revolving around powergrid operations. The internal engine state representing the simulation world, including all physical dynamics and structures, is fully exposed to external applications, such as control algorithms or time series analysis tools. Thus, BIFROST can be employed as a virtual testbed for Smart Energy System installations, allowing for the evaluation of scenarios which would be hard or impossible to stage in-field.

Keywords: Smart energy system · Simulation · UX · Scenario planning Reactive systems

1 Introduction

Infrastructure automation and ubiquitous networking increasingly empower residents to migrate into active “prosumer” roles in the context of the “Energiewende” [1]. In parallel, municipal governments push towards increasingly data- and statistics driven decision making, fueled by distributed sensing of environmental parameters, materiel and power flows, traffic and sociological contexts. The *Smart City* manifests the utopian vision of a well-oiled, rational machine affording all the benefits of dense urban living while mitigating its negative effects through algorithms.

In this context, manufacturers of software and hardware components often face difficult environments for testing and demonstration: critical infrastructure is affected; the well-being and comfort of residents directly impacted; operator network topologies must be navigated; and, especially in Europe, stringent privacy laws limit data availability and resolution [2]. Testbeds for isolated functionality are readily available. But to fully realize the vision of a holistic integration of components in the Smart City context, comprehensive testing and demonstration often mandates scenario planning and simulation across domains.

W. Karwowski and T. Ahram (Eds.): IHSI 2019, AISC 903, pp. 217–222, 2019.
https://doi.org/10.1007/978-3-030-11051-2_33

In this paper, we present BIFROST, a framework for modelling and simulation of Smart Cities. BIFROST takes cues from game engines to present a consistent *state* as the single source of truth for the city model. A variety of APIs allow *actions* to manipulate the state, for example by introducing new buildings, wiring powerlines or manipulating a buildings' AC settings. External software can interface with BIFROST to provide models or affect the simulation directly, affording testing as IOT-in-the-loop approach. BIFROST ships with a powerful, browser-based interface for constructing *settlements* and controlling the simulation. In the following, we will proceed to outline the BIFROST backend as an open hub for Smart City modelling, present the client interface and illustrate a use case in the powergrid domain.

2 The Backend

BIFROST is designed as a single-instance, concurrent multi-user environment, which means that clients can edit and view its structure and manipulate the simulation of the same settlement in real-time. The single source of truth for a settlement is its representation in the *state*. A theoretically unlimited number of settlements can be maintained in parallel in the state, each outfitted with its own *simulation control*. Internally, the state is maintained as a plain JSON object. The BIFROST *engine* uses Redux [3] as a modelling/control flow paradigm/library. Redux mandates that the state be immutable. The control flow is unidirectional: *actions* (themselves plain JSON objects) introduce changes to the state via *reducers*, which return a novel state object manifesting the requested changes. In BIFROST, actions can come from a variety of sources, including the simulation (e.g., to update the current in-world time), the browser client (e.g., to build a new structure) or external software (e.g., to affect the position of a power switch).

2.1 State Shape

Top-level entries in the state all have a similar, normalized structure, as seen in the listing below (quotation marks omitted for brevity).

```
{
 settlements: {
   byId: {
     solheim: {
                 name: solheim,
                 hash: cc82bfc5b69adf534469dc0,
                 owner: rmos,
                 ...
     }, ...
   },
   allElements: [...]
}
```

The state fragment above holds meta-information about all settlements, such as their name and *hash*. The state hash of a settlement is used in synchronization: all actions dispatched by clients must include the currently valid state hash or be rejected, the assumption being that the client tries to manipulate an out-of-date version of the world. Whenever the state updates, the hash is recalculated and the state broadcast to all interested parties, most notably the web clients.

Aside from a top-level sub-state containing information about the simulation, such as the current in-world time and simulation step, the central element of the state are the *domain layers*. A domain is characterized as

1. grouping elements under the same ontological "type"
2. having a tangible representation on the client interface and
3. holding references to *dynamics*, *affordances* and *models* for its elements.

Examples of domains include *architecture* (the bulk of all man-made physical structures, such as housing and roads) and *powergrid* (all electrical elements and their connections). *Dynamics* are physical quantities relating to the objects' behavior: the voltage level of a grid node; the ambient temperature and precipitation; or the structural integrity of a building. *Affordances* are entry points for manipulation of an entities' behavior in the simulation: a switch position; the mode of an e-car charging pole; a modifier for the cloud cover. *Models* affect entity dynamics in the context of the simulation (see below). Domain elements only reference these quantities, which are held in separate parts of the state. A description of which elements incorporate which affordances, dynamics and models is kept in separate *directories*. This loose coupling allows for painless extension, addition and modification of domain elements. Adding an air-condition control for a building is as simple as modelling it in the affordance directory, and attaching it to the appropriate building types in the architecture directory.

2.2 Simulation Control

The simulation loops for each settlement all follow the same basic steps:

1. Advance the in-world time by one t_{step}.
2. Invoke all active element models and update dynamics according to their results.
3. For each domain that experienced changes due to model updates, run any attached network *solvers*.
4. Delay the loop to achieve t_{loop}.

For an (accelerated) real-time simulation, t_{loop} will generally remain at 1 s (barring longer calculation times for models and solvers) while t_{step} can be adapted to conform to the model resolution, or be identical to t_{loop} for a true real-time simulation.

While models are expected to only manipulate the dynamics of their respective element, solvers influence an entire domain state. For example, the currently implemented powergrid solver will use topological information, switch states and building power demand (as provided by their load models) to calculate node voltages on any point in the grid.

3 Usage Scenario

BIFROST is furnished with a powerful, browser-based interface for construction of arbitrary settlements, powered by React [4] and WebGL. A lush variety of building models – both for visual as well as behavioral diversity – are provided. Roads, terrain and power lines can be freely picked from a selector and placed on the regular 2.5D grid. Figure 1 shows the construction interface, with the building variant picker and a large settlement already established.

Fig. 1. BIFROST main construction interface with open building selector

On the upper left corner, the simulation control is visible as a large "play" button. Once the simulation is started, it is surrounded by an arc control for adjusting the simulation step. This design element is highlighted in Fig. 2.

Fig. 2. Design elements of the BIFROST client, from left to right: simulation control with start/stop button, speed control, weather and time information; building information overlay with dynamics, affordances and models, including an active voltage limit monitoring; search interface

During simulation, it is possible to highlight powergrid structures, as shown in Fig. 3. This is a snapshot of a typical scenario simulation: a power switch near the bottom center on the rightmost feeder has been disabled, rendering several buildings powerless, as indicated by the magenta icons. These icons are the result of enabling dynamics monitoring in the building overlay, shown in the center of Fig. 2.

Fig. 3. BIFROST during a running simulation, with cloud visualization and powergrid highlight view enabled

In this exemplary scenario, the event information – building power outage – was relayed to an external event collection and visualization system [5], which proceeded to present relevant time series data pulled from the state. This is an example where testing/demonstrating this secondary system in a real in-field scenario would have not been possible without risking the ire of the building residents.

4 Conclusion

In this paper, we gave a brief overview of our co-simulation/demonstration system for Smart Energy System environments. In a first practical application BIFROST will be used as virtual testbed for reactive operation scenarios representing a real "Smart Grid Village" within the European research project DECAS [6, 7].

BIFROST is built as loosely coupled, open architecture which allows for flexible integration of new elements, models, and even entire simulation domains. In its current implementation, it is specifically suited for powergrid simulation and scenario analysis. New areas of application are planned as part of the Smart City research projects in Seestadt Aspern (Austria), as well as in the area of new operating concepts for local energy communities.

Acknowledgements. Project DECAS has received funding in the framework of the joint programming initiative ERA-Net Smart Grids Plus, with support from the European Union's Horizon 2020 research and innovation program.

References

1. Kühne, O., Weber, F.: Bausteine der energiewende – einführung, übersicht und ausblick. In: Bausteine der Energiewende, pp. 3–19. Springer VS, Wiesbaden (2018)
2. EU Regulation 2016/679 (General Data Protection Regulation), 25 May 2018
3. ReduxJS. https://redux.js.org/. Accessed 2 Oct 2018
4. React. https://reactjs.org/. Accessed 2 Oct 2018
5. Mosshammer, R., Diwold, K., Einfalt, A., Groiss, C.: Reactive operation: a framework for event driven low voltage grid operation. In: IHSI 2018 International Conference, Dubai (2019)
6. DECAS. http://www.decas-project.eu/. Accessed 15 Oct 2018
7. Uebermasser, S., et al.: Requirements for coordinated ancillary services covering different voltage levels. In: CIRED, Glasgow (2017)

Stochastic Drop of Kernel Windows for Improved Generalization in Convolution Neural Networks

Sangwon Lee and Gil-Jin Jang(✉)

School of Electronics Engineering, Kyungpook National University, 80 Daehakro Bukgu, Daegu 41566, South Korea
{lsw0767,gjang}@knu.ac.kr

Abstract. We propose a novel dropout technique for convolutional neural networks by redesigned Dropout and DropConnect methods. Conventional drop methods work on the individual single weight value of the fully connected network. When they are applied to convolution layers, only some kernel weights are removed. However, all the weights of the convolutional kernel windows together constitute a specific pattern, so dropping part of kernel window weights may cause change of the learned patterns and may model completely different local patterns. We assign the basic unit of drop method for convolutional weights to be the whole kernel windows, so one output map value is dropped. We evaluated the proposed DropKernel strategy by the object classification performance on CIFAR10 in comparison to conventional Dropout and DropConnect methods, and showed improved performance of the proposed method.

Keywords: Convolutional neural networks · Dropout · Object recognition

1 Introduction

Artificial neural network (ANN) models are suitable to dealing with large scale dataset because the model complexity can be simply extended by adding more layers or number of nodes in the layer, and Dropout [1] and DropConnect [2] are very simple but effective methods to deal with overfitting problems. To handle large-scale images, Convolution neural network (CNN) has been proposed [3], but Dropout and DropConnect are not applicable because they were designed for fully connected layer.

In this paper, we propose novel dropout methods CNNs by redesigned Dropout and DropConnect. We apply Dropout and DropConnect on 2-dimensional representation; in other words, we change the unit of Dropout and DropConnect to be suited to convolutional layers. As the original Dropout and DropConnect methods can deal with 0D value in 1D-input and 1D-output layer like fully connected layer, our proposals deal with 2D value in 3D-input and 3D-output layer such as 2D convolution layer.

This paper is organized as follows: we briefly interpret Dropout and DropConnect in Sect. 2. And we describe the details of proposal in Sect. 3, followed by experiment and result in Sect. 4. In Sect. 5, we analyze experiment results and summarize them.

W. Karwowski and T. Ahram (Eds.): IHSI 2019, AISC 903, pp. 223–227, 2019.
https://doi.org/10.1007/978-3-030-11051-2_34

2 Related Work

2.1 Dropout

Dropout was proposed by Hinton [1] for generalization method in fully connected layer. Dropout keep or remove each element of layer's output with probability constant p. Many other papers and extensive experiment successfully show its generalization effect. In general fully connected layer, the output can be described as follow:

$$\boldsymbol{y} = \boldsymbol{W}\boldsymbol{x} \tag{1}$$

where $\boldsymbol{y}$ and $\boldsymbol{x}$ denotes current and previous layers' output of size $n \times 1$ and $m \times 1$, and $\boldsymbol{W}$ denotes weight parameters of size $m \times n$. When dropout is applied to this layer, we can write Eq. 1 with binary mask $\boldsymbol{M}$ as:

$$\boldsymbol{y} = \boldsymbol{M} \otimes (\boldsymbol{W}\boldsymbol{x}) \tag{2}$$

where $\otimes$ is element wise product. With some activation function such relu and tanh and the property that $f(0) = 0$, Dropout can be applied both input and output of activation.

2.2 DropConnect

DropConnect was proposed by Wan [2], which is generalization of dropout. DropConnect works similarly with Dropout, but it is applied to weight parameters, rather than output of layer. So the size of binary mask in DropConnect is the same as the size of the weight parameters. When it is applied to fully connected layer, the output can be described as:

$$\boldsymbol{y} = (\boldsymbol{M} \otimes \boldsymbol{W})\boldsymbol{x} \tag{3}$$

The reason why DropConnect is generalization of dropout is that, from Eqs. 2 and 3, DropConnect can cover more cases than Dropout. Dropout drops a node directly, which is equal to drop all of connections linked to that. But if only a part of connections linked to same node is dropped, Dropout cannot represent this situation.

3 Redesigned Dropout and DropConnect

3.1 Convolution with Conventional Dropout

Conventional drop methods work on 0-dimension, each single mask drop a single value. When they are applied to convolution layer, only some part of channels or kernel will be dropped like Fig. 2. But in CNN, a channel has more information than a single value in FCN, such as spatial information. So dropping individual value in CNN equals adding random noise, not generalizing. Instead we changed the basic unit of drop method. In other word, we redesigned drop methods by changing dimensionality of them (Fig. 1).

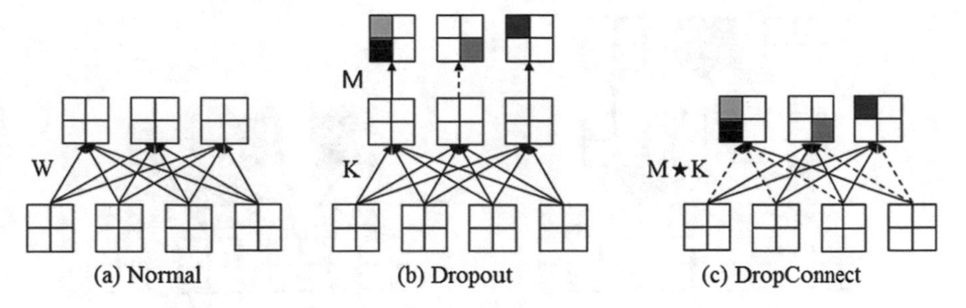

Fig. 1. Conventional Convolution with Drop methods.

3.2 Expansion to 2D

Fully connected layer is usually described as vector-in and vector-out operation. In this form, every nodes and connections' weight parameter are zero dimensional value. Also convolution layer can be described like this zero dimensional form, even though it is not a vector-in and vector-out operation. In this case it has sparse connection than fully connected one and shows weight sharing with different shape on different weights. It is the reason the Drop methods are not suitable to convolution layer; they were designed to work in dense connections, not in spares connection.

But if we bind some nodes which are in same channel, the connections between bundles become dense. where a node represents a channel, and a connection between channels represents a kernel. And these values can be described in 2D as:

$$\boldsymbol{y}_j = \boldsymbol{W}_{ij} * \boldsymbol{x}_i \tag{4}$$

where * denotes convolution operation, and $\boldsymbol{W}_{ij}$ represent a 2-dimensional kernel between input channel i and output channel j. Then we adapt Dropout and DropConnect methods over it. Whit expanded dimension, Drop methods work little differently. In 2D, a node is extended to a channel. Like it, a weight parameter is extended to a convolution kernel. Then, Dropout randomly selects channels to deactivate as:

$$\boldsymbol{y}_j = \boldsymbol{M}_{ij} * (\boldsymbol{W}_{ij} * \boldsymbol{x}_i) \tag{5}$$

where $\boldsymbol{M}_{ij}$ denotes binary mask of channel j. Likewise, DropConnect randomly selects kernels to drop as:

$$\boldsymbol{y}_j = (\boldsymbol{M}_{ij} * \boldsymbol{W}_{ij}) * \boldsymbol{x}_i \tag{6}$$

where $\boldsymbol{M}_{ij}$ denotes binary mask for a kernel between input channel i and output channel j In this form, each channel is dropped independently.

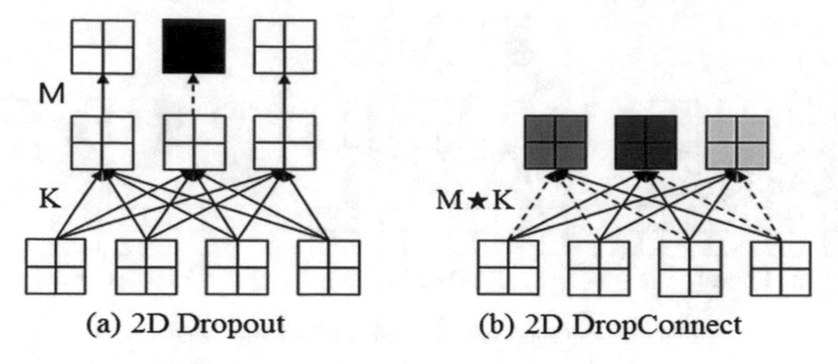

Fig. 2. Dropout and DropConnect comparison in 2D space.

4 Experiments

To analyze the effect of proposal methods, we implemented CNN models with and without them, and compare the classification accuracy of each model on CIFAR-10 dataset. We used WRN-10-10, kind of Wide ResNet [4], to maximize the effect of drop method. This model has only 10 convolution layers, but has 10 times more kernels than ResNet so Drop methods affect to training well. Also we used bacth normalization [5] and L2 normalization for higher test accuracy. All the results are averages of three independent models with same architecture. We built total 4 WRN-10-10 models and trained them in Tensorflow [6]. One is original model for baseline, 2 of others are original Dropout and DropConnect adapted, and the others are proposal Dropout and DropConnect adapted as Fig. 4. Dropout and DropConnect are only used to first convolution layer in every residual block as Fig. 3. We used momentum optimizer with 0.9 of momentum and 0.1 of initial learning rate. We trained the models on Cifar10 dataset for 100 epochs, with dividing learning rate by 10 at every 30, 60 and 80 epochs.

The result of CIFAR-10 classification test is described on Table 1. Most of all, proposal methods show better accuracy than conventional methods. The gap between conventional and proposal Dropout is larger than between the others. Which means, our modification is more suitable for Dropout than DropConnect.

Table 1. Test accuracy on CIFAR-10 database.

Drop Method	Accuracy
None	91.48%
Dropout	93.44%
Proposed Dropout	93.56%
DropConnect	93.32%
Proposed DropConnect	93.08%

5 Conclusion

In this paper, we propose redesigned Dropout and DropConnection for convolution layer and compare these with conventional ones. We modify binary mask units of Dropout and DropConnection from 0D to 2D, which is suitable to convolution operation. The experimental results show that the proposed modification successfully improves Dropout and DropConnection to be suitable to convolution layers. Future work includes generalization of the proposed methods.

Acknowledgments. This work was supported by Institute for Information and communications Technology Promotion (IITP) grant funded by the Korea government (MSIT) (No. R7124-16-0004, Development of Intelligent Interaction Technology Based on Context Awareness and Human Intention Understanding, 50%) and by the National Research Foundation of Korea (NRF) grant funded by the Korea government (MSIP) (No. NRF-2017M3C1B6071400).

References

1. Srivastava, N., Hinton, G.E., Krizhevsky, A., Sutskever, I., Salakhutdinov, R.: Dropout: a simple way to prevent neural networks from overfitting. J. Mach. Learn. Res. **15**(1), 1929–1958 (2014)
2. Wan, L., Zeiler, M., Zhang, S., Cun, Y. L., Fergus, R.: Regularization of neural networks using DropConnect. In: the 30th International Conference on Machine Learning (ICML 2013), pp. 1058–1066 (2013)
3. Krizhevsky, A., Sutskever, I., Hinton, G.E.: ImageNet classification with deep convolutional neural networks. In: Advances in Neural Information Processing Systems, pp. 1097–1105 (2012)
4. Zagoruyko, S., Komodakis, N.: Wide residual networks. arXiv:1605.07146 (2016)
5. Ioffe, S., Szegedy, C.: Batch normalization: accelerating deep network training by reducing internal covariate shift. In: the 32nd International Conference on Machine Learning (ICML 2015) (2015)
6. Abadi, M., Agarwal, A., Barham, P., et al.: TensorFlow: large-scale machine learning on heterogeneous distributed systems. arXiv:1603.04467 (2016)

Challenge of Tacit Knowledge in Acquiring Information in Cognitive Mimetics

Pertti Saariluoma[1(✉)], Antero Karvonen[1], Mikael Wahlstrom[2], Kai Happonen[2], Ronny Puustinen[1], and Tuomo Kujala[1]

[1] Cognitive Science, University of Jyväskylä, Jyväskylä, Finland
ps@jyu.fi
[2] VTT Technology Research Center, Vuorimiehentie 3, Box 1000, Espoo, Finland

Abstract. Intelligent technologies are rising. This is why methods for designing them are important. One approach is to study how people process information in carrying out intelligence demanding tasks and use this information in designing new technology solutions. This approach can be called cognitive mimetics. A problem in mimetics is to explicate tacit or subconscious knowledge. Here, we study a combination of thinking aloud in ship simulator driving and focus group commenting the solutions of subjects. On the ground of these early experiments, a multiple method combination seems to be the best way forward to solve problems of tacit or subconscious knowledge.

Keywords: Design science · HTI · Mimetics · AI

1 Introduction

Intelligent technologies are in focus. Robotics, Artificial intelligence, cognitive automation or technologies and autonomous systems are under intensive development [1, 2]. The unifying factor of these new technical openings is the systems' capacity to carry out intelligence demanding tasks. Consequently, human technology interaction specialists have to meet the problem of how to design technologies with intelligent capacities.

The first serious example of a machine that could take human role in intelligence requiring task was presumably Turing machine [3]. Of course, computing machines were not new, but it was possible for a universal Turing machine to be programmed to perform very different intelligent processes from mathematics to chess. The Turing machine created a frame which could be applied in very different types of intellectual tasks [3].

The next step forward was invented by Herbert Simon and his colleagues [4]. They understood that Turing's model of the human mind was too intuitive [3]. It was possible, for example, to have unlimited number of machines which could solve some problem. So, it was impossible to tell which one of these processes would be the true model of the human mind. Thus, it was a good idea to study how people really processed information.

W. Karwowski and T. Ahram (Eds.): IHSI 2019, AISC 903, pp. 228–233, 2019.
https://doi.org/10.1007/978-3-030-11051-2_35

It was important to adopt the empirical research methods of modern psychology of thinking and thus study how people process information when they carry out intelligent tasks such as chess playing [4]. The empirical research in human thinking enabled on the one hand researchers to study the best possible models for programming intellectual machines and on the other to think of the limits of computational machines as models of the human mind.

The intimate relations of computational models and human thinking has been problematic and under intensive research for decades [1]. Recent advancement in developing intelligent technologies brings a new aspect to this discussion. Modern intelligent machines are designed to take care of tasks, which have earlier been done by people. Consequently, to improve their design, designers have to understand how people process information in such tasks.

2 Multiple Realizability and Cognitive Mimetics

For several thousands of years, people have been able to steer boats and ships from one place to another because of their sufficient cognitive capacities. They have been able to perceive and categorize things, create spatial mental representations, remember these representations, and manipulate mental representations or think. Consequently, ships have found right places, thanks to human information processing capacities.

People have been components in shipping (and also in other machine systems) because of their capacity to process information [4]. No ape, bear, or ant, for example, would be able to steer ships from one port to another. Thus, human information processing has made seafaring possible. However, today, it is possible to construct machines that can take care of many parts of these complex steering tasks with minimum human involvement, and, for this reason, a new goal for designers has emerged, which is to design machines that can replace people in tasks requiring human-like information processing.

A prerequisite for successful replacement of human work in intelligence demanding tasks is that machines can take care of the same things that people do when they take care of these same tasks. Intelligent machines need not do the same thing in the same way as people do. The main thing is to get the same outcome. Chess playing machines do not think like people, but they perform as well as or even better than people do.

Turing machines illustrate quite well that it is possible to carry out same information processing tasks in different ways. Different physical systems can carry out same tasks. People can drive industrial trucks and move goods from one place to another, but this can be done by automatic systems also. Thus, one can carry out intelligence demanding tasks with different types of physical "platforms". This property of intelligent performance can be called *multiple realizability*.

Originally, multiple realizability referred to fact that human brains need not be in precisely the same physical state to represent something. Different physical states can carry out have same information states. The idea can be generalized over many types of physical objects and information states. Thus, computers can solve the same mathematical problems as people can. Such a general multiple realizability is the ground idea of intelligent technologies.

The generalized multiple realizability has an important consequence. Because people have been able to process information and because information processes can be realized by different physical systems, it is natural to ask if we can use human information processing as a model for designing intelligent machines. In fact, this was what Turing did. He used his intuitive idea about how mathematicians compute to model their minds. Similarly, modern AI and autonomy designers could mimic human information processes in constructing intelligent artifacts. Such approach to intelligence design can be called cognitive mimetics. Here we study how to develop the methods of cognitive mimetics, and for this reason our focus in is methodology of knowledge acquiring rather than results. The latter can be reported separately.

3 The Acquisition of Knowledge

Here, we made a two-stage design simulation experiment to analyze how to collect tacit information. In the first stage, we made a simulator driving experiment with thought aloud protocols [5]. In the latter stage, we asked sea captains as a focus group to comment the presented driving solutions. Thus, we could get additional information about subconscious information processing.

3.1 Simulation Method

The first part of our study investigated the use of simulator in collecting knowledge about subjects thinking.

Data Collection and Analysis: Our chosen method aims at analyzing how captains think when steering a ship. Simulator setting was applied data collection, because, in contrast to real at-the-sea setting it allows to, controlled collecting data in challenging tasks.

Subjects: The study subjects were experienced maritime pilots (highly experienced ship-handlers, former ship captains) (n = 6) and other professionals (n = 2), all being high-grade licensed mariners. They had on average 20 years of experience (between 4 and 40 years) of ship steering.

Equipment: The simulator features a 180-degree view from the bridge, playing a 3D simulation of the environment (Fig. 1). Other sources of information were a simulated radar, an electronic chart display system (ECDIS), and a screen displaying information of the ship, such as speed, rudder angle, bearing, etc. The ship was controlled by a console which had a mouse operated (heading) autopilot, manual rudder, throttle, and front propeller. The console also held a conference telephone used for simulating radio communications. The experiments were recorded using two video cameras, one placed behind the subject and the other to his or her left. The audio recordings collected by the cameras were transcribed into text.

Fig. 1. Simulation environment

Tasks: After an introduction to the system and a chance to rehearse for a while in the simulator, the participants were instructed to navigate the ship to the harbor (and back during the second run) and informed that VTS (vessel traffic service) would be (unrealistically) quiet but would answer when asked. The participants were asked to voice their thoughts as they observed their environment and navigated the ship and were reminded to do so during the experiment if they forgot to speak.

The data collection and the simulator scenarios were designed by a multidisciplinary team consisting of engineers, maritime professionals (including a maritime pilot), and behavioral experts (with background in social psychology and cognitive sciences). The main idea was to simulate a challenging but not unrealistically difficult journey. The setting chosen was a specific real-life Finnish harbor (that of Vuosaari). The simulations took 20–30 min one way (Fig. 2).

3.2 Results – The Ontology of Steering

Description of human thinking is normally qualitative. Time and other numerical parameters do not normally give a clear idea about what people think and, for this reason, the most natural way of investigating thinking is to concentrate on the information contents of the thoughts. A good way of expressing is to use ontologies [6, 7]. Ontologies provide frameworks in which not only the contents of a particular domain of knowledge but also the conceptual lenses for exploration and research can be expressed. Thus, the main outcome of our empirical work here is ontology for describing thinking of steering captains. This ontology contains: (1) situational information; (2) chosen action; and (3) why the action was chosen. The last component was called explanation.

On the ground of protocol analysis, we abstracted an ontological schema which gives an idea about the content structure of steering a ship through a regatta.

On the ground of protocols we suggested an ontology of three categories (Fig. 3).

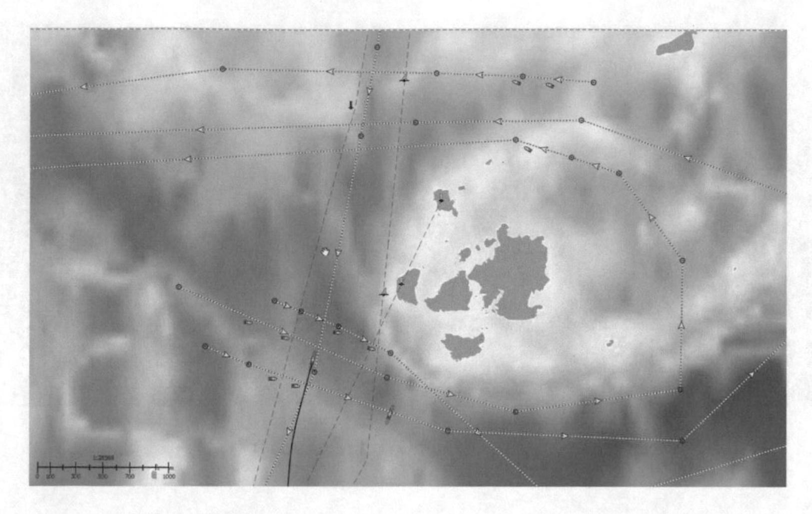

Fig. 2. The regatta problem: the ship of the participant can be seen with a black line trailing behind it. The ship is in the middle of crossing the regatta which can be seen around the ship to the front, back, and right and left (six in total). The route of the sailing boats can be seen circling the island, and was designed to re-cross the path of the participants? vessel again from the right

INFORMATION ACQUISITION	ACTION	EXPLANATION

Fig. 3. Ontology of steering

The analysis of protocols lead to a three component model of steering activity: (1) Information acquisition (2) The actual steering actions. (3) Finally, the explanations i.e., the reasons and justification subjects can give to their actions.

3.3 Focus Group Method

In the second part, we asked five experienced sea farers to tell by means of forms with the given ontology how they would operate in the task situation. One of them was a pilot, one was engineer, three were captains or co-pilots. All of them had over 10 years of professional experience. Our main methodical problem was, if the presented ontology can be used in collecting additional knowledge.

3.4 Results

The main additional information with focus groups was explication of reasons. As they were specifically asked in the form subjects were explicit with them. All the presented actions were motivated. Reasons mostly concerned the control of the ships, but there were also a number of ethical and economic reasons such as avoidance of lost life or damage, or avoidance of breaking the ship engine. Clearly, it makes sense to ask explicitly people for reasons for their actions.

4 General Discussion

Our example illustrates how designers must meet the problem of tacit or subconscious or tacit information. Subconscious processes are everyday life in clinical counseling. People are not aware of what happens in their mind. For this reason, clinical processes often take long times. Logically, one would think that applying multiple methods in combination and thus investigating human information processing in several ways and from several points of view could be used to get a clearer picture. Consequently, we decided to pilot a multimethod approach to improve human information processing. The target information process is autonomous shipping.

Our pilot study focusses on only one problem in designing intelligent technologies, which is the problem of tacit or tacit or subconscious knowledge. In any case, it is vital to use multiple information collection methods in order to get a complete picture of how people process information. Here, we used a focus group to get extra information, but it is not the only method. The main goal is to get as complete a picture of information as is possible and the reasons why people act as they do.

In sum: The investigation of human information processing and thinking provides an important source of knowledge for developing intelligent technical systems. A problem in this research is tacit or subconscious knowledge, which is not available to experts themselves though it affects their thinking. One can explicate tacit or subconscious knowledge by means of a multimethod approach and present the results by means of action ontologies [7]. They can be used to collect and communicate results.

References

1. Boden, M.: Artificial Intelligence and Natural Man. Basic Books, New York (1988)
2. Saariluoma, P.: Four challenges in structuring human-autonomous systems interaction design processes. In: Williams, A., Sharre, P. (eds.) Autonomous Systems. NCI, The Hague (2015)
3. Turing, A.M.: On computable numbers, with an application to the entscheidungs problem. In: Proceedings of the London Mathematical Society, vol. 42, pp. 230–265 (1936–1937)
4. Newell, A., Simon, H.: Human Problem Solving. Prentice-Hall, Englewood Cliffs (1972)
5. Ericsson, K.A., Simon, H.A.: Protocol Analysis. MIT-press, Cambridge (1993)
6. Chandrasekaran, B., Josephson, J.R., Benjamins, V.R.: What are ontologies and why do we need them? IEEE Intell. Syst. **14**, 20–26 (1999)
7. Saariluoma, P., Cañas, J., Leikas, J.: Designing for Life. PalgraveMacmillan, London (2016)

Population Healthcare AI (PopHealthAI)—The Role of Geospatial Infused Electronic Health Records in Creating the Next Generation Preventive HealthCare

Chandrasekar Vuppalapati[1(✉)], Anitha Ilapakurti[1], Sharat Kedari[1], Rajasekar Vuppalapati[1], Jayashankar Vuppalapati[2], and Santosh Kedari[2]

[1] Hanumayamma Innovations and Technologies, Inc., Fremont, CA, USA
{cvuppalapati, ailapakurti, sharat, raja}@hanuinnotech.com

[2] Hanumayamma Innovations and Technologies Private Limited, Hyderabad, India
{jaya.vuppalapati, skedari}@hanuinnotech.com

Abstract. Geospatial data is a location-specific data. The data contains natural geographical markers and man-made changes. For instance, natural markers include geolocation perimeter and man-made changes include global warming trends & pollution indexes. We strongly suggest that interweaving geospatial data, especially pollution index, with outpatient electronic health records can lead into detection of critical health markers and can make it possible to shift from reactive to preventive health care, thereby saving billions of dollars worldwide and improve overall health outcomes to outpatients. In this research paper, we propose an integration of geospatial data with the EHR and aim to solve one of the most important issues in outpatient healthcare "on-set of life-threatening diseases due to changes in geospatial". Finally, the paper presents a prototyping solution design as well as its application and certain experimental results.

Keywords: Electronic health records · Geospatial · Asthma attack
Preventive healthcare · Sanjeevani electronic health records · Outpatient

1 Introduction

1.1 Asthma and Outdoor Air Pollution

Asthma is a serious and sometimes life-threatening chronic respiratory disease that affects almost 25 million Americans and costs the nation $56 billion per year [1]. Every year more than 2 million emergency department visits [2] are due to asthma related

W. Karwowski and T. Ahram (Eds.): IHSI 2019, AISC 903, pp. 234–240, 2019.
https://doi.org/10.1007/978-3-030-11051-2_36

disease and each ED visit would cost $1502.[1] The Geospatial Determinants of Health Outcomes Consortium[2] (GeoDHOC) study has investigated and clearly concluded that intra-urban air quality variations are related to adverse respiratory events in Detroit, Michigan, USA and Windsor, Ontario, Canada. As part of the study NO2, SO2, and volatile organic compounds (VOCs) were measured at 100 sites, and particulate matter (PM) and polycyclic aromatic hydrocarbons (PAHs) at 50 sites for two 2-week sampling periods in 2008 and 2009 [1]. The Unites States Environmental Protection Agency (EPA) advises[3] that "Air pollution can make asthma symptoms worse and trigger attacks". The Fig. 1 lists asthma capitals of USA.

These cities have the highest asthma-related emergency department visits:

Emergency Department Visits Ranking	Metropolitan Area	Overall Asthma Capital National Ranking
1	Springfield, MA (highest in U.S.)	1
2	Virginia Beach, VA	22
3	Omaha, NE	13
4	Dayton, OH	3
5	Greensboro, NC	9
6	Richmond, VA	2
7	Youngstown, OH	7
8	Winston-Salem, NC	17

Source: Asthma and Allergy Foundation of America asthmacapitals.com

Springfield, MA

Fig. 1. Asthma capitals [2]

It's evidently clear that asthma is costly and more prevalent diseases across the nation and has affected millions of Americans. The best antidote for asthma is prevent triggers of asthma attacks. The triggers could vary from person to person but on a macro level the air quality & pollution indexes dominate. Common asthma triggers include[4]: Environmental Tobacco Smoke, Dust Mites, Outdoor air pollution, Cockroach Allergan, Pets, and Mold [3].

One can control their asthma [4] provided take preventive measures. For instance, Environmental Tobacco Smoke trigger can be mitigated by not smoking or by not exposing to second hand smoke. Similarly, Dust Mites bugs are every home [3] and one way to prevent the trigger is to use mattress covers and pillowcase covers to make a barrier between dust mites and yourself. The only trigger that is dynamic and have least control upon is air pollution. Air pollution can change anytime, or air quality could change due to seasonal and geolocations. One way of mitigating the risk is pay

[1] Asthma capitals 2018 - http://www.aafa.org/asthma-capitals-emergency-department-visits/.

[2] Geospatial relationships of air pollution and acute asthma events across the Detroit–Windsor international border: Study design and preliminary results - https://www.nature.com/articles/jes201378.

[3] EPA - https://www3.epa.gov/airnow/asthma-flyer.pdf.

[4] You can control your asthma - https://www.cdc.gov/asthma/pdfs/asthma_brochure.pdf.

attention to air quality forecasts on radio, television, and the Internet and check your newspaper to plan your activities for when air pollution levels will be low.

The other way is your Electronic Health Records(EHR) provide timely, geolocation & contextual based recommendations that can work for you to eliminate asthma triggers permanently. We propose an intelligent Population AI that works for us, Humans. Simply put, EHR triggers notifications to Outpatients whenever there is change in air quality.

1.2 Air-Quality Data

The following air quality parameters data available and data of pollutants can be integrated to EHR (Fig. 2):

- (CO, Pb, NO2, Ozone, PM10, PM2.5, and SO2)
- PM2.5 Chemical Speciation Network monitors
- IMPROVE (Interagency Monitoring of PROtected Visual Environments) monitors
- NATTS (National Air Toxics Trends Stations)
- NCORE (Multipollutant Monitoring Network)
- Nonattainment areas for all criteria pollutants

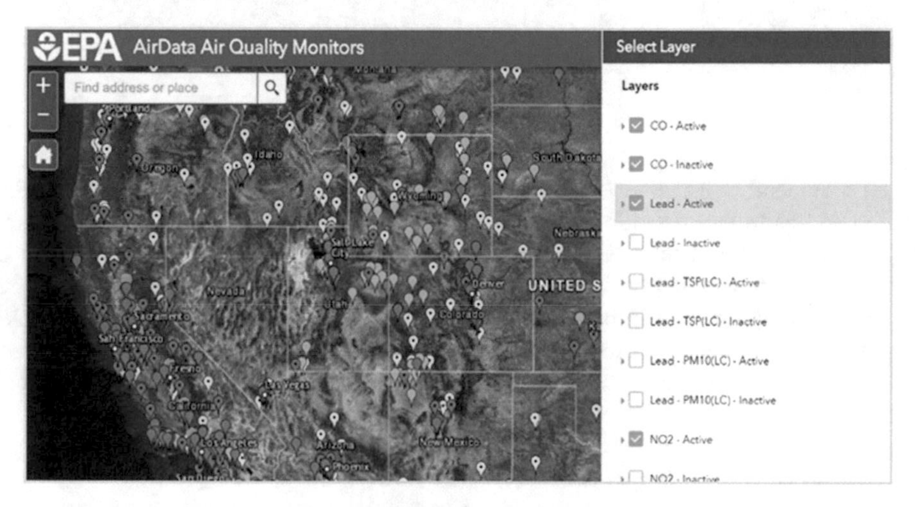

Fig. 2. EPA air quality data (Air Quality Data - https://epa.maps.arcgis.com/apps/webappviewer/index.html?id=5f239fd3e72f424f98ef3d5def547eb5&extent=-146.2334,13.1913,-46.3896,56.5319)

The structure of this paper is presented as: Sect. 2 discusses the basic concepts and methods about Machine Learning Algorithms. Section 3 presents our Population AI architecture, and Sect. 4 shows a case study.

2 Understanding Machine Learning Algorithms Population AI

2.1 Geo Location Data

In order to recommend air quality to asthma recommendations, we need to get Geo location of the User. In this regard, we can use Android Maps API and iOS Map Kit. We can apply forward Geocoding and reverse Geocoding. Reverse-geocoding requests take a latitude and longitude value and find a user-readable address. Forward-geocoding requests take a user-readable address and find the corresponding latitude and longitude value [5] (Table 1).

Table 1. Geo location APIs (iOS & Android).

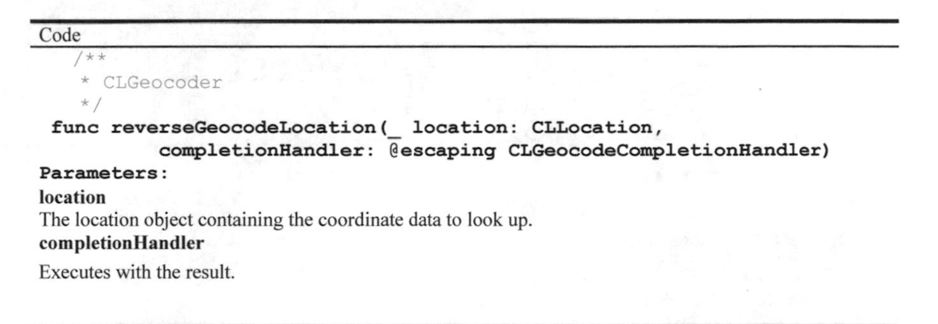

Code

```
/**
 * CLGeocoder
 */
func reverseGeocodeLocation(_ location: CLLocation,
        completionHandler: @escaping CLGeocodeCompletionHandler)
Parameters:
```

location

The location object containing the coordinate data to look up.

completionHandler

Executes with the result.

2.2 Clustering

To recommend based on the locations, Geo clustering is needed. Clustering could be Hierarchical (agglomerative or Divisive). This is achieved through Similarity or K-Means clustering (Fig. 3).

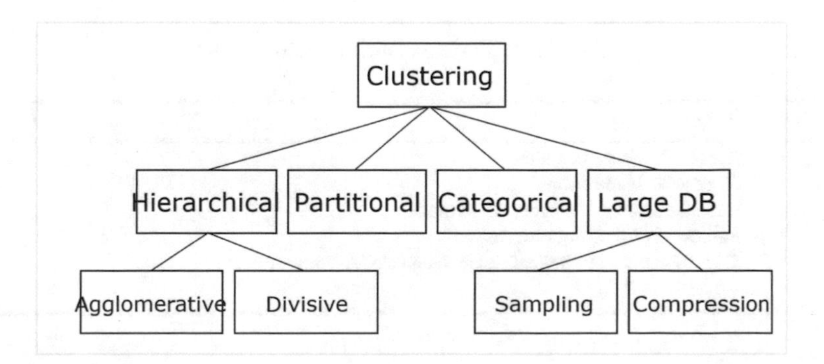

Fig. 3. Clustering [6]

3 Architecture

The system architecture consists of four major parts: (1) Mobile Layer (2) Cloud EHR Layer, (3) Real-Time Geolocation Data, and (4) Recommendation System.

Mobile Layer. The Mobile Layer Uploads Geo Location details of the User periodically. When there is significant change in the User location, the most recent Latitude/Longitude values are posted to Cloud Layer (Fig. 4).

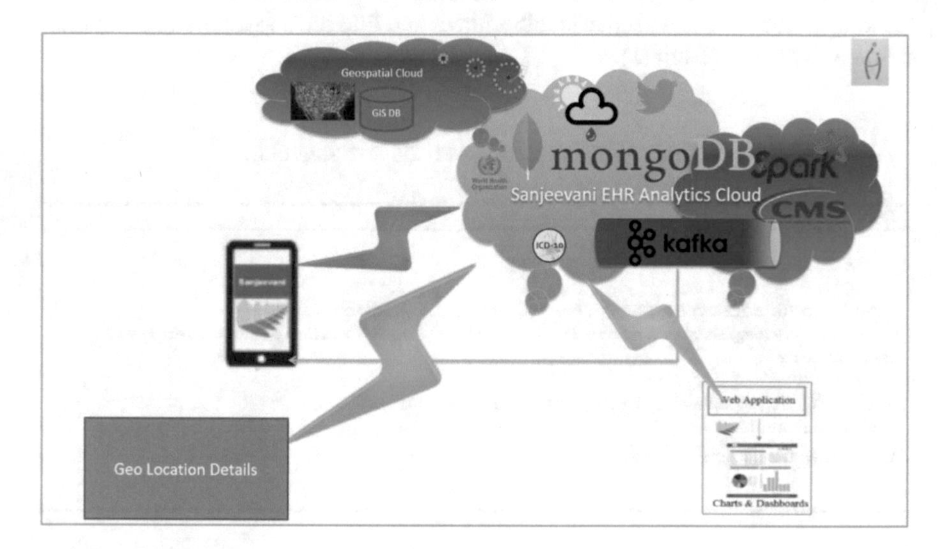

Fig. 4. Architecture

Cloud Layer. The location analytics and combination of Electronic Health Records data with air quality mapping is performed in the Cloud layer. Cloud layer only computes air pollution to Outpatient if the user exhibits high risk stratification for Asthma (Table 2).

Table 2. High risk stratification to asthma

Stratification Rules
• Recent Asthma Encounters • High Risk Medication • Recent changes in country of domicile • Fall events for senior citizens • Twitter analysis (* indicating prevalence of Asthma disease)

Real-Time Integration with Air Quality Databases. The systems constantly get real-time data of air quality of the User locations. Based on the location air quality parameters, the system calculates the risk score for any asthma event. The Data can be retrieved from EPA Database[5].

The Carbon Monoxide Data attributes include (Fig. 5):

Date	AQS_SITE_	POC	Daily Max 8-hour CO Concentration	UNITS	DAILY_AQ	DAILY_OB	PERCENT_	AQS_PARAMETER_CODE	AQS_PARAMETER_DESC	CBSA_CO	CBSA_NAME
1/1/2017	60010009	1	0.3	ppm	3	18	75	42101	Carbon monoxide	41860	San Francisco-Oakland-Hay
1/2/2017	60010009	1	0.3	ppm	3	24	100	42101	Carbon monoxide	41860	San Francisco-Oakland-Hay
1/3/2017	60010009	1	0.4	ppm	5	24	100	42101	Carbon monoxide	41860	San Francisco-Oakland-Hay
1/4/2017	60010009	1	0.4	ppm	5	24	100	42101	Carbon monoxide	41860	San Francisco-Oakland-Hay
1/5/2017	60010009	1	0.7	ppm	8	24	100	42101	Carbon monoxide	41860	San Francisco-Oakland-Hay
1/6/2017	60010009	1	0.5	ppm	6	24	100	42101	Carbon monoxide	41860	San Francisco-Oakland-Hay
1/7/2017	60010009	1	0.4	ppm	5	24	100	42101	Carbon monoxide	41860	San Francisco-Oakland-Hay
1/8/2017	60010009	1	0.4	ppm	5	24	100	42101	Carbon monoxide	41860	San Francisco-Oakland-Hay
1/9/2017	60010009	1	0.4	ppm	5	24	100	42101	Carbon monoxide	41860	San Francisco-Oakland-Hay
1/10/2017	60010009	1	0.4	ppm	5	24	100	42101	Carbon monoxide	41860	San Francisco-Oakland-Hay
1/11/2017	60010009	1	0.4	ppm	5	24	100	42101	Carbon monoxide	41860	San Francisco-Oakland-Hay
1/12/2017	60010009	1	0.4	ppm	5	24	100	42101	Carbon monoxide	41860	San Francisco-Oakland-Hay
1/13/2017	60010009	1	0.9	ppm	10	24	100	42101	Carbon monoxide	41860	San Francisco-Oakland-Hay
1/14/2017	60010009	1	0.7	ppm	8	24	100	42101	Carbon monoxide	41860	San Francisco-Oakland-Hay
1/15/2017	60010009	1	0.7	ppm	8	24	100	42101	Carbon monoxide	41860	San Francisco-Oakland-Hay
1/16/2017	60010009	1	0.6	ppm	7	24	100	42101	Carbon monoxide	41860	San Francisco-Oakland-Hay
1/17/2017	60010009	1	0.6	ppm	7	24	100	42101	Carbon monoxide	41860	San Francisco-Oakland-Hay
1/18/2017	60010009	1	0.5	ppm	6	24	100	42101	Carbon monoxide	41860	San Francisco-Oakland-Hay
1/19/2017	60010009	1	0.4	ppm	5	24	100	42101	Carbon monoxide	41860	San Francisco-Oakland-Hay
1/20/2017	60010009	1	0.3	ppm	3	24	100	42101	Carbon monoxide	41860	San Francisco-Oakland-Hay
1/21/2017	60010009	1	0.3	ppm	3	24	100	42101	Carbon monoxide	41860	San Francisco-Oakland-Hay

Fig. 5. CO data for California [7]

Recommendation System. The purpose of the recommendation system is to cluster Outpatient with GEO location and provide notifications on the mobile about changes in the air quality index.

4 A Case Study

This paper presented a novel approach Population Healthcare AI (PopHealthAI) - the role of geospatial infused Electronic Health Records in creating the next generation Preventive HealthCare. We staunchly believe that integrating geospatial index with EHR would provide new pathways to preventive healthcare, one of the most important being Asthma.

We strongly believe that Population Healthcare AI will not only reduce the cost factor for outpatients but also saves the lives.

[5] EPA Daily Data Download - https://www.epa.gov/outdoor-air-quality-data/download-daily-data.

References

1. Barnett, S.B., Nurmagambetov, T.A.: Costs of asthma in the United States: 2002–2007. https://www.ncbi.nlm.nih.gov/pubmed/21211649
2. Asthma and Allergy Foundation of America, Asthma Capitals 2018: Asthma-Related Emergency Department Visits. http://www.aafa.org/asthma-capitals-emergency-department-visits/, http://www.aafa.org/asthma-capitals-emergency-department-visits/
3. Center for Disease Control (CDC), "Common Asthma Triggers". https://www.cdc.gov/asthma/triggers.html
4. Air Quality Monios. https://epa.maps.arcgis.com/apps/webappviewer/index.html?id=f239fd3e72f424f98ef3d5def547eb5&extent=-146.2334,13.1913,-46.3896,56.5319
5. Apple, Core Location Service. https://developer.apple.com/documentation/corelocation/clgeocoder
6. Leskovec, M.J., Rajaraman, A., Ullman, D.: Mining of Massive Datasets. ISBN 9781107077232
7. Han, J.: Data Mining: Concepts and Techniques (2000). ISBN-13 978-0123814791

Correlation of Driver Head Posture and Trapezius Muscle Activity as Comfort Assessment of Car Seat

Alberto Vergnano[1(✉)], Francesco Pegreffi[2], and Francesco Leali[1]

[1] Department of Engineering Enzo Ferrari, University of Modena and Reggio Emilia, Via P. Vivarelli, 10, 41125 Modena, Italy
{alberto.vergnano, francesco.leali}@unimore.it
[2] School of Pharmacy, Biotechnology and Motor Sciences, University of Bologna, Via dei Mille, 39, 47921 Rimini, Italy
f.pegreffi@unibo.it

Abstract. Car design must very care comfort and driving pleasure. Nonetheless, the design choices are tested with subjective evaluations. In the present research, an objective measurement equipment for driving comfort assessment is proposed. The muscles activity of the driver in different maneuvers is considered the gauge of her/his feeling with the car. The activity of trapezius muscles of both shoulders is monitored by electromyography (EMG), through electrodes applied to her/his skin. The driver posture is monitored with a robust device for head tracking, using two 9-axis orientation sensors, including gyroscope. Real driving experiments are performed both with a luxury SUV and a high end car. As expected, the first resulted more comfortable. The proposed equipment proved to be effective in assessing the driving comfort for different seat designs and car layouts.

Keywords: Comfort · Driveability · Electromyography · Head tracking

1 Introduction

Driving comfort is a key factor for marketability of a cars, especially for Euro segments C and uppers. Research investigated comfort more as long driving fatigue, rather than car driveability or driving pleasure, [1]. However, a customer usually evaluates a car for making the final decision about its purchase or not with a more or less short test drive. Long before, the car manufacturer assessed the possible design choices affecting car driveability with subjective evaluations by their experienced test drivers, thus hazily considering the actual expectations of average final customers. The objective measurement of driving comfort would better assess the car development process.

Higher driving comfort is perceived in case of easy handling the driving wheel and maneuvering the car. The muscles response to their nerves stimulation can be monitored through electromyography (EMG), which is commonly used in patient rehabilitation, [2]. The electrodes must applied to the test driver skin. Then the EMG oscilloscope displays information about the muscles activation during an exercise by measuring their electrical activity.

W. Karwowski and T. Ahram (Eds.): IHSI 2019, AISC 903, pp. 241–247, 2019.
https://doi.org/10.1007/978-3-030-11051-2_37

Driver upper body and head can be monitored through artificial vision and machine learning techniques, but with limited robustness to illumination and pose changes during driving maneuvers. Seat pressure monitoring enables to estimate the body gross motions, but not the fine ones for neck and head, [3]. A robust head tracking system can be achieved with a sensor device, [4].

The present research aims at quantitatively monitoring the muscle activity in relation with driver posture in driving maneuvers. The activity of trapezius muscles of both shoulders is assumed as predicting factor of car seat comfort. A head tracking system, triggered by an output signal of the EMG equipment, monitors the driver movements relative to the moving car.

The paper is organized as follows. Section 2 presents the electromyographic measurements, while its integration with the head tracking device is discussed in Sect. 3. Preliminary results of driving experiments are reported and discussed in Sect. 4, while the concluding remarks are drawn in Sect. 5.

2 Trapezius Muscles Activity Monitoring

The muscles activity is monitored with SHoW Motion 3D kinematic tracking system (NCS Lab®, Carpi, Italy). This system enables to measure the joint movements and the related muscles activity through the synchronization of the signals from superficial Wave Plus Wireless EMG (Cometa®, Bareggio, Italy) and WISE Magnetic Inertial Measuring Unit (NCS Lab®, Carpi, Italy). The system implements a biomechanical model based on the ISEO protocol, [5]. The anatomical coordinate system is defined acquiring a static reference trial with the subject standing upright, the humerus positioned alongside the body and the elbow flexed at 90°. Several repeatability studies showed the consistency of data provided by this protocol, [6].

The EMG signals are sampled at 2000 Hz with 16 bits of resolution. Two surface electrodes Ag/AgCl, 24 mm diameter ARBO Kendall (Covidien®, Gosport Hampshire, United Kingdom), are placed in single differential configuration over the fibers of both superior trapezius muscles, at 2 cm distance, as shown in Fig. 1. The electrodes are

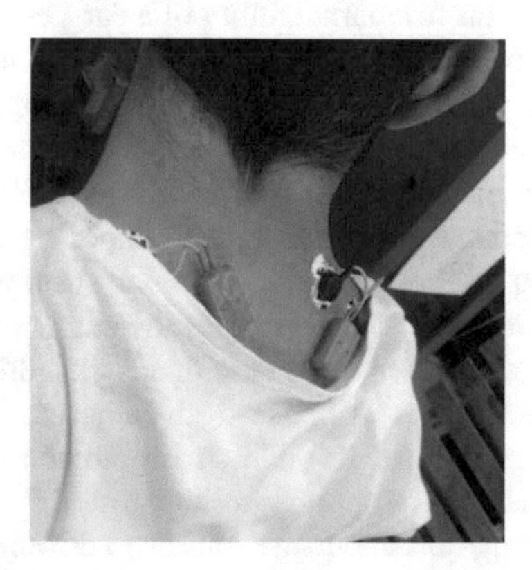

Fig. 1. EMG electrodes placed over both superior trapezius muscles.

placed on the center of the muscle belly, in the direction of the muscle fibers, according to the European recommendations for surface electromyography [7]. According to [8], the acquired signals must be subtracted by their average values and digitally filtered. The output of these elaborations represents the linear envelopes. The envelopes are normalized and expressed as a percentage of the EMG peak value. The maximum value of the normalized envelope in a 0.5 s window is calculated for each task.

The SHoW Motion system provides upper body motions, monitoring position and orientation of thorax, left scapula, right scapula, right humerus, left humerus, right forearm and left forearm. Head tracking would be fundamental for posture recognition in the maneuvers. So, an additional equipment is set up in the present research, and triggered by an output signal from SHoW Motion system.

3 Driver Head Tracking Device

The head tracking equipment uses 9-axis BNO 055 sensors (Bosh®, Reutlingen, Germany) as in [4]. The internal microcontroller, running Bosch Sensortec sensor fusion software, integrates an accelerometer, a gyroscope and a geomagnetic sensor, each being a triaxial component. The BNO055 is simply mounted through a headband, as shown in Fig. 2(a). In this application we are interested in the driver movements as relative to a moving platform. So, a second BN055 is expected to be mounted on the car in order to compensate for noise and acceleration, as shown in Fig. 2(b). For the present work, the developed software simply integrates the gyroscope vectors over time in order to get the head and car angular positions. This would be a design and verification tool, thus it must be user friendly and error proof. So, the car model serves as aiding a first mounting reference of the equipment, in order to make robust the sensor alignment to the real car through the software.

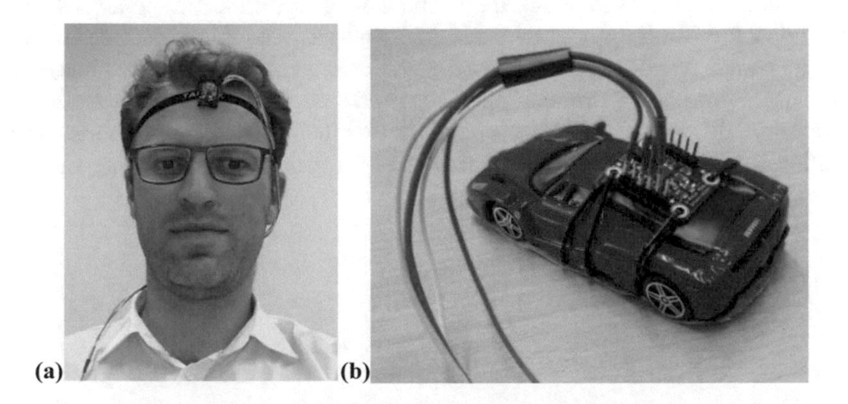

Fig. 2. Head (**a**) and car (**b**) mounted BNO055 sensors.

The two BNO055s are controlled through an Arduino Mega controller (Arduino®, Turin, Italy). The BNO055s are interfaced with the controller through I2C communication, doubled by a TCA9548A I2C multiplexer board (Adafruit Industries LLC®,

New York, USA). The complete system includes also LCD screen, LED and push-button serving as HMI, SD card slot for data logging and 9 V battery to enable a stand alone device. Figure 3 shows the signals and energy connections in the system. The complete wiring is omitted for clarity of presentation. The cycle time is 27 ms.

4 Driving Experiments

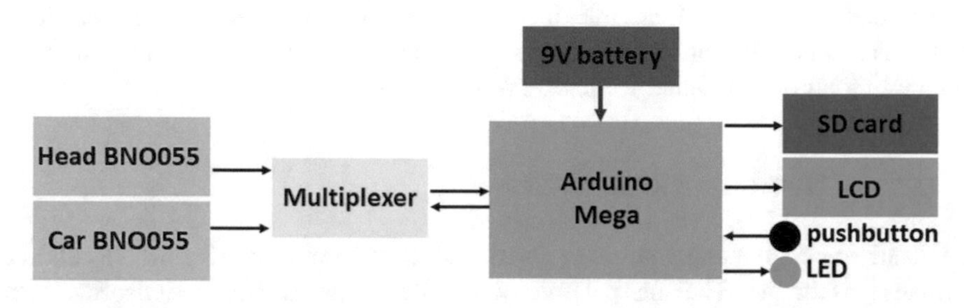

Fig. 3. Configuration of the sensor based head tracking system.

Before seating on the car, the driver is checked to exclude cervical pain, muscles discomfort, previous neck and/or upper-harm surgery, dismetabolic conditions and eye disorders. Before the electrodes application, a proper skin preparation enables to reduce many artifacts from the muscles.

The driving experiment consists in a 8 track: short straight road, about 270° left turn, back straight, about 270° right turn, final short straight. The maneuver is first driven with a Levante Maserati® (Modena, Italy), as shown in Fig. 4, producing the results shown in Fig. 5. Then, the same maneuver is driven with a Quattroporte Maserati®, with results shown in Fig. 6. The graphs report the turning angles around a vertical direction for the car, black solid line, and the head, black dotted line, and the electromyograph responses for left (red dotted line) and right (blue dotted line) superior trapeziuses. From the experiment, the 8 maneuver results in a major activation of the superior trapezius while driving a Levante than while driving a Quattroporte. Also, the activation is generally greater for the first left turn than for the following right one.

Fig. 4. Experiment setup on the Levante car.

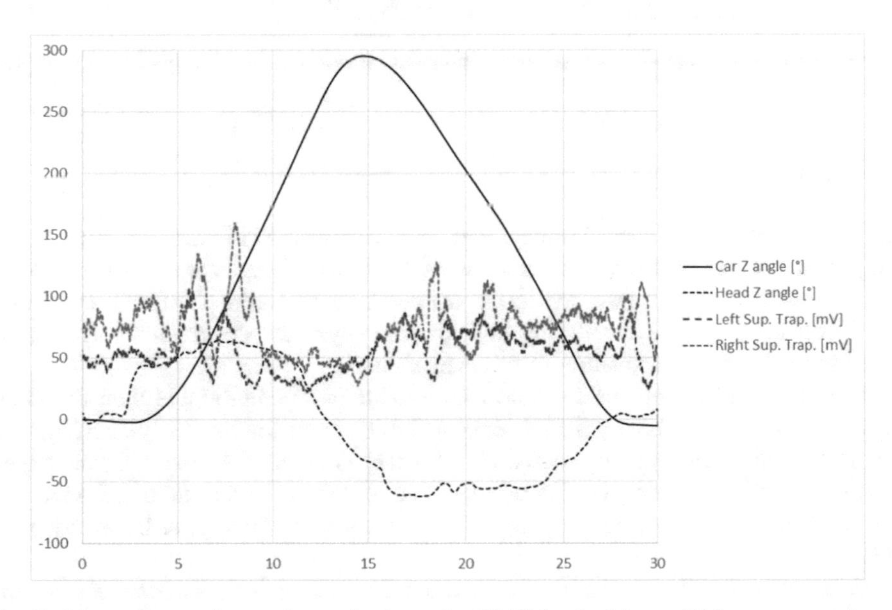

Fig. 5. Maneuver rotations and superior trapezius EMG for the Maserati® Levante experiment.

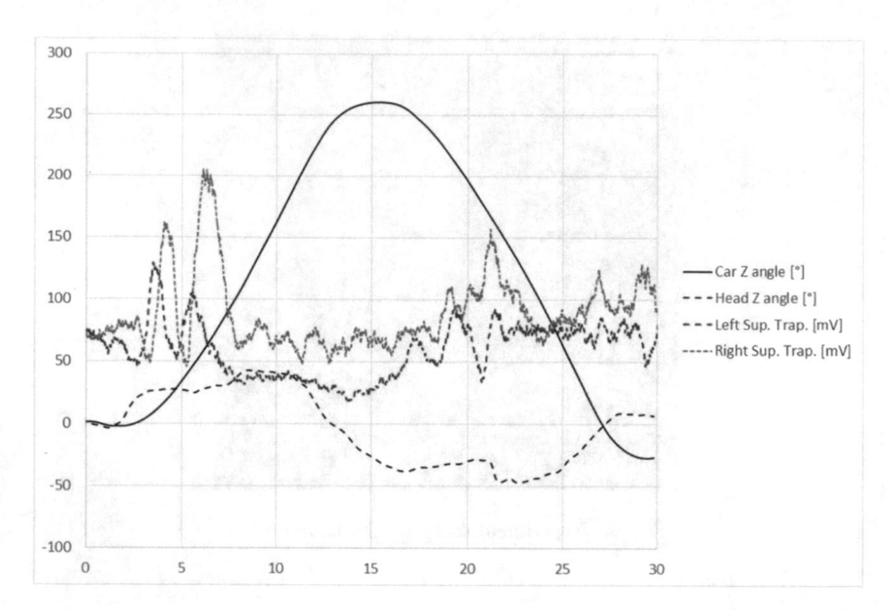

Fig. 6. Maneuver rotations and superior trapezius EMG for the Maserati® Quattroporte experiment.

5 Conclusions

The SHoW Motion system is integrated with a driver head tracking device. The equipment is tested in real driving maneuvers with two different car models in order to validate the system with two comfort concepts. The synchronous sEMG acquisition and driver head posture monitoring enables to analyze the driver movements, characterizing the muscles activation in the maneuvers.

For the tested driver, the Levante car resulted more comfortable than the Quattroporte one. Indeed, a luxury SUV is expected to be more comfortable than a high-end car. Then, a discussion can be started about what the different segment drivers expect from their cars. So, the proposed equipment can be used to fit the steering feedback to as more comfortable for cars intended to provide a smooth driving, or harder for sport cars.

Future works will test other drivers and, for the same drivers, different seat setups.

Acknowledgements. The authors gratefully acknowledge NCS Lab Srl, Carpi, Italy, for supporting the research with an effective test system and with precious knowledge.

References

1. Grujicic, M., Pandurangan, B., Xie, X., Gramopadhye, A.K., Wagner, D., Ozen, M.: Musculoskeletal computational analysis of the influence of car-seat design/adjustments on long-distance driving fatigue. Int. J. Ind. Ergon. **40**(3), 345–355 (2010)
2. Schüldt, K., Ekholm, J., Harms-Ringdahl, K., Arborelius, U.P., Németh, G.: Influence of sitting postures on neck and shoulder EMG during arm-hand work movements. Clin. Biomech. **2**(3), 126–139 (1987)
3. Vergnano, A., Leali, F.: Monitoring driver posture through sensorized seat. In: 1st International Conference on Human Systems Engineering and Design: Future Trends and Applications, Reims (2018)
4. Vergnano, A., Leali, F.: Out of position driver monitoring from seat pressure in dynamic maneuvers. In: 2nd International Conference on Intelligent Human Systems Integration: Integrating People and Intelligent Systems, San Diego (2019)
5. Cutti, A.G., Giovanardi, A., Rocchi, L., Davalli, A., Sacchetti, R.: Ambulatory measurement of shoulder and elbow kinematics through inertial and magnetic sensors. Med. Biol. Eng. Comput. **46**(2), 169–178 (2008)
6. Cutti, A.G., et al.: Prediction bands and intervals for the scapulo-humeral coordination based on the Bootstrap and two Gaussian methods. J. Biomech. **47**(5), 1035–1044 (2014)
7. Hermens, H.J., Freriks, B., Disselhorst-Klug, C., Rau, G.: Development of recommendations for SEMG sensors and sensor placement procedures. J. Electromyogr. Kinesiol. **10**(5), 361–374 (2000)
8. Konrad, P.: The ABC of EMG. A Practical Introduction to Kinesiological Electromyography. Noraxon U.S.A. Inc., Scottsdale (2006)

Significance of Technology Factors in the Context of Development of Health Care Sector in Latvia

Daiga Behmane[1], Henrijs Kalkis[1(✉)], Anita Villerusa[1], Uldis Berkis[1], and Didzis Rutitis[2]

[1] Riga Stradins University, Dzirciema 16, Riga 1007, Latvia
{Daiga.Behmane,Anita.Villerusa,Uldis.Berkis}@rsu.lv, henrijs.kalkis@gmail.com

[2] BA School of Business and Finance, Kr. Valdemara 161, Riga 1063, Latvia
Didzis.Rutitis@ba.lv

Abstract. The aim of the study is to evaluate significance of technology factors in the context of development of health care sector in Latvia in order to improve the external environment and performance of the health sector from the perspective of increasing the opportunity to provide medical service exports in Latvia. The study evaluates external national level conditions affecting the development of Latvian health care industry by applying the PEST analysis framework. The ranking of the PEST components was carried out using the expert method and implementing structured interviews for data collection. The study justifies that there are opportunities for increasing competition, and technological progress is indicated as the leading technology factor in the Latvian health care market.

Keywords: Health care industry · Competitiveness · External factors Technological factors

1 Introduction

Health care industry experiences increasing importance in the context of modern socioeconomic processes, globalization and technological progress worldwide. Competitiveness of national health care providers has turned out to be of similar strategic importance to the traditional manufacturing, energy and service industries. Competitiveness can be characterized by capacity of health care providers to supply services to foreign patients, thus, revealing the capacity and attractiveness of the sector. Increasing demand for health services due to the population ageing and technological progress, as well as the citizens' mobility and demand for social convergence determines the growing proportion of health care in national economies and potentially creates new segments of the international health care market.

The processes related to medical services export growth are influenced by the global technological progress. However, they need to be carefully analysed within the

W. Karwowski and T. Ahram (Eds.): IHSI 2019, AISC 903, pp. 248–254, 2019.
https://doi.org/10.1007/978-3-030-11051-2_38

context of country's ability to maintain efficiency of public administration, current economic and social factors, actual business environment and political agenda.

The aim of the study is to determine the significance of technological factors to improve the external environment and performance of the health sector from the perspective of facilitating the medical service exports from Latvia.

2 Methodology of Research

This paper reflects results from a study aimed to evaluate external national level conditions affecting the development of Latvian health care industry by applying the PEST analysis framework in the context of technological factors. In order to create the PEST matrix, international reviews and publications, including the WHO, WB, OECD, Eurostat, WEF publications and studies were analyzed.

The ranking of the PEST components was implemented by applying expert method and by using structured interviews for data collection. Twenty experts representing ambulatory and hospital care from institutions of different ownership form (state, municipality and private) were identified from the Registry of Medical Institutions Offering Treatment Services to Foreign Patients run by the Health Inspectorate of Latvia. To compare the actual performance of the factor with the desired result, experts were asked to rank the PEST factors according to two aspects: (1) significance of the factor and (2) actual performance of the factor by using the Likert scale (from 0 to 5). Significance and performance indicators for each of the factors were calculated and by further application of the GAP analysis the difference in the rank of performance from the desired indicator for each factor was calculated.

Data were collected between July and November 2017.

3 Background

3.1 The Importance of Technological Factors in the Development of Health System

External factors are important determinants of successful operation by the health sector service providers to deliver efficient and high-quality health services. They also encourage countries and respective economies to maximize their comparative advantage by improving productivity. Medical tourism industry is dynamic and volatile, and a range of factors including the economic climate, domestic policy changes, advertising practices, geo-political shifts, and innovative forms of treatment may all contribute towards shifts in patterns of consumption and production of domestic and overseas health services [1].

As recognized by researchers, technology drives health care more than any other force, and in the future, it will continue to develop in dramatic ways. Technological advance in health care relate to the progress in medical technologies, the ways of service production, readiness of support and communication systems as well as company's ability to operate in the global medical knowledge exchange environment.

It is also commonly known practice that health care organizations compete by setting up equipment from well-known producers and brands that are not always economically justified [2].

Porter has identified importance of health care technology providers within the chain of value creation and their influence towards provision of value-based health care. Regarding the role of IT systems, Porter indicates that State-of-the-art information technology cannot fix a broken health care system. Instead, if technology is treated as a tool, not a solution, health care professionals can stay focused on the primary goal of health care reform: providing value for patients. Moving to better IT systems can enable a move toward a value-based health care delivery model that improves patient outcomes. While some health care professionals state that technology often escalates costs, in a value-based competitive environment where costs and outcomes are accurately measured, technology is viewed as a strategically employable component to be applied when it's most efficient and effective [3].

3.2 Selection of Technological Factors for the PEST Analysis

PEST analysis covers 4 types of factors - *political factors* that are linked to the impact and opportunities provided by government attitudes towards the industry, changes in political institutions and the direction of political processes, legal issues and the general legislative environment; *economic factors* that relate to the economic structures of the society, the country's economic policy and capacity, tax and investment policies; *social factors* that relate to shared values, cultural attitudes, ethical beliefs, demographics, educational levels, and *technological factors* that are linked to changes in technology that can change the provider's competitive position, improvement of current products and process innovations that can reduce production costs [4].

The development of medical care increasingly depends on the technological progress.

Technology drives healthcare more than any other force, and in the future it will continue to develop in dramatic ways [5]. More often health care organizations usually compete for prestige medical equipment, even not always economically justified. A key priority for regulators is the identification of relevant systems to support the need for holistic governance approach in health sector [6].

The authors have derived the following technological factors within the PEST matrix to be evaluated by experts:

- Technological progress
- Threats from competing technology
- Innovation in service provision
- Research funding by government
- ICT support
- High standards for health information
- International knowledge transfer

The authors have applied the conceptual approach that technological factors relate not only to the medical technological progress itself, but also to innovation in service provision and organization, environment for knowledge and research based solutions supported by information exchange and communication platforms and other ICT support as well as company's ability to operate in the global medical knowledge exchange environment.

4 Research Results

The results of external factor evaluation by experts (see Table 1) reveal that national economic and technological external environment factors have the greatest significance in the health sector development, ranked 4.36 and 4.35 respectively, followed by social factors (4.23) and political factors (3.97).

Table 1. Ranking of PEST factor groups by significance and performance, mean values (0–5, Likert scale)

		Significance (0–5)	Performance (0–5)	Performance %
1	Political factors	3.97	2.18	54.86
2	Economic factors	4.36	2.62	60.16
3	Social factors	4.23	3.02	71.45
4	Technological factors	4.35	2.86	65.68

In relation to the difference between the assigned values for the significance and real performance, there is the smallest gap for the technological factor group. This justifies that health care managers evaluate the progress of the technological development of the sector relatively high, at the level at 65.68% from what they expect. In comparison, the performance of political factors has been rated the lowest – only at the level of 54.86%.

The expert evaluation of certain technological dimension factors (see Table 2) confirms hypothesis that technological advancement of the sector is of crucial importance for ensuring a competitive edge. Factor *technological progress* itself is ranked as most important technological factor (4.80) among all, followed by *ICT support* (4.60), *international knowledge transfer* (4.55) and *innovation in service provision* (4.50). Somewhat lagging behind is the ranking of the significance factor - *threats from competing technology* (3.95) characterizing the modest competition level between health care providers in Latvia.

Performance of the factor *technological progress* is ranked high (3.85) and at the level of 80.21% from expected value, suggesting that the provision of health care services is ensured by high technological support in Latvia. The high evaluation of the availability of advanced medical technologies can be justified by the situation of significant investment in medical technologies from the European Structural Funds in recent years.

Factor *innovation in service provision* is also ranked rather high – at the level of 75.6% from the expected value, stating that health care institutions are ready for an innovative approach to patient care. As indicated by respondents, patient-oriented

Table 2. Ranking of technological factors, mean values (0–5, Likert scale)

Technological factors	Significance (0–5)	Performance (0–5)	Performance, %
Technological progress	4.80	3.85	80.21
Threats from competing technology	3.95	2.85	72.15
Innovation in service provision	4.50	3.40	75.56
Research funding by government	3.90	1.65	42.31
ICT support	4.60	2.80	60.87
High standards for health information	4.15	2.35	56.63
International knowledge transfer	4.55	3.10	68.13
Mean value	4.35	2.86	65.68

solutions and clustering possibilities in service provision are explored to increase their competitiveness.

In the performance dimension the lowest rank (the level of 42.31%) is attributed to the factor - *available research funding* (1.65), although the significance of the factor is ranked as moderate (3.90), which points to some limited funding opportunities for research as a whole at the national level. Although medical technologies are international in nature and their introduction by health care providers is not considered to be a narrative for research, the need for more research in the area of implementation and application to the local context to serve the needs of the population in the values context potentially will highlight in future.

The performance of the factor *high standards for health information* is ranked rather low, at the level of 56.6%, which can be explained by the recent remarkable international attention to the protection of patient data and the processing of information, which is in the implementation phase in Latvia. Although a common approach is being developed at national level, the implementation is highly relevant to the understanding and capabilities of each service provider.

Although the significance of the factor *ICT support* is ranked high (4.6), it's real performance is evaluated to be very moderate (60.87%), which points to the need for service providers to have much better common IT solutions in place to serve the needs for the information exchange between providers. The underestimation can be attributed to the long-lasting but complex and controversial implementation of e-health solutions at national level.

The importance of the factor *international knowledge transfer* (4.55) is appreciated by health care providers highlighting the nature of the health sector as a strong international environment. The performance evaluation at the level of 3.1 (68.13%) shows the positive fact that the development of the health sector in Latvia to a large extent is based on the transfer of international knowledge to the local system.

5 Conclusions

The research results indicate that the success of health care organizations to be competitive at international level, highly depend on external environment – mainly economic and technological factors.

The study proves that the comprehensive approach to the evaluation of technological factors. The study highlights the most important external technological factors for the competitive development - medical technological progress is evaluated as the main driving factor for the sector, followed by the need for ICT support, innovations in service provision and international knowledge transfer.

The study reveals the main gaps in the performance of the technological factors in the Latvian health care industry, which are attributed to the overall limited research environment in the health sector, gradual implementation of overall standards of patient information processing and the limited support by integrated ICT tools. The slow introduction of national e-health system is considered as a threat for competitive development of the sector. There are still opportunities for rising a competitive development between service providers thus contributing to the general competitiveness of the health sector itself.

References

1. Lunt, N., Smith, R., Exworthy, M., Green. S.T., Horsfall, D., Mannion, R.: Medical tourism: treatments, markets and health system implications: a scoping review. Directorate for Employment, Labour and Social Affairs, OECD (2014)
2. Competition among health care providers: Investigating policy options in the European Union, expert panel on expert panel on effective ways of investing in health (2015). https://ec.europa.eu/health/expert_panel/sites/expertpanel/files/008_competition_healthcare_providers_en.pdf
3. Porter, M.: Harvard Business School. https://www.isc.hbs.edu/health-care/vbhcd/Pages/information-technology-platform.aspx
4. Wiley Encyclopedia of Management. Wiley. https://onlinelibrary.wiley.com/doi/book/10.1002/9781118785317
5. Thimbleby, H.: Technology and the future of healthcare. J. Public Health Res. **2**(3), e28 (2013)
6. Bodolica, V., Spragon, M., Tofan, G.: A structuration framework for bridging the macro–micro divide in health-care governance. Health Expect. **19**(4), 790–804 (2016)
7. Ketels, C.: Review of Competitiveness Frameworks. An Analysis Conducted for the Irish National Competitiveness Council (2016)
8. Health 2020: the European policy for health and well-being. http://www.euro.who.int/en/health-topics/health-policy/health-2020-the-european-policy-for-health-and-well-being
9. van den Ven, W.P., Beck, K., Buchner, F., Schokkaert, E., Schut, F.T., Shmueli, A., Wasem, J.: Preconditions for efficiency and affordability in competitive healthcare markets: are they fulfilled in Belgium, Germany, Israel, the Netherlands and Switzerland? Health Policy **109**, 226–245 (2013)
10. Rechel, B., Wright, S., Edwards, N., Dowdeswell, B., McKee, M.: Investing in Hospitals of the Future. Observatory Studies No 16. European Observatory on Health Systems and Policies, Copenhagen (2009)

11. Nunes, R., Brandao, C., Rego, G.: Public accountability and sunshine regulation. Health Care Anal. **19**, 352–364 (2011)
12. Health Policy in Latvia. OECD Health Policy Overview. http://www.oecd.org/els/health-systems/Health-Policy-in-Latvia-March-2017.pdf

The Influence of Illuminance and Color Temperature on Target Dragging Task

Yingwei Zhou[1], Tuoyang Zhou[1], Yuting Zhao[2], Haixiao Liu[2], Chi Zhang[1], Dan Wang[2], Jinshou Shi[1], Chuang Ma[2], Xin Wang[1], Xiai Wang[2], and Jianwei Niu[2(✉)]

[1] China Institute of Marine Technology & Economy, Beijing 100081, China
[2] School of Mechanical Engineering, University of Science and Technology Beijing, Beijing 100083, China
niujw@ustb.edu.cn

Abstract. Eye control, a human behavior that is often overlooked in daily life, has long been considered a desktop technology that can be used in human-computer interaction interfaces to control operating conditions. Moreover, eye tracking methods are often used in cognitive research and can understand individuals' intention, habits and how they use their knowledge and skills when performing tasks by operating computer. The purpose of this study was to investigate the effects on color temperature and illumination of the use of computer screen manipulation through eye control human-computer interaction interface, and to obtain design factors that affect human body performance through experimental results. Multivariate analysis of variance showed that under different illumination and color temperature conditions, different levels of illuminance had a significant effect on the accuracy of the operation. However, color temperature has no significant effect on any of them.

Keywords: Eye control · Illumination · Color temperature · Human-computer interaction interface

1 Introduction

During the past decade, eye tracking has developed in a variety of directions, and individuals' cognition and behavior through eye control have developed in various fields [1]. The pioneer work can be traced to 1995, Kathmann N and Friedman L focus on eye cognition and healthy people's cognition through eye tracking [2, 3]. Nowadays, more attention is paid to eyesight tracking or eye movement trajectory for cognitive processing according to the needs of the moment [4].

However, there are very few eye movement control interactive interfaces that explore user manipulation capabilities. Although Cáceres' study shows that 30 test subjects prefer to use the mouse to control the interface pointer [5], in order not to let the mouse control time response error affect the experimental time, it also needs eye movement control to move the pointer, so that the intuitive response observer's attention force and cognitive response.

W. Karwowski and T. Ahram (Eds.): IHSI 2019, AISC 903, pp. 255–259, 2019.
https://doi.org/10.1007/978-3-030-11051-2_39

2 Materials and Methods

2.1 Experimental Facility

The equipment for monitoring the eye movement process and interface operation of the participants in the experiment was Tobii Eye Tracker 4C. Dimensions of Tobii Eye Tracker 4C was 0.7 in. × 0.6 in. × 13.2 in. (W × D × H), Wight 3.35oz, OS Compatibility Windows 7, 8.1 and 10 (64-bit only), PC Windows Hello, System Recommendation 2.0 GHz, Intel i5 or i7, 8 GB RAM, Max Screen Size 27 in. with 16:9 Aspect Ratio or 30 in. with 21:9 Aspect Ratio. Operating Distance 20–37″/50–95 cm, Track Box Dimensions 20–37″/50–95 cm. Frequency 90 Hz, Illuminators Near Infrared (NIR 850 nm) Only. Tracking Population 97%.

The human-computer interaction screen used a resolution of 2560 × 1440, Processor Up to 8th Generation Intel® Core™ i5-8550U Processor (1.80 GHz, up to 4.0 GHz with Turbo Boost, 8 MB Cache). Graphics was Integrated Intel® UHD Graphics 620, Dimensions (W × D × H) is 12.59″ × 8.89″ × .74/320 mm × 226 mm × 18.8 mm.

2.2 Experimental Design

One hypothesis was that light has an effect on the accuracy of the target drag and divided the light into illuminance and color temperature into three levels. According to the recommendations of optical research experts, the illuminance and color temperature were equally divided within a certain range, and the illuminance level was divided into 125, 200, and 275. The color temperature level was divided into 3500, 4500, and 5500. The trial was repeated 6 times at each different level, and the order appeared randomly. The eye control replaced the mouse operation to perform target click, drag and so on the human-computer interaction interface, and monitored the trial's eye activity. During the trial, control other variables, such as the target objective was a rectangle, the size was 60px, and the color was gray. The distance between the eyes and the operating screen during the participant's trial was about 700 mm.

2.3 Experimental Procedure

In the trial, the participant's main task was to drag the rectangular target objective in the interface into the specified circular range under different levels of color temperature and illumination environment. The relative distance between the target objective and the specified range was fixed, both 1280px. The position of the target objective was random. Each participant needed to repeat 6 times. The main recorded dependent variables are dragging time, dragging numbers and accuracy (Fig. 1).

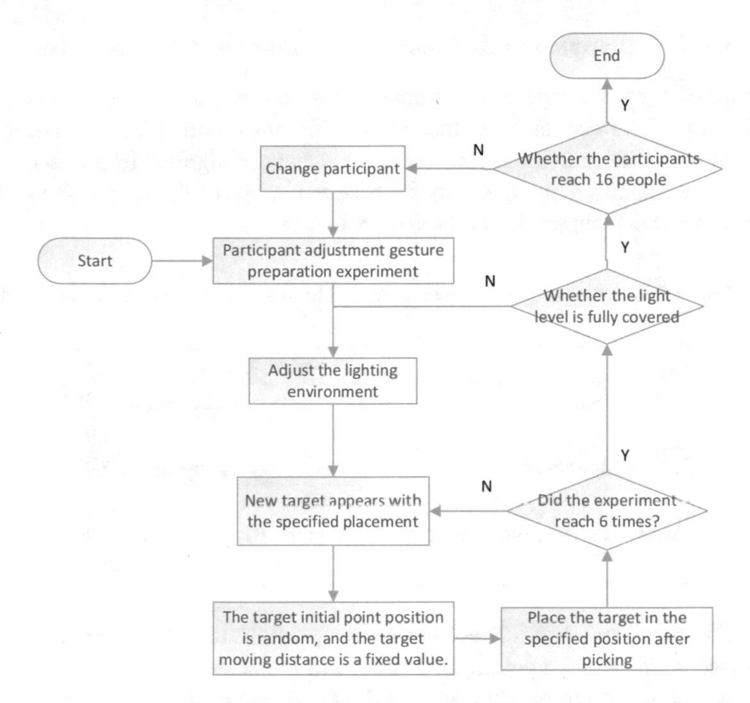

Fig. 1. Flow chart of human operation under different illumination and color temperature levels

2.4 Data Analysis

In statistics, multivariate analysis of variance (MANOVA) was an analytical method that compares the mean of multiple dependent variable samples, and usually follows the significance test of each dependent variable [6]. Participants used the effects of eye control on interactive interface operations in multiple dependent variables (different levels of illuminance and color temperature), used multivariate analysis of variance to explore the significant effects between individual dependent and dependent variables, and designed for subsequent user manipulation Elemental paving.

3 Results

3.1 Participants

Sixteen individuals (eight graduate students and eight experts engaged in human-computer interaction research) were randomly distributed to participate in the study. The ratio of male to female participants was 3:5 (6 males and 10 females), and the average age of students was 23 years old (range 21–24), the average age of the experts was 35 years (range 29–40), and their vision was normal after correction.

3.2 Results in Different Levels Color Temperature and Illuminance

The average of 6 repeat experimental data performed by each participant was used to reduce the error. The results show that under different illumination and color temperature conditions, different levels of illuminance have a significant effect on the correctness of the operation. In the tests of between-subject effects, the P value of the illuminance for the accuracy is 0.039 < 0.05 (Table 1).

Table 1. Tests of between-subject effects under different illumination and color temperature conditions

Source	Depend variable	Sig.
Illumination	Average dragging time	0.961
	Accuracy	0.039
Color temperature	Average dragging time	0.973
	Accuracy	0.950
Illumination × Color temperature	Average dragging time	0.999
	Accuracy	0.993

Regarding the interaction between color temperature and illuminance in estimated marginal means of the average dragging time and the accuracy, when the illumination is 200lx and the color temperature is 3500K, the average dragging time of the target objective is significantly lower than the dragging time under other interaction levels, but it is not operated correctly under this condition (Fig. 2 shows the estimated marginal means of dragging time).

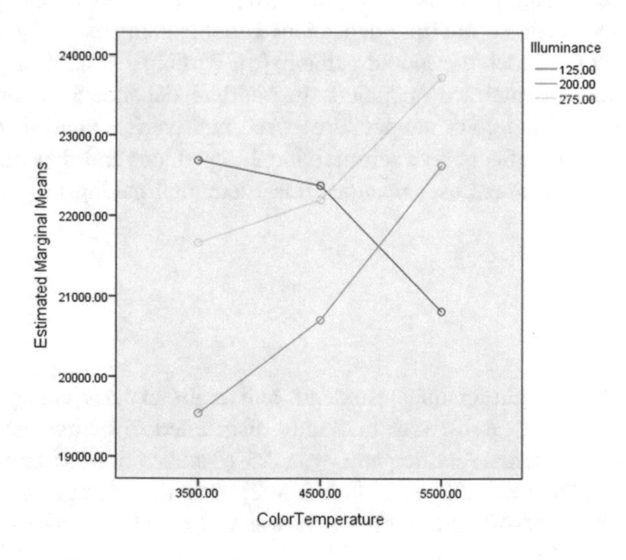

Fig. 2. Estimated marginal means of dragging time

4 Discussion

The perception of light by the eyeball is the result of the interaction of columnar and cone cells on the retina. The action of columnar cells becomes more and more prominent as the level of brightness decreases. However, the cone cells become less and less sensitive. While cone cells are more sensitive to color, the study is based on the effect of sensitivity on eye movement control pointers in human-computer interaction. These experimental results support the original hypothesis that different levels of illuminance had a significant effect on the accuracy of the operation.

In future studies, the relationship between the eye movement control and the operation time should be discussed. Moreover, this paper concludes with a discussion of future research consideration in section that split color temperature and illumination into more levels for experimentation and adding more dependent variables.

References

1. Celine, G., Cho, V., Kogan, A., Anthonappa, R., King, N.: Eye-tracking in dentistry: what do children notice in the dentist? J. Dent. **78**, 72–75 (2018)
2. Kathmann, N., Wagner, M., Rendtorff, N., Schöchlin, C., Engel, R.: Information processing during eye tracking as revealed by event-related potentials in schizophrenics, alcoholics, and healthy controls. Schizophr. Res. **16**(2), 145–156 (1995)
3. Friedman, L., Kenny, J.T., Jesberger, J.A., Choy, M.M., Meltzer, H.Y.: Relationship between smooth pursuit eye-tracking and cognitive performance in schizophrenia. Biol. Psychiat. **37** (4), 265–272 (1995)
4. Huddleston, P.T., Behe, B.K., Driesener, C.: Inside-outside: using eye-tracking to investigate search-choice processes in the retail environment. J. Retail. Consum. Serv. **43**, 85–93 (2018)
5. Cáceres, E., Carrasco, M., Ríos, S.: Evaluation of an eye-pointer interaction device for human-computer interaction. Heliyon **4**(3), e00574 (2018)
6. Warne, R.T.: A primer on multivariate analysis of variance (MANOVA) for behavioral scientists. Pract. Assess. Res. Eval. **19**(17), 1–10 (2010)

Applying Random Forest Method to Analyze Elderly Fitness Training Routine Data

Chia Hsuan Lee[1], Tien-Lung Sun[2], Diana Eloisa Roa Flores[2], and Bernard C. Jiang[1](✉)

[1] National Taiwan University of Science and Technology, 1 No. 43, Section 4, Keelung Road, Da'an District, Taipei 106, Taiwan
iebjiang777@gmail.com
[2] Yuan-Ze University, 135 Yuan-Tung Road, Chung-Li, Tao-Yuan, Taiwan

Abstract. This study used the random forest algorithm to predict Senior Fitness test results on the execution of Synchronized Monitoring Analysis Record Care (SMARC) programs with the aim of aiding healthcare professionals in modifying patients' training routines to improve their effectiveness. Twenty-three subjects in a community center performed a fitness training routine using the SMARC series of equipment and training modes, and took timed "Up and Go" tests before and after their performances. The 74 combined features (categorical + numerical) of the series were used as input features, and performance was measured by the Timed Up and Go (TUG) score. The results show that the top five features ranked with the highest importance were associated with Machines F (16.5%), D (15.4%), E (13.9%), H (13.9%), and B (12%), with 35% unassignable. The results can aid healthcare professionals in planning and adjusting more targeted health-promotion exercises programs using assistive devices for the elderly.

Keywords: Random forest · Fitness test · Functional training equipment Timed up and go · Synchronized Monitoring Analysis Record Care (SMARC)

1 Introduction

Longevity and quality of life are important topics of concern to an aging society. As the global population ages, more knowledge and evidence are needed to support policy-making to promote the independence and dignity of the elderly, maintain sustain-able development of societies, and ensure high quality of life. As part of the ten-year long-term care program, the Health Promotion Administration (HPA) has integrated local resources through the health departments and community medical institutions. In 2013, health stations in 22 cities and counties, and 438 medical institutions partnered with 1,672 Community Care Sites to hold health promotion activities, increasing the partnership rate at care sites to over 80% [1]. While an elderly employment resource center was launched in October 2014 to bring employable senior citizens into the job market. A total of NT$10 billion (US$329.27 million) has been earmarked to build elderly day care centers in 368 townships by 2016 [2].

W. Karwowski and T. Ahram (Eds.): IHSI 2019, AISC 903, pp. 260–264, 2019.
https://doi.org/10.1007/978-3-030-11051-2_40

Moreover, the program highlights health promotion issues such as healthy diet, exercise, prevention of falls, drug use safety, prevention of chronic diseases, health examinations and blood pressure measurement. The establishment of a long-term care system generates more possibilities and opportunities for an emerging health industry in order to meet the growing long-term care demands. Since long-term care services have become one of the fastest growing segments in the healthcare industry, developing industrial cooperation related to long-term care services through participating in innovation of assistive devices for the elderly and for active aging research. In this study, it is hoped that the Senior Fitness test results on the execution of "SMARC" programs can be predicted in order for the healthcare professionals be able to modify the patients training routine and improve their efficiency.

2 Methods

Twenty-three subjects in a community center performed fitness training routine using "SMARC" which is a complete series of 8 functional training equipment (machine A-H), with 5 training modes and 67 performance and difficulty features. In this study, we collected and analyzed the data of a fitness training routine that involves the use of different fitness equipment, in order to understand their Senior Fitness test results, and to determine which features collected from the fitness equipment with different training mode affect the Senior Fitness test results. The methodology applied to reach the research objective is the use of the Random Forest Algorithm, for this the inputs and parameters were defined. Some functional fitness tests were (such as Timed Up and Go test) also taken before and after the performance of the fitness training routine.

2.1 SMARC

The Taiwan Health Promotion Administration [1] has entrusted groups of academics and invited experts in various fields to form health promotion teams. On the health promotion activities for the elderly, "SMARC" which is a complete series of 8 functional training equipment, with integrated hardware and software combination, in order to relief syndrome, improve functional capacity, quality of motion and cognitive level. The fitness training routine of the present study was done using SMARC Machine, as shown in Fig. 1.

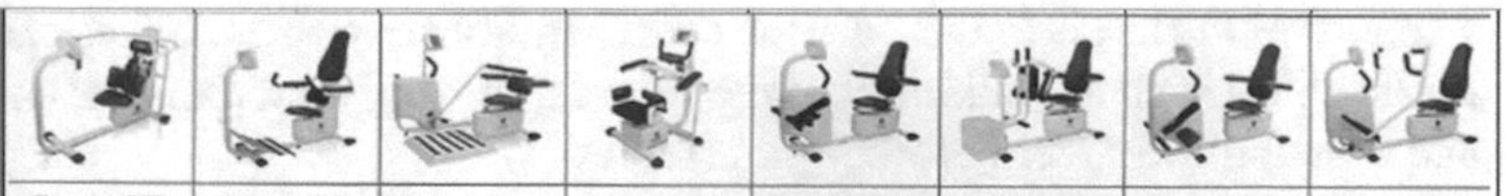

Machine	Smart US	Smart Tilt	Smart Arc	Smart Core	Smart Psoas	Smart Coxa	Smart LS	Smart CS
Machine Code	A	B	C	D	E	F	G	H

Fig. 1. SMARC machine

2.2 Fitness Test

The progress of the fitness training routine for the elderly group was evaluated using the Senior Fitness test. According with Jones and Rikli [3] the Senior Fitness test is a set of test items assessing the functional fitness of older adults that fulfill scientific standards for reliability and validity in a safe ambiance. These items are:

1. Maximum grip strength test (Kg). This test requires a hand dynamometer calibrated to 100 kg of force that is squeezed as hard as possible with the preferred hand while standing and with the other arm down. The goal of this test is to measure the upper body muscular strength.
2. Chair Sit and Reach or Hamstring Stretch (cm +/−): The test consists in sit down in a chair, extend both legs straight out until knees are as straight as the person can achieve, then reach for the toes, hinging from the hips and slightly rounding the back. The aim of this test is to measure the body lower flexibility, specifically your hamstring flexibility. This flexibility plays a role in the balance, posture, in fall prevention, and in your gait, or walking.
3. Back Scratch Test (cm). In the test both hands are placed behind the back. The purpose of the examination is measuring the flexibility of the upper body. This flexibility affects the ability to reach items that may be high or to require arm and/or shoulder movement.
4. Chair Stand (number of stands). The test involves stand up and sit down in a chair during 30 s. The objective of the test is to measure the strength of the lower body. This is important for activities such as getting out of a chair, on the bus, out of the car.
5. Arm Curl during in 30 s (number of repetitions). For this test is asked to flex and extend arms. The target of the test is to measure the strength of the upper body, which is important for activities such as carrying laundry, groceries, and luggage.
6. Two Minute Step Test (number of steps). The test requires to lift the legs during 2 min, setting the minimum knee height for person who take the test. The aim of the Two-Minute Step Test is to measure the endurance or physical stamina. The endurance affects the ability to performance daily activities.
7. One-Legged Balance Test (seconds). The position for this test is to stand in one leg with the opposite foot lifted halfway up the calf of the supporting leg. Timing is stopped if the supported foot leaves its original position on the floor or the other foot touches the floor. The objective of the test is to measure the agility and balance.
8. Up and Go (seconds). On this evaluation is required to walk 2.44 m and sit down. The goal of the test is to measure the speed, agility and balance. These are important for activities such as walking through crowds, moving on unfamiliar environments, and crossing the street before light changes.

2.3 Random Forest

A random forest is an ensemble classifier that first constructs a number of decision tree classifiers [4] and averages their predictions to improve the accuracy of variable estimation and control overfitting [5]. Each tree in the ensemble is built from a sub-sample drawn with replacement (i.e., a bootstrap sample) from an original dataset or training set that contains a collection of features [6]. Each tree depends on the values of a random vector (X) that is sampled independently and with the same distribution for all of the trees in the forest [5]. Thus, according to the classifier concept, random vector $X = (X1, \ldots, Xp)^T$ represents the real-valued input or features and a random feature γ represents the real-valued response, assuming an unknown joint distribution P_{XY} (X, Y). The goal is to identify a function f(X) for accurately predicting γ [7].

The following approaches were set for the inputs:

- *Random vector (X) approach:* this approach considered all combined features (1394 features by 23 samples) that have an effect on the Timed Up and Go test: D, E, F, G, H.
- *Random feature (y) approach:* This input measured the time variation between the initial TUG test and the same test's results after the subjects had performed the training routine. A decrease of at least 10% in test completion time was defined as an improvement (1). Otherwise, the score denoted no improvement' (0).
- *30%–70% split selection:* The split was determined to be 70% for the training set (16 samples) and 30% for the test set (7 samples). To obtain this split, we applied the Scikit-Learn function random split on the Python code.
- *Leave-One-Out (LOO):* One sample was used for the test set and the other 22 samples were used for training until all samples were tried. To obtain the split, we applied the Scikit-Learn function LeaveOneOut on the Python code.

The approaches for X, y, and the split were combined to generate a total of eight models, as shown in Table 1. Once the random forest models were defined, the random forest classifier algorithm from the Scikit-Learn implementation was used to obtain the classifications. All models were run for six forests of different sizes with 50, 100, 1000, 5000, 10000, and 50000 trees.

Table 1. Random forest models

Models	
X_1-y_1-30/70	X_1-y_1-LOO
X_2-y_1-30/70	X_2-y_1-LOO
X_1-y_2-30/70	X_1-y_2-LOO
X_2-y_2-30/70	X_2-y_2-LOO

3 Results

The five features that the X1-y1-30/70 model ranked with the highest importance was associated with machines F (16.5%), D (15.4%), E (13.9%), H (13.9%), and B (12%). Figure 2 illustrates the features grouped by machine and shows their rank according to the model. The selection of features of the F, D, E, H, B machines among the top five is consistent with what is advertised about the impact of those SMARC machines on TUG test performance (D, E, F, G & H) [8]. However, machine B (Smart Tilt) had not been previously known to have an effect on that test. As this machine was chosen by the model, it might be that the manufacturers consider it to have an effect on test performance. Indeed, the manufacturer's description of that machine emphasizes its usefulness in pre-gait training and reducing lower back pain [8]. In addition, this effect aligns well with machine B's training objectives (mobility, symmetric performance, dynamic balance and functional performance).

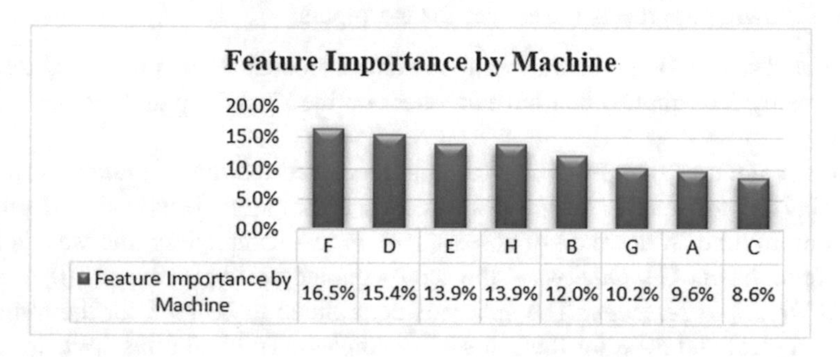

Fig. 2. Features importance per machine

4 Conclusion

The Random Forest Algorithm was effectively implemented to analyze the incidence of a fitness training routine data performed, over their functional fitness test results. The results were useful for the healthcare professionals in grasping and following-up of a health-promotion exercises program process.

References

1. Taiwan Health Promotion Administration, Ministry of Health and Welfare. http://www.hpa.gov.tw/English/file/ContentFile/201502140514171717/2014_Health_Promotion_Administration_Annual_Report.pdf
2. Department of Information Services, Executive Yuan. Health & Welfare. http://www.ey.gov.tw/en/cp.aspx?n=F85CABCA09695756
3. Jones, J., Rikli, E.: Measuring functional fitness of older adults. J. Act. Aging **1**, 24–30 (2012)
4. Breiman, L.: Random forests. Mach. Learn. **45**(1), 5–32 (2001)
5. Pedregosa, F., et al.: Scikit-learn: machine learning in Python. J. Mach. Learn. Res. **12**, 2825–2830 (2011)
6. Geurts, P., Irrthum, A., Wehenkel, L.: Supervised learning with decision tree-based methods in computational and systems biology. Mol. BioSyst. **5**(12), 1593–1605 (2009)
7. Cutler, A., Cutler, D.R., Stevens, J.R.: Random forests. In: Zhang, C., Ma, Y. (eds.) Ensemble Machine Learning, pp. 157–175. Springer, Boston (2012)
8. United Lifestyle. SMARC (2018). http://unitedlifestyle.com/smarc/. Accessed 26 Oct 2018

The Influence of Target Layout on Dragging Performance Based on Eye-Control Technique

Tuoyang Zhou[1], Yingwei Zhou[1], Dan Wang[2], Yuting Zhao[2], Haixiao Liu[2], Chi Zhang[1], Jinshou Shi[1], Chuang Ma[2], Xin Wang[1], Xiai Wang[2], and Jianwei Niu[2(✉)]

[1] China Institute of Marine Technology & Economy, 100081 Beijing, China
[2] School of Mechanical Engineering, University of Science and Technology, 100083 Beijing, China
niujw@ustb.edu.cn

Abstract. Eye-control technique can achieve convenient and rapid real-time operation through the movement of the eyes and reduce unnecessary manual operations, which has been a hot topic in human-computer interaction research. Due to the layout determines the location orientation, organizational complexity, cognitive consistency, and predictive ability of the information display, the interface layout design affects the user's perception of information intensity, complexity, and logic. The purpose of this study is to investigate the influence of target layout on dragging performance based on eye-control technique. We proposed a hypothesis that the target layout has a significant impact on dragging performance. The results indicate that there was no significant effect of target layout on dragging time and numbers, which did not support our hypothesis. The reason may be related to the setting of experimental conditions (e.g. lighting level and screen resolution). More affecting variables could be considered in future studies.

Keywords: Eye-control technique · Target layout · Human-computer interaction · Dragging performance

1 Introduction

The eye-tracking technology has widely spread in the last decade. Eye-control system is an advanced application in human-computer interaction research, which can achieve convenient and rapid real-time operation through the movement of the eyes and reduce unnecessary manual operations. With the deepening of research, this technology will be more widely used in medicine, military and entertainment.

Spatial attributes (such as topology, geometry, spatial relationships, etc.) are mapped to functional attributes (such as causality, hierarchical relationships, associations, etc.). For example, Liu et al. pointed out that stable covariant structural information can reduce the complexity of the scene while increasing its predictability [1]. Due to the layout determines the location orientation, organizational complexity,

W. Karwowski and T. Ahram (Eds.): IHSI 2019, AISC 903, pp. 265–270, 2019.
https://doi.org/10.1007/978-3-030-11051-2_41

cognitive consistency, and predictive ability of the information display, the interface layout design affects the user's perception of information intensity, complexity, and logic. In the 1950s, Fitts et al. studied the series of eye movements in the pilot's landing process to find out the effective method of assessing the importance of the instrument, the difficulty of instrument reading and the instrument layout design by eye-moving techniques [2, 3]. Eye movements are thought to provide an indication of the amount of cognitive processing a display requires [4] and eye tracking can be a tool for the assessment of usability [5]. Pušnik et al. studied how layout, typeface use, position of titles and/or text, color combination etc. draw attention and affect the recall of presented content by using the eye-tracking technology [6].

This paper is aimed to investigate the influence of target layout on dragging performance based on eye-control technique. We proposed a hypothesis that the target layout has a significant impact on dragging performance and designed an eye-controlled human-computer interaction experiment to verify the hypothesis. Through the study of visual perception and information processing mode, we explored whether the position in the screen would affect the operation process. The results could provide some suggestions for human-machine interface designing that are more suitable for eye-control system.

2 Method

2.1 Experiment Design

We performed a repeated-measures experiment to verify the hypothesis that the target layout has a significant impact on dragging performance. The independent variable was the initial position of target, which had five levels: screen center, top left, bottom left, top right, and bottom right. The dependent variables were the dragging time (the time of dragging the target into the specified range successfully) and the dragging numbers (the numbers of dragging the target into the specified range successfully), which were considered as variable indicators of dragging performance.

2.2 Task Design

The square target has five positions: screen center, top left, bottom left, top right, and bottom right. A set of a square target (one of the five positions) and a circular target (around the square target) are randomly displayed on the screen. The interactive interface is shown in Fig. 1. The participants were asked to use eye ball movement to drag the target into the specified circular range. The operation of putting the square target into the circular range successfully is shown in Fig. 2.

Fig. 1. The interactive interface

Fig. 2. The operation of putting the square target into the circular range successfully

2.3 Participants

Sixteen participants were recruited to participate in this experiment. All the subjects were required to have normal vision without glasses, which could prevent the lens from affecting the capture of eye movement. There were no restrictions on age and gender. The participants were asked to use eye ball movement to drag the target into the specified circular range. Each participant needed to drag the target 5 times and repeat for 4 groups of experiments, i.e. 20 trials in total. All the participants had finished the experiments.

2.4 Experimental Facility

We chose the Tobii Eye Tracker 4C for eye movement control and a notebook with 15.6-inch display for connecting the eye tracker in the experiment. The resolution of the notebook is 1920 × 1080. The Tobii Eye Tracker scans at a frequency of 90 Hz and supports head tracking. The effective distance is 20–37″/50–95 cm. The tracking range is 40 × 30 cm, and the preparation efficiency is up to 97%. It works by emitting infrared rays and then using the receiver to detect the direction of the eye for fine tracking. The Tobii Eye Tracker 4C only needs to be fixed under the experimental computer screen, which is a good solution to reduce the experimental error caused by the uncomfortable side-effect of a long-time experiment. It provides a natural environment for collecting multi-channel data, such as voice, motion, etc. It comes with Clear View data analysis software to analyze eye movement data and actual interface, sound and video of user's action.

2.5 Experiment Procedure

The participants were asked to complete a profile questionnaire (basic information) and an informed consent form first. Then, we explained the experimental content to the participants. Each participant was asked to sit in front of the notebook, created a new eye movement information and proofread by the eye tracker. After a set of a square target and a circular target were displayed on the screen, the participants fixated on the square target for 1 s to pick up the square target. Then, he/she needed to drag the square target to the circular range and put down it by fixating for 1 s. This was a complete dragging operation. Participants could drag and drop multiple times until the square target was placed in the circular range successfully. Three seconds after the completion of one operation, the target would randomly appear in the next position. The operation in five locations is one group of experiments, and the experiments was repeated for four groups, i.e. 20 trials per participant in total.

2.6 Data Analysis

We performed repeated-measures analysis of variances (ANOVAs) on the dependent variables. Multivariable analysis, which includes Pillai's Trace, Wilks' Lambda, Hotelling's Trace and Roy's Largest Root, was used to test the effect of the increase of experiment time and operation repeating numbers on the experimental results. Then, according to the result of Mauchly's test of spherical ($P < 0.05$), we analyzed the corrected part of the unary analysis for studying the effect of target layout on dragging time and numbers.

3 Results

The repeated-measures analysis of the dragging time and numbers is shown in Table 1. Multivariable analysis, which includes Pillai's Trace, Wilks' Lambda, Hotelling's Trace and Roy's Largest Root, was used to test the effect of the increase of experiment time and operation repeating numbers on the experimental results. The result ($P > 0.05$) showed that the experiment time and operation repeating numbers did not make a difference to the five levels of the target's initial position, which indicated that the increase of experiment time and operation repeating numbers did not affect the dragging performance (dragging time and numbers).

Table 1. The repeated-measures analysis of the dragging time and numbers

Effect	Method	Value	F	*P*
Dragging time	Pillai's Trace	0.010	0.445[b]	0.721
	Wilks' Lambda	0.990	0.445[b]	0.721
	Hotelling's Trace	0.010	0.445[b]	0.721
	Roy's Largest Root	0.010	0.445[b]	0.721
Dragging time * screen position	Pillai's Trace	0.063	0.746	0.706
	Wilks' Lambda	0.938	0.745	0.707
	Hotelling's Trace	0.065	0.744	0.708
	Roy's Largest Root	0.053	1.839[c]	0.125
Dragging numbers	Pillai's Trace	0.010	0.479[b]	0.697
	Wilks' Lambda	0.990	0.479[b]	0.697
	Hotelling's Trace	0.010	0.479[b]	0.697
	Roy's Largest Root	0.010	0.479[b]	0.697
Dragging numbers * screen position	Pillai's Trace	0.069	0.821	0.628
	Wilks' Lambda	0.933	0.815	0.635
	Hotelling's Trace	0.071	0.808	0.642
	Roy's Largest Root	0.041	1.418[c]	0.231

According to the result of Mauchly's sphericity test ($P < 0.05$, See Table 2), we should analyze the results of corrected tests (See Table 3). The results indicated that the position of target had no significant effect on the dragging time ($P = 0.211$) and dragging numbers ($P = 0.211$).

Table 2. Mauchly's sphericity test[b]

Within subjects effect	Mauchly's W	Approx. Chi-Square	df	Sig.	Epsilon[a]		
					Greenhouse-Geisser	Huynh-Feldt	Lower-bound
Dragging time	0.770	36.226	5	0.000	0.883	0.959	0.333
Dragging numbers	0.770	36.226	5	0.000	0.883	0.959	0.333

Table 3. Tests of within-subjects

Source	Measure	Type III Sum of squares	df	Mean square	F	Sig.
The position of target	Dragging time	3,499,074,782.022	4	874,768,695.506	1.480	0.211
	Dragging numbers	3,499,074,782.022	4	874,768,695.506	1.480	0.211

Tests the null hypothesis that the error covariance matrix of the orthonormalized transformed dependent variables is proportional to an identity matrix.

(a) May be used to adjust the degrees of freedom for the average tests of significance. Corrected tests are displayed in the Tests of Within-Subjects table.
(b) Design: Intercept + The position of target.

Within Subjects Design: Dragging time + Dragging numbers.

4 Conclusion

In this experiment, each subject performed four repeated experiments on the target initial position. Multivariable analysis was used to test the effect of the increase of experiment time and operation repeating numbers on the experimental results. The result indicated that the experiment time and operation repeating numbers did not make a difference to the five levels of the target's initial position. Then, according to the result of Mauchly's test of spherical ($P < 0.05$), we analyzed the results of corrected tests. The results indicate that there was no significant effect of target layout on dragging time and numbers, which did not support our hypothesis. The reason may be related to the setting of experimental conditions (e.g. lighting level and screen resolution). In order to continue to investigate the influence of target layout on performance based on eye-control technique, more affecting variables could be considered in future studies.

References

1. Liu, L.L., He, X.Z., Fu, C.W.: Study on usability of agricultural product web page layout based on eye tracker. In: Proceedings of the International Conference on Automatic Control and Information Engineering. pp. 78–82. Atlantis Press, Paris (2016)
2. Fitts, P.M., Jones, R.E., Milton, J.L.: Eye fixations of aircraft pilots. III. Frequency, duration and sequence fixations when flying air force ground-controlled approach system (GCA), vol. 10, pp. 1–25. Air Materiel Command (1949)
3. Fitts, P.M., Jones, R.E., Milton, J.L.: Eye movements of aircraft pilots during instrument landing approaches. Aeronaut. Eng. Rev. **9**, 1–5 (1950)
4. Raynor, K., Pollatsek, A.: The Psychology of Reading. Lawrence Erlbaum Associates, Mahwah (1994)
5. Renshaw, J.A., Finlay, J.E., Ward, R.D., Tyfa, D.: Designing for visual influence: An eye tracking study of the usability of graphical management information. In: Proceedings of the IFIP Conference on Human-Computer Interaction, pp. 144–151. IOS Press, London (2003)
6. Pušnik, N., Tihole, K., Možina, K.: Testing magazine design with eye-tracking technology. In: The 8th International Symposium on Graphic Engineering and Design, Novi Sad, Serbia, 3–4 November (2016)

A Platform for Assessing Physical Education Activity Engagement

Rafael de Pinho André(✉), Alberto Barbosa Raposo, and Hugo Fuks

Department of Informatics, Pontifícia Universidade Católica do Rio de Janeiro, Rua Marquês de São Vicence 225, Rio de Janeiro, Rio de Janeiro, Brazil
{rafael.andre,alberto.raposo, hugo.fuks}@inf.puc-rio.br

Abstract. Physical activity is an important part of the healthy development of children, improving physical, social and emotional health. One of the main challenges faced by physical educators is the assembling of a physical education program that is compelling to all individuals in a diverse group. Recent advances in Human Activity Recognition (HAR) methods and wearable technologies allow for accurate monitoring of activity levels and engagement in physical activities. In this work, we present a platform for assessing the engagement of participants in physical education activities, based on a wearable IoT device, a machine learning HAR classifier and a comprehensive experiment involving 14 diverse volunteers that resulted in about 1 million data samples. Targeting at a replicable research, we provide full hardware information and system source code.

Keywords: Physical education · Human Activity Recognition
Wearable computing and wearable sensing · Healthcare systems

1 Introduction

Physical education and school sport (PESS) aims at (i) developing children's cognitive capacity and motor skills, (ii) teaching about health and the benefits of physical activity and (iii) fostering emotional intelligence [1]. Recent research on successful PESS programs shows a strong correlation between physical activity and learning performance, school attendance and academic success of children and young people [2]. Bevans [3] discusses that an adequate exposure to PESS during school day increases children energy expenditure and allow for the maintenance of a healthy weight, and [4] suggests that physically active children have reduced chances of experiencing chronic disease factors and becoming obese throughout adolescence. However, while studies such as [5] consider that an adequate exposure to PESS for school-age children is at least 60 min per day, other studies such as [6] and [7] point out that the average physical activity level of children worldwide is low and decreasing, and there is a correlated increase in childhood obesity.

This work addresses one of the main challenges of PESS, which is the engagement of school-age children in the physical activities. By proposing a foot-based wearable IoT device and a Human Activity Recognition (HAR) classifier that can assess the

W. Karwowski and T. Ahram (Eds.): IHSI 2019, AISC 903, pp. 271–276, 2019.
https://doi.org/10.1007/978-3-030-11051-2_42

activity level of an individual during a planned physical activity, we aim at offering an alternative for PESS teachers to appraise their programs and tailor them to meet the needs of each class. The main contributions of this work are an experiment with 14 diverse individuals that resulted in about 1 million data samples for analysis and, above all, a wearable IoT device and a HAR classifier that can assess the activity level of an individual during a planned physical activity. The prototyping of the wearable IoT device, sensors deployment and replication information are shown on section Building the HAR Classifier, along with the details of the experiment conducted to develop the activity model, collect ADL data and build the HAR classifier. Section Assessing Activity Level shows the assessment of the activity level and discusses the results, and Section Conclusion presents the findings and future work.

2 Literature Review

This section presents a literature review of HAR research based on feet movement and posture information focused on health and sports activities.

2.1 Recognition of Common Movement Activities

Common movement activities recognition is the most common type of research found in this literature review. Many works, such as [8], [9] and [10] rely on plantar FSR pressure sensors to classify user activity according to a previously elaborated activity model. Other works, such as [11] and [9], rely on inertial motion units (IMUs) located on user's feet for that purpose. Sensor fusion - FSRs and IMUs - is employed by works such as [12], [13] and [14] achieving good overall results. Only a few of the surveyed works used sensors other than ground contact force (GCF) sensors and IMUs, such as infrared sensors [15] or capacitive sensing technology [16] and [17]. Some positioned extra sensors in other places beyond the user's feet, such as [18]. They all use very similar activity models comprising sitting, standing, running, walking and slope-walking activities, with the main difference being the machine learning algorithms applied and the context of the experiments.

2.2 Recognition of Specific Activities

Many of the surveyed studies were conducted in the recognition of activities related to healthcare well-being, such as (i) the research presented in [19], that aims at recognizing caregiver's patient handling activities (PHA) and movement activities to help prevent overexertion injuries, (ii) the work presented in [20], that measures activity in people with stroke, (iii) the work presented in [21], that recognizes activities and postures to provide behavioral feedback to patients recovering from a stroke, and (iv) the research proposed by [22], in which researchers present a pair of shoes that offer low-cost balance monitoring outside of laboratory environments and uses features identified by geriatric motion study experts. The lightweight smart shoes are based on the MicroLEAP wireless sensor platform [23], that uses an IMU and FSR pressure sensors embedded inside each insole for data acquisition.

Some other shoe-based wireless sensor platforms, such as the SmartStep [24], were used by many different healthcare related works. In [25], the former platform was used to develop an Android application to capture data from the wearable device and provide real time recognition of a small set of activities. In [26] and [27], the SmartShoe platform is used for energy expenditure estimation after the classification of the activities performed by the user, and in [28] it is further used to predict body weight. The same platform is then used by [29] and [30] to identify activity levels and steps in people with stroke.

2.3 Literature Review Discussion

The measuring of GCFs is the most prevalent approach used by the surveyed works for the task of recognizing user activity, followed using IMUs and sensor fusion. No work thoroughly addresses the challenge of adequately positioning the GCF sensors, although studies such as [14] and [19] recognize that this is a very important factor for HAR. Considering the wearable devices presented in the literature, two characteristics impair their reproducibility: (i) the lack of information about sensor positioning and orientation and the (ii) absence of sensor model information or specification. Most of the works analyzed provided detailed information regarding the activity model of its HAR classifiers, but few studies detailed the validation techniques used for building the activity classifiers. The success rate of activities classification, with one notable exception, fell into the 80%–100% range. It was also observed that although most works informed the number of participants, only a few informed the dataset size. Detailed knowledge of datasets is especially important to assess (i) works that use similar activity models and sensor placement and (ii) machine learning classifiers results. As discussed in [18], dataset disclosure is crucial for benchmarking purposes, given that classification algorithms rely heavily on datasets.

The prevailing suggestions for future works and contribution found in the literature follows the conclusion of the review presented in [31]: (i) increase the data set through longer data collection intervals and the diversification of participant's profiles, (ii) improve the classifier algorithms and (iii) adapt the activity model to a specific challenge, such as helping patients to avoid falls.

3 Building the HAR Classifier

On this Section, we describe the stages followed to develop the HAR classifier – prototyping the wearable device, conducting the experiment and building and validating the model.

The wearable device comprises two components: an insole that houses the plantar pressure sensors and an external protective case that houses the microcontroller and the other sensors. The insole employs four GCF sensors for monitoring plantar pressure distribution, following the recommendations found in works such as [14], in addition to the lessons learned from the prototype presented in [31]. The main component of the external protective case is a WIFI enabled microcontroller that collects and transmits sensor data to the database. The ABS 3D printed external protective case also holds the accelerometer, gyroscope, magnetometer, barometer and range finder sensors. The prototype is powered by a 2,200 mAh lithium ion battery pack.

The first experiment, aimed at building the HAR classifier, was conducted with twelve volunteers carefully selected for their diverse characteristics. We collected 12 h of activity data - 1 h of feet posture and movement data from each volunteer. The activity model we developed for the experiment comprises 10 activities: walking straight (2 km/h), walking slope up (2 km/h), walking slope down (2 km/h), slow jogging (6 km/h), slow jogging slope up (6 km/h), slow jogging slope down (6 km/h), hopping, ascending stairs, descending stairs and sitting. The experiment was conducted in 4 distinct sessions, where participants performed a subset of the planned activities. During the data acquisition stage, a stream of raw, unprocessed signals of the combined sensors was stored in the microcontroller in JSON format and periodically sent to the application server. The same data acquisition, data processing, feature extraction and feature selection pipeline proposed in [31] was employed. Different strategies were then experimented to build the classifier, and the Random Forest Algorithm with Leave-one-out Cross Validation was selected for classification achieving an average accuracy of 91.26%.

In the second experiment, our goal was to use the prototyped wearable IoT device and the HAR classifier to assess activity level of different individuals during a planned physical activity - as a proof of concept that both could be used for investigating engagement in PESS programs. Considering the most commonly practiced PESS activities of country in which the study was conducted - soccer, basketball and volleyball -, basketball was chosen to avoid exposing the foot-based device to direct physical contact in the case of football or falls to ground in the case of volleyball. Two volunteers were selected for this experiment, both without professional experience in basketball. All sessions were performed on a basketball inside a private condominium. The experiment was conducted in one session that lasted for 10 min, and data collecting followed the first experiment model. The two participants were asked to play against each other in a friendly game, without any reward for the winner.

This activity level of both participants were successfully assessed, even when we account for the error of the classifier. This result indicates that both the wearable IoT device and the HAR classifier can be used to measure activity levels of individuals and groups of individuals during physical activities. Those activity levels can be used by a qualified PESS teacher to (i) understand how an individual responds to an activity or PESS program when compared to other individuals or to his own past records, (ii) assess how a group of individuals - i.e. a class - responds to a particular activity or PESS program when compared to other groups, thus allowing for continuous improved based on this real time feedback and (iii) experiment on different PESS programs with the support of quantitative data.

4 Conclusion

In this work, we conducted two experiments: (i) the first with 12 volunteers, to evaluate the recognition of 10 different activity classes through a machine learning HAR classifier based on feet movement and posture information and (ii) the second with 2 volunteers, as a proof of concept of an alternative to investigate the engagement of individuals in PESS and validate the feasibility of our model. We were also able to expand the activity model by more than 65% - when comparing to the 6-activity classes model presented in [31] - with a drop of only 2.12% in the overall accuracy. This result suggests that the proposed wearable IoT prototype can be used for further investigation of HAR-related challenges and employed by other researchers in PESS studies.

Currently, we are employing the proposed wearable IoT device prototype in a study that aims to assess group and individual engagement in basketball PESS activities at a technical high school. We are now performing tests and reworking the protective case to allow for the experiment to commence, since the original case was not resilient enough to be used in prolonged games.

References

1. Kohl III, H.W., Cook, H.D.: Educating the student body: taking physical activity and physical education to school. In: Committee on Physical Activity and Physical Education, Institute of Medicine, pp. 199–199. National Academies Press (2013)
2. Rasberry, C.N., et al.: The association between school-based physical activity, including physical education, and academic performance. In: Preventive Medicine, pp. 10–20. Centers for Disease Control and Prevention (2011)
3. Bevans, K.B., et al.: Physical education resources, class management, and student physical activity levels. J. Sch. Health **80**, 573–580 (2010)
4. Carlson, S.A., et al.: Physical education and academic achievement in elementary school. Am. J. Public Health **98**, 721–727 (2008). American Public Health Association
5. Sollerhed, A.C.: Young today – adult tomorrow! Ph.D. thesis, Lund University, Sweden (2006)
6. Tester, G., et al.: A 30-year journey of monitoring fitness and skill outcomes in physical education. Published online at Scientific Research, 7 (2014)
7. Morgan, P., Hansen, V.: Classroom teachers' perceptions of the impact of barriers to teaching physical education on the quality of physical education programs. Res. Q. Exerc. Sport. **79**, 506–516 (2008). National Institutes of Health
8. De Santis, A., et al.: A simple object for elderly vitality monitoring. In: Mechatronic and Embedded Systems and Applications (MESA), ASME, pp. 1–6. IEEE (2014)
9. Drobny, D., et al.: Saltate!: a sensor-based system to support dance beginners. In: CHI 2009 Extended Abstracts on Human Factors in Computing Systems, pp. 3943–3948. ACM (2009)
10. Doppler, J., et al.: Variability in foot-worn sensor placement for activity recognition. In: International Symposium on Wearable Computers, 2009. ISWC 2009, pp. 143–144. IEEE (2009)
11. Ghobadi, M., Esfahani, E.T.: Foot-mounted inertial measurement unit for activity classification. In: Engineering in Medicine and Biology Society, EMBC, pp. 6294–6297. IEEE (2014)
12. Tang, W., Sazonov, E.S.: Highly accurate recognition of human postures and activities through classification with rejection. IEEE J. Biomed. Health Inform. **18**(1), 309–315 (2014)
13. Sazonov, E.S., et al.: Monitoring of posture allocations and activities by a shoe-based wearable sensor. IEEE Trans. Biomed. Eng. **58**(4), 983–990 (2011)
14. Lin, F., et al.: Smart insole: a wearable sensor device for unobtrusive gait monitoring in daily life. Trans. Ind. Inform. **12**(6), 2281–2291 (2016)

15. Jiang, X., et al.: AIR: recognizing activity through IR based distance sensing on feet. In: International Joint Conference on Pervasive and Ubiquitous Computing: Adjunct, pp. 97–100. ACM (2016)
16. Haescher, M., et al.: CapWalk: a capacitive recognition of walking-based activities as a wearable assistive technology. In: International Conference on Pervasive Technologies Related to Assistive Environments, p. 35. ACM (2015)
17. Matthies, D.J.C., et al.: CapSoles: who is walking on what kind of floor? In: Proceedings of 19th International Conference on Human-Computer Interaction with Mobile Devices and Services (2017)
18. Zhu, C., Sheng, W.: Multi-sensor fusion for human daily activity recognition in robot-assisted living. In International Conference on Human Robot Interaction, pp. 303–304. ACM (2009)
19. Lin, F., et al.: Sensing from the bottom: smart insole enabled patient handling activity recognition through manifold learning. In: Connected Health: Applications, Systems and Engineering Technologies (CHASE), pp. 254–263. IEEE (2016)
20. Zhang, T., et al.: Using decision trees to measure activities in people with stroke. In: Engineering in Medicine and Biology Society (EMBC), pp. 6337–6340. IEEE (2013)
21. Edgar, S.R., et al.: Wearable shoe-based device for rehabilitation of stroke patients. In: Engineering in Medicine and Biology Society (EMBC), pp. 3772–3775. IEEE (2010)
22. Noshadi, H., et al.: HERMES: mobile system for instability analysis and balance assessment. ACM Trans. Embed. Comput. Syst. **12**:57 (2013)
23. Au, L.K. et al.: MicroLEAP: energy-aware wireless sensor platform for biomedical sensing applications. In: Biomedical Circuits and Systems Conference. BIOCAS, pp. 158–162. IEEE (2007)
24. Hegde, N., Sazonov, E.: Smartstep: a fully integrated, low-power insole monitor. Electronics **3**(2), 381–397 (2014)
25. Hegde, N., et al.: Development of a real time activity monitoring android application utilizing smartstep. In: Engineering in Medicine and Biology Society, pp. 1886–1889. IEEE (2016)
26. Sazonov, E., et al.: Posture and activity recognition and energy expenditure estimation in a wearable platform. IEEE J. Biomed. Inform. **19**, 1339–1346 (2015)
27. Sazonova, N., et al.: Accurate prediction of energy expenditure using a shoe-based activity monitor. Med. Sci. Sports Exerc. **43**(7), 1312–1321 (2011)
28. Sazonova, N.A., et al.: Prediction of bodyweight and energy expenditure using point pressure and foot acceleration measurements. Open Biomed. Eng. J. **5**, 110 (2011)
29. Fulk, G.D., Sazonov, E.: Using sensors to measure activity in people with stroke. Top. Stroke Rehabil. **18**(6), 746–757 (2011)
30. Fulk, G.D., et al.: Identifying activity levels and steps in people with stroke using a novel shoe-based sensor. J. Neurol. Phys. Ther. **36**(2), 100 (2012)
31. De Pinho André, R., et al.: Bottom-up investigation: human activity recognition based on feet movement and posture information. In: Proceedings of the 4th International Workshop on Sensor-based Activity Recognition and Interaction, pp. 10:1–10:6 (2017)

Adaptive Learning for Robots in Public Spaces

Xiaohua Sun(✉), Jan Dornig, and Shengchen Zhang

Tongji University, Shanghai, China
{xsun, seanzhang}@tongji.edu.cn, jandornig@gmail.com

Abstract. Proper functioning of robots deployed in public spaces often require extensive knowledge of its environment of use, which is completely unknown prior to deployment. The methods for acquiring and utilizing such knowledge also varies depending on the nature of the public space and the tasks the robot needs to perform. This calls for development and application of adaptive learning methods specifically designed to take into consideration the nature and key properties of various public spaces and robotic tasks. In this paper, we study typical types of public spaces for deployment of robots, and analyze robotic tasks required in each type of space to derive common capabilities that the robots need to have. We then consider three adaptive learning methods: (1) autonomous learning, (2) unsupervised learning from real-time on-site data, and (3) guided learning. Applicability of the methods to improve each common capability and possible means of application are further discussed.

Keywords: Human robot interaction · Robotic and design · Artificial intelligence and design

1 Introduction

After decades of experimental research and tentative applications, robots are still a rare sight in public spaces. Applications like performance [1], working with homeless people [2], delivery [3, 4] and entertainment [5] are mostly still experimental, but have shown promising results. One obstacle hindering wider application is the complexity and dynamic nature of the environment that the robot operates in. In this case, simulations and lab environment tests are often inadequate to prepare for real world scenarios. In the recent paper "A Berkeley View of Systems Challenges for AI", the researchers present this problem and highlight *Acting in dynamic environments* and especially *Continual Learning* as one of the most promising research opportunities for AI. They continue to state how especially in the field of robotics, the ability to perform online learning will be crucial for robotic applications. And point specifically to methods like reinforcement learning and simulated reality to enable robots to make the right choices in dynamic, mission critical situations [6].

Other efforts include work like multiple "network robot system" frameworks, which show a concerted network of multi-robot and components systems and how they

W. Karwowski and T. Ahram (Eds.): IHSI 2019, AISC 903, pp. 277–285, 2019.
https://doi.org/10.1007/978-3-030-11051-2_43

use technologies to deal with their environment. Though as Glas et al. present their extensive system architecture, little room was given so far to consider the differences between types and demands of public spaces and scenarios are mostly focused on indoor places like shopping malls, exhibition spaces or building lobbies [7, 8].

Some studies are also concerned with exploring strategies for the robot to use adaptive capabilities to deal with social situations that the robot encounters in public spaces. Social norms and multi-party situations present complex hierarchies and decisions which can change depending on participants, children, adults, staff or visitor, who also can have varying expectations dependent on the space [9, 10]. [11] used a co-creation approach and facilitated workshops with the stakeholders to understand their expectations, concerns and how success would be defined.

In the light of this, this paper looks at the possibilities of using adaptive capabilities in robots to facilitate better functionality in robots deployed in public space. We outline types of public spaces, areas of adaptation and their possible influence on robot capabilities, with the goal to facilitate designing and building robotic systems that are able to adapt to and learn from the environment.

2 Analysis of Public Spaces and Corresponding Robotic Capabilities

Typical functionalities like autonomous navigation, natural language interaction and screen-based interaction are what allows the robot to function as desired in complex environments like public spaces. However, such functionalities are often hindered by the rapid changing environment and a lack of knowledge of its immediate surroundings. These problems can largely be attributed to the nature of each individual public space, whose details are completely unknown to a robot prior to deployment. This calls for development and application of adaptive learning methods specifically designed to take into consideration the nature and key properties of the robot's environment of use. In this chapter, we begin by analyzing possible types of public spaces for robot deployment and proceed to derive the tasks required for the robot to perform and common capabilities needed for proper functioning.

2.1 Types of Public Spaces

Depending on the type of space, the responsible institution and ongoing activities, each type of public space harbors different robotic tasks. Different sets of challenges are also posed, which requires the application of adaptive learning techniques. In this section, we analyze categories of public spaces by robotic tasks and challenges posed. The resulting correspondence is shown in Table 1. Robotic tasks and challenges are referred to by their indices for brevity. Example spaces are also given for each category.

Detailed definition of each type of robotic tasks and challenges are further expanded in Tables 2 and 3.

Table 1. Categories of public spaces, corresponding robotic tasks and challenges.

Category	Examples	Robotic tasks	Challenges
S1	Restaurants, cafes	T1, 3, 6	Ch1, 2, 3, 6
S2	Malls, plazas, markets	T1, 2, 3, 4	Ch1, 3, 4, 6, 7
S3	Train stations, airports, subway stations	T1, 2, 3, 8, 9	Ch1, 3, 4, 5, 7
S4	Hotels, hospitals	T1, 2, 3, 7	Ch1, 2, 6, 7
S5	Museums, exhibitions, tours	T1, 2, 3	Ch6, 7
S6	Libraries, town halls	T1, 2, 3, 7	Ch1, 2, 6, 7, 9
S7	Community areas, campuses, offices	T1, 2, 3, 4, 7	Ch1, 2, 6, 7, 8
S8	Lobbies (in theaters, banks, office buildings)	T1, 2, 5, 8	Ch2, 6, 7
S9	Stores	T2, 5, 7	Ch1, 2, 3

Table 2. Main robotic tasks.

Robotic tasks	Description
T1. Physical guidance	Guiding users to a specific destination or through certain space
T2. Information/help	Responding to user inquiries with corresponding information
T3. Maintenance	Repairs, janitorial services, installation, etc.
T4. Entertainment	Performance, interactive games, etc.
T5. Sales	Performing retail duties, providing information, selling products/service, or advertising the brand
T6. Waiting	Serving as waiter at restaurants, cafes, etc.
T7. Delivery of goods	Performing longer distance delivery duties
T8. Security	Surveillance, patrol and alarm, etc.
T9. Transportation	Transporting goods or users to a specific location in the same space

Table 3. Types of challenges posed by public spaces, and corresponding demands for adaptive learning techniques.

Challenges	Demands for adaptive learning techniques
Ch1. Demand for recognizing and handling unknown objects	The robot's task may require handling objects that is specific to the space, which cannot be pre-trained
Ch2. Demand for recognizing and tracking previously unknown individuals.	The robot's task may require recognizing and tracking human users that may exhibit traits and behaviors not included in factory settings
Ch3. Rapidly changing environment	The objects in the environment or the structure of the environment may undergo rapid modification
Ch4. Possibility of damage/vandalism	The robot may be damaged due to intentional/unintentional mistreatment. This situation could be improved by enabling detection of human intentions
Ch5. High density pedestrian flow	High density pedestrian flow cannot be taken in to account a priori, which causes great difficulty in the robot's navigation
Ch6. Demand for natural interaction	The robot's role is to the public, which may require natural and easy-to-use interaction that adapts to the user
Ch7. Demand for detailed general knowledge of environment	The robot's role may require answering inquiries about events and things in its surroundings, which is unknown prior to deployment
Ch8. Demand for detailed personal knowledge	The robot's role may require personalized interaction using knowledge of regular users
Ch9. Limited modality	Certain modality is unusable either due to regulation, norms or due to specific conditions of the space

2.2 Task-Specific Robotic Capabilities

We proceed to analyze the robotic capabilities required for improving the performance of the robotic tasks mentioned above. Common capabilities among them are what we argue will contribute the most to overall performance improvement, if improved by applying adaptive learning methods (Table 4).

Table 4. Possible robotic tasks and capabilities needed.

	T1	T2	T3	T4	T5	T6	T7	T8	T9
C1. Navigation/wayfinding	√		√			√	√	√	√
C2. Communication	√	√		√	√	√	√	√	√
C3. Carrying humans and luggage									√
C4. Activity & identity recognition	√	√				√		√	
C5. Object recognition & handling	√		√			√	√	√	√
C6. Relevant knowledge	√	√			√	√			
C7. Cleaning			√						
C8. Payment processing					√	√			
C9. Performance				√					

3 Adaptive Learning for Common Capability Improvements

In the process of adapting the robot to its environment, the system has to go through different stages towards the improvement. As starting status, the robot will have the pre-programmed baseline setup. Once deployed, it will start gathering data as discussed before. Crucial for success is the successful analyses of any such data. Shortcomings and room for improvement have to be recognized. We identify three broader aspects of learning. They are not exclusive to each other and might be used as different parts of a holistic process: (1) autonomous learning, (2) unsupervised learning from real-time on-site data, and (3) guided learning.

In any case, limits have to be set to how the performance of the robots can vary and how much it can change its behavior autonomously. We define this as “freedom of adaptation”, it determines how much leeway a robot can be given to autonomously adapt to its environment. These margins would be used to test different possibilities and behaviors before refining the parameters to adapt to the current situation and ideally would routinely test for ongoing changes in the environment that would warrant new adjustments in the robot. These limits can define the limit for overall deviation from the baseline but also the maximum variation in parameters each time, which is to ensure that though the robot is testing possible improvements, it continues to display a continuous performance without seeming random.

Below we further describe each type of adaptive learning method.

1. *Autonomous learning* is a process that the robot self-initiates. This can be achieved by incorporating either continuously running systems that self evaluate the robot’s situation or trigger situations that the robot would react to. In certain situations, the robot asks for help from humans and utilizes the help to improve the situation.
2. *Unsupervised learning from real-time on-site data* can use an automated system utilizing data mining and unsupervised learning algorithms to learn from real-time onsite data. Methods like reinforcement learning and knowledge graph embedding-based reasoning could also prove helpful depending on the challenges. If multiple high-level options are being considered, methods like A/B testing can be employed on a local or broader scale, testing if options can be provable advantageous for the robot and if it applies only locally. Similarly, robots might employ a multi-armed bandit approach to test between different possibilities.
3. *Guided learning* is an alternative that involves humans. The gathered data might be used for analysis by an expert in the loop and certain situations like a lack of requested information might lead to clear gaps, which can be filled by the human counterpart. The robot might request human assistance, or the human initiates it. The human on-site and after being informed of the problem, or having seen the robot misbehaving, steps in to teach the robot the correct way in a choreographed interaction. This can happen through learning by demonstration or other interactions which should be designed based on the robot’s tasks specifically to support a swift and productive process between local personal and robot.

In this chapter, we study methods for applying adaptive learning methods to the common capabilities we analyzed in the former chapter for improved performance. To further simplify the vast possibilities of features, we assume that the very elementary functions like motor and sensor control are already given. We then look for possible areas of improvements and adaptations in the respective capabilities.

3.1 Navigation/Wayfinding

The awareness of the current physical position in a larger context of varying degree and the ability to navigate the space without collision, based on different forms of data like stored maps as well as sensor input.

Adaptation: Unsupervised learning from real-time on-site data apply to this kind of problems. The robot could profit from continuous updating of maps with multidimensional information, like the inclusion of time-sensitive data with respective algorithms predicting the likeliness of path obstruction during specific times in the day. The robot might update the map with temporary and permanent changes to the physical environment. Trial and error approaches to learn from and continuous updating with temporary information, to predict the likeliness to encounter an obstacle at a specific point in the map, with time sensitivity if possible, might also help.

3.2 Recognizing Individual Human Activity

The ability to recognize an approaching user, predict future paths and recognize human activity from skeletal or image data. This can be used to avoid collisions and foster natural human-robot interaction.

Adaptation: Unsupervised learning from real-time on-site data applies. Using simulation and prediction methods based on data captured on-site will be key for robots to act in the right ways without needing to much time on actual trial and error actions in the real world, which are time consuming. [12] studied capability of an algorithm trained to predict human movements and actions from their current body movement. It is not far-fetched to reason that people in public spaces exhibit similar behaviors that the robot can be trained on. Similarly, the robot might learn to differentiate who in this particular space exhibits behavior that means they are about to approach the robot.

3.3 Communication

For a social robot, the need for communication is very high and might include multimodal communication means like speech, gesturing as well as interface input/output. Robots that perform non-social tasks still require certain communication functionalities. They must be able to communicate about their status in situations like initial deployment, task failure, or regular maintenance. These involve mostly trained personal and minimum multimodality.

Adaptation: Autonomous learning methods could be applied to this situation. In terms of communication, parallels can be drawn from other media that performs communication optimization. The methodology of A/B testing can be adjusted to fit the context of robots. With every interaction, the robots can gather feedback on the performance.

Depending on the task and goals, the metrics have to be structured accordingly. For entertainment robots, it might be about keeping the human engaged as long as possible, while a sales robot should rather be judged on resulting products sold. Factors like phrases used, structure of a conversation as well as how multimodal means are used can all be tested this way.

At the same time, unsupervised learning can also be performed on gathered onsite data. Technologies like Natural Language Processing can benefit from the collected interactions and language samples and used to further increase diversity in datasets.

3.4 Relevant Knowledge

The robot might provide a visitor with knowledge about the space similar as what is expected from an information booth. This requires the robot to have an up-to-date database and the ability to understand and answer user input through various possible ways like voice or screen-based interaction or other expressive behavior.

Adaptation: Autonomous learning and guided learning both fit for this situation. Though any preparation will be done to avoid missing information, it is only natural that people would ask the robot things he does not currently know. A knowledge graph could enable it to not only structure its current and newfound knowledge but also infer new relations. [13] While the robot probably would get irrelevant questions that don't fall into his responsibility, recurring questions from different people might signify missing information and the robot could have a process in place to let staff or the servicing company know about this lack of knowledge and receive fitting updates. A more natural interaction could be to use a "smart" process that lets the robot get the help of human staff for unknown customer requests and let it witness and learn from the interaction between human staff and customer. Similarly, the robot might need to be updated with current information when changes in the environment happen. The design of a specific human-robot interaction that enables local staff to brief the robot on changes, like it is already doing with traditional staff, would be beneficial too.

Unsupervised learning methods can also be applied to data gathered in the process. Apart from the actual knowledge, the robot could learn about frequently asked questions and either restructure information to automatically include often requested information in initial answers and adjust dialog to provide it directly. In case of a shopping mall, it might learn that people enquire different things about different businesses for example and provide these more directly instead of a general answer for all businesses alike.

3.5 Object Recognition and Handling

Actual environments of use may present new objects that is not the robot has no prior knowledge for methods of recognition and handling. The robot needs to adaptively learn such knowledge.

Adaptation: This case especially calls for the application of autonomous and guided learning methods. The recognition of objects will need continuous updating in most environments. The shape, color and location of objects might change frequently in different locations, especially shops. The robot needs to be able to express what it knows and what it doesn't, so that interactions can be facilitated which allow the humans to teach the robot about the objects it has to work with. The ability to handle the objects might need a specific training preparation, preferably can be achieved by demonstration. As it is a time-consuming task to teach a robot to handle objects, it would be necessary to design a natural process.

Unsupervised method like simulated reality would be one way to bridge the gap. As the robot perceives real world data like shape, material, weight of an object, it would relay this to a digital simulation which attempts to learn and predict the best course of action.

4 Conclusion

Proper functioning of robots deployed in public spaces often require extensive knowledge of its environment of use, which is completely unknown prior to deployment. The methods for acquiring and utilizing such knowledge also varies depending on the nature of the public space and the tasks the robot needs to perform. In this paper, we addressed these problems by presenting adaptive learning methods to enhance performance of public space robots. We studied typical types of public spaces for deployment of robots, and analyzed robotic tasks required in each type of spaces. We derived five types of common capabilities that the robots need to have: (1) navigation/wayfinding, (2) recognizing individual human activity, (3) communication, (4) relevant knowledge and (5) object grasping and handling. Three adaptive learning methods are then considered: (1) autonomous learning, (2) unsupervised learning from real-time on-site data, and (3) guided learning. We discussed the applicability of each adaptive learning method for the common capabilities and proposed possible ways of implementation. The results provide a basis for further research into the implementation of each method in specific field settings.

References

1. Sone, Y.: Futuristic spectacle: robot performances at expos. In: Japanese Robot Culture. Palgrave Macmillan, New York (2017)
2. Security robot Knightscope article. http://www.businessinsider.com/security-robots-are-monitoring-the-homeless-in-san-fraccisco-2017-12. Accessed 31 Mar 2018
3. Delivery robots article. https://www.cnbc.com/2015/11/02/forget-delivery-drones-meet-your-new-delivery-robot.html. Accessed 31 Mar 2018
4. Domino´s delivery robot article. https://www.dominos.com.au/inside-dominos/technology/dru. Accessed 31 Mar 2018
5. Pepper Robot. https://www.ald.softbankrobotics.com/en/robots/pepper. Accessed 31 Mar 2018
6. Stoica, I., et al.: A Berkeley view of systems challenges for AI. CoRR abs/1712.05855 (2017)
7. Liu, Y., Yang, J., Wu, Z.: Ubiquitous and cooperative network robot system within a service framework. Int. J. Humanoid Robot. **8**(01), 147–167 (2011)

8. Glas, D.F., Satake, S., Ferreri, F., Kanda, T., Hagita, N., Ishiguro, H.: The network robot system: enabling social human-robot interaction in public spaces. J. Hum.-Robot Interact. **1** (2), 5–32 (2013)
9. Mussakhojayeva, S., Kalidolda, N., Sandygulova, A.: Adaptive strategies for multi-party interactions with robots in public spaces. In: International Conference on Social Robotics, pp. 749–758. Springer, Cham (2017)
10. Mussakhojayeva, S., Sandygulova, A.: Cross-cultural differences for adaptive strategies of robots in public spaces. In: 2017 26th IEEE International Symposium on Robot and Human Interactive Communication (RO-MAN), pp. 573–578. IEEE (2017)
11. Niemelä, M., Heikkilä, P., Lammi, H.: A social service robot in a shopping mall: expectations of the management, retailers and consumers. In: Proceedings of the Companion of the 2017 ACM/IEEE International Conference on Human-Robot Interaction (HRI 2017), pp. 227–228. ACM, New York (2017)
12. Schydlo, P., Rakovic, M., Jamone, L., Santos-Victor, J.: Anticipation in human-robot cooperation: a recurrent neural network approach for multiple action sequences prediction. arXiv:1802.10503 [cs.HC]. 28 February 2018
13. Nickel, M., Murphy, K., Tresp, V., Gabrilovich, E.: A review of relational machine learning for knowledge graphs. Proc. IEEE **104**(1), 11–33 (2016)

How to Design Assembly Assistance Systems

Sven Hinrichsen[1(✉)] and Manfred Bornewasser[2]

[1] Industrial Engineering Lab, Ostwestfalen-Lippe University of Applied Sciences, Liebigstr. 87, 32657 Lemgo, Germany
sven.hinrichsen@hs-owl.de
[2] Institute of Psychology, University of Greifswald, Greifswald, Germany

Abstract. Empirical research shows that the informational design of manual assembly systems is becoming increasingly important in the light of growing complexity. Assembly assistance systems supply employees with information according to their needs and individual situation. This article aims to present important principles for the design of informational assembly assistance systems. The empirical basis for these principles is formed by projects involving the introduction of informational assistance systems for assembly work. The trends and design recommendations are explained using a model, which illustrates important associations between the complexity of assembly tasks, the demands on the mental capacity of employees, work productivity and the use of assembly assistance systems.

Keywords: Assistance systems · Manual assembly · Cognitive ergonomics

1 Introduction

The higher the complexity level of manual assembly from the viewpoint of the assembly worker, the greater the focus on the informational design of the assembly system should be. Complexity mainly results from a large number of product and component versions, from small batch sizes, and from a rapid rate of change and development in the products themselves [1]. Besides the trend toward mixed model assembly systems [2], other developments can be observed here. For example, an ever-growing number of additional functions are being incorporated in products, so increasing the number of components for assembly [3, 4].

Empirical research concerning the supply of information in the manual assembly environment shows that—in addition to the work of Hollnagel [5]—five categories of deficits can be identified [6]: (1) Necessary information is missing in the assembly system. (2) Unnecessary information is displayed. (3) Information is supplied at the wrong time and in the wrong quantity. (4) Information is not up to date and/or (5) not prepared in such a way that it is easy for the assembly worker to comprehend and process. This results, for example, in interruptions in work, search operations or reworking, leading to poor acceptance of information management. The consequence of these deficits in conjunction with new technological possibilities has been the development and implementation of various assistance system technologies for assembly in recent years.

W. Karwowski and T. Ahram (Eds.): IHSI 2019, AISC 903, pp. 286–292, 2019.
https://doi.org/10.1007/978-3-030-11051-2_44

The aim of this article is to present important principles for the design of informational assistance systems for manual assembly tasks. The empirical basis of these principles is formed by projects involving the introduction of informational assistance systems for assembly work. The design principles are thus above all based on the requirements of industry and its employees, who, as the customer, define the demands of each project. These industrial companies mostly belong to the mechanical engineering sector. The trends and design recommendations are explained using a model, which illustrates important associations between the complexity of assembly tasks, work productivity, mental strain of employees, and the use of assembly assistance systems.

2 Fundamentals

The informational design of a manual assembly system is a sub-area of ergonomic assembly system design [7], as shown in Fig. 1. This sub-area investigates how information should be displayed to ensure the effective and efficient comprehension and processing of this information in sensory terms. Efficiency means here that the process of comprehension and processing of the information by humans should be as short as possible. By effectiveness we understand that the subsequent assembly operations ensuing from processing of the information lead to the desired result: A product that has been flawlessly assembled. At the same time, the demands on the mental resources of the assembly workers, which have become more demanding due to the trend toward increasing complexity, should be reduced to an appropriate degree.

As shown in Fig. 1, informational assembly system design can be either explicit by means of work instructions, or implicit through the design of assembly system elements. Design is implicit, for example, when component containers, tools and devices are provided with color coding [8]. Such color codes tell the assembly worker, for instance, that objects with the same colored markings belong to a specific assembly set or product type. Informational aspects can also be taken into consideration via product design. For example, the risk of confusing parts that closely resemble each other in terms of their geometry can be reduced during assembly by designing the product with special tactile or visual features [7]. The explicit communication of information can take place via paper-based work documents [9] or in digital form. The digital delivery of information can in turn be either static or dynamic as regards the volume of information and the time of supplying individual items of information. The static delivery of information takes place where assembly instructions exist for a product—e.g. in the form of a pdf document—and they are displayed in their entirety—for instance, on a monitor. The dynamic delivery of information takes place when information is supplied to assembly workers according to their current needs and situation. Technical systems for the dynamic information delivery are known as assistance systems. Assistance systems capture data via sensors and inputs and process it to supply employees with the right information ("what"), at the right time ("when"), and in the required form ("how") [2, 5, 10].

While automation aims to entirely replace the work of man by machines, assistance systems combine the special skills of humans with the positive properties of technical systems. The relevant literature also refers to assembly assistance systems as worker guidance systems [11], cognitive information systems [2] or worker information systems [12]. Assembly assistance systems can help to make up for functional limitations of humans (compensation), protect human health (prevention), and support the learning processes of humans (capability) [13]. In addition, informational assistance systems contribute to reducing training times for new staff, boosting work productivity and avoiding assembly errors, thereby assuring product quality [14]. Informational assistance systems moreover make a significant contribution to ensuring continuous digital value chains within a company. They also boost productivity in terms of administration, as the design and management of assembly instructions is simplified or partially automated [15].

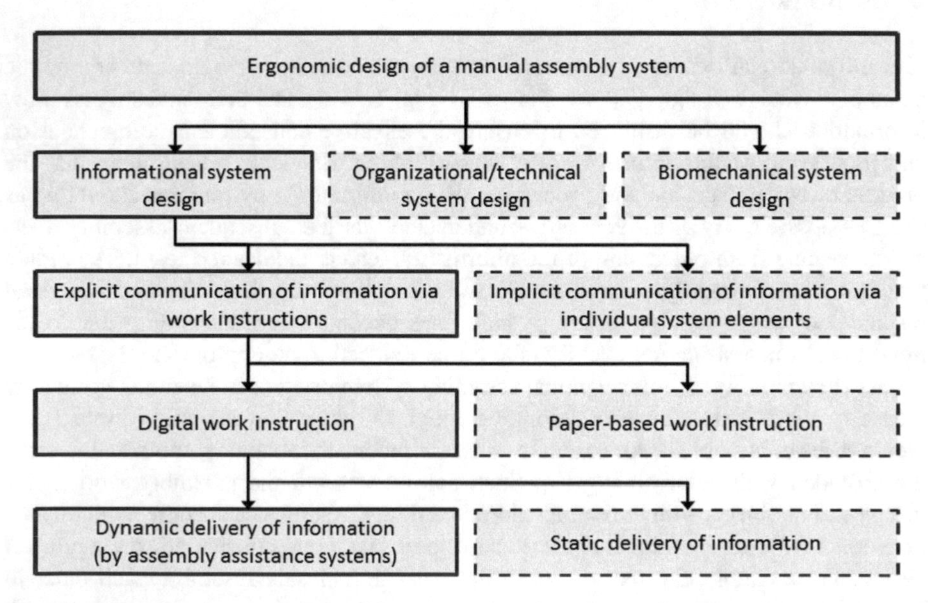

Fig. 1. Dynamic delivery of information as part of the ergonomic design of a manual assembly system

3 Recommendations

The main design recommendations for assembly assistance systems are explained using the model shown in Fig. 2. The model is based on the assumption that a high informational complexity as an independent variable results in high informational strain, which have an adverse effect on work productivity [4]. Such impacts can be mitigated or eliminated by moderating variables—shown in gray in Fig. 2—by means of the following:

- application of mental resources and information processing effort can be reduced for employees with an assistance system featuring user-centered design
- developing employee skills in dealing with work system-specific complexity

– incorporating assistance systems in existing technical systems to avoid, e.g. media discontinuity, so making additional demands on workers.

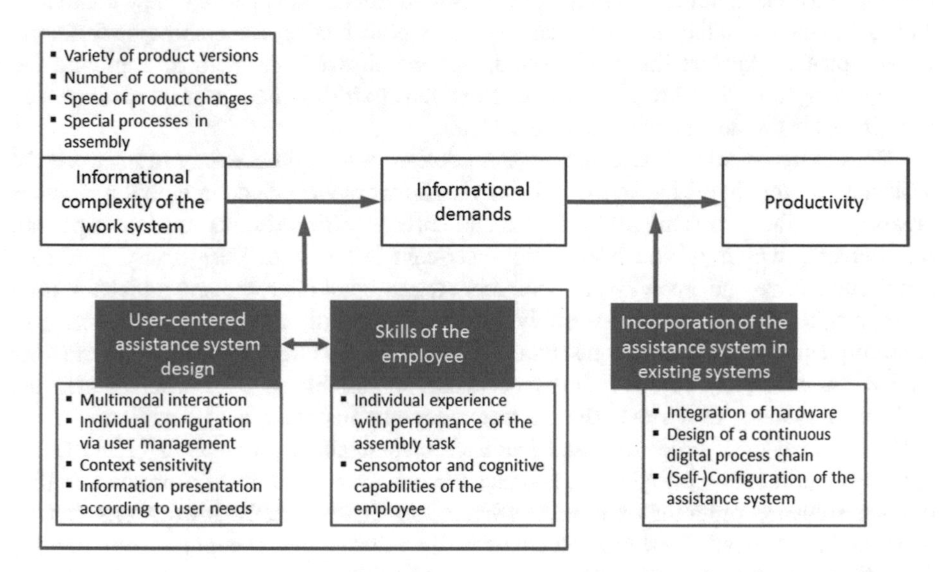

Fig. 2. Impact of assembly assistance systems on informational demands and productivity [4]

User-centered assistance systems act as a moderating variable, as shown in Fig. 2. They can reduce the perceived level of complexity, offer guidance to employees and thus help to improve work productivity. Key features of such an assistance system include the possibility of multimodal interaction, the creation of individual system configurations, the context sensitivity of the system, and the preparation of the assembly information according to user needs. The trend in the development of assembly assistance systems is clearly away from unimodal toward multimodal man-machine interaction. The employee can use voice, gesture or key-based inputs to control the system and can switch between these inputting modes in any way required during assembly without having to change the system settings. To date, assembly assistance systems have been frequently employed to create standardized assembly instructions, which are then used by all workers in the assembly system, regardless of their individual characteristics and needs. In future, employees will increasingly enjoy support from individualized assembly instructions and work conditions, for example taking into account the individual level of training. Assistance systems are described as context-adaptive when the system detects changes in the working environment, in performance of the task or in the person carrying out the work, and reacts to such changes in real time [16]. In association with the scope of an employee's experience, context sensitivity means that the assistance system is capable of assessing on the basis of historical data to what extent a worker needs support with individual steps. As regards the preparation of assembly information, various design principles should be taken into account. An overview of these principles is offered by [17]. In line with the model shown here, there is close interaction between the perceived level of complexity, the functionality of the assistance system and the skills of the employee [4]. The more experience workers have with an assembly task, the less complex they will consider it to be. The assistance system is thus felt by the employee as less helpful and may—if the system fails to adapt to the new context or cannot be adapted to the individual case—then be perceived as bothersome, so forfeiting its acceptance. Against this background, options allowing workers to configure the assistance system themselves or the context sensitivity of the system constitute key principles for the design of assistance systems.

Besides user-centered assistance system design, it is also necessary to integrate the assistance system in the hardware and software infrastructure. Such integration includes creating interfaces to standard hardware and software. It represents another important requirement, which is consistent with user-centered system design. By hardware infrastructure we understand work equipment (e.g. digital micrometers) which are used by employees during manual assembly tasks. Besides designing interfaces to standard assembly equipment, it is also necessary to implement a direct or indirect interface to an ERP system [15]. A direct interface makes it possible to read order data to the assistance system to retrieve the correct assembly instructions. In addition to the automatic read-in of order data and feedback of assembled products or orders to the ERP system, it may be logical to connect the assistance system to a product configuration software program or a PLM system, so that assembly instructions can be generated dynamically, and product changes can be taken into account in the assembly system automatically. Such integration of the assistance system in the existing IT systems make an important contribution to designing continuous digitally supported value chains as a vision of the so-called Industry 4.0 [15].

The software for assistance systems is being increasingly designed in such a way that industrial users can adapt the assistance system to the operational requirements with a configurator and develop assembly instructions. This involves taking decisions on information output devices (e.g. projector, monitor), information input devices and/or sensors (e.g. touch display, microphone for voice input, optical sensor for gesture recognition), and the hardware required (e.g. digital micrometer). The objective must be for the assistance system software to directly detect and integrate connected devices in terms of self-configuration (“plug & produce”).

4 Conclusion

Assistance systems make a significant contribution to the informational design of manual assembly systems. They can help to reduce the perceived complexity of work in mixed model assembly systems and with it, the demands made on the mental capacity of employees by offering them guidance via the system. There is close interaction here between the perceived level of complexity, the functionality of the assistance system and the development of employee skills, whereby these aspects need to be adapted to each other dynamically. In the past there was a tendency to assign only small-scale tasks to employees in order to take advantage of specialist skills and to guarantee high work productivity. In the future assistance systems will offer new opportunities to give assembly workers more extensive tasks without an increase in errors during assembly or any decline in productivity.

References

1. Stork, S., Schubö, A.: Human cognition in manual assembly: theories and applications. Adv. Eng. Inform. **24**(3), 320–328 (2010)
2. Claeys, A., Hoedt, S., Soete, N., Van Landeghem, H., Cottyn, J.: Framework for evaluating cognitive support in mixed model assembly systems. IFAC-PapersOnLine **48**(3), 924–929 (2015)
3. Brecher, C., Kolster, D., Herfs, W.: Innovative Benutzerschnittstellen für die Bedienpanels von Werkzeugmaschinen. ZWF Zeitschrift für wirtschaftlichen Fabrikbetrieb **106**(7–8), 553–556 (2011)
4. Bornewasser, M., Bläsing, D., Hinrichsen, S.: Informatorische Assistenzsysteme in der manuellen Montage: Ein nützliches Werkzeug zur Reduktion mentaler Beanspruchung? Zeitschrift für Arbeitswissenschaft **73**, 264–275 (2018)
5. Hollnagel, E.: Information and reasoning in intelligent decision support systems. Int. J. Man-Mach. Stud. **27**(5–6), 665–678 (1987)
6. Hinrichsen, S., Bendzioch, S.: How digital assistance systems improve work productivity in assembly. In: Nunes, I.L. (ed.) Advances in Human Factors and Systems Interaction. Proceedings of the AHFE 2018 International Conference on Human Factors and Systems Interaction, 21–25 July 2018, Orlando, pp. 332–342. Springer, Cham (2018)
7. Luczak, H: Manuelle Montagesysteme. In: Spur, G., Stöferle, T. (eds.) Handbuch der Fertigungstechnik, Fügen, Handhaben und Montieren, vol. 5, pp. 620–682. Hanser, Munich (1986)
8. Takeda, H.: LCIA–low Cost Intelligent Automation: Produktivitätsvorteile durch Einfachautomatisierung. mi-Fachverlag, Landsberg am Lech (2006)
9. Thorvald, P., Backstrand, G., Hogberg, D., De Vin, L.J., Case, K.: Information presentation in manual assembly: a cognitive ergonomics analysis. In: Proceedings of 40th Annual Nordic Ergonomics Society Conference (NES 2008): Ergonomics is a Lifestyle, Reykjavik, Iceland (2008)
10. Hinrichsen, S., Riediger, D., Unrau, A.: Assistance systems in manual assembly. In: Villmer, F.-J., Padoano, E. (eds.) Proceedings of 6th International Conference on Production Engineering and Management, pp. 3–14 (2016)
11. Wiesbeck, M.: Struktur zur Repräsentation der Montagesequenzen für die situationsorientierte Werkerführung. Dissertation, TUM, Herbert Utz, Munich (2014)
12. Lušić, M.: Ein Vorgehensmodell zur Erstellung montageführender Werkerinformationssysteme simultan zum Produktentstehungsprozess: Bericht aus dem Lehrstuhl für Fertigungsautomatisierung und Produktionssystematik. Meisenbach, Bamberg (2017)
13. Apt, W., Bovenschulte, M., Priesack, K., Weiß, C., Hartmann, E.A.: Einsatz von digitalen Assistenzsystemen im Betrieb. Forschungsbericht 502 des Bundesministeriums für Arbeit und Soziales, Berlin (2018)
14. Alexander, U., Hinrichsen, S., Riediger, D.: Development of projection based assistance system for manual assembly. In: 6th International Ergonomics Conference on Ergonomics 2016 – Focus on Synergy, Zadar, Zagreb, Croatia, pp. 365–370 (2016)
15. Hinrichsen, S., Riediger, D., Unrau, A.: Montageassistenzsysteme – Begriff Entwicklungstrends und Umsetzungsbeispiele. Betriebspraxis & Arbeitsforschung **232**, 24–27 (2018)

16. Stoessel, C., Wiesbeck, M., Stork, S., Zaeh, M.F., Schuboe, A.: Towards optimal worker assistance: investigating cognitive processes in manual assembly. In: Mitsuishi, M., Ueda, K., Kimura, F. (eds.) Manufacturing Systems and Technologies for the New Frontier, pp. 245–250. Springer, London (2008)
17. Mattsson, S., Li, D., Fast-Berglund, Å.: Application of design principles for assembly instructions – evaluation of practitioner use. Procedia CIRP **76**, 42–47 (2018)

How to Increase Crane Control Usability: An Intuitive HMI for Remotely Operated Cranes in Industry and Construction

Felix Top(✉), Michael Wagner, and Johannes Fottner

Technical University of Munich, Boltzmannstraße 15, 85748 Garching, Germany
{felix.top,michi.wagner,j.fottner}@tum.de

Abstract. Current interfaces for operating cranes are non-intuitive and lack a user-centered design since the user constantly needs to determine the joint velocities that are required to generate the desired velocity of the crane's hook. The presented research aimed to make crane operation more intuitive by using an innovative HMI which includes inverse kinematics and several controllers. The user directly specifies the desired direction the load should move in. Inverse kinematics was used to calculate necessary joint velocities, while controllers ensured that the real movement and the desired movement matched closely.

Keywords: Crane control · Inverse kinematics · Human machine interface

1 Introduction

Cranes are essential tools in numerous areas of industry and construction. Load manipulation is realized by the direct actuation of individual crane joints, such as linear moving hydraulic cylinders or rotary electric drives. Hence, the final load movement results from the superposition of all individual joint movements. Additionally, many cranes can be operated with a remote control, which allows the user to carefully watch both load and crane and thus maintain process and personal safety. Also, it guarantees high operator comfort and flexibility.

To determine the necessary commands for all crane drives, the operator needs to convert the desired hook movement into the crane's joint movements, where he not only has to consider the current crane position and the system's kinematics but also his personal orientation relative to the crane. This results in a considerable cognitive burden, as there is a need for continuous internal conversions due to the lack of compatibility between the coordinate system of the crane, the load and the operator. Hence, current interfaces for operating a crane cannot be considered to be intuitive and show a lack of usability and user-centered design. This, in turn, leads to unnecessary operator fatigue and high error probability, where the latter causes inefficiencies due to avoidable incorrect movements and a corresponding high damage risk.

Different investigations show that there is a high potential for improving the human machine interface [1–3]. Modifications lead to significant improvements with respect to usability, operator fatigue, occurring errors and operation time [4–6]. Nevertheless, present improvement strategies lack the combination of both high usability and remote

W. Karwowski and T. Ahram (Eds.): IHSI 2019, AISC 903, pp. 293–299, 2019.
https://doi.org/10.1007/978-3-030-11051-2_45

operation. Hence, the remote human machine interface is the subject of further research. There exists particular potential when adapting the control strategy to the current operator's position [7]. However, that requires a solution for directly connecting the operator's input to a hook movement parallel to the input command, and therefore an approach to determine all crane joint movements such that the desired hook velocity results in a superposition of all joint velocities.

2 Problem Statement

Within the framework of the presented research, an innovative human machine interface is proposed. The main goal is to increase both usability and intuitiveness for all crane types without lowering safety standards or process quality.

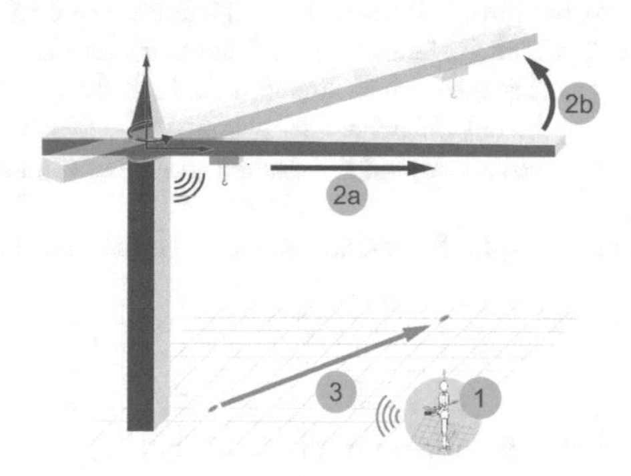

Fig. 1. Increased intuitiveness: direct load movement specification.

As illustrated in Fig. 1, to manipulate a load with the new intuitive remote control, the user specifies the direction of the hook movement by deflecting a joystick (1) in the desired direction. Using sensors, the system determines the current orientation of the operator, the position of the crane and the orientation of all joints. The system then calculates the desired load movement (3) in machine coordinates based on the operator's input (1) in his own coordinate system. Subsequently, inverse kinematics determines all necessary joint movements (2a) (2b) (2n), such that the resulting hook movement (3) is parallel to the user's input (1). Furthermore, controllers are added to the system such that the real load movement closely meets the operator's desire.

We face the problem that we have to find inverse kinematics and a control strategy that allows the calculation of all joint velocities to achieve the desired hook velocity, which, in turn, is parallel to the user's input at the joystick.

3 Proposed Method

Research was done on both a real loader crane of type Palfinger PK 7.501 SLD 5 (Fig. 2, left) and a simulation in MATLAB of the same crane model (Fig. 2, right). Loader cranes have a high number of joints that have to be operated simultaneously by often unskilled users, which is why the proposed solution was implemented exemplarily on such a crane. However, the research is not limited to a certain crane type.

The loader crane can be seen as a robotic manipulator. Consequently, it is described in the field of robotics as having three revolute (q_1, q_2, q_3) and one prismatic joint (q_4), and thus has four degrees of freedom (DOF). The crane's hook is located at the end of the extension system and constitutes the end effector. The cartesian position of the end effector is denoted by $h \in R^3$.

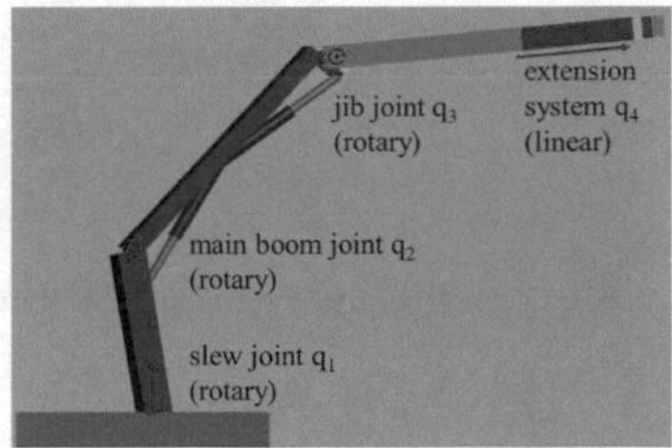

Fig. 2. Hydraulic loader crane (left); simulation model with joints $q_1 - q_4$ (right).

3.1 Inverse Kinematics

Inverse kinematics is a procedure to find joint velocities $\dot{q}_1 \ldots \dot{q}_4$ that result in the desired end effector velocity $\dot{h}$. As the system is underdetermined (four DOFs, but three constraints $\dot{x}, \dot{y}, \dot{z}$), there exist infinite solutions to satisfy the equation

$$\dot{h} = \begin{pmatrix} \dot{x} \\ \dot{y} \\ \dot{z} \end{pmatrix} = J_e \dot{q}, \tag{1}$$

where $\dot{h}$ is the hook speed, $\dot{q} = \begin{pmatrix} \dot{q}_1 & \ldots & \dot{q}_4 \end{pmatrix}^T$ the joint velocity vector and

$$J_e = \left[\frac{dh}{dq} \right] \tag{2}$$

is the Jacobian Matrix of the end effector.

An overview of approaches to solve Eq. 1 can be found in [8]. In this case, the inverse kinematics needs to solve Eq. 1 and also should fulfill three additional demands:

a. Joint limits have to be avoided, thus penalizing unfavorable joint velocities.
b. The inverse kinematics must prefer velocity contributions of joints q_2 and q_3 over q_4, as experiments have shown that joints q_2 and q_3 are faster than q_4.

c. The inverse kinematics must limit velocities, to prevent crane damage.

Furthermore, inverse kinematics should avoid singularities, consume little computing time and also produce meaningful solutions for all joint velocities and positions.

Inverse kinematics that proves to be capable of handling all requirements is configuration control (CC) as described in [9]. After it is modified slightly to satisfy all additional demands, it reads as follows:

$$\dot{q} = (J_e^T W_e J_e + J_l^T W_l J_l + J_p^T W_p J_p + W_v)^{-1} (J_e^T W_e \dot{h}), \tag{3}$$

where $W_e = 30 \cdot I_3$ and $J_l = I_4$, $W_v = 0.01 \cdot I_4$, with I_n indicating the n x n identity matrix. Additionally, $J_p = \text{diag}(0\ 1\ 1\ 1)$ and $W_p = \text{diag}(0\ 0.1\ 0.1\ 2)$.

H(q) is the performance criterion as stated in [10]:

$$H(q) = \sum\nolimits_{i=1}^{n} \frac{1}{4} \frac{(\overline{q_i} - \underline{q_i})^2}{(\overline{q_i} - q_i)(q_i - \underline{q_i})}, \text{ with} \tag{4}$$

$$W_l = \text{diag}(\omega_1 \omega_2 \omega_3 \omega_4) \text{ and } \omega_i = \begin{cases} 1 + \left|\frac{\partial H}{\partial q}\right|, & \text{if } \Delta\left|\frac{\partial H}{\partial q}\right| > 0 \\ 1, & \text{else} \end{cases}, \tag{5}$$

with $\overline{q_i}$ being the upper joint limits and $\underline{q_i}$ being the lower joint limits, respectively. All numeric values were estimated and proved by simulations.

3.2 Control Strategy

Additionally, to generate suitable joint velocities, it was necessary to generate commands for all drives in such a way that the real joint velocities matched the desired velocities. The control strategy consisted of two controllers and one rejector.

Joint-Velocity Controller. Two nested PID controllers were used to keep the deviation between desired and actual velocities small and also to deal with the system's response times. Additionally, a feed forward look-up table described the non-linear relationship between joint and hydraulic piston velocities.

Path Controller. A path reconstructor in combination with a PID controller was proposed for path tracking in order to prevent the crane's hook from leaving the desired plane of movement when reacting to the inverse kinematics' inputs. This controller estimated both the intended path based on user inputs (reconstructor) and kept the deviation between operator intention and resulting movement small (PID controller).

Inner Disturbance Rejector. Inner disturbances do not originate from external sources but from sources within the control loop. That could be the case, e.g., when the operator accidentally hits an actuator because he is inattentive. Since unintended commands usually differ greatly from previous inputs in direction, extent and duration, they can be detected and compensated for by statistical outlier compensation. In this case, a comparison between the present user input and the median of the previous user inputs was used to determine whether a current signal could be considered as an outlier.

The proposed median filter was parametrized by a window length which specified the number of samples from before the outlying input that was used to calculate the median. In this case, the window length was set to 0.5 s, which should be able to reject wrong commands shorter than 0.25 s while keeping filtering time delay low.

4 Results

4.1 Inverse Kinematics

Testing was conducted by discretizing the crane's workspace in steps of 5° (q_1, q_2, q_3) and 5 cm (q_4), respectively, and by subsequently conducting a simulative performance study. In this study, firstly, every point within the workspace was examined with respect to reachability. The second step consisted of joint velocity calculations for moving the hook from one point to the next in every movement direction, thus proving the correct calculation of needed joint velocities. Simulation experiments and analytical investigations (e.g., with respect to singularities) revealed the following performance results of the proposed inverse kinematics (Table 1):

Table 1. Performance criteria of inverse kinematics.

Criteria	Result
Deviation: end-effector velocity vs. desired velocity	Average deviation: 4%
Joint limit avoidance (a)	Success rate: 100%
Preferring q_2 and q_3 over q_4 (b)	Success rate: 97.4%
Avoidance of improper velocities (c)	Given, if input is limited.
Singularity avoidance	Success rate: 100%

4.2 Control Strategy

Examining the velocity controller in a simulation setup showed small relative errors e for q_2 and q_3 when controlling one joint movement at a time ($e < 5\%$). The relative error while controlling q_1 or q_4 exceeded 5% with values up to 40%. Additionally, when actuating all four joints simultaneously, the model became unstable due to the inaccurate parametrization of the dynamics of the counterbalance valves, which in turn lead to piston speed oscillations at joints q_2 and q_3.

Evaluation of the path controller showed that there were considerable deviations between the intended path and the actual path, even when using the path controller. However, the reconstructed path kept both the shape and the plane of the intended path. The disturbance rejector absorbed disturbances with a duration up to 0.2 s, which was found to be suitable for unintended inputs. Delays were as low as 0.2 s.

4.3 Resulting Framework for Intuitively Controlling a Loader Crane

The resulting control framework is shown in Fig. 3.

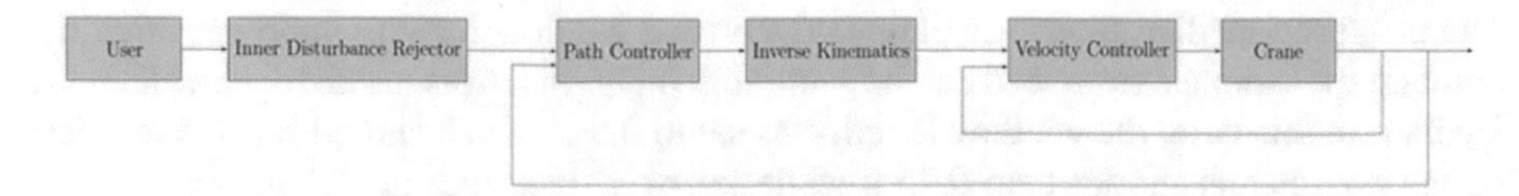

Fig. 3. Resulting framework.

The user commands a certain movement of the hook, which is first filtered by the inner disturbance rejector. The filtered input is passed on to the path controller, which ensures that the desired hook movement corresponds to the intended path.

Subsequently, the inverse kinematics takes the desired hook movement and computes the needed joint velocities, which are then passed to the velocity controller. The velocity controller finally actuates the crane in such a way that the hook moves in the direction specified by the user.

5 Discussion

The proposed inverse kinematics fulfilled all tasks sufficiently well. The small deviation between actual and desired velocities was very acceptable, as the inverse kinematics fulfilled all other requirements. Furthermore, considering that there exists a control strategy to further decrease the observed deviation, the proven performance seems sufficient. Still, the velocity controller should be the subject of further research, as relative errors are not yet fully satisfactory when controlling q_1 and q_4 and, in addition, it shows instabilities when controlling all joints simultaneously. The path controller kept both the plane the and shape of the intended movement and therefore worked as desired. The inner disturbance rejector solved the problem satisfactorily.

6 Conclusion

With the presented intuitive remote HMI and the corresponding control strategy, crane operation becomes easier and faster for both occasional and expert users. Furthermore, incorrect movements that occur in present systems as a result of demanding internal conversions are minimized, which leads to reduced manipulation times and a lower damage risk. The presented control strategy allows for a direct crane hook manipulation and hence contributes to a successful implementation of the intuitive remote HMI. However, further performance in real-life studies need to be conducted to fully prove that the observed deviation does not interfere with the user's expectations.

References

1. Peng, K., Singhose, W.: Crane control using machine vision and wand following. In: International Conference on Mechatronics. IEEE, Piscataway (2009)

2. Peng, K., Singhose, W., Gessesse, S., Frakes, D.: Crane operation using hand-motion and RFID tags. In: International Conference on Control and Automation. IEEE, Piscataway (2009)
3. Sorensen, K., Spiers, J., Singhose, W.: Operational effects of crane interface devices. In: Conference on Industrial Electronics and Applications, pp. 1073–1078. IEEE, Piscataway (2007)
4. Kim, D., Singhose, W.: Performance studies of human operators driving double-pendulum bridge cranes. Control Eng. Pract. **18**, 567–576 (2010)
5. Manner, J., Gelin, O., Mörk, A., Englund, M.: Forwarder crane's boom tip control system and beginner-level operators. Silva Fennica **51**(2), 1717 (2017)
6. Kivila, A., Porter, C., Singhose, W.: Human operator studies of portable touchscreen crane control interfaces. In: International Conference on Industrial Technology, pp. 88–93. IEEE (2013)
7. Kivila, A., Singhose, W.: The effect of operator orientation in crane control. In: Dynamic Systems and Control Conference. ASME (2014)
8. Siciliano, B.: Kinematic control of redundant robot manipulators: a tutorial. J. Intell. Robot. Syst. **3**, 201–212 (1990)
9. Fahimi, F.: Autonomous Robots—Modeling, Path Planning, and Control. Springer, Boston, (2009)
10. Chan, T., Dubey, R.: A weighted least-norm solution based scheme for avoiding joint limits for redundant joint manipulators. IEEE Trans. Robot. Autom. **11**, 286–292 (1995)

Warehouse Storage Assignment by Genetic Algorithm with Multi-objectives

Chi-Bin Cheng(✉) and Yu-Chi Weng

Department of Information Management, Tamkang University, New Taipei City, Taiwan
cbcheng@mail.tku.edu.tw,
emily091131027@gms.tku.edu.tw

Abstract. The proper assignment of storages to stocks prior to their picking is critical to reduce order picking costs. Appropriate storage assignments can also shorten storing time, improve storage utilization, and facilitate inventory management. Three objectives are considered: minimizing the routing length of storing stocks, maximizing the future chance of adjacent stocks to be picked together, and minimizing the storage distance to the access point for popular stocks. This study devises a genetic algorithm to find feasible solutions and uses a method to determine the final storage assignment from a set of Pareto solutions.

Keywords: Storage assignment problem · Distribution center · Genetic algorithm · Multi-objective decision making · Order picking

1 Introduction

The cost of the picking task accounts for up to 55% of the total warehouse operating costs [1]. The aim of this study is to improve the efficiency of the order picking of a distribution center under the constraints of its current layout, routing method, and picking policy. The case study distribution center is an auto parts seller who owns a few warehouses where facility layout was just renovated. The distribution center adopts a discrete picking policy and strictly follows a first-in-first-out rule that for the same items the earliest stored ones are always picked first. The S-shape routing strategy is used when picking an order. This strategy leads to a route in which the aisles to be visited are totally traversed while aisles without items to be picked are skipped, as shown in Fig. 1 [2]. The order picker thus enters an aisle from one end and leaves the aisle from the other end. This strategy is frequently used due to its simplicity and explicitness.

With the above constraints imposed by the current operations of the distribution center, this study focus on the design of the storage location assignment method to improve the order picking efficiency of the distribution center. The current storage assignment method of the distribution center is similar to a random storage policy. An additional rule by the case study distribution center is that receiving items are assigned to storage frames where the same items have been stored previously, if such storage

W. Karwowski and T. Ahram (Eds.): IHSI 2019, AISC 903, pp. 300–305, 2019.
https://doi.org/10.1007/978-3-030-11051-2_46

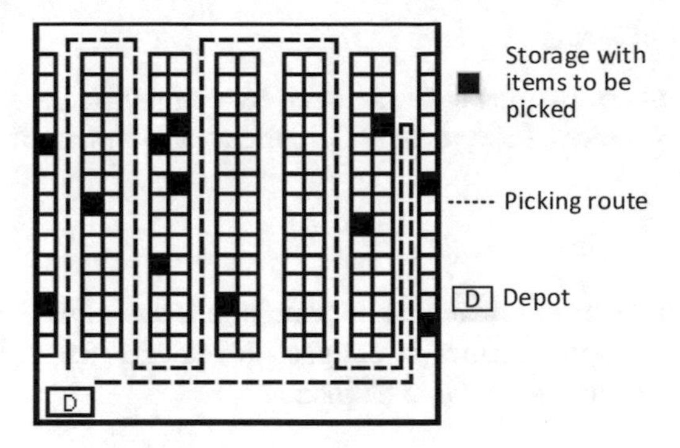

Fig. 1. The S-shape routing strategy [2]

locations are available; otherwise, to save space, the receiving items are assigned to other storage frames with different items in a random fashion.

The use of random storage policy may create inefficiency in the future order picking operations, since, in many cases, the demand frequencies of items are not unique but often vary greatly among them. With this consideration, this study attempts to solve the storage assignment problem by taking into account the demand frequencies of items; that is, the frequently demanded items ought to be stored near the depot. Furthermore, since the case study distribution center adopts a discrete picking policy where the items in the same order are picked together, we also consider the possibility that the receiving item will be picked with certain items in the future and prioritize their locations for storing the receiving item. Finally, the items are received in batch and generally the receiving amount exceeds the available capacity of a single storage frame, and thus, it is required to seek multiple available storage frames in such a case. This study also considers the traverse distance among the storage frames when assigning storage locations for the receiving item. Consequently, the storage assignment model of this study has three objectives: minimizing the distance from the storage location to the depot, maximizing the possibility of the item to be picked with items in its adjacent locations, and minimizing the traverse distance among storage frames when storing the item.

2 Storage Assignment with Multiple Objectives

The three objectives mentioned in the previous section are all related to moving distances. Estimates of such distances based on probability distributions of orders might be performed to obtain an expected measure, but are difficult due to the great uncertainties of future orders. Instead, we determine the storage assignment via optimizing a set of indexes corresponding to the three objectives.

2.1 Correlation of Items

The correlation between two items is measured by computing the support of the two items from historical orders. Let c_{ij} denote the correlation measure of items i and j, then

$$c_{ij} = \frac{n_{i \cap j}}{N} \tag{1}$$

where $n_{i \cap j}$ is the number of orders that contain both items i and j, and N is the total number of orders. Suppose item i is assigned to storage frame k, the correlation between the item and the frame, f_{ik}, is defined as

$$f_{ik} = \frac{\sum_j c_{ij} q_{jk}}{\sum_j q_{jk}} \tag{2}$$

where q_{jk} is the amount of item j in storage frame k. The objective is to maximize an overall correlation index, CI, over all the storage frames assigned for item i:

$$CI = \sum\nolimits_{k \in K} f_{ik} \tag{3}$$

where K is the set of storage frames assigned for item i.

2.2 Frequency-Distance Index

The items demanded more frequently are stored nearer to the depot, while items demanded less frequently are stored farther away from the depot. Let $FI \in [0, 1]$ be an index that reflect the above intention, and the greater the FI the more satisfaction of the intention. Such a tendency is analog to the fuzzy XOR operator by Mela and Lehmann [3] is adopted by this study, and thus FI is defined as:

$$FI = \overline{\phi}_l + \overline{d}_l^{\max} - 2\overline{\phi}_l \cdot \overline{d}_l^{\max} \tag{4}$$

where $\bar{\phi}_l \in [0, 1]$ is the normalized demand frequency of item i and is computed by

$$\bar{\phi}_l = \frac{\bar{\phi}_l - \phi^{\min}}{\phi^{\max} - \phi^{\min}} \tag{5}$$

with $\phi^{\min}$ and $\phi^{\max}$ being the minimum and the maximum among the demand frequencies of all items; $\bar{d}_l^{\max} \in [0, 1]$ is the normalized distance of the farthest storage frame assigned for item i and is obtained by

$$\bar{d}_l^{\max} = \frac{d_i^{\max} - d^{\min}}{d^{\max} - d^{\min}} \tag{6}$$

with $d_i^{\max}$ as the distance (to depot) of the farthest frame that stores item *i*, and $d^{\min}$ and $d^{\max}$ and the smallest and the greatest distances among all frames in the warehouse to the depot, respectively.

2.3 Traverse Distance Among Frames

The receiving item often requires multiple storage frames to accommodate the entire batch. This objective is to find a set of storage frames that minimize the distance required to visit these frames. Due to the storage layout and aisle structure, the distance between two frames is a Manhattan distance. There are numerous routes to reach a storage frame from another one. Here, we apply the Dijkstra shortest path algorithm to determine the distance between two storage frames. The traverse distance *t* among a set of storage frames assigned for item *i* is again normalized by

$$T = \frac{t - t^{\min}}{t^{\max} - t^{\min}} \tag{7}$$

where $t^{\max}$ is the possibly longest distance predetermined based on experience, and $t^{\min}$ is the possibly shortest distance, i.e. $t^{\min} = 0$ when the entire batch of item *i* is stored in a single frame.

3 Solution Procedure with Genetic Algorithm

3.1 Solution Encoding

The solution to the problem is a sequence of storage frames to be visited and store the item. A chromosome is used to represent such a sequence as shown in Fig. 2, where each cell of the chromosome indicates the frame number. The frames are visited by the orders of cells and the traverse distance among frames is thus computed. The chromosomes are generated randomly or through genetic operations. Unlike regular GA, the length of chromosome is not fixed in our algorithm. If the chromosome is not long enough to accommodate the entire batch, it is discarded; otherwise, it is truncated if too long. For example, suppose the batch is 100 units, and the available capacities of the frames in the chromosome of Fig. 3 are 50, 20, 15, 25 and 30 units in turn, then the last cell is discarded since the first four frames are abundant already to take the entire batch.

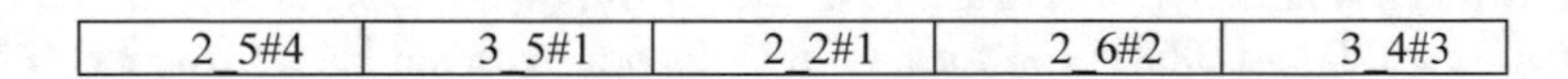

Fig. 2. An exemplar chromosome

3.2 Fitness Function

Desirability function [4] is a classical approach for solving multiple objective optimization problems by combining individual objectives into a single index/measure.

The difficulty of the desirability function approach is the specification of weights associated with objectives, which relies upon prior information regarding the objectives. This study adopts the weighted sum method to approximate the Pareto front, where the weighted sum of the three objectives serves as the fitness function of the posed GA and is defined as:

$$FT = w_{CI} \cdot CI + w_{FI} \cdot FI + w_T \cdot (1 - T) \tag{8}$$

where w_{CI}, w_{FI} and w_S are weights of the three objective respectively and $w_{CI} + w_{FI} + w_T = 1$. After the Pareto frontier is obtained or at least approximated, the final solution is selected within the set represented by the Pareto frontier using the empiric rule suggested by Bortolini et al. [5]:

$$\max_p \left\{ \frac{CI_p}{CI^*} \cdot \frac{FI_p}{FI^*} \cdot \frac{(1 - T_p)}{(1 - T^*)} \right\} \tag{9}$$

where CI_p, FI_p and T_p are the p-th Pareto solution, and CI^*, FI^* and T^* are the respective single objective optimal solutions.

3.3 Solution Procedure

The procedure begins with a setting of the weight vector used in the fitness function. This fitness function is sent to the GA to obtain a storage assignment solution. The solution is recorded to approximate the Pareto frontier of the multi-objective problem. If all the weight vectors have been enumerated, the algorithm determine the final solution from all Pareto optimum.

4 Performance Evaluation

To evaluate the performance of our method, we simulate the procurement decision and customer order arrival of the case study distribution center based on its inventory policy and historical orders. Performance evaluation is carried out by comparing the storage assignment decisions for receiving items by the principle of current operations and by the proposed method.

The weights used in the fitness function (9) are changed from 0 to 1 with an increment of 0.1, which results in 66 sets of weights. Part of the computational results are shown in Table 2. The individual best value of single objective is 0.76 for $(1 - T)$, 0.541666 for CI, and 0.6285714 for FI. From the approximate Pareto frontier we find $G^* = 0.8749984$. It is noted that at G^* it is also the best solutions for the two objectives, T and CI. The moving distance produced by G^* is compared to that by the current operations. Figure 3 shows the comparison, where our method outperforms the current method for both item storing and order picking.

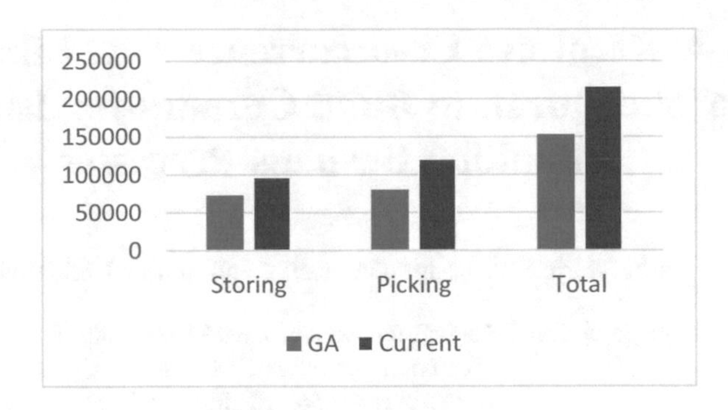

Fig. 3. Comparison of moving distances by two methods

References

1. Tompkins, J.A., White, J.A., Bozer, Y.A., Tanchoco, J.M.A.: Facilities Planning. Wiley, Hoboken (2010)
2. Material Handling Forum: Routing strategies. https://www.erim.eur.nl/material-handling-forum/research-education/tools/calc-order-picking-time/what-to-do/routing-strategies/
3. Mela, C.F., Lehmann, D.R.: Using fuzzy set theoretic techniques to identify preference rules from interactions in the linear model: an empirical study. Fuzzy Sets Syst. **71**, 165–181 (1995)
4. Derringer, G., Suich, R.: Simultaneous optimization of several response variable. J. Qual. Technol. **12**, 214–219 (1980)
5. Bortolini, M., Faccio, M., Ferrari, E., Gambori, M., Pilati, F.: Fresh food sustainable distribution: cost, delivery time and carbon footprint three-objective optimization. J. Food Eng. **174**, 56–67 (2016)

A Recursive Co-occurrence Text Mining of the Quran to Build Corpora for Islamic Banking Business Processes

Farhi Marir[1](✉), Issam Tlemsani[2], and Munir Majdalwieh[1]

[1] College of Technological Innovation, Zayed University, Duabi, UAE
farhi.marir@zu.ac.ae
[2] IE Business School, Madrid, Spain
i.tlemsani@tcib.org.uk

Abstract. As the holy Quran text is time and place based and has "blockchain" like-structure where stories are spread like pieces of puzzles and linked to each other among 6,236 verses from 114 chapters, we present a new recursive co-occurrence text mining algorithm to mine the Holly Quran to build corpora which can be further processed to develop Islamic banking business processes complying which Islamic Sharia' law. First, we create a list containing all the business process action terms and their synonyms in Arabic. Then for each term, we run the recursive co-occurrence algorithm to parse the holy Quran and extract all verses that contain the business process action term and all the verses where the action term co-occurs with any other term in the list. The retrieved verses along their chapter number and their verse number are saved and then compiled into business process corpora using. The resulting business processes could further be processed to generate Islamic Business processes that comply with Islamic Sharia Law.

Keywords: Holy Quran · Blockchain · Recursive co-occurrence
Text mining · Corpora

1 Introduction

The Islamic banking market has grown rapidly from about US$200 billion in 2003 to about US $1.8 trillion in 2013 and with expectations of market size to be $3.4 Trillion by end of 2018. It has evolved from a niche offering, limited to banks in the Middle East to part of the mainstream financial services landscape globally. In principle, Islamic financial institute are governed by Sharia law, which includes a combination of sources such as the Quran, the Hadiths and fatwas (the rulings of Islamic scholars). Islamic financial services institutions' products and services must be Sharia compliant. However, Sharia law lacks standardization and is interpreted differently in each

---

The original version of this chapter was revised: The affiliation of the author "Issam Tlemsani" has been updated. The correction to this chapter is available at https://doi.org/10.1007/978-3-030-11051-2_141

W. Karwowski and T. Ahram (Eds.): IHSI 2019, AISC 903, pp. 306–312, 2019.
https://doi.org/10.1007/978-3-030-11051-2_47

country, as it is dependent on their individual Sharia. Islamic financial institute's clients are becoming more educated about Islamic products and services and are seeking convincing evidence of the sharia compliance of products and services that the institute is offering [1, 2]. Various bodies in several countries and regions are formed to address the creation of a coherent set of Islamic banking standards. This puts more pressure on the managers of Islamic Financial Institutes (IFIs) to not only maximize the value of their investments, but also to achieve these objectives in a Sharia compliant way [3]. Sharia law compliance has not been researched or explored from a business process management perspective and the current literature has shown a lack of a well-defined methodology for integrating Sharia compliance controls into Islamic Sharia business processes [4]. [1] indicated that the lack of Islamic financial standards would affect the ability of the banks to implement Islamic products and services.

This paper presents a new text mining technique based on recursive co-occurrence of synonym words to build corpora for Islamic banking business processes based on the Quran stories related to Islamic finance. This text mining technique is proposed to fit the recursive structure of the Quran, in which verses are partially inter-connected but spread like puzzles all over the holy Quran 6,236 verses from 114 chapters.

2 Literature Review

As Arabic language a challenging task when applying machine learning and artificial intelligence techniques, limited research has considered the Arabic text of the Quran. Recent research work on analyzing the Holy Quran includes the research paper by [5] in which the authors initiated a series of statistical analysis and information retrieval techniques to show a variety of characteristics of the Holy Quran. The most recent reported by [6] measures the strength of the semantic relations of the AND conjunction between two semantically same terms in different position around the AND conjunction in the Holy Quran. [7] used text mining for analyzing religious texts.

As for text mining, the co-occurrence based methods look for concepts that occur in the same unit of text typically a sentence or verse in Quran, but sometimes as large as an abstract and suggest a relationship between them as reported in [8] used basic text mining approach based on terms co-occurrence and transitive relationship (A $\rightarrow$ B $\rightarrow$ C). In [9] co-occurrence, based meta-analysis of scientific texts is used for retrieving biological relationships between genes.

In line with these successful texts mining, we are proposing in this paper a recursive co-occurrence approach of text mining approach to reflect the blockchain like structure of the holy Quran. The proposed algorithm will be run to build a corpus for *Murabaha* business process (shown in Fig. 1 below).

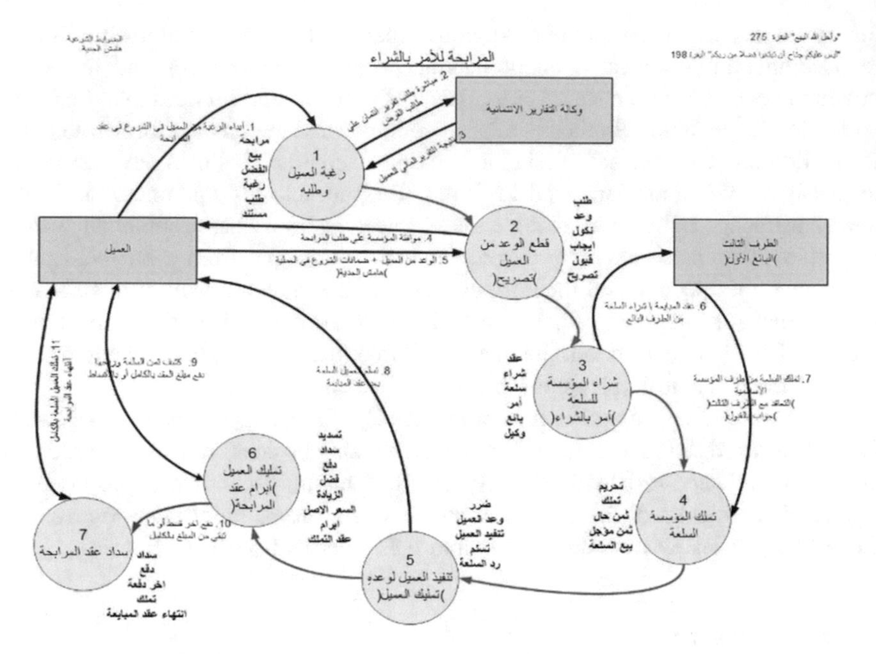

Fig. 1. The seven actions in Murabaha business process

3 The Recursive Co-occurrence Algorithm

The next step is to mine Quran to find out how the Murabaha business process shown (Fig. 1) is interpreted in holy Quran text. Text mining techniques to analyze and identify patterns through given free text is used in different domains, for instance medical science. In this research, we will use recursive co-occurring text mining for each action of the business process (see Algorithm in Fig. 2 below) to extract recursively linked verses based on the same term or its synonym in the same way a blockchain elements are linked. This retrieval of semantically linked Quran verses will be saved in a tree like structure which later will parsed to extract all verses and compile them into a sub-corpus. The process will be repeated for each action term of the business process until a comprehensive corpus of verses related to that business process is built. This business process corpus will be ready for further analysis to design a related Sharia' law business processes. The proposed Algorithm is as follows:

```
Create Bpr Corpus (IN: Quran Text, OUT: BprCorpus)
   {Gather all the action terms e.g. transaction, sign
   pledge agreement, etc. of the business process
   adding their Arabic synonyms t_i to make the tBpr
   list}
    tBpr = {t_1, t_2,..., t_m}
 1. While (i<=m)
    - Use Sketch Engine to extract all verses
      containing t_i and all verses containing any of
      terms {t_i+1, t_2,..., t_m} occurring with t_i
    - Save the retrieved verses along their ID numbers
      in a new vfile_i
    - i=i+1 {move to the next term in tBpr}
 2. End While
 3. Compile all vfile_i into a BprCorpus business process
    related corpus
 4. Undertake further text mining analysis on the
    BprCorpus to generate a sequential of action terms
    that will compose the Sharia compliant business
    process
```

Fig. 2. Recursive co-occurrence text mining algorithm

3.1 Preparing the Holy Quran Texts

The Arabic version of the Holy Quran has been downloaded from Tanzil project website (http://tanzil.net/download) which represents an authentic verified source of the holy Quran text as shown in Fig. 3 below. It is composed of 114 chapters and 6,236 which were compiled into a corpus using Sketch Engine text analysis tool.

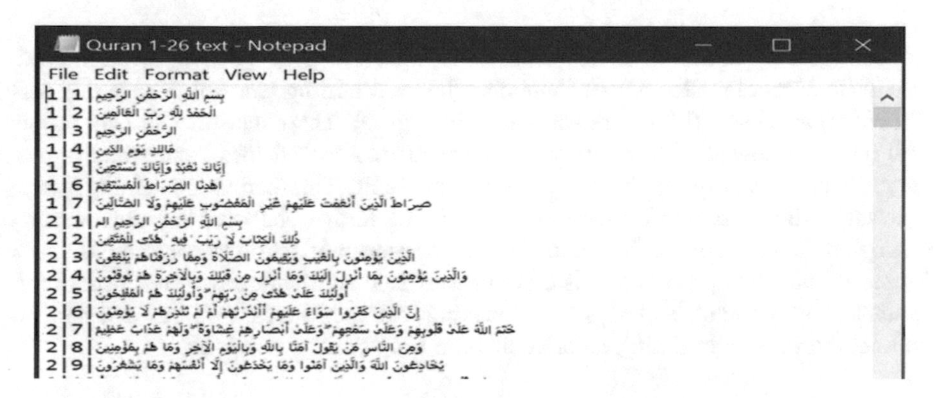

Fig. 3. Portion of Tanzil Holy Quran text along chapter and verses numbers

3.2 Running the Co-occurrence Algorithm

In the first step, we browse through Murabaha business process of Fig. 1 to collect the initial terms that defines tasks or action as: tBpr = {مضاربة (Mudaraba); مرابحة (Murabaha); بيع (Sale); شراء (Purchase); عقد (Contract); … etc.}. Then we used the process co-occurring terms (word sketch operation in the Sketch engine) to extract terms within initial co-occurring together (shown inside red rectangles) and manually collect the new observed terms (shown in blue rectangle) e.g. observed = {رهن (mortgage); قرض (loan) … etc.} relevant to the business process in hand as shown (see Fig. 4 below). These new observed terms would be used to perform another recursive iteration to discover new relevant terms to the Murabaha business process. Based on the co- occurrence of terms, a list will be built to explicitly reflect these relationships to be used as an indexing to verses texts of the Quran.

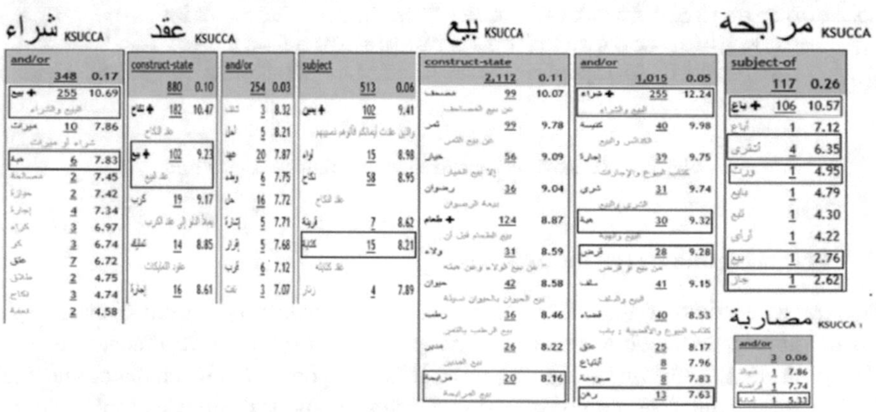

Fig. 4. Extracting terms that co-occur with Murabaha action terms

Using the hyper link of co-occurring terms shown in (Fig. 4 above), we drill down into the indexes of related verses containing the co-occurring terms like (بيع (Sale); شراء (Purchase) and مرابحة (Murabaha) as shown in (Fig. 5). These indexes will help further drill down to download new verses to be saved into special files from which a new business process corpus will be compiled and built. This corpus containing verses relevant to the Murabaha business process will be further analyzed to extract orderly one by one action terms to be assembled in a sequence of action terms to build a new Murabaha business process that is compliant to Islamic Sharia Law. To illustrate this example, some Islamic banks instead of making loan to buy a car they buy and sell cars without the interest by fixing its price and the time limit to pay back.

Fig. 5. Files indexes generated from the co-occurring process of (بيع (Sale); شراء (Purchase) and مرابحة (Murabaha)

4 Conclusion

Several challenges to mine the holy Quran; the complex structure of the Quran as stories are spread in different chapters and verses and the Arabic version of the Holy Quran text is the most challenging natural languages in natural in machine learning. In this research work we achieved the first results by developing a new text mining approach that takes each action term of the existing business process recursively to discover new terms through co-occurrence process. This recursive and co-occurrence approach reflects the Quran structure where each story is spread like piece of puzzle all over the different chapters and verses. The text mining approach is run successfully on Murabaha (financial participatory) business process and a corpus composed of all verses containing business action terms, their synonyms and all other financial terms related to Murabaha. Future work will focus on further analyzing this Murabaha business process based corpus to build its equivalent Islamic Sharia complaint business process by assembling sequentially one by one term (verses) action from the corpus. This process will be repeated for each Banking business process to replace all banking business processes that involve interest rate by other formula compliant with Islamic Sharia Law that abolish interest rate.

References

1. Kuppusamy, M., et al.: A perspective on the critical success factors for information systems deployment in Islamic Financial Institutions. EJISDC **37**(8), 1–12 (2009)
2. Chapra, M.U., et al.: Corporate Governance in Islamic Financial Institutions, Islamic Development Bank. Islamic Research and Training Institute, Periodical Doc. No. 6 (2002)

3. Archer, S., et al.: Financial contracting, governance structures, and the accounting regulation of Islamic banks: An analysis in terms of agency theory and transaction cost economics". J. Manag. Gov. **2**, 149–170 (1998)
4. Vayanos, P., Wackerbeck, P., Golder, P.T., Haimari, G.: Competing Successfully In Islamic Banking. Booz & Co, Retriev (2008)
5. Alhawarat, M., et al.: Processing the text of the Holy Quran: a text mining study. March Int. J. Adv. Comput. Sci. Appl. **6**(2), 262–267 (2015)
6. Bentrcia, R., Zidat S., Marir F.: Extracting semantic relations from the Quranic Arabic based on Arabic conjunctive patterns. J. King Saud Univ. – Comput. Inf. Sci. http://dx.doi.org/10.1016/j.jksuci.2017.09.004 (2017)
7. McDonald, D.: Text mining analysis of religious texts. J. Bus. Inq. **13**(1 (Special Issue)), 27–47 (2014). http://www.uvu.edu/woodbury/jbi/articles/
8. Swanson, D.R.: Fish oil, Raynaud's syndrome, and undiscovered public knowledge. Perspect. Biol. Med. **30**, 7–18 (1986)
9. Yang, H., Spasic, I., Keane, J.A., Nenadic, G.: A text mining approach to the prediction of disease status from clinical discharge summaries. J. Am. Med. Inform. Assoc. **16**(4), 596–600 (2009)

Digital Television in EU in Terms of Increasing Demographic Fragmentation of the Audience (A Cross-National Research)

Nadezhda Miteva(✉)

Faculty of Journalism and Mass Communication, The St. Kliment Ochridsky Sofia University, 49 Moskovska Street, 1000 Sofia, Bulgaria
nmiteval@uni-sofia.bg

Abstract. The coordinated transition from analogue to digital terrestrial broadcasting in the European Union (except in Romania) is completed process. One of the positive results of the digitalization of television is the allocation of valuable frequency resources (digital dividend) for future development of modern telecommunications services, including distribution of video content. The ageing of the population in Europe and the entering of the Millennials generation the labour-market define the balance in consumption of traditional and contemporary television services. The future of digital terrestrial television in Europe is guaranteed until 2030 and it depends on the ability to find hybrid solutions to watch live, diverse, multiplatform and convenient media content.

Keywords: Digital transition · Digital dividend · Hybrid television · Viewing habits · Millennials

1 Introduction

As a whole the European single market remains one of the most lucrative ones for investors from media, entertainment and telecommunications industries due to its numerous population, high living standard (mostly of the people from Western and Northern Europe), as well as the existence of adopted by the European institutions common rules and regulations for governments, businesses and users. Meanwhile, every country has media and telecommunications infrastructure of its own which differs from the ones in other countries while being marked by geographical, economical and cultural particularities and traditions.

Aspirations for harmonized action while keeping in sight national characteristics continue to exist and are present during the transition from analog to digital terrestrial broadcasting in the European Union countries. The process is amazing by scope (511.8 million people in 28 countries), as well as by time span (it began in the end of the 20 century and continues nowadays), irregularity and coordination.

The analog switch-off and the building of infrastructure for digital terrestrial broadcasting become part of the digital transformation of the European single market. This transformation coincides to a great extent with a period of significant demographic

W. Karwowski and T. Ahram (Eds.): IHSI 2019, AISC 903, pp. 313–318, 2019.
https://doi.org/10.1007/978-3-030-11051-2_48

changes (related mostly to age groups) in European societies, which set the pattern for changes in supply and demand of audiovisual media content.

The suggested research puts forward the following research questions:

1. Is the transition from analog terrestrial broadcasting to digital one successful, respectively, which are the successful models in the European Union countries?
2. To what extent characteristics related to the age of population in European countries influence the development of digital television systems?

The main findings are:

- All EU member states succeed to get certain positives and advantages from introducing digitalization of television broadcasting, although to various extent for each and every country.
- The ratio linear (live) television/internet tv consumption depends directly from the ratio *adults/young adults & children*. This determines the persistent interest to traditional television, typical for Europe, projected on the background of growing consumption of Internet based TV services.

2 Methodology

Digitalization of television broadcasting in EU is a many-sided process and it is difficult for one researcher alone to study its influence on society, economy and culture. That is the reason why for the purposes of measuring the "success" of this process, the suggested research focuses on the basic results: market positions of digital terrestrial broadcasting in the respective EU member countries; perspectives for the development of digital terrestrial television (hybrid solutions, launch of DVB-T2); usage of the allocated radio frequency spectrum (digital dividend), which has most favourable features and is priority means for the development of 4G connectivity in the EU. The results can be described by using the following formula:

$$S = DTT_{max} + 4G_{max} + P \quad (1)$$

where:

DTT_{max} is the level of consumption (% of households) thus turning digital terrestrial broadcasting into a leading platform for every respective country. Maximum possible value = 100.

$4G_{max}$ is a peak value of 4G coverage with networks and services (% of households), which brings each and every respective country among the top ten in the EU, based on EU Digital Economy and Society Index (DESI) 2018. Maximum possible value = 100.

P is the evaluation of the perspective for development of digital terrestrial television (DTT) platform, based on the existence or the absence of hybrid solutions and/or opportunity to migrate to DVB-T2. Maximum possible value = 2.

S is the basic achievement of the transition from analog to digital terrestrial broadcasting that was carried out. $S_{max} = 202$.

The combination of the three indicators' peak values represents the cluster of the EU countries, which benefit to the fullest from the transition to digital television and as a result of it they have well shaped consumption and perspective for development of digital terrestrial broadcasting as well as high level of 4G connectivity penetration, which places them in the group of the EU most successful economies [1]. The formula used describes only basic direct and collateral consequences of the analogue to digital transition. The formula does not pretend to provide comprehensive description of effects from this process.

3 Data Analysis

Digitalization of television broadcasting (wholly or in part) is outstanding achievement of its own and as such it has important consequences for the digital economy and the digital society. That is the reason why, it would be wrong to speak about a failed model in the EU.

Nevertheless, the countries which to a maximum extent succeed to benefit most from the digital transition and the allocated frequency spectrum are Finland, Great Britain, Spain and the Czech republic. The first three countries mentioned are among the most developed digital economies, while the Czech republic ranks among top five EU countries in terms of best 4G coverage. Besides, digital terrestrial broadcasting is a leading distribution platform with developed accompanying services (internet based applications, HbbTV etc.) and perspective to migrate to the more perfected terrestrial standard DVB-T2 (Fig. 1).

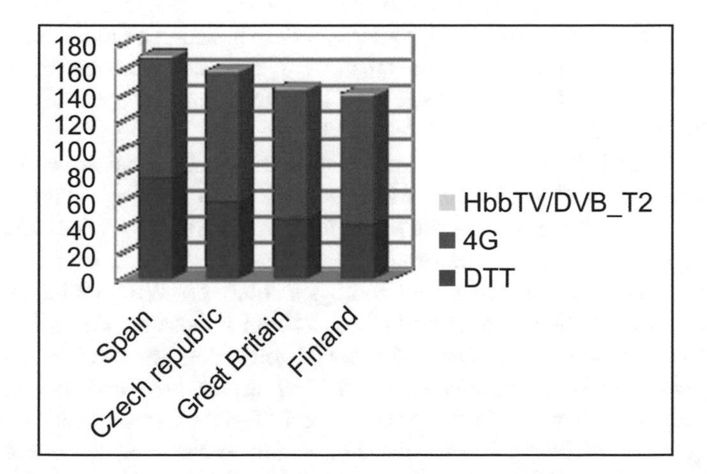

Fig. 1. The most successful models of analog to digital switchover in EU

Bulgaria, Slovakia and Romania are the member-states with the least usage of digital television. In Bulgaria and Slovakia digitalization of television broadcasting is accompanied by decrease of popularity, absence of interactive applications and no intentions to move on to DVB-T2. Romania is the only country within the EU which has not completed the switch over. These three countries are among the least developed digital economies with the worst 4G coverage in the EU (Fig. 2).

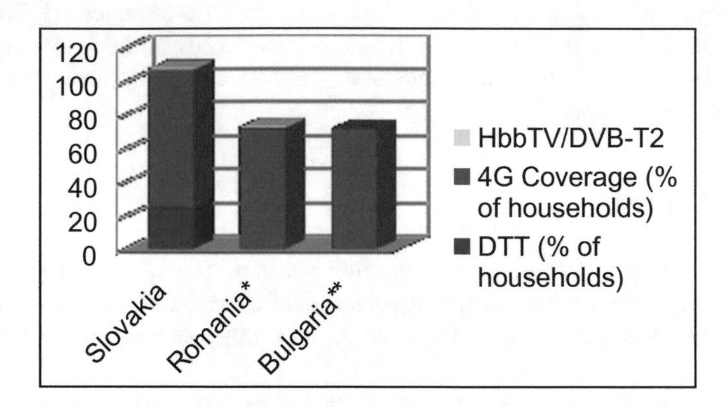

Fig. 2. The less successful models of analog to digital switchover in EU (**: Digital switchover not yet completed; **: No current data of DTT usage available*)

The other 22 countries are part of the group of states that benefited partially from the transition to digital television with varying combinations of indicators.

The proposed ranking shows that to various extent all EU countries manage to get some advantages for their respective societies in the process of transition from analog to digital terrestrial broadcasting. Still, terrestrial broadcasting is among the dominant tv platforms in the EU with penetration rate of 27.7% among EU 27 [2]. Following successfully carried out transition to digital and the allocation of radio frequencies for broadband and radio services, European citizens now enjoy fastest high-speed 4G networks in the world. European states rank 1–9 out of top ten countries in the world ranking with the fastest broadband internet [3].

As the ITU and the European commission look for harmonious co-existence of terrestrial television and broadband wireless services in Europe, some authors and professionals approach the issue of the principle of the two technologies excluding each other by suggesting a possible future without terrestrial television services [4, 5].

Although new generations of technologies such as WiB (Wideband Reuse) appeared, the Digital Video Broadcasting consortium (whose standards for terrestrial, satellite, cable and mobile platforms for digital television are used in the EU) also decided to terminate future research in this field of development and accepted that, for the time being, the Shannon limit is reached and "further improvement could not be expected" [6]. The main reasons for the consortium to focus on the development of hybrid solutions for combining broadcast DTV with Over-The-Top services (DVB-I) are insufficient spectrum efficiency of current technologies (DVB-T2, DVB-S2X) as well as upcoming changes in consumers` habits within different generations. The bearers of these changes are predominantly the young adults (Millennials) and children, although new online viewing habits are being registered throughout all age groups [7].

Two main demographic trends characterize EU population: uneven, however generally valid ageing, as well as Millennials, who are more highly ICT skilled than the population as a whole, entering the labour-market en masse [8].

The latter is the reason for the decrease in television watching globally. The decrease is measured by several minutes on an annual basis and is visible especially in Asia, the USA, Great Britain, while in South America, Africa and most European countries (Germany, France, Bulgaria etc.) it remains high. More and more end devices for watching are at hand—television sets, computer, smart phone, tablet. Audiences frequently reach out and migrate to the online environment. Trends indicate development of OTT television services, increased demand for content such as web series, streaming of video, live sports events, virtual reality and video 360 [9].

Although slowly, elderly people (65+) improve their media and digital skills [10]. The prognosis is that during the next several decades the relative share of this social group of the population in Europe will increase [11]. This provides reasonable ground to expect that the sustainable persistent consumption of linear television services will remain in place.

4 Discussion

The ascending trend of audiences acquiring new online viewing habits is irreversible. The dynamics of fading of linear television services and the rise of online television services is both technology as well as generation driven process. The prognosis that the demand for linear television services in the EU will continue even after 2030 are well-founded by demographic reasons [12]. This is the reason why digital terrestrial television is looking for hybrid solutions by combining terrestrial broadcasting with the advantages of Internet in order to counter competition from mobile technologies as well as to attract the attention of young active and participative consumers (hybrid television). However, users would be interested in watching terrestrial television to the extent it meets their needs to use hybrid services, watch "live TV", watch diverse and multiplatform content and—last, but not least, to the extent digital terrestrial broadcasting provides economically viable and profitable alternative to other platforms and content distribution channels.

Acknowledgments. This text has been developed within the framework of the COST Action of the European Commission: CA 16211: Reappraising Intellectual Debates on Civic Rights and Democracy in Europe, supported by the National Scientific Fund of Bulgaria: DCOST-01.25-20.12.2017.

References

1. European Commission Digital Economy and Society Index (DESI) (2018). https://ec.europa.eu/digital-single-market/en/desi
2. Digital Video Broadcasting. https://www.dvb.org/news/why-is-switzerland-switching-off-dtt

3. Tidey, A.: Europe dominates global broadband speed ranking. In: News. http://www.euronews.com/2018/07/12/europe-dominates-global-broadband-speed-ranking. Accessed 12 July 2018
4. Zarri, M: Bringing broadcast to mobile. In: Future Networks. https://www.gsma.com/futurenetworks/digest/bringing-broadcast-mobile/. Accessed 26 Sept 2016
5. Ala-Fossi, M., Lax, S.: The short future of public broadcasting: replacing digital terrestrial television with internet protocol? Int. Commun. Gaz. **78**(4), 365–382 (2016). ISSN 1748-0485. orcid.org/0000-0003-3469-1594. https://doi.org/10.1177/1748048516632171
6. Siebart, P.: A Word from the DVB project office. DVB Scene (52), 2 (2018). https://www.dvb.org/resources/public/scene/dvb-scene52.pdf
7. Eurodata TV Worldwide. https://www.digitaltveurope.com/2017/10/03/eurodata-unveils-european-viewing-trends/
8. Eurostat: being young in Europe. In: Statistic Books, pp. 12–48 (2015)
9. Eurodata TV Worldwide. http://www.mediametrie.com/eurodatatv/
10. Raycheva, L., Miteva, N., Peycheva, D.: Overcoming the vulnerability of older adults in contemporary media ecosystem (international policies and Bulgarian survey). In: Zhou, J., Salvendy, G. (eds.) ITAP. LNCS 10926, Part I, pp. 118–134. Springer, Heidelberg (2018)
11. Eurostat. https://ec.europa.eu/eurostat/statistics-explained/index.php?title=Population_structure_and_ageing/bg. (in Bulgarian)
12. Broadcast Network Europe: the digital terrestrial television facts&figures. In: Google Drive (2016). https://drive.google.com/

Understanding Ordinary and Disruptive Events Discussion in Twitter: Barbados Environmental Health Hazard as a Use Case

Adel Alshehri[1,2](✉) and Aseel Addawood[3,4]

[1] University of South Florida, Tampa, USA
Adelalshehri@mail.usf.edu
[2] King Abdulaziz City for Science and Technology, Riyadh, Saudi Arabia
[3] University of Illinois at Urbana Champaign, Champaign, USA
Aaddaw2@illinois.edu
[4] Al Imam Mohammad Ibn Saud Islamic University, Riyadh, Saudi Arabia

Abstract. Online users may utilize social media to discuss issues related to their environment. Understanding such discussions can help with predicting early warning signs for crisis situations and to enhance situational awareness and emergency preparedness. Over a period of more than four years, we collected and adopted a filtering method to obtain 30,358 tweets concerning environmental health risks in Barbados. In this study, we implemented an unsupervised machine learning algorithm to discover and understand how social media is used in the discussion of environmental health situational awareness. Moreover, what other topics online users are exposed to when engaging in such conversations. Our results show that there is a distinction between disruptive and ongoing events by exploring the number of tweets at a certain point of time and the sentiment of each event.

Keywords: Text mining · Topic model · Social media · Disaster
Sewage · LDA

1 Introduction

Communities are at risk of suffering from natural disasters or the outbreak of diseases; some are more defenseless than others. For example, Barbados and other small island developing states (SIDS) are susceptible to hurricanes, floods, and the expanded dangers of waterborne and foodborne illness in addition to other mosquito-borne diseases. Social media, such as Facebook and Twitter, let users share different types of content and communicate interpersonally. Due to the nature of social media, challenges can be found in tracking, monitoring and evaluating environmental health conversations to separate between disruptive and ongoing events.

Throughout this paper, the term event can be referred to "An incident that caused an increase in the number of text data that addresses the associated topic at a particular time." In [1], a disruptive event can be defined as "An event that opposes another event or disrupts an ordinary event. It may happen during a day or multiple days, causing troubles and may result in anxiety, sadness, and discontinuity. "Lastly, an ongoing

W. Karwowski and T. Ahram (Eds.): IHSI 2019, AISC 903, pp. 319–325, 2019.
https://doi.org/10.1007/978-3-030-11051-2_49

event can be defined as "An event that has been occurring for quite a long time and is expected to remain for some time in the future."

In this work we try to address two research questions: (1) How social media, mainly Twitter, is used in the discussion of environmental health situational awareness? (2) What type of topics online users get exposed to when engaging in an environmental health discussion in social media? We apply unsupervised machine learning technique (topic modeling) to gain a better understanding of what people are talking about in such health issues.

The remainder of this paper is organized in the following. In Sect. 2 we give a literature review on adopting social media during environmental disaster events. Next, Sect. 3 we discuss the framework components. In Sect. 4 we present our results and highlight key findings. Finally, we conclude with a review and directions for future research in Sect. 5.

2 Literature Review

To date, several studies have used social media platforms for the exploration of public health issues and tracking environmental health risks. In 2010 [2] a group of researchers collected tweets about the Deepwater Horizon Oil Spill crisis to evaluate how members of the society came to understand the potential consequences of the tragedy. The results revealed that Twitter users desired to cooperate in and contribute to response efforts. In [3] the researchers collected 206,764 tweets during the tornado that struck Joplin, Missouri (USA) and 140,000 tweets during Hurricane Sandy. Then, they sorted them into different dimensions, e.g., warning and advice, losses and destruction, donation and offer, and information source.

Much of the topic modeling and latent Dirichlet allocation (LDA) research has focused on classification [4], dataset search [5], and recommendation [6]. Analyses of the topic models show that disaster preparedness is an integral part of disaster risk reduction by improving solid waste management and evacuation preparation. LDA and its alternatives generally use statistical modeling approach implemented in event identification tasks [7, 8]. Further, classification-clustering method along with textual, temporal and geolocation features have been used in [1] to provide a way to detect events. To this aim, we chose the environmental health risks such as wastewater (sewage) and mosquito-borne diseases in Barbados as a case study to understand what people say during crises. Our study may contribute important considerations for decision-makers to prepare for disasters and save time in choosing where to focus their limited resources.

3 Method

In our study we concentrated on Twitter. It has been widely used for sharing news, activities during crises and natural emergencies. Our proposed framework is presented in Fig. 1, where its components are split into five stages:

Tweets Collection. We used Crimson Hexagon [9] to collect a corpus of 3,897,789 English posts from January 1, 2014, to May 31, 2018 with the keyword Barbados. Then, since our main interest was understanding the different environmental issues, we adopted a filtering method with a list of keywords containing the following: crisis, wastewater, sewage, Water Authority, water quality, Climate Change, Zika, mosquito, West Nile, disease, and health risk. The total sample size was 30,358 tweets.

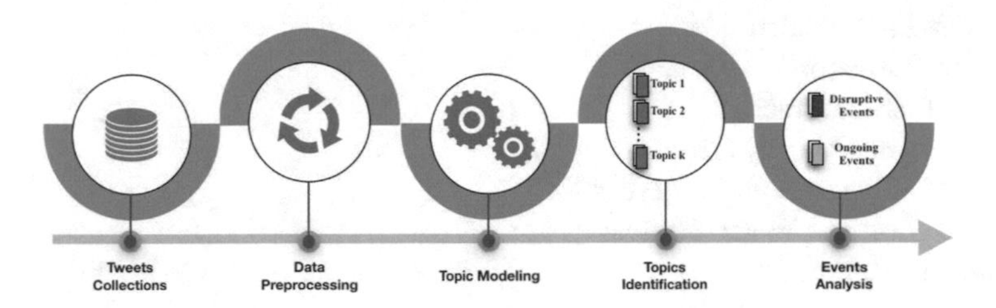

Fig. 1. Twitter topic modeling and event detection framework.

Data Preprocessing. Twitter users tend to use idioms, abbreviations, and grammatical errors in their posts. We applied text-processing methods like stopwords removal (the, a, an, in, etc.), punctuation, Lemmatization (converting a word to its root), lowercased all characters, and removed unnecessary white spaces using the Natural Language Toolkit library available in Python programming language.

Topic Modeling. In this step, we applied an unsupervised learning where topic modeling is a typical task. Topic model technique is based on two basic assumptions: (1) Each dataset consists of a mixture of topics, and (2) each topic consists of a collection of words. As a result, the goal of topic modeling is to uncover these hidden topics that shape the meaning of our dataset. From a machine learning perspective, Latent Dirichlet allocation (LDA) [10] is a Bayesian probabilistic model used for topic modeling. It uses term frequency and inverse document frequency TF-IDF approach, which treats each document as a vector of word counts. LDA has three important parameters: number of documents (M), number of topics (K), and number of words per each document (N). In our study, [M = 30,358 tweets], [k = 9], and finally [N = 10].

Topics Identification. Once the topics were extracted, we identified the top keywords that describe each topic. From there, we analyzed each topic and its relation to environmental health issues such as zika.

Events Analysis. We hypothesized that a disruptive event can be characterized by set of features: temporal, spatial and sentiment features. The temporal features are related to the diffusion of tweets counted over time frames. The spatial features include location inside Barbados or outside Barbados (international). We analyzed sentiment features (positive or negative) to understand whether the sentiment expressed in each topic can help with identifying disruptive events. We assumed neutral sentiment will not add any impact so we excluded it.

4 Results and Discussion

4.1 Topics Analysis

The (nine) topics are shown in Table 1 with a summary of some statistics:

1. Barbados on the Water
2. Environment in Barbados
3. Holiday in Barbados
4. Zika in Barbados
5. Head of Environment
6. Environment Agency Boss
7. Barbados Water Authority-BWA
8. Barbados on the Water Festival
9. Project in Barbados.

Table 1. Environmental topics in Barbados with some statistics

Summary	Water	Environment	Holiday	Zika	Head Envi	Envi boss	BWA	Water festival	Project
Min	89	3	0	0	0	0	0	0	0
1st Qu	205	9	0	0	0	0	9	0	0
Median	239	15	0	9	0	3	15	3	0
Mean	274	126.7	53.77	53.04	48.02	30.45	18.85	15.34	13.77
3rd Qu	303	37	6	31	3	6	28	15	6
Max	706	2176	1390	933	1897	926	73	284	517
Total Tweets	14520	6715	2850	2811	2545	1614	999	813	730

Each topic is a combination of weightage and keywords that contribute to the topic. Such as Topic 5 "Head of Environment" is a represented as: (0.046, head) + (0.045, environmental) + (0.043, environ) + (0.043, room) + (0.043, parliament) + (0.043, Barbados) + (0.043, walking) + (0.043, unit) + (0.042, state) + (0.042, nation). As shown in Table 1, the topic "Barbados on the Water (water)" has the highest number of tweets while the topic project got only 730 tweets. Also, we can see that topics such as "zika" and "BWA" have fewer tweets in the 1st quarter compared to the topic "water." These results help with identifying ongoing topics.

4.2 Event Analysis

To identify the difference between ongoing and disruptive event, we analyze three factors, (a) Temporal, (b) Sentiment (c) Spatial.

(a) Temporal Analysis. We keep collecting data for a long duration because some valuable posts in the past may not be as critical today or in the future. The results are presented in Fig. 2. A topic like "Barbados on the water" was discussed

14,520 times which might be assigned as an ordinary event. On the contrary, zika in Barbados was visible as a disruptive event that has been discussed 2,811 times overall but mostly in 2016 when the World Health Organization announced the Zika outbreak in 2016. The variation of the time (horizontal axis) per each topic is displayed in Fig. 2 where it was demonstrated that Barbados on the water continuously appears since the first day of data collection. Conversely in 2016, zika curve rises dramatically and then falls the same way.

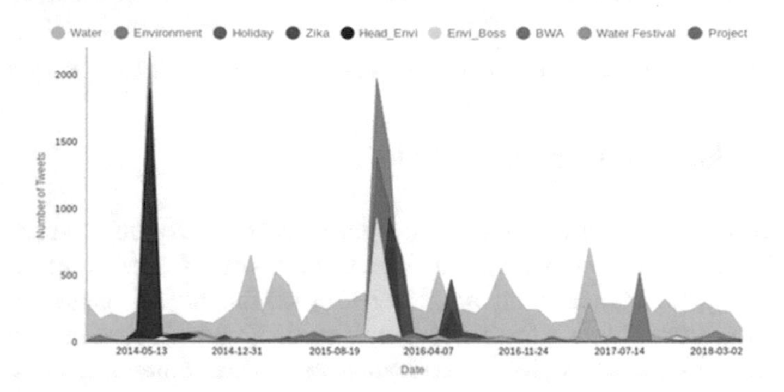

Fig. 2. Topic waves illustrates the most important topics have been discussed during the data collection period.

(b) Sentiment Analysis. The sentiment expressed in each topic can also be a factor in identifying the type of an event. Interestingly, in Fig. 3, in 2016, a disruptive topic such as zika has high negative sentiment (361 tweets) compared to its positive sentiment (33 tweets). Positive sentiment analysis for an ongoing topic such as "Barbados on the Water" was continuously higher than negative sentiment except for the year 2016 where positive tweets were higher with 812 tweets and 1338 tweets with negative sentiment. To sum up, negative tweet can be a strong predictor in identifying disruptive events which can be used by governmental agencies and policymakers to understand which topics users engage with the most. Moreover, a disruptive event can occur during ordinary events. This finding is consistent with previous research [8] where negative emotion tweets have a good confirmation rate for reporting disruptive events.

(c) Spatial Analysis. The Crimson Hexagon tool can estimate the location of the user or the event based on various pieces of information such as their profile information. Once the location has been extracted from each tweet, we aggregate them to determine two groups of tweets that originated from Barbados and tweets from abroad. The country with highest number of tweets is the US with 5,609 posts, Barbados comes next with 4,783 posts followed by the United Kingdom (UK) with 3,423 tweets. The engagement of countries such as the UK, USA, and Canada can be justified because sewage mess drove these countries to advise their citizens to avoid such affected area.

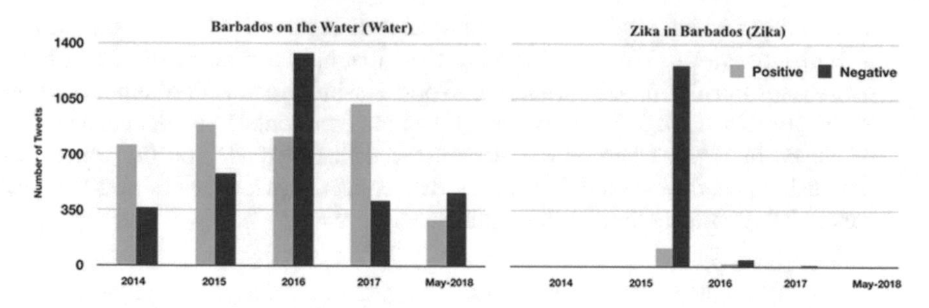

Fig. 3. Sentiment comparison "Barbados on the water" vs. "zika in Barbados."

5 Conclusions and Future Work

This empirical study contributes to this domain by reviews-public interaction with social media platforms due to environmental health concerns. The results indicate that it is not enough to consider temporal, spatial, or sentiment in isolation. Instead, the aggregate of features leads to a better ability to distinguish between events. In the future, we plan to use multi-class classification to predict community engagement on Twitter during environmental health hazards.

References

1. Alsaedi, N., Burnap, P.: Feature extraction and analysis for identifying disruptive events from social media. In: Proceedings of the 2015 IEEE/ACM International Conference on Advances in Social Networks Analysis and Mining, pp. 1495–1502. ACM (2015)
2. Starbird, K., Dailey, D., Walker, A.H., Leschine, T.M., Pavia, R., Bostrom, A.: Social media, public participation, and the 2010 BP Deepwater Horizon oil spill. Hum. Ecol. Risk Assess.: Int. J. **21**(3), 605–630 (2015)
3. Imran, M., Elbassuoni, S., Castillo, C., Diaz, F., Meier, P.: Practical extraction of disaster-relevant information from social media. In: Proceedings of the 22nd International Conference on World Wide Web, pp. 1021–1024. ACM (2013)
4. Mcauliffe, J.D., Blei, D.M.: Supervised topic models. In: Advances in Neural Information Processing Systems, pp. 121–128 (2008)
5. Rosen-Zvi, M., Griffiths, T., Steyvers, M., Smyth, P.: The author-topic model for authors and documents. In: Proceedings of the 20th Conference on Uncertainty in Artificial Intelligence, pp. 487–494. AUAI Press (2004)
6. Wang, C., Blei, D.M.: Collaborative topic modeling for recommending scientific articles. In: Proceedings of the 17th ACM SIGKDD International Conference on Knowledge Discovery and Data Mining, pp. 448–456. ACM (2011)
7. Vavliakis, K.N., Symeonidis, A.L., Mitkas, P.A.: Event identification in web social media through named entity recognition and topic modeling. Data Knowl. Eng. **88**, 1–24 (2013)
8. Pan, C.C., Mitra, P.: Event detection with spatial latent Dirichlet allocation. In: Proceedings of the 11th Annual International ACM/IEEE Joint Conference on Digital Libraries, pp. 349–358. ACM (2011)

9. Etlinger, S., Amand, W.: Crimson hexagon [program documentation] (2012). Accessed 15 September 2016
10. Blei, D.M., Ng, A.Y., Jordan, M.I.: Latent Dirichlet allocation. J. Mach. Learn. Res. **3**(Jan), 993–1022 (2003)

Integrating Hydrodynamic Models and Satellite Images to Implement Erosion Control Measures and Track Changes Along Streambanks

Mohamed Elhakeem[1,2(✉)], A. N. (Thanos) Papanicolaou[2], and Evan Paleologos[1]

[1] Abu Dhabi University, Abu Dhabi, United Arab Emirates
mohamed.elhakeem@adu.ac.ae
[2] University of Tennessee, Knoxville, USA

Abstract. Dike structures are used to restore the original bankline by redirecting the flow away from the banks towards the center of the stream. 2D-hydrodynamic models integrated with satellite images proved to be a powerful tool to enhance the design of dike structures in meandering streams and access dikes overall performance in controlling stream-bank erosion. In this study, the integrated 2D-hydrodynamic model with satellite images approach was used to design and evaluate the performance of dike structures at two meandering stream reaches along the Raccoon River, Iowa, USA. Analysis of the Google Earth pictures for the sites showed that the banklines almost recovered back in their original profiles and that the dikes became part of the restored banks, a strong indication of the success of the proposed dikes design.

Keywords: Stream-Bank erosion · 2D-Hydrodynamic models
Satellite images

1 Introduction

Dike structures, or barbs, have been used in many streams to control streambank erosion by redirecting the flow away from the banks towards the center of the channel (Fig. 1). Design and evaluation of dikes performance in controlling streambank erosion require often the use of physical and/or numerical models due to site specific aspects, which vary widely from one location to another [1, 2]. Numerical models are more adaptable to different environmental and geomorphologic domains than physical models [3].

Numerous, commercially available hydrodynamic models (either 1D, 2D, or 3D) exist for simulating flow around structures [4]. 1D-hydrodynamic models cannot easily simulate flow around structures. 2D-hydrodynamic models deemed to be ideal for simulating flow conditions in meandering stream reaches containing dike structures because they can resolve the large-scale turbulent flow structure and provide spatially varied information regarding the flow around the dikes [2]. In addition, 2D models require less data than 3D models for model calibration and verification [4].

W. Karwowski and T. Ahram (Eds.): IHSI 2019, AISC 903, pp. 326–331, 2019.
https://doi.org/10.1007/978-3-030-11051-2_50

Fig. 1. Dike structures along river bends.

The main objective of this study is to demonstrate how 2D hydrodynamic models integrated with satellite images can be used to design and evaluate the overall performance of a series of dike structures in mitigating streambank erosion and their interaction with the stream reach. The study was performed at two sites along the Raccoon River, Iowa, USA, where the Iowa Department of Transportation deemed these sites as in critical need for remediation due to the occurrence of large amounts of streambank erosion at both sites. The study was performed in the period from 2006 to 2010, where a series of dikes were installed in two meandering reaches located on the Raccoon River to control stream-bank erosion. The two sites are located on the Raccoon River between US Highway 35 and US Highway 169, where site 1 is located right upstream US Highway Bridge 169 and site 2 is located near Booneville right upstream US Railroad Bridge R22 (Fig. 2). The bridges at both sites have accelerated the development of the meander bends upstream of them causing the excessive bank erosion. The bridges act as control structures that do not allow the downstream portions of the reaches to adjust to the upstream changes.

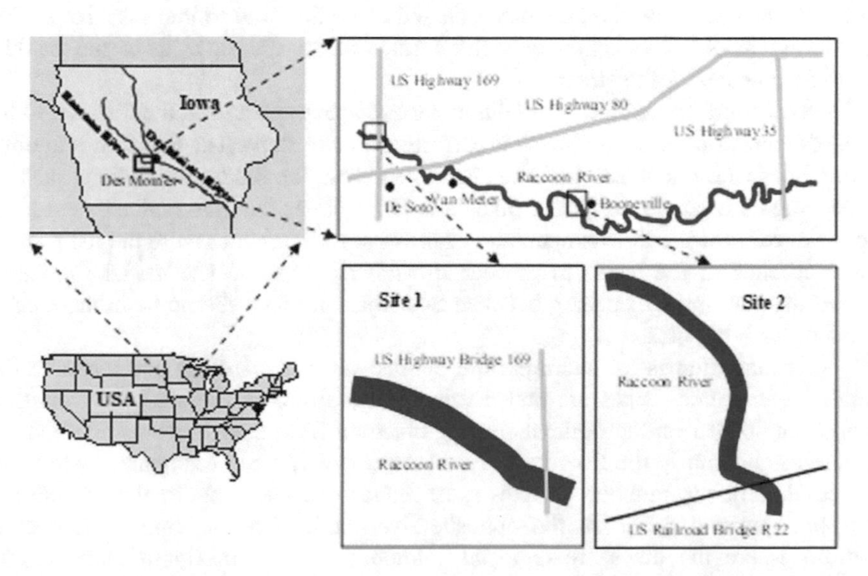

Fig. 2. Study site locations on the Raccoon River, Iowa, USA.

2 Model Description

The 2D Finite Element Surface Water Modelling System (FESWMS) was chosen to simulate the flow conditions around the dike structures at the two sites (Fig. 2). FESWMS is part of the commercially available Surface water Modelling System (SMS) software package (version 12.1) with a graphical interface that combines a series of hydrodynamic/sediment codes developed by the Federal Highway Administration [5]. FESWMS solves the differential forms of the continuity and the momentum equations in the stream wise and transverse directions using the Galerkin method of weighted residuals providing water depth and depth-averaged velocity magnitude in x and y directions at each node in the grid [5].

FESWMS inputs for model calibration are the Manning's coefficient of roughness, n, and the eddy viscosity, v. The Manning's n is an empirical coefficient that accounts for the total flow resistance from interactions with the boundary, while the eddy viscosity v accounts for flow resistance due to the internal shear stresses, or the Reynolds' stresses of the fluid incorporating the added energy dissipation due to turbulence in the flow [2].

3 Results

Dikes of different shapes and geometry (e.g., barbs, spurs, bendway weirs) have been proposed for streambanks erosion control. The hydrodynamic model, FESWMS, was used in this study to obtain the optimal number and spacing between the dikes, as well as the riprap lining length along the river-bank. Figure 3(a) shows the exact number, spacing and arrangement of the dikes at site 1. A series of alternating spurs and bendway weirs were used for protecting the river-bank at both sites. The major advantage of this alternating pattern is the correction of the incoming flow. Spurs protect the bank line on the structure side and force the flow to the inner bank of the meandering reach. The bendway weirs further direct the flow from the bankline towards the center of the stream.

The procedure to find the optimal number and spacing between the dikes as well as the dikes' effect on the stream reach is performed in two steps. The first step is to obtain the optimal spacing between the dikes iteratively from the model simulations such that the spacing between two adjacent dikes must be smaller than the reattachment length (the distance from the downstream end of the dike to the point where the main stream flow reattaches to the bank) to prevent streambank erosion. The model simulations showed that the optimal spacing between two adjacent dikes should be in the range of 45–50 m for both sites.

The second step is to determine the optimal number of dikes and evaluate their overall performance. Dikes are first distributed equally over the study reach with no riprap lining based on the optimal spacing obtained from step 1. If this configuration increases significantly the flow velocity on the opposite bank (compared with the no dike condition), the number of dikes must reduced in the next simulations until the dikes has minimal effect on the opposite river-bank. Riprap lining is used at the locations where the dikes are removed. Although the use of riprap lining is more economic than using dikes, it does not help in recovering the bankline by gaining sediment like the dikes effect.

The overall effect of the dikes on the stream reaches was evaluated by performing numerical simulations with and without the presence of the dikes. The no-dikes simulated flow condition was used as a reference for comparison. The change in the flow condition due to the presence of the dikes reflects their effect on the stream reach, either positive or negative. The model inputs are the following: bathymetry, Manning's coefficient $n = 0.03$, eddy viscosity $\nu = 0.1$ m^2/s, bank-full flow rate at the upstream section ($Q = 312$ m^3/s), and corresponding water depth at the downstream section ($y = 4.3$ m). The bank-full flow condition was used to evaluate the overall performance of the dikes at controlling bank erosion in the two stream reaches.

The model simulations at site 1 showed that the optimal number of dikes are 9 (5 bendway weirs and 4 spurs) with a spacing of 45–50 m between them (Fig. 3a). No-riprap lining was needed here because the last dike was close to the bridge abutment. Compared to the condition with no-dikes present (Fig. 3b), it can be seen in Fig. 3(c) that the dikes reduced the velocity considerably near the north bank where the dikes were constructed and increased the velocity in the core of the stream reach. Figure 3(d) shows the Froude number which was also within the accepted range. The model simulations for site 2 showed that the optimal number of dikes are similar to site 1 (5 bendway weirs and 4 spurs) with a spacing of 45–50 m between them. However, a 275 m riprap lining was required in the bend section downstream of the last dike. Similar trends of velocity to site 1 were also observed at site 2.

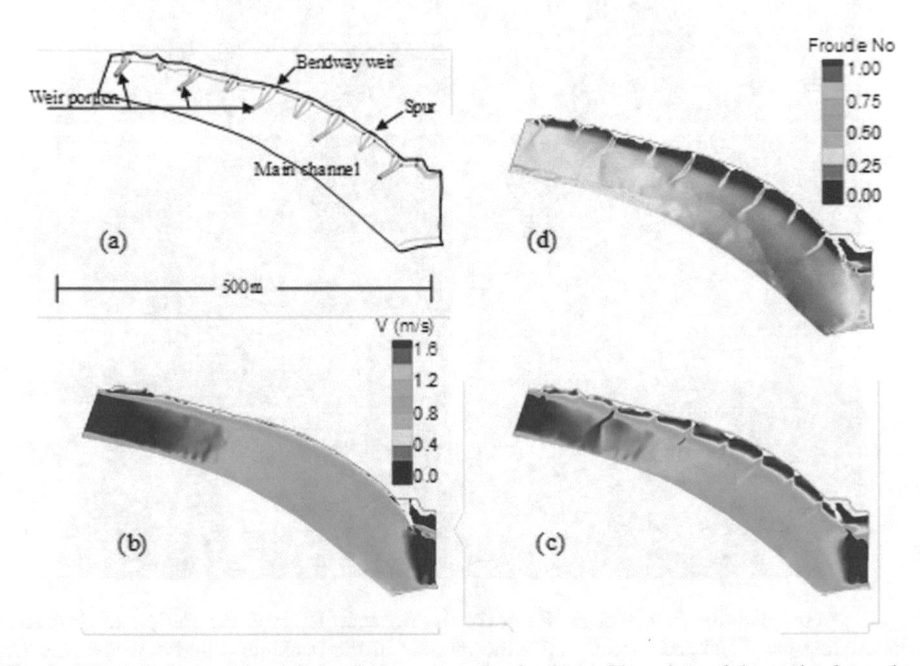

Fig. 3. Final design and model simulations at study site 1: (a) Plan view of the optimal number of dikes, arrangement and spacing; (b) flow velocity with no dikes; (c) flow velocity with the present of dikes, (d) the Froude number with the present of dikes.

Tracking of the bankline development after construction of the dikes from consecutive Google Earth pictures over the past 10 years shows that the banklines at the two study sites are almost recovered. Figure 4 shows a comparison between the banklines just after constructing the dikes and recent pictures of the banklines taken from Google Earth on 2016 for both sites. The recent Google Earth picture of study site 1 (Fig. 4 b) shows that the bankline recovered back to its original profile and that the dikes are completely buried and became part of the restored bank. However, the recent Google Earth picture of study site 2 (Fig. 4 d) shows that the river-bank is still building up and the dikes are partially impeded in the river-bank. The complete development of the river-bank at site 1 compared to site 2 which is still building up is attributed to the fact that the dikes at site 1 were constructed 4 years before the dikes at site 2. Figures 4 b and d show also that the dikes succeeded in moving the channel thalweg toward the center of the river. From these observations and inspection of the Google Earth pictures, we can conclude that the bankline took nearly 8 years for it to recover back to its original profile. This period is considered short in terms of morphological time scale indicating the success of the proposed dikes design in restoring the original river-banks.

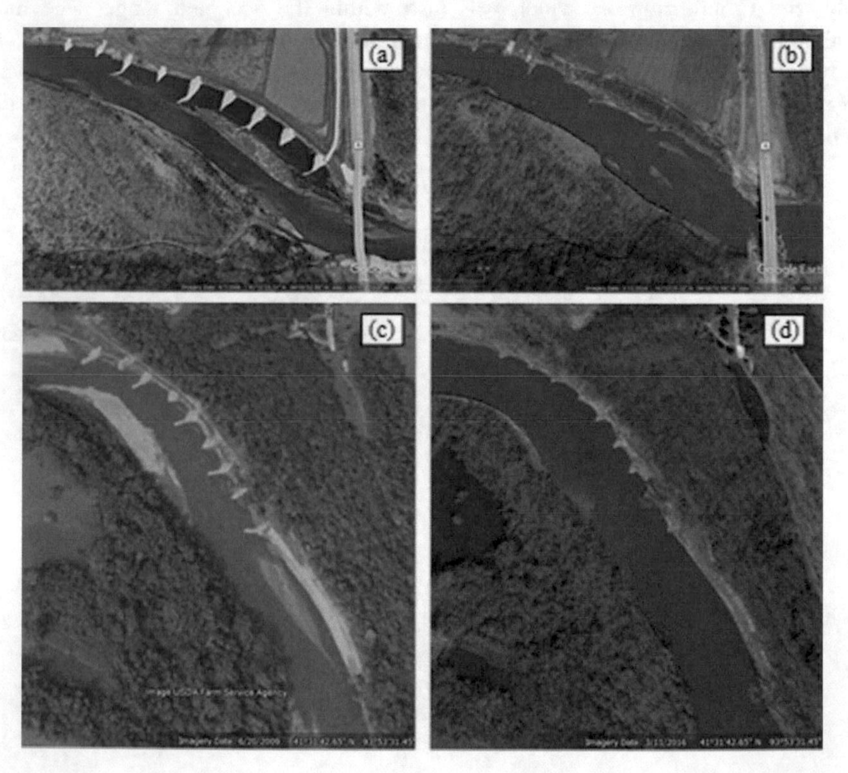

Fig. 4. Aerial pictures from Google Earth for the Raccoon River at the study sites just after constructing the dikes and recently after the recover of the bankline: (a) site 1 upstream US highway bridge 169, April 2006; (b) site 1 upstream US Highway Bridge 169, March 2016; (c) site 2 upstream US railroad bridge R22, Sept. 2009; (d) site 2 upstream US railroad bridge R22, March 2016.

4 Discussion and Conclusion

Recommended maximum velocity in erodible boundary streams with silt-clay banks similar to the two investigated sites should be in the range of 1.0 m/s to avoid bank erosion [6]. In addition, the Froude number should be in the range of 0.3 so that the water surface does not become rough, especially downstream of obstructions and in meandering sections [6]. From Fig. 3(c) and (d), it can be seen that the velocity and the Froude number in the stream reach after the construction of the dikes are well below the recommended values for the stability design of erodible boundary channels. The modified hydraulic conditions due the present of dikes provided an ideal situation for the incoming sediment transport to settle and redistribute around the dikes and along the bankline.

The 2D-hydrodynamic model FESWMS is used in this study to design dike structures to control river-bank erosion in two meandering stream reaches located on the Raccoon River, Iowa, USA. FESWMS was used to find the optimal number and spacing between the dikes and to access their effect on the study reaches. The model results showed that the proposed dikes structure, which consist of alternate bendway weirs and spurs successfully reduced the flow velocity along the outside bank and increased the velocity in the center of the stream, thereby successfully increased the conveyance towards the core of the river. The estimated velocity and Froude Number values along the outside river-banks where the dikes exist were within the recommended values for erodible channel stability design. The recent Google Earth pictures for the sites showed that the banklines almost recovered their original profiles and that the dikes are covered with sediment and become part of the restored banks. It was concluded from tracking the Google Earth pictures of the two sites over the past 10 years that the bankline took nearly 8 years to recover its original profiles and the stream reach took nearly 10 years to stabilize.

References

1. Papanicolaou, A.N., Elhakeem, M., Dermisis, D., Young, N.: Evaluation of the Missouri River shallow water habitat using a 2D-hydrodynamic model. River Res. Appl. **27**(2), 157–167 (2011)
2. Papanicolaou, A.N., Elhakeem, M., Wardman, B.: Calibration and verification of a 2D-hydrodynamic model for simulating flow around bendway weir structures. J. Hydraul. Eng. **137**(1), 75–89 (2010)
3. Elhakeem, M., Papanicolaou, A.N., Wilson, C.G.: Implementing streambank erosion control measures in meandering streams: design procedure enhanced with numerical modelling. Int. J. River Basin Manag. **15**(3), 317–327 (2017)
4. Papanicolaou, A.N., Elhakeem, M., Krallis, G., Prakash, S., Edinger, J.: Sediment transport modeling review - current and future developments. J. Hydraul. Eng. **134**(1), 1–14 (2008)
5. Froehlich, D.: User's manual for FESWMS Flo2DH: two-dimensional depth-averaged flow and sediment transport model. Release **3** (2002)
6. Chaudhry, M.H.: Open Channel Flow, 2nd edn. Springer, New York (2008)

A Systematic Literature Review About Quantitative Metrics to Evaluate the Usability of E-Commerce Web Sites

Ediber Diaz(✉), Juan Jesús Arenas, Arturo Moquillaza, and Freddy Paz

Pontificia Universidad Católica del Perú, San Miguel, Lima 32, Lima, Peru
{diazr.e, amoquillaza, fpaz}@pucp.pe, jjarenas@pucp.edu.pe

Abstract. Usability has become an essential aspect for the success of the software products. In the E-Commerce domain, the relevance of this quality attribute is even more significant. For this reason, there are methods that can be used to determine whether a software product meets the appropriate level of usability. In the category of quantitative techniques, the metrics are the most used by the specialists to establish the usability degree of a software system. However, the traditional metrics fail to cover aspects that the new emerging types of applications present. This fact leads to the postulation of new metrics that can measure the degree to which these aspects affect the usability. In this paper, the authors present a systematic review about the metrics that are reported in the literature as an assessment tool for the usability of E-Commerce websites.

Keywords: Systematic review · E-Commerce applications · Quality software Usability evaluation · Software metrics

1 Introduction

Usability is a software quality attribute that nowadays represents one of the main concerns of the software developers since this aspect is extremely related to the satisfaction degree of the final users [1]. Considering techniques and methods that guarantee the obtaining of an intuitive and easy-to-use product is essential to the project success and to meet the real expectations of the users. In the Web-specific domain of E-Commerce, the usability becomes even a more important characteristic, since can make a difference with respect to the competitors which can be found just a click away. In this sense, the user-centered design approach (UCD) provides a set of mechanisms and tools that allow software developers to evaluate the design proposals against the actual user and organizational requirements and verify if the software product satisfies the expectations [2]. The usability degree of an E-Commerce Web site, although being a non-functional aspect, can be the decisive factor for the users to choose the Web application of an explicit company from several options that are available on the Internet [3]. The current context of the Web, forces software development teams to consider the usability as a significant advantage in a market that is highly competitive.

W. Karwowski and T. Ahram (Eds.): IHSI 2019, AISC 903, pp. 332–338, 2019.
https://doi.org/10.1007/978-3-030-11051-2_51

The importance of the usability has led to the emergence of numerous methods that allow specialists to determine systematically through a well-defined and structured process if a Web site is usable, understandable and easy-to-use [4]. These techniques can be classified into qualitative or quantitative methods [5], in concordance with the result that is obtained from the application of them. Most of the methods are qualitative since these approaches allow to identify design issues from the graphical user interface to be subsequently described and cataloged as usability problems. On the contrary, the qualitative methods are focused on obtaining a numeric value about the usability degree of the Web site [6]. In some scenarios, quantifying the user experience as well as the usability can be more advantageous because allow consultants to perform comparisons with the closest competitors of a company. By establishing the usability degree in numerical values is possible for companies to determine how far they are from their competitors or notice to which extent they exceed to their main referents online.

The user tests with software metrics are one of the most used quantitative approaches to evaluate the usability of a software product [7]. This method involves the use of a set of measurement scales that are applied during a test scenario, in which potential users of the system are requested to interact with the software product, to determine according to certain mathematical formulas, whether the appropriate level of usability is achieved. The ISO 9126 standard [8] provides a series of metrics for the quantitative usability assessment of software products, and it is the main reference regarding this evaluation method. However, the standard fails to cover relevant aspects that the current software categories present nowadays. The E-Commerce websites at present are embedded with new features that previously did not implement, such as real-time processing, sophisticated designs, complex components and excessive functionality, that affect the usability of the web applications directly. In this sense, it becomes important to determine the metrics that are being used and the new proposals that have been developed to solve the gap in research.

In this study, the authors present a systematic review to identify the quantitative metrics that are reported in the literature to evaluate the usability of E-Commerce web applications. The review covers a spectrum of five years (from 2014 to 2018), and it is intended to propose as future work, a consolidated approach together with an evaluation process, that can be used to assess this type of software products accurately. In this way, the authors provide to the industry, the academia, and the specialists in HCI, with a tool that can improve remarkably the quality of the web applications.

2 Systematic Literature Review

The systematic literature review (SLR) is a structured methodological process that allows researchers to identify all the relevant studies that have been developed in relation to a specific subject of interest [9]. Unlike the traditional search process, the SLR provides an unbiased method to find research papers, is a replicable process and establish mechanisms to verify if the results are veridical. In the current section, the authors detail the protocol that guided the review process to find the existing metrics, that are reported in the literature, for the usability evaluation of E-Commerce websites. The process was based on the guidelines provided by Kitchenham [10] for the execution of SLRs in the area of software engineering (SE). In this study, Kitchenham reorganizes the traditional procedure that is commonly used in Medicine [11], for the application of the method in the software engineering domain.

The systematic review was conducted with basis on the following research question:

- What metrics have been reported in the literature during the last five years for the usability evaluation of E-Commerce websites?

Likewise, the elaboration of the search string was based on the PICOC method (Population, Intervention, Comparison, Results, and Context) which allows the identification of the general concepts involved in the systematic review [10]. The definitions for each item in relation to the research topic are shown in Table 1. Given that the purpose of this review was not focused on comparing different techniques that are employed for the usability evaluation of software products, the comparison criterion was not considered for the definition of PICOC.

Table 1. Definition of the general concepts using PICOC

Criterion	Description
Population	E-Commerce web applications.
Intervention	Quantitative metrics to evaluate the usability.
Outcomes	Study cases which report the use of a set of quantitative metrics to evaluate the usability degree of E-Commerce web applications.
Context	Academic context, software industry and all kinds of empirical studies.

According to the research question, we determined a list of terms that could contribute to the search of relevant studies: *e-commerce*, *website*, *commercial*, *transactional*, *metrics*, *measurements*, *quantitative*, *usability*, *heuristic*, *evaluation*, *methods*, *assessment*, *inspection*, *E-Commerce*, *principles,* and *guidelines*. The terms were grouped in accordance with the results of the PICOC method. The terms used in the search were linked with logical connectors (AND/OR) to achieve an appropriate portfolio of results. For the search, we only considered studies whose year publication was from 2014 onwards, to avoid outdated proposals and to determine only tendencies and current approaches to be recognized as future standards. After evaluating different viable options, we defined the following search string:

- *C1*: (“E-Commerce” OR “e-commerce website” OR “commercial website” OR “transactional website” OR “web application”)
- *C2*: (“metric” OR “measurement” OR “quantitative metric” OR “usability” OR “usability metric” OR “user experience”)
- *C3*: (“evaluation” OR “heuristic” OR “heuristic evaluation” OR “quantitative evaluation”)
- *C4*: (publication year > 2013)

The complete string used in our search was:

$$C1\ AND\ C2\ AND\ C3\ AND\ C4$$

The search process was performed using four recognized databases in the field of software engineering: IEEE Xplore, SCOPUS, SpringerLink and ISI (considering Web of Science & Web of Knowledge). The search string was adapted to be executed according to the particular guidelines of each search engine. No secondary search was performed. Gray literature was excluded because it is not peer-reviewed.

We defined the following inclusion criterion: *the paper describes a study case in which one or some metrics have been used to evaluate quantitatively the usability degree of an E-Commerce website*. If the study was not framed in this specific scenario, then the research paper was excluded from the review. For instance, studies related to the usability evaluation of physical devices were discarded.

3 Search Results

After an analysis of the relevant papers [8, 12, 13], we were able to determine thirty-five metrics that are currently reported in the literature during the period 2014-2018 for to assess the usability of transactional websites. The results are presented in Table 2.

Table 2. List of quantitative metrics to evaluate the usability of E-Commerce websites

Name	Formula
M1. complete description	X = A/B where A is the number of functions that are understood by the user and B is the total of functions.
M2. accessibility demonstration	X = A/B where A is the number of demonstrations/tutorials that the user can access, and B is the total number of demonstrations/tutorials available.
M3. demonstration of accessibility in use	X = A/B where A is the number of satisfactory cases in which the user achieves to watch the demonstration and B is the number of cases in which the user is requested to watch the demonstration.
M4. demonstration effectiveness	X = A/B where A is the number of functions correctly performed after the tutorial and B is the total number of demonstrations or tutorials reviewed by the user.
M5. understandable functions	X = A/B where A is the number of interface functions correctly described by the user and B is the number of functions available in the interface.

(*continued*)

Table 2. (*continued*)

Name	Formula
M6. understandable inputs and outputs	X = A/B where A is the number of input and output data elements that the user understands successfully and B is the number of input and output data elements available from the interface.
M7. easy of learning to perform a task	T where T is the sum of the user's operating time until the user is able to perform the specified task within a period of time.
M8. help accessibility	X = A/ B where A is the number of tasks that are found in the help system and B is the total number of tasks that were evaluated.
M9. operational consistency	X = 1 – A/B where A is the number of functions that are inconsistent with what the user expected and B is the total number of functions.
M10. error correction	T = Tc – Ts where Tc is the time that takes the user to complete the corrections of the specified errors and Ts is the start time of the correction of errors.
M11. attractive interaction	Questionnaire to evaluate the attractiveness of the interface for users after the interaction.
M12. proportion of functional elements with the appropriate name	X = A/B where A is the number of functions with a correct name and B is the total number of functions.
M13. proportion of functional elements used without errors	X = A/B where A is the number of functions that were used correctly and B is the total number of functions.
M14. proportion of exceptions that are correctly understood	X = A/B where A is the number of exceptions that were used correctly and B is the total number of exceptions.
M15. Proportion of returned values that are correctly understood	X = A/B where A is the number of returned values that are understood by the user and B is the total number of returned values.
M16. proportion of arguments that are correctly understood	X = A/B where A is the number of arguments that are correctly understood and B is the total of arguments.
M17. average time of component use	T = Hu – Hi where Hu is the end time after the use of the component and Hi is the initial time of component use.
M18. average time to master the component	T = Hd – Hi, where Hd is the end time after mastering the component, and Hi is the initial time of the test.
M19. error message by the density of functional elements	X where X is the total number of errors that appear due to overload.
M20. proportion of error messages that are correctly understood	X = A/B where A is the number of errors that are understood by users and B is the total number of errors.

(*continued*)

Table 2. (*continued*)

Name	Formula
M21. operation interface density	X where X is the number of functions found in a GUI.
M22. chars typed	X where X is the number of characters that are typed.
M23. cursor move time	T where T is the time in which the cursor was moved.
M24. cursor speed	X = A/B where A is the length of the movement of the cursor and B is the time the cursor was moved.
M25. cursor speed X	X = A/B where A is the length of the movement of the cursor in X direction and B is the time the cursor was moved in the X direction.
M26. cursor stops	X where X is the number of cursor stops.
M27. cursor trail	X where X is the distance the cursor has moved.
M28. hovers	X where X is the number of suspensions of a component.
M29. hover time	T where T is the total time of suspension of a component.
M30. time of use of the website	T where T is the total time of use of the website.
M31. changes in the scroll direction	X where X is the number of times the scroll bar has been moved.
M32. scroll maximum distance	X where X is the maximum distance the scroll bar has moved.
M33. scroll pixel amount	P where P is the number of pixels in the displacement.
M34. text selections	S where S is the number of selected texts.
M35. text selection length	T where T is the total length of the selected texts.

4 Conclusions and Future Works

Based on the initial research question, it is possible to conclude that there are studies in which several metrics have not only been proposed but also validated in practice. These measurements are focused on assessing different usability aspects of the E-Commerce websites, allowing companies to provide a better service to their customers. The benefits of providing to the specialists with a tool to measure this quality attribute in numerical values, is the possibility to perform comparisons with the main competitors, or even to decide between different available design proposals of graphical user interfaces. The list of metrics that has been identified based on the literature can be used as a preliminary approach for the future elaboration of a new set of usability evaluation metrics.

References

1. Quiñones, D., Rusu, C., Rusu, V.: A methodology to develop usability/user experience heuristics. Comput. Stand. Interfaces **59**, 109–129 (2018)
2. Cayola, L., Macías, J.A.: Systematic guidance on usability methods in user-centered software development. Inf. Softw. Technol. **97**, 163–175 (2018)
3. Paz, F., Paz, F.A., Pow-Sang, J.A.: Experimental Case Study of New Usability Heuristics, pp. 212–223. Springer International Publishing, Cham (2015)
4. Paz, F., Pow-Sang, J.A.: A systematic mapping review of usability evaluation methods for software development process. Int. J. Softw. Eng. Appl. **10**, 165–178 (2016)
5. Fernandez, A., Insfran, E., Abrahão, S.: Usability evaluation methods for the web: a systematic mapping study. Inf. Softw. Technol. **53**, 789–817 (2011)
6. Sauro, J., Lewis, J.R.: Quantifying the User Experience: Practical Statistics for User Research. Morgan Kaufmann, Cambridge (2016)
7. Paz, F., Pow-Sang, J.A.: Current trends in usability evaluation methods: a systematic review. In: Proceedings of the 7th International Conference on Advanced Software Engineering and Its Applications, pp. 11–15 (2014)
8. ISO: Software engineering – Product quality – Part 3: Internal metrics. Geneva, Switzerland (2003)
9. Kitchenham, B.A.: Systematic review in software engineering: where we are and where we should be going. In: Proceedings of the 2nd International Workshop on Evidential Assessment of Software Technologies, pp. 1–2. ACM, Lund (2012)
10. Kitchenham, B., Pearl Brereton, O., Budgen, D., Turner, M., Bailey, J., Linkman, S.: Systematic literature reviews in software engineering – a systematic literature review. Inf. Softw. Technol. **51**, 7–15 (2009)
11. Whiting, L.S.: Systematic review protocols an introduction. Nurse Res. (through 2013) **17**, 34–43 (2009)
12. Speicher, M., Both, A., Gaedke, M.: Ensuring Web Interface Quality through Usability-Based Split Testing, pp. 93–110. Springer International Publishing, Cham (2014)
13. Santos, C., Novais, T., Ferreira, M., Albuquerque, C., Farias, I.H.d., Furtado, A.P.C.: Metrics focused on usability ISO 9126 based. In: 2016 11th Iberian Conference on Information Systems and Technologies (CISTI), pp. 1–3 (2016)

Environment-Factor-Intellectual Concentration (EFiC) Framework: Method for Deriving Mechanism for Improving Workplace Environment

Kyoko Ito[1(✉)], Daisuke Kamihigashi[2], Hirotake Ishii[2], and Hiroshi Shimoda[2]

[1] Office of Management and Planning, Osaka University, Osaka, Japan
ito.kyoco@gmail.com

[2] Graduate School of Energy Science, Kyoto University, Kyoto, Japan
{kamihigashi,hirotake,
shimoda}@ei.energy.kyoto-u.ac.jp

Abstract. In this paper, a framework has been examined for quantitatively analyzing the relation between the workplace environment and intellectual concentration, through "factors" that connect between them, in order to improve intellectual concentration in the office. Specifically, "human characteristics" have been focused on and the factors affecting intellectual concentration was categorized into two groups. Using the factors, the measurement method and the quantification method have been considered and EFiC framework (Environment-Factor-intellectual Concentration) has been proposed for deriving the mechanism of intellectual concentration affected by the workplace environment. In order to confirm the effectiveness of EFiC framework, it was applied to the measurement data acquired in a past experiment of the intellectual concentration affected by lighting environment. As a result, concrete suggestions to improve the operating environment based on the characteristics of people were obtained. By applying the framework to measurement experiments of various intellectual concentration, it is expected that effective suggestions for improving intellectual concentration will be obtained.

Keywords: Intellectual concentration · Human characteristics
Factors · EFiC (Environment-Factor-intellectual concentration) framework
Covariance structure analysis

1 Introduction

Intellectual concentration and intellectual productivity have been actively studied to improve the work efficiency of office workers by controlling the office environment in the office in recent years [1]. In order to evaluate the influence on the intellectual concentration by the workplace environment, it is necessary to design "factors" that mediate the workplace environment and intellectual concentration [2]. In this study, a mechanism model is defined as a model that the mediated factors influence the

W. Karwowski and T. Ahram (Eds.): IHSI 2019, AISC 903, pp. 339–344, 2019.
https://doi.org/10.1007/978-3-030-11051-2_52

intellectual concentration from the workplace environment, and the derivation method of the mechanism is focused on. The mechanism is a relationship that connects what kind of factors the individual workplace environment concretely mediates and quantifies what impacts intellectual concentration. There is no research that quantifies the "factors" that connect the workplace environment and intellectual concentration and that clarified the mechanism and a method therefor. In this study, the method for deriving the mechanism is called a framework.

In the mechanism model, when influencing intellectual concentration by mediating factors from the workplace environment, "factor" is the characteristics of individuals and the state of individuals. The workplace environment such as temperature and humidity influence individual condition such as fatigue and then the change of the individual state affects intellectual concentration. The influence from the workplace environment to the state of the individual is the individual difference. For example, the temperature that you feel comfortable and the brightness that you feel bright are different as others'. In this study, individual condition and difference are taken as human characteristics. By deriving a mechanism based on human characteristics, it is expected to lead to the improvement of work efficiency by selecting of an office environment for improving intellectual concentration considering individual differences.

The purpose of this study is to provide a framework for quantitatively analyzing the relation between the workplace environment and intellectual concentration, through "factors" that connect between them, in order to improve intellectual concentration in the office.

As related researches, there are some researches to improve intellectual productivity. There are two main categories: one category of researches [3–5] that investigate the influence of workplace environment on intellectual productivity and another category of research [1] focusing on factors that affect intellectual productivity. In these researches, tasks are used for knowledge processing among intellectual productivity. In addition, some methods for evaluating intellectual productivity have been proposed.

2 Method

The "human characteristics'" have been focused on and the factors affecting intellectual concentration is categorized into two groups. They are dynamic factors (arousal, mood, fatigue, stressor assessment) [6, 7] which vary during work and static factors (reference value such as demographics, environmental sensitivity, etc.) which do not change. Using the factors, the measurement method and the quantification method have been considered, and EFiC framework (Environment-Factor-intellectual Concentration) has been proposed for deriving the mechanism of intellectual concentration affected by the workplace environment.

The relationship between intellectual concentration and factors is shown in Fig. 1. A mechanism model based on Fig. 1 for change of intellectual concentration is shown in Fig. 2. First, the workplace environment influences factors. Among the factors, static factors affect dynamic factors. Dynamic factors are four: arousal [8], mood [9], fatigue [10], and stressor evaluation, and the four factors are expected to affect each other. And the factors as a whole affect the intellectual concentration.

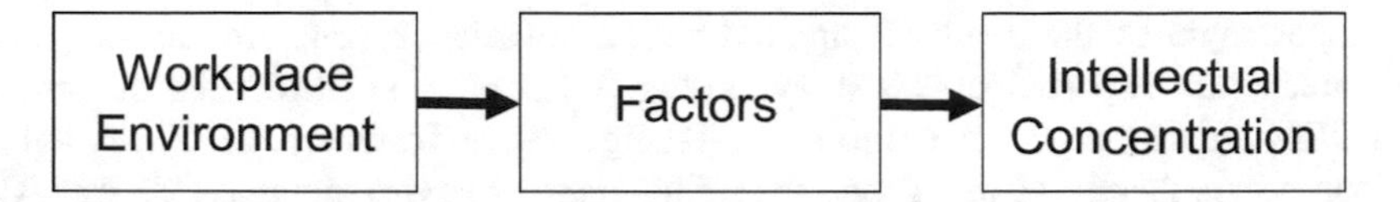

Fig. 1. An outline of a mechanism model for the change of intellectual concentration

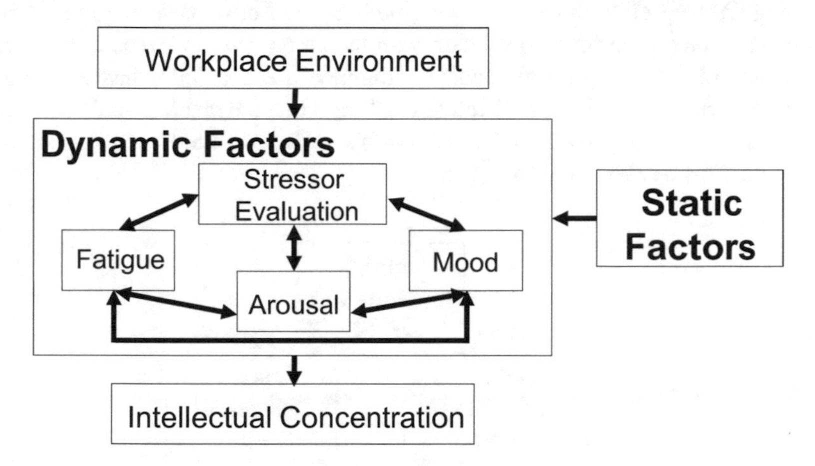

Fig. 2. A mechanism model with dynamic and static factors for changes in intellectual concentration.

When using measurement data obtained from large-scale measurement experiments or measurement experiments for participants with different attributes, a new mechanism may be derived. Also, if we can use data obtained from measurement experiments that change the workplace environments such as air quality, thermal environment, sound environment, etc., we may derive a new mechanism for the change of intellectual concentration.

3 Verify the Effect of EFiC Framework: A Case Study of Lighting Environment

In order to confirm the effectiveness of EFiC framework, it was applied to the measurement data acquired in a past experiment of the intellectual concentration affected by lighting environment.

The two target environments are normal ceiling lighting and New Task and Ambient (N-TA) lighting which adjusts illuminance and color temperature around the office worker and of the whole room. N-TA lighting uses higher color temperature lighting than ceiling lighting. The number of participants in the experiment was twenty-four. Also, there are some data losses, and "arousal" has no data measured. Thus, the mechanisms statistically analyzed within the data are derived.

An example of the results is applied with a covariance structure analysis, and a mechanism for the participants is shown in Fig. 3. In order to consider the change from the ceiling lighting environment to the N-TA lighting environment, the N-TA lighting environment is quantified as "1" and the ceiling lighting environment as "0". And, CTR means intellectual concentration. Figure 3 shows the N-TA lighting environment reduces "bright-good-pleasant environment" and "refreshed environment" compared to the ceiling lighting environment. "bright-good-pleasant environment" and "refreshed environment" have a positive correlation with the other. Then, CTR is reduced due to improvement of "bright-good-pleasant environment". Also, "refreshed environment" reduces "refreshing", "refreshing" improves "enjoyment concentration" and "enjoyment concentration" improves CTR. To improve CTR, two kinds of paths are shown from the lighting environment in Fig. 3.

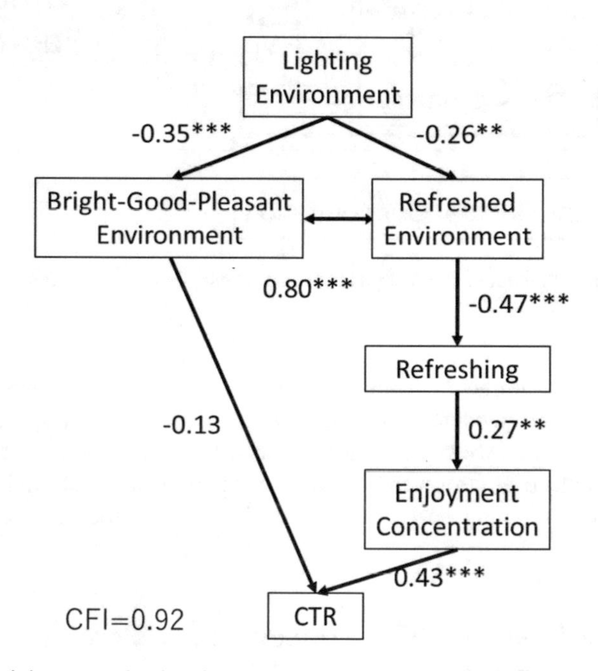

Fig. 3. Deriving a mechanism by covariance structure analysis for the participants.

As a result, some concrete suggestions to improve the environment with the basis on the characteristics of people were obtained. By applying the framework to measurement experiments of various intellectual concentration, it is expected that effective suggestions for improving intellectual concentration will be obtained.

4 Conclusion

In this study, we have considered a framework for quantitatively analyzing the relationship between the workplace environment and intellectual concentration via "factors" connecting them. Specifically, focusing on "human characteristics", we have classified the factors that affect intellectual concentration into two categories: dynamic factors (arousal, mood, fatigue, stressor assessment) which changes during work and static factors (reference value such as demographics, environmental sensitivity, etc.) which do not change. Using the factors, we have examined the measurement method and the quantification method and proposed EFiC framework (Environment-Factor-intellectual Concentration) for deriving mechanisms for the change of intellectual concentration by the change of the workplace environment. In order to verify the effectiveness of EFiC framework, it was applied to a measurement data acquired in a past experiment of the intellectual concentration by the lighting environment. It is a case study. The target of the case study is for environmental changes in lighting. By applying EFiC framework to changes in the air conditioning environment, for example, similarities and differences may be considered in comparison with the mechanisms obtained by the change of lighting environment. Based on the results, we can analyze the influence on intellectual concentration in detail. By analyzing using EFiC framework, it is expected to provide guidelines for appropriate combinations of environmental factors.

From now on, by applying EFiC framework to various measurement experiments of intellectual concentrations, it is expected that effective guidelines for improvement of intellectual concentration can be obtained. In addition, in order to expand the framework in general purpose, it is conceivable to consider more effective measurement method and analysis method.

Acknowledgements. This work was supported by JSPS KAKENHI Grant Number 17H01777.

References

1. Tanabe, S., Nishihara, N.: Productivity and fatigue. Indoor Air **2004**, 126–133 (2004)
2. Bosch-Sijtsema, P.M., Ruohomaki, V., Vartiainen, M.: Knowledge work productivity in distributed teams. J. Knowl. Manag. **13**(6), 533–546 (2009)
3. Wargocki, P., Wyon, D.P., Fanger, P.O.: Productivity is affected by the air quality in offices. Proc. Healthy Build. **2000**, 635–640 (2000)
4. Fisk, W.J., Price, P.N., Faulkner, D., Sullivan, D.P., Dibartolomeo, D.L., Federspiel, C.C., Liu, G., Lahiff, M.: Worker performance and ventilation: analyses of time-series data for a group of call-center workers. Indoor Air **2002**, 784–795 (2002)
5. Kroner, W.M., Stark-Martin, J.A.: Environmentally responsive workstation and worker productivity. ASHARE Trans. **100**, 750–755 (1994)
6. Sensharma, N.P.: An extension of a rational model for evaluation of human responses, occupant performance, and productivity. healthy buildings 2000. In: Workshop, vol. 9 (2000)
7. Swain, A.D., Guttmann, H.E.: Handbook of human reliability analysis with emphasis on nuclear power plant applications, pp. 51–76. NRC, U.S.A (1983)

8. Matthews, G., Jones, D.M.: Refining the measurement of mood: the UWIST mood adjective checklist. Br. J. Psychol. **81**(1), 17–42 (1990)
9. Terasaki, M., Kishimoto, Y., Koga, A.: Construction of a multiple mood scale. Jpn. J. Psychol. **62**(6), 350–356 (1991–1992)
10. Working Group for Occupational Fatigue of Japan Society for Occupational Health : Handbook of Industrial Fatigue. Labor Standards Committee (1988)

Code of Breakthrough Innovations

Anda Batraga, Jelena Salkovska, Liga Braslina, Henrijs Kalkis(✉),
Daina Skiltere, Aija Legzdina, and Girts Braslins

Faculty of Business, Management and Economics, University of Latvia, Asp
Aspazijas blvd. 5, Riga, Latvia
{anda.batraga,jelena.salkovska,liga.braslina,henrijs.kalkis,daina.skiltere}@lu.lv, aija.legzdinal@gmail.com, girts.braslins@inbox.lv

Abstract. Innovations make a significant contribution to economies, while at the same time 85% of innovations in the market are unsuccessful. This is a record high number of unsuccessful innovations, considering the high macro-level support of innovations and researches, devoted to innovations. The study analyzes in detail the technological essence of innovation process, its constituent components, the elements of commercially viable innovations and their inter-relations with the aim to find the common thread of successful innovations in the market. The result is the conclusion that there is a specific algorithm that is repeated in the context of commercially successful innovations. The study offers the list of innovative components of innovations and the analysis of the regularities of commercially successful innovation process technology.

Keywords: Innovation components · Innovation process technology
Breakthrough innovations

1 Introduction

Commercially successful innovations provide for an average of 45% of market growth in the world largest economy, the United States [1]. In Europe, the average new SKU[1] hits € 160,000 in its first year, which is about 60% of the average amount generated by new and established SKUs [1]. Within the framework of innovative economy there is a lot of researches and methods for the successful development of innovations. However, 85% of innovations launched in the market are removed as unsuccessful [1]. In 2014, in the US market, only 7 out of 12,000 launches combine the strength of relevance, endurance and distinction to deliver superlative performance [1]. The high rate of failure of innovations sometimes indicates to companies that the commercial success of innovations is rare and unpredictable. The long-term researches in innovation discipline by the authors provide the evidence that innovation is the same economic discipline like others with its

[1] SKU – stock keeping unit.

W. Karwowski and T. Ahram (Eds.): IHSI 2019, AISC 903, pp. 345–351, 2019.
https://doi.org/10.1007/978-3-030-11051-2_53

certain regularities. This author's study provides an in-depth analysis of the components of commercially successful innovations and conclusions that there is a specific algorithm that is repeated in the context of commercially successful innovations. Within the study 32 commercially successful innovations in fast moving consumer goods (FMCG) market have been analyzed, which in the period from 2014 to 2016, the year after launching, achieved the following sales results, which are the highest in innovation sector in each world region: *Europe:* Generated a minimum of €7.5 million in first-year sales per market for Western Europe and € 5 million for Eastern Europe, according to Nielsen's data. Achieved at least 90% of the year-one sales in year[2] two. *U.S.:* Generated a minimum of $50 million in a year-one U.S. sales, according to Nielsen data. Achieved at least 90% of the year-one sales in year two. *Australia:* Generated a minimum of $4.5 million in annualized first year sales. Achieved at least 90% of the year-one sales in year two. Asia: Generated significant year-one sale according to Nielsen. Achieved at least 90% of the year-one sales in year two. Narrowing of the research is the analysis of innovative products for the FMCG market without considering process, service, marketing or organizational innovations. The article reviews innovations from the standpoint of business environment companies. The study analyzes product innovation components without consideration of marketing mix elements – price, placement and promotion. The study does not cover or analyze the micro- and macroeconomic environment processes of innovation environment.

2 Innovation Essence and Components

The definitive essence of innovations is to promote socioeconomic growth in the society. In capitalistic society, innovation is a dynamic and multilateral concept that relates to new products and/or processes in society. General classification of innovations is broad and there are different dimensions for specific sectors, categories and segments, which may be not representative or relevant for separate innovation classifications. To make a classification of innovation components, the authors of the study have analyzed in detail more than 25 definitions of innovations [2] and performed an in-depth analysis of the interpretations, provided Schumpeter [3], Drucker [4], the company Booz Allan [5], Nielsen, and Praude [6]. The OECD definition [7] is used in the research to specify and identify innovation. Specifically, the common characteristics and components of innovations are relatively little researched, as they apply to a wide variety of industries and different products and processes. The authors of the study have carried out an extensive analysis by comparing academic-related researches with data of research companies. The bibliography from the OECD Directive Oslo Manual [8] and the researches by Schumpeter and Mensch [8], Kondratiev [9], Chan

---

[2] This measure confirms a sustained level of consumer demand while allowing for some drop in revenue during the transition from trial to adoption.

and Mauborgne [10] and Praude [6], has been analyzed as well as the components of the unique innovation method approach InnoMatrix [11] have been investigated. An in-depth analysis of the key dimensions of commercially successful innovations of the research company Nielsen has been carried out. Based on the studied literature of innovations and commercially successful innovations, the authors have summarized the components of innovations with the aim to analyze its commercially successful innovations in the next section (Table 1).

Table 1. Innovation components and their classification. Source: table composed by the authors of the study, based on literature review.

Innovation components	Classification
1. Company power in category	Leading, existing, new
2. Brand power in category	Leading, existing, new
3. New consumer experience	Existing experience or new experience
4. Compensating effect	Yes, no
5. Innovation toxification type	Brand extension, package extension, flavor extension, functional intensity extension, category merge, subcategory merge, new product shape
6. Consumer experience	Existing or new experience
7. New claim	Yes, no
8. New consumption ritual	Yes, no
9. New flavor/fragrance	Yes, no
10. Value dimensions	Functional, emotional and/or social
11. Price segments	Lux, premium, upper middle, middle, lower middle, economy
12. Scientific discovery	Yes, no
13. Technology progress	Scale
14. Claimed eco friendly	Yes, no
15. Claimed health supportive	Yes, no
16. New external packaging form	Yes, no
17. New product form	Scale
18. Outstanding design	Yes or no
19. Innovation development strategies	InnoMatrix©classification

This study does not analyze marketing communication components of commercially successful innovations, but their detailed analysis is significantly different from product components. Based on the analyzed components of innovation, the authors of the study distinguish the complex of elements of commercially successful innovations and the approach to their creation – the strategy, collectively defined as a commercially successful innovation code.

3 Content Analysis the Elements of Commercially Successful Innovations

In 2014–2016, a half (52%) of the world's breakthrough innovations in retail trade of FMCG were created by world-leading companies, one-third – by leading local companies while 19% comprise innovations of existing local, non-leading companies. One-fifth of breakthrough innovations come from local companies that are not leaders in their markets, that is the evidence of the capacity of a powerful innovation to break through the market also from the starting positions of a local company. Almost 90% (87%) of all breakthrough innovations are innovative extensions of existing brands. This shows the high importance of the brand in the context of commercially successful innovations, also if the innovation is distinctively technologically functional. However, it should be noted that 13% of the breakthrough innovations are the innovations of local brands, not dominating in the market, that managed to achieve a significant importance in the market relatively fast within the last year. It also points out that a local company can create commercially successful innovations that affect the consumption patterns of a particular category over the course of a year. An important component of commercially successful innovations is their ability to generate a new consumption experience. However, an empirical study indicates that only 2/3 of commercially successful innovations bring new experiences. 30% of breakthrough innovations do not create new consumer experiences, rather using the existing experience. Most (70%) of commercially successful innovations provide some other process economy or other process enhancement. However, one-third of commercially successful innovations do not provide a compensating effect. In terms of functionality, one-half (48%) of commercially successful innovations are the innovations that go beyond their usual category by combining two categories of promise to the consumer (2 categories merge). Another commercially successful innovation approach applied is brand strength transfer to another category (16%) by adding some new element. One of the consumer behavior phenomena the most difficult to influence is the change in consumer behavior. It is relatively difficult to achieve even with the most successful innovations - only 1/5 (19%) of commercially successful innovations have succeeded. However, most commercially successful innovations (81%) fail to change consumer habits. A similar Pareto principle is also applicable to new flavors and fragrances component in the assessment of commercially successful innovations. A new flavor or fragrance component is found only in 1/5 of commercially successful innovation assessment. It should be noted that even in cases where innovation has a new flavor or fragrance, it is not the only element of innovation complex of commercially successful innovations. An essential component in innovation assessment is the presence of a new functional, emotional or social value provided by innovation. In an empirical study, all commercially successful innovations turned to have a functional innovative value, a half (47%) had both functional and emotional value, while 16% had a social value. Most often, in 55% of cases, within one innovation one innovative value added was used - either functional, emotional or social. In 32% of cases 2 values were used, and in 13% of cases all 3 consumer values were used. As regards the price segment, the index for commercially successful innovations is 3.7, where 1 is economic price and 6 - luxury price, indicating in general that 63% of commercially successful innovations relate to the upper middle price segment. Only 28% of commercially successful innovations include a scientific component, although the impact on the market of all selected innovations is significant, according to Nielsen classification. This confirms P. Drucker's approach that innovation is not so much associated with technological advancement as with the impact on consumer behavior. Only 9% of breakthrough innovations are eco-related and only 19% promise a healthier lifestyle. Considering the megatrend of consumer health, the authors of the study have concluded that these are relatively small numbers compared to the total amount of commercially successful innovations and the trend of the new principle "health is new wealth" in the economy. 25% of commercially successful innovations include new packaging, and only 12% have a distinctively different (outstanding) design, which is significantly different from other products in the category. New product form plays the key role among innovation components: 31% of all commercially successful innovations have a new form.

Summarizing the analysis of innovation elements and evaluating it compared to the results provided by InnoMatrix, it turns out that application of two different approaches in innovation analysis leads to surprisingly similar conclusions, differing essentially in percentage sharing of innovation codes applied. See Fig. 1.

In essence in 80% of cases 3 approaches to innovation technology process are applied. The most commonly used approach to the creation of commercially successful innovations is product hybridization with another category, thus expanding the category and getting new consumers in the category. The second most successful approach in innovation development is the maximization or reduction of functional values of existing components of a product in relation to comparable products in different graduations. The third most successful approach is the transfer of brand strength from one category to another by adding some innovative component. These conclusions

Commercially successful innovations analyse by 2 approaches-components detailed analyse and InnoMatrix approach			
I Commercially successful innovations analyse of packed food sector by InnoMatrix *(scope- 100 food innovations)*		**II Commercially successful innovations analyse of all FMCG sector by components analyse** *(scope- 32 FMCG innovations)*	
Innovation development code- Category expanded by merging or hibernating			
35%		**53%**	
Hybridization between the largest and the fastest growing category	14%	Merging 2 related categories (2CM)	47%
Hybridization between a major industry category or segment and an innovative component	8%	Hybridize with leading components from other related categories (HYB with LC)	6%
Hybridization without definite analysis, between previously unmatched categories	7%		
Transfer of the fastest growing category or segment component to another segment	6%		
Innovation development code-Brand power transferred to new category with some innovative component			
3%		**9%**	
Existing brand transfer to the largest or fastest category or segment of another category	3%	Brand awareness moved to new related category+ new functional value (BA+NC+FC)	6%
		Leader brand moved to category leading format with new emotional value (L+LP)	3%
Innovation development code- product components reduced or maximized			
40%		**13%**	
Maximum geolocalized - either globalized or localized product	8%	Reduce or Maximize product components (MinMax)	13%
Maximum reduced or propagated components	8%		
Maximum reduced or enlarged product size	7%		
Color and perfume maximum and zero gradations	7%		
Maximum or reduced product exposure (power)	6%		
Personalization gradations	4%		
Other innovation development code			
22%		**25%**	
A time-saving approach to achieving the desired end position	5%	Leader brand complemented with some scientific discovery component (L+Sc)	13%
Copy / paste to fast growth leader	4%	Change shape of leading format (LF FCh)	9%
High-speed product segment transfer to highly low or high price segments	4%	Stand Out by diametrically opposite approach vs mainstream (S-Out)	3%
Hybridized product packaging mixing different physical states	3%		
A simpler approach to achieving the desired end position	3%		
Components that physically change the states of the ego consciousness	2%		
Parodical	1%		

Fig. 1. Analysis of commercially successful innovation development approaches by comparing the analysis of innovation elements with the results of InnoMatrix © model *Source*: Content analysis by the study authors, based on Nielsen Breakthrough Innovation reports (US, Europe, Asia, Australia) data 2014–2016 and InnoMatrix© model reported results.

confirm P. Drucker's approach that technological process of a commercially successful innovation relates not so much to the category of novel, innovative components but rather to the extension of the existing consumer category between the categories, providing to the consumer a compensating effect and a new consumption experience.

References

1. Nielsen breakthrough innovation reports. European Edition (2014, 2016), U.S. Edition (2014), Australia Edition (2016), Asia Edition (2016)
2. Dubra, I.: Innovation in the Baltic States and its Influencing Factors (Inovācija Baltijas valstīs un to ietekmējošie faktori). University of Latvia, Riga (2014)
3. Durlauf, N.S., Blume, E.L.: The New Palgrave Dictionary of Economics. Palgrave Macmillan, London (2008)
4. Drucker, P.F.: Classic Drucker: Essential Wisdom of Peter Drucker from the Pages of Harvard Business Review. Harvard Business Review Press (2006)
5. Booz Allen, H.: New Products Management for the 1980s. Booz Allen Hamilton Inc., New York (1982)
6. Praude, V., Salkovska, J.: Integrated Marketing Communication (Integrētā Mārketinga Komunikācija). Burtene, Riga (2015)
7. Organisation for Economic Co-operation and Development (OECD) Staff: Oslo Manual: Guidelines for Collecting and Interpreting Innovation Data. Oecd Publishing (2005)
8. Mensch, G., Coutinho, C., Kaasch, K.: Changing capital values and the propensity to innovate. Futures **13**, 276–292 (1981)
9. Kondratieff Cycles - A Thumbnail Sketch. Building Prosperity, pp. 7–11. Business Source Complete, EBSCOhost (2000)
10. Chan, W.K., Mauborgne K.R.: Value innovation -the strategic logic of high growth. Harward Bus. Rev. **82**, 172–180 (2004)
11. Batraga, A., Salkovska, J., Braslina, L., Legzdina, A., Kalkis, H.: New innovation identification approach development matrix. In: Advances in Human Factors, Business Management and Society, vol 783, pp. 261—273. Springer (2018)

Application of the Usability Metrics of the ISO 9126 Standard in the E-Commerce Domain: A Case Study

Freddy Paz[1(✉)], Ediber Diaz[1], Freddy A. Paz[2], and Arturo Moquillaza[1]

[1] Pontificia Universidad Católica Del Perú, Lima, Peru
{fpaz, diazr.e, amoquillaza}@pucp.pe
[2] Universidad Nacional Pedro Ruiz Gallo, Lambayeque, Peru
freddypazsifuentes@yahoo.es

Abstract. Quantitative metrics are one of the few tools that allow specialists in the field of Human-Computer Interaction (HCI) to obtain numerical values about the level of usability of a software system. Although the advantages that software metrics provide, there is not enough evidence about their use in the literature. In this paper, we describe the evaluation process and the application of the usability metrics proposed by the ISO 9126 standard to evaluate a specific software product in the E-Commerce domain. The purpose of this study is to establish a guide that can be used by professionals in Software Engineering and related areas, to determine the correct level of usability of a set of graphical user interfaces that are designed as part of the front end of a system. In the same way, we describe the example of an application in which the entire inspection process is followed.

Keywords: Human factors · Human-systems integration · Systems engineering

1 Introduction

In the e-commerce domain, the usability has become in an extremely relevant aspect that impacts directly on the user's decision of using a specific website [1]. According to several authors [2–4], if a web application is difficult to use or fails to clearly establish the products or services that are offered, the users leave the site in search of other available alternatives on the online market. In this sense, the usability can be determinant and for this reason, a factor to be considered during the software development process.

Given the importance of the usability, several evaluation methods have been developed in order to provide specialists with a formal and systematic procedure that allows them to determine how usable the graphical user interfaces of a software system are [5]. Most of these methods are focused on the identification of usability problems through subjective approaches, and only a small percentage of them establish the protocol to obtain a numerical value about the usability of the software product. However, in many scenarios and situations, obtaining the usability degree of the system

W. Karwowski and T. Ahram (Eds.): IHSI 2019, AISC 903, pp. 352–356, 2019.
https://doi.org/10.1007/978-3-030-11051-2_54

in quantitative terms is necessary, especially, if the company requires to compare itself with the competition, or if there is required to determine the most suitable interfaces from multiple design proposals. For companies, it would be useful to have a methodological procedure that allows them to know the usability degree of their websites and at the same time perceive how far they are from the competition [6]. The software metrics, besides providing quantitative measurements about the usability degree of a software product, also establish the aspects in which the application must improve to be more intuitive and functional. In contrast to qualitative methods such as the heuristic evaluation and the cognitive walkthrough [7], which only allow the identification of issues that are present on the interface design, with the metrics is possible to know precisely the effort that e-commerce enterprises must perform to reach and even overcome their biggest competitors.

The number of usability problems provided by the qualitative methods is not a reliable calculation that can describe the usability level of a system, given that a website can have multiple issues, but if these only represent cosmetic aspects, the scenario is not as unfavorable as if they were catastrophic. A web application with a fewer number of usability issues can be a more critical scenario if the problems are severe. The metrics, on the contrary, provide a more objective perspective since the results of the evaluation are based on human actions performed by real users in interaction with the software system [8]. However, although the valuable information they can offer, there is little evidence in the literature that of their application as a usability evaluation tool [9]. In this paper, we describe a case study in which the metrics are used to assess the usability degree of an e-commerce web application. Likewise, an evaluation protocol is presented as well as an analysis of the aspects that are covered by the standard ISO 9126 for the inspection of software products in this specific domain.

2 Usability Metrics

The software metrics and the questionnaires are the primary techniques to quantify the usability degree of a software product. Both methods are time-consuming and require a previous planning session that allows specialists to formulate the usability tests correctly with users [6]. The questionnaires are focused on gathering subjective data related to the satisfaction of the users through pre-defined questions with Likert scale for subsequent analysis of the numerical values. On the contrary, the software metrics are objective measures related to the user's primary-task performance. Thus, the users are requested to interact with the software product according to a list of tasks, at the same time that an observer is taking notes about the performance of the users. The results of these metrics are based on direct actions that can be easily observed during the interaction, such as the success rate of achievement of certain requested tasks, the time that it takes the users to complete a specific workflow or the number of attempts that it takes the users to reach their goals.

The usability metrics are mathematical formulas that allow specialists to measure the usability degree of a software product in numerical values. According to the ISO 9126 standard, the usability is the capability of a software system to be understood, learned, operated, attractive and compatible with regulations of usability [10]. In this sense, the metric must be aligned to these aspects that represent sub-characteristics of this quality attribute. An example of a metric can be appreciated in Table 1. Although the metrics are difficult to use, are an objective way to compare design proposals from the user's direct interaction and determine how far the companies are from their main competitors.

Table 1. Definition of the metric: "completeness of description".

Aspect	Description
Name	Completeness of description
Purpose	What proportion of functions (or types of functions) is understood after reading the product description?
Formula	X = A/B A = Number of functions (or types of functions) understood B = Total number of functions (or types of functions)
Interpretation	0 <= X <= 1 The closer to 1.0 is the better
Metric scale type	Absolute
Measure type	A = Count B = Count X = Count/Count
Input to measurement	User manual operation (test) report
Target audience	User
Usability aspect	Understandability

3 Assessment Protocol

The usability evaluation process through the use of quantitative metrics is not formally established by the specialists. There is little evidence about how to apply these metrics for the evaluation of software products. However, a previous systematic literature review about evaluation methods [11] has allowed the authors of this research, to elaborate a proposal of the activities that can guide the process:

1. The supervisor of the evaluation process must develop a test plan, composed of a set of tasks based on the usability metrics that will be evaluated.
2. The supervisor of the evaluation process must select the participants. The selection must be performed through a questionnaire, to corroborate if the people meet the profile of potential users of the system.
3. The evaluation team must prepare a set of materials: confidential agreement, instructions, pre-test questionnaire, post-test questionnaire, test plan, and observation sheet, to document the results of the usability assessment.
4. The usability test begins with a brief explanation to the participants about the tasks they must perform in interaction with the software product.

5. The observer must take notes about the interaction between the user and the software product, emphasizing on the quantitative metrics and according to the mathematical formulas.
6. The evaluators can request optionally the completion of a user satisfaction questionnaire by which the measurement results can be contrasted.
7. Given the results of all participants involved in the experiment, the numerical results for each metric are averaged.
8. The supervisor proceeds to document all findings in a report.
9. Finally, the assessment team identifies opportunities to improve the graphical user interface based on the results.

4 Case Study: Usability Evaluation of *Alibaba.Com*

The authors proceeded to evaluate the usability of an E-Commerce website using the software metrics proposed by the ISO 9126 standard [10]. For this study, the website of a world company specializing in e-commerce and retail was selected: *Alibaba.com*. In this experiment, four participants with similar experience and background in online transactions were requested to interact with the software application. In Table 2, the authors present the results of the usability evaluation for each metric.

Table 2. Results of the Usability Evaluation Using Metrics.

Metric	Results per participant					Average
	P1	P2	P3	P4	P5	
Ease of function learning to perform a task in use	2	7	120	3	60	38.4
Ease of function learning	2	7	120	3	60	38.4
Attractive interaction	4	4	4	4	3	3.67
Completeness of description	1	1	1	1	0.75	0.95
Understandable input and output	1	1	1	1	1	1
Result of the usability of the website						97.5%

5 Conclusions and Future Works

The metrics are an approach that allow specialists in HCI to quantify the level of usability of a software product. This tool can be useful in the sense that allow companies to perform comparisons with their main competitors or in a development process to select the best option from different design proposals. The ISO 9126 standard establishes metrics that are clearly observed during a user test and that can be apply to any kind of software product. However, the aspects that are considered in this norm are general and nowadays, the E-Commerce websites present features that are not covered for the evaluation. In this sense, it is necessary to develop a new approach that can be used to evaluate effectively the usability of the software products in this domain.

Acknowledgements. The authors would like to thank all the participants involved in the preliminary experiments, especially the members of "HCI – DUXAIT" (HCI, Design, User Experience, Accessibility & Innovation Technologies). HCI – DUXAIT is a research group of the *Pontificia Universidad Católica del Perú* (PUCP).

References

1. Paz, F., Paz, F.A., Arenas, J.J., Rosas, C.: A perception study of a new set of usability heuristics for transactional web sites. In: Karwowski, W., Ahram, T. (eds.) Intelligent Human Systems Integration, IHSI 2018. Advances in Intelligent Systems and Computing, vol. 722, pp. 620–625. Springer, Cham (2018)
2. Nielsen, J.: Usability Engineering. Morgan Kaufmann, San Francisco (1993)
3. Alarcón, C., Medina, F., Villarroel, R.: Finding usability and communicability problems for transactional web applications. IEEE Latin Am. Trans. **12**(1), 23–28 (2014)
4. Díaz, J., Rusu, C., Collazos, C.A.: Experimental validation of a set of cultural-oriented usability heuristics: e-commerce websites evaluation. Comput. Stand. Interfaces **50**, 160–178 (2017)
5. Paz, F., Pow-Sang, J.A.: A systematic mapping review of usability evaluation methods for software development process. Int. J. Softw. Eng. Appl. **10**(1), 165–178 (2016)
6. Paz, F., Paz, F.A., Sánchez, M., Moquillaza, A., Collantes, L.: Quantifying the usability through a variant of the traditional heuristic evaluation process. In: Marcus, A., Wang, W. (eds.) Design, User Experience, and Usability: Theory and Practice, DUXU 2018. Lecture Notes in Computer Science, vol. 10918, pp. 496–508. Springer, Cham (2018)
7. Nielsen, J.: Usability inspection methods. In: Conference Companion on Human Factors in Computing Systems, CHI 1994, pp. 413–414. ACM, New York, USA (1994)
8. Harrati, N., Bouchrika, I., Tari, A., Ladjailia, A.: Exploring user satisfaction for e-learning systems via usage-based metrics and system usability scale analysis. Comput. Hum. Behav. **61**, 463–471 (2016)
9. Paz, F., Pow-Sang, J.A.: Current trends in usability evaluation methods: a systematic review. In: 7th International Conference on Advanced Software Engineering and Its Applications, ASEA 2014, pp. 11–15. IEEE, New York, USA (2014)
10. International Organization for Standardization: ISO/IEC 9126-3:2003 Software engineering – Product quality – Part 3: Internal metrics. Switzerland, Geneva (2003)
11. Paz, F., Pow-Sang, J.A.: A systematic mapping review of usability evaluation methods for software development process. Int. J. Softw. Eng. Appl. **10**, 165–178 (2016)

Design of Smart City Evaluation Based on the Theory of "White Bi"

Xinying Wu(✉), Minggang Yang, and Xinxin Zhang

School of Art, Design and Media, East China University of Science and Technology, M. Box 286 NO. 130, Meilong Road, Xuhui District, Shanghai 200237, China
{XinyingWu, MinggangYang, XinxinZhang, 82150987}@qq.com

Abstract. With the continuous development of information and communication technology, smart cities have become the dominant slogan for building cities. Smart cities are new plans created to solve common problems in cities. The current perspective focuses on the integration of smart cities and technologies, and believes that smart cities are a combination of digital and technology, information and communication, ignoring the function of cities. The aesthetics of "White Bi" theory focusing on function and essence is precisely the guiding ideology missing in the construction of contemporary smart cities. This article takes the ancient Chinese "White Bi" philosophy as the starting point and guides the development concept of the smart city, so that the smart city construction pays attention to the essence of the city and restores the city function. In view of the construction of the current smart city, an effective design theory reference is proposed. Let the smart city form a new model of "integral", "intelligent", "harmonious" and "functional" in the future.

Keywords: Smart city · Theory of "White Bi" · Sustainable design

1 Introduction

When the wave of industrialization, electrification, and informatization swept the world and gradually receded, humanity ushered in an era of wisdom. The massive investment in resources and energy has led to a rapid advancement in urbanization. The accelerated accumulation of population and rapid economic development and the rapid expansion of the city have brought about many "urban diseases" such as energy shortage, waste of resources, ecological damage, and traffic congestion [1]. The emergence of urban diseases has seriously affected the quality of urban development and the experience of living. The smart city construction is a good medicine to solve urban diseases in the era of wisdom. As a result, smart cities have generated a lot of attention around the world. In the academic and urban planning fields, the development of smart cities is highly valued. The academic community has extensive research and discussion on smart cities. In the urban planning and construction of the world, there are many cases of actual construction of smart cities. However, the actual construction of smart cities involves a lot of relevant knowledge, not the combination of the Internet and the Internet of Things is the core strength of building a smart city. More is devoted to the

W. Karwowski and T. Ahram (Eds.): IHSI 2019, AISC 903, pp. 357–362, 2019.
https://doi.org/10.1007/978-3-030-11051-2_55

development of smarter cities, to meet the needs of the sustainable development of urban cycles. When society does not meet the requirements of a unified definition of smart cities, it will produce multi-faceted cognition.

In recent years, information and communication technology (ICT) has gradually become the main perspective of smart city construction. ICT is the main carrier for building a smart city information platform and one of the most important infrastructures for smart cities. However, building a smart city from the ICT alone will bring about one-sidedness in the construction of a smart city. The fundamental purpose of smart city construction is not only to achieve the wisdom of urban information, but also to solve the most fundamental problems of the modern city itself, rather than building an information platform, the smart city becomes an individual who can think. The research and planning related interest groups of smart cities should be considered and discussed from the urban ontology.

Zhouyi is an important classic in ancient Chinese philosophy. Confucius believes: "Yijing is a rich book containing infinite truths between heaven, earth and people. [2] " Yijing has had a profound influence on ancient Chinese culture, showing the mysteries and laws of the universe. The philosophical truth contained in the Book of Changes is the law of survival of human wisdom. "White Bi" represents the traditional Chinese philosophy and has important significance. After thousands of years of interpretation, the theory of White Bi has been widely used in related fields. Ancient Chinese philosophical thinking holds that everything has its roots and that returning to its origin can achieve a balance between man and nature. This paper studies the smart city from the perspective of "White Bi" as a theoretical entry point. Under the guidance of the simple concept, natural concept, harmonious concept and ecological concept of "White Bi", the "Wen" and "Zhi" of the smart city are discussed, so as to find the reason and direction of restoring the function of the city.

2 Problems in Smart Cities

The development of smart cities has achieved remarkable results so far. Smart cities have become the core guiding ideology of urban planning in various countries. However, the city does not give it the wisdom of the label to become intelligent, but gives it a smart platform and intelligent planning. The smart city began to experience the wisdom of the earth proposed by IBM in 2009, and has gradually become an urban planning system. There are also many problems in the development. The author has drawn the following related questions through the investigation of the construction of smart cities in China.

2.1 Cognitive Bias of Smart City Concept

The main body of smart city construction planning is the local government. Local government perceptions of smart city concepts affect the construction of smart cities in the region. Due to the different economic developments of different cities, the limitations of urban development are not the same [3]. Smart city development commonality: the more developed the economy, the more comprehensive the transformation of wisdom. Compared with the region of economic development, the transformation of wisdom is very limited. A smart city is not a slogan. It is tailored to the local conditions. Smart transformation within limited conditions requires practical application to be put into action. This requires local government interest groups to understand the concept of smart city and translate it into practical action.

2.2 Lack of Evaluation Criteria and Top-Level Design

The construction of a smart city is inseparable from the standard system and precise evaluation standards. Although there was a spontaneous exploration in the early stage of the construction of a smart city, the system gradually formed in the later period with certain standards. Throughout the construction of smart cities around the world, there are more or less the lack of standard systems and top-level planning, resulting in ambiguous responsibilities of smart city management, and overall planning at the national level is not comprehensive. There are also differences in the infrastructure construction of smart cities in various regions.

2.3 Lack of Core Technology, Lack of Urban Function

The core technology is the guarantee of intelligent operation of smart cities. There are common problems in the investigation of the construction of smart cities in China. There is a widespread lack of core technology. In the planning of building a smart city, rushing to build and ignoring the upgrading of core technologies led to a lack of intelligent experience in the post-construction planning case, thus weakening the urban intelligence function. The construction of smart cities involves many aspects of technology. The lack of technologies such as important components, data processing, and high-precision chips will cause problems in urban operation.

2.4 Lack of Autonomy, Low Participation of Citizens

Among the many problems facing urban development, the fundamental shortcoming is the lack of comprehensive urban autonomy. The lack of autonomy also reflects the real problem of low citizen participation. The resume of a smart city is not just a smart information exchange platform, but more about the improvement of the citizen's living experience [4]. This requires the public to participate in the construction. As a public place, the improvement of the sense of citizenship will bring about a fundamental improvement in the city. In the final analysis, smart cities are people-oriented. Therefore, it is necessary for residents to experience feedback and participation.

3 White Bi Ontology

Divinatory symbols of Bi refers to the meaning of decoration, but it contains the truth of the harmonious unity of human civilization and natural laws. The birth and progress of civilization is a certain law, and it is also in line with the development of nature. Divinatory symbols of Bi contains upper and lower parts. The upper part represents the mountain. The lower squat represents the fire. It meaning the scene of the fire shining through the mountains [5]. This bright sight means social civilization. The combination of fire and mountains reflects the harmonious unity of culture and nature, and it represents the philosophy of harmony between man and nature. Divinatory symbols of Bi contains White Bi. White Bi represents the philosophical concept that is decorated to the extreme and will be attributed to calm. It embodies the process of things from simple to simple, and it is also the process of returning things to their origins. Zhou Yi believes that all things in the universe are cyclical, unified, and mutually different, resulting in a rich form. Therefore, returning to the origin of things can find the fundamental way to solve the problem.

3.1 The Application of the Concept of White Bi in Smart Cities

The continuous advancement of the smart city has blurred the function of the city's origin. The city is ultimately the home of human habitation and the carrier of human beings and nature. A successful urban planning case will lead to imitation, and the phenomenon of a thousand cities will become more and more serious. The concept of "returning to the truth" contained in "White Bi" represents the meaning of returning to the source. Since the city is a carrier connecting humans and nature, there is a corresponding language in the city. Smart cities bring both human and intellectual experiences, but they also bring problems. The rules of the day are in accordance with the laws of nature and focus on cultural heritage. The increase in cultural elements in smart cities helps to form features. The opposition between city and nature has always been a topic for urban planners and their troubles. Combining the concept of White Bi to return urban planning to pay attention to the function of the ontology, as a breakthrough point, it can establish a new relationship between the city and nature.

3.2 Sustainability

The destruction of ecology has created a sense of crisis in urban expansion. The natural ecological environment has a mutually restrictive relationship with urban expansion and urbanization. Soil, water, atmosphere and vegetation in the ecological environment are the basic elements of urban development. The concentration of urban population and the continuous advancement of social civilization have led to the expansion of urban land use, resulting in the reduction of ecological basic elements. When the basic elements are reduced to a critical point, the city will not be able to continue to expand. White Bi advocates the principle of survival of natural harmony in line with the concept of sustainable development of smart cities. White Bi advocated pay attention to people's feelings, as well as smart cities.In the smart city PSF evaluation model, the human factor ranks first, firstly with the human needs, experience, residence and life as the core target layer, indicating the importance of people in the smart city.

The development concept of China's smart cities: innovation, coordination, green, open, and sharing. The concept of White Bi contains: all things follow, the laws of nature, the unity of nature and man, the coexistence of symbiosis. It can be seen that the concept of "White Bi" has a guiding role in the development of smart cities in China. White Bi originated from the oldest philosophical thoughts in ancient China, and is a "theory" based on China's actual conditions. Therefore, the concept of White Bi has an important guiding role for the development of China's smart city.

3.3 "Wen" and "Zhi" Unified Smart City

There is also a relationship between "Wen" and "Zhi" in the smart city system (Fig. 1). The urban natural system is the basic system of a smart city, equivalent to "Zhi". The urban social system contains both "Wen" and "Zhi". The urban economic system is purely "Wen". In this system diagram, "Wen" and "Zhi" have the same weight, and the system works to bring the best urban living experience. The living experience of the citizens and the green development concept are the new topics that the smart city needs to consider in the future.

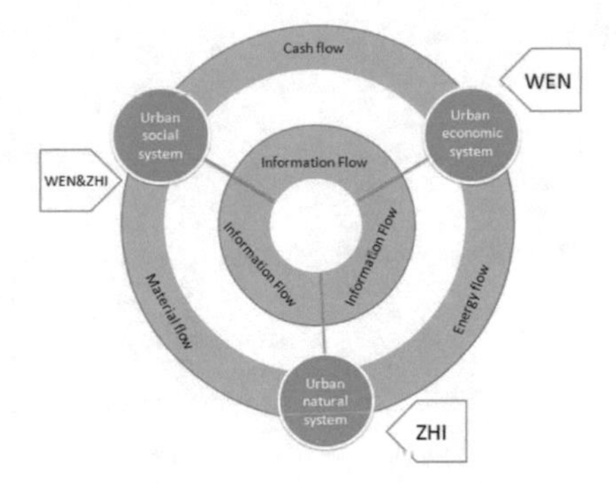

Fig. 1. Smart city system

4 Conclusion

The theory of "White Bi" originated from ancient Chinese philosophical thoughts represents China's most traditional theoretical system. Therefore, the theory of "White Bi" is the development theory that best fits the actual situation in China. The construction of a smart city needs to be adapted to local conditions in order to play a sustainable role. The effective use of the theory of "White Bi" can make the development of smart cities more effective. Therefore, in the follow-up study, we can start from the most fundamental function of the city and carry out reasonable and effective planning according to the characteristics of regional development.

References

1. Xi, G., Zhen, F.: The spatial organization and planning of smart cities based on the sustainable development goals. J. Urban Develop. Stud. 102–105 (2014)
2. Anan Y.: The Multi-dimensional enlightenment of Zhou Yi's concept of aesthetics. J. Soc. Sci. Yunnan, 171–174 (2014). (in Chinese)

3. Guo, M., Liu, Y., Yu, H., Hu, B., Sang, Z.: An overview of smart city in China. J. China Commun. 203–211 (2016)
4. Xu, Z., Liu Y.: Research on the development of smart city based on the thought of "urban brain". J. Reg. Econ. Rev. (2018)
5. Huang, L.: On the aesthetic value of divinatory symbols of Bi in the book of changes. J. Jishou Univ. 55–56 (2006). (Social Sciences Edition)

Building on Water: The Use of Satellite Images to Track Urban Changes and Hydrodynamic Models to Simulate Flow Patterns Around Artificial Islands

Mohamed El Amrousi(✉), Mohamed Elhakeem, and Evan Paleologos

Abu Dhabi University, Abu Dhabi, United Arab Emirates
mohamed.elhakeem@adu.ac.ae

Abstract. Dubai's interest to attract investors through exclusive urban enclaves is best exemplified by the artificial islands that reshaped much of its shoreline. Several artificial islands such as Palm Jumeirah, Palm Jebel Ali and the World Islands reinforce the trend of fostering extensive urban development along shorelines. Creating artificial islands is not only limited to Dubai a much larger scale artificial island "Forest City" is emerging in Johor Bahru on the southern tip of Malaysia fronting Singapore. Forest City advocates a multi-layered city built on reclaimed land to offer luxury homes with vertical gardens to wealthy investors. This paper juxtaposes the use of satellite images to track urban changes in these emerging artificial islands and 2D-hydrodynamic models to simulate the flow patterns around the Palm Jumeirah as an example.

Keywords: The Palm Jumeirah-Dubai · Forest City-Johor Bahru
2D-hydrodynamic surface water modeling

1 Introduction

Dubai is one of the cities that is constantly reshaping its urban image and brand value to diversify its economy, cater to investors and attract tourism as part of its strategy to reduce dependency on oil revenues. Mega projects target aspirations of the middle-class seeking lifestyles that mark them as successful entrepreneurs in a competitive global economy [1]. There is no doubt that today the Arab world, and especially Gulf state cities are becoming urbanized, at diverse rates and with comparable characteristics. They have become perhaps more connected to a network of global cities and economies beyond their regional and culture [2]. Academics and urban planners resort to satellite images from google-earth to track such rapid urban changes in cities because of their increased geometric capabilities and applications such as street map updating, shoreline modification updates and emerging building patterns [3]. Satellite imagery has many advantages over an aerial photography in terms of its availability to all, surface coverage and lower image distortion [4]. Intensifying the urbanization of Dubai's artificial islands through the addition of high-rise buildings, viewing tower and mega malls reflects an understanding that waterfronts are precious resources with unique potential for investment and diversified opportunities for economic

W. Karwowski and T. Ahram (Eds.): IHSI 2019, AISC 903, pp. 363–369, 2019.
https://doi.org/10.1007/978-3-030-11051-2_56

development and public enjoyment. Although faced by construction challenges developments on this reclaimed land have been further extended beyond initial design to add facilities and spaces of gathering. Artificial islands as secluded dreamscapes for the elite are becoming more popular. The mega project "Forest City" in Johor Bahru on the southern tip of Malaysia fronting Singapore illustrates this relationship between simultaneous increase in urbanization, land reclamation and popularity of artificial island development. As artificial island cities grow in population, economic and political importance, they expand further into the water too as construction technologies develop; their adjacent mainland also expands in size as in the case of the Dubai Marina. Land reclamation has become a favored strategy for coastal and island city development in part because it is a means by which elite actors can create new urban spaces while bypassing cultural and stylistic existing urban heritage [5]. The idea of building a new city in front of a present one has existed since the 19th century as seen in colonial urban additions to Algiers-Algeria and Casablanca-Morocco. Colonial architects evaded the existing medina with its dense urban fabric and pre-dominantly Muslim culture to create new building typologies and spaces of entertainment along the coasts of the Mediterranean. Artificial islands as new assemblages of exclusion and inclusion offer new spaces of leisure for wealthy investors while physically still remaining in proximity to their own country [6].

2 Artificial Islands, Reclaimed Land and Engineered Urbanism

Dubai is constantly pushing the limits of its offshore projects, the Palm Jumeirah is the first of a series of artificial islands located off the coast of Dubai, manifesting clear geometries to be seen from above as a "palm tree" with a protective crescent-shaped offshore breakwater partially enclosing it. The Palm Jumeirah (Fig. 1) started in 2001 and was built according to a tight schedule. The project comprises of a central trunk and 16 fronts that were created by dredging and spraying 100 million cubic meters of sand in a rainbow-like arch using GPS technology [7]. To protect the island against wave attack, an offshore crescent breakwater surrounding the island with a total length of 11 km was constructed at the same time [8]. Satellite images indicate that Palm Jebel Ali surpasses in scale Palm Jumeirah, and another Palm "Deira," is under construction to offer more villas and high-rise buildings separated by interconnected channels and marinas [9]. This new form of upscale urbanism is further manifested in Forest City-Malaysia yet on a larger scale. The four artificial islands of "Forest City" are a 100 billion dollar project that is being developed by the Chinese developer "Country Garden" and is advertised as a car-free urban enclave that promotes sustainable strategies and utilizes the latest technologies. Forest City with its high-rise greened towers built on reclaimed land questions the relation of such projects to nature, as land is reclaimed and concrete towers are planted, favoring artificial dreamscapes and the influx of new settlers/investors is bound to inflate the property market and change morphologies.

Fig. 1. Palm Jumeirah and Forest City Johor Bahru-Malaysia.

Artificial islands brand their city(s) as a financial powerhouses capable of creating complex projects, and highlight that they are cities were investment in real-estate is plausible. Cultural clustering is most visible and characterized by an increase of the spaces of leisure that revive Disneyland characters that are well represented by post-modern architecture such as the Atlantis Hotel in the Jumeirah Palm. Urban settings staged for tourists in the Palm and Forest City include monorails that travel from one end of the island(s) lagoons to centers of leisure activities with restaurants, cinemas, and fitness complexes to create epic clusters. The lack of public spaces in artificial islands can be understood as actively reconstituting beach-front property in relation to ownership an alternative narrative to socio-cultural habits of local communities [6]. Forest City (Fig. 2) secluded spaces for its investors with virtually self-contained shopping centers and theme parks on a mega scale that collage post-modern pastiches that transgress national boundaries. Forest City and the Palm Jumeirah (Fig. 3) highlight that artificial islands development is taking place on a global scale therefore a balance needs to be negotiated with the environment especially with respect to shoreline modifications. Intensifying construction of residential and commercial estates in order to cover construction and post-construction cost contradicts initial perceptions of islands as secluded low-density spaces, escapes from the modern metropolis. Increased population on artificial islands may also have substantial effects on lagoon water quality trapped within the breakwater resulting in increased sediment, organic and inorganic materials [9].

Fig. 2. Forest City-Malaysia; Entrance pavilion, city model, fake seals and beach.

Fig. 3. Palm Jumeirah-Dubai; Breakwater promenade and Atlantis Hotel.

3 Simulation of the Flow Around the Palm Jumeirah Island

The Palm Jumeirah Island (PJI) was built on the Arabian (Persian) Gulf in front of Dubai City and is surrounded with a crescent shape offshore breakwater to damp the wave energy and control erosion of the island shoreline (Fig. 1). Due to the complex geometry of the Palm Jumeirah Island, a 2D hydrodynamic model was used to investigate the flow pattern of the water surrounding the island. In this study, we chose the 2D Finite Element Surface Water Modelling System (FESWMS), which is part of the commercially available Surface water Modelling System (SMS) software package (version 12.1) developed by the Federal Highway Administration [10]. FESWMS solves the differential forms of the continuity and the momentum equations in the stream wise and transverse directions using the Galerkin method of weighted residuals providing water depth and depth-averaged velocity magnitude in x and y directions at each node in the grid [10].

The model inputs are: the island bathymetry, Manning's coefficient $n = 0.025$, eddy viscosity $v = 0.3$ m^2/s, flow rate entering the island from the breakwater openings $Q = 3000$ m^3/s, and flow depth of the water surrounding the island $y = 10$ m. Figure 4 shows the simulations of the water depth and the velocity vectors around the island. It can be seen from the figure while complex flow circulation patterns developed around the island, the water has almost a constant flow depth of 10 m. It should be pointed out here that these circulation patterns are important because they help in controlling sediment deposition, aquatic growth, and sulphide formation. Nonetheless, it can be seen from Fig. 4 that there is no circulated flow in the inner part of the island and the water is almost stagnant. This can increase salinity, sediment, organic and inorganic materials in this area, which may affect water quality and marine life. An artificial flow circulation mechanism may be needed in this area to improve water quality. Figure 5 shows the water velocity simulation, which is in the range of 0.03 to 0.5 m/s. Thus, the offshore breakwaters damped the wave energy effectively by reducing the velocity around the island and hence, preventing erosion of the island shoreline.

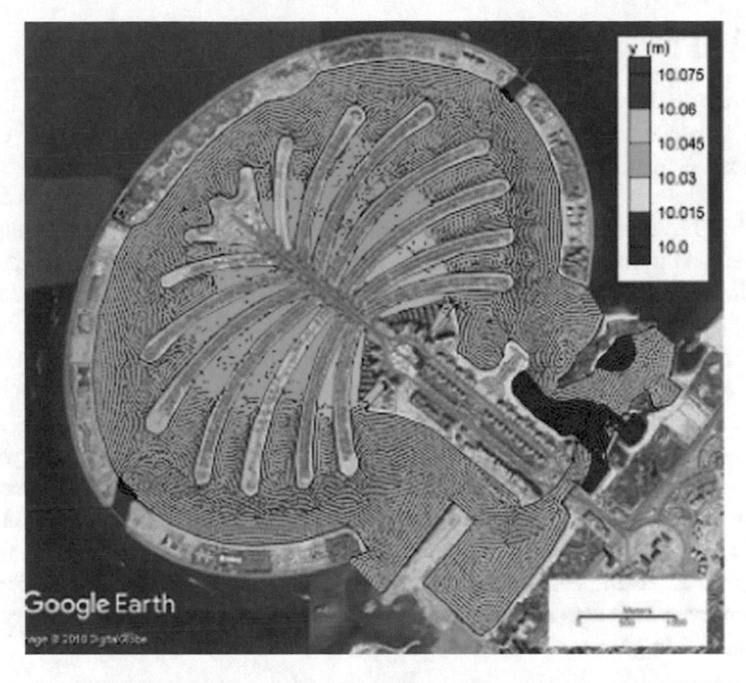

Fig. 4. Simulation of the flow depth and circulation pattern around PJI, UAE.

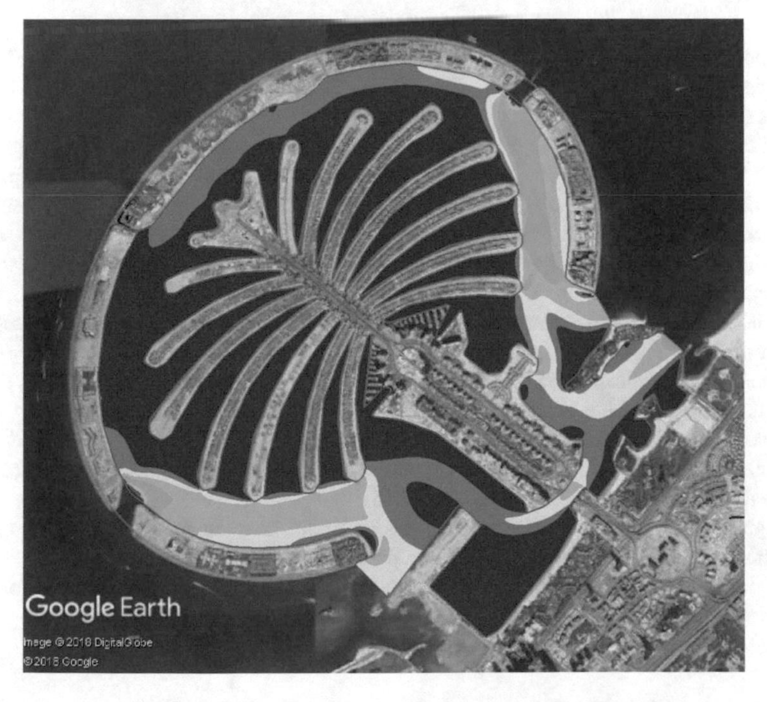

Fig. 5. Simulation of the flow velocity around the PJI, UAE.

4 Conclusion

Palm Jumeirah-Dubai and Forest City-Malaysia represent new forms of urbanism that manifest financial power and connect cities as part of a global network of emerging economies. As artificial islands their designs challenge and construct new relationships between offshore urban enclaves and existing shorefronts of the city. We studied through a 2D hydrodynamic model the flow patterns around the Jumeirah Palm Island. The model predictions showed that while complex flow circulation patterns developed around the island, the water around the Jumeirah Palm has almost a constant flow depth of 10 m. These circulation patterns are important because they help in controlling sediment deposition, aquatic growth, and sulphide formation. It was observed that there was no circulated flow in the inner part of the island and the water was almost stagnant, which can increase salinity, sediment, organic and inorganic materials in this area affecting water quality and marine life. The water velocity around the island was in the range of 0.03–0.5 m/s. Thus, the offshore breakwaters damped the wave energy effectively by reducing the velocity around the island and hence, preventing erosion of the island shoreline. To conclude satellite imagery is helpful to see the development of artificial islands during construction and to view the shapes of such dreamscapes from above, however, it is also necessary to experience actual space and how it evolved the post-construction phase.

References

1. Haines, C.: Cracks in the façade: landscapes of hope and desire in Dubai. In: Worlding Cities: Asian Experiments and the Art of being Global, pp. 160–181. Wiley-Blackwell, UK (2011)
2. Malkawi, F.: The new Arab metropolis: a new research agenda. In: Elsheshtawy, Y. (ed.) The Evolving Arab City: Tradition, Modernity and Urban Development, pp. 27–36. Routledge, New York (2008)
3. Pillay, D.L.: The use of medium to high resolution satellite imagery for urban mapping applications. In: Proceedings of the 21st International Cartographic Conference (ICC), Durban, South Africa, pp. 1247–1252 (2003)
4. Meinel, G., Lippold, R., Netzband, M.: The potential use of new high resolution satellite data for urban and regional planning. In: ISPRS Commission IV Symposium on GIS - Between Visions and Applications, Stuttgart, Germany, vol. 32/4, pp. 375–381 (1998)
5. Grydehøj, A.: Making ground, losing space: land reclamation and urban public space in island cities. Urban Island Stud. **1**, 96–117 (2015)
6. Gupta, P.: Futures, fakes and discourses of the gigantic and miniature in 'The World' islands Dubai. Island Stud. J. **10**(2), 181–196 (2015)
7. Higgins, K.: Engineering Challenges of Dubai's Palm Jumeirah. Coastal and Ocean Engineering, Undergraduate Student Forum, ENGI.8751 (2013)
8. De Jong, R.E., Lindo, M.H., Saeed S.A., Vrijhof, J.: Execution Methodology for Reclamation Works Palm Island 1. Terra et Aqua. Report Number 92 (2003)

9. Cavalcante, G.H., Kjerfve, B., Feary, D.A., Bauman, A.G., Usseglio, P.: Water currents and water budget in a coastal megastructure, Palm Jumeirah Lagoon, Dubai, UAE. J. Coast. Res. **27**(2), 384–393 (2011)
10. Froehlich, D.: User's Manual for FESWMS Flo2DH: two-dimensional depth-averaged flow and sediment transport model. Release **3** (2002)

An Intelligent Tool to Facilitate Home Building: U-Design

Neda Khakpour, Edwina Popa, Jose Luis Lamuno, Syeda Anam Zaidi, Bao Truong, and Ebru Celikel Cankaya(✉)

Department of Computer Science Richardson, University of Texas at Dallas, Richardson, TX, USA
{nxk144430, exp150030, jxl162731, saz160330, bqt140030, exc067000}@utdallas.edu

Abstract. Considering the need for a fully automated tool, this study aims at filling a gap by introducing a comprehensive design that facilitates home building from inside out. The system we propose provides a user friendly intelligent interface by parameterizing the entire process of designing home elements including furniture and structural components, as well as an ability to navigate and contact real estate agents for further intents including rents or purchases of homes. We present a preliminary design for the U-Design tool that promises to be a comprehensive framework for home design in its entirety: From submitting requirements for home structure preferences to working with a consultant to rent or purchase the dream home. The performance measurements of our tool yield promising and comparable values.

Keywords: Intelligent systems · Home design · Real estate automation

1 Introduction

In a current market dominated by Zillow and Trulia, U-Design aims to add an extra layer of customization in terms of the home building process rather than home finding. In addition, U-design gives users the ability to fully design their home inside and out with access to furnishing retailers. We intend on creating a software that differs from those on the current market.

In this work, we introduce an application that allows future homeowners to create the house that they want, from the size and style down to the cabinetry and flooring. We also provide a price and mortgage calculator. An inherent customer database allows the customers to save their preferences and build a service that compares prices across all house builders to get the best price.

Our motivation is to create an easy-to-use platform that any user can access to construct the home of their dreams in an affordable manner in comparison to the other homes on the market, avoiding extra consulting. It is currently a difficult process for home buyers to design a home of their desire. Therefore, this software will guide users through the steps, having answers to frequently asked questions along the way. Real estate across the nation is on the rise, leading to a high demand in homes. By creating

W. Karwowski and T. Ahram (Eds.): IHSI 2019, AISC 903, pp. 370–377, 2019.
https://doi.org/10.1007/978-3-030-11051-2_57

this internet-based software, our goal is to support this demand and add a new dimension to the home buying process. Our design is able to handle any use case involving the customization, purchase, and comparison of a home.

2 Background and Related Work

In [1] Hromada introduces mapping of real estate prices data mining technique, where the author employs the use of data mining to analyze changes in the real estate market. Their application looks at price quotations of apartments, houses, business properties and building lots. This was done to prevent manipulation of information (such as when done by real estate companies, developers, etc.). This work relates to our own software in the sense that to give the customer the best price a similar algorithm would have to be employed. A data mining approach would allow us to consistently provide the customer with the best prices available on the market regardless of external factors. They can then use this information to their advantage to negotiate for the properties they are interested in.

Authors in [2] introduces the use of GIS for the collection, storage, management and analysis of housing space. Their application is interesting as it is a system capable of assembling, storing, manipulating and displaying geographically referenced information. Such information would be something like an unfavorable neighborhood, areas prone to flooding, foundation problems due to geographic shifting, etc. This relates to our own software in the sense that customers always want to make informed decisions. By providing users of our software with complete information such as this we are allowing them to choose more freely than if they were to use a real estate software and also provide more of the information needed to successfully design their homes.

In another work in [3], authors integrate virtual reality with a mobile device in order to design architectural models and interiors of a house. Their design provides an easy to change software, and a wide variety of environments can be created without the hassle of traveling back and forth from the home improvement store. Also instant feedback is received as the software works extremely fast. This idea applies to our software as we want to provide our users with an interface which allows them to design their home as well as preview it. It would be extremely beneficial to set up a virtual reality scenario in a mobile device where the user can walk through their desired home to preview any changes before making them in the real world. Users with a virtual reality headset would obviously benefit more from the experience but the software would not be limited by this.

A similar work employs a virtual reality software environment for evaluation and support of interior design [4]. The focus of this software is not only in the use of virtual reality for design, but also in making it more accessible for users not versed in CAD systems. This would in turn make design using software easier for non-expert users. This work pertains to our software because we want to make it as accessible as possible. Implementing a CAD system for house design has been done before but this still does not make it easier for the average person to use. A virtual reality software based on ease of use would be of great benefit to our software as we want to make it available to as many people as possible.

Another scholar work integrates hedonic pricing models for evaluation purposes [5]. The price of a house is determined by the characteristics of the property itself (e.g.

its size, appearance, features like solar panels or state-of-the-art faucet fixtures, and condition). Similar parameters are included in U-design, thus the users can predict the prices after they finish a blueprint of their houses.

3 Our Design

We wanted to create a product that simplified the process of designing and building homes on the customer end. The system will be developed using React JS platform and compatible to any kind of Web Browsers on the market. Our design works as follows.

First, the customer will create an account like in Fig. 1 or click "Log In" to access your account. Afterwards, the user can continue onward to design their floor plan as shown in Fig. 2. Their design will be true to life and after they are done, they can save it to their computer or book a consultation with one of the realtors or architects. The customer can also choose preset floor plans like in Fig. 3 that they can send to their realtor or architect.

Fig. 1. Creating an account

Fig. 2. Custom designing the home.

Our application provides a flexible platform to help users custom design their dream home. A user typically can work on each room of the home individually by custom designing windows such as flooring, painting, tile, carpet, doors, etc.

Additionally, our application supports extra features for home and room size (square footage) selection as demonstrated in Fig. 3.

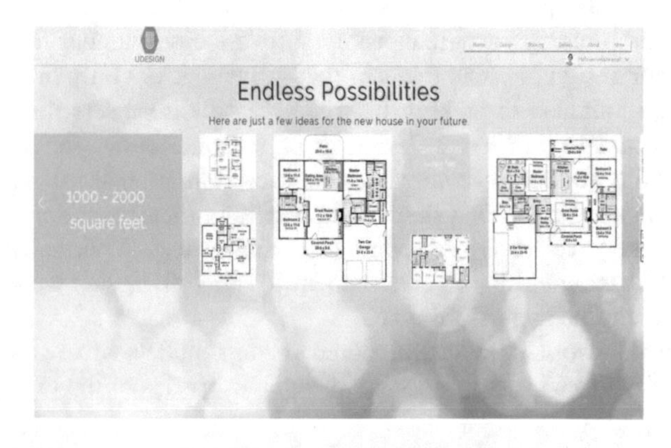

Fig. 3. More ideas

Next, the customer can book a consultation with either an architect or a realtor via the booking page. The higher prices per hour reflect the architects. The customer can then choose, depending on their region, a Texas, California, or New York realtor or architect as in Fig. 4.

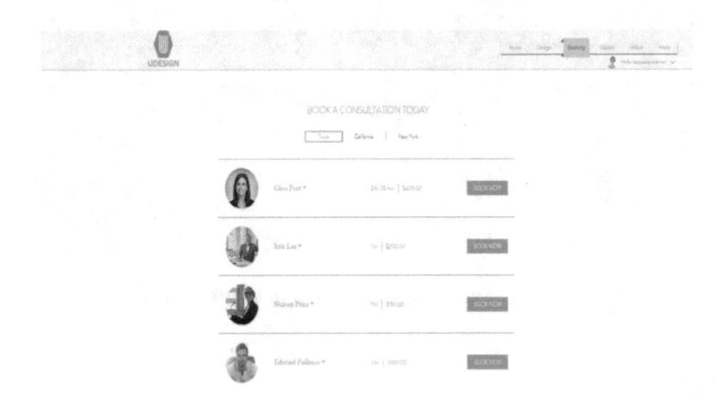

Fig. 4. Booking in Texas.

Lastly, the customer can then go back into their accounts and view their general information like in Fig. 5 or look at their previous and current bookings like in Fig. 6. In Fig. 7, we see that the customer can also upload their photos and it will be sent into the database from where the realtors and architects can view it.

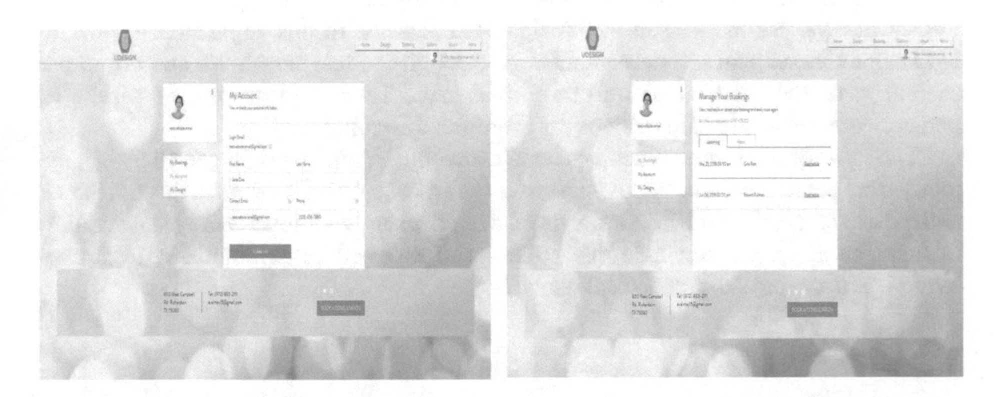

Fig. 5. User account info

Fig. 6. User account bookings

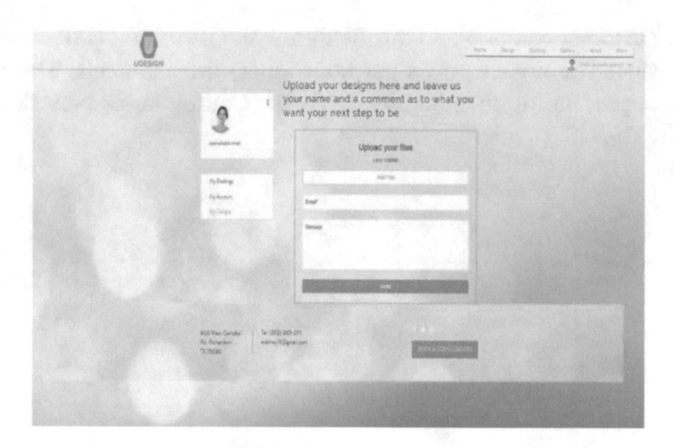

Fig. 7. User account design submission

3.1 Performance

We are required to execute performance testing on different scenarios to examine the effectiveness of U-Design and possibility of errors before it is fully integrated. We tested for the system's responsiveness, stability and reliability. For the testing purposes, we used the functional requirements of creating or editing the design and booking for consultation. We were able to measure the throughput, response time, and errors per second for these cases based on one virtual user. Table 1 illustrates the results of performance testing for the stated functional requirements.

To evaluate the efficiency of our system, we conducted a performance test for a similar home designing software called Homebyme and compared results with U-Design. We use the scenario of designing a room with this application similar to U-Design. Once again the measurement of throughput, response time, and errors per second were recorded based on one virtual user. Table 2 illustrates the results of performance testing for Homebyme software for designing a room.

The data for the testing outcomes demonstrated a significant difference between the total errors per second and throughputs for both software. The reason for this difference and possible result could be further investigated using additional functional requirements. Therefore, we conclude that U-Design performed better compared to Homebyme based on provided statistical analysis.

Table 1. Performance testing results for U-design

Label	#Samples	Average	Min	Max	Std. Dev.	Error %	Throughput	Received KB/sec	Avg. Bytes
Home page	1	2000	2000	2000	0	0.00%	0.5	514.72	1054146
Design page	1	88	88	88	0	0.00%	11.36364	89.68	8081
Add table	1	719	719	719	0	0.00%	1.39082	3.56	2619
Rotate table	1	194	194	194	0	0.00%	5.15464	13.57	2696
Add plant	1	174	174	174	0	0.00%	5.74713	15.15	2699
Move chair	1	166	166	166	0	0.00%	6.0241	15.71	2671
Delete chair	1	154	154	154	0	0.00%	6.49351	16.44	2592
Add wall	2	185	182	188	3	0.00%	5.39084	13.67	2597.5
Adjust wall	1	172	172	172	0	0.00%	5.81395	14.7	2589
Add door	1	169	169	169	0	0.00%	5.91716	15.32	2651
Delete door	1	180	180	180	0	0.00%	5.55556	13.96	2574
Booking page	1	262	262	262	0	0.00%	3.81679	14.15	3796
Consultation for California	1	461	461	461	0	0.00%	2.1692	121.15	57189
Select Consultant	1	252	252	252	0	0.00%	3.96825	221.54	57169
Consultation fees	1	274	274	274	0	0.00%	3.64964	203.83	57189
Enter Client email	1	290	290	290	0	0.00%	3.44828	193.75	57535
Book Appointment	1	71	71	71	0	100.00%	14.08451	8.43	613
Booking Confirmation	1	52	52	52	0	100.00%	19.23077	11.39	633
Checkout more services	1	207	207	207	0	0.00%	4.83092	271.53	57555
Consultants for Texas	1	205	205	205	0	0.00%	4.87805	274.18	57555
Total	26	248	0	2000	382.01	26.92%	4.54069	246.36	55559.3

Table 2. Performance testing results for Homebyme

Label	#Samples	Average	Min	Max	Std. Dev.	Error %	Throughput	Received KB/sec	Avg. Bytes
Homebyme home page	1	1087	1087	1087	0	100.00%	0.91996	1.41	1573
Login	4	0	0	0	0	100.00%	0	0	1359
Context	1	604	604	604	0	0.00%	1.65563	1.3	803
User	1	552	552	552	0	100.00%	1.81159	0.89	504
New project	1	129	129	129	0	100.00%	7.75194	3.82	504
Log interaction	1	206	206	206	0	100.00%	4.85437	3.44	725
Platform	3	137	131	147	7.12	0.00%	3.75	46.6	12725
Cursor-drag-wall	1	127	127	127	0	0.00%	7.87402	5.18	674
Cursor-drag-vertex-white	1	127	127	127	0	0.00%	7.87402	8.15	1060
Products	1	129	129	129	0	100.00%	7.75194	3.82	504
Brands	1	130	130	130	0	100.00%	7.69231	3.79	504
Search	1	155	155	155	0	100.00%	6.45161	3.18	504
Search sofa	1	129	129	129	0	100.00%	7.75194	3.82	504
Add sofa	1	132	132	132	0	100.00%	7.57576	3.73	504
Delete sofa	1	130	130	130	0	0.00%	7.69231	95.59	12725
Total	20	202	0	1087	253.78	65.00%	8.88099	28.06	3234.9

4 Conclusion and Future Work

We propose a new scheme that provides a novel framework to facilitate home designing as well as seeking consultants for rental and purchasing purposes. Our design is naive yet, still is unique in combining multiple features in the same application. As is, it promises to outperform those tools available on the market, which focus on single or limited features of home building or searching.

Our design is fast, reliable, and practical with reasonable response time. We plan on expanding our design to include more features for a fully comprehensive design. For this, we collect user feedback so as to adopt our work to their requirements. Moreover, we are planning to add enhanced security features to our design so as to ensure user privacy while also satisfying secrecy concerns for users' design ideas.

References

1. Eduard, H.: Mapping of real estate prices using data mining techniques. Proc. Eng. **123** (2015), 233–240 (2015). ScienceDirect
2. Lin, C., Meng, L., Pan, H.: Applications and research on GIS for the real estate. School of Remote Sensing Information Engineering, [online document], November 2001
3. Mudliyar, P., Ingale, Y., Bhalerao, S., Jagtap, O.: Virtual reality for interior design. Int. J. Res. Advent Technol. **2014**, 260–263 (2014)
4. Vosinakis, S., Azariadis, P., Sapidis, N., Kyratzi, S.: A virtual reality environment supporting the design and evaluation of interior spaces. Department of Product and System Design Engineering (2007)
5. Monson, M.: Valuation using hedonic pricing models. Cornell Real Estate Rev. **7**, 62–72 (2009). Cornell University

Toward Clarifying Human Information Processing: A Case Study of Big Data Analysis in Education

Keiko Tsujioka(✉)

Institute for Psychological Testing, 1-4-1 Higashimachi Shinsenri, Toyonaka, Osaka, Japan
keiko_tsujioka@sinri.co.jp

Abstract. Our purpose of this paper is to make language processing more clearly by finding out differentiation among individual traits, depending on various kinds of materials, such as context (emotional and nonemotional) and media (sound voice and text). We have carried out comparative experiments, using questionnaires of personality testing and measured the decision time of ninety-eight female university students. The result of those data analysis has shown that the members of Type X have sustained learning more effectively than Type Y ones. It is implied that the relationship between higher order and language information processing might be implemented depending on individual differences and diversity of materials. Those views of this study would continuously help teachers to predict students' behavior and plan their instructions so that they can make teaching and learning effectively.

Keywords: Human information processing · Individual traits · Sound voice and text · Emotional and non-emotional context · Big Data analysis

1 Introduction

Along with developing technology, it has been revealing language information processing [1]. On the other hand, it has been increasingly used text message in education, however, the differences between sound voice and written language has not been made clearly. For example, while we are reading text message, we do not seem to have noticed reading, but as if we are hearing senders' voice, and it looks like face to face interaction. It has not been found the reason clearly why we behave or feel like that, because it has not been explained enough about the language information processing yet, especially concerning with individual differences, diversity of materials, the function of a high order and so on. In Japanese case, especially, it is supposed that we do not always read written sentences, transforming from sound [2], because there are two types of characters, Kanji and Kana; an ideogram and a phonogram.

From those points of view, our purpose of this paper is to make language processing more clearly by finding out differences among individual traits and conditions of contexts (emotional and nonemotional context) [3]. We have carried out comparative experiments between sound voice and letters which are presented one hundred-twenty

W. Karwowski and T. Ahram (Eds.): IHSI 2019, AISC 903, pp. 378–383, 2019.
https://doi.org/10.1007/978-3-030-11051-2_58

questionnaires each to 98 female students who have participated in. The questionnaires are used psychological testing which is a measurement of personality consisted by mainly two types of traits. In other words, questionnaires are divided into two kinds of materials, emotional context and nonemotional contexts. We have also measured the decision time from the beginning of presenting questionnaires to replying by students so that we can analyze the data, comparing between two kinds of contexts.

In addition to them, we have also collected students' various data which are related to collaborative learning, for example, observations of interaction among team members (four students each) during both face to face in class and distance communication (LMS: Learning Management System), their performance of low and high stakes assessments and so on. Our aim of this study is to improve collaborative education and learning by providing the results of those Big Data analysis, such as relevance between their performance and individual traits so that instructors can predict learners' behavior and attitude in class. Moreover, if we can find out those individual differences, it might be applied to machine learning, like AI for teaching.

2 System

2.1 Concept of Education and Learning System

From those previous points of views, we have investigated information processing by gathering variety of students' data and analyzing their learning processing. Learning is considered that it comes effectively between instructor-learner or learner-learner interactive communication with mainly language information. It means that learners are thinking with cognitive management during language information processing at the same time (Fig. 1). The information (problems) is input ① to learners from teachers. Next, learners engage to solve problems ② with their existing knowledge ③, ④ and make their decision ⑤ to feedback their progress ⑥, ⑦, and then output their results ⑧.

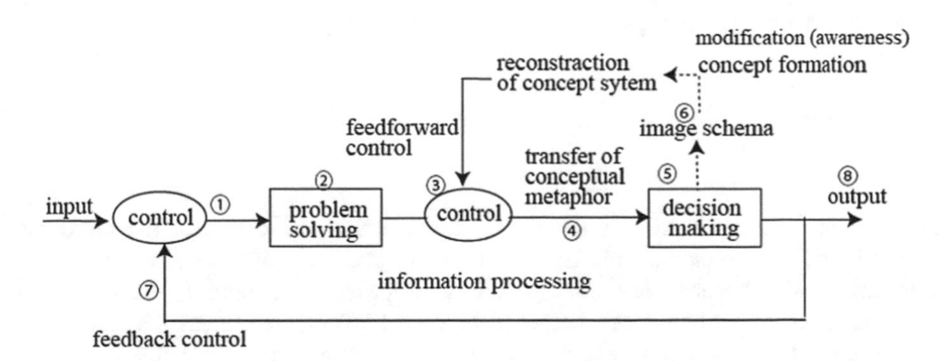

Fig. 1. Model of human information processing (learning processing)

2.2 Hypothesis

We have remarked autographic memory which is one of every day memory, and schema theory. Among them, autobiographic memory is considered as one of specific episodic memory and recalled easier than others. Especially, it is reported that relating with ones' characteristic words tend to be remembered and recalled easier. Reisberg [4] reported that many authors proposed emotion memory promotes episodic memory. He also said that personal events were vividly recalled and were more emotional than the public events. In addition to this, he introduced Buchanan has argued that emotional and nonemotional contexts may be differently operated by encoding, consolidating and retrieving memories. Buchanan has explained it from neuroanatomy and emphasized the role of the amygdala with hippocampus, which tends to be effect by phoneme. From those perspective on information processing, we supposed that learners' differences while they are studying, such as problem solving and decision making, might be found out by measurement of personality with various materials of questionnaires, comparing between sound voice and text, emotional and nonemotional contexts and so on.

In Japanese language case, two kinds of characters are used for writing, kanji and kana. Kanji is an ideogram which is understood meaning from its shape. On the other hand, kana is a phonogram which shows pronunciation how it is read phonologically and its function as an auxiliary to a main word is grammatically important. From those aspects, we have generated a series of hypotheses. When students input information from instructors, different strategies of reading are selected in order to solve problems and decision making; (1) kanji is changed to sound phonologically and understood like sound voice, (2) kanji is not changed to sound but understood with symbol which has its meaning, (3) both strategies are mixed, depending on context.

If the attitude of reading (1), the decision time which is measured from the beginning of presenting one short sentence of a questionnaire to the answer selected by a student, should be correlated strongly with duration of reading while the questionnaire is presenting as a condition of sound voice. In the case of (2), those correlation coefficients might be low. Moreover, depending on contexts, students might select their reading strategies variously (3). Next, we will examine those hypotheses by gathering empirical data of psychological testing in a standardized manner.

3 Design

3.1 Method

A preliminary experiment was designed procedure for verification of the prototype model in order to prepare for the implementation of practical nursing classes and verify this case study, because our final goal is to improve collaborative education and learning. Both experiments of measurement were implemented under the same condition and nature, such as environment, proceedings, age and sex of subjects, and so on. Materials are applied on YGPI (Yatabe-Guilford Personality Inventory) questionnaires. (1) YGPI is composed 120 items related to personality. (2) Those items are divided into twelve factors and assessed highly reliability. (3) Its method of operation is standardized how to present questionnaires (auditory for subjects) so that subjects can reply in a time limit. (4) Written questionnaires are also standardized for executors so that they can present them auditory to subjects. (5) It is possible to divide 120 items into two contexts, emotional and nonemotional conditions [5]. Each questionnaire is consisted by one short sentence and their average number of words and duration of reading presentation are not significantly different between emotional and nonemotional conditions. We have operated an informed consent procedure and obtained approval from all students and their affiliations.

3.2 Procedure

Preparatory Experiment. (1) Purpose: confirmation and implementation of verification procedure. (2) Participants: 28 female freshmen of university. (3) Duration: from January to March in 2015. (4) Method: under the same condition, verification experiments for repeatability of the prototype equipment. (5) Implementation: dividing into two groups for a counterbalance to the order of auditory and visual experiments and implementation of the same way twice in January and March. (6) Data: gathering data of responses to questionnaires and decision time.

Practical Research. (1) Purpose: calibration of the prototype equipment and review of results by compering between measurements and students' performance in practical class, identifying individual differences through language information processing. (2) Participants: 98 female freshmen of university. (3) Duration: from April in 2015 to March in 2016. (4) Method: formation of teams referring the results of the measurement; gathering other results of students' performance. (5) Implementation: orientation to students and instructors, practice with team members, observation in practical classes. (6) Data: gathering data of responses to questionnaires and decision time, students' performances, observations and reports.

3.2.1 Methods of Analysis

(1) Comparison of correlations between decision time and duration of presenting standard sound voice. (2) Identifying individual differences from strength of correlations. (3) Investigating the relations between individual traits and performances.

4 Results

4.1 Results of Preparatory Experiment

(1) Repeatability; validation of correlation coefficient between decision time and duration of standard sound voice presentation: Results of comparisons between pre and post correlation coefficient of both sound voice and text are not shown significantly different.
(2) Comparison of Individual traits: Decision time of types depending on strength of the correlation coefficient is shown the same feature at the pre and post experiments. Decision time of Type S (auditory type: $r > 0.5$) was significantly longer than Type V (visual type: $r < 0.3$). Both results of pre and post experiments were appeared the same tendency.

4.2 Results of Practical Research

(1) Comparison of the average decision time between TypeS and TypeV: In the case of sound voice presentation, both types of decision time were not significantly different. On the other hand, in the case of text presentation, there were significantly different between them (df. = 5154, $t = -21.05$, $p < .001$). The average of decision time of TypeS was much longer than those of TypeV. This result was the same as those of the preparatory experiment.

(2) Investigating the relations between individual traits and performances; The comparison of the correlation coefficient in emotional context between TypeX ($\overline{X} = > \mu + \sigma$) and TypeY ($\overline{X} = > \mu - \sigma$) is visualized by scatter diagram methods in Fig. 2, which is their distribution of decision time at text presentation. It helps us to compare the correlations to the standard duration of sound voice (reading aloud) presentation. In the case of SubB-1 (TypeX), the correlation coefficient of emotional context is stronger than nonemotional context. In addition to this, the decision time of emotional context is faster than those of nonemotional context. In contrast, the decision time of SubB-2 (TypeY) between two kinds of context are not significantly different.

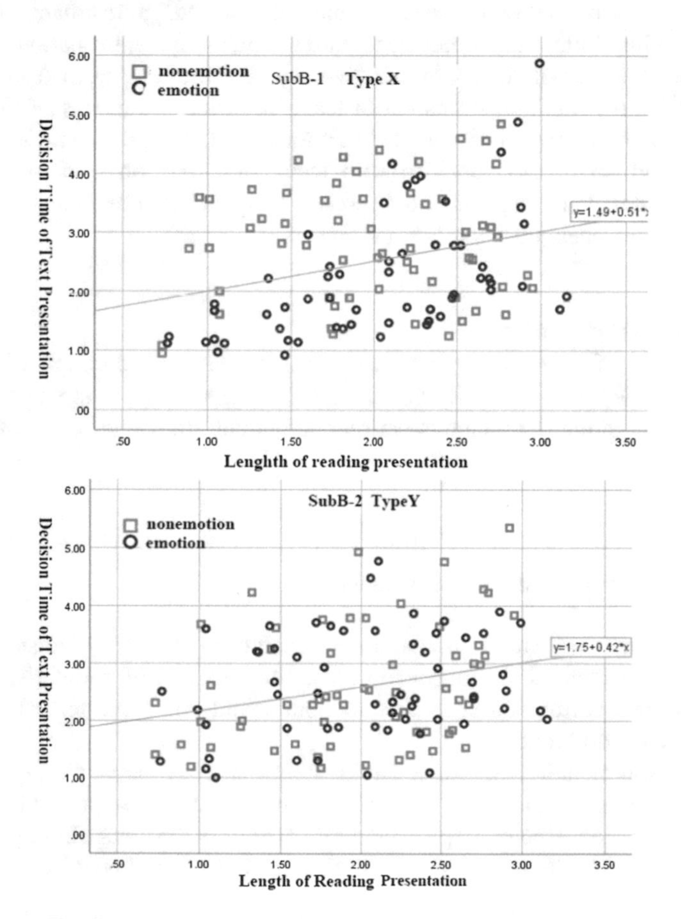

Fig. 2. Distribution of 120 decision time at text presentation

5 Discussion and Conclusion

Hypothesis (1) kanji is changed to sound phonologically and understood like sound voice: high correlation Group TypeS need decision time longer than Group TypeV. This suppose that TypeS might have taken time for phonologization when they understand sentences and decision making. In contrast, Hypothesis (2) kanji is not changed to sound but understood with symbol which has its meaning: TypeV might not do such behaviors, however, understand letters directly. In this case, their attitude of information processing might be explained positive feedback loop in Fig. 1⑦. On the other hand, TypeS might tend to process information through negative feedback loop with schema (Fig. 1⑥) for their prediction as feedforward control (Fig. 1③).

In SubB-1(TypeX) case, however, she responses to emotional context with high correlation but faster than nonemotional context. This means her attitude of information processing is opposite to others. From this point of view, TypeX might be able to adjust their behavior of information processing unconsciously by Hypothesis (3) both strategies mixed, depending on context; positive feedback in emotional context and negative feedback for forecasting by feedforward control in nonemotional context. This tendency is appeared by other TypeX students and the results of their performances become higher than TypeY students (including SubB-2 who is also TypeV) in second semester. It implies that negative feedback loop might bring out conceptual reconstruction through schema during information processing [6] and might have effects on deep learning.

In conclusion, we found out some individual differences of information processing behavior and attitude. It must be useful to understand our information processing by visualizing of Big Data analysis [7] and would contribute to sustainable educational effect and imply how to apply to AI and machine learning.

References

1. Yamada, A., Sakai, L.K.: Syntactic processing in Broca's area: Brodmann areas 44 and 45. Brain Nerve **69**(4), 479–487 (2017)
2. Hashimoto, R., Sakai, L.K.: Learning letters in adulthood: direct visualization of cortical plasticity for forming a new link between orthography and phonology. Neuron **42**(2), 311–322 (2004)
3. Buchanan, T.W., et al.: Recognition of emotional prosody and verbal components of spoken language: an fMRI study. Cogn. Brain Res. **9**, 227–238 (2000)
4. Buchanan, T.W., Adolphs, R.: The neuroanatomy of emotional memory in humans. In: Reisberg, D., Hertel, P. (eds.) Memory and Emotion. Oxford University Press (2003)
5. Tsujioka, K.: A study of the impact of audio or visual media on decisions in Japanese: comparing response and reaction times in tasks between emotionality and introversion-extroversion. Osaka University, Ann. Educ. Stud. (16), 33–44 (2011)
6. Baldwin, M.W.: Relational schemas and the processing of social information. Psychol. Bull. **112**(3), 461–484 (1992)
7. A case study of ICT used by big data processing in education: discuss on visualization of RE research paper, pp. 160–164. Association for Computing Machinery (2018)

Computational Modeling and Simulation

Simulation-Based Planning and Programming System for the Assembly of Products with a Wide Range of Variants in Collaboration Between Worker and Robot

Werner Herfs, Simon Storms(✉), and Simon Roggendorf

Laboratory for Machine Tools and Production Engineering (WZL), Chair of Machine Tools, Steinbachstraße 19, 52074 Aachen, Germany
{w.herf,s.storms,s.roggendorf}@wzl.rwth-aachen.de

Abstract. In the field of robot-based assembly especially for products with a high number of variants, robot programming causes a large part of the product costs. The use of automation solutions for specialized production steps is particularly unprofitable for such small production batches. Human-robot collaboration (HRC), where tasks are allocated according to the respective skills, provides a good start-up point in automation of manual processes. However, robot programming remains a complex topic with many new programming possibilities during the last years, especially in the HRC sector. The concepts presented in this paper are intended to present a solution for a flexible assembly system, which is able to create a robot program and path automatically based on given information. In particular, this will be based on comprehensive product modeling and automatic process simulation.

Keywords: Human robot collaboration · Robot assembly · Simulation Automatic programming

1 Introduction

According to IFR estimates, the number of industrial robots used worldwide will double between the years 2014 and 2020 [1]. Small and medium-sized enterprises are also increasingly benefiting from this trend, even if the benefits and opportunities are related to high investment costs. The downside of this development is that robots are one driving factor of technologically based unemployment. One trend that can work against the initial risk for automation and the increasing danger of unemployment is collaborative workplaces in which humans and robots work hand in hand [2]. However, collaborative scenarios pose challenges, in particular to worker acceptance, occupational safety and economic efficiency (due to reduced robot speeds). In addition, the problem remains that the robot-based assembly system must always be reprogrammed, especially for products with many variants, which requires experts who are often not available in small and medium-sized companies.

W. Karwowski and T. Ahram (Eds.): IHSI 2019, AISC 903, pp. 387–392, 2019.
https://doi.org/10.1007/978-3-030-11051-2_59

On the basis of comprehensive and standardized data modeling, linked to a collaborative workstation consisting of a robot, tools, safety and vision technology, methods for task distribution and automated machine programming are presented in this paper. The focus is on the collaborative assembly of products with many variants. In a first step, resources, components and process steps are modeled in a suitable way to illustrate the integrated planning process. Based on this, an automated skill-based task assignment for humans and robots takes place, as described in the preliminary work [3]. This also represents the detailed planning of the work steps. On this basis, the assembly system now automatically plans the further steps necessary for assembly. In particular, this includes the selection of suitable tools (e.g. grippers, screwdrivers), the use of these tools (e.g. gripper planning, screwdriving programs) and path planning. A simulation environment, extended by specific algorithms and interfaces, is used to implement these planning steps. Planning can be carried out both in equipment engineering and work preparation as well as just-in-time during assembly.

2 State of the Art

Programming Methods for Collaborative Robots. In the development of robot-supported applications, robot programming takes up an elementary part of development costs and must always be adapted to new products and variants. Industrial robots of different kinematics generally have large overlaps in their range of functions, including general logics, loops, conditional statements and IO interaction, as well as drive strategies such as point-to-point, linear or circular. Robot manufacturers have usually developed their own programming languages for this purpose, mostly text-based, such as KRL (KUKA), TPE (Fanuc) or VAL3 (Stäubli). New simple programming methods have been developed, in particular through the introduction of collaborative robots, which allow hand-guided teach-in of positions (play back). An example of this is the Panda Robot from Franka Emika, which can be programmed with the use of apps, which can be arranged by drag and drop and parameterized by simple input on the robot [4]. A similar concept is used with the Sawyer Robot from rethink robotics. On the ROS-based operating system Intera, function blocks can be easily arranged and parameterized in a kind of petry net [5]. In this and similar systems, only rudimentary robotics knowledge is required, allowing the production worker to make program changes.

A further simplification of robot programming is the standardization of languages and interfaces. The idea is almost as old as industrial robotics (e.g. Industrial Robot Language from 1992 [6]), but is still the subject of research today. The industrial communication standard OPC-UA [7] is currently one of the most advanced technologies in the field of standardization for production plants. Within this standard there are current researches on a non-proprietary interface for industrial robots [8].

Modeling as Base for Resource Planning and Programming. Within this work, the assembly cell will independently select resources (tools) based on a qualified basis (modeling). For this purpose, a suitable modeling of tasks, resources and components is essential. The state of the art already offers multi-domain modeling languages and standards that can be used as a basis for the required data. Unified Modeling Language (UML) offers a particularly graphically possibility of modeling by describing systems with 14 different language units [9]. However, the modeling of physical properties is only possible to a limited extent [10], which means that the language cannot be used optimally for the planning of collaborative applications with special needs. The Systems Modeling Language (SysML) is based on UML, but does not use specific parts of the standard, but in return extends it to describe across domains and over the entire lifecycle. A current research approach to modeling languages is the Interdisciplinary Modeling Language (IML), which tries to combine the advantages of UML and SysML, but the number of diagram types used is more manageable [11]. In particular, IML allows the tools used in the development process of production machines to be combined with one another. Similar goals are pursued by the Standard Automation Markup Language (AutomationML), which in turn consists of various standards (like COLLADA, CAEX or PLCopen) and is very popular in industry-related research [12]. The open graphics format JT [13], which can store product and manufacturing information (PMI) in addition to pure geometry data, should also be mentioned in particular.

Robotic-Based Path Planning. Even though the use of collaborative robots allows collisions between the robot and humans within the limits of defined forces and conditions (see [15]), it increases worker acceptance if the robot avoids these collisions. Therefore a collision free robot path has to be generated repeatedly during the automatic planning and programing. This is known as a mathematical problem called the Pianos Mover Problem [14], which is solved by sampling based path planning in the very most cases. Due to changing environmental conditions for real-time path planning, single-query models (e.g. Exploring Random Tree RRT) must be used. Multiple query models (e.g. Probabilistic Roadmap PRM), which create a graph from valid paths in advance, are thus omitted.

3 Concept

The concept presented in this paper is intended to enable automatic programming of a robotic cell with a focus on collaborative assembly. In a certain way, it thus represents an alternative to the state of the art methods for simple programming of collaborative robots using predefined apps. As Fig. 1 shows, the concept can be roughly divided into four main modules, which are linked together by a Product Lifecycle Management (PLM) including defined workflows. In the first module (modeling), the necessary components imported to the system and described with meta information. This provides the basis for the second module (planning). Here, tasks are allocated to resources. In particular, the allocation of tasks between humans and robots is also realized here. The planning is validated in the third module (simulation). Finally, the work plan is executed on the collaborative workstation. All the modules are working together, so that a work plan is generated within iteration cycles.

Modeling. In the course of modeling, the geometric model is first imported into the system as a JT format and provided with manufacturing information. The focus is on the tasks of gripping (fitting), screw application and media application (e.g. adhesives or greases). For this purpose, gripping points, screwdriver programs or quantities must be defined in the model. These definitions are stored independently of resources. For the automatic allocation of resources, tools and robots are also described semantically. The individual components are related to each other with the help of a work plan in AutomationML format. References are made to the PMIs of the JT format.

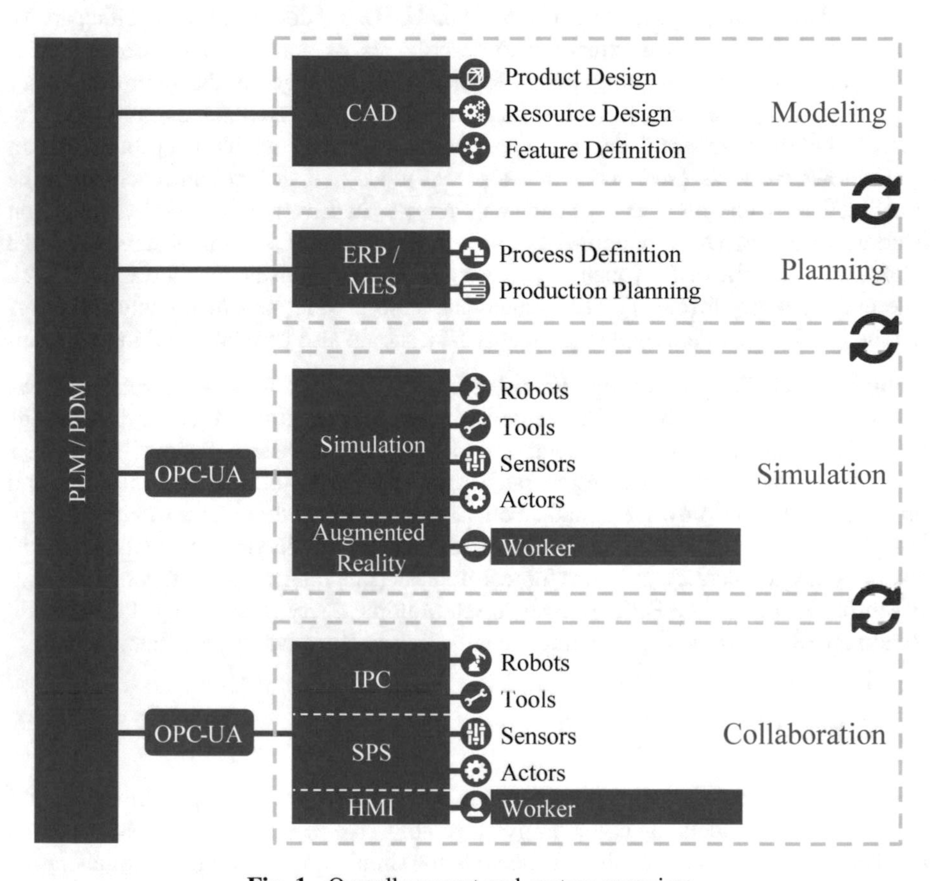

Fig. 1. Overall concept and system overview

Planning. In the course of planning, tasks are now allocated to resources on the basis of skills. This also involves the division of tasks between humans and robots, which can be done using machine learning (see an earlier published paper [3]). The information stored in the modeling allows the planning module to select suitable tools for each task and to use them appropriately. The executing robot should be able to change its tool autonomously at the tool center point (TCP). After planning has checked which resources are generally suitable for a task, detailed planning is done with the final assignment based on capabilities and availability. During execution, the planning unit acts as a Manufacturing Execution System (MES) and controls the corresponding physical units.

Simulation. The simulation works closely together with the planning so that a work plan is generated in the sequence of several iterations. As soon as the planning module has assigned a task to the robot, this assignment is verified with the help of the simulation environment. Component positions are either known through defined storage locations or are determined using cameras of the assembly cell. The positions relative to the component can be taken from the PMIs stored in the JT file. In several steps it is now checked whether the robot is actually able to perform the task with its allocated tool. The fact, that the tool is basically capable on the basis of its characteristic data, was already ensured during the planning phase. The first step is to determine whether the positions can generally be reached. In this case, it is verified whether these positions can also be reached collision-free and whether a collision-free path between the positions can be generated. The result of the validation is sent back to the planning module including the estimated times. Based on the simulation results the planning module will finally decide which resource should be used or whether the task must be delegated back to the human being.

Collaboration. Finally, the simulated machine program is executed in collaborative process mode. Both, the simulation and the physical production cell are connected to the overall system via OPC-UA. With the described robot standard it will be possible to change not only between the simulation and the real cell but also between robots of different manufactures. The program on the robot accepts commands from the planning system and has no predefined programs, so that the robot program can be changed in real time. The cell includes sensors and cameras that can recognize the component positions and obstacles. This information is sent back to the simulation, which can adjust the paths (or resource allocation) if necessary. The human being receives his instructions via monitor or VR glasses.

4 Validation Summary

The concept was validated within the use case of the assembly of a pump with many variants and small batches. The assembly process consists of mating and crimping parts, screwing together and applying mediums (water, different oils, and glues) and was made completely manually in the past. The manual workstation was extended by a collaborative robot and all tools where modified so that they can be used by the robot as well as by the human. One human and one robot are working on two pumps in parallel. They alternate between the two pumps, so the workload respectively the time has to be the same to effort the best equipment efficiency. The result of the planning process is that the worker is better qualified in mating parts, especially non-rigid ones (e. g. wires and gaskets). The robot assume recurring tasks like screwing and applying mediums, which also increases the quality.

With the help of the presented method, it is possible to apply a new kind of robot programming. It therefore represents an alternative to the novel programming methods of collaborative robots. A standardized integration of resources creates the advantage of being able to work independently of manufacturers. The late and situational assignment of working steps to resources enables the simple transfer of new products to an existing assembly system or known products to a new system. The simulation can be used to determine process times in order to quickly make statements about cycle times and profitability.

References

1. Gemma, J., Wyatt., Litzenberger, G.: How robots conquer industry worldwide. In: IFR Press Conference, 27 September 2017, Frankfurt (2017)
2. Jun Li, R.: Top 8 Trends for Robotics, Shanghai, 21 December 2016
3. Storms, S., Roggendorf, S., Stamer, F., Obdenbusch, M., Brecher, C.: PLM - supported automated process planning and partitioning for collaborative assembly processes based on a capability analysis. In: Schmitt, R., Schuh, G. (Hrsg.) 7. WGP-Jahreskongress Aachen, 5–6 Oktober 2017, S. 241–249. Apprimus Wissenschaftsverlag, Aachen (2017)
4. Franka Emika Gmbh (Hrsg.): User Manual. Panda Research, München, 06 October 2017
5. Rethink Robotics Inc.: Rethink robotics: Intera. http://mfg.rethinkrobotics.com/intera. Accessed 09 Oct 2018
6. DIN DE 66312: Industrial Robot Language (IRL) (withdrawn), September 1996
7. IEC 62541-100: OPC Unified Architecture Specification - Part 100: Device Interface, March 2015
8. VDMA 40010-1: OPC UA Companion Specification Robotics – Part 1: Condition monitoring, asset management, predictive maintenance, vertical integration, November 2018
9. BS ISO/IEC 19505-1:2012: Information technology. Object Management Group Unified Modeling Language (OMG UML). Infrastructure, 10 November 2012
10. Barbieri, G., Fantuzzi, C., Borsari, R.: Tools for the development of a design methodology for mechatronic systems. In: 2013 IEEE 18th Conference on Emerging Technologies & Factory Automation (ETFA), S. 1–4. IEEE (2013)
11. Brecher, C., et al.: Interdisciplinary specification of functional structures for machine design. In: 2016 IEEE International Symposium on Systems Engineering (ISSE), S. 1–8. IEEE (2016)
12. Din Norm DIN EN 62714-1: Engineering data exchange format for use in industrial automation systems engineering - Automation markup language - Part 1: Architecture and general requirements, February 2018
13. ISO 14306:2017: Industrial automation systems and integration – JT file format specification for 3D visualization, November 2017
14. Siciliano, B., Khatib, O.: Springer Handbook of Robotics. Springer, Berlin (2007)
15. ISO 15066: Robots and robotic devices – Collaborative robots, February 2016

VR Experience from Data Science Point of View: How to Measure Inter-subject Dependence in Visual Attention and Spatial Behavior

Pawel Kobylinski[1(✉)], Grzegorz Pochwatko[2], and Cezary Biele[1]

[1] Laboratory of Interactive Technologies, National Information Processing Institute, al. Niepodleglosci 188b, 00-608 Warsaw, Poland
{pawel.kobylinski, cezary.biele}@opi.org.pl
[2] Virtual Reality and Psychophysiology Lab, Institute of Psychology, Polish Academy of Sciences, ul. Jaracza 1, 00-378 Warsaw, Poland
grzegorz.pochwatko@psych.pan.pl

Abstract. Any Virtual Reality (VR) immersive experience inherently allows its subjects to choose their own paths of visual attention and/or spatial behavior. If a VR designer employs any system of attentional cues, they might be interested in measuring the system's effectiveness. Eye tracking (ET) time series data can be used as a visual attention trail and positional time series data can be used as spatial behavior trails. In this paper we are addressing the issue of measuring inter-subject dependence in visual attention and spatial behavior. We are arguing why recently developed distance correlation coefficient [1, 2] might be both a proper and convenient choice to either measure the inter-subject dependence or test for the inter-subject independence in visual and behavioral data recorded during a VR experience.

Keywords: Virtual reality · Narration · Attention · Behavior · Eye tracking Positional tracking · Data science · Applied statistics · Energy statistics Distance correlation · Distance variance · Human-Technology Interaction User experience · Research methodology · Social sciences · Psychology

1 Introduction

Though the concept of Virtual Reality (VR) occupies our minds for decades [3] and the history of Head-mounted Displays (HMD) integrated with a computer reaches back to the 1960s [4], it is only recently that we have a new wave of interest in VR technology and its applications. Current wave of technological, intellectual, financial, and marketing efforts in the VR area is facilitated by the promising fact that processing power of recent consumer computer devices allows for a credible and rich immersion. Companies, academia, designers, and artists provide solutions (HMDs, sensors, graphic cards, software) and content (simulations, games, experiences, and last but not least, Cinematic VR). Social sciences recognize the availability of methodology that proposes not only controlled experiments, but also fully controlled environment [5].

W. Karwowski and T. Ahram (Eds.): IHSI 2019, AISC 903, pp. 393–399, 2019.
https://doi.org/10.1007/978-3-030-11051-2_60

Any VR immersive experience inherently allows its subjects to choose their own paths of visual attention and/or spatial behavior. Contrary to non-immersive, linear, two-dimensional means of expression (e.g. 2D video), VR experience cannot force its participant's attention to follow any linear narration intended by a VR experience designer. If a VR designer employs any system of attentional cues (e.g. spatial sounds), they might be interested in measuring the system's effectiveness. ET time series data (if available) can be used as a visual attention trail and positional (position of a HMD and of other sensors) time series data can be used as spatial behavior trails. It is reasonable to assume that if there exists any intended or unintended effective system of attentional cues in a VR experience, then there is statistical dependence between visual and/or behavioral data trails left by two or more subjects. In this paper, we are addressing the issue of measuring inter-subject dependence in visual attention and spatial behavior. We are arguing why recently developed distance correlation coefficient [1, 2] might be both a proper and convenient choice to either measure the inter-subject dependence or test for the inter-subject independence in visual and behavioral data recorded during a VR experience.

We believe that our paper may serve as a worthy inspiration to teams of cooperating designers of VR experiences and researchers on human factor in VR.

2 Motivation

In the course of our professional activities as VR-related Human-Technology Interaction researches, we have been being approached by VR content designers, e.g. educational and artistic Cinematic VR creators, who seek our advice regarding problems with guiding visual attention and spatial behavior of people participating in VR experiences. This boils down to the problem of narration in VR experiences.

2.1 The Problem of Narration in VR Experiences

While we may define narration broadly as a human communication paradigm [6], for purposes of this paper we treat narration more narrowly as an actual way of conveying a particular story in the VR medium [7].

It is only natural that a participant of a VR experience can choose freely where to look and/or in which direction to move. This means they are probable to miss an element of a story intended as important to the plot by an author. This issue is especially important in the case of Cinematic VR experiences, where cinematic language well established for traditional 2D video or movies simply does not work. This is why VR experience designers devise systems of attentional cues based on gaze, motion, sound, context, and perspective [8].

In this paper we are neither proposing nor analyzing ways to guide attention during a VR experience. Instead we assume that if a VR designer employs any system of attentional cues (e.g. spatial sounds), then they might be interested in measuring the system's effectiveness. This is where data science and statistics are ready to help.

2.2 Inter-subject Dependence Between Attentional and Behavioral Trails

Data trails reflecting visual attention and spatial behavior of a VR experience participant can be recorded and analyzed. We assume that if two or more people participating in a VR experience pay attention to similar elements of the experience in similar time and/or move in similar way, they leave data trails that change in a similar way. This similarity in change (similarity in data patterns) can be measured.

In this paper we are putting stress on the inter-subject dependence in visual attention and spatial behavior. Nevertheless, the mathematical method addressed here can be used for intra-subject situations (to investigate similarity in data trails left by the same person during a repeated VR experience).

3 Mathematical Method

In mathematics, statistics, and data science the general notion of dependence in data patterns is operationalized usually in terms of correlation coefficients. The choice of a correlation coefficient proper for a given task depends to a great extent on the properties of available data.

3.1 Data Trails Recorded During VR Experiences

Several types of data reflecting the human factor can be recorded during a VR experience. Table 1 summarizes briefly data sources, data types, and general types of information the data may represent [9].

Table 1. Typical data sources, data types, and general types of information obtainable for recording during a VR experience.

Data source	Data type	Indicatum
HMD	Position	Behavior
HMD	Rotation	Behavior/attention
Eye tracker	Gaze position	Attention
Handheld controller	Position	Behavior
Handheld controller	Rotation	Behavior
Handheld controller	Input	Behavior/attention
Wearable sensor	Position	Behavior
Wearable sensor	Rotation	Behavior

Positional and rotational data obtained during a VR experience is by its nature multivariate. At any given time $t \geq t_0$ (where t_0 denotes the time at which a VR experience starts) three values should be recorded for every single positional and rotational variable, including gaze position determined by an eye tracker, if one is integrated with a HMD. Data trails detected by separate sensors can be treated separately, yielding three-dimensional, multivariate time series variables. Alternatively, the trails can be combined into more-dimensional time series (e.g. HMD position + gaze position would yield a six-dimensional time series). It is worth to remember that the VR data trails, especially the ET data, may require processing, e.g. filtering and/or categorization [9].

3.2 Correlation and Dependence Between Data Trails

I order to effectively measure the inter-subject dependence or test for the inter-subject independence between data trails, we must apply an association measure (a correlation coefficient) which matches the properties of recorded data, maximizes the scope of associations that can be detected, and last but not least, allows to detect lack of dependence. For VR data trails we need a coefficient that: (1) handles multivariate data, (2) detects non-linear associations, (3) has a value that implies independence between variables, (4) can be statistically tested for null hypothesis (5) can be used in the case of time series data. Table 2 compares traditional and novel association measures [10, 11]. Recently developed distance correlation coefficient [1, 2] happens to have the properties desired for correlating VR and ET data trails. Distance correlation coefficient can be used to both measure non-linear dependence and test for independence between two matrices of the same number of rows. Considerable work has been also done to prove the coefficient can be applied to time series data [12].

Table 2. Comparison of traditional and novel association measures. Distance correlation has the properties required to investigate dependence between VR data trails.

Coefficient	Variables	Association	Zero value
Pearson's r	Univariate	Linear	Does not imply independence
Hoeffding's D	Univariate (ranks)	Non-linear	Does not imply independence
Maximal information	Univariatc	Non-linear	Implies independence
Distance correlation	Multivariatc	Non-linear	Implies independence

3.3 Distance Correlation Coefficient

Let us address a situation in which at any given time $t \geq t_0$ (where t denotes a discrete time series variable and t_0 denotes the time at which a VR experience starts) m values are recorded for every VR experience participant. Thus, throughout a VR experience, for every participant we obtain at least one m-dimensional variable recorded n times. This variable can be represented as a matrix T with n rows and m columns. After the filtering/categorization of the matrix rows we obtain a corrected matrix T', which in the next step is transformed into matrix of Euclidean distances between rows treated as m-dimensional vectors of observations (square matrix D, with n rows and n columns). Next, D is double-centered (i.e. the row and column means are subtracted and the grand mean added) in order to obtain final matrix D'.

Let D' denote the transformed distance matrix for a VR experience participant p and let E' denote the transformed distance for a participant r. The distance covariance $dCov$ is defined as follows [1, 2, 13, 14]:

$$dCov_{pr} = \frac{1}{n}\sqrt{\sum_{i,j=1}^{n} \boldsymbol{D}'_{ij}\boldsymbol{E}'_{ij}}. \tag{1}$$

The empirical version of distance correlation coefficient [1, 2, 13, 14] is defined on the basis of *dCov*:

$$dCor_{pr} = \begin{cases} \dfrac{dCov_{pr}}{\sqrt{dCov_{pp}dCov_{rr}}}, & dCov_{pp}dCov_{rr} > 0 \\ 0, & dCov_{pp}dCov_{rr} = 0 \end{cases}. \tag{2}$$

dCor lies in the interval [0, 1] and takes zero value only in case of independence between matrix variables T'_p and T'_r. In our specific VR case, $dCor_{pr} = 1$ implies maximal inter-subject dependence between data patterns left by VR experience participants p and r, whereas $dCor_{pr} = 0$ implies independence between the data patterns.

Research problems may naturally involve $N > 2$ VR experience subjects. Since *dCor* is calculated pairwise, for $N > 2$ the empirical distribution of the $N(N-1)$ *dCor* values has to be analyzed in order to gain insight into the dependence between multiple variables. The median of the obtained *dCor* values can serve as a measure of the central tendency relatively resistant to potential outliers and skewness of the distribution. The percentiles may add some easily interpretable information about variability of the pairwise dependences. For tasks requiring a more synthetic procedure for investigating the mutual inter-subject dependence between $N > 2$ data trails, a generalized version of the distance correlation may be considered [15].

3.4 Distance Variance

The distance variance $dVar_p = dCov_{pp}$ should not be treated only as a byproduct of the *dCor* procedure. Distance variance [1, 2, 13, 14] itself can serve as a comparative measure of variability in attentional and/or behavioral data trail left by a single VR experience participant. For multiple participants the median and other percentiles of the empirical distribution of the N *dVar* values can be employed to sum up information about inter-subject differences in spatial dispersion of visual attention and behavior during a VR experience.

3.5 Energy Statistics

dCor, *dCov*, and *dVar* are based on Euclidean distance matrices. These statistical measures generalize to a wider class of functions of distances between statistical observations in metric spaces. The class is named energy statistics by analogy to Newton's gravitational potential energy, which is also a function of distances between physical objects [13, 14]. Among other interesting implications, distance based approach paves the way to clustering procedures allowing a researcher to identify subgroups of similar observations in data.

4 Conclusion

We believe that our paper may help teams of cooperating designers of VR experiences and researchers on human factor in VR to investigate narration-related attentional and/or behavioral processes. We also hope the sound methodology described in our paper may inspire social science researches. Practical, real-life application of the statistical procedures is convenient, as their authors took effort to implement them into R packages [16–18]. Distance based energy statistics may be applied to data trails recorded during single- or multi-user VR experiences. It is also worth to note that the scope of research in which the method may find its applications is not limited to the VR field. Research involving Augmented Reality (AR), standalone and mobile ET (including parallel ET), and positional tracking in real space setups may benefit from usage of the methodology described in this paper.

References

1. Szekely, G.J., Rizzo, M.L., Bakirov, N.K., Nail, K.: Measuring and testing dependence by correlation of distances. Ann. Stat. **35**, 2769–2794 (2007)
2. Szekely, G.J., Rizzo, M.L.: Brownian distance covariance. Ann. Appl. Stat. **3**, 1236–1265 (2009)
3. Lowood, H.E.: Virtual Reality (2018). https://www.britannica.com/technology/virtual-reality
4. Sutherland, I.E.: A head-mounted three dimensional display. In: AFIPS Fall Joint computer Conference, pp. 757–764. ACM, New York (1968)
5. Fox, J., Arena, D., Bailenson, J.N.: Virtual reality: a survival guide for the social scientist. J. Media Psychol. Ger. **21**, 95–113 (2009)
6. Fisher, W.R.: Narration as a human communication paradigm: the case of public moral argument. Commun. Monogr. **51**, 1–22 (1984)
7. Aylett, R., Louchart, S.: Towards a narrative theory of virtual reality. Virtual Real. **7**, 2–9 (2003)
8. Godde, M., Gabler, F., Siegmund, D, Braun, A.: Cinematic narration in VR - rethinking film conventions for 360 Degrees. In: Chen, J., Fragomeni, G. (eds.) VAMR 2018. LNCS, vol. 10910, pp. 184–201. Springer, Cham (2018)
9. Duchowski, A.T.: Eye Tracking Methodology: Theory and Practice. Springer, Berlin (2007)
10. Clark, M.: A comparison of correlation measures. Technical report, University of Notre Dame (2013)
11. de Santos, S.S., Takahashi, D.Y., Nakata, A., Fujita, A.: A comparative study of statistical methods used to identify dependencies between gene expression signals. Brief. Bioinform. **15**, 906–918 (2014)
12. Davis, R.A., Matsui, M., Mikosch, T., Wan, P.: Applications of distance correlation to time series. Bernoulli **24**, 3087–3116 (2018)
13. Szekely, G.J., Rizzo, M.L.: Energy statistics: a class of statistics based on distances. J. Stat. Plan. Infer. **143**, 1249–1272 (2013)
14. Szekely, G.J., Rizzo, M.L.: The energy of data. Ann. Rev. Stat. Appl. **4**, 447–479 (2017)
15. Fan, Y., Lafaye de Micheaux, P., Penev, S., Salopek, D.: Multivariate nonparametric test of independence. J. Multivar. Anal. **153**, 189–210 (2017)

16. Rizzo, M.L., Szekely G.J.: energy: E-statistics: multivariate inference via the energy of data. R package version 1.7-5 (2018). https://CRAN.R-project.org/package=energy
17. Lafaye de Micheaux P., Bilodeau, M.: Software: R Package, IndependeceTests, Version 0.2 (2012). https://CRAN.R-project.org/package=IndependenceTests
18. R Core Team: R: A Language and Environment for Statistical Computing, Vienna, Austria (2018). https://www.R-project.org

Augmented Reality in Order Picking—Boon and Bane of Information (Over-) Availability

Ralf Elbert and Tessa Sarnow(✉)

Technische Universität Darmstadt,
Hochschulstraße 1, 64289 Darmstadt, Germany
{Elbert,Sarnow}@log.tu-darmstadt.de

Abstract. Nowadays, 75% of warehouses are operated manually with more than half of the costs caused by the process of order picking. To enhance efficiency various technical support systems are used. Recently, augmented reality applications gain great attention in operational processes in logistics. While most of them focus on the different process steps, there is a lack of research on the question how to employ augmented reality under consideration of cognitive ergonomics and related design principles. Making use of literature as well as empirical data gained in qualitative interviews with eleven order pickers, this research proposes designs for the user interfaces for three primal levels of order picking support to make the amount of available information controllable. Furthermore, three calibration options used to control the visualization of information are evaluated regarding usability and necessity. The results indicate that individual calibration of the interface is seen as crucial. Additionally, requirements for the primal support levels were directed towards a clear line-of-sight and presentation of additional order-related information.

Keywords: Augmented reality · Order picking · Logistics
Cognitive ergonomics

1 Introduction

The story of augmented reality (AR) goes back to the 1960's when first military applications in aviation were developed and continues to the present even more diversified [1]. The main advantages are the increased situation awareness and the reduction in workload. The technology's strengths become apparent when considering the wide variety of application scenarios of which logistics is only one. Applicability of AR in logistics in general is the research subject of Schwerdtfeger et al. [2] and others. Klinker et al. [3] show the numerous processes in logistics where AR can be of supportive character. Most of the identified use cases belong to intralogistics, as holds for further research, too [4, 5]. In the aforementioned works, the order picking process is the main area of interest. The term describes the retrieval of items in a warehouse according to customer orders and can be designed as parts-to-picker or picker-to-parts system with the names reflecting the configuration. Other research is closely related to application-oriented developments as a pick-order visualization tool [2]. Possibilities to train employees via AR for order picking are discussed in the literature, too [6].

W. Karwowski and T. Ahram (Eds.): IHSI 2019, AISC 903, pp. 400–406, 2019.
https://doi.org/10.1007/978-3-030-11051-2_61

Manual order picking in a picker-to-parts system provides good opportunities for application of the AR technology. This is because the usage for information transmission is of special interest in this process, as 55% of warehousing costs stem from it and it is the most common configuration of an order picking system. Thus, an efficiency increase leads to a decrease of warehousing costs and hence in total expenditure [7]. This is possible due to the fact, that cognitive and motoric processes are executed sequentially instead of parallel. Through improved provision of content, AR is able to reduce the time between presentation of content and reaction of the user. Thereby, effort of the order picker as well as performance can be improved [8]. Additionally, when employing smart glasses, the order picker can work with both hands and best performance of human-machine interaction can be expected [9].

Related research far less regarded than the aforementioned is the scientific examination of different approaches to present information under ergonomic considerations. Generally, the human factor or human-machine interaction is analyzed and assessed when designing AR systems without taking findings from the field of cognition into account on a large scale. However, especially the consideration of findings from cognition research is important for the development of AR applications. The importance of a well-considered design of an AR system is underlined by Hegarty [10]. In her review on works from the field of cognitive science, focusing on design of visual-spatial displays she says that although the content of two displays is the same, the resulting task performance is not. Hereinafter she names numerous works covering a broad time range that reinforce this statement with empirical evidence. Nonetheless, is can be observed, that there is a multitude of works on application of AR but far less focusing on reasonable design of the AR interface. The results are interfaces that are confusing, misunderstood and often come with an information overload, which in turn increases the cognitive load for the user. Therefore, we see a research gap in the diligent development of the visual presentation of information on an AR device. In the context of manual order picking, we aim at a decrease of cognitive task load and thereby reduction of the gap between cognitive abilities of the user and the requirements of the task. During the development, feedback from future users is considered to identify such information that is crucial during the picking process but to avoid information overload. The research question is the following: How is an AR application for manual order picking with smart glasses to be designed under consideration of cognitive ergonomics and related design principles? To answer this research question, we have to analyze, which factors have the largest impact on cognitive task load. We assume that there are two main factors to be considered. The first one tackles individual preferences in settings e.g. font size of the user. We address this factor by introducing calibration options, leading to the following first research proposition P1: *"Usage of calibration options for individual adjustment leads to reduction in task load."* Secondly, we see the necessity to consider the difference in need for information depending on experience and skills of the user. For example, an experienced order picker might feel information overload when using the same support level as a new colleague. Consequently, the second research proposition P2 is formulated as follows: *"Usage of different support levels depending on the current process step leads to reduction in task load."*

As a first step to test the propositions, we reviewed literature on cognition for identification of basic design principles. The rather abstract results were then transferred into design proposals and evaluated during semi-structured interviews with order pickers. The interviews took place in august 2018 at the work place of the pickers. They lasted between 20 and 30 min each. We terminated the interviews when no new insights could be given by the interviewees, which was the case after ten interviews. One additional interview was conducted for control. During the interviews, the order pickers were shown the implementation of the developed picking support and asked to imagine the support to be used at their workplace. This can be described as a multi-aisle small parts warehouse with several shelves and shelf levels. Pick orders are to be completed manually in a picker-to-parts manner. For the development, Unity and the Microsoft 10 SDK was used as software and employed on Microsoft's HoloLens.

In this paper the results of the literature review as well as the interviews is given structured as follows: In section two, the theoretical foundation is laid by review of findings from cognition research in literature. These findings are then applied to the use case manual order picking, by designing three calibration options and assistance levels, each. The results are given in section three, followed by their discussion in section four.

2 Cognitive Ergonomics and Assistance Levels for AR in Order Picking

As we proposed before, AR poses an opportunity for the order picking process as it can relieve the picker to a notable extent. This is because task load is reduced significantly compared to conventional picking support systems e.g. handhelds. According to the multiple resource theory, this reduces dual task interference. Nonetheless, there is the need to consider design directives based on findings from cognition when developing an AR order picking support system, as motoric-cognitive interference is at a low level but still influences the resulting performance. According to Funk et al. [11] the increase in use of support systems enhancing information exchange by employing technologies as AR is inevitable. At the same time, it influences cognitive load on the user largely [12]. Ruffieux et al. [13] claim that under normal circumstances, simple motoric processes as walking are not significantly influenced when performed simultaneous to cognitive demanding tasks. This is because motoric tasks are dependent on cognitive resources only to a very small proportion. Still, there might be situations requiring bringing all available cognitive resources into play, thus restricting execution of motoric tasks. Another factor influencing the performance are the cognitive skills of the user. This underlines the influence of the relation between a task's cognitive load and the user's cognitive abilities on resulting cognitive and motoric performance. As both are crucial for an efficient picking process, a prudent development of an AR application for manual order picking is of major interest.

General design principles and recommendations on the configuration of visual displays are sufficiently available. The issue at hand becomes clearer when looking at the first "principle of effective graphics" compiled by Hegarty [10]. It emphasizes the dependency between application context of a display and its design. Consequently, besides consideration of design principles on an abstract level, one has to take into account the specific context in which a display is used and adapt its design accordingly. According to Hegarty [10], the most important design principles are the following:

1. Display only such information as is needed to perform the next step.
2. Take into account the abilities of the user.
3. Make sure the content is understood as intended.
4. Differences (e.g. in size or orientation) must be perceptible.
5. Form and content must match.
6. Highlight the most important.
7. Offer a help menu.
8. When using multiple instances of a display, make sure that overlapping aspects are consistent across all instances.

In this first approach, we consider the order picker to have reached the right aisle, searching for the right box to withdraw the right amount of items. Therefore, the information needed (1) is about the location of this specific box and the ordered amount. To meet the abilities of order pickers (2), the information is to be plain and rather graphical than textual. Correct understanding (3) can be assumed as the design is based on the proposals of the future users. As long as no packing instructions are targeted, the consideration of sizes and orientation (4) is negligible. Alignment of form and content (5) can cause problems e.g. when thinking about displaying a picture of the item to pick and the amount. Strictly speaking, the picture has to show not only one item but also the right quantity, which would require a database with photos of each item in various quantities from which the right one is chosen and shown to the user. Highlighting of important aspects (6) is easily realized but could be a question of subjective feeling. A help menu (7) shall always be accessible during usage of the application. The demand for consistency across different instances (8) is of major importance as we plan to create different levels of support that can be changed at run time. Thus, overlaps of content across support levels are in the same position and design in all levels of support.

From these findings, we came to a twofold approach for the application in manual order picking as is also shown by the two research propositions mentioned before. First calibration options shall be given to enable individual adjustment of basic settings. The calibration options that were developed, proposed to the order pickers during the interviews and discussed are font size, font color, and highlighting (see Fig. 1 a–c). Secondly, different levels of support shall be available to account for different experience and skill levels of the users (see Fig. 1 d–f). In this first approach, the support levels are oriented towards the existing support systems for order picking. Thus, the lowest level is a plain textual presentation of the pick list without additional support or control functions. Secondly, highlighting of the respective withdraw location referring to a pick-by-light system is realized. The last proposed support level was an extension of the aforementioned with the function of additional quantity control.

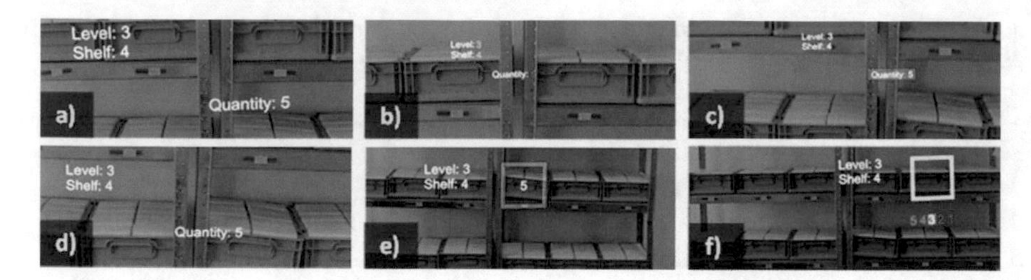

Fig. 1. Calibration options: (a) Font size, (b) Font color, (c) Highlighting Support level: (d) Pick list, (e) Withdraw location, (f) Quantity control

3 Empirical Results

All of the interviewed order pickers agreed that individual setting of the font size as shown in Fig. 1(a) is necessary to meet the varying needs of the users. Nonetheless, a limitation in expansion of the size should be implemented to prevent obstruction of a too large field of vision, which could cause safety issues. Only one order picker liked the option to change the font color individually as shown Fig. 1(b). More valuable than letting the order pickers decide for their favorite color may be the approach to adapt the font color to the lighting conditions. The option to highlight certain lines of the pick order by coloring the background of the text as shown in Fig. 1(c) was appreciated by three of the interviewees. Additionally, they suggested transferring the highlighting option from the personal calibration to the context-based picking support. This means, that depending on the status of an order, only the relevant is colored. During the interviews, several additional supporting functionalities were mentioned by the order pickers including for example inventory control as well as warnings and hints.

Regarding the proposed support levels there is consensus that presenting the picklist on an AR device as shown in Fig. 1(d) has slight advantages to an analogous picklist but that the cost-benefit ratio has to be enhanced further by making use of additional supportive features. The interviewees also agree completely, that highlighting of the withdraw location as can be seen in Fig. 1(e) is a reasonable feature which can be used in all kinds of situations occurring in a warehouse without presenting a distraction. In contrast to this, opinions differ when it comes to quantity control like in Fig. 1(f). Only five of the interviewed order pickers found it helpful to have access to this feature and many requirements in this context were brought forward. They include primarily the consideration of differences in packaging units as well as batches. We find the former to be an issue mainly of the database but see the relevance to prepare for difficulties related to this topic. When it comes to recognition of specific batches, we drift towards the field of identification of items rather than performing a simple count for quantity control. Nevertheless, both issues have to be considered. Rather than adding the item identification functionality to the quantity control feature, we propose its assignment to a new support level having item identification as the aim and enabling scanning and related bookings in the warehouse management system.

4 Discussion and Outlook

With this first qualitative assessment of the propositions, a first step towards answering the research question was taken. The feedback from the interviewed order pickers is promising and encourages further investigation. Overall, the following was revealed: Font size shall be adjustable for each user individually, while font color should be based on ambient conditions rather than individual taste. Highlighting is seen as a valuable option to draw ones attention towards the most important information. This means it is removed from the calibration options and added at the support levels. Thereby, the only resulting calibration option is font size. For the support levels, we got the feedback, that investment in AR technology is justified only when functionality goes beyond static displaying of the picklist. Highlighting of the currently relevant information in a stepwise manner is seen as a possibility for a support level 1. Support level 2 shall be the highlighting of the withdraw location, while use of quantity control is to be discussed. The research at hands is limited as only order pickers from one warehouse were interviewed which may reduce direct transferability. Furthermore, the assessment of the presented calibration options and support levels was done based on individual opinions of the order pickers with no operational assessment. This can be overcome by testing the options in an operative context and evaluating resulting performance and task load. Thereby the next step for testing of the propositions and thus answering of the research question can be taken. In the future, further enhancement can be realized by implementing identification of items to allow comparison to a target item specified by the database in the warehouse management system. This would enable a multitude of sub-functionalities as quantity control and booking.

References

1. Cirulis, A., Ginters, E.: Augmented reality in logistics. Proc. Comput. Sci. **26**, 14–20 (2013)
2. Schwerdtfeger, B., Reif, R., Günthner, W.A., Klinker, G.: Pick-by-vision: there is something to pick at the end of the augmented tunnel. Virtual Reality **15**, 213–223 (2011)
3. Klinker, K., et al.: Structure for innovations: a use case taxonomy for smart glasses in service processes. In: Multikonferenz Wirtschaftsinformatik Lüneburg, Deutschland (2018)
4. Sarupuri, B., Lee, G. A., Billinghurst, M.: Using augmented reality to assist fork-lift operation. In: Proceedings of the 28th Australian Conference on Computer-Human Interaction, pp. 16–24. ACM (2016)
5. Ginters, E., Martin-Gutierrez, J.: Low cost augmented reality and RFID application for logistics items visualization. Proc. Comput. Sci. **26**, 3–13 (2013)
6. Elbert, R., Knigge, J., Sarnow, T.: Transferability of order picking performance and training effects achieved in a virtual reality using head mounted devices. In: Preprints of the 16th IFAC Symposium on Information Control Problems in Manufacturing (2018)
7. Petersen, N., Stricker, D.: Cognitive augmented reality. Comput. Graph. **53**, 82–91 (2015)
8. Wang, X., Dunston, P.S.: Compatibility issues in augmented reality systems for AEC: an experimental prototype study. Autom. Constr. **15**, 314–326 (2006)
9. Renner, P., Pfeiffer, T.: Augmented reality assistance in the central field-of-view outperforms peripheral displays for order picking: results from a virtual reality simulation study (2017)

10. Hegarty, M.: The cognitive science of visual-spatial displays: implications for design. Topic Cogn. Sci. **3**, 446–474 (2011)
11. Funk, M., Mayer, S., Nistor, M., Schmidt, A.: Mobile in-situ pick-by-vision: order picking support using a projector helmet. In: Proceedings of the 9th ACM International Conference on Pervasive Technologies Related to Assistive Environments (2016)
12. Kretschmer, V., Eichler, A., Spee, D., Rinkenauer, G.: Cognitive ergonomics in the intralogistics sector. In: Interdisciplinary Conference on Production, Logistics and Traffic, Darmstadt (2017)
13. Ruffieux, J., Keller, M., Lauber, B., Taube, W.: Changes in standing and walking performance under dual-task conditions across the lifespan. Sports Med. **45**, 1739–1758 (2015)

Use of Virtual and Augmented Reality as Tools for Visualization of Information: A Systematic Review

Alexandre Cardoso(✉), Gabriel F. Cyrino, Jose C. Viana, Mauricio J. A. Junior, Pedro A. M. T. Almeida, Edgard A. Lamounier, and Gerson F. M. Lima

Federal University of Uberlandia, Uberlandia, Brazil
{alexandre, lamounier}@ufu.br,
gabrielcyrino@hotmail.com, joscorvi@gmail.com,
drmurisystem@gmail.com, pedro.aml4ll@gmail.com,
gersonlima@ieee.org

Abstract. Visualization of Information aims to present methodologies to optimize the cognition of the agent that seeks to identify, segment and learn from information that can be presented in various forms. Based on that, this study aims to identify the availability of information through virtual environments with a focus on Virtual Reality and Augmented Reality as a support for Visualization of Information. Thus, a Systematic Literature Review (SLR) at IEEE Xplore, ScienceDirect and ACM Digital Library databases, from September 20, 2016 to November 18, 2016. Of the 174 studies surveyed, 22 met the inclusion criteria. As an analysis, this article briefly presents the contributions of each of the articles, and a discussion is made of the applicability and research opportunities that can still be made in this area.

Keywords: Systematic review · Information visualization · Virtual reality Augmented reality

1 Introduction

Society finds itself in an era full of information emerging from various different sources, and that information comes in all shapes and forms. Today, the Internet has made access to many sources of information a lot easier, and for the a wide array of purposes, such as entertainment, research, work advise, seeking psychological help and even in diagnosing diseases and evaluating symptoms, which today even counts with professional medical supervision to maintain such information available to the public in the most reliable and trustworthy fashion possible.

According to the 2016 paper *"How Big is the Internet, Really?"*, available on the website *Live Science* and based on a study published at the journal *Supercomputing Frontiers and Innovations* in 2014, it is estimated that the Internet has a data storage capacity of 10^{24} bytes, in other words: one million exabytes. However, there is great challenge when one seeks to define, handle and analyze data, and to be able to identify

W. Karwowski and T. Ahram (Eds.): IHSI 2019, AISC 903, pp. 407–417, 2019.
https://doi.org/10.1007/978-3-030-11051-2_62

techniques that simplify data interpretation, to optimize cognitive processing, is a pressing necessity.

Information Visualization (InfoVis) can have many meanings. In general terms, it is used in the field of psychology to represent events, objects, processes and procedures in a visual-spatial manner. In terms of technology, there are techniques and procedures to support segmentation, handling and presentation of information, such as graphics, virtual environments, tables and text. Thus, this paper presents studies that use technology as an attempt to contribute with and enhance cognition through the use of computer-based visual-interactive representation [1].

Virtual Reality (VR) allows user interaction and navigation in computer-maintained 3D environments, utilizing mapping channels and user behavior analysis, making information exchange between the virtual environment and the user possible, thus affecting one or many of the human senses [2]. Augmented Reality (AR) has as a goal the creation of virtual elements to complement the world, and offers, just as with VR, the stimulus of one or more of the human senses to allow user immersion through applied technology [3].

A Systematic Literature Review (SLR) is the execution of a secondary and retrospective study, with the goal of synthesizing scientific studies in a thorough manner, with the use of explicit and systemized methods of search, extraction and quantitative analysis of the results obtained [4]. This research technique was initially employed in the realm of healthcare for the identification and execution of procedures based on evidence, seeking to lay solid foundations for research concerning quality of life [5].

The use of an SLR is a helpful technique in many areas of knowledge, for instance in [6], an SLR is carried out to investigate previous works that apply gesture communication in interactive smart city environments applied to healthcare.

Having the considerations made in mind, this piece presents an SLR being performed where Virtual and/or Augmented Reality were adopted to enhance Information Visualization in various fields of study and applications between the years of 2011 and 2016, with the stated goal of making a contribution to the research community about the intents of the research carried out in this time frame.

2 Methods

An SLR used for identification, classification and evaluation of research in a replicable and thorough manner, is that which reduces bias for the identification, and even selection, of papers. Bias is something that the SLR aims to avoid, but even still seems to always make itself present in an SLR. In the section Threats to Validity, criteria that will focus on paper selection bias will be described as it concerns the papers chosen for the construction of the SLR.

For the development of the SLR, the model proposed by [6] will be used as an archetype. Figure 1 presents the tasks that, according to the authors, allow for adaptation according to the necessity and reality of your application in order to develop an SLR.

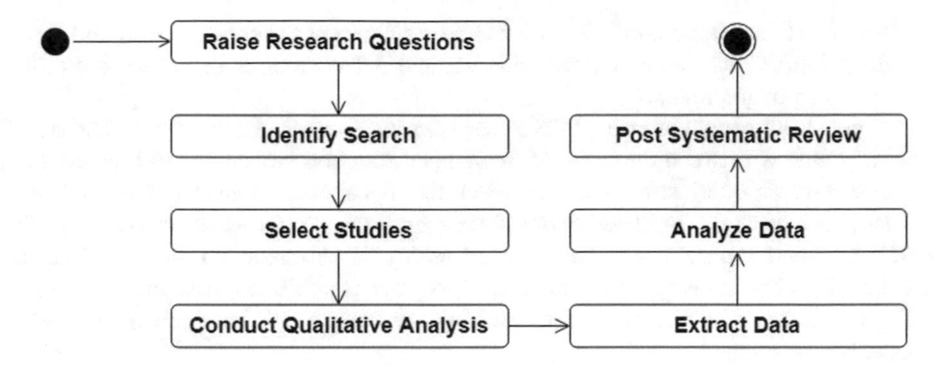

Fig. 1. Flowchart for Systematic Review [6]

The main tasks have responsibilities and subprocesses each, that guide the execution of the SLR, those being:

– Raise Research Questions: this activity aims to represent which questions must be answered after the application of systematic review. Three questions of research will be raised in consideration, those being:

    1. *Q.1: Which studies present case studies on the use of Virtual and Augmented reality for IV?* Presents the fields of study in which research applied VR and AR in InfoVis systems, regardless of the field at hand, as long as either VR or AR is integrated in aiding Information Visualization.
    2. *Q.2: What are the InfoVis evaluation techniques in AR and VR environments?* To make possible the identification of studies that present techniques used in evaluating the Visualization of Information when either VR or AR are employed.
    3. *Q.3: Which InfoVis techniques are employed in developing a VR or AR proposal?* It asks if the studies raised present or are based in techniques to justify how the environment or data are presented, furthermore, it asks if the means through which this data is made available to the user is justified.

– Identify Search: this activity is carried out through the outlining of research strategy, selection of publishing channels and documentation of the research done.
– Select Studies: this task outlines study selection criteria, which encompasses both inclusion and exclusion criteria.
– Conduct a Qualitative Analysis: in this activity the instruments of quality and quality thresholds of previously selected studies are defined.
– Extract Data: in this activity data from the selected studies is defined, extracted, and handled.
– Analyze Data: this activity is carried out by performing a descriptive and qualitative synthesis of the selected research.
– Post the Systematic Review: lastly, this activity has the goal of defining and formalizing the SLR structure for presentation and publishing.

As exclusion criterion, the abstract and introduction of papers were read and those in which InfoVis was not correlated with VR and AR were discarded at the outset for the next step of the review.

Papers were sought after in *IEEE Xplore Digital Library*, *ScienceDirect* and *ACM Digital Library*, between the dates of September 20th and November 18th of 2016.

The research query string used to search the repositories, written in the English language, was: *((“visualization of information”) OR (“information visualization”)) AND ((“virtual reality”) OR (“augmented reality”))*. A restriction was imposed on these digital libraries as to narrow the results to projects published between the years of 2011 and 2016, with the goal of identifying the most recent research in the fields previously mentioned.

3 Results

There were 174 matches to the database query. The information presented in Table 1 shows the number of papers presented and the results applied to each step of the classification process.

Table 1. Number of studies selected on each step

Digital Library	Step 1	Step 2	Step 3
IEEE Xplore	50	27	7
ScienceDirect	106	31	14
ACM Digital Library	18	12	1
Total	174	70	22

The first step was the selection of papers based on the query string which returned a total of 174 matches. In the second classification step, selection based on exclusion criteria was carried out. Lastly, the third step presents the results correlating inclusion criteria, those being the three research questions raised. In Table 2 the studies elected after the second step are presented.

Table 2. Study classification based on inclusion criteria

Research question raised papers		Quantity
Q1	[7], [8], [9], [10], [11], [12], [13], [14]	8
Q2	[15]	1
Q3	[16], [17], [18], [19], [20], [21], [22], [23], [24], [25], [26]	11
Q1, Q3	[27]	1
Q1, Q2, Q3	[28]	1
Total	–	22

For the execution of steps two and three, the papers were divided between the researchers. This decision may infer a higher bias during paper selection. This statement will be further discussed in the Threats to Validity section.

4 Discussion

The paper [7] presents, through the use of AR, how to carry out equipment assembly by presenting an accompanying user interface that facilitates error detection and correction, besides the ease of use due to the real environment being integrated with feedback-generating instructions. The authors expect to execute comfort tests on HMD after prolonged continuous use.

The work [8] executes a case study comparing physical and immersive virtual environments with the goal of planning office environments. A research with 112 participants was carried out in order to compare physical and virtual environments, and from that, it was possible to verify that the combination diversity with virtual environments was greater than with real ones. The authors concluded that this tool might help with profiling users and decision making, besides being helpful when it comes to identifying risks in civil construction.

The paper [9] presents the development of an application that makes information available in different levels of detail, and even different kinds of information, depending on the position the image was taken or on the object contained in it. This allows for different information to be made available when observing the same object at different levels of detail.

The study [10] deliberates on the topic of using Virtual Environments for visualization of complex information geared towards the field of factory automation systems. The authors compare different kinds of virtual environments (headsets, cave automatic virtual environments and personal computers) and present examples for each one of the aforementioned kinds.

The work [11] presents the use of AR in order to view and interact with machines, presenting a case study for hydraulic systems. The application proposes problem solving by analyzing the logical model and converting said model to a virtual one. From the actions carried out by the user it was possible to carry out a documentation process of the machinery through multimedia reports.

The study [12] brings forward a proposal of using Smart Glasses to visualize manoeuvre information in a port. This application makes not only information about the vessel available (such as length, width, speed and position) but also compares the distance between the port. The authors reported a good acceptance rate among users that tested the proposal.

The work [13] presents the use of AR to carry out sanitary control of vegetation that protects the river banks through a mobile monitoring application. As a future work, the authors presented the use of autonomous robots to monitor the river banks facilitating mapping and classification of risk cases.

In the paper [14] three case studies related to construction projects are presented, and special attention is given to their methods of data registry and visualization of civil construction activities. The results show that important construction information related to security and ongoing onsite operations can be automatically monitored and viewed in real time, thus offering great benefit to the workers, equipment operators or decision makers.

The paper [15] presents an evaluation methodology to explore the effects when making use of various virtual environments rich in many kinds of information, and then collect the results through a series of studies on the users. This methodology is presented as a set of design and layout guidelines to support developers of new visual interfaces. One of the references used by the authors was the Gestalt laws, applied in virtual environments. In the pursuit of efficient and effective human-computer interaction, the authors aimed to reduce the cognitive distance between user and system, increasing the information absorption rate. The conclusion was that it is possible, with this methodology, to enhance comprehension and absorption of information in users in virtual environments.

In the paper [16] an AR system for mobile devices is presented, which aims to present points of interest and their information through recognition and map positioning (such as Google Street View), reducing also confusion in information already presented. Having as main goal to encompass AR design considerations with experience in discovery and interaction with the world, through the application it is possible to view and identify a wide array of information about buildings, streets, monuments, among others, spread throughout the city. According to the authors, the interactive system presented brings some innovative elements that can facilitate discussion about the construction of AR experiences focused on better final user experience.

In the paper [17] the development of an AR prototype for the visualization of nutritional information was proposed. Two techniques were applied in the creation of said information: detection, through digitalization of the object's image without needing an AR marker, and tracking, using a smartphone camera. The data was represented in a visual fashion through a pointer gauge with colours being used as markers, and a panel that provides information about the amount of calories to contribute to the user's dietary plan. According to the authors, the application will help users in the process of regulating their diet plan, providing visual nutritional information for them to improve their health, especially for patients with diabetes that need to control the amount of calories in their bloodstream.

The paper [18] presents a CARMMI approach, which aims to integrate information derived from *CAx* tools, Mixed and Augmented Reality tools and incorporated Intelligent Maintenance Systems. It has as a goal to support technicians and operators during maintenance tasks through mixed reality, allowing for easier information access and comprehension of different systems. The information about where, when and which data will be presented on the interface are all dictated by CARMMI. The paper present three test cases which were carried out employing the proposed concepts and infrastructure. According to the authors, the concepts and tools proposed allow for not only a reduction in the time it takes for the operator to execute their tasks, but also increases the reliability of their actions, as well as improving them.

In the paper [19], AR was used alongside Building Information Modeling (BIM) in order to develop an industrial solution in the liquefied natural gas industry, to support efficient project progress guidance and control. With that it is possible, for example, to view architectural projects generated in AR with their many layers of information, and beyond that, it can also make a projection of components and their information in work environments. According to the authors, this solution could solve problems with low productivity in the recovery of information, tendencies to make mistakes in the assembly process and low communication efficiency.

In the study of [20], the techniques of Visualization of Information have as goal to represent data sets aiming to enhance comprehension, allowing for better analysis. Content-Based Image Retrieval (CBIR) has as a goal the recovery of images using characteristics extracted through algorithms representing them numerically. In order to contribute to minimize this problem, this paper has the goal of defining and implementing visualization techniques for the data generated by two characteristic extracting algorithms. The results obtained were evaluated by CBIR specialists, having been verified that the visualization contributed to validate the extractors for convex models as well as non-convex models. The paper demonstrates the results obtained with elaborate visualization methods for the Distance Histogram and ETH3D extractors in 3D models.

The paper [21] presents that temporal coherence of annotations is an important factor in AR user interfaces and for visualization of information. In this paper, four different annotation techniques were evaluated. The results show that presentation of annotations in the object space or in the image space brings to a significant difference in task performance. Beyond that, there is significant interaction between the rendering space and the annotation update frequency.

The study [22] presents an ongoing research about generic architecture of sensor fusion and their application in maritime surveillance. The importance of information fusion for various kinds of sensors and specialized sensor fusion systems are discussed in various fields of study. Traditionally, the visualization of information limited itself through 2D representations, mainly due to the prevalence of 2D exhibition and report format. However, there has been a recent growth in the popularity of HMDs that allow for 3D visualization. The ubiquity of such monitors allows for the possibility of stereoscopic and immersive visualization environments. While techniques that deploy such immersive environments have been extensively explored for scientific and spatial visualizations, very little has been explored for Visualization of Information.

The paper [23] presents the authors' very own considerations about layout, rendering and interaction methods to view graphics in an immersive environment. A user study was performed in order to evaluate their techniques in comparison to the traditional 2D graphical visualization. The results show that participants responded significantly faster with a lower number of interactions when utilized techniques proposed by the authors, especially for more difficult tasks. Despite the global correction rates not being significantly different, the authors discovered that participants had significantly better responses when using the techniques for greater graphics.

The paper [24] poses that, despite VR having an enormous success in enhancing quality of scientific visualization applications, there is a considerable gap in development of similar applications when it comes to Visualization of Information. Some researchers in Visualization of Information claim that the 2D representations are sufficient when it comes to data analysis. However, in the case of multidimensional data sets, other researchers state that the simultaneous study of multiple dimensions is advantageous.

The research carried out in [25] defends that self-localization in large environments is a vital task for visualization of information registered accurately in outdoor AR applications. In this work, a system of self-localization in mobile phones using a GPS prior and an online-generated panoramic view of the user's environment is presented. The approach is adjusted to run entirely on current generation mobile devices, such as smartphones. According to the authors, it was possible to make a 3D point reconstruction allowing for self-localization and real time registration in in large scale environments. It was concluded that the positioning estimating algorithm allowed for highly accurate results in case of a successful run. When detailing reprojection error, the traditional precision indicator was omitted, seeing as it was considered to be an inadequate measurement in this case.

The authors of [26] informs that the use of Head-Up Displays (HUD) in automobiles to visualize multiple kinds of navigational information on the windshield has risen rapidly over the past years. This paper introduces a new generation of AR automotive systems to support the driver during nighttime, detecting 3D positions of potential collision partners, as well as performing eye tracking on the driver in order to exhibit warnings and information in the exact position of the full-windshield (FWS) the driver is looking at. With the goal of viewing graphics on the windshield, the concept of indirect projection, due to its integration convenience, was applied. The concept of emissive indirect graphical projection for the windshield using laser projectors was employed. This method allows for low-cost windshield projection, without needing to make significant modifications in the vehicle. Finally, it is important to note that the graphics emit upwards and not in the direction of oncoming traffic.

The study [27] presents an analysis of different learning pathways within a group of 227 children aged 7 through 10. The learning profile of these children was constructed through the use of statistical mapping to identify which variables had more weight in the learning process of these groups. Considering factors such as information necessity, media interface and their affection state, the authors verified that: 1) the children favored interfaces for entertainment over problem solving; 2) The text-sound interface showed itself to be more user friendly, and finally 3) the children that searched for information as a way of entertainment were more susceptible to dealing with uncertainty than those that sought information for assigned tasks when problem solving.

The paper [28] presents the use of AR to develop a real time immersive analysis tool. The authors present a simulation in which the user can go shopping in a mall an in which it is possible, for instance, to view the properties of a product, compare products and even to identify a specific product. Furthermore, the authors also present a new form of Visual Analysis based on the proposed system.

5 Threats to Validity

The main goal of the SLR is the possibility of doing research in certain fields of study for the identification of evidence around a specific defined theme and, furthermore, to have that be carried out in a thorough and systemic fashion, in order to mitigate existing bias when selecting a base of research for analysis.

However, there are some items that, for the making of this paper, should be taken into account to justify, that even when applying the SLR, some activities in its execution may contain bias. The first one being in relation to the publishing channels. Three channels were selected to choose papers from, if a larger group of publishing means were to be selected, it would help to reduce study bias.

Another criterion and likely the most important are the criteria used for paper selection. Seeing as there was division of labor among researchers, the selection for all three steps was subjected each to the outlook of the individual researcher. An alternative to mitigate said problem is to rotate visualization between selected projects through the selection criteria, seeing as all works would go through an evaluation from all researchers, thus it would be possible for weighting in the selection of works to be carried out.

This section makes itself relevant in order to emphasize some tasks carried out during the execution of this paper that could be improved, for this very research as well as to readers that come to read it and desire to carry out an SLR.

6 Conclusions

This paper presents the result of carrying out an SLR related to the use of VR and AR in IV, contemplating case studies, applications of InfoVis concepts or even criteria to make a qualitative evaluation of the manner in which information is presented to the end user in many business domains.

Based on the studies that were selected for step three, it was possible to observe that the preoccupation in presenting information, focusing on enhancing cognition, is common ground within the analyzed works and that it is an existing concern in many different fields of study and across various age groups. It was possible to observe also that few works use InfoVis evaluation techniques before the development of applications or environments supporting user information availability.

As for future works, a tertiary study (or Systematic Review of Systematic Reviews) in the same field of study as the one presented in this paper is intended to be carried out aiming to comprehend and seek to identify new research opportunities in this promising looking field, especially concerning evaluation techniques for InfoVis methods.

References

1. Keller, T., Tergan, S.O.: Knowledge and information visualization (2005). https://doi.org/10.1007/s13246-010-0014-8
2. Burdea, G.C., Coiffet, P.: Virtual reality technology. Presence Teleoperators Virtual Environ. (2003). https://doi.org/10.1162/105474603322955950
3. Malbos, A.N., Rochadel, W., De Lima, J.P., Da Silva, J.B.: Aplicação da realidade aumentada para simulação de experimentos físicos em dispositivos móveis. In: Proceedings of 2014 11th International Conference on Remote Engineering and Virtual Instrumentation, REV 2014 (2014). https://doi.org/10.1109/REV.2014.6784263

4. Higgins, J.P.T., Green, S., (eds.): Cochrane Handbook for Systematic Reviews of Interventions Version 5.1.0 (2011). https://doi.org/10.1088/0004-637X/699/2/L76
5. Sampaio, R., Mancini, M.: Estudos de revisão sistemática: um guia para síntese criteriosa da evidência científica. Revista Brasileira de Fisioterapia (2007). https://doi.org/10.1590/S1413-35552007000100013
6. Fontana, E., Biduski, D., Marchi, A.C.B.D., Rieder, R.: Smart environments using gesture-based interactions for health: a systematic review. In: 2015 XVII Symposium on Virtual and Augmented Reality (2015). https://doi.org/10.1109/SVR.2015.32
7. Mura, M.D., Dini, G., Failli, F.: An integrated environment based on augmented reality and sensing device for manual assembly workstations. Proc. CIRP **41**, 340–345 (2016). https://doi.org/10.1016/j.procir.2015.12.128
8. Heydarian, A., Carneiro, J.P., Gerber, D., Becerik-Gerber, B., Hayes, T., Wood, W.: Immersive virtual environments versus physical built environments: a benchmarking study for building design and user-built environment explorations. Autom. Constr. **54**, 116–126 (2015). https://doi.org/10.1016/j.autcon.2015.03.020
9. Kim, M., Lee, J.Y.: Interactive lens through smartphones for supporting level-of-detailed views in a public display. J. Comput. Des. Eng. **2**, 73–78 (2015). https://doi.org/10.1016/j.jcde.2014.12.001
10. Ulewicz, S., Pantförder, D., Vogel-Heuser, B.: Interdisciplinary communication and comprehension in factory automation engineering - a concept for an immersive virtual environment. IFAC-PapersOnLine (2016). https://doi.org/10.1016/j.ifacol.2016.10.529
11. Neges, M., Wolf, M., Abramovici, M.: Secure access augmented reality solution for mobile maintenance support utilizing condition-oriented work instructions. Proc. CIRP **38**, 58–62 (2015). https://doi.org/10.1016/j.procir.2015.08.036
12. Ostendorp, M.C., Lenk, J.C., Lüdtke, A.: Smart glasses to support maritime pilots in harbor maneuvers. Proc. Manuf. **3**, 2840–2847 (2015). https://doi.org/10.1016/j.promfg.2015.07.775
13. Pierdicca, R., et al.: Smart maintenance of riverbanks using a standard data layer and augmented reality. Comput. Geosci. **95**, 67–74 (2016). https://doi.org/10.1016/j.cageo.2016.06.018
14. Cheng, T., Teizer, J.: Real-time resource location data collection and visualization technology for construction safety and activity monitoring applications. Autom. Constr. **34**, 3–15 (2013). https://doi.org/10.1016/j.autcon.2012.10.017
15. Polys, N.F., Bowman, D.A., North, C.: The role of depth and gestalt cues in information-rich virtual environments. Int. J. Hum. Comput. Stud. **69**, 30–51 (2011). https://doi.org/10.1016/j.ijhcs.2010.05.007
16. Fedosov, A., Misslinger, S.: Location based experience design for mobile augmented reality. In: Proceedings of the 2014 ACM SIGCHI symposium on Engineering interactive computing systems - EICS 2014 (2014). https://doi.org/10.1145/2607023.2611449
17. Bayu, M.Z., Arshad, H., Ali, N.M.: Nutritional information visualization using mobile augmented reality technology. Proc. Technol. **11**, 396–402 (2013). https://doi.org/10.1016/j.protcy.2013.12.208
18. Espíndola, D.B., Fumagalli, L., Garetti, M., Pereira, C.E., Botelho, S.S., Ventura Henriques, R.: A model-based approach for data integration to improve maintenance management by mixed reality. Comput. Ind. **64**, 376–391 (2013). https://doi.org/10.1016/j.compind.2013.01.002
19. Wang, X., Truijens, M., Hou, L., Wang, Y., Zhou, Y.: Integrating augmented reality with building information modeling: onsite construction process controlling for liquefied natural gas industry. Autom. Constr. **40**, 96–105 (2014). https://doi.org/10.1016/j.autcon.2013.12.003

20. Bergamasco, L.C.C., Campos, H.B., Nunes, F.L.: Interactive Visualization of Three-Dimensional Descriptors Using Virtual Reality. In: 2015 XVII Symposium on Virtual and Augmented Reality (2015). https://doi.org/10.1109/SVR.2015.40
21. Madsen, J.B., Tatzqern, M., Madsen, C.B., Schmalstieg, D., Kalkofen, D.: Temporal Coherence Strategies for Augmented Reality Labeling. IEEE Transactions on Visualization and Computer Graphics (2016). https://doi.org/10.1109/TVCG.2016.2518318
22. Gunasekara, C., et al.: Sensor information fusion architecture for virtual maritime environment. In: Proceedings of International Conference on Advances in ICT for Emerging Regions, ICTer 2012 (2012). https://doi.org/10.1109/ICTer.2012.6422832
23. Kwon, O.H., Muelder, C., Lee, K., Ma, K.L.: A study of layout, rendering, and interaction methods for immersive graph visualization. IEEE Trans. Vis. Comput. Graph. **22**, 1802–1815 (2016). https://doi.org/10.1109/TVCG.2016.2520921
24. Garcıa-Hernandez, R.J., Anthes, C., Wiedemann, M., Kranzlmuller, D.: Perspectives for using virtual reality to extend visual data mining in information visualization. In: IEEE Aerospace Conference Proceedings (2016). https://doi.org/10.1109/AERO.2016.7500608
25. Arth, C., Klopschitz, M., Reitmayr, G., Schmalstieg, D.: Real-time self-localization from panoramic images on mobile devices. In: 2011 10th IEEE International Symposium on Mixed and Augmented Reality, ISMAR 2011 (2011). https://doi.org/10.1109/ISMAR.2011.6092368
26. Hosseini, A., Bacara, D., Lienkamp, M.: A system design for automotive augmented reality using stereo night vision. In: Proceedings of IEEE Intelligent Vehicles Symposium (2014). https://doi.org/10.1109/IVS.2014.6856484
27. Wu, K.C.: Affective surfing in the visualized interface of a digital library for children. Inf. Process. Manag. **51**, 373–390 (2015). https://doi.org/10.1016/j.ipm.2015.02.005
28. ElSayed, N.A., Thomas, B.H., Marriott, K., Piantadosi, J., Smith, R.T.: Situated analytics: demonstrating immersive analytical tools with augmented reality. J. Vis. Lang. Comput. **36**, 13–23 (2016). https://doi.org/10.1016/j.jvlc.2016.07.006

Visual System Examination Using Synthetic Scenarios

Robert Manthey[1(✉)], Rico Thomanek[2], Christian Roschke[2], Tony Rolletschke[2], Benny Platte[2], Marc Ritter[3], and Danny Kowerko[1]

[1] Junior Professorship Media Computing, Faculty of Computer Science, Technical University of Chemnitz, 09107 Chemnitz, Germany
{robert.manthey, danny.kowerko}@informatik.tu-chemnitz.de
[2] Faculty Media Sciences, University of Applied Sciences, Technikumplatz 17, 09648 Mittweida, Germany
{rthomane, roschke, rolletsc, platte}@hs-mittweida.de
[3] Faculty Applied Computer Sciences and Biosciences, University of Applied Sciences, Technikumplatz 17, 09648 Mittweida, Germany
ritter@hs-mittweida.de

Abstract. Many systems use visual devices to detect, inspect and analyze persons, scenes, and properties of objects. Often, they use samples to learn relevant indicators to reach a high level of quality of the appropriated operation. Nevertheless, collecting samples and annotate the relevant parts may be a hard, expensive and error prone task in same fields of use. To overcome this problem we create a system to generate synthetic scenarios based on predefined and exact definitions of the content as well as the sample production process. To demonstrate the usability we apply a scenario with a humanoid with known activity and with various environment objects to different systems for visual detection and analysis.

Keywords: System evaluation · System inspection · Dataset generation Human detection · Human activity recognition · Usability testing

1 Introduction

In the modern world cars detect pedestrians with small cameras, check the condition of the driver and follow the lane of the highway[1], quality inspection systems search for faulty products and mobile phones identify the user by his face[2]. Robots assist humans at work [1] or guide them through museums [2].

To realize this behavior, huge amounts of data showing the expected content is required as well as the enhancement with annotations. The volume, diversity and quality of them affect the correctness and conductivity of the systems, and are required during creation as well as for verification of the functionality [3]. Acquire them from

[1] https://www.engadget.com/2016/11/19/tesla-self-driving-demo-shows-car-view/
[2] https://www.apple.com/iphone/#face-id

W. Karwowski and T. Ahram (Eds.): IHSI 2019, AISC 903, pp. 418–422, 2019.
https://doi.org/10.1007/978-3-030-11051-2_63

real world is a time consuming, error-prone and expensive task which need human work for annotation.

However, synthesis the data as result of well-known definitions removes the need of annotations as well as undesired influences of the real world environment. In addition, the creation of scenarios showing dangerous situations or a huge amount of different facets of the same scene may become feasible [4].

2 System Architecture and Technical Realization

To realize a system being able to synthesis the desired scenarios we use the open-source 3D modeling software Blender[3] which supply the possibility to create arbitrary virtual objects and virtual environments, as well as compositions. With humanoid modeling tools, arbitrary persons can be created to represent humans. Captured or self-defined activity data provide well-known actions the humanoid may execute. Combined in mostly any desired combination, arbitrary scenarios can be generated by Blender, as shown in Fig. 1, with already known data as ground truth.

The examined visual system process the generated data, create their results and allow a comparison with the ground truth.

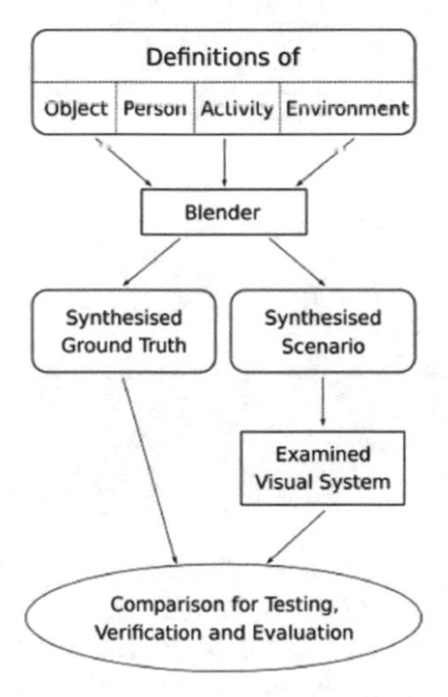

Fig. 1. Schema of the workflow to generate synthetic scenes and to analyze existing visual detection systems. The abstract definitions of objects, persons, activities and the environment representing the scene, is used by Blender to combine the components and to render visual data for a input of the examined system as well as ground truth. A later comparison of both set shows the properties of this system.

[3] https://www.blender.org

3 Application and Results

We create a virtual environment comparable to the setup of our laboratory with a scaffolding holding our sensors and effectors like cameras, microphones and audio boxes. Some further objects were modeled as templates to fill the environment like bikes, chairs. With the open-source modeling software MakeHuman[4] the creation of humanoids, each with 163 bones, were realized and allow the definition of their properties like gender, age, body structure and clothing. Recorded and custom-build activities were prepared combined together by Blender to form scenarios as shown in Figs. 2 and 3.

For each defined camera, the scenario is rendered with $1920 \times 1080@25$ and the generated video is processed by the desired visual detection system. The results are explored and conspicuous are shown in Figs. 2 and 3.

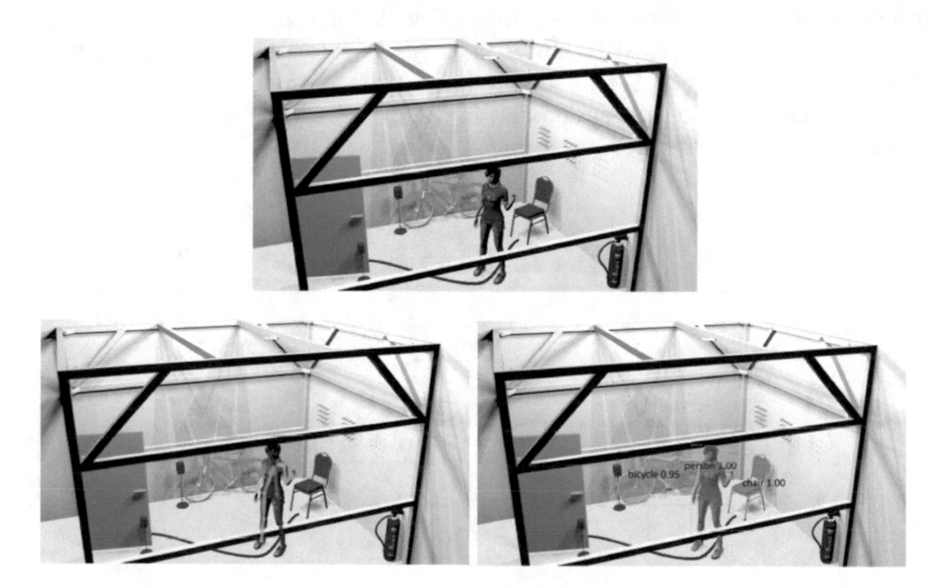

Fig. 2. (*Top middle*) show a synthetic scene modeling a setup of the stage in our laboratory and some sample objects like a door, sound boxes, bike, chair, fire extinguisher and three microphone arrays as well as a female person in a wave pose. The red line on the ground represents a visualization of the path of the humanoid through the stage. (*Down Left*) showing the detection result of Openpose (https://github.com/CMU-Perceptual-Computing-Lab/openpose) as color-coded skeleton from the humanoid at correct location. (*Down Right*) showing the Detectron (https://github.com/facebookresearch/Detectron) results with correct classification of the bike, humanoid and chair with high confidence and good shape matching.

The results from Fig. 2 show that the proposed workflow is applicable and the visual detection system produce expected results. A more detailed exploration with

[4] https://www.makehuman.org

multiple camera positions and perspectives, shown in Fig. 3, confirm this, but also show false detections as well as missing detections as real world data would reveal. Due to the absence of noise and similar effects, the confidences are higher and fewer detection problems occur. Therefore, examinations with easier to produce and favorable synthetic data represent a step detection systems should pass before working with real world data.

Fig. 3. Multiple cameras at various locations and perspectives capture the scene of Fig. 2. (*Top*) show the ability of Openpose to detect the person, but with a false detection in the upper left. (*Down*) show the results of Detectron with the missing of the bike down left.

4 Summary and Future Work

The inspection of visual systems requires well-known data to detect problems and to verify known constraints. To produce this data we build a system being able to synthesis them based on exact definitions of objects, persons, motions etc. We process a sample by different visual detection systems showing the usefulness of our solution. Further developments will focus on automatic variations of surface textures and motions to simplify further scene generations.

Acknowledgements. This work was partially accomplished within the project localizeIT (funding code 03IPT608X) funded by the Federal Ministry of Education and Research (BMBF, Germany) in the program of Entrepreneurial Regions InnoProfile-Transfer.

References

1. Hägele, M., Schaaf, W., Helms, E.: Robot assistants at manual workplaces: effective cooperation and safety aspects. In: Proceedings of the 33rd International Symposium on Robotics, pp. 1–6 (2002)
2. Germa, T., Lerasle, F., Danes, P., Brethes, L.: Human/robot visual interaction for a tour-guide robot. In: Proceedings of IEEE/RSJ International Conference on Intelligent Robots and Systems, pp. 3448–3453. IEEE Press (2007)
3. Bilenko, M., Kamath, B., Mooney, R.J.: Adaptive blocking: learning to scale up record linkage. In: Universal Access in Human-Computer Interaction, pp. 460–467. Springer, Heidelberg (2016)
4. Manthey, R., Conrad, C., Ritter, M.: A framework for generation of testsets for recent multimedia workflows. In: Proceedings of the 6th IEEE International Conference on Data Mining, pp. 87–96. IEEE Press (2006)

Ubiquitous and Context-Aware Computing Modelling: Study of Devices Integration in Their Environment

Francisco Gregório(✉) and Vítor Santos

NOVA Information Management School, Campus de Campolide, 1070-312 Lisbon, Portugal
{m2016531, vsantos}@novaims.unl.pt

Abstract. In an almost imperceptible way, ubiquitous and context-aware computing make part of our everyday lives, as the world has developed in an interconnected way between humans and technological devices. This interconnectedness raises the need to integrate humans' interaction with the different devices they use in different social contexts and environments. In the proposed research, it is suggested the development of new scenario building based on a current ubiquitous computing model dedicated to the environment context-awareness. We will also follow previous research made on the formal structure computation model based on social paradigm theory, dedicated to embed devices into different context environments with social roles. Furthermore, several socially relevant context scenarios are to be identified and studied. Once identified, we gather and document the requirements that devices should have, according to the model, in order to achieve a correct integration in their contextual environment.

Keywords: Context-aware computing · Ubiquitous computing
Pervasive computing · Social paradigms · Organization theory

1 Introduction

At the present time, technological advances in computation, sensing, storage, and communications are changing the current mobile devices' state and transforming it from near-ubiquitous devices into a global sensing device that is aware of the environment, as studied by [1]. Visions of context-awareness are a reality in everyday products, with devices being equipped with sophisticated capabilities such as sensing, motion state, location and environmental state capture. However, this has not been overpassed from a single user level to a multiple and large-scale networked system within several users' communities [2].

Therefore, the emerged concept of a device that is aware of its environment starts to play an interesting role on how technology can aid human beings to perform their activities of daily life, thus benefiting their social and technological relationships. As in research by [3, 4], it was developed Java based frameworks and application based interfaces that represented context-awareness in clinical/hospital settings, and [5] developed the AWARE architecture to support context awareness and management

W. Karwowski and T. Ahram (Eds.): IHSI 2019, AISC 903, pp. 423–428, 2019.
https://doi.org/10.1007/978-3-030-11051-2_64

regarding a user's working context. Therefore, this signals the applicability of context-aware and ubiquitous computing, being that already exist frameworks to support posterior developments onto the area of ubiquitous and context-aware technology. Moreover, a technique was developed by [6] that enabled automatic situation recognition and its performance evaluation in real user situational tests and perceived contextual information, distinguishing itself by its ontology-based nature.

Besides the development of the above mentioned technological applicability, emerged a newly conceptualization of the importance that social paradigms have in the connection between human interaction with technology. By including a device to do specific tasks' execution in a determined environment, it becomes viable to conceptualize new types of self-conscious applications of their role in a specific environment.

Henceforth, [7] provide a detailed analysis on the applicability of the social paradigms theory and underlying concepts for Smart Cities accessibility improvement, thus being a very relevant contribution to a challenge faced at the present time.

Additionally, research made by [7–9] references ontology and social paradigms' application to systems' design. The used ontology and social paradigms concepts combine social reasoning, ontology models, and the organization theory notions in a context-aware behavior of mobile devices in a computing system with a cooperative structure having several responsibilities for a device. There exists a need for advancements in research on how to develop a device capable to adapt to the different environment and social context it is at a certain moment. Hence, this need can both be in terms of device's requirements and specifications but also in terms of the different purposes and usability that it can infer from different contexts, with the latter requiring social context interpretation, and where social paradigms theory might be of extreme importance. Henceforth, our research will directly impact their work and provide specific insights into how the concepts used in the research can be applied in real-life situations.

This research focuses on ubiquitous computing and context-aware computing with the center on social paradigms theory, and with the aim to study the current models of the subject matter and the advancements made so far.

It is also proposed to conceptualize new applicable scenarios for a relevant model in ubiquitous and context-aware computing, using social paradigms theory. The scenarios identified are intended to be realistic and relevant to society.

Moreover, it is also proposed to study, discover and recommend the requirements a device should have in these different scenarios so that it is apt for embedding itself on the environment with other devices, considering that all can collaborate among each other.

2 Scenario Building Based on Social Paradigms

Under the model revised by [8], it is addressed the need to supply the device with information about the different roles, being proposed a web interface that does this through a database with the full context formal structure and the device registration skills, being organized as web services. This way, it is possible to convey the device competence and skills, receiving the requirements needed to execute a task.

Considering this, and once a device owns a specific role, it is admitted that the device implements itself all the functions required for the correct performance of the role(s).

Scenario Identification. Below we present the social scenarios identified and built, constructed upon the Social Paradigms ontology and model developed by [8]. The reasoning behind these conceptualized model and ontology lies in the need for a device to acquire the specific context it is at. Hence, this can be achieved either through the usage of diverse technology available for this purpose or to include the option on a user's device to selectively manage and identify which is the current environment and context. After this step, the device requests to download the ontology related to that context and to the user's device role in it. As researched by [8], this can be achieved through the ontologies' upload to a system database (DB) and user interface managed by an administrator. Then, the device can connect to this specific DB, and the context selection process can start, with the context data packages made available to public download, being then possible to download to all users' systems and, in a next step, actuate per the selected context and its specific role in the ontology.

The chosen scenarios identified were 1. School and 2. Gymnasium.

Scenario Building and Device Technological Requirements' Specification. To build the presented below scenarios it was followed an organizational structure as seen in the originally developed model [9].

For the school scenario, the technology that is proposed to be used is the AWARE middleware, P2P, and WiMax.

The AWARE middleware is proposed as it enables devices in a network to possess a set of capabilities that allow them to divulge the contextual environment they are at. This is achieved through personal settings and communications' definitions, allowing for each of the device's data to be shared with the applications that can make use of it, i.e.: geo-location. Thus, by having an application that can load it, the device can be totally monitored and served with useable information regarding the context it informs it is at.

Once a device is setup to the different environmental contexts where it can be present, it is possible for it to download the set of data packages loaded with the ontologies related to those specific contexts. Once downloaded, it is enabled with the capabilities to assume the different socio-contextual roles of the environment and participate actively in it, interacting both with the other context's system devices and user controlled devices.

Additionally, having P2P technology implemented, and being aware of its roles in the environment, a users' device is empowered with the ability to connect to the other peer nodes in the network to, cooperatively or not, access or allow access to specific sets of data. It is then possible to make the request of access to specific data to be shared, and to make it available in the network, constituting a fundamental requirement in the automation of information sharing in a setting where users are required to be given certain types of information or files related to the type of environment they are at. This is a situation that could be already made available in a theoretical school scenario. Once in operation using AWARE middleware, the network should be enabled with a high range span signal, so that every user could be reached and connected. For this,

WiMax seems to be the right technology to use, due to its high range actuation. One requirement would be the maintenance of transmitters to accommodate a capacity of 500 users per transmitter.

Thus, with this technology, users are able to be connected through the network and communicate easily between each other, exchanging data in the process.

For the Gymansium scenario, it is proposed to be used the BLE, RFID, and WINDware middleware.

The BLE technology usage is proposed considering the existence of a WiFi network. This way, it can take advantage of the supposed existent WiFi network already spread around a building, enabling a new layer of connectivity between all the network environment of devices that can be connected on it. This way, devices can communicate with the master node, transmitting information and receiving instructions and contextual data. Thus, it enables the connection between all the network devices, providing a powerful method for interaction between users and a user-device relationship. Moreover, BLE has the capacity to make a device detect the environment it is at, through geo-fencing technology, making it a powerful context identifier if the user's device already has predefined data of geographical context information of the specific areas it wants to be connected in. Hence, and due to its decreased power consumption, it is a viable solution for scenarios where communication is made through wireless networks and confined to the inside of a single building, such as a Gymnasium. This way, the user's device can download the specific identified context, and the related social ontology, providing afterwards to its users the option to select their social role in that specific context. From that moment on, the user is employing its role(s) in the current system, interacting with the other devices available in the network, being this either another person's device or also the specific several different devices available in the gym for its usage.

Moreover, in situations where one intends to grant access to people, by granting access through the validation of their devices, the RFID technology is of easy and adequate usage. Considering the developed middleware of WINDware, one can further make use not only of the access granting in one single location but also in the whole grid where people bring their devices to. The requirement is the implementation of RFID sensors across the locations requiring access granting, and the possession of the WINDare middleware from both the receiver and transmitter, which, on the receiver side, can be achieved by the download of a simple data package with the middleware code.

This way, it is achievable a perfect integration of the users' devices in the environment it is at.

3 Results and Discussion

After a scenario simulation, it was possible to demonstrate that the developed model is adequate to the GYM and SCHOOL real-life scenarios. The main reasons to the possibility of such are the fact that currently existent technology make it possible to create a viable technological infra-structure and environment by the application of non-intrusive hardware in combination with diverse mobile and non-mobile devices. It was also proved, by the use of the previously modelled social-paradigms theory, that devices can have the capacity to detect the current environments they are at, and to interact, within specific contexts, with each other. Furthermore, the social paradigms theory construct makes the communication between devices with roles a reality, and enables multiple interactions between diverse context actors with the supposed roles and tasks each need to attain and provide to each other. A device can thus be part of a context, having several roles that entitle or enforce it to conduct certain tasks within a ruled contract.

The researched and presented technology is on par with the most recent advancements in the field of Information and Communication systems, being of high knowledge, reputation and actual applicability in a vast range of existent infrastructures. The aim was to try to discover what were the theoretical possibilities of implementation and to gather the set we found to be the most appropriate for the desired model application. Thus, all the specified and selected technology requirements were chosen having this in consideration and the current public usage.

Henceforth, it is easy to admit that the model would also be utile and applicable in other contexts/scenarios, applying the same technological constructs and proposed options, with the known need of adjustments in terms of the specific and different contexts' requirements.

4 Conclusion

Within the research, it was proposed and achieved the study, analysis and gathering of aspects and characteristics in existent models, following with a conceptualization of new applicable scenarios for a relevant model in ubiquitous and context-aware computing, evaluating its functioning using social paradigms theory. Social paradigms theory comprehends that devices and users are aware of their social role within an environment, being the owner or the executor of a specific task or any other social role applicable. The main challenges include the knowing and definition of the information structure, social roles possible for devices and methods to include in each device that could enable it to integrate and interact in social environments.

As a strong advancement had already been made recently by [7–9], this was the one model we decided to provide a further study and contribution, but also due to the fact that it expresses in a new socio-contextual form the dynamic integration of a device into a socially enabled computing system, where devices exist to aid in.

With our research, we intended to directly impact their ongoing work, which includes the supply of information to the device about the different existent roles, being proposed a web interface through a database connection that has the full context formal structure and device registration skills, organized as web services, and provide insight and support on specific implementation and how the concepts used in their research could potentially be applied to real-life situations.

5 Future Work

The model used for our research, in our opinion, and as already expressed in the past by Kamberov [8], is sturdy enough to be executed but should be revisited once the website and database structure are put in practice, since some IT engineering layers might need to be constructed to make the model work seamlessly in a IOT context.

As exposed in the limitations area, it would be relevant if a general logical framework could be used to adapt to every possible scenario, enabling the model to be dynamic to every context. However, we think it is of great feasibility and application through web-based services.

It was addressed communication between devices and support from one device to another, but we think it is relevant to study collaboration between multiple devices in a subsequent research.

It would be of interest, as well, to start implementing the model in real-life, since it seems to be of applicability to the introductory and validated scenarios. This requires technology use, adaptation and development, and in a subsequent phase, would allow for beta-testing and the public/private use of it in a real-life setting.

References

1. Campbell, A.T., et al.: The rise of people-centric sensing. IEEE Internet Comput. **12**(4), 12–21 (2008)
2. Lukowicz, P., Pentland, A., Ferscha, A.: From context awareness to socially aware computing, pp. 32–39. IEEE CS Press (2012)
3. Bardram, J.: Applications of context-aware computing in hospital work: examples and design principles. In: Proceedings of the 2004 ACM Symposium on Applied Computing (2004). http://dl.acm.org/citation.cfm?id=968215
4. Bardram, J.: The Java context awareness framework (JCAF)-a service infrastructure and programming framework for context-aware applications. In: Pervasive Computing, pp. 98–115, April 2005
5. Bardram, J., Hansen, T.R.: Context-based workplace awareness. In: Computer Supported Cooperative Work (CSCW), vol. 19, pp. 105–138 (2010)
6. Attard, J., Scerri, S., Riviera, I., Handschuh, S.: Ontology-based situation recognition for context-aware systems. In: Proceedings of the 9th International Conference on Semantic Systems – I. SEMANTICS 2013, p. 113 (2013)
7. Santos, V., Santos, C., Cardoso, T.: Use of sociology concepts as the basis for a model for improving accessibility in smart cities. In: Procedia Computer Science (DSAI), vol. 67, pp. 409–418 (2015)
8. Kamberov, R., Granell, C., Santos, V.: Sociology paradigms for dynamic integration of devices into a context-aware system. J. Inf. Syst. Eng. Manag. **2**(1), 2 (2017)
9. Santos, V.: Use of social paradigms in mobile context-aware computing. Procedia Technol. **9** (2012), 100–113 (2012)

Dreaming Mechanism for Training Bio-Inspired Driving Agents

Alice Plebe[1], Gastone Pietro Rosati Papini[2], Riccardo Donà[2(✉)], and Mauro Da Lio[2]

[1] Department of Information Engineering and Computer Science, University of Trento, Trento, Italy
alice.plebe@unitn.it

[2] Department of Industrial Engineering, University of Trento, Trento, Italy
{gastone.rosatipapini,riccardo.dona, mauro.dalio}@unitn.it

Abstract. This paper addresses one of the key ideas embraced by the European funded H2020 research project Dreams4Cars: to borrow resemblances from the way humans learn to drive. We developed an artificial driving agent, called "Co-driver", characterized by an architecture mimicking the fundamental components of the human brain involved in the learning of complex sensorimotor abilities, like driving. We implemented a dream-like mechanism to train and test the self-driving agent, and we will show two case studies demonstrating the effectiveness of such approach.

Keywords: Human systems integration · Artificial neural networks · Autonomous driving · Dreaming mechanism

1 Introduction

The ability to drive a car is certainly outside our natural endowments, what human beings have instead is an extraordinary capability to learn highly specialized sensorimotor behaviors. Still, a complete picture of how the human mind comes to learn such complex behaviors is missing.

Our project relies on a set of theories that currently appears as the most promising in explaining such advanced sensorimotor learning. The core theory is proposed in [1], according to which cognitive activity is performed, at least in some cases, as a simulated interaction with the environment. An exemplification of this principle is *perceptual imagery*, the mechanism by which a simulation of what is acquired during the initial phases of perception is present, but the sensory stimulus is not actually being perceived. Perceptual imagery has a strong connection with the phenomenon of *dreaming* from which the acronym of the project derives.

There are several evidences leading to claim that, during the first childhood, dreaming is fundamental to build up simulative skills, for the purpose of learning new sensorimotor control schemes. Under this perspective, Dreams4Cars aims to develop a dream-based environment to train and test the self-driving agent we propose, called "Co-driver".

W. Karwowski and T. Ahram (Eds.): IHSI 2019, AISC 903, pp. 429–434, 2019.
https://doi.org/10.1007/978-3-030-11051-2_65

1.1 Introducing the Co-driver

The Co-driver has a brain-inspired architecture with three main components: dorsal stream, basal ganglia and cerebellum. Such architecture takes inspiration from one of the most promising interpretation of the human brain functionality, proposed by Cisek and Kalaska [2]. The driving agent receives sensory inputs and computes via optimal control theory a space of possible actions, in terms of longitudinal jerk and lateral steering rate. This process is tantamount to the function of the *dorsal stream* activating/inhibiting the biological *motor cortex*. Then, a *basal ganglia*-based action selection mechanism chooses the optimal motor output among the feasible actions encoded in the motor cortex. Lastly, the cerebellum carries out the dreaming mechanism, providing simulated perceptions given efferent copies of motor commands.

1.2 Paper Contribution

This work focuses on the integration of the dreaming mechanism with the Co-driver and shows two case studies demonstrating the effectiveness of such approach. The imaginary process is realized by means of autoencoder neural network, whose generative capabilities represent one of the closest surrogates of the biological dreaming mechanism. We implemented a first autoencoder extracting the salient features of real road profiles. The decoding part of the network is then exploited to "dream" novel realistic roads. Similarly, a second autoencoder generates imaginary braking maneuvers of other preceding vehicles.

We used such generated data to effectively train the driving skill of the Co-driver, learning human-like steering actions and safe vehicle following maneuvers, in a way resembling the sensorimotor learning process occurring in infants.

2 How Humans Learn to Drive

To be sure, nobody knows how a person becomes a car driver. A justified knowledge of the changes in the brain of a person, when acquiring driving skills, is not yet available.

This project does not claim to reproduce the same mechanisms used by the human brain for driving. Our aim is, instead, to find guidance in a number of general principles, discovered so far, for sensorimotor learning in the brain [3, 4].

The ability to drive is just one among the many highly specialized sensorimotor behaviors performed by humans. The key common aspect is that the sequencing and coordination of all these behaviors is not programmed in advance in the brain, its impressive power is in the ability of *learning* by experience and training.

Learning comprises a number of interacting tasks, including selection of perceptual information relevant to an action, selection of motor sequences performing an action, and the coordination of predictive and reactive processes. In this section, we will pinpoint a key mechanism found in theories of sensorimotor learning that will be exploited in the project, and then describe the available computational architectures able to approximate this mechanism. A more comprehensive description of the neurocognitive theories of sensorimotor control grounding this project can be found in [5].

2.1 Simulation and Dreaming

The account of sensorimotor learning embraced by the Dreams4Cars project is the so-called *simulation theory of cognition,* more precisely in the account reviewed by [1]. Hesslow argues that thinking in general is explicated by simulating perceptions and actions involved in the thought, without the need of actually executing the actions, or perceiving online what is imagined. Simulation, conceived as a general principle of cognition, is explicated, according to Hesslow, in at least three different components: actions, perception, and anticipation. Dreaming found an interpretation in this theory as an extreme form of simulation at perceptual and action levels. In fact, dreaming shares features with the routine phenomenon of mental imagery. It happens when a representation of the type created during the initial phases of perception is present, but the stimulus is not actually being perceived; such representations preserve the perceptible properties of the stimulus [6]. For a long period, it was debated whether mental imagery, in the case of vision, do actually involve the same cortical areas engaged in direct perception, of higher cortical areas only. The former hypothesis proved correct [7], and today the same has been assessed for dreaming. It is even possible to infer which category of objects have been visualized in dream, by analyzing fMRI voxels in visual areas [8]. While it is easy to experiment mental imagery when deliberately thinking on a specific object, its more pervasive role is in the continuous generation of predictions, based upon past experience. Martinez-Conde [9] has forcefully shown how predictive imagery is at the basis of several allegedly "magic" tricks. For example, the trick of a tossed coin disappearing through the air, is a consequence of visual areas predicting – and so visualizing – the trajectory of the flying coin based on the hand movements of the magician, who in fact surreptitiously holds the coin in his palm and stops it from flying. It has been argued [10] that dreaming too has a role in refining predictive mechanisms. In infants through to early childhood, dreaming is a safe way to trigger and exercise the capacity of simulating.

3 Imagery and Autoencoders

For primary sensorial areas to be activated both by online stimuli or by perceptual imagery, there should be a bidirectional stream of processes from lower to higher areas, and the other way around. A prominent proposal for this sort of mechanism is formulated in term of convergence-divergence zones (CDZs) [11]. CDZs receive convergent projections from the early sensorimotor sites and send back divergent projections to the same sites. This arrangement has the first purpose to record the combinatorial organization of the knowledge fragments coded in the early cortices, together with the coding of how those fragments must be combined to represent an object comprehensively. CDZ records are built through experience, by interacting with objects. The CDZ framework can explain perceptual imagery, as it proposes that similar neural networks are activated when objects or events are processed in perceptual terms and when they are recalled from memory.

In the attempt to capture key principles of the simulation theory of sensorimotor control, the best computational framework is certainly that of artificial neural networks, because the first and foremost lesson from the theory is that sensorimotor skills derive from a learning process. Within Artificial Intelligence neural networks are the methods mostly based on learning, in contrast with rule-based methods. Within neural networks, one of the best candidate for simulating imagery, following the CDZs hypothesis, is the *autoencoder*. It has been the cornerstone of the evolution from shallow to deep neural architectures [12, 13]. The crucial issue of training neural architectures with multiple internal layers was initially solved associating each internal layer with a Restricted Boltzmann Machine [12], so that they can be pre-trained individually in unsupervised manner. The adoption of autoencoders overcame the crucial energy-based training of Boltzmann Machine: each internal layer is trained in unsupervised manner, as an ordinary fully connected layer. The key point is to use the same input tensor as target of the output, training therefore the layer to minimize the reconstruction of the input [14]. In the first layer the inputs are that of the entire neural model, for all subsequent layers the hidden units' outputs of the previous layer are now used as input. The overall result is a regularization of the entire model similar to the one obtained with Boltzmann Machine [15], or even a better one [13]. Soon after, refinement of algorithms for initialization [16] and optimization [17] of weights, made any type of unsupervised pre-training method superfluous. However, autoencoders find a new role for capturing compact information from visual inputs [18]. In this kind of models the task to be solved by the network is to simulate as output the same picture fed as input. The advantage is that, during learning of the task, the model develops a very compact internal representation of the visual scene.

4 Applications of the Dreaming Mechanism

This section describes the detailed implementation of the dreaming mechanism by means of episodic simulations in two cases studies. Commonly to both applications there is the architecture of the autoencoder, used as a generative tool to model both imaginary roads and other road users velocity profiles.

4.1 Dreaming Roads

In the first case study, we generated realistic road profile starting from digital maps data. The salient features of the roads were then encoded in the kernel of the autoencoder during the learning phase of the network.

The trained decoding part of the autoencoder was then implemented to generate realistic road profiles by the injecting of random signals in the kernel of the network.

Then, we built an interface for running the agent on the "dreamt" roads via the open-source driving simulator OpenDS (https://opends.dfki.de). In the simulation environment, the agent could drive the vehicle along the generated lanes while simultaneously recording driving actions. The offline investigation of the data provided us the relevant information regarding the driving skills of the Co-driver. For instance, we could evaluate the compliance of both GG-diagram and velocity in curves to human common driving behavior. The same structure could alternatively be used in order to fine tune the driving skills by providing a "dreamt" gym for the agent.

4.2 Braking Dreams

The second case study is what we refer to as the "Braking dreams". In this context, the main purpose was to investigate the behavior of the Co-driver in the task of following a vehicle moving along a straight road while maintaining the desired time headway. In this case study, we used the industry state-of-the-art IPG CarMaker software as simulation environment.

The training set was obtained from real log data where the vehicle was driven by multiple (human) drivers along several roads. The collected data were then clusterized in order to find the braking actions which corresponded to the full stop maneuvers. Such batch of data was then implemented for the training process of the autoencoder which compressed the 10 s time-series data into a 10-feature size kernel. The trained decoding part can then be used for generating a virtual infinite number of realistic velocity profiles modeling the behavior of other road users.

In the next step, we automated the execution of the experiments such that for any generated velocity profile a new simulation is engendered and the Co-driver automatically started. Eventually, we were able to compute the distribution of *time to collision* and *time headway* to the front vehicle and assessing that the agent behaved as expected.

5 Conclusions

This paper presented a biologically-based approach to engineer and validate the internally developed driving agent Co-driver. The agent was designed by emulating real brain functionalities such as reported in [2] while the quality assessment process was based on the dreaming mechanism.

For the time being, only two simple examples of such closed-loop realizations of the mechanism were demonstrated. We plan to introduce more complicated and realistic "dreams" in the ongoing activity of the Dreams4Cars project to show the effectiveness of the overall structure.

References

1. Hesslow, G.: The current status of the simulation theory of cognition. Brain **1428**, 71–79 (2012)
2. Cisek, P., Kalaska, J.F.: Neural mechanisms for interacting with a world full of action choices. Ann. Rev. Neurosci. **33**, 269–298 (2010)
3. Wolpert, D.M., Diedrichsen, J., Flanagan, R.: Principles of sensorimotor learning. Nat. Rev. Neurosci. **12**, 739–751 (2011)
4. Hardwick, R.M., Rottschy, C., Miall, C., Eickhoff, S.B.: A quantitative meta-analysis and review of motor learning in the human brain. NeuroImage **67**, 283–297 (2013)
5. Da Lio, M., Plebe, A., Bortoluzzi, D., Papini, G.P.R., Donà, R.: Autonomous vehicle architecture inspired by the neurocognition of human driving. In: International Conference on Vehicle Technology and Intelligent Transport Systems, pp. 507–513. Scitepress (2018)
6. Kosslyn, S.M.: Image and Mind. Harvard University Press, Cambridge (1980)

7. Kosslyn, S.M.: Image and Brain: The Resolution of the Imagery Debate. MIT Press, Cambridge (1994)
8. Horikawa, T., Kamitani, Y.: Hierarchical neural representation of dreamed objects revealed by brain decoding with deep neural network features. Front. Comput. Neurosci. **11** (2018). Article 4
9. Macknik, S., Martinez-Conde, S., Blakeslee, S.: Sleights of Mind: What the Neuro-Science of Magic Reveals About Our Everyday Deceptions. Henry Holt and Company, New York (2010)
10. Thill, S., Svensson, H.: The inception of simulation: a hypothesis for the role of dreams in young children. In: Carlson, L., Hoelscher, C., Shipley, T.F. (eds.) Proceedings of the 33rd Annual Conference of the Cognitive Science Society (2011)
11. Meyer, K., Damasio, A.: Convergence and divergence in a neural architecture for recognition and memory. Trends Neurosci. **32**, 376–382 (2009)
12. Hinton, G.E., Salakhutdinov, R.R.: Reducing the dimensionality of data with neural networks. Science **28**, 504–507 (2006)
13. Vincent, P., Larochelle, H., Lajoie, I., Bengio, Y., Manzagol, P.A.: Stacked denoising autoencoders: learning useful representations in a deep network with a local denoising criterion. J. Mach. Learn. Res. **11**, 3371–3408 (2010)
14. Larochelle, H., Bengio, Y., Louradour, J., Lamblin, P.: Exploring strategies for training deep neural networks. J. Mach. Learn. Res. **1**, 1–40 (2009)
15. Bengio, Y.: Learning deep architectures for AI. Found. Trends Mach. Learn. **2**, 1–127 (2009)
16. Glorot, X., Bengio, Y.: Understanding the difficulty of training deep feedforward neural networks. In: International Conference on Artificial Intelligence and Statistics, pp. 249–256 (2010)
17. Kingma, D.P., Ba, J.: Adam: a method for stochastic optimization. In: Proceedings of International Conference on Learning Representations (2014)
18. Krizhevsky, A., Hinton, G.E.: Using very deep autoencoders for content-based image retrieval. In: European Symposium on Artificial Neural Networks. Computational Intelligence and Machine Learning, pp. 489–494 (2011)

The Construction of Agent Simulations of Human Behavior

Roger A. Parker(✉)

AirMarkets Corporation, Seattle, WA, USA
rap@airmarkets.aero

Abstract. A general definition of the minimum required structure of computer agents that are capable of accurately representing human behavior in agent-based models is offered. Included is the abstract definition of the Environment, the State Vector, the Perceptor, the Actor, and the Ratiocinator components of the agent structure and their interaction. The synthetic population and associated incident distributions are then defined. Finally, practical considerations of time, accuracy and path dependency are examined.

Keywords: Human agent model · Synthetic population · Narrative Ratiocinator · Memory state vector · Virtual markets · Path dependency Structural limits

1 Introduction

A general definition of "agent" has been presented by this author and many other individuals over the past several years. For example, consider the presentations in Parker and Bakken [1], in Parker and Perroud [2], and in Parker [3]. This discussion describes the minimum set of components required for the representation of viable human activity in an agent-based, computer context. It is based on the realization and implementation of the fundamental concept of narrative that must be recognized if the integration of human and artificial intelligence is to be successful. The concept of narrative and its implications for agent-based modeling are the topics of Parker [4].

This simple agent structure consists of four components, each of which is examined separately. Corresponding to the memory for a human being, the *state vector* is an array of variables which describe the current state of the agent in the simulation. Further, the agent implementation has code that allows it to receive information about the current conditions in its environment through the *perceptor*. Opposite of receiving information about the environment through the perceptor is taking actions on the environment, managed by the *actor* component of the agent. The actions of the agents are created and managed by the *ratiocinator* component. The ratiocinator fulfills the role of choice maker and adaptation mechanism. But choice mechanisms are always stochastic, and how these mechanisms are implemented are part of the agent's ratiocinator.

A collection of agents is referred to as a *synthetic population*, and one of the powers of agent-based modeling is that the collections of agents can be heterogeneous. Since the state space carries data necessary for the representation of the underlying narrative

W. Karwowski and T. Ahram (Eds.): IHSI 2019, AISC 903, pp. 435–441, 2019.
https://doi.org/10.1007/978-3-030-11051-2_66

supporting the agent's behavior, the *distribution* of the state space variables in the synthetic population is an important datum for the representativeness and calibration of the simulation. Such distributions are referred to as *incidence distributions*.

2 Agent Structure

An *agent* is a computer-based object representing the behavior of a human being. The *environment* for one agent usually contains other agents, either like itself or with different capabilities. Within the agent are mechanisms for perceiving the environment, making choices about what to do given the information from the environment in the context of the agent's narrative, and taking actions that advance the agent's narrative. This simple agent structure consists of four components. See Fig. 1.

Corresponding to the memory for a human being, the *state vector* is an array of variables which describe the current state of the agent in the simulation. These can be very simple variables, the existence or absence of some condition, or very complex, such as an array of parameters defining the probability distribution of some choice protocol, or the history of the agent in the simulation. Structures more complex than simple numbers can be specified. For example, it can contain the probability distributions of the outcome sets relevant to the events the agent will encounter during the simulation. The state vector is maintained as appropriately defined data structure within computer memory.

Like the ability of humans to receive, filter and understand information, the agent includes code that allows it to receive information about the current conditions in its environment through the *perceptor*. The purpose of the perceptor is to 'observe' the

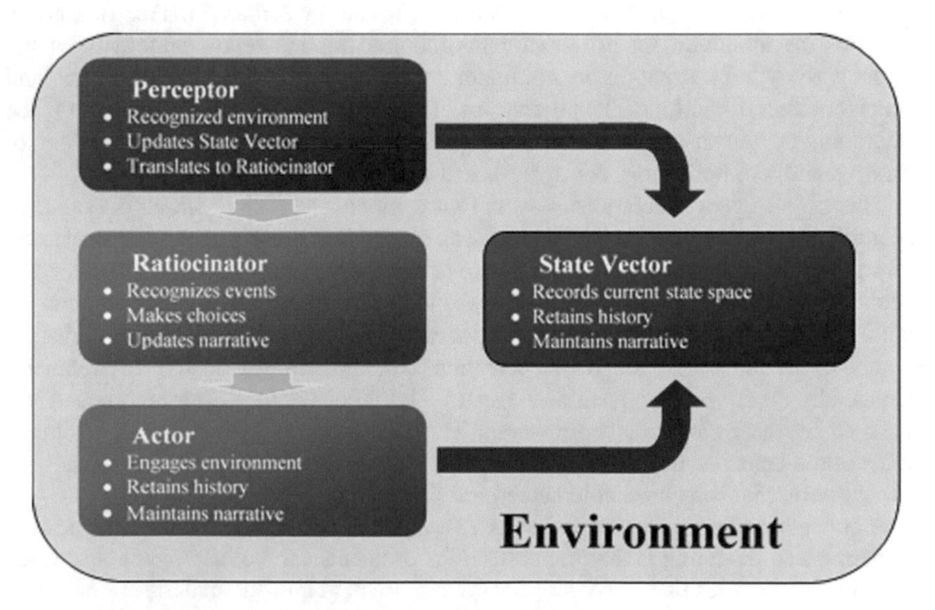

Fig. 1. The general structure of an agent

current state of the agent's environment, and filter and translate that information into a form that can be used by the choice-making component of the agent in accordance to the narrative by converting the messages from the environment into an internal form of use to the agent. This internal translation can be unique to each agent, which thus allows for the design of agents that interpret the same external message differently. This is important, for instance, if different agents had different narrative events that are triggered by the same external environmental conditions.

Opposite of receiving information about the environment through the perceptor is taking actions on the environment, such as implementing a choice. Such actions are managed by the *actor* component of the agent. The activity carried out by actor is very often the notification to the environment (more particularly, specific agents in it) of a choice that has been made by the agent. This is done by means of a messaging structure, quite similar to the messaging operations handled by the perceptor.

The actions of the agents to external messages, both those which are put into the environment as output messages by the agent actors and those required for changes in the internal state vector of the agent, are managed by the *ratiocinator* component. The ratiocinator is the choice maker and adaptation mechanism. This is the part of the agent which replicates how the human intersects with the simulated world in accordance with the then-active narrative structure. In simple simulations, the narratives of interest are atomic narratives, dealing with a single narrative event. For example, in the word-of-mouth agent-based simulation suggested by Conway [5] or Epstein and Axtell [6], the atomic narrative is a choice to buy, wait or ignore, depending (stochastically) on the input about the product the agent receives from its immediate physical neighbors. In the AirVM airline passenger virtual market examined in detail in Parker [7], the atomic narrative is the selection of a particular itinerary from those available at the time of purchasing a ticket for a flight. For even the simplest case of the atomic narrative with a single event, the agent components must contain significant data. The perceptor needs to be designed to recognize the occurrence of the event and the perceived state of the environment at the time of the occurrence. The ratiocinator must be programmed to execute one of a set of choice protocols the agent uses to exercise the choice required by the event, including the availability and allocation conditions of the resources at the agent's disposal. And the actor must have, in its repertoire of possible actions, those that are suitable given the event perception and protocol requirements. Furthermore, the connections between the component events must also be precisely represented in order to reliably represent the agent's actions.

3 An Example: Virtual Markets

A useful example of an agent-based model using this definition of agent is a *virtual market*. A virtual market is a simulation of the market for a defined set of products being offered for sale by a collection of sellers and available for purchase by a collection of customers. It sharply differs from other forms of modeling used in marketing analysis, such as hierarchical Bayes (Rossi et al. [8]) or diffusion theory (Bass [9]).

There must be at least two kinds of agents in a virtual market – *customers* and *sellers*. Both need to be agents with perceptors, actors, ratiocinators and state vectors.

Customer agents must have sufficiently detailed representations of the narratives which are the focus of the simulation, that is the decision context of the customer when considering alternative products to fulfill their narrative-based needs. And the choice protocol for the customer agent is an essential element of the agent. Very often this can be extremely simple: the perceptor recognizes the availability of two or more products and the ratiocinator employs a simple weighted random choice mechanism to make the choice. In such a simple case, the state vector is simply temporary memory for the variables which describe the available product options, and minimal internal agent parameters, such as weights, that need to be applied. On the other hand, it is also possible to build agents which utilize quite complex molecular narratives, where one decision leads to a course of additional choices, while another takes the agent down a different path. In all cases the ratiocinator's choice protocol must be stochastic, but it does not have to be rational. Agents are representations of how people actually behave, not how we might think they should behave.

For an agent-based model to be a virtual market there must also be agents that are sources of products for the customer agents to consider – seller agents. These agents also have narratives which motivate their behavior, but they are distinguished from customer agents in that generally such narratives are merely ways to make a profit. But it is conceivable that much more complex seller interaction, such as competition using various economic game choice protocols, could be represented in a virtual market. Perhaps the most obvious contribution of agent-based models to marketing science is that of a *laboratory*. Experiments can be done in a virtual market that would be impossible otherwise. Of equal importance, given the inherently stochastic nature of a virtual market, Monte Carlo methods can be employed to yield probability distributions for variables of interest that validly represent the probability structures found in the actual market place.

Humphries [10] asserts that computational science is an extension of humanity's ability to observe and understand the world. This function is an attribute of all of computational science and is thus a property of virtual markets. A virtual market can be expected to reveal aspects of the market place it represents that would otherwise not be visible or apparent. The sharper view of the dynamics of the virtual market creates the opportunity for forecast mechanisms of greater validity, and for specifying realistic confidence bounds on such forecasts. Furthermore, where such mechanisms cannot exist – such as in the stock exchange – the reason they cannot exist may well be more convincingly demonstrated through the virtual market instrument.[1] Much of the motivation for the development of a virtual market comes from firms and institution that participate in that market.

Airlines want to use an application called the AirMarket Simulator (Parker and Holmes [11]), for example, because they want to understand the behavior of passengers when presented with travel opportunities, they are considering offering. In this case, the seller agent in the simulation is a representation of a client. Moreover, the ratiocinators

[1] A wonderful experiment would be to use a virtual market representation of a stock market to test the hypothesis that the confidence interval on any market forecast is so wide as to make moot the forecast itself. In other words, the forecast can never be better than a random guess.

of the client seller agents are incomplete in that the client institution expects to provide choice protocols, which reflect its narrative within the simulation and which requires access to the agent ratiocinator. When a decision is called for the agent doesn't execute the ratiocinator process, the client participating in the simulation does.[2] In this situation, the virtual market offers an *avatar* for the simulation client. In addition, other seller agents often represent competitors which react in some way based on choices made by the client agent. The AirMarkets Simulator virtual market has client agents in this sense.

4 Synthetic Populations and Incidence Distributions

Every agent outside of any individual agent is part of that individual agent's environment. For many consumer markets, the number of customer agents can rise into the millions. And many markets worthy of simulation have a number of seller agents as well. A collection of agents of the same type (e.g. customer agents) is referred to as a *synthetic population* of those agents. In most virtual markets, for example, there should be at least two synthetic populations – one for the customers and one for the sellers.

The collection of agents in a synthetic population can be heterogeneous. Agents of the same type have similar state space structures (which, basically, forms the definition of agent "type"), and heterogeneity is reflected in different values for variables maintained in the state vector. For example, if gender is an important variable in a purchase decision, (e.g. if the product were cosmetics) then the individual customer agents would have to be assigned a gender when they are created in the simulator. Therefore, since the state space carries data necessary for the representation of the underlying narrative supporting the agent's behavior in the virtual market, the *distribution* of the state space variables in the synthetic population is an important datum for the representativeness and calibration of the simulation. When the agents are created this distributional data is used to set the relevant state space values of the agents in the synthetic populations. The distributions are referred to as *incidence distributions*.

5 Practical Issues: Time, Accuracy and Path Dependency

The time aspect of a simulation highlights one of the practical issues that arise in the design and construction of simulation models. Grimm and Railsback [12] provide an excellent reference on how to build agent-based models (or individual-based models, IBMs, as they put it), with an important emphasis on the practical problems that must be addressed and resolved. Although their subject matter is ecology, their observations and advice extend to any agent-based development program. However, Grimm and Railsback do not include *model accuracy* as one of their criteria. The first intuition is to

[2] Agent-based models are finding some interesting uses in this context, where one or more of the agents in the simulation are actually people. It's obvious that this is the case in many electronic simulation games, where the player manages an avatar which behaves under the direction of the player and has no independent volition. .

compare the model's output with observations to the real world and see if the two differ. But no model is one hundred percent accurate, for models are always simplifications of what is being studied, and aspects of the phenomenon being modeled that are simplified away can have some effect on real world observations. Indeed, they had better have some effect, for otherwise they are irrelevant. So rather than measure accuracy against real-world observations, because that will always be wrong, it is wiser to measure the accuracy with respect to other models.

The situation is often exacerbated by the nature of the thing being modeled. When applied to complex adaptive systems, for example, each execution of an agent-based model represents one dependent path of the evolution of the system being modeled. Moreover, each independent observation of the complex system is also the result of a single dependent path. It is highly unlikely that the simulation and the real phenomenon would match, purely because of the path dependency. (In fact, with systems that have an uncountable number of dependent paths, the probability that two paths are identical is zero.) This situation can be detected by measuring the actual phenomenon multiple times. If path dependency is at work, it will be very difficult, if not impossible, to get two measurements that are identical. Repeated measurements of even simple variables – say the load on a flight from point A to point B – rarely yield equal values. But, one might suggest, take the average or some other statistical summary of the real observations, and compare that with a similar statistic derived from the simulation. That only begs the question. It is difficult in most cases to get *independent* observations from a complex system (path dependency is a wicked thing), so the probability distribution of the observed values is not knowable, and thus the probabilities associated with the statistical comparison, whatever it is, is also not knowable.

References

1. Parker, R.A., Bakken, D.: Predicting the unpredictable: agent-based models in market research. In: Mouncey, P., Wimmer, F. (eds.) Best Practices in Market Research. Wiley, Hoboken (2007)
2. Parker, R.A., Perroud, D.: Exploring markets with agent-based computer modeling. In: Proceedings of the 61st ESOMAR International Congress, Montreal, Canada (2008)
3. Parker, R.A.: An agent-based simulation of air travel itinerary choice. In: Proceedings of the ABMTRANS 2017 Conference, Madeira, Portugal (2017). https://doi.org/10.1016/j.procs.2017.05.419
4. Parker, R.A.: The concept of narrative as a fundamental for human agent-based modeling. In: Karwowski, W., Abram, T. (eds.) Intelligent Human Systems Integration. Advances in Intelligent Systems and Computing, vol. 722 (2018). https://doi.org/10.1007/978-3-319-73888-8_59
5. Conway, J.: Mathematical games: the game of life. In: Gardner, M. (ed.) Scientific American (1970)
6. Epstein, J., Axtell, R.: Growing Artificial Societies: Social Sciences from the Bottom Up. MIT Press, Cambridge (1996)
7. Parker, R.A.: A demonstration of AirVM. In: Proceedings of the AGIFORS Schedule and Strategic Planning Study Group, Lausanne, Switzerland (2010)

8. Rossi, P., Allenby, G., McCulloch, R.: Bayesian Statistics and Marketing. Wiley, Hoboken (2005)
9. Bass, F.: A new-product growth model for consumer durables. Manage. Sci. **15**, 215–227 (1969)
10. Humphreys, P.: Extending Ourselves: Computational Science, Empiricism, and Scientific Method. Oxford University Press, Oxford (2004)
11. Parker, R.A., Holmes, B.: Using predictive analytics for corporate shuttle decisions. In: Proceedings of the NBAA Convention, Las Vegas, Nevada, USA (2015)
12. Grimm, V., Railsback, S.: Individual-Based Modeling and Ecology. Princeton University Press, Princeton (2005)

Effects of Simulator Sickness and Emotional Responses When Inter-pupillary Distance Misalignment Occurs

Hyunjeong Kim and Ji Hyung Park(✉)

Center of Robotics Research, Korea Institute of Science and Technology,
Seongbuk-gu, Seoul, Republic of Korea
{jeong, jhpark}@kist.re.kr

Abstract. The aim of this study was to empirically investigate the effects of inter-pupillary distance (IPD) misalignment on simulator sickness and emotional responses. Twenty participants were recruited from an advertisement. The experiment has five conditions to change the differences between IPD and head-mounted display (HMD) optical systems. As results, simulator sickness significantly decreased by aligning the HMD optical systems based on the IPD. Extending previous studies, the results demonstrated the different patterns in emotional arousal and valence. For future study, we are conducting a second experiment using eye-tracker to investigate whether anomalies occur in pupil movements. This study might help to inform the important of IPD misalignment problem by measuring visual discomfort quantitatively.

Keywords: Human factors · Virtual reality · Inter pupillary distance
Simulator sickness · Head mounted display · IPD misalignment

1 Introduction

Virtual reality (VR) provides higher senses of immersion and presence; however, it causes simulator sickness with symptoms such as dizziness, fatigue, blurred vision, headache and nausea [1]. Above all, inter-pupillary distance (IPD) misalignment that incorrect between IPD of a user and the HMD optical systems is one of the important causes [1]. The IPD misalignment is a fundamental problem because it can lead to simulator sickness independently of the other factors (e.g. sensory conflict and postural instability) in the event of an optical misalignment.

There are few studies on the relationship between the IPD misalignment and the simulator sickness in VR. Howarth investigated the heterophoria effects when people used different types of HMDs that optical systems created three different inter-ocular distance (IOD) [2]. As the result, the differing IODs induced heterophoria with outward or inward. However, they did find significant correlations between the degree of IPD disparity and the measured heterophoria, and did not compare the SS and the IPD directly. Kolasinski et al. found the correlation between the simulator sickness and the IPD using HMDs with fixed IOD and the results showed that converging eye movement caused less ocular stress and eyestrain than diverging eye movements [3]. There is the serious

W. Karwowski and T. Ahram (Eds.): IHSI 2019, AISC 903, pp. 442–447, 2019.
https://doi.org/10.1007/978-3-030-11051-2_67

limitation in these studies that experimental environments are inconsistent because HMDs have the different configuration including the location of lenses and screen.

To overcome this issue, we proposed an experimental protocol that can examine the IPD misalignment empirically. The experiment has five conditions that can adjust the HMD optical systems according to the individual IPD. In addition, we investigated emotional arousal and valence, and the symptoms of simulator sickness. We verified the hypothesis that: (1) the greater differences between IPD and HMD optical systems (IPD misalignment) enhanced more simulator sickness and negative emotion; (2) the B + 4 condition elicited greater emotional arousal and lower valence than other conditions.

2 Methods and Materials

2.1 Participants

Twenty participants (M = 28.4, SD = 3.7) were recruited through advertisements. Wc measured each participant's IPD, and it was assigned to baseline (B) condition value.

2.2 Hardware and Software

A High-end graphics card (NVIDIA GeForce) and an HTC Vive HMD with high resolution was used in this experiment. The HMD can adjust the distance between lenses according to the IPD values. We developed the main program using Unity3D (Unity Technologies SF, California, USA).

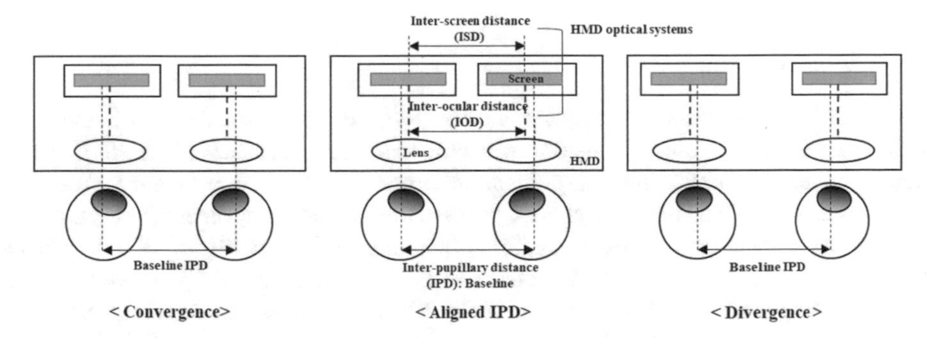

Fig. 1. Concept images according to the conditions. Left: HMD optical systems are decreased compared to the IPD (B-2 and B-4 condition). Mid: optical systems are same as the IPD (B condition). Right: optical systems are increased compared to the IPD (B + 2 and B + 4 condition).

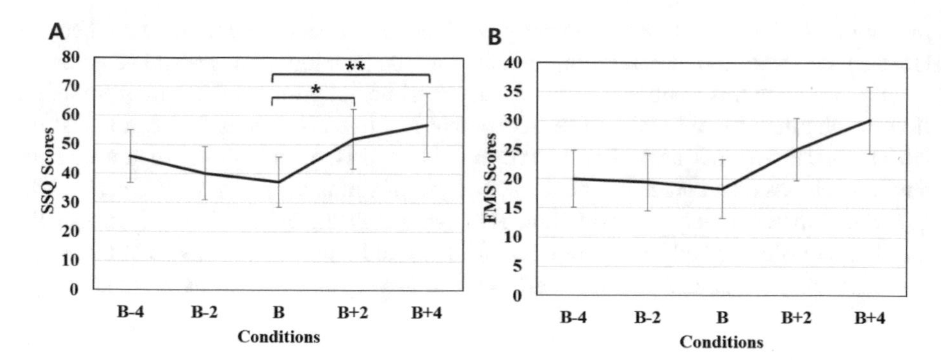

Fig. 2. Results of the differences between the five conditions. (A) Total scores of simulator sickness questionnaire (SSQ). (B) Average scores of fast motion sickness (FMS); ** $p < .001$, ** $p < .001$.

2.3 Task Explanation

We measured each participant's IPD and assigned it to the baseline. Therefore, all the participants have different individual baseline value. This study has five conditions (Fig. 1): Baseline − 4 mm (B − 4), Baseline − 2 mm (B − 2), Baseline (B), Baseline + 2 mm (B + 2), and Baseline + 4 mm (B + 4). HMD optical systems were adjusted according to the five conditions. The intervals of the HMD optical systems were respectively 2 and 4 mm greater than the participants' IPD under the B + 2 and B + 4 conditions, whereas the intervals were decreased by the same magnitudes under the B − 2 and B − 4 conditions.

2.4 Measurements

We used the simulator sickness questionnaire (SSQ) to assess the degree of simulator sickness including oculomotor distortion, disorientation, and nausea [4]. The SSQ has 16-item symptoms checklist and four-point scale format (0 = None, 3 = Severe). To assess the degree of perceived motion sickness, we used the fast motion sickness (FMS) that use a Likert scale (0 = no sickness at all, 20 = frank sickness) verbally [5]. Every minute during each condition, the experimenter asks the participant to rate perceived motion sickness using the FMS rating system verbally. The Self-Assessment Manikin (SAM) is visual descriptions of a manikin with the various emotional expression used a Likert Scale (1–9) to measure emotional arousal (1 = Extremely calm, 9 = Extremely excited) and valence (1 = Extremely unpleasant, 9 = Extremely pleasant) [6].

2.5 Procedure

Prior to the task, participants received to complete a brief questionnaire containing a name, age, gender, eyesight, and SAM. For a baseline, the experimenter helped the participants to measure their IPD accurately. After describing the experimental process, the experimenter adjusted the HMD optical systems based on the IPD every condition and helped them to wear the HMD to fit their heads. The participants performed the tasks under five conditions that made a movie-clip for 7 min in counterbalanced orders. During the task, they were asked to rank the FMS verbal ratings from zero to twenty every minute. After finishing each condition, they immediately filled in the SSQ and SAM, and then rested for approximately 10 min to prevent fatigue effect after every condition. They could rest longer if they encountered a high degree of sickness. After finishing all conditions, the experimenter debriefed the aim of this study and gave them a class credit.

3 Results

3.1 Simulator Sickness

We conducted repeated measures of analysis of variance (ANOVA) to test the simulator sickness relationship between five conditions (B, B $\pm$ 2, and B $\pm$ 4). As shown in Fig. 2, the SSQ total scores were significantly different by conditions ($F[4, 76] = 4.741$, $p < .005$, and $\eta 2 = .200$). The B + 4 condition elicited the highest degree of simulator sickness, followed by B + 2, B − 4, B − 2, and B condition (Table 1).

We also found statistically significant differences in all sub-scores; oculomotor distortion ($F[4, 76] = 4.694$, $p < .005$, and $\eta 2 = .198$), disorientation ($F[4, 76] = 3.840$, $p < .01$, and $\eta 2 = .168$), and nausea ($F[4, 76] = 2.646$, $p < .05$, and $\eta 2 = .122$). Moreover, oculomotor distortion showed highly significant effect between five conditions.

3.2 Fast Motion Sickness

The average of FMS scores ($F[4, 76] = 3.471$, $p < .05$, and $\eta 2 = .154$) and peak scores ($F[4, 76] = 4.441$, $p < .005$, and $\eta 2 = .189$) showed a significant main effect between the five conditions. As shown in Table 1, the total score of B + 4 condition was the highest among five conditions, and B condition elicited the lowest score.

Table 1. Univariate main effects of questionnaire.

Scale	Conditions					*p*-value
	B − 4	B − 2	B	B + 2	B + 4	
SSQ total	46.0 (9.1)	40.2 (9.1)	37.2 (8.6)	52.0 (10.3)	56.8 (11.0)	**
FMS average	21.5 (4.9)	19.3 (4.9)	17.2 (5.0)	23.1 (5.4)	26.7 (5.7)	**
Arousal	5.5 (0.4)	5.5 (0.5)	5.8 (0.4)	5.7 (0.5)	5.3 (0.5)	>.062
Valence	3.0 (0.5)	2.6 (0.5)	2.6 (0.4)	3.0 (0.5)	3.0 (0.5)	>.072

Note. Mean [$\pm$ standard error (SE)]; ** $p < .001$; SSQ: simulator sickness questionnaire; FMS: fast motion sickness

3.3 Emotional Reactivity

There were no significant main effect in emotional arousal ($p > .062$) and valence ($p > .072$). In addition, we found the participants felt the highest score of arousal most unpleasant emotions in the B + 4 and the pleasant in B condition.

4 Discussion

The primary purpose of this study was to investigate whether IPD misalignment elicits simulator sickness and emotional responses. We performed an experiment with five different conditions changing the differences between the HMD optical systems and the IPD. As results, reducing the IPD misalignment induced smaller simulator sickness and more pleasant and stable emotion. During the current experiment (Fig. 2), the participants reported that their eyes abnormally centered (convergence) or opened outward (divergence), and it was supported by related studies [2, 3]. This study strongly suggests designers adjust the configuration in HMDs so that users can adjust both the lenses (IOD) and the screen display (ISD).

We experimentally confirmed the occurrence of heterophoria phenomenon. This study suggests that HMD manufactures should produce optical systems are consistent with the users' IPDs to avoid discomfort and unusual eye movements particularly children. Previous studies have shown that young children can be affected by becoming strabismus (vergence and accommodation) if they use stereoscopic images for long durations [7]. In addition, severe oculomotor discomforts (visual fatigue, blurred vision, and double vision) can occur when experiencing a virtual environment with misalignment IPD [2].

Emotional arousal means the state of being awake psychologically, and emotional valence means the intrinsic attractiveness. As results of arousal and valence, the B condition evoked the highest emotional arousal and the lowest emotional valence. Therefore, people felt more sense of stability and pleasant when alignment condition between their IPD and HMD optical systems than IPD misalignment condition. According to the participants' report, the emotional responses were not only caused by the simulator sickness, but also by the inconvenience of the HMD optical systems. These results anticipated that significant differences would be obtained with more participants; therefore, we will conduct the second experiment with more participants.

IPD misalignment induced significantly greater symptoms of oculomotor distortions, disorientation, and nausea to HMD user especially eye fatigue. Accurate binocular alignment might is one of the crucial contributing factor causing simulator sickness and negative emotion, therefore HMD users should be aware that they are aligning the HMD optical systems with their IPD. We are conducting a second experiment using eye-tracker to investigate whether anomalies occur in pupil movements. This study might help to inform the important of IPD misalignment problem by measuring visual discomfort quantitatively.

Acknowledgments. This research was supported by the Global Frontier Program through the National Research Foundation of Korea (NRF) funded by the Ministry of Science, ICT & Future Planning (NRF-M1AXA003-2011-0028362).

References

1. Kolasinski, E.M.: Simulator Sickness in Virtual Environments (No. ARI-TR-1027): Army research Inst for the behavioral and social sciences, Alexandria, VA (1995)
2. Howarth, P.A.: Oculomotor changes within virtual environments. Appl. Ergon. **30**, 59–67 (1999)
3. Kolasinski, E.M., Gilson, R.D.: Simulator sickness and related findings in a virtual environment. Proc. Hum. Factors. Ergon. Soc. Ann. Meet. **42**, 1511–1515 (1998)
4. Kennedy, R.S., Lane, N.E., Berbaum, K.S., Lilienthal, M.G.: Simulator sickness questionnaire: an enhanced method for quantifying simulator sickness. Int. J. Aviat. **3**, 203–220 (1993)
5. Keshavarz, B., Hecht, H.: Validating an efficient method to quantify motion sickness. Hum. Factors **53**, 415–426 (2011)
6. Bradley, M.M., Lang, P.J.: Measuring emotion: the self-assessment manikin and the semantic differential. J. Behav. Ther. Exp. Psy. **25**, 49–59 (1994)
7. Tsukuda, S., Murai, Y.: A case report of manifest esotropia after viewing anagryph stereoscopic movie. Jpn. Orthoptic J. **16**, 69–72 (1988)

Mixed Reality-Based Platform for Smart Cockpit Design and User Study for Self-driving Vehicles

Xiaohua Sun[1(✉)], Shiyu Wu[1], Shengchen Zhang[1], and Hanlin Wang[2]

[1] College of Design and Innovation, Tongji University, Shanghai, China
{xsun, seanzhang}@tongji.edu.cn, shinywsy@gmail.com
[2] Philips Design Center, Shanghai, China
294619413@qq.com

Abstract. The change of the role people plays in autonomous driving leads to different use scenario and user behavior, which should be considered in the design of smart cockpit to enable better experience. However, the smart cockpit design and user study under autonomous driving conditions are difficult to achieve by existing approaches due to high cost and the limitation current methods pose on designer imagination. Mixed reality incorporates the ability of virtual reality to simulate immersive scenarios and events, and combines both virtual and real worlds, thus enables further natural user interaction and intuitive experience. We present in this paper a platform for designers to build virtual cockpit and validate their design based on mixed reality. In innovation and design phase, the platform provides basic modularized components which can be assembled to various usable space and HMI. It helps designers to explore the possibility of cockpit space of self-driving vehicles unlimitedly and build it conveniently according to different scenarios and functions. In user validation phase, designers can do user study for designed cockpit in virtual potential use scenarios, integrating a series of driving condition and events. With the support of mixed reality, the platform can provide more natural experience of some components (seats, screens, etc.) to better validate and optimize the design.

Keywords: Self-driving vehicles · Smart cockpit · Modular components
Design validation · User experience

1 Introduction

Self-driving vehicles is developed by the combination of technologies of artificial intelligence, automatic control, vision algorithm [1], monitoring and navigation system [2]. The driving operation from drivers has been replaced by vehicle, which liberates more space and time and become safer [3, 4]. So, drivers and passenger can get away from some safety based behavior, and have more working and entertainment behaviors. Self-driving vehicles also generates new scenarios of driving, resting or working as well as human-vehicle authority transition [5, 6]. Besides, the application of display

W. Karwowski and T. Ahram (Eds.): IHSI 2019, AISC 903, pp. 448–459, 2019.
https://doi.org/10.1007/978-3-030-11051-2_68

technologies (HUD [7], VRHUD, holographic projection [8], etc.), as well as interaction technologies (gesture interaction [9], voice recognition, etc.) will greatly enrich the operating modes and experience in cockpit of self-driving vehicles. In this way, human, vehicle, environment and information will be interrelated and coordinated by these high-tech technologies, and form a more interactive cockpit system.

The change of space and behavior possibilities and interaction medium will be a great stimulation for the innovation of self-driving vehicle cockpit. A comfortable, interactive and entertaining experience is as important as the performance of autonomous driving algorithms. Deeply analyzing and discussing these two changes is important in the design of self-driving vehicle cockpit.

However, physical model making using traditional methods like clay modeling is usually labor extensive and time consuming. When it comes to driving scenarios, due to the technology limitation and security problem of self-driving, the test is difficult to realize. Most of the tests can only be carried out on roads under severely limited conditions, such as those found in auto-test site and speed limited roads [10], which will also be the limitation to designers' innovation.

Virtual reality can simulate the autonomous driving scenarios and events, and be able to provide more convenient creation and iterative approaches compared with traditional solid model. So it's a good tool for the evaluation of user experience in Human–Product Interaction studies [11]. Only virtual reality cannot provide users with natural and intuitive experience because of its unreal operation feeling. Mixed reality makes a world that physical and digital objects coexist and interact with each other [12], which solve the demands of experience authenticity and study effectiveness.

Some studies have been made for the application of virtual reality and mixed reality in interior or cockpit design. Pacheco used virtual reality in the design of indoor wall colors [13]. Purschke applied it in vehicle styling and design based on CAD [14]. Immersive virtual reality has also been used to eliminated the time-consuming of qualitative assessment for vehicle design in normal CAD environment [15, 16]. Sportillo described a fully immersive simulator for semi-autonomous vehicles and focused on its control recovery and its most appropriate interface [17]. These studies mainly focus on styling or parts design and modification of vehicle, or some certain interfaces. More study need to be done for the usage of virtual reality in space and HMI design in self-driving vehicle cockpit.

The paper proposes the application of virtual reality and mixed reality in the study of self-driving vehicles, and constructs a platform for smart cockpit design. The platform provides approaches for designers to explore the design possibilities of self-driving vehicle cockpit, and ensures the intuitive and nature experience during the validation process at the same time.

2 Platform Structure

The platform of self-driving vehicle cockpit design and validation is based on Unity 3D, but also contains varieties of devices and program as the support. It can be divided into three main layers: client, host and application, MR/VR system (Fig. 1).

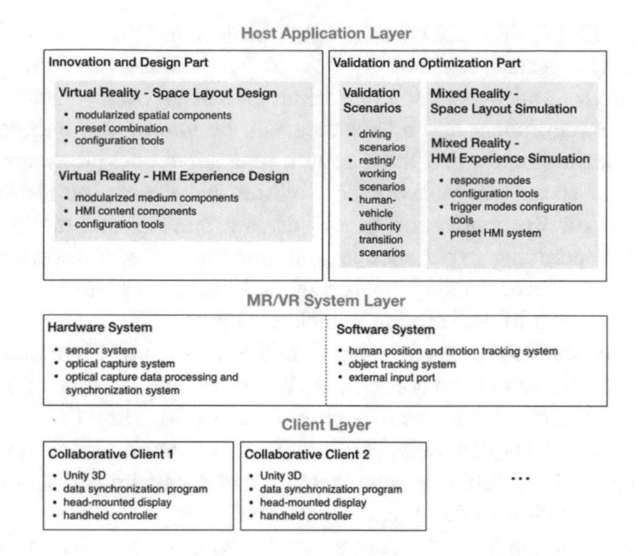

Fig. 1. Platform system

Client Layer. The client layer consists of one (single platform) or multiple collaborative (multiple collaboration platform) clients, which is the foundation of the whole platform. Each of the clients adopts HTC Vive and its handheld controller as the main devices. There is also a data synchronization program in each host to make multi-users cooperating possible.

MR/VR System. The MR and VR system layer consists of hardware and software systems that support the implementation of the host application layer. The hardware includes sensor system, optical capture system, optical capture data processing and synchronization system. The software includes human position and motion tracking system, object tracking system and external input port. The host application layer can obtain real-time user operation and behavior data through series of hardware and software systems, and make visual feedback according to the data results.

Hardware devices involved in the platform, including VR head-mounted display, handheld controllers, and tools used in mixed reality, are based on infrared-reflection markers and high-speed camera (Fig. 2). The location of the hardware devices is determined by infrared cameras and particular combination of these reflective markers which can be recognized into unique identities. So the platform can get the spatial coordinates of each tool according to the spatial coordinates of markers' combination [18]. When users operate the tool, it can present the model and the operation result in virtual environment as well. Besides, external input port is set to deal with the data from sensors on physical components.

Host Application Layer. The core function of this platform exists in the application layer, which is also the main layer displayed to the user. This layer mainly includes the smart cockpit application of innovation and design as well as validation and optimization. The two applications can be quickly switched between each other to ensure the flexibility of the design and iteration process.

The natural human-computer interaction of virtual reality and mixed reality runs through the whole layer and provides good user experience for the platform users. The interaction methods in the platform are designed through considering the characteristics and advantages of interface and operation in three-dimension environment [19]. In addition, it also takes possibility of cooperative operation into account.

3 Innovation and Design Stage

The stage can be roughly divided into three steps: defining the application context and getting guidance, using modularized space components to explore possibilities of cockpit space, using modularized medium and HMI content components to explore possibilities of. The latter two can be switched with each other (Fig. 3).

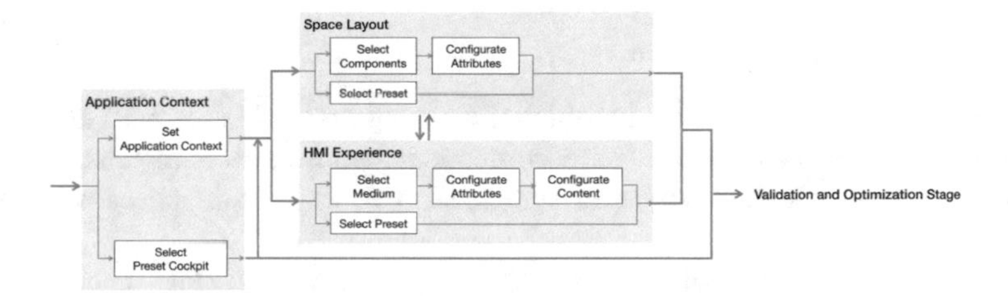

Fig. 2. The logic flow of innovation and design stage

Leaving out control system of self-driving vehicle, the design of smart cockpit is mainly composed of internal parts (tables, chairs, walls, lights, etc.) and HMI parts (internal control, dashboard, rear seat infotainment, etc.). Components provided by the platform has comprehensively considered designers' operation simplification as well as creation freedom. Designers use modularized components cannot only fully demonstrate their imagination, but also easily and conveniently complete the cockpit construction.

3.1 Application Context Guidance

The design of self-driving vehicle cockpit depends on the application context and its behavior demands. It provides more space and time for users that is possible for more kinds of behavior demands. And as the features of self-driving vehicle make the interaction of passengers more frequent, the relationship and number of passengers will also affect the behavior demands of users to some extent. Therefore, the definition of application context contains parameters like persons capacity, space size, usage scene. The definition affects design direction of cockpit interior space and HMI in the innovation and design stage, and can be seemed as essential filter criterion for events priority and importance especially in validation and optimization stage.

In order to better guide designers to consider the application context, the platform provides video and image references from the perspective of different application context. They are not only based on cockpit scene, but also various possible scene with the behaviors and spatial layout, which will inspire designers to think outside the box and create more possibilities.

The platform also provides preset cockpits based on different user behaviors in common contexts. In addition to directly creating new cockpit through setting the parameters, designers can also choose to load one of preset cockpits, which gives guidance to designers. Preset cockpits consist of personal cockpit, family cockpit, commercial cockpit, shuttle cockpit, etc.

After setting up the application context, the platform will generate the corresponding model frame according to the parameters. The frame is just a scale reference of suitable cockpit in such the application context, which can be adjusted again to meet the demands. Following operation will be carried out in the frame.

3.2 Space Layout Exploration

The cockpit space is set as a grid in which the size of one cell is 50 * 50 cm. Platform provides a series of modularized space components whose size is in cells, and guides designers to use their creativity and imagination to explore layout possibilities of spatial components.

The platform disassembles what likely appears in cockpit space into four component-based modular: horizontal board, vertical board, solid box, hollow box (Fig. 4). Because of the demands of less limitation for creativity, the components are given no explicit name like table or seat. Horizontal board can support the design of folding seat and table, vertical board of seat back and partition, solid box of central control and seat, hollow box of storage space and so on. The size of four basic components can be adjusted, but each has its own limits that can't be converted to each other freely. It gives clearer definition for designers so that they can operate the components more logical and clearer.

Next step is to set the components' attributes including the definition of functions and the configuration of operations. For example, designers can set whether the component can be placed with heavy items, folded, raised or lowered, and whether it has sensors and safety system. The different combination of components and their attributes gives different usage semantics to each of them.

As the combination possibilities of horizontal board, vertical board, solid box and hollow box can be thousands, the process might be confusing especially for some amateurs or designers that use the platform for the first time. Therefore, the platform provides a series of typical preset combinations of the four basic components (Fig. 4). Each combination has some certain functions, such as a seat with folding table or available for temporary private space. There is also an animation for each combination to show the usage approach and process. It gives combination guidance for designers to innovate their own cockpit for application demands.

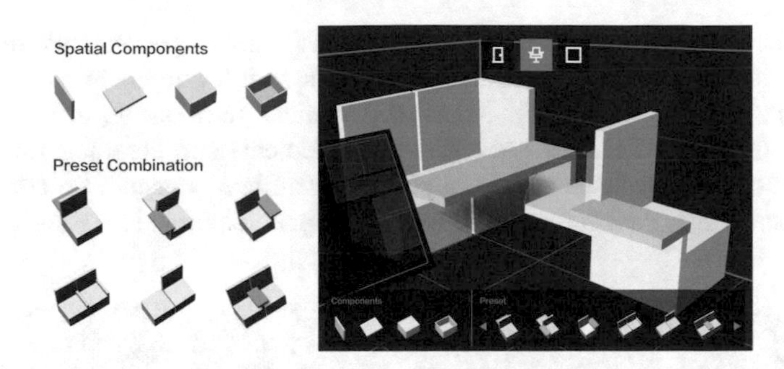

Fig. 3. Modular spatial components and typical preset combination

3.3 HMI Experience Exploration

Multi-channels interaction adopts various interaction devices and technologies to enable users to make natural, parallel and collaborative man-machine conversation. The design of HMI for self-driving vehicle should set type and position of mediums at first, and then arrange the function for each medium based on its features. Therefore, the HMI experience exploration part can be divided into two main contents: modularized medium components setting and HMI content components setting.

Modularized Medium Components Setting. The platform provides medium components for HMI design except for the space component. In self-driving vehicle cockpit, the mediums might be physical screen, the attached LCD, HUD, projector and attached touchpad, so to meet virtual reality design requirements. Three-dimensional medium can support the design of solid screen and suspended projection, two-dimensional medium can support HUD and attached screen, while speaker can support aural response or control system (Fig. 5). Designers attach mediums on any space components and adjust its position and size to build medium system for HMI design. The system emphasizes sensory equipment but ignores the projector, camera, voice recognition and synthesis sensor and other auxiliary equipment which is not necessary to display in concept design phase. In this way, the designers can focus on exploration and innovation instead of those details.

The same as space components, each medium component also has its own attributes. Designer can set whether it is normally open, choose one activation mode from touch, button, position and speech, and associate one medium to another. Medium have no connection with each other in default settings, but will synchronize the operation to another medium when having set connected.

In the context of self-driving, although there are varieties of different medium settings, the existence of basic medium such as central control and dashboard is still necessary. The platform preset several basic mediums with necessary functions, which can be directly use to ensure the normalization of the whole cockpit design.

HMI Content Setting. In general, the process of HMI development will go through product definition, user research, human-computer interaction design, UI design, evaluation, final development, on-line and iteration, and will usually last for about a year. In order to make the whole process smoother to meet the demands of following validation, and considering the difficulty of design and development in virtual reality environment, the platform pre-develops HMI content components to use directly in stage of design [20, 21]. They package visual elements and common functions into reusable ones. What designers do is to place content components into corresponding mediums, and adjust their information hierarchy, position and size (Fig. 5).

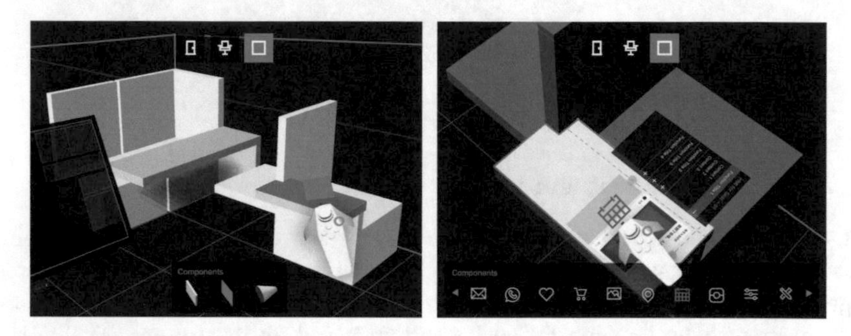

Fig. 4. Modular medium components and HMI functions

Information Architecture. The same as designing on traditional interface, there is also information architecture of functional content in HMIs. It includes the adjustment of hierarchy and display mode. There is a list of all content beside the current interface. The content component can be dragged over the list to adjust its hierarchy, and the display mode can be switched to underlying background or modal card.

Interface Layout. The HMI interface is designed as a grid, too, with cells of 200 * 200 px. According to the common layout of HMIs, the interface is divided into five regions: bottom part that often used in cockpit state control, top part that often used to display the current state, left and right parts that often used for function switching, and central part that often used for core content. The size of each part can be adjusted by moving the parting line cell by cell.

Component Size. There are six sizes of different display forms based on the size of basic cell, which are applicable to different position of interface. The limitation of available range of sizes is in accordance with where the component is.

Although component-oriented HMI design might limit the designer's freedom, and might not provide satisfying HMI GUI style, it does help them quickly iterate their design and make preliminary study of the user experience in self-driving vehicle cockpit.

4 Validation and Optimization Stage

The design and validation of self-driving vehicle cockpit consists of two phases of rules setting and validation. What designers do is to contextualize the design content to make it appear in the corresponding events, and to validate all design results naturally and intuitively to optimize them for better experience (Fig. 6). During the validation phase, the cockpit will be optimized to detailed model (Fig. 7), which runs less smoothly but provides better experience. And also, the phase is based on mixed reality, with which testers can have real experience of operating physical objects.

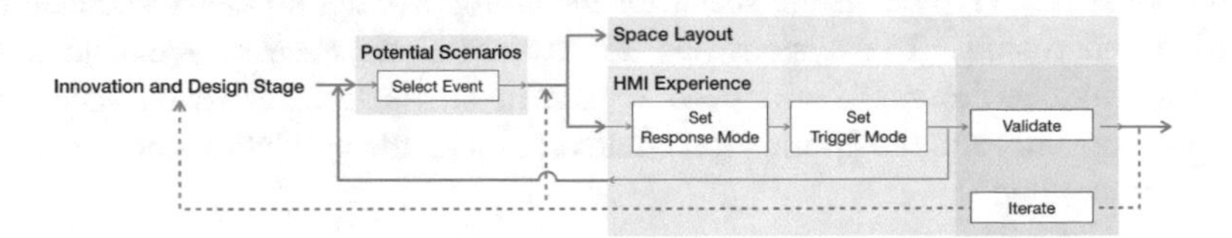

Fig. 5. The logic flow of validation and optimization stage

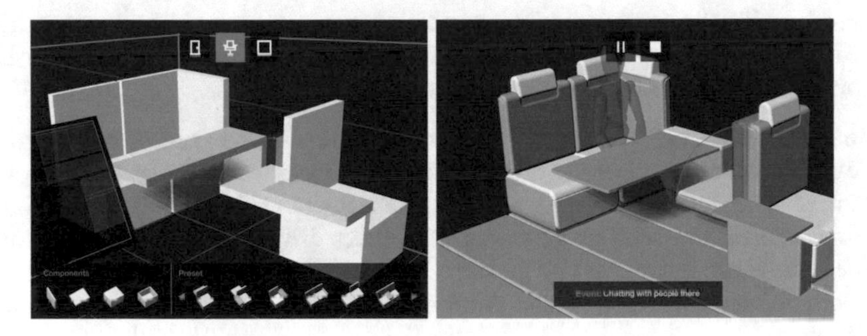

Fig. 6. From simple model to detailed model

4.1 Validation Scenarios Setting

Considering the context of self-driving, exterior scenarios will affect interaction system of vehicle in speed, pause time, drivers' attention, alert frequency, etc. Therefore, the classification of exterior scenarios can be considered from these features above, and turns out with four typical types: central city scenario, extreme mountain scenario, way-finding scenario, highway scenario. A series of unexpected accidents caused by external scene change are set based on these scenarios. At this point, the system will automatically send out warning or voice alerts, and display the vehicle status and location information. They remind users to do further judgement and operation like switch driving modes to drive by themselves.

The interior scene mainly refers to the interactive space composed of modularized space and HMI components. Classifying from space functions and user behaviors, it can be divided into three kinds of scenarios: space construction, information exchange and vehicle control. Space construction scenarios usually occurs in the process of constructing specific space, or transforming between two types of spaces. For example, there will be events of communication or children's entertainment space construction in the activity space. Information exchange scenarios focus on the concern and feedback for all kinds of information, which includes the information from mobile phones, third party applications and records in smart cockpit, such as medium entertainment, television, video chat, surrounding facilities promotion, photograph. Vehicle control scenarios are the active and passive control over the vehicle during driving process. The events here are the control of vehicle driving status, vehicle operation status [22], route planning, etc., such as driving data query, congestion reminder, and power reminder.

Therefore, combining the exterior and interior scene and events above, the platform already builds three typical usage scenarios including driving, human-vehicle authority transition and resting. The system has already set some necessary events in these scenarios, but the designers still need to screen and select scenarios according to validation demands and the application context to make the validation and optimization stage more targeted.

Driving Scenarios. Driving scenario refers to the process of driving by users. The exterior scenes are mainly mountains, highway or other extreme ones. Central city or business district will also appear in some situation. In the scenario, the user behaviors are similar with the ones in ordinary vehicle, where users need to operate driving related components like steering wheel, brake and accelerators. There is little information exchange and vehicle control events in such a scenario.

Resting and Working Scenarios. The scenario refers to the process of self-driving. The exterior scenes are highway or central city with clear way-finding objective. Users don't need to operate the driving related components, but help make decisions through real-time information such as vehicle status, route, driving plan and accident alarm. Besides, due to the role played by the users has been changed from driver to passenger, there will be more entertainment and working behaviors. It is a peculiar situation during self-driving.

Human-Vehicle Authority Transition Scenarios. It is the conversion process of manual driving and self-driving. It happens when accidents, such as equipment failure, vehicle accident, external scene change, etc., occur, and also when users temporarily add events that effect driving process. Besides, users' active operation, for example, push the steering wheel (ID BUZZ, as an example), can also make human-vehicle authority transition. In the scenarios, the driving status and warnings will present automatically and frequently. It is also peculiar in self-driving. In some cases, human-vehicle authority transition will happen many times.

4.2 Space Layout Validation

The validation and optimization of spatial components combination and space layout is mainly through the field experience of before and after the layout change of the cockpit. In this stage, the events will occur as tasks or instructions, according to which testers can actively operate the content on mediums. For example, there is an event of space construction in resting scenarios that requires users to create a face-to-face communication space. In this case, testers need to turn the seat back and put down the table according to the instruction of the event task.

The platform has added mixed reality on the basis of virtual reality in order to achieve better operation and space experience. It includes not only virtual scenarios and virtual interaction components, but also some physical objects associate with virtual ones. The physical objects in the physical environment can be mapped into the virtual environment in real time, through installing infrared reflected markers on target devices and binding this unique identity of markers to virtual components. In addition, to meet the operation demands of driving scenarios and human-vehicle authority transition scenarios, there are sensors on necessary driving-related objects such as steering wheel, braking, etc. The sensors will sense the operation by users, and transfer the data to processor to make judgement. As a result, users can really operate and experience physical objects with simpler shapes to make change of the position and status of virtual components with more complex and real shapes. Mixed reality provides more natural space feeling and operation experience, thus to increase the feasibility of the validation.

4.3 HMI Experience Validation

Events is the core content in the process of HMI experience validation, which guides testers to validate the responsiveness and operability medium by medium, function by function. It includes both users' active command in the context and passive response to the context. Events of active command refer to what are only tasks and need to be actively operated to complete, the same as that in the space layout validation phase. Events of passive response refer to what have already occurred and need to be checked, or need users' timely operation, to complete.

There has already been a mapping of preset HMI system validation scenarios in the default setting of the platform. The default function in HMI will respond in a specific form when the specific scenario occurs, and users can execute the event through default mode (screen operation). For example, when there is a call-in event, all HMI mediums with call-in function will response in screen notices until the testers putting through or hanging up the call.

However, for new components add by users, or components with special verification requirements, designers should customize response modes and performing logic to make expected relevance effect between HMI content and scenarios, so that they can make targeted optimization of the cockpit.

The modes of medium-event response consist of all-medium notification, current-medium or nearest-medium activation and notification as well as voice notification. Medium-event response usually appears in the notification events of exterior scene change. Designers can set the response modes in the platform. When selecting an event of a scenario, it can be set whether to take on voice notification and Do Not Disturb (only present on current medium or nearest medium) or not. At the same time, the potential responsive medium components are marked to be set switch status (Fig. 7). The default settings are all-medium response and voice notification on.

In some cases, the completion of events needs actively command and operation to the corresponding components. Active command often occurs in scenarios and events of space construction, information exchange, vehicle status control. Similar with activation modes of mediums, the command modes consist of touching, button control, positioning, voice control, etc. When selecting an event of a scenario as what have done in setting response modes, it can also be set whether to take voice control on. For marked components, their specific command modes can also be set (Fig. 7). The default settings are touch trigger and voice control on.

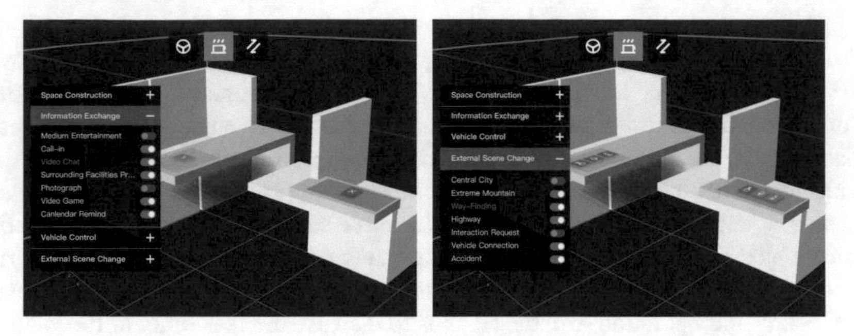

Fig. 7. Response (left) and command (right) modes setting

5 Conclusion and Limitation

The paper introduces a mixed reality platform for design and user study of self-driving vehicle cockpit. It presents modularized spatial components, medium components and HMI function components to assist designers make free exploration of space layout possibilities and HMI design possibilities of self-driving vehicle cockpit. It also provides mapping mechanism for the HMI components and different events to help validate and optimize them. The attempts above might contribute to the research of design and user study platform for self-driving vehicle.

As the validation stage is based mainly on the designers' or testers' subjective feelings of the usability and feasibility now, the next step is to add objective numerical measurement channel. Setting up objective monitoring and data recording system, as well as the evaluation index system is necessary in the further study of the platform. The data and scenarios will also be integrated into database for further statistics and inspection. The work is helpful for more targeted optimization.

References

1. Sotelo, M., Rodriguez, F., Magdalena, L., Bergasa, L., Boquete, L.: A color vision-based lane tracking system for self-driving on unmarked roads. Self-driving Robots. **16**, 95–116 (2004)
2. Levinson, J., et al.: Towards fully self-driving: Systems and algorithms. In: Intelligent Vehicles Symposium (IV), pp. 163–168. IEEE (2011)
3. Martens, M.H., Wilschut, E.S., Pauwelussen, J.: Semi-autonomous driving: do drivers still respond to unexpected events?. In: 15th World Congress on Intelligent Transport Systems and ITS America's 2008 Annual MeetingITS AmericaERTICOITS JapanTransCore (2008)
4. Reimer, B., et al.: Behavioral impact of drivers' roles in automated driving. In: Proceedings of the 8th International Conference on Automotive User Interfaces and Interactive Vehicular Applications, pp. 217–224. ACM (2016)
5. Walch, M., Lange, K., Baumann, M., Weber, M.: Autonomous driving: investigating the feasibility of car-driver handover assistance. In: Proceedings of the 7th International Conference on Automotive User Interfaces and Interactive Vehicular Applications, pp. 11–18. ACM (2015)

6. Schieben, A., Temme, G., Köster, F., Flemisch, F.: How to interact with a highly automated vehicle. Generic interaction design schemes and test results of a usability assessment, pp. 251–267. DLR (2011)
7. Park, H.: In-vehicle AR-HUD system to provide driving-safety information. ETRI J. **35**, 1038–1047 (2013)
8. BMW unveiled its BMW i Inside Future sculpture at CES (2017). http://www.bmwblog.com/2017/01/04/bmw-unveiled-its-bmw-i-inside-future-sculpture-at-ces-2017/
9. Sun, X., Li, T.: Survey of studies supporting the use of in-air gesture in car HMI. UXPA, China (2014)
10. Schmidt, R., Weisser, H., Schulenberg, P., Goellinger, H.: Autonomous driving on vehicle test tracks: overview, implementation and results. In: Proceedings of the IEEE Intelligent Vehicles Symposium 2000, IV 2000, pp. 152–155. IEEE (2002)
11. Rebelo, F., Noriega, P., Duarte, E., Soares, M.: Using virtual reality to assess user experience. Hum. Factors **54**(6), 964–982 (2012)
12. Desai, S., Blackler, A., Popovic, V.: Intuitive interaction in a mixed reality system. In: 2016 Design Research Society 50th Anniversary Conference. Brighton (2016)
13. Pacheco, C., Duarte, E., Rebelo, F., Teles, J.: Using virtual reality in the design of indoor environments: selection and evaluation of wall colors by a group of elderly. In: Kaber, D.B., Boy, G. (eds.) Advances in Cognitive Ergonomics. Advances in Human Factors and Ergonomics Series, pp. 784–792. CRC Press, Miami (2010)
14. Purschke, F., Schulze, M., Zimmermann, P.: Virtual reality - new methods for improving and accelerating the development process in vehicle styling and design. In: Computer Graphics International, p. 789. IEEE Computer Society (1998)
15. Noon, C., Zhang, R., Winer, E., Oliver, J., Gilmore, B., Duncan, J.: A system for rapid creation and assessment of conceptual large vehicle designs using immersive virtual reality. Comput. Ind. **63**(5), 500–512 (2012)
16. Meyrueis, V., Paljic, A., Leroy, L., Fuchs, P.: A template approach for coupling virtual reality and CAD in an immersive car interior design scenario. Int. J. Prod. Dev. **18**(5), 395–410 (2017)
17. Sportillo, D., Paljic, A., Boukhris, M., Fuchs, P., Ojeda, L., Roussarie, V.: An immersive virtual reality system for semi-autonomous driving simulation: a comparison between realistic and 6-DoF controller-based interaction. In: The International Conference on Computer and Automation Engineering, pp. 6–10 (2017)
18. Bajana, J., Francia, D., Liverani, A., Krajčovič, M.: Mobile tracking system and optical tracking integration for mobile mixed reality. Int. J. Comput. Appl. Technol. **53**, 13 (2016)
19. LaViola, J., Kruijff, E., McMahan, R., Bowman, D., Poupyrev, I.: 3D User Interfaces. Addison-Wesley, Boston (2017)
20. Component driven design and development. https://events.drupal.org/barcelona2015/sessions/component-driven-design-and-development
21. Dery-Pinna, A.M., Fierstone, J., Picard, E.: Component model and programming: a first step to manage human computer interaction adaptation. In: International Conference on Mobile Human-Computer Interaction, pp. 456–460. Springer, Heidelberg (2003)
22. Koo, J., Kwac, J., Ju, W., Steinert, M., Leifer, L., Nass, C.: Why did my car just do that? Explaining semi-autonomous driving actions to improve driver understanding, trust, and performance. Int. J. Interact. Des. Manuf. (IJIDeM) **9**, 269–275 (2014)

Development of a Scanning Support System Using Augmented Reality for 3D Environment Model Reconstruction

Yuki Harazono[1(✉)], Hirotake Ishii[1], Hiroshi Shimoda[1], and Yuya Kouda[2]

[1] Graduate School of Energy Science, Kyoto University, Yoshida Honmachi, Sakyo-Ku, Kyoto-Shi, Kyoto, Japan
{harazono, hirotake, shimoda}@ei.energy.kyoto-u.ac.jp
[2] Fugen Decommissioning Engineering Center, Japan Atomic Energy Agency, Myojin-Cho, Tsuruga-Shi, Fukui, Japan
kouda.yuya@jaea.go.jp

Abstract. 3D reconstruction models are useful for many situations in maintenance and decommissioning work at nuclear power plants (NPPs). There is a method to make the models from color and depth images. When using the method, it is necessary to scan a target environment without missing in order to make detailed and precise models. However, work sites at NPP are very complicated, and it is difficult to scan without missing. In this study, we aim to develop a scanning support system that enables users to make 3D reconstruction models without missing even in a very complicated environment such as NPPs. The system reminds and encourages users to scan work sites by visualizing unscanned area using an algorithm extended truncated signed distance function.

Keywords: Augmented reality · Model reconstruction · Environment scanning support

1 Introduction

Recently, by spreading of a RGB-D camera, which can get color and depth image, it becomes easier to make 3D reconstruction models which reflect situations of scanned objects. 3D reconstruction models made by scanning work sites in nuclear power plants (NPPs) using a RGB-D camera have some merits. These models reflect even small facilities that do not exist in computer-aided design (CAD) models made by hand. Once they have been made, workers can verify the work site situation at any time without visiting the site. A work support system which enables workers to refer work-related information using 3D reconstruction models is expected to increase safety and efficiency in maintenance and dismantling work [1].

There are some methods to make 3D reconstruction models, such as using color and depth images obtained by RGB-D camera [2]. On the other hand, work sites at NPP are very complicated. Therefore, even if the users scan the environment carefully, some parts of the environment are tend to remain unscanned. As the results, some areas

W. Karwowski and T. Ahram (Eds.): IHSI 2019, AISC 903, pp. 460–464, 2019.
https://doi.org/10.1007/978-3-030-11051-2_69

in the 3D reconstruction model are ended up missing. These reasons cause some problems when scanning to make detailed and precise reconstruction models. Furthermore, when using 3D reconstruction models in actual maintenance and decommissioning works at NPPs, not IT experts but the workers have to scan work sites. Therefore, even those unfamiliar users with scanning need to be able to scan easily.

In this study, we aimed to develop a scanning support system that enables users to make 3D reconstruction models without missing even in a very complicated environment such as NPPs. In order that even unfamiliar users with scanning can scan easily, the system presents information, such as unscanned areas, using augmented reality (AR).

2 Requirements of Scanning Support System

This section explains a method to make 3D reconstruction models and information to present to users.

2.1 Method to Make Reconstruction Models

There are some methods to scan the environment and make 3D reconstruction models, such as the methods using drones or mobile robots [3] and the methods using camera or laser scanner [4]. Considering the costs, the methods using drones or mobile robots are expensive. Work sites at NPP are very complicated, and it is dangerous to use flying devices such as drones in work sites. Therefore, it is suitable that users move around and scan. If users use laser scanner, it will need some laser scanners to scan without missing. The method using RGB-D camera needs the users to scan by moving around with a handheld RGB-D camera. This method suits the situation at work sites. Therefore, in the system, we employed a method that users move around and scan with a handheld RGB-D camera.

2.2 Information to Present to Users

In the method that users move around and scan with RGB-D camera, 3D reconstruction models are made from color and depth images obtained by RGB-D camera. However, if a user moves RGB-D camera quickly when scanning, obtained color and depth images include blurs. Because of this, users can make only 3D reconstruction models which have collapsed shapes. Therefore, it is necessary to present a current speed of moving RGB-D camera and a limit of acceptable speed when scanning.

Work sites at NPP are very complicated, and there are many areas hidden by other facilities as seen Fig. 1. Even if the users scan the environment carefully, some parts of the environment are tend to remain unscanned. Therefore, it is necessary to visualize unscanned areas in order to make 3D reconstruction models without missing.

In addition, high texture resolution is also necessary in order to make detailed 3D reconstruction models which workers at NPP can verify work site situations. As seen Fig. 2, areas scanned from far or oblique position have low texture resolution.

Therefore, it is necessary to present areas with low texture resolution and encourage users to rescan from proper position and direction.

Thus, we decided that a speed of moving RGB-D camera, unscanned areas, and areas with low texture resolution are the information to present to users in order to make more precise and detailed 3D reconstruction models. In the study, we developed the scanning support system which presents these three information to users.

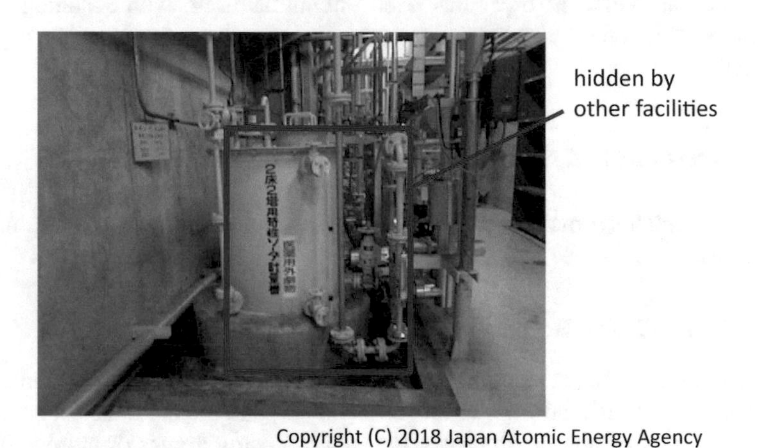

Fig. 1. An example of the area hidden by other facilities.

Fig. 2. An example of an area with low texture resolution.

3 Development of Scanning Support System

3.1 Design of Scanning Support System

The system calculates a speed of moving RGB-D camera from the amount of the camera movement between each consecutive frame. The system visualizes it using a gauge bar. Using a gauge bar, the extent of a speed can be presented intuitively. In the system, unscanned areas and areas with low texture resolution are detected by the algorithm extended truncated signed distance function (TSDF) [5]. The system visualizes them using AR. However, it is thought difficult to understand when visualizes the whole unscanned areas using AR simultaneously. Therefore, the system visualizes only boundaries between scanned areas and unscanned areas. Visualizing only the boundaries is expected to present areas simply where users should scan. The system visualizes areas with low texture resolution by superimposing the figures, such as arrows, which enable the users to understand both positions and directions the users should scan.

Figure 3 presents the overall of the system. The system calculates these information from scanned color and depth images and superimposes them on the camera view using AR.

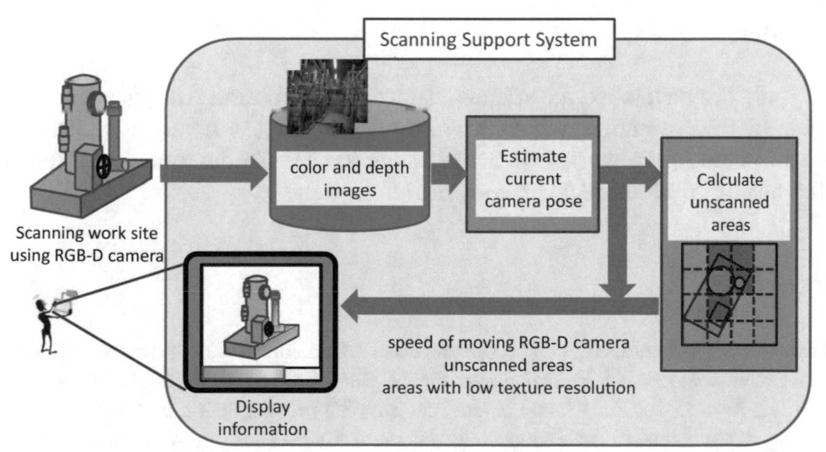

Fig. 3. An overview of scanning support system.

3.2 Calculation of Unscanned Areas

In the system, we employed the algorithm extended TSDF to detect unscanned areas and areas with low texture resolution. TSDF is an algorithm to make depth map of a scanning environment from scanned depth images. In TSDF, scanning environment is divided into certain size voxels. Distances between each voxels and nearest surfaces from each voxels are calculated and stored from every depth image. In the voxels inside objects, negative values are stored. In contrast, in the voxels outside objects, positive values are stored. From these values stored in each voxels, the system classifies each voxels into scanned areas, unscanned areas, borders between them, or surface areas. The algorithm considers surface areas scanned from far or oblique positions as areas with low texture resolution.

Figure 4 shows an example when applied the algorithm to synthesized computer graphics images, and we confirmed the algorithm calculates and presents the borders between scanned areas and unscanned areas.

Fig. 4. Examples of visualization when applying developed algorithm to synthesized computer graphics images provided in [6].

4 Future Works

In this study, we developed the scanning support system using AR which enables users to make 3D reconstruction models without missing even in a very complicated environment. As the future works, we are planning to evaluate the usability of the system and the quality of obtained 3D reconstruction models.

References

1. Harazono, Y., Kimura, T., et al.: Development of an information reference system using reconstruction models of nuclear power plants. Nucl. Eng. Technol. **50**(4), 606–612 (2018)
2. Izadi, S., Kim, D., et al.: KinectFusion: real-time 3D reconstruction and interaction using a moving depth camera. In: Proceedings of the 24th Annual ACM Symposium on User Interface Software and Technology, pp. 559–568 (2011)
3. Remondino, F., Barazzetti, L., et al.: UAV photogrammetry for mapping and 3D modeling current status and future perspectives. In: International Archives of the Photogrammetry, Remote Sensing and Spatial Information Sciences, pp. 25–31 (2011)
4. Xiong, X., Adan, A., et al.: Automatic creation of semantically rich 3D building models from laser scanner data. Autom. Constr. **31**, 325–337 (2012)
5. Curless, B., Levoy, M.: A volumetric method for building complex models from range images. In: Proceedings of the 23rd Annual Conference on Computer Graphics and Interactive Techniques, pp. 303–312 (1996)
6. Choi, S., Zhou, Q., Koltun, V.: Augmented ICL-NUIM Dataset. http://redwood-data.org/indoor/dataset.html. Accessed 15 Oct 2018

The Effects of Situational Context on Information Valuation

Justine P. Caylor[1(✉)], Robert J. Hammell II[2], Timothy P. Hanratty[1], Eric G. Heilman[1], and John T. Richardson[1]

[1] U.S. Army Research Laboratory, Aberdeen Proving Ground, MD, USA
{justine.p.caylor.ctr,timothy.p.hanratty.civ,eric.g.heilman.civ,john.t.richardson7.civ}@mail.mil
[2] Towson University, Towson, MD, USA
rhammell@towson.edu

Abstract. A commonality between simple, everyday tasks and complex, military operations is that they both are dependent on decision-making. In this era of big data, successful decision-making is reliant on the effective *contextual exploitation* of information. Understanding information and its value within context is a complicated task. For this research *context* is defined as the macro environment surrounding a decision space. With that understanding, the U.S. Army Research Laboratory (ARL) in collaboration with Towson University is investigating how humans perceive and judge information value within varying context. This paper highlights the experimental design and presents the early findings of an Amazon Mechanical Turk (MTurk) experiment where a human population judged the effects of varying information sources on three different online purchase contexts. Preliminary results indicate that information and situational context play a significant role in discerning information value.

Keywords: Situational context · Information value · Decision-Support agents Amazon Mechanical Turk

1 Introduction

From artificial intelligence (AI) applications that assist in everyday tasks to more impactful uses of AI in military operations [1, 2], advanced technologies are changing the way decisions are made. Information affects decision-making by the way it is perceived and how valuable it is to the contextual situation. However, this can be a daunting and difficult task for a human to accomplish [3–5]. In a move toward combating information overload and assisting timely decision-making, ARL has developed a Value of Information (VoI) metric. The VoI metric reflects information value based on the characteristics of the source of the information, the content of the information, and its latency [6–8]. This paper extends the Value of Information research and examines how humans perceive and judge information value within varying context.

On the battlefield, information drives action [3]. Military operations are defined by an unprecedented volume, velocity, variety, veracity, and value of information on the battlefield [9]. While information is critical, the large amount of available information

W. Karwowski and T. Ahram (Eds.): IHSI 2019, AISC 903, pp. 465–470, 2019.
https://doi.org/10.1007/978-3-030-11051-2_70

can be problematic; it is often full of uncertainty and contradiction [4, 9]. Recent ARL research in the VoI domain focuses on evaluating the role context plays in the value of information. Towards this end, ARL participated in the U.S. Army Research Institute's Umbrella Weeks (UWs) by performing an experiment that included surveying Soldiers on how different context affects the valuation of information in a military environment [10]. Analysis from the UWs data revealed that situational context plays a role in discerning information value for battlefield scenarios. The work described herein seeks to evaluate how the general population perceives and interprets pieces of information and its characteristics.

This paper highlights the experimental design and discusses a preliminary observation of trends and patterns that emerged from the data related to how situational context affects information valuation. Results from the experiment and subsequent data analysis will be used to create a foundation for comparison with on-going ARL research investigating context in military decision-making. The overall goal is the creation of computational models that support intelligent human-agent collaboration.

2 Crowdsourced Approach

To collect a large population of diverse people, the experiment was run using the crowdsourcing platform Amazon Mechanical Turk. The experiment casts participants in the role of a customer shopping for various products for different reasons. Each participant is presented three scenarios that feature three reviews of a product conveying various levels of sentiment. Reviews can provide customers with an insight that affects a products perceived value [11]. Using a Likert rating scale from zero to 10 (zero represents low value, and 10 represents high value), participants judge the value of these reviews and their influence towards purchase decisions. In the study, three scenarios are measured:

- Vignette A high importance (situation that affects you and or high cost involved), the shopping scenario consisted of the participant needing to purchase a new roof due to significant damage (NEED purchase).
- Vignette B low-medium importance (situation that affects someone that is not you and or low cost involved), the shopping scenario consisted of the participant purchasing a cake to bring to a housewarming party (WANT purchase).
- Vignette C unknown-low importance (no established importance), the shopping scenario consisted of the participant seeing an online advertisement for foot massaging slippers for a low price of $19.99 (IMPULSE purchase).

The preliminary results in this paper focus on the analysis of Vignettes A vs C to gain an understanding of how the extreme cases are perceived and valued.

Decision making typically involves the dissemination of various pieces of information. These pieces of information can have varying degrees of agreement, from contradictory to complementary. To measure how value and decision making are affected by additional pieces of information, participants were given an initial review and then two additional reviews: one that was positive (complementary) to the initial review and one that was negative (contradictory) to the initial review. Vignette Groups (VGs) #1 and #2 present the participant with the initial review, a positive additional review, and then a negative additional review. VGs #3 and #4 present the participant with the initial review, a negative additional review, and then a positive additional review.

For the experiment, four information source categories are utilized (listed from "best" to "worst"): *usually helpful* (UH), *fairly helpful* (FH), *usually not helpful* (UNH), and *unverified* (UV). The reviews (initial, first additional, second additional) provided for each vignette group were as follows:

- VG #1 – FH, FH (+), FH (-)
- VG #2 – UV, FH (+), FH (-)
- VG #3 – UH, UH (-), UH (+)
- VG #4 – UNH, UH (-), UH (+)

3 Preliminary Results

In total 304 participants were collected from the experimental run. After removing distinct outliers, which attributes to participants interpreting the directions incorrectly or "fast-clicking" through the experiment, 278 participant results remained for analysis. The participants ranged from the ages 18+, and varied in occupation and highest level of education attained.

For the investigation, the contexts were broken into separate groups by the priority of purchase score, which was captured by participants on a Likert scale ranging from one to five. Vignette A had the expectation of being a "5" (need) priority and Vignette C had the expectation of being a "1" (impulse) priority. For our study, "expectation" is defined as the "correct" value based on the researchers' opinion. In relation to the priority level, it is considered that the higher the priority, the higher the value (and vice versa). Breaking the contexts into separate groups allows analysis related to how participants who matched to the expectation valued the information, and allows comparison with the other groups that did not match the expectation.

3.1 Evaluation of Vignette A Value

Vignette A for every VG was the representation of a high importance/high priority scenario. 63% of participants interpreted that the priority of purchase is a 5, thus matching our expectation.

These 175 (63%) responses were further examined as shown in Table 1. The median value score given after the initial review ranged from 5 to 8 across all four vignette groups, which is above the decision space of "likely to purchase/medium value" and under the decision space of "will purchase/high value". After Review #1 (a positive review) was provided to VGs #1 and #2, the value scores rose; for VGs #3 and #4 the negative Review #1 decreased the value scores. The fourth column highlights the median delta change between the value after Review #1 and the Initial Value. The fifth column in Table 1 shows that the negative Review #2 for VGs #1 and #2 decreased the value scores while the positive Review #2 for VGs #3 and #4 increased the value scores. The final column highlights the median delta change between the values after all reviews versus the value after Review #1.

The pattern observed after the subsequent reviews followed our expectation that when the participants are faced with a positive review, they would increase their perceived value score, and decrease when faced with a negative review. Specifically speaking, 161 out of 175 participants within this priority group performed expectedly after reading Review #1, and 168 out of 175 participants performed expectedly after reading Review #2.

Table 1. Vignette A information valuation

Vignette group	Initial value	Value after review #1	Delta change after review #1	Value after all reviews	Delta change after review #2
VG #1	7	8(+)	1	6(−)	−2
VG #2	5	7(+)	2	4(−)	−2
VG #3	8	5(−)	−2	8(+)	2
VG #4	5	5(−)	−3	7(+)	4

An F-test statistic was calculated to compare the participants initial value scores of the different priority groupings and information sources to see if there was any significant differences. After calculation, it was determined that the way the participants gave value to Vignette A across priority groupings and information source was significantly different. This indicates that differences in priority and information sources have an effect on perceiving value in high importance situations.

3.2 Evaluation of Vignette C Value

Vignette C for every VG was the representation of an unknown-low importance/low priority scenario. Only 43% of participants interpreted that the priority of purchase is a 1, thus matching our expectation.

These 119 (43%) responses were further examined as shown in Table 2. The average value score after the initial review ranged from 1 to 4, which is slightly above the decision space of "will not purchase/low value" and slightly under the decision space of "likely to purchase/medium value". After Review #1 (a positive review) was provided to VGs #1 and #2, it had a marginal effect on the value score; for VGs #3 and #4 the negative Review #1 decreased the value scores. The fourth column highlights the median delta change between the value after Review #1 and the Initial Value. The fifth column in Table 2 shows that the negative Review #2 for VGs #1 and #2 decreased the value scores while the positive Review #2 for VGs #3 and #4 marginally increased the value scores. The final column highlights the median delta change between the values after all reviews versus the value after Review #1.

The pattern observed after the subsequent reviews followed our expectation partially: there was a minimal increase when faced with a positive review and decrease when faced with a negative review. Specifically speaking, 95 out of 119 participants within this priority group performed expectedly after reading Review #1, and 106 out of 119 participants performed expectedly after reading Review #2.

Table 2. Vignette C information valuation

Vignette Group	Initial value	Value after review #1	Delta change after review #1	Value after all reviews	Delta change after review #2
VG #1	4	4(+)	0	2(−)	−2
VG #2	2	2(+)	0	1(−)	−1
VG #3	5	1(−)	−3	2(+)	0
VG #4	2	1(−)	−2	2(+)	1

Similar to the calculation in Sect. 3.1, an F-test statistic was calculated to compare the participants initial value scores of the different priority groupings and information sources to see if there was any significant differences. After calculation, it was determined that the way the participants gave value to Vignette C across priority groupings and information source was not significantly different. This indicates that differences in priority and information sources did not have an effect on perceiving value in unknown-low importance situations.

4 Conclusion and Future Works

For both Vignette A and Vignette C, the initial value ratings show that source does influence perceived value, where the highest rated source (UH/VG #3) is correlated with the higher initial value scores and low rated source (UV/VG #2 and UNH/VG #4) is correlated with the lower initial value score. This pattern was observed across all priority groupings. A trend of priority score emerged, which shows that value scores where similar for vignettes with the same priority scores (i.e. the value scores of those who identified Vignette A as a 3 is similar to those who identified Vignette C as a 3). This indicates that priority or perception of situation importance has an influence on information valuation.

The effect on the participants' decision-making with respect to how the new reviews were introduced was also examined. Across all priority groupings and sources for Vignette A and C, when participants were given a negative review (regardless of whether it was presented first or second relative to the positive review), it substantially decreased their value score. The positive additional review did not bring as much of a notable impact to the value score; however, it was observed to have a stronger impact when presented after the negative review, rather than before it.

While more analysis that is detailed is needed, this preliminary analysis of the data indicates that the decision-making thought processes among both military and general population groups are similar, even though the specific value changes and ranges vary. Further analysis of other dependent/independent variables and a closer examination of the data to discover other trends will be done in the future. This paper presents preliminary research and experimental trends and validates that further examination of data along the lines of these experiments is warranted.

Acknowledgements. The Research Associateship Program at USARL administrated by ORAU supported this research.

References

1. Chen, J.Y., Barnes, M.J.: Human–agent teaming for multirobot control: a review of human factors issues. IEEE Trans. Hum.-Mach. Syst. **44**(1), 13–29 (2014)
2. Barnes, M.J., Evans III, A.W.: Soldier-robot teams in future battlefields: an overview. In: Human-Robot Interactions in Future Military Operations, pp. 29–50. CRC Press (2016)
3. Hanratty, T., Hammell, R.J., Heilman, E.: A fuzzy-based approach to the value of information in complex military environments. In: International Conference on Scalable Uncertainty Management, pp. 539–546. Springer, Heidelberg (2011)
4. Richardson, J.T., Caylor, J.P., Heilman, E.G., Hanratty, T.P.: ACT-R modeling to simulate information amalgamation strategies. In: International Conference on Applied Human Factors and Ergonomics, pp. 326–335. Springer, Cham (2018)
5. Alberts, D.S., Garstka, J.J., Hayes, R.E., Signori, D.A.: Understanding information age warfare. Assistant Secretary of Defense (C3I/Command Control Research Program), Washington DC (2001)
6. Hanratty, T., Heilman, E., Dumer, J., Hammell II, R.J.: Knowledge elicitation to prototype the value of information. In: Proceedings of the 23rd Midwest Artificial Intelligence and Cognitive Sciences Conference (MAICS 2012), Cincinnati, OH, pp. 173–179 (2012)
7. Hanratty, T., Dumer, J., Hammell II, R.J., Miao, S., Tang, Z.: Tuning fuzzy membership functions to improve value of information calculations. In: Proceedings of the 2014 North American Fuzzy Information Processing Society Conference (NAFIPS 2014), Boston, MA, 24–26 June 2014
8. Michaelis, J.: Requirements for value of information (VoI) calculation over mission specifications. In: SPIE Next Generation Analyst, Orlando, FL, April 2017
9. Hanratty, T., Heilman, E., Richardson, J., Caylor, J.: A fuzzy-logic approach to information amalgamation. In: FUZZ-IEEE, Naples, Italy, 9–12 July 2017
10. Hanratty, T., Heilman, E., Richardson, J., Mittrick, M., Caylor, J.: Discerning the role context plays in the value of information, a key to effective human-agent teaming. In: Proceedings of 2019 Hawaii International Conference on System Sciences (HICSS), Maui, HI, January 2019
11. Hu, N., Liu, L., Zhang, J.J.: Do online reviews affect product sales? The role of reviewer characteristics and temporal effects. Inf. Technol. Manage. **9**(3), 201–214 (2008)

Aligning Teams to the Future: Adapting Human-Machine Teams via Free Energy

Adam Fouse[1(✉)], Georgiy Levchuk[1], Nathan Schurr[1], Robert McCormack[1], Krishna Pattipati[2], and Daniel Serfaty[1]

[1] Aptima, Inc., 12 Gill Street Suite 1400, Woburn, MA 01801, USA
{afouse, georgiy, nschurr, serfaty}@aptima.com, rmccormack@ucon.edu
[2] University of Connecticut, Storrs, CT, USA
Krishna.pattipati@aptima.com

Abstract. Future hybrid human-machine teams will need to optimize their performance in uncertain environments by adapting their team structure. To address this need, we have developed a framework based on minimization of variational free energy, an information theoretic measure that has been shown to account for a variety of biological self-organizing phenomena. This paper proposes a novel approach to balance team structure by adapting roles and relationships based upon this framework. We apply this approach to evaluate possible structures for an infantry squad of human soldiers and autonomous systems. Using our STATES team simulation environment, we simulate mission performance for these teams and demonstrate that this approach enables a 12-person team to achieve performance results on par with a 15-person traditional team in terms of mission execution time. We argue that these results indicate that the free energy approach will lead to better hybrid team adaptations and improved performance.

Keywords: Human-machine teaming · Team structure adaptation
Adaptive teams · Free energy · Active inference

1 Introduction

Teams of the future will include both humans and increasingly complex and capable systems. To enable effective teamwork in these hybrid teams, we must understand the mechanisms that allow humans to team with multiple intelligent entities to effectively explore, execute, and learn in changing and unpredictable environments. In addition, we argue that these human-system teams must optimize their performance and efficiency in the face of adversity and uncertainty in the environment and themselves by adapting through changes in team structure.

Approaches to this must not only maintain enough intelligence to perceive and act locally, but they must also possess team-level collaboration and adaptation processes to improve overall performance. We posit that such processes embody energy-minimizing mechanisms found in all biological and physical systems and operate over the objectives and constraints that can be defined and analyzed locally without the need for

W. Karwowski and T. Ahram (Eds.): IHSI 2019, AISC 903, pp. 471–477, 2019.
https://doi.org/10.1007/978-3-030-11051-2_71

global centralized control. In this paper, we present a continuation of our work on adaptive self-organizing teams using active inference, a model based on the iterative minimization of variational free energy that encodes task performance and team structure complexity.

1.1 Application of Free Energy to Team Design

We have developed a generalizable mathematical abstraction for the design of adaptive human-machine teams, based on application of the *free energy principle* to hybrid teams of human and machine actors. The free energy principle tries to explain how self-organizing systems, such as the brain, restrict themselves to a small number of attracting states to avoid disorder in uncertain environments and minimize surprise [1]. Free energy optimization has been shown to account for a variety of phenomena in sensory, cognitive, and motor neuroscience, and provides useful insights into structure-function relationships in the brain [2]. These unifying ideas about how the brain processes information can now be applied to the design teams of machines and humans.

True surprise minimization requires marginalization over all possible states of the world, which is infeasible. Instead, the free energy principle argues that adaptive self-organizing systems minimize the bounds on surprise, called *variational free energy* [3]. Our framework postulates that hybrid human-machine teams can efficiently operate in dynamic uncertain environments by minimizing this variational free energy, producing *bounded rational behavior.*

Application of this framework takes the form of an iterative perception-action loop, with a learning process to update knowledge of team capabilities, a perception process to estimate mission execution, and a control process to change team structure [4]. This formalism can be applied for both evaluating among pre-defined team structures as well as for generating new team structures [5]. Evaluation of team structures computes values of free energy through simulation or human-in-the-loop experimentation. Generative application requires definition of modular elements of team structure that can be modified through the control process to adapt teams to expected mission distributions.

2 Team Structure and Simulation

2.1 Team Structure and Mission Definition

Modeling the application of free energy to the design of teams requires definition of team structures and expected tasks of the team. Our team structure abstraction consists of the team members, the capabilities of those team members, and the communication and task assignment relationships between those members. Our abstraction uses an attributed graph to represent that structure. In this representation, the nodes of the graph are human or robot team members (agents), with attributes representing the capabilities of each member to perform a task. The links between nodes represent functional roles and relationships, and task assignment authority.

Missions are defined probabilistically, with each mission consisting of a set of events that occur with a defined frequency across a set of geospatial areas. For each event, a set of tasks are defined with probabilities that each task is required to complete the event. Events have pairwise transition probabilities defined to determine the order of events that unfold during a mission. Each task has defined relationships with other tasks. Combined, the event-task and task-task connections define a space of possible task networks for each event.

2.2 Simulation of Task Assignment for Teams to Evaluate Structure (STATES)

To test the application of free energy to the team design, we have developed a mechanism for simulating the performance of teams, which we call the Simulation of Task Assignment for Teams to Evaluate Structure (STATES). As shown in Fig. 1, STATES takes as input team structures and mission scenarios. The core of STATES consists of probabilistic generation of task assignment vectors based on the ranges, probabilities, and locations defined in the mission. Through multiple simulations, it generates instantiated event/task networks based on the probabilities defined by the mission, and simulates task assignment and performance to generate metrics of the teams' processes, outcomes, and energy.

The STATES outputs include measures related to outcomes, including time-based measures such as total mission time and task delay, measures related to process, such as task workload and team balance, and measures related to energy, including expended energy, entropy, and free energy. By looking both at the evolution of task assignment over the span of a single mission simulation, and at the range of task assignment vectors generated over many simulations, the output of the simulation can inform both the overall quality and adaptability of the team, as well as provide insight into the specific tasks or team members that create bottlenecks or team weaknesses.

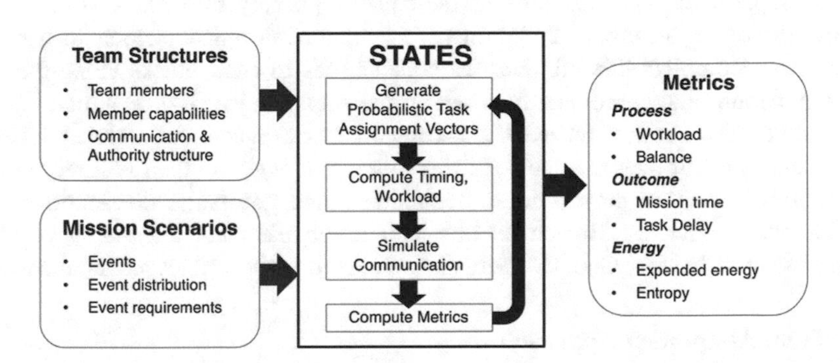

Fig. 1. STATES simulation architecture

3 Simulation of Team Adaptation

3.1 Military Infantry Squad Domain

We have used these definitions to apply the free-energy team design formalism to the design of military teams. Specifically, we have focused on the structure of infantry squads, such as are found in the US Army or US Marine Corps (USMC). A squad typically consists of nine to fourteen members, led by a squad leader, and further organized into fire teams of 3–5 members. The structure of squads, and how that structure supports the integration of automated systems, is under active consideration. For example, the USMC announced in 2018 that in future squads, fire teams would be reduced from four to three members, and a command element would be created that include the squad leader, an assistant squad leader, and a squad systems operator charged with managing and operating automated systems [6].

3.2 Simulated Team Structures and Scenario

Within the infantry squad domain, we have focused our squad structure considerations on questions of team size (total size and fire team size), integration of autonomous systems, and role definition. For the results reported here, we were primarily interested in the effect of total squad size, with a design goal of mitigating the loss of resources when reducing the number of human team members from 15 to 12.

Each squad has a command element with a squad leader, assistant squad leader, and squad systems operator. Each fire team has a fire team leader and a mix of riflemen, automatic riflemen, and systems operators, depending on the fire team size. Each squad also has two Unmanned Aerial Vehicles (UAVs) and one Unmanned Ground Vehicle (UGV). An example of a 15-person squad is shown in Fig. 2.

We tested these squads with a scenario with a heterogeneous set of events and tasks. In support of a platoon assault, the squad's primary mission is to establish a cordon position in an urban environment, to support an allied platoon in a nearby district. To accomplish this mission, the squad needs to clear blocks along the way, seize key terrain, and coordinate both within the squad and with other units.

In coordination with a retired U.S. Army Master Sergeant, we defined 13 event types that may occur over the course of this mission, along with their task requirement distributions, transition probabilities, frequencies, and geospatial distributions. This includes events such as: Breach an obstacle, Enter and clear a building or room, Conduct support by fire, Conduct tactical movement, and Conduct reconnaissance.

3.3 Team Adaptation Approach

While the free energy formalism suggests several possible approaches to team adaptation, here we focus on "local" adaptation by using free energy to balance the workload among team members by adapting their roles and relationships on the team. This balance across both human and agent team members aims to increase the number of good future options when performing inter-related tasks. This is enabled by agents that reason and infer about the capabilities of their teammates and make decisions about team adaptation in a decentralized manner.

In our initial simulations, 12-person squads had shown significantly worse performance than 15-person squads. The initial simulations also showed that squads with larger fire teams performed worse than smaller fire teams. Our initial interpretation was that these differences were due to overall decreased resources within squad and an inability to create effective task assignments, but further analysis based on the agent-task distribution indicated that tasking was unbalanced and that the team could generate better options with shifted roles, delegation, and change in technology.

Inspection of the free energy, workload, and task delay metrics revealed a set of tasks that were connected to high agent-task loads or contributed to significant task delay. By reconfiguring the roles of both humans and autonomous systems, we created new team structures (shown in Fig. 2 for a 15-person squad) designed to increase the number of good *task assignment options* at any given time during the mission. We created a new team structure with the following four changes:

- Technology Capabilities: Move *Communication Relay* task from UGV to UAV
- C2 Delegation: Delegate the *Report to Headquarters* task from the squad leader to the team lead of the support team
- Tactical Delegation: Move the *Tactical Questioning* tasks from the squad leader to the team lead of the support team
- Role Restriction: Remove *Overwatch* responsibilities from the command team

These same adaptation techniques were applied to the 12-person squad with 3-person fire teams, and to the 15-person squad with 6-person fire teams, with minor differences in the specific fire teams that took on shifted roles.

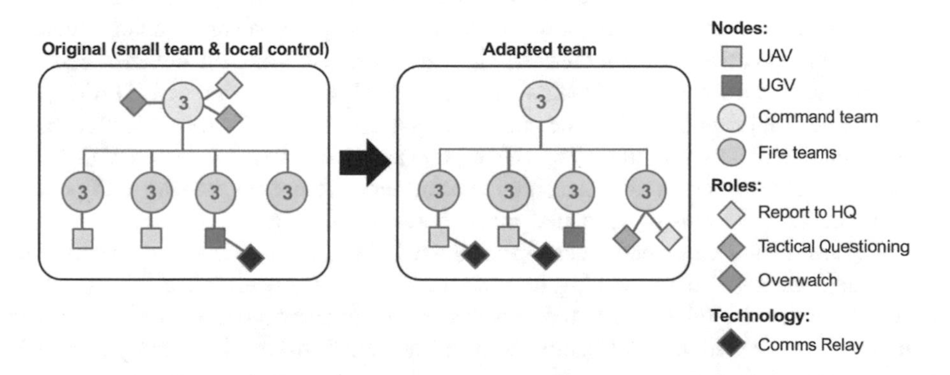

Fig. 2. Team structure adaptation for infantry squad for 15-person squad

4 Results

Overall, the results from our simulations demonstrate improvement in mission performance through the application of the adaptations. This is primarily seen in the overall mission time metric, which is shown for three of the base and adapted team structure pairs in Fig. 3.

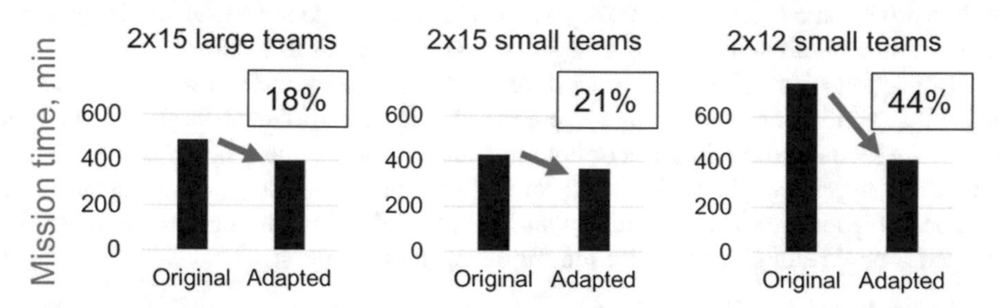

Fig. 3. Simulated performance for original and adapted teams. Performance is measured in time (lower is better). Percentage improvement shown in top right

As seen in the figure, all teams showed a decrease in total mission time with the adapted teams. The 12-person squad in particular showed the largest improvement in performance, with a 44% improvement. This new 12-person squad is on par with the performance of the original 15-person squad.

5 Discussion and Future Work

Our aim with this work is to enable heterogeneous teams of humans and agents to perform more efficiently in the face of uncertain environments and events. In this paper, we have demonstrated the application of the free energy formalism to team structure design, with a particular focus on balancing the workload among team members by adapting their roles and relationships on the team. We have leveraged STATES to illustrate how application of the free energy principles can identify how a team should be structured and how it should adapt. Using this approach, we have demonstrated that we can enable a 12-person team to achieve performance results on par with a 15-person traditional team in terms of simulated mission execution time.

We have identified several areas of future work. In addition to the work reported in this paper, we are also researching how application of free energy can inform adaptation dynamics, including when and how teams should adapt so as to balance the cost and benefit of adaptation. Finally, our claims of the benefits of the free-energy approach will be strengthened through ongoing work to test team structures with human participants performing simulated missions.

Acknowledgements. This work was supported by the Defense Advanced Research Projects Agency through contract N66001-17-C-4053. We would like to thank Dr. John Paschkewitz for his guidance of this work. Any opinions, findings, and conclusions or recommendations expressed in this paper are those of the authors and do not necessarily reflect the views of the sponsors.

References

1. Friston, K.: The free-energy principle: a unified brain theory? Nat. Rev. Neurosci. **11**(2), 127–138 (2010)
2. Friston, K., Schwartenbeck, P., FitzGerald, T., Moutoussis, M., Behrens, T., Raymond, R.J., Dolan, J.: The anatomy of choice: active inference and agency. Front. Human Neurosci. **7**, 598 (2013)
3. Friston, K., Thornton, C., Clark, A.: Free-energy minimization and the dark-room problem. Front. Psychol. **3**, 130 (2012)
4. Levchuk, G., Pattipati, K., Fouse, A., Serfaty, D.: Application of free energy minimization to the design of adaptive multi-agent teams. In: Disruptive Technologies in Sensors and Sensor Systems, vol. 10206, p. 102060E. International Society for Optics and Photonics (2017)
5. Levchuk, G., Pattipati, K., Fouse, A., Serfaty, D., McCormack, R.: Active learning and structure adaptation in teams of heterogeneous agents: designing organizations of the future. In: Next-Generation Analyst VI. Proceedings of SPIE, vol. 10653, p. 1065305, 21 May 2018
6. South, T.: 12-man rifle squads, including a squad systems operator, commandant says. Marine Corps Times, 3 May 2018. https://www.marinecorpstimes.com/news/your-marine-corps/2018/05/04/12-man-rifle-squads-including-a-squad-systems-operator-commandant-says/. Accessed 18 Oct 2018

An Agent-Based Model of Plastic Bags Ban Policy Diffusion in California

Zining Yang(✉) and Sekwen Kim

Claremont Graduate University, Claremont, USA
{zining.yang, se-kwen.kim}@cgu.edu

Abstract. This paper uses an agent-based model to study plastic bags ban policy adoption in California. By simulating the policy diffusion among counties in California with close-to-reality data, this study seeks to identify the mechanism of policy diffusion and interaction among individuals as well as counties and between individuals and counties. This work models each individual with his/her own attributes, including education, preference, and wealth. Individuals may either influence or be influenced by interacting with others based on preference difference from others. Preferences of individuals are also affected by neighboring counties' preference changes, while aggregated individual preference change determines the policy adoption and the change of preference of counties. By understanding the mechanism of policy diffusion from interconnected multi-level interaction, this work insights applicable to different areas of policy diffusion.

Keywords: Simulation · Agent based model · Public policy · Policy diffusion
Environmental policy

1 Introduction

Before California adopted the policy banning the single-use plastic bags usage in 2016 and became the first state adopted this environmental regulatory policy, there has been similar policy adoptions of similar policy in municipal legislature, such as Malibu, Fairfax, and others [1]. This study simulates innovative policy diffusion in California in 2016 by using agent-based model (ABM) incorporating policy diffusion model and social judgement theory. By using ABM, we capture the emergent behavior from multi-level interaction and identify mechanism of policy diffusion, including interaction among individuals, among counties, and between individuals and counties.

2 Literature Review

Our work utilizes diffusion theory of innovation to identify significant socio-economic attributes of agents on diffusion and social judgement theory to identify interaction and persuasion mechanism between individuals, micro level agents, to draw a plausible explanation on the pattern of the innovative policy diffusion.

W. Karwowski and T. Ahram (Eds.): IHSI 2019, AISC 903, pp. 478–483, 2019.
https://doi.org/10.1007/978-3-030-11051-2_72

In the policy diffusion model, regional proximity and socio-economic homogeneousness play a significant role [2, 3]. Owing to flexible and frequent mobility and political and demographic similarity, neighboring legislatures adopt similar policy to reduce the risk of innovative policy adoption [4–8]. In the policy diffusion, a leader, who is risk taker and socio-economically developed initiates and a follower, who is risk aversive and tries to mitigate the risk of policy adoption, follows adopting innovative policy [9]. However, the diffusion model states that the changes occur similarly at all system levels, while ABM emphasizes a scale-free network where micro level changes influences the meso and macro level changes [10]. Incorporating both models allows us to gain a comprehensive understanding of the diffusion.

The social judgment theory seeks to capture mechanism of micro level persuasion by using preference distance, determined by latitude of acceptance, rejection, and non-commitment [11]. The theory formulates the persuasive communication mechanism as a function of preference distance between persuaders and message recipient [11, 12]. However, the social judgement theory only focuses on micro-level communication and persuasion, but does not consider different level of communication, such as between micro and meso- levels [10, 13]. By using ABM, we capture mechanism of inter-level effect on changes of micro-level agent.

3 The Model

ABM simulates action and interaction between autonomous agents and system, leading to complex network synchronization and emergent behavior adaptive to the environment [14, 15]. ABM, as a tool of complex adaptive systems, can capture non-linear relationship between the cause and the effect in a non-linear manner, which is the characteristic of real world situation [15, 16].

Our ABM consists of micro-level agents like individuals in each county, meso-level agents representing 9 counties in California including Los Angeles, and a macro-level agent as the aggregation of 9 counties.

Figure 1 portrays the interactive and feedback mechanism between micro-level agents, meso-level agents, and between the two levels. When initialized, individuals are randomly categorized as either initiators or followers with different preference, wealth and education, toward policy in each county. The number of individuals in each county is calibrated based on actual population.

In the first iteration, initiators select, interact with, and persuade proximate followers to accept initiator's preference toward policy within each county. Persuasion is determined by education and wealth level between initiator and followers. If initiators succeed in persuading followers, followers will change to initiators, which select, interact with, and persuade other proximate followers. If initiators fail to persuade, followers maintain their preference unchanged. At each iteration, both initiators and followers repeat the same process. After aggregating individual preference, which determines the median preference level of a county, a county would adopt the policy if its median preference level is higher than cut-off point of 50. Once a county adopts the policy, the policy adoption would increase the probability of initiators persuading followers in neighboring county. The model stops when all counties adopt the policy.

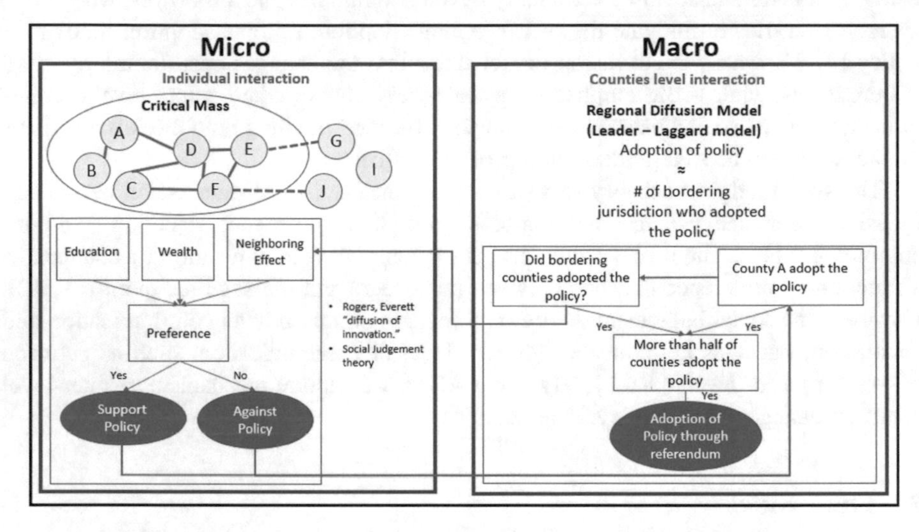

Fig. 1. Flow chart of interaction between Micro and Macro agents. While micro level agents determine the policy adoption of meso level agents, the policy adoption of neighboring meso level agents influence the micro level agents as well.

4 Parameterization and Simulation

By dividing monitors into 9 squares each representing 9 counties, policy adoption can be displayed in the interface by the changes of colors from white to yellow. Sliders to manipulate the attributes of each county, wealth, education, and population based on US Census Bureau [17] are created to adjust the input. The model considers each iteration t as a week and each run has 624 t equivalent to 3 presidential election, 12 years.

4.1 Baseline – Los Angeles

The baseline simulation is the reflection of real world policy adoption in California initiated by Los Angeles County with the most population. Los Angeles adopted environmental policy after 57 t by achieving the median preference above 50. It took 231 t to achieve all 9 counties to change their preference and adopt (Fig. 2).

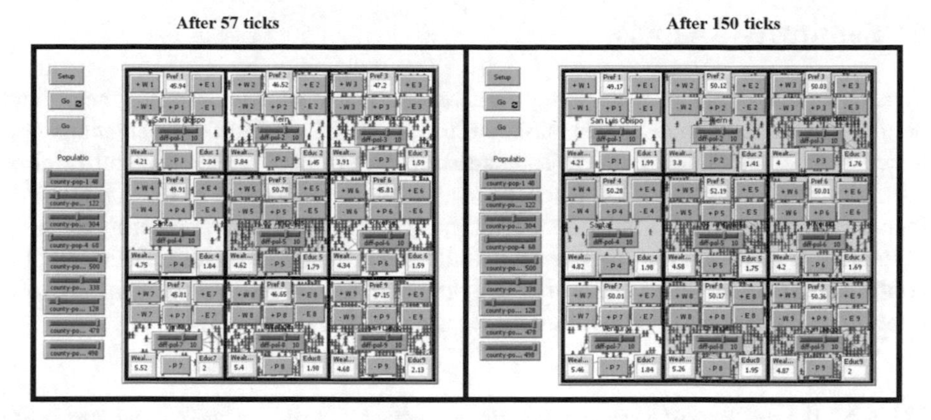

Fig. 2. Interface after 57 ticks and 150 ticks. After 57 ticks, Los Angels adopted the policy and after 150 ticks every neighboring counties except San Luis Obispo and Ventura adopted the policy.

4.2 Scenario - Adoption Initiated with San Luis Obispo

The scenario aims to test whether diffusion will be delayed if the least populated county at the peripheral initiates the policy adoption. San Luis Obispo is made the first county to adopt the policy at 52 *t*, but policy adoption was diffused to many other counties even after 150 *t*, as the diffusion first affected Los Angeles and 9 counties at 277 *t* (Fig. 3).

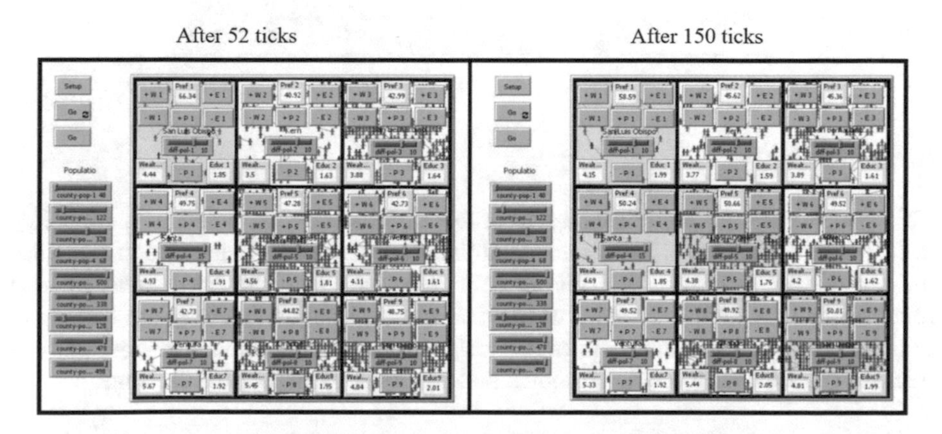

Fig. 3. Interface after 52 ticks and 150 ticks. After 52 ticks, San Luis Obispo adopted the policy and after 150 ticks, Santa Barbara, Los Angeles, and San Diego adopted the policy.

5 Sensitivity Analysis

Sensitivity test explores how policy preference, education level, wealth mean, and neighbor policy adoption impacts policy diffusion based on 24,960 observations collected from simulation result. While increase in mean policy preference and level of education would negatively influence the median policy preference by 0.21 and 0.61%, increase in mean wealth and number of neighbor county policy would positively influence the median policy preference by 0.02 and 0.29% respectively. The result also confirms the neighborhood effect of policy diffusion, though model fit is relatively low, due to simplification of real-world non-linear complexity (Table 1).

Table 1. The Heat map of interaction analysis is congruent with design concept and mechanism of interaction between attributes, with darker color indicating stronger correlation. Wealth and policy diffusion are positively correlated, while education shows second-degree polynomial relationship with policy diffusion, as U-shape curve. The relationship between education level and policy diffusion can be translated as low and high level of education have stronger policy diffusion effect then moderate level of education, confirming the persuasion mechanism

		Policy			
		Isolation	**Low**	**Moderate**	**High**
Wealth	**Low**	2.9	3.6	3.6	4.0
	Moderate	1.4	2.2	3.0	2.4
	High	2.9	4.0	2.7	4.7

		Policy			
		Isolation	**Low**	**Moderate**	**High**
Education	**Low**	2.1	3.1	3.1	4.7
	Moderate	2.0	2.9	3.6	2.7
	High	2.7	3.2	2.4	3.5

6 Conclusion

This paper identifies significant socio-economic attributes of individuals of policy diffusion by using ABM. It explores how the loop of inter-level interaction influence the lower-level agents and diffusion of policy adoption. The simulated environmental regulatory policy diffusion guides us to a feasible explanation that there are within not only microscopic interaction influencing policy diffusion, but also interaction between microscopic and macroscopic attributes that significantly impact the outcome.

Acknowledgements. The authors would like to acknowledge Mathew Gomes and Jingyu Wang for their contributions to previous policy diffusion ABM work.

References

1. Californians Against Waste: Single-Use Bags Ordinances in CA (2016). https://static1.squarespace.com/static/54d3a62be4b068e9347ca880/t/583f1f57e4fcb5d84205b330/1480531800415/LocalBagsOrdinances1Pager_072815.pdf
2. Berry, F.S., Berry, W.D.: Innovation and diffusion models in policy research. In: Theories of the Policy Process, p. 169 (1999)
3. Rogers, E.M.: Diffusion of innovations. Simon and Schuster (2010)
4. Shipan, C.R., Volden, C.: The mechanisms of policy diffusion. Am. J. Polit. Sci. **52**(4), 840–857 (2008)
5. Volden, C., Ting, M.M., Carpenter, D.P.: A formal model of learning and policy diffusion. Am. Polit. Sci. Rev. **102**(3), 319–332 (2008)
6. Boehmke, F.J., Witmer, R.: Disentangling diffusion: the effects of social learning and economic competition on state policy innovation and expansion. Polit. Res. Q. **57**(1), 39–51 (2004)
7. Mooney, C.Z.: Modeling regional effects on state policy diffusions. Polit. Res. Q. **54**(1), 103–124 (2001)
8. Bennett, C.J.: What is policy convergence and what causes it? Br. J. Polit. Sci. **21**(2), 215–233 (1991)
9. Crain, R.L.: Fluoridation: the diffusion of an innovation among cities. Soc. Forces **44**(4), 467–476 (1966)
10. Rogers, E.M., Medina, U.E., Rivera, M.A., Wiley, C.J.: Complex adaptive systems and the diffusion of innovations. Innov. J.: Public Sector Innov. J. **10**(3), 1–26 (2005)
11. Sherif, C.W., Sherif, M., Nebergall, R.E.: Attitude and Attitude Change: The Social Judgment-Involvement Approach. Greenwood Press, Westport (1981)
12. Griffin, E.M.: A first look at communication theory. McGraw-Hill, New York (2006)
13. Gilbert, N., Troitzsch, K.: Simulation for the social scientist. McGraw-Hill, New York (2005)
14. Bankes, S.C.: Agent-based modeling: a revolution? Proc. Nat. Acad. Sci. **99**(suppl 3), 7199–7200 (2002)
15. Railsback, S.F., Grimm, V.: Agent-based and individual-based modeling: a practical introduction. Princeton University Press, Princeton (2011)
16. Berry, B.J., Kiel, L.D., Elliott, E.: Adaptive agents, intelligence, and emergent human organization: capturing complexity through agent-based modeling. Proc. Natl. Acad. Sci. **99** (suppl 3), 7187–7188 (2002)
17. US Census Bureau (2017). https://www.census.gov/quickfacts/fact/table/TX,NV,WA,CA,OR/PST045216. Accessed 22 Nov 2017

The Impact of Analogic, Digital and Hybrid Representations in the Ideation Phase of an Artifact Design: An Educational Perspective

Vasco Santos[1,2(✉)], Ana Ferreira[1,2], and Eduardo Gonçalves[1,2]

[1] Universidade Europeia, IADE, Av. D. Carlos I, 4, Lisbon 1200-649, Portugal
{vasco.milne, ana.margarida.ferreira, eduardo.goncalves}@universidadeeuropeia.pt

[2] UNIDCOM/IADE – Unidade de Investigação em Design e Comunicação, Av. D. Carlos I, 4, Lisbon 1200-649, Portugal

Abstract. The present study focuses on the understanding of the relationship effect between analogical and digital representation forms upon the reflective act and consequently with the creative result in product design. The action field is characterized by the operative constituents of the design process. Within three decades, we watched the influence of the digital age on project practice [1] without new procedures about the way which was integrate in design project curricula, but the reality is that technologies are developing fast. Based on this paradigm, we need to restructure the project habits, using new semantics to describe and materialize our concepts. The starting question is: are we articulating and using better the representation tools in the ideation phase of design project? With this research, we seek to quantify the semantics reflection process, using the synergistic of analogical and digital modelling, to create best creative results.

Keywords: Design process · Creativity · Analogical methods
Digital methods · Creative performance · Innovation

1 Introduction

In 1986, Manzini [2] focused a paradigmatic theme that would change the contexts of the way of thinking and doing the project, alerting to the phenomenon of our perceived differences when we are subject to natural environments or simulated by digital communication. The transition of new social and economic experiences would come to be pondered by Oxman [1] referring that the digital media came to contribute to a new domain of knowledge and creation, mentioned "Information has become a new material for the designers". The new digital technologies brought the transformations to the level of the concepts representing practice, they changed the cognitive processing in problem framing reflection and in solution discovery. In the design and architecture

W. Karwowski and T. Ahram (Eds.): IHSI 2019, AISC 903, pp. 484–490, 2019.
https://doi.org/10.1007/978-3-030-11051-2_73

perspectives, investigations are divided by the positive and negative influences of the integration of the digital or analogical media, in which we highlight the works of Zhu et al. [3], Walther et al. [4], Dorta et al. [5]. Through bibliography review, it's possible to state that the analog representation process (drawing and physical models) support the ambiguity and allow flexibility or change, the versatility of hypothesis management, guarantee the spatial and real visualization of the concepts, trigger direct experimentation and decision making. The negative aspects respect the requirement of technical knowledge, materials and techniques of representation knowledge, manipulative skills, time and costs. Digital representation process is positively described by the possibility of anticipating the production processes, allowing the exploration of complex geometries (free-forms), evolving in the new stylistic qualities as can be seen in the works of Zaha Hadid, predictability, performance and realistic visualization of concepts. The negative aspects described are the creation of a circumscribed or partial thought of the problem, limitation of hypothetical thought, generation of premature concept fixation, illusory formation effect of good performance, matter is amorphous regardless of the material to be used, geometries sometimes interfere with the established problem requirements. Ibrahim and Rahimiam [6] notify the importance of knowing how to decide and use the inherent tools and processes of the project because they represent a space of 80% of the decisions made. Analogical and digital processes are pointed as not being the best to communicate the ideas and they concluded that inexperienced designers (students) have great difficulty in transforming tacit knowledge into explicit knowledge given the lack of capacity for visualization, technical capacity of representation and thinking with the tools. We considered to have negative effects for creative development, and it was verified in a survey that we disseminated in 2016, for two hundred students of higher education at the IADE -European University Portugal, Faculty of Architecture - University of Lisbon, European University of Madrid and Polytechnic of Bari-Dicar Italy. According to our field observation, we are experiencing a paradox in design education, because the students describe having great difficulties in drawing by three-dimensional sketch (associated with the preparation process and generation of ideas). The digital 3D representation (associated with the project communication) is also referred to the high difficulty due to the time it takes to learn and how to work with the software. According to Goldschmidt [7], the difficulty in media using compromises the visual and spatial images, essentially to the reflection process, and consequently the creative performance. Another problem pointed out that the projective teaching programs in Portugal are affected by a strong sketch culture in the ideation phase, and there is a heterogeneous allocation of the analogical modeling tools in the initial phase of the project and the digital tools in the final phase. The integration of the curricular units is not very evident and drawing and 3D digital teaching modeling are based on a formative techniques and tools knowledge. However, they are not applied as project reflection tools as discussed by Duarte et al. [8]. Our main concern is the understanding of how the cognitive and creative act of the project is processed in the designer and tools relationship.

2 Framework

Design research highlights have determined that the evaluation of creative performance is analyzed in the cognitive ability of the designer, in the process or in the resulting product. The creative assessment is a complex process and the difficulty of finding good evaluations lies in the fact that the studies subjects are abstract, analyze personal value, social aspect, psychological state, concept of novelty, originality, utility, psychological state and influencing or limiting factors. Many studies used Comparative Creativity Assessment and the Multi-point Creativity Assessment [9], Creative Product Semantic Scale [10], Consensual Assessment Technique [11], Innovative Characteristics [12], Quality Scale [13], SAPPhIRE Method [14], Linkography [15]. The methods of data collection are also several and the think aloud method of protocol analysis is the most used method, consisting of an empirical and introspective process that captures the sequence of thinking while performing a task. Data collection is done by video and audio recording and the verbalization of the participants can be collected by the concurrent data method (currently) or retrospective data. Questioning its reliability of the data collection, Chi [16] reformulated the method and made it more trustworthy by asking participants to verbalize not what they were doing, but what they were thinking and intended to do, making the process even more introspective and rich in the reproduction of cognitive action.

3 Hybrid Representation Model Enhances the Creative Process? Research Methodology

The research was based on three moments. First, the state of art, based on the literature review, direct field observation (curricular programs analysis of three portuguese universities, a class exercise with a report and a survey disseminated in four national and international institutions). The second moment focused on the model of experimental analysis and implementation (portable toothbrush concept for the adults). We used two groups, a control group that used the analogical process (drawing + physical model executed as they did in the project classes), a control group that used the digital process (3D modeling in software 3D MAX) and an experience group that used the hybrid process with a methodology of alternation between analog and digital modeling. The second moment of preparation and implementation of the case study used the data collection by the concurrent data and the verbal analysis of Chi [16]. The third stage focused on data processing, through the application of descriptive and inferential statistical processes, to validate the hypothesis that the synergistic use of analog and digital modeling in the design phase of the design project enhances the creative results. The study analyzed 24 cases and the exercise briefing consisted of three distinct phases, the preparation (students structure the first principles), incubation (looking for associated elements), generation (hypotheses to solution discover). Based on the literature, the study variables were fluency, flexibility, coevolution, divergent context and novelty. The study formula applied consisted in: CQ (Creativity quantification) = RP (Reflexive Process) [F (Fluency) + F (Flexibility) + Ce (Coevolution)] + PP (Practical

Process /artifact)] = or ≠ N (Novelty). Fluency and flexibility evaluated through the number quantification of the productions made and the diversification or the solution proposals change (number of inferences generated in the problem definition and the diversity of thought contexts. The identification of the inferences was conceived by protocol checking with a semantic matrix of problems categories (e.g. morphology, size, usability, production, etc.), that correspond to the heuristics synthesis discovered by Yilmaz [17]. The actual creep score was quantified in the following formula: IQm (Inference Quantity per minute) = IQGP (Inference Quantity Generation Phase) − IQPP (Inference Quantity Preparation Phase or Incubation): GTU (Generation Time Used). The Flexibility or context changing score was quantified in the following formula: DI (Diversity of Inferences) = DCFI (Diversity Categories Lighting Phase) − DCFP (Diversity of Categories Preparation Phase). To quantifying the coevolution with problem and solution framing [18], we found that there are partial correlations when the creative agent fits the problem and then thinks of subproblems and in the solution, and the direct correlations when there is the problem framing and then the pre-solution. The coevolution score used the following formula: Ce (Coevolution) = NCP (Number of Partial Correlations) − NCD (Number of Direct Correlations). In the practical field, we used the CAT (Consensual Assessment Techniques) with ten designers with more than ten years of industrial design experience. Intending to evaluate divergent thinking (represents an ambiguous, abstract thinking), its quantification becomes complex and sensitive in reliability. We found the way to perceive the amount of divergent thinking, accounting for the amount of convergent, more pragmatic, technical and related thinking about the viability of concepts. According to Tschimmel [19], divergent and convergent thoughts are always implicit in the formation of concepts, but ideation is more related to divergent thinking. In order to quantify the novelty degree, we used the SAPPhIRE method in which the metric evaluates the creative products by S (State of Change), A (Action), P (Parts), Ph (Phenomenon Physical), I (Inputs), R (Organs), E (Effect). To analyze the novelty degree, we segmented the universe of portable toothbrushes, into four typologies and established for each one a framework of analysis under the SAPPhIRE metric. The 24 products were quantified by the method of comparing their characteristics in relation to the protocol metric.

4 Data Analysis and Results

Under a descriptive statistical analysis, we could see that there is a difference of the results obtained in the experimental group (hybrid condition) in relation to the control groups (analog and digital). The independent variables showed that hybrid condition had a higher value and the result was reinforced by SAPPhIRE method. Under the inferential statistical analysis, the one-way Anova (single factor) was used to determine if there was a statistically significant difference in the score of the potential of the creative process in the three conditions and in the relation between the variables of the cognitive action. The Levene test indicated that there was no homogeneity of variance ($p = .073$) and the score was numerically higher (mean and standard deviation) in the hybrid condition (32.821 ± 16.582), followed by the analog condition (21.993 ± 8.264) and the digital condition ($21{,}150 \pm 8{,}916$). The numerical differences between groups did not present evidence against the null hypothesis, $F (2,21) = 2,405$, $p = 0,115$. The observed power value was low (.431), which for Hoenig and Heisey [20] shows to be an indicator that the result could have been affected by sample size. Post hoc tests, such as Games-Howel, showed that between the hybrid and digital conditions there is a difference $p = .044$, but not between the hybrid and analog conditions $p = .060$. Thus, based on Kirk [21], we intend to do a clinical analysis of effect size observation with the Glass Delta, whose results indicated between the hybrid and analog condition the value of (1,31), and between the hybrid and digital condition the value of (1,30). According to Cohen's *d* scale, the expression of magnitude is great. Measurement of effect size in common language indicated that there is a 71.4% likelihood of a random hybrid condition case being greater than in the analog group, and 68.9% hybrid group being greater than in the digital group. The verification of the practical action (divergent thinking) it was used the same statistical tests and the results presented a superior numerical score in the hybrid condition ($2,300 \pm 0,427$), followed by the digital condition with ($2,012 \pm 0,574$) and the analogous condition with (1.985 ± 0.516). As in the results of the cognitive domain, the practical domain presented no significant evidence against the null hypothesis, $F (2.21) = 0.928$, $p = 0.411$. By verifying the effect size with the Glass Delta test, the results indicated between the hybrid and analog condition, the value of (0.61) and between the hybrid and digital condition, the value of (0.50), a size of Cohen's *d* scale average effect. Measurement of the effect in common language indicated that there is 61% chance that a random case of the hybrid condition will be greater than in the analog condition and 51% probability that a random case of the hybrid condition will be greater than in the digital group. Considering the extra sensitivity given to the somewhat small sample (case study nature), it became apparent that the descriptive and inferential results clearly point to a considerable numerical difference in the score of creative potential and novelty, demonstrating that the creative process in the ideation phase, can be powered by the hybrid condition.

5 Conclusions

Study highlights reveal some worrying aspects in the design teaching, debating the representative difficulty of university students of design. Understanding the importance of analogical and digital representations in the reflexive action between the problem framing and, in the solutions search, we believe that the creative action is compromised by the inability, conditioning the concepts simulation. The curricula programs of the analog and digital units of design education in Portugal also demonstrate a formative concern about technical knowledge, teaching to work with the tools, but not to know how to reflect with these tools, realizing their potential. Creative action has two major types of thinking, the divergent and the convergent, and the analogical and digital modulations provide each other, with different proportions. About this issue, we disagree with the heterogeneous representation media allocation in the project, feeling that there is an emerging need to create a greater articulation between the curricular units of analog and digital contents, and to allow an openness to other forms of representation, not just drawing. The present study, having an empirical nature, demonstrates that there is a different reflexive and practical dynamics, when the two modellings (hybrid condition) are used in synergy. However, the results achieved should not be understood as absolutes, but as being capable of being improved or refuted. This is the reason for the studies that integrate the human subjective thought and creation. We aim to open the spectrum for the problems, raised to better understand the design action.

References

1. Oxman, R.: Theory and design in the first digital age. Des. Stud. **27**(3), 229–265 (2006)
2. Manzini, E.: A Matéria da Invenção. Centro Português de Design, Lisboa (1993)
3. Zhu, Y., Dorta, T., De Paoli, G.: A comparing study of the influence of CAAD tools to conceptual architecture design phase. In: Proceedings of EuropIA'2011: 11th International Conference on Design Sciences and Technology, pp. 29–43 (2011)
4. Walther, J., Robertson, B., Radcliffe, D.: Avoiding the Potential Negative Influence of CAD Tools on the Formation of Students' Creativity. The University of Melbourne, Department of Computer Science and Software Engineering (2007)
5. Dorta, T., Perez, E., Lesage, A.: The ideation gap: hybrid tools, design flow and practice. Des. Stud. **29**(2), 121–141 (2008)
6. Ibrahim, R., Rahimian, F.P.: Comparison of CAD and manual sketching tools for teaching architectural design. Autom. Constr. **19**(8), 978–987 (2010)
7. Goldschmidt, G.: The dialectics of sketching. Creat. Res. J. **4**(2), 123–143 (1991)
8. Duarte, J.P., Celani, G., Pupo, R.: Inserting computational technologies in architectural curricula. In: Computational Design Methods and Technologies: Applications in CAD, CAM and CAE Education, pp. 390–411. IGI Global (2012)
9. Oman, S.K., Tumer, I.Y., Wood, K., Seepersad, C.: A comparison of creativity and innovation metrics and sample validation through in-class design projects. Res. Eng. Des. **24** (1), 65–92 (2013)
10. Besemer, S.P., Treffinger, D.J.: Analysis of creative products: review and synthesis. J. Creat. Behav. **15**(3), 158–178 (1981)
11. Amabile, T.M.: Social Psychology of Creativity: A Consensual Assessment Technique. J. Pers. Soc. Psychol. **43**(5), 997 (1982)
12. Saunders, M.N., Seepersad, C.C., Hölttä-Otto, K.: The Characteristics of Innovative, Mechanical Products. J. Mech. Des. **133**(2), 021009 (2011)
13. Linsey JS: Design-by-analogy and representation in innovative engineering concept generation. A dissertation for degree of Doctor of Philosophy (Engineering), Department of Mechanical Engineering, University of Texas, Austin, TX (2007)
14. Chakrabarti A, Khadilkar P: A measure for assessing product novelty. In: DS 31: Proceedings of ICED 2003, the 14th International Conference on Engineering Design, pp. 159–160. Stockholm (2003)
15. Goldschmidt, G.: Linkography: Assessing Design Productivity. In: Cyberbetics and System'1990, Proceedings of the Tenth European Meeting on Cybernetics and Systems Research, pp. 291–298. World Scientific (1990)
16. Chi, M.T.: Quantifying qualitative analyses of verbal data: a practical guide. J. Learn. Sci. **6** (3), 271–315 (1997)
17. Yilmaz, S.: Design heuristics. A dissertation for degree of Doctor of Philosophy (Design Science). In: The University of Michigan (2010). https://deepblue.lib.umich.edu/handle/2027.42/77845

18. Cross, N., Dorst, K.: Co-evolution of problem and solution spaces in creative design: observations from an empirical study. In: Gero, J.S., Maher, M.L., (eds.) Computational Models of Creative Design IV, University of Sydney, New South Wales (1998)
19. Tschimmel, K.: Sapiens e Demens no Pensamento Criativo do Design. A dissertation for degree of Doctor of Philosophy (Design Science), University of Aveiro, 2010, Departamento de Comunicação e Arte. http://ria.ua.pt/bitstream/10773//1/2010000838.Pdf
20. Hoenig, J.M., Heisey, D.M.: The abuse of power: the pervasive fallacy of power calculations for data analysis. Am. Stat. **55**(1), 19–24 (2001)
21. Kirk, R.E.: Practical significance: a concept whose time has come. Educ. Psychol. Measur. **56**(5), 746–759 (1996)

The Civil Affairs Information Matrix: Designing Context-Aware Visual Analytics Enabling Mission Planning with Ensemble Learning

Ryan Mullins[1(✉)], Benjamin Ford[1], Lynndee Kemmet[2], and Shana Weissman[1]

[1] Aptima, Inc., 12 Gill Street Suite 1400, Woburn, MA, USA
rmullins@aptima.com
[2] United States Military Academy, West Point, NY, USA

Abstract. With the pivot towards grey zone operations, the United States Army must engage, influence, and partner with local populations in order to achieve the military objectives of their foreign policy. However, the knowledge that enables successful partnerships is split between the Civil Affairs and Intelligence staff elements. Here, we present the Civil Affairs Information Matrix, a context-aware visual analytics capability enabling automated gap analysis for grey zone operations using fused Civil Affairs and Intelligence data. The Civil Affairs Information Matrix uses semi-supervised and unsupervised learning techniques to fuse and annotate data (for example, surveys and news reports), generate potential actions from this data to address gaps and leverage strengths, and visualize the annotated data and rank-ordered potential actions to planners for consideration. Our ensemble approach enables more collaborative planning processes for grey zone operations that will ensure appropriate and dynamic responses to local population interests while achieving the planned outcomes.

Keywords: Visual analytics · Context-aware systems · Ensemble modeling Civil-military operations · Mission planning · Intelligent decision support

1 Background

Future United States military conflicts will be characterized as grey zone operations in which adversarial forces engage with or otherwise leverage local populations to achieve the military objectives of their foreign policy [1]. U.S. success in grey zones will require the ability to achieve desired ends while respecting the affected populations [2]. Within the U.S. Army, the essential knowledge for grey zone planning is split between the Intelligence and Civil Affairs (CA) staff elements [3]. Intelligence is charged with knowing the adversary and CA is charged with knowing the local population. Effective grey zone operations require a planning capability that enables critical collaborative planning functions via data fusion, analysis, and visualization across these differing perspectives.

W. Karwowski and T. Ahram (Eds.): IHSI 2019, AISC 903, pp. 491–496, 2019.
https://doi.org/10.1007/978-3-030-11051-2_74

Gap analysis is one function in the planning process that provides an opportunity to improve data fusion, and thereby collaboration, between the Intelligence and CA staff elements. As currently practiced, gap analysis is a cognitively intense task that identifies weaknesses in the staff's understanding of the battlespace [3]. Often, this analysis takes the form of a cross-walk, where planners manually create a matrix into which information is binned and from which weaknesses are identified to inform information collection requirements. The rows and columns of this matrix are typically the mission and operational variables used across the Army Mission Command and Intelligence communities [4]. This process offers opportunities for improvement in terms of automation and augmentation.

In this paper, we introduce the Civil Affairs Information Matrix (CAIM), a context-aware visual analytics capability under development that enables gap analysis across a broad spectrum of grey zone operation planning needs. The CAIM augments traditional planning tools by integrating automated gap analysis directly into the search functions used to retrieve and compile information for plans. It further augments the process by enhancing the traditional gap analysis process with ensemble learning to derive strengths, in addition to weaknesses, in the staff's understanding of the battlespace, and recommend actions that can be taken to leverage those strengths and weaknesses during engagements with the local population.

2 The Civil Affairs Information Matrix

The Civil Affairs Information Matrix (CAIM) is an adaptive user interface that augments traditional search capabilities used in military planning processes with the addition of automated gap analysis and action recommendation. In this section, we discuss the three components of the CAIM – the visualization, the data fusion, and the action generation and recommendation.

The capability arose from the need to translate between the CA and Intelligence perspectives on the command staff. As described by expert participants in focus group discussions, this translation is very similar to a tactical-to-operational crosswalk [4]. However, instead of communicating solely about the direct effects of engagements with adversaries, this translation must communicate how our engagements with adversaries can directly and indirectly impact the local population, as well as how our engagements with the local population can directly or indirectly affect, deter, or inhibit actions of adversarial forces.

For the purposes of this research, our user is a Civil Affairs staff officer. Their function is to generate a list of potential courses of action (COAs) that will be sent to the Intelligence staff element for collation, assessment, and recommendation to the commander. The CAIM must provide for two needs in their workflow. First, it must help the user assess relative strengths and weaknesses in their understanding of the civil component of the operational environment to inform additional data collection requirements. Second, it must support action identification and integration as part of COA development, where individual actions are sequenced to achieve an effect.

2.1 Matrix Visualization

The centerpiece of the CAIM is the matrix visualization, shown in Fig. 1. The rows of the matrix represent the CA perspective on information categorization and the columns represent the Intelligence perspective. Labels for rows and columns are the name of the respective category. Each dot represents the subset of data at the category intersection. The total count of items (e.g., news reports, surveys) at that intersection is shown in the center of the dot, and it is possible for one item to be referenced at the intersection of multiple dots. A dot's hue is determined by the row in which it sits, and its saturation is based on the automated gap analysis, described below. This creates a heatmap, informed by the gap analysis, of strengths and weaknesses in the data and acts as the basis for further exploration.

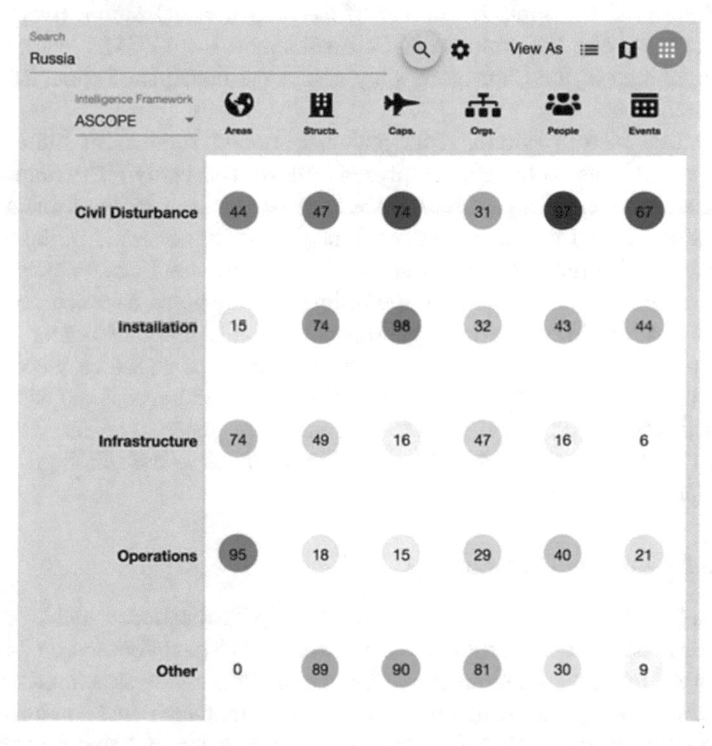

Fig. 1. The Civil Affairs Information Matrix presented in the context of a mission planning tool and displaying the highest-level Civil Affairs categories and the Intelligence categories related to tactical planning.

Army Civil Affairs uses a hierarchical categorization method driven by the structure of the ground truth information they collect when deployed [5]. Ground truth CA data is collected using surveys that cover a range of topics from civil engineering to descriptions of social organizations that are important in a community. More recently, there has been a push to integrate diverse, heterogeneous data sources such as news media and Intelligence data to supplement the surveys and fill the long temporal collection gaps between deployments. We have scoped this research to focus specifically on open source news media. Survey data is structured such that the labels exist at the survey question level. News data, however, is not labeled, and we thus rely on natural language processing techniques to assign these labels probabilistically at the document, paragraph, and sentence level, as described below. The CA hierarchy can be explored and exploited by the user when viewing the matrix. When the user clicks on a CA category's row label, and that category has subcategories, all other CA categories in the current view will be hidden and new rows will be added for the sub-categories associated with the selected category.

Army Intelligence uses a different method for categorization: a flat set of categories for each of the "levels of war", from tactical to operational to strategic [4, 6]. The user can choose which categorization scheme to view from the "intelligence framework" drop-down menu in the upper-left corner of the matrix visualization. Every combination of categories can be explored and visualized in the CAIM, which presents a cognitive challenge to the user: thoroughly assess the contents of some 2,600 hierarchically nested dots.

Our solution is to generate, rank, and recommend actions for these bins. The process by which these actions are analyzed is described below. The output of those analytics is a list of high-impact actions that can be assessed by the user. Actions are displayed in contextual menus accessible through a sticker overlaid in the upper-right corner of the dot. To minimize the number of attentional conflicts, we present at most ten stickers containing no more than three actions for any given parameterization of the matrix visualization. When reviewing potential actions, the staff officer has several interaction possibilities. They may: (1) accept the action as-is, which incorporates the action into their other planning documents; (2) modify-and-accept the action; (3) decline the action outright; or (4) decline and manually generate an alternative action. These four types of interaction provide valuable feedback to the action generation and recommendation analytics to learn from.

2.2 Data Fusion and Gap Analysis

As discussed above, the CAIM is an abstract visualization capable of support heterogeneous data. For the purposes of this research, we have focused on the fusion of two types of information, semi-structured CA survey data and unstructured open source news reporting. Surveys are structured in such a way that each question-answer pair is labeled according to the Civil Affairs and Intelligence categories that drive the matrix visualization shown above. News reports are uncategorized, and therefore must be labeled in order to be integrated with the survey data. Our method uses a semi-supervised approach to assign CA and Intelligence category labels using probabilistic keyword matching and topic modeling approaches. Civil Affairs and Intelligence subject matter experts—those responsible for data quality assurance—have the ability to inspect, update, and delete the keywords and topics associated with each category. Individual data elements can also be manually added to each of the category intersections. Reinforcement learning monitors these user interactions and updates the associated probabilities for matching.

- Automated gap analysis is performed through a semi-supervised machine learning capability. We define gap analysis as a value-of-information (*VOI*) function that takes input in the form of a user query (q), a geographic area of interest (g), and an intended time of engagement (t). For every bin in the matrix, the *VOI* for each datum-entity pair (de) (e.g., a news article talking about construction of a new soccer stadium) in the bin is computed. Then, those *VOI* scores are aggregated into that bin's score. The goal of the bin score is to assess the bin's information sufficiency, which we define as the distribution of that information.
- For each de, a baseline score (S_{de}) in the range of [0, 1] is computed that represents the completeness of de (e.g., how many questions in a survey are answered). To compute the *VOI*, the baseline score is augmented by the recency (r) and trustworthiness (tr) functions (Eq. 1). Recency uses a decay function defined by the difference between t and the publication/collection time for the scored de. Trustworthiness uses the doctrinally-defined scales for reliability (assessing sources) and credibility (assessing accuracy and precision) [7].

$$VOI_{de}(q,g,t) = S_{de}(q,g) \times r_{de}(q,g,t) \times tr_{de}(q,g) \quad (1)$$

$$S_b(q,g,t) = \frac{\sum_{de \in b} VOI_{de}(q,g,t)}{|b|} \quad (2)$$

- Augmented *VOI* scores for data-entity pairs are averaged to create bin scores (S_b) (Eq. 2), which define the hue of the bin's corresponding dot in the matrix. The value of S_b is a number in the range of [0, 1] where values greater than or equal to 0.5 represent sufficient information to enable engagement and values less than 0.5 represent gaps requiring data collection. The value of S_b also defines the types of actions that are generated and recommended to the user.

2.3 Action Generation and Recommendation

Following gap analysis, the information space is assessed to identify actions that could be executed by Army units. We define an action in the "who-did-what-to-whom" fashion associated with event data [8], expressed as the tuple $actor, event, target$. The event is derived from a curated list of potential actions germane to CA operations [9], for example a key leader engagement, a veterinary clinic, or a survey collection. Action events are divided into two classes based on the bin score (see Eq. 2). Gap-filling actions, such as data collections, are derived when $S_b < 0.5$ and are ranked based on the predicted change to the aggregate S_b. Engagement actions are derived when $S_b \geq 0.5$ and are ranked based on their predicted movement towards the commander's stated intent. The actor is an appropriately-sized Army unit such as a Squad, Platoon, or Company. Specific units will be assigned by the appropriate staff element during the planning process. The target is derived from the available entities in the information space.

Regardless of event class, actions are aggregated into a single list from which the top actions are recommended to the user. Recommendations are updated in response to user interactions. For example, if a staff officer declines an action, that action is removed from the pool and the next most-valuable action is displayed.

3 Discussion and Future Work

In this paper, we introduced the Civil Affairs Information Matrix and its constituent elements for visualization, data fusion, and analytics. The visualization component provides an approachable, familiar mechanism for exploring complex, many-to-many relationships in a semi-structured dataset used across Army Civil Affairs and Intelligence staff elements. The analytic components support the cognitively intense task of gap analysis at the scale of the modern military information enterprise, while providing sufficient human-in-the-loop interactions to enable learning, honing, and adaptation of staff officer skills over time. We believe our approach will enable more collaborative planning for grey zone operations. End-user evaluations regarding usability, utility, and performance are being planned and executed with operational users. We intend to publish our findings in the future.

Acknowledgements. This work was conducted in connection with contract W56KGU-17-C-0039 with the U.S. Army Communications-Electronics Research, Development, and Engineering Center. We would like to thank Mr. Raymond McGowan, Dr. Lynn Copeland, Mr. Tim Strong, and Ms. Stacy Pfautz for their contributions to this work as thought partners.

References

1. Kapusta, P.: The Gray Zone. United States Special Operations Command, Tampa (2015)
2. Echevarria, A.I.: Operating in the Gray Zone: An Alternative Paradigm for US Military Strategy. Army War College-Strategic Studies Institute, Carlisle (2016)
3. Army, U.S.: Field Manual 6–0: Commander and Staff Organization and Operations. Department of the Army, Washington, D.C. (2014)
4. Army, U.S.: Army Doctrine Reference Publication 5–0: The Operations Process. Department of the Army, Washington, D.C. (2012)
5. Army, U.S.: Army Techniques Publication 3-57.60: Civil Affairs Civil Information Management. Department of the Army, Washington, D.C. (2013)
6. Stolberg, A. G.: Making National security policy in the 21st century. In: Bartholomees, J. B. (Ed.) The US Army War College Guide to National Security Issues: National Security Policy and Strategy, vol. 2, pp. 29–46. United States Army War College Strategic Studies Institute, Carlisle (2010)
7. Army, U.S.: Army Techniques Publication 2-22.9: Open Source Intelligence. Department of the Army, Washington, D.C. (2012)
8. Bagozzi et al.: Political language ontology for verifiable event records event, actor, and data interchange specification. Technical Report, Open Event Data Alliance (2018). https://github.com/openeventdata/PLOVER
9. Army, U.S.: Field Manual 3-57: Civil Affairs Operations. Department of the Army, Washington, D.C. (2014)

Modified Baum Welch Algorithm for Hidden Markov Models with Known Structure

Kim Schmidt(✉) and Karl Heinz Hoffmann

Chemnitz University of Technology, Straße der Nationen 62, 09111 Chemnitz, Germany
{kim.schmidt,hoffmann}@physik.tu-chemnitz.de

Abstract. Hidden Markov Models (HMMs) are widely used in speech and handwriting recognition, behavior prediction in traffic, time series analysis, biostatistics, image and signal processing, and many other fields. For some applications in those real world problems, a-priori knowledge about the structure of the HMM is available. For example the shape of the state transition matrix and/or the observation matrix might be given. We might know that some entries in these matrices are equal and others are zero. For training such a model, we have two options: use the common Baum Welch Algorithm (BWA) and enforce the given structure after training or modify the BWA to enforce it during training. This paper shows several approaches for modifying the BWA and compares the results of all training methods.

Keywords: Hidden Markov Model · HMM · Baum Welch Algorithm · Multiple sequence learning · A-priori knowledge

1 Introduction

Hidden Markov Models (HMMs) are already widely distributed in speech and handwriting recognition [1, 2], behavior prediction in traffic [3, 4], time series analysis [5, 6], biostatistics [7, 8], image and signal processing [9, 10], and many other fields. In some of these applications, the structure of (parts of) the HMM is given. We might know that some entries of the matrices describing a HMM are equal and others are zero. While training those HMMs with the Baum Welch Algorithm (BWA) we could ignore that given structure or enforce it in the end or even during learning.

In this paper, we want to analyze and compare the results for these different approaches: training with BWA ignoring structure, ignoring structure during learning but enforce it after training, and enforcing the given structure during learning and thus after training. The BWA ignoring structure during learning is the common BWA (see [11]). If the structure is ignored during learning but enforced after training, some matrix entries of the trained model can be replaced by their properly scaled mean value to satisfy the structure. This method is called BWA+structure. The BWA enforcing structure during learning is a newly developed variant, which replaces entries that have to be equal by their properly scaled mean value and so recovers the given structure in each iteration step of the BWA. This algorithm is called structure enforcing BWA (SBWA).

W. Karwowski and T. Ahram (Eds.): IHSI 2019, AISC 903, pp. 497–503, 2019.
https://doi.org/10.1007/978-3-030-11051-2_75

2 Hidden Markov Model

A HMM is a stochastic process that consists of two parts, a hidden Markov process and observations generated by the hidden part. For the hidden Markov process, the initial state probability distribution at time $t = 1$ is stored in the start vector π. The transitions from state i to state j (a_{ij}) are defined in the state transition matrix A as a real square matrix with size $N \times N$, where N is the number of possible states. All passed states are stored as the state sequence $Q = \{q_1, \ldots, q_T\}$, where T is the length of the sequence. Each state q_t generates an observation out of M possible observations in a probabilistic way according to observation matrix B's entry b_{jk} (size $N \times M$). The observation o_t at time t is added to the observation sequence $O = \{o_1, \ldots, o_T\}$. The parameters A, B and π form the HMM λ which is depicted in Fig. 1 with its Trellis structure.

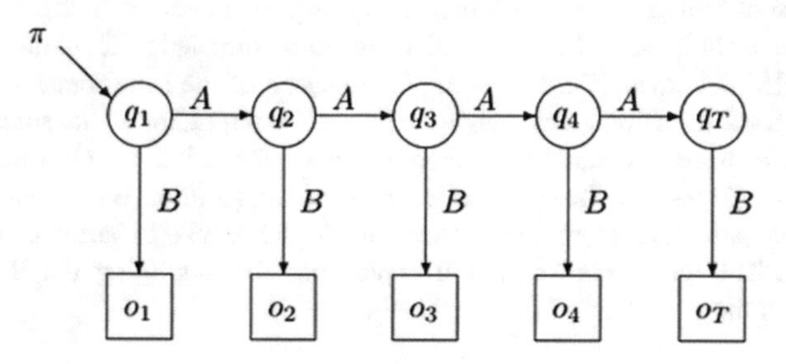

Fig. 1. Trellis structure of a Hidden Markov Model

In the following, we use a discrete and homogeneous HMM. Therefore time, states, and observations are discretized and neither A nor B are time dependent. The HMM includes a first order hidden Markov chain that fulfills the Markov property: The probability for the next state depends only on the current state.

The HMM is trained with multiple sequences. There are H observation sequences given for learning whereby $O^{(h)}$ is a single sequence and T_h is the individual length of sequence h. The final solution of the training is called λ^*.

For detailed algorithms, proofs and derivations of HMMs and multiple sequence learning see [1–12].

3 BWA+Structure

We now turn to the subclass of HMM of interest here. For these we have a particular structure of elements of the HMM given: We know that some entries of a matrix are equal, where those are located and which of them are zero. This information can be given for one or more of the HMM parameters. If a-priori knowledge exists about the HMM that shall be estimated, it should be used for a better representation of the true process that generated the observation sequences. Considering that, this structure has to be included in the trained model.

The information which entry is equal to zero is processed in the initialization of the BWA. An entry equal to zero stays zero during training. This is handled by the common BWA.

After the BWA estimated the HMM λ^* with the training sequences, the whole given structure is enforced. That means entries of the matrices of the trained model of which we know they should be equal are replaced by their properly scaled mean value. This has to be followed by a renormalization to make sure that the constraints of the HMM (right stochastic matrices, range of values between 0 and 1) are fulfilled.

This method will be called BWA+structure in the following.

4 SBWA

The structure enforcing BWA is also using the common BWA with a little modification. In contrast to BWA+structure, the known structure is not enforced in the end of the training process, but in each iteration step of the BWA. This is also done by replacing entries that should be equal by their properly scaled mean value and renormalization.

We expect that SBWA leads to better results than BWA+structure because of including the structure in the optimization (training) process.

5 Experimental Set-up

Two HMMs λ_x (x =1, 2) of different sizes were generated. The special structure in the experimental set-up is only known for A_x. B and π have fix values. In the following A_1 and A_2 are shown.

$$A_1 = \begin{pmatrix} a_1 & 0 & a_3 & 0 \\ a_1 & 0 & a_3 & 0 \\ 0 & a_2 & 0 & a_4 \\ 0 & a_2 & 0 & a_4 \end{pmatrix}, \quad A_2 = \begin{pmatrix} a_1 & 0 & 0 & 0 & a_3 & 0 & 0 & 0 \\ a_1 & 0 & 0 & 0 & a_3 & 0 & 0 & 0 \\ 0 & a_1 & 0 & 0 & 0 & a_3 & 0 & 0 \\ 0 & a_1 & 0 & 0 & 0 & a_3 & 0 & 0 \\ 0 & 0 & a_2 & 0 & 0 & 0 & a_4 & 0 \\ 0 & 0 & a_2 & 0 & 0 & 0 & a_4 & 0 \\ 0 & 0 & 0 & a_2 & 0 & 0 & 0 & a_4 \\ 0 & 0 & 0 & a_2 & 0 & 0 & 0 & a_4 \end{pmatrix} \tag{1}$$

For these example systems, we generated n = 50 training sets of H =10 observation sequences $O^{(h)}$ with length T =100 each. Each model λ_x was trained with BWA, BWA +structure, and SBWA. The logarithmic probabilities that show how probable the training sequences were generated by the final trained models ($\log P(O|\lambda_x^*)$) have been used for the comparison in the next section. A higher value of $\log P(O|\lambda_x^*)$ means the model suits the training sequences $O^{(h)}$ better.

6 Results and Discussion

The first intuition for training a HMM is using the BWA. For a common HMM without a-priori knowledge about the structure we can get a good estimation. As Table 1 shows, we can find also a good estimation for a HMM with particularly given structure. However, the training result λ_x* does not display that a-priori knowledge.

Consequently, the next step is to train the model with BWA+structure. The training results of this method are also shown in Table 1 for four exemplary samples. For each model λ_x log $P(O|\lambda_x{*})$ of BWA is higher than log $P(O|\lambda_x{*})$ of BWA+structure. For testing whether this observation is significant, we use the one-tail Wilcoxon Signed Rank test (WSR) [13]. The result of this test is, that log $P(O|\lambda_x{*})$ is significantly higher for BWA than for BWA+structure (WSR: $W < W^* = 249$, $z < z^*{=}4.265$, $p < 0.00001$, $n = 50$). The strength of this effect is $r > r^* = 0.7$ and corresponds to a strong effect. The test values W^*, z^*, and p^* can be found in the appropriate tables in [14–16]. The results of W, z, and r can be found in Table 2 (BWA vs. BWA+structure). Consequently, the BWA fitted the model significantly better than the BWA+structure.

As depicted in Fig. 2, there is an obvious difference in the data points and the medians of all n samples for training with BWA (black dots and line) and BWA +structure (blue squares and dashed line). It is obvious that BWA+structure is not the optimal way to include the a-priori knowledge.

We then developed a different method to enforce structure: SBWA. The structure is not enforced after training, but during training. In Table 1 for each model λ_x log $P(O|\lambda_x{*})$ of SBWA is higher than log $P(O|\lambda_x{*})$ of BWA+structure. The result of the significance test is, that log $P(O|\lambda_x{*})$ is significantly higher for SBWA than for BWA (WSR: $W < W^* = 249$, $z < z^* = 4.265$, $p < 0.00001$, $n = 50$). The strength of this effect is corresponded again to a strong effect. The results of W, z, and r can be found in Table 2 (BWA+structure vs. SBWA). That means, the SBWA fitted the model significantly better than the BWA+structure.

Table 1. Exemplary results of log $P(O|\lambda_x{*})$ for training with BWA, BWA+structure, and SBWA.

HMM	Run	log $P(O\|\lambda_x{*})$		
		BWA	BWA+structure	SBWA
λ_1*	1	−583.871	−645.525	−597.272
	2	−585.154	−661.832	−608.624
	3	−544.539	−626.142	−566.406
	4	−552.517	−668.649	−579.957
λ_2*	1	−542.678	−647.977	−604.844
	2	−539.708	−640.928	−610.436
	3	−508.764	−605.754	−560.203
	4	−508.490	−612.994	−582.080

Table 2. Results of the one-tail Wilcoxon Signed Rank test.

HMM	Method	W	z	r
λ_1*	BWA vs. BWA+structure	0	−6.154	0.870
	BWA+structure vs. SBWA	0	−6.154	0.870
	BWA vs. SBWA	0	−6.154	0.870
λ_2*	BWA vs. BWA+structure	0	−6.154	0.870
	BWA+structure vs. SBWA	0	−6.154	0.870
	BWA vs. SBWA	0	−6.154	0.870

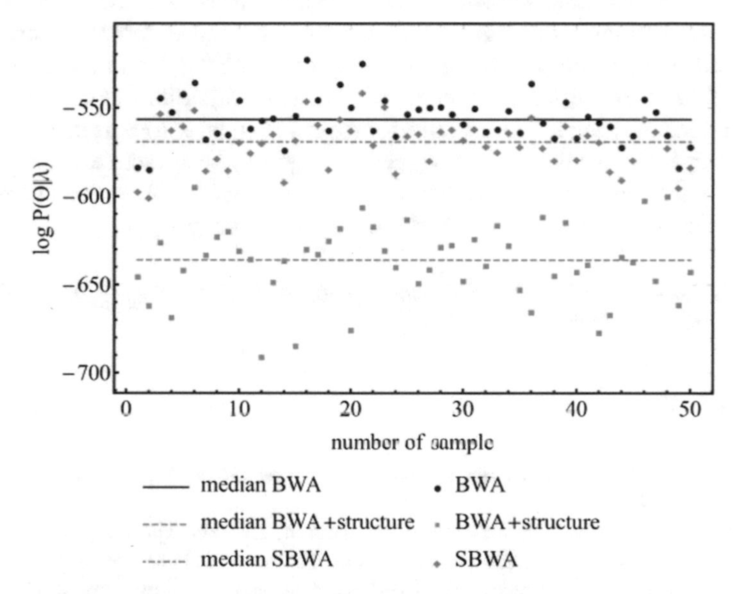

Fig. 2. log $P(O|\lambda_x)$ for all n = 50 samples for λ_2* after training with BWA, BWA+structure, and SBWA as dots and their medians as lines.

Nevertheless, training with SBWA results in a less good outcome than training with BWA (WSR: $W < W^* = 249$, $z < z^* = 4.265$, $p < 0.00001$, $n = 50$, $r > r^* = 0.7$, see Table 2, BWA vs. SBWA). This is not surprising, because by enforcing the structure we have less values for fitting and so we reduce the degrees of freedom of the matrix A_x. It must be stressed that even though the log $P(O|\lambda_x*)$ of the BWA is lower than that of SBWA, thus does not mean that training the model with the BWA is better than the model trained with SBWA. Only the latter model effects the known structure of the original problem.

7 Conclusions

In this paper, we presented the newly developed SBWA for training HMMs with known structure. This method was compared to previously used approaches: BWA and BWA+structure.

A typical finite set of observation sequences does not exactly represent the known structure of the HMM generating it. As a consequence the classical BWA captures that "faulty" information. This is successfully suppressed by our new SBWA.

We tested the three different algorithms on a HMM with a sparse transition matrix A. In general we can say, that learning with BWA results in the best outcome for log $P(O|\lambda_x)$, though it is no good representation of the true process. If the given structure is enforced in the final trained model, SBWA is significantly better then BWA+structure and quite close to the results of training with common BWA.

A next step will be to study the results for given structure for B and π and also combinations up to a-priori knowledge for the whole HMM. Another interesting point will be investigations with less sparse matrices and more complex dependencies between the several values in a matrix. Furthermore, the SBWA will be optimized by a better integration of the structure enforcement in the algorithm. Additionally, other methods instead of using the mean value for replacing equal entries as using the highest or lowest value, the median, or a weighted average can be attractive.

The results of this paper and further improvements of SBWA can be interesting for special application as for example in chemistry and biology.

Acknowledgments. This research was supported by the European Social Fund and the Free State of Saxony under Grant No. 100269974.

References

1. Wendemuth, A.: Grundlagen der stochastischen Sprachverarbeitung. Oldenbourg Verlag, München (2004)
2. Fink, G.A.: Markov Models for Pattern Recognition, 2nd edn. Springer, London (2014)
3. Kuge, N., Yamamura, T., Shimoyama, O., Liu, A.: A driver behavior recognition method based on a driver model framework. SAE Technical Paper, Technical report (2000)
4. Streubel, T., Hoffmann, K.H.: Prediction of driver intended path at intersections. In: Intelligent Vehicles Symposium Proceedings, 8–11 June 2014, pp. 134–139. IEEE, Dearborn (2014)
5. Knab, B.: Erweiterung von Hidden-Markov-Modellen zur Analyse ökonomischer Zeitreihen. Ph.D. dissertation, Universit ̈at zu Köln (2000)
6. Hassan, M.R., Nath, B.: Stock market forecasting using hidden Markov model: a new approach. In: 5th International Conference on Intelligent Systems Design and Applications (ISDA 2005), pp. 192–196 (2005)
7. Schliep, A., Schönhuth, A., Steinhoff, C.: Using hidden Markov models to analyze gene expression time course data. Bioinformatics **19**(1), i255–i263 (2003)
8. Holzmann, H., Munk, A., Suster, M., Zucchini, W.: Hidden Markov models for circular and linear-circular time series. Environ. Ecol. Stat. **13**(3), 325–347 (2006)
9. Li, J., Najmi, A., Gray, R.M.: Image classification by a two-dimensional hidden Markov model. IEEE Trans. Signal Process. **48**(2), 517–533 (2000)
10. Dash, D.P., Kolekar, M.H.: Epileptic seizure detection based on EEG signal analysis using hierarchy based Hidden Markov Model. In: 2017 International Conference on Advances in Computing, Communications and Informatics (ICACCI), pp. 1114–1120 (2017)
11. Rabiner, L.R., Juang, B.-H.: An introduction to hidden Markov models. IEEE ASSP Mag. **3** (1), 4–16 (1986)

12. Li, X., Parizeau, M., Plamondon, R.: Training hidden Markov models with multiple observations - a combinatorial method. Trans. PAMI **22**(4), 371–377 (2000)
13. Wilcoxon, F.: Individual comparisons by ranking methods. Biometrics Bull. **1**(6), 80–83 (1945)
14. McCornack, R.L.: Extended tables of the Wilcoxon matched pair signed rank statistic. J. Am. Stat. Assoc. **60**(311), 864–871 (1965)
15. Pearson, E.S., Hartley, H.O.: Biometrika Tables for Statisticians, vol. 1. Cambridge University Press, New York (1966)
16. Owen, D.B.: Handbook of Statistical Tables. Addison-Wesley, Reading (1962)

Input Data Dimensionality Reduction of Abnormality Diagnosis Model for Nuclear Power Plants

Jae Min Kim, Gyumin Lee, Suckwon Hong, and Seung Jun Lee(✉)

Ulsan National Institute of Science and Technology, 50, UNIST-Gil, Ulsan 44919, South Korea
sjlee420@unist.ac.kr

Abstract. Nuclear power plants are diagnosed by operators according to the alarms and plant parameters that can be identified in the main control room. The operators are trained to conduct tasks in any cases by following an operating procedure. When a component has in malfunction, the operator must choose the appropriate abnormal operating procedures to stabilize the plant. However, the operators take a high burden because this task requires complex judgement with large amounts of information in a short time. To support the operators, this paper studied the data preprocessing methods to develop the nuclear power plant abnormal state diagnosis system using deep learning algorithms. A nuclear power plant simulator was used to produce training data which includes more than 2800 variables recorded in the given time. It is necessary to reduce the dimensionality of the generated data to achieve the best estimation of the training. There are two ways to reduce the dimensionality of the data: feature selection and feature extraction methods. Abnormal operating procedures of the advanced pressurized water reactor 1400 were analyzed to select parameters related with abnormal events. On the other hand, principal components analysis was used as one of the feature extraction methods. Preprocessed data through two methods were trained by the same deep learning algorithm, gated recurrent unit. The data selected by humans and the data extracted by considering the relationship among the variables showed different performance for diagnosing the plant state. The results showed that it is advantageous for the developing diagnosis model to learn and judge through the feature extraction method.

Keywords: Principal components analysis · Deep learning · Gated recurrent unit · Nuclear power plant

1 Introduction

Nuclear power plants (NPPs) are subject to strict safety standards to prevent severe effects on surroundings when they fail. It is required that NPPs must be operated by operating procedures based on design specifications and operational experience for maintaining the plant safety. The diagnosis of the NPP was the responsibility of a human operator, there have been attempts to develop NPP diagnosing systems on behalf of the operators [1–3].

W. Karwowski and T. Ahram (Eds.): IHSI 2019, AISC 903, pp. 504–509, 2019.
https://doi.org/10.1007/978-3-030-11051-2_76

In this paper, abnormality diagnosis model for NPPs is suggested to support the operators. An advanced pressurized reactor 1400 (APR 1400), the latest model operating in Korea, has 82 kinds of abnormal operating procedures (AOPs). When some components have malfunction, the operator should diagnose the plant state as soon as possible and stabilize the plant with an appropriate AOP. However, considering the number of stages of subdivision in all AOPs, it is high burden for the operators to compare lots of information in a short period of time. The suggested model used a deep learning algorithm to comprehend the relationship among the plant parameters to detect any abnormalities. The framework of this work is shown in Fig. 1.

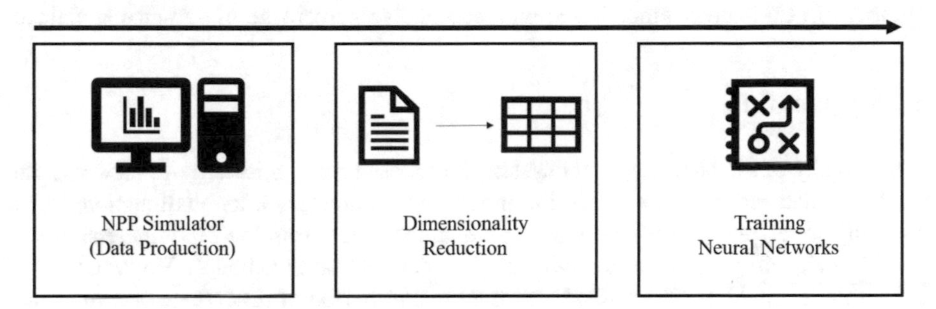

Fig. 1. Framework to develop abnormal state diagnosis system

The data was produced from an NPP simulator. Before training the deep learning algorithm, it is necessary to reduce the dimensionality of the generated data to achieve the best estimation of the training. This paper covers data preprocessing methods divided into feature selection method and feature extraction method. The performance of the NPP state prediction results was compared for the two methods.

2 Data Preprocessing

To obtain the best prediction result, it is necessary to reduce the dimensionality of the data to prevent the problem so called as 'the curse of dimensionality' [4]. Feature selection method is one of the methods for it by using domain knowledge by the user. In this case, AOPs can provide the information currently used to determine abnormal conditions. On the other hand, feature extraction method extracts latent features from the original data without the user intervention. Principal components analysis (PCA) was used as one of the methods of feature extraction.

2.1 Analysis of Abnormal Operating Procedures

An AOP includes stages indicating the detailed causes of abnormal events. Although it was not the same environment as the real NPP, the variables were selected including the ones that can be confirmed in the simulator and related with the safety critical function. In addition, by analyzing the AOPs, it was possible to set up the examples associated with them.

2.2 Principal Components Analysis

PCA is a method to convert the original data into a dimensionally reduced one. PCA calculates a linear transformation matrix [5]. Once the correlation matrix is calculated, it is decomposed by its eigenvectors and eigenvalues through the process of the PCA. The eigenvectors represent the principal components (PCs), and the eigenvalues represent the percentage of variances described by the corresponding eigenvectors. The number of PCs is determined based on a cumulative percentage of variance explained.

3 Application

Abnormality of the charging water system was chosen for application of this work. The produced data contained records for monitored parameters after malfunctions for a charging water pump, a flow control valve and leakage from the pipe line occurred.

RNN algorithm is one of the most used models for data prediction. However, as the algorithm learns high dimensional data over a long period of time, there is a vanishing gradient problem that the learning result becomes inaccurate. Long short-term memory (LSTM) solves this problem by using three function gates to handle variables among cells. In advance, gated recurrent unit (GRU) has the same advantage of LSTM with update and reset gates [6]. It allows GRU to reduce computational complexity and enable more efficient training. In this work, GRU was used for the deep learning algorithm to diagnose the given data states (Fig. 2).

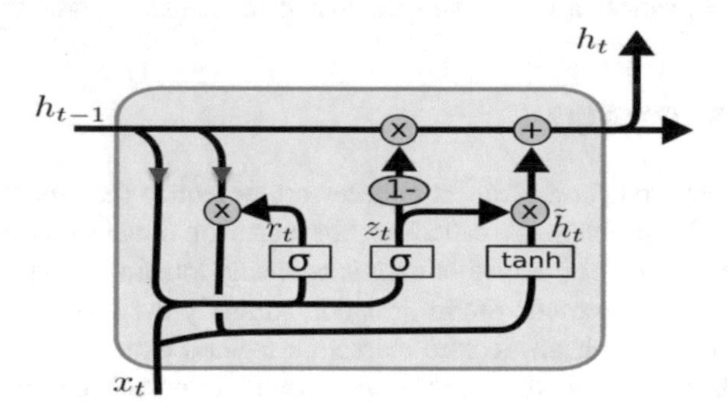

Fig. 2. Schematic diagram of gated recurrent unit

4 Result and Discussion

4.1 Feature Selection Method

Figures 3 and 4 show the training results of GRU using the 62 variables based on the AOP.

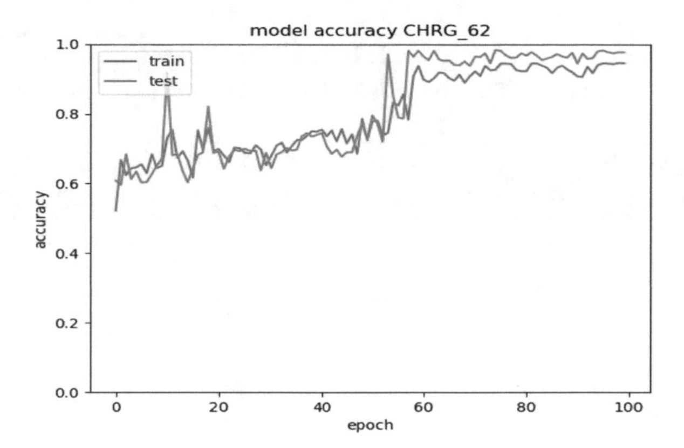

Fig. 3. Model accuracy of the feature selection method

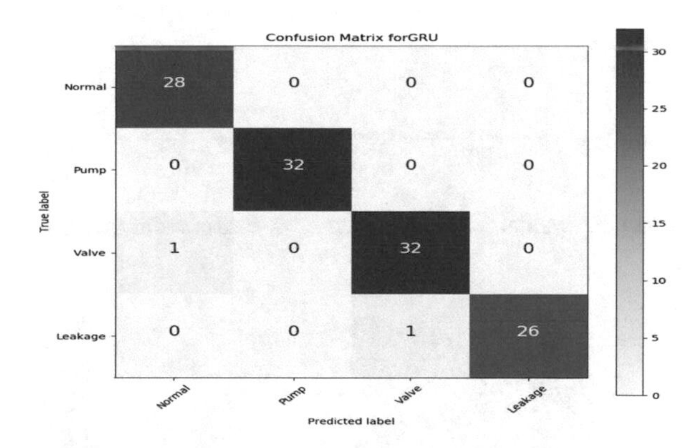

Fig. 4. Confusion matrix of the feature selection method

Although the simulator monitored the variables based on the AOP of the different plant type, training results showed high prediction level as training epochs increasing. There were only two inaccurate prediction out of 120 cases for the malfunction on the valve and leakage from the pipe line. It might be difficult to judge the selected stages because they cause similar symptoms in the same system.

4.2 Feature Extraction Method

Table 1 shows the PCA result from 2829 parameters to 10 PCs. Figures 5 and 6 show the training results of GRU using 10 variables through the PCA.

Table 1. PCA results with 10 PCs

PC	1	2	3	4	5
Eigenvalue	0.802586	0.101177	0.032922	0.014469	0.014016
Cum. value	0.802586	0.903763	0.936684	0.951153	0.965169
PC	6	7	8	9	10
Eigenvalue	0.008008	0.00452	0.003231	0.002993	0.002308
Cum. value	0.973177	0.977698	0.980928	0.983921	**0.986229**

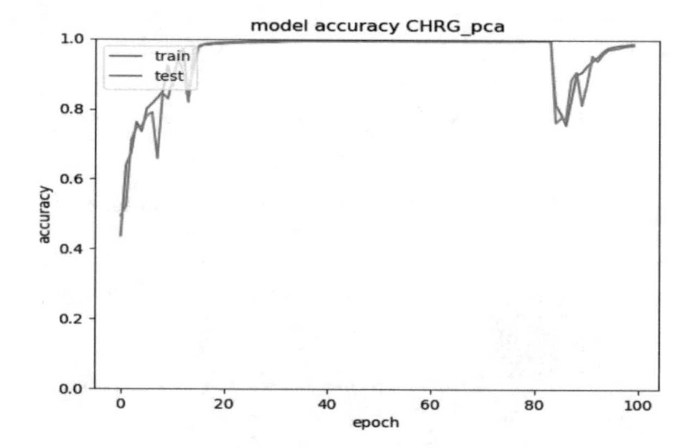

Fig. 5. Model accuracy of the feature extraction method

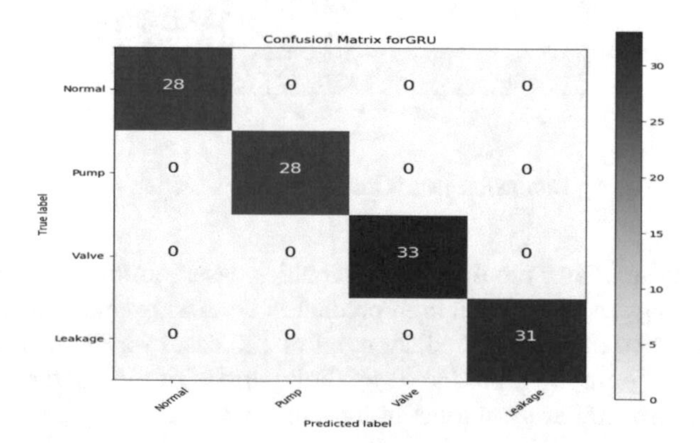

Fig. 6. Confusion matrix of the feature extraction method

With only 10 PCs, the preprocessed data could preserve total 98.6229% of information from the raw data. Even only two PCs could contain about 90% of information. The data from the PCA made accurate prediction possible through the GRU. It is noticeable that there was no need to have many epochs to achieve high level of prediction after about 20 epochs. At the end of the training, model accuracy was dropped and recovered. When compared to the feature selection method, since the original data covered whole plant parameters, the results of the feature selection method could be improved if an NPP expert adds variables that are not currently considered. In addition, another reason is that the number of variables were much less than that of the feature selection method, which makes the density of the data high.

5 Conclusion

In this paper, data preprocessing methods was studied as a part of developing abnormality diagnosis system for NPPs. GRU algorithm leaned the data in the same environment by preprocessing in different ways. Based on analysis of AOPs, feature selection method showed high prediction level except for two missing in the test case. However, with 10 PCs passed through the PCA, GRU algorithm predicted all test cases. Moreover, feature extraction method has another advantage of reducing the time required to analyze a new operating procedures. Therefore, further study to develop the diagnosis model will use PCA for the data preprocessing.

Acknowledgements. This work was supported by the Korea Institute of Energy Technology Evaluation and Planning (KETEP) and the Ministry of Trade, Industry & Energy (MOTIE) of the Republic of Korea (No. 20171510102040).

References

1. Miller, D.W., et al.: Experience with the hierarchical method for diagnosis of faults in nuclear power plant systems. Reliab. Eng. Syst. Saf. **44**(3), 297–311 (1994)
2. Horiguchi, M., Fukawa, N., Nishimura, K.: Development of nuclear power plant diagnosis technique using neural networks. In: Proceedings of the First International Forum on Applications of Neural Networks to Power Systems, Seattle, WA, USA, pp. 279–282 (1991)
3. Lee, S.J., Seong, P.H.: A dynamic neural network based accident diagnosis advisory system for nuclear power plants. Prog. Nucl. Energy **46**(3–4), 268–281 (2005)
4. Zimek, A., Schubert, E., Kriegel, H.: A survey on unsupervised outlier detection in high-dimensional numerical data. Stat. Anal. Data Min. **5**, 363–387 (2012)
5. Jolliffe, I.: Principal Component Analysis, 2nd edn. Springer, pp. 1–59 (2002)
6. Chung, J., Gulcehre, C., Cho, K., Bengio, Y.: Empirical evaluation of gated recurrent neural networks on sequence modeling, arXiv preprint arXiv:1412.3555 (2014)

Personalized Product Recommendation for Interactive Media

Hal James Cooper[1(✉)], Garud Iyengar[1], and Ching-Yung Lin[2]

[1] Columbia University, New York, USA
{hc2683,gi10}@columbia.edu
[2] Graphen, Inc., New York, USA
cylin@graphen.ai

Abstract. The video game industry is larger than both the film and music industries combined yet has received scant academic attention. We explore recommendations that makes use of interactivity, arguably the most distinguishing feature of video game products. We show that implicit data that tracks user-game interactions and levels of attainment (e.g. Microsoft Xbox Achievements) has high predictive value when making recommendations. Furthermore, we argue that the characteristics of the video gaming hobby (low cost, high duration, socially relevant) make clear the necessity of personalized, individual recommendations that can incorporate social networking information. We tackle this problem from the viewpoint of graph querying and demonstrate the foundation of a new approach for learning structured graph queries from data.

Keywords: Social recommendation · Human-centered computing

1 Introduction

Recommendation systems and the techniques developed for them have historically focused on music [1], film [2], and other media for which consumption is a predominantly passive process. These services (see [3] for a recent review of the field) primarily employ "explicit" ratings data, wherein a user directly and deliberately inputs some form of rating for a product. In contrast, implicit data, i.e. data that may *implicitly* indicate user preferences, but that does not involve the user explicitly designating a score or writing a review, is often seen as less causally informative, more biased, and less precise. The implicit data collected for the purposes of making recommendations is often basic (e.g. "user *i* watched movie *j*"). We argue that there exists a class of products for which implicit data is disproportionately detailed and relevant; interactive media products for which interaction is the primary method of consumption. In this paper, we use the videogame industry (specifically, Steam [4] and Xbox Live [5]) as our source of interactive media products. Steam is the largest digital distribution service for video games on the PC, Mac, and Linux platforms, with more than 18,000 available video game products. Xbox Live is a similar digital distribution service for Windows PCs and Microsoft's Xbox line of home video game consoles. Products on Steam and Xbox Live are actively marketed to users running each service, with regular discounted sales events and pop-ups of recommended purchases. Though Steam and Xbox Live do allow users

W. Karwowski and T. Ahram (Eds.): IHSI 2019, AISC 903, pp. 510–516, 2019.
https://doi.org/10.1007/978-3-030-11051-2_77

to post reviews of purchased products, the use of such data in making recommendations could be called into question due to the frequency of review bombing incidents [6, 7] and the relative sparsity of the review datasets. Fortunately, there exists a wealth of additional data less susceptible to manipulation that we demonstrate is very informative for the purposes of making recommendations. Though video game digital distribution service datasets have been analyzed previously [8–10], this is the first work to use achievement data for social recommendation beyond simple friendship networks.

2 Attainment Ratings

Xbox Live was explicitly designed as a social network. Users have avatars that represent their likenesses, and add other users, known in real-life or encountered during online play, to their list of friends. The concept of an "Xbox Live Achievement" (hereafter simply referred to as an "achievement") was created, "gamifying" the play of video games [11]. Achievements are awarded when players complete certain (potentially difficult) tasks in a game. These achievements are publicly visible, giving users a form of bragging rights that demonstrates their skill. Digital marketplaces for other consoles [12] and computers (e.g. Steam) have since followed suit.

In this paper we argue that achievements are highly informative implicit data that can be used to determine a score representing how much a user likes a game. The rationale is simple; the more a user enjoys a game, the more they will attempt to complete everything it has to offer. We compute an overall achievement score for a user-game pair by combining each individual achievement weighted by a difficulty level computed using global achievement rates. Total time played (hereafter referred to as "play times") has been proposed in the literature [13] as a metric to approximate ratings. However, there are significant issues with the use of simple play times. For instance, the length of games is not uniform across genres, and many highly acclaimed games are very short (such as the Playstation title, Journey). Though we can "normalize" play times (e.g. by uniform scaling, using z-scores, or taking quantiles), the process of normalization requires explicitly comparing all users, a computationally expensive operation. We propose to eliminate these shortcomings by using achievement data to define what we call an attainment rating. These attainment ratings can be computed without explicitly comparing users yet results in a score that is globally consistent across all users.

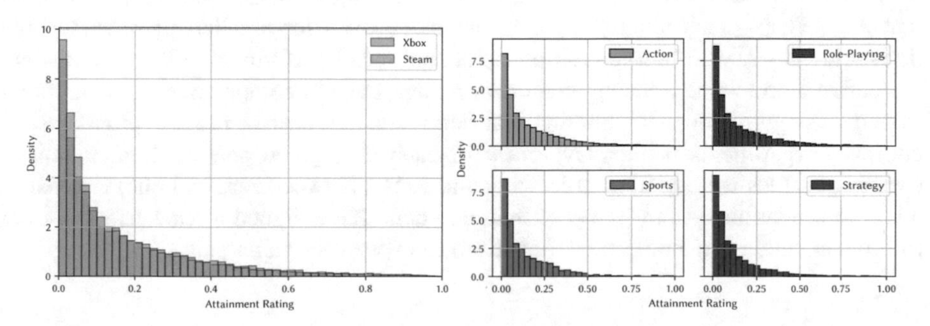

Fig. 1. Density histograms of attainment ratings, split by dataset (LHS) and genre (RHS).

The attainment rating is computed by combining each individual achievement weighted by a difficulty level computed using global achievement rates. For a game g we denote its set of achievements as $A_g = \{A_{g1}, A_{g2}, \ldots, A_{gN_g}\}$, where N_g is the total number of achievements of game g. Each $A_{g_i} \in A_g$ is a binary vector of length $\| P_g \|$, where P_g is the set of players who own game g, such that A_{g_is} indicates whether or not player s has achieved achievement number i in game g. Clearly, the proportion of players who have achieved A_{g_i} can be calculated as $C_{g_i} = \sum_{p \in P_g} A_{g_ip} / \| P_g \|$, with an attainment score A_{gs} of player s for game g then calculated as follows:

$$A_{gs} = \sum_{A_{g_i} \in A_g} \frac{A_{g_is} \cdot \left(1 - C_{g_i}\right)}{N_g}. \quad (1)$$

Although C_{g_i} compares all players, these values are available a priori due to gamification; they are precalculated by the services to allow users to understand their rarity. We can therefore calculate Eq. (1) without the need for explicit user comparisons. We also note that a rare achievement, i.e. one with $C_{g_i} \approx 0$, makes a large contribution to Eq. (1). Conversely, extremely common achievements with $C_{g_i} \approx 1$, contribute very little to the attainment rating. From Eq. (1), we have well-defined bounds on the range of possible attainment scores, $0 \leq A_{gs} \leq 1 - 1/ \| P_g \|$. In a traditional rating system, different users may have entirely different standards for what constitutes a given score. In contrast, the attainment ratings in Eq. (1) are precise and represent the same degree of attainment for every user. In Fig. 1, we show density normalized histograms of the attainment ratings calculated using Eq. (1), split by both dataset and genre.

3 Attainment Recommendations

In Fig. 2, we demonstrate the use of attainment ratings in a collaborative filtering (specifically, SVD++ [14] using default parameters of Surprise [15] due to the implicit nature of the attainment ranking data) based recommendation system.

We observe that precision and recall both improve as the threshold increases but note that the universe of games that are "relevant" or recommended becomes vanishingly small as the threshold increases (consider the proportion of ratings indicated by Fig. 1 as greater than a given threshold), resulting in performance measurements that are perhaps misleadingly high. Nonetheless, even for smaller, more reasonable threshold values, we observe high precision and recall performance. These results are indicative of the value of using attainment ratings, even in comparatively simple rating systems using collaborative filtering that don't consider user or product properties. In contrast, experiments using play times for each user game pair as implicit ratings (where quantiles were used in order to bound values between zero and one) resulted in a recommendation system no better than a system that assigned a relevance score 0.5 (the mean, under a quantile transformation) to every user game pair.

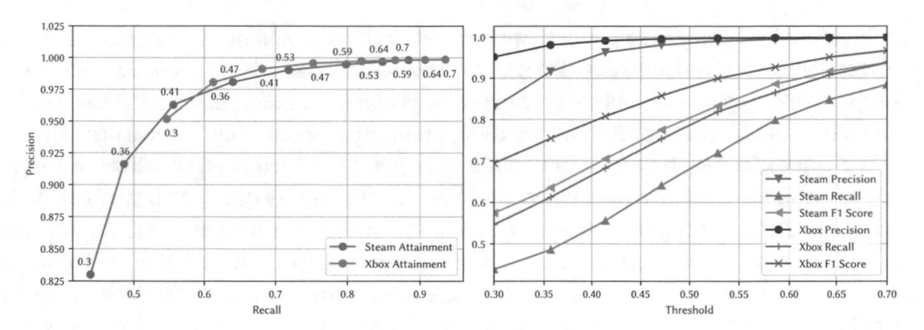

Fig. 2. Precision vs. Recall (LHS), and Precision, Recall, and F1 scores with respect to relevance thresholds (RHS) for SVD++ attainment recommendation.

4 Graph Recommendations and Future Work

The field of social recommender systems [16, 17] has become increasingly important in recent years, growing alongside the size and complexity of social networks. In its most basic form, the problem of social recommendation can be viewed as an example of link prediction on graphs with many vertex (e.g. person, product, organization, news item) and edge types (friends, owns, belongs to, read). In both the Steam and Xbox Live datasets, we designate player, game, developer, and genre as vertex types, and friendship, ownership, developed by, and genre association as edge types. The social aspect of using ratings data from actual friends (as opposed to other users designated as "similar" via some clustering or similarity measure) is reasonable in the context of video games, where online multiplayer is often a big part of a product's appeal.

The average playtime of a video game is considerably longer than many other forms of entertainment, with games in the action genre having playtimes typically in the dozens of hours, and role-playing games often extending into hundreds of hours. Given the long length of video games (in comparison to other media) and the large number of products in the dataset, the ability to recall even a small fraction of relevant products is enough for users for the duration of their use of the platform. This encourages the repeated use of more personalized queries, where a user can narrow these general recommendations to match their specific desires of the moment. For example, such a request might be "for a certain user, find the names and purchase costs of five games that are owned by friends, that are developed by companies who have developed games already owned by the user, and order responses according to how the friends 'rated' them with respect to attainment", which can be interpreted as a graph query. Here we set about learning such an interpretable graph query from data, which could be easily personalized for a user in accordance with their custom preferences.

The problem of learning queries from data is alternatively referred to as "reverse engineering", the "reachability problem", or "query by example" [18–20]. To the best of the authors' knowledge, existing solution approaches for this problem are all combinatorial in nature [21]. Here, we instead outline a method for learning a query that generates recommendations from the graph datasets and the calculated attainment ratings by solving a continuous optimization problem, where personalization can be achieved by additional constraints, or by appropriately modifying the learned query.

We approximate a query on the graph by N parameterized random walks of length M, where θ_{ijm} are learnable parameters defining the probability of traversing from a node of type i to a node of type j on the m th step of the traversal, where $m = \{1, \ldots, M\}$ and $i, j \in \{player, game, developer, genre\}$. For the m th step of traversal n, we draw a sample $A_{nm}(\theta)$ as the largest element of a concrete distribution [22] with category probabilities defined by θ (where $\sum_j \theta_{ijm} = 1 \forall i, m$, and each traversal chooses the appropriate θ variables corresponding to current step and vertex type). We denote by x_{nm} the vertex of the n th traversal after the m th step, and let $avgAttainment(\cdot)$ be a function that takes a vertex and returns its average attainment score (and is therefore non-zero only if the vertex is a game vertex). Then we solve the following optimization:

$$\begin{aligned} &\max_{\theta} \sum_{n=1}^{N} \prod_{m_1=1}^{M} A_{nm_1}(\theta) \sum_{m_2=1}^{M} avgAttainment(x_{nm_2}) \\ &s.t. \sum_{j} \theta_{ijm} = 1 \forall i, m \\ &\theta \geq 0 \end{aligned} \tag{2}$$

For sufficiently small values of the temperature λ of the Concrete distribution, each $A_{nm}(\theta) \approx 1$, yet is a function of the parameters θ. In this manner, random walks that traverse games with high average attainment will back-propagate rewards through these parameters that guided the walks on their paths. Here we have used the reparameterization trick [23] such that the discrete choices of the random walk made during each traversal can propagate the gradients of the continuous parameters with minimal effect on the actual value of the objective function. In Fig. 3, we observe very fast convergence to the globally optimal solution, an interpretable query of "traverse repeatedly from users to owned games, to other users who own that game, to games that they own".

The results of this paper clearly demonstrate the worth of attainment ratings when making recommendations in the video game context. This suggests the need for future work to explore recommendation in the context of other interactive products that likewise generate implicit data. Even in the context of video game recommendation systems alone, there is room for development in the use of more advanced recommendation systems, such as those that explicitly use product and user properties. Furthermore, the method of using differentiable parameterized query families to solve and represent graph analytics has clear implications beyond generating product recommendations.

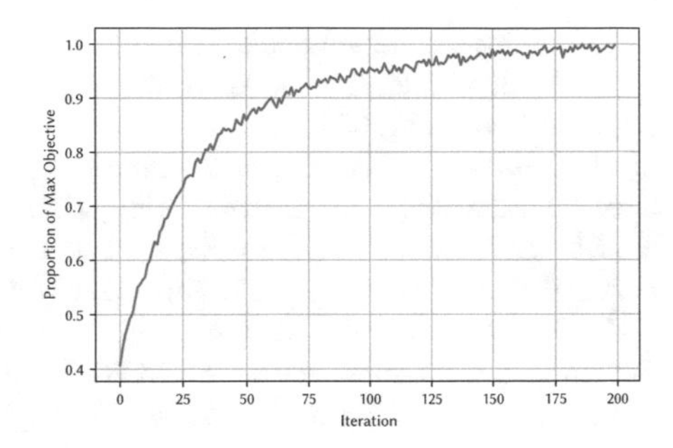

Fig. 3. Training of Eq. (2) using SGD with a learning rate of 0.01.

References

1. Song, Y., Dixon, S., Pearce, M.: A survey of music recommendation systems and future perspectives. In: 9th International Symposium on Computer Music Modeling and Retrieval (2012)
2. Gomez-Uribe, C.A., Hunt, N.: The netflix recommender system. ACM Trans. Manag. Inf. Syst. **6**, 1–19 (2015)
3. Aggarwal, C.C.: Recommender Systems. Springer International Publishing (2016)
4. Welcome to Steam. http://store.steampowered.com/
5. Microsoft: Xbox Live—Xbox. https://www.xbox.com/en-US/live
6. Kuchera, B.: The Anatomy of a Review Bombing Campaign – Polygon. https://www.polygon.com/2017/10/4/16418832/pubg-firewatch-steam-review-bomb
7. Grayson, N.: Total War Game Gets Review Bombed On Steam Over Women Generals. https://steamed.kotaku.com/total-war-game-gets-review-bombed-on-steam-over-women-g-1829283785
8. Becker, R., Chernihov, Y., Shavitt, Y., Zilberman, N.: An analysis of the steam community network evolution. In: 2012 IEEE 27th Convention of Electrical and Electronics Engineers in Israel (IEEEI), pp. 1–5 (2012)
9. Blackburn, J., et al.: Cheaters in the Steam Community Gaming Social Network. ArXiv e-prints (2011)
10. O'Neill, M., Vaziripour, E., Wu, J., Zappala, D.: Condensing steam: distilling the diversity of gamer behavior. In: Proceedings of 2016 ACM Internet Measurement Conference, pp. 81–95 (2016)
11. Jakobsson, M.: The achievement machine: understanding Xbox 360 achievements in gaming practices. Game Stud. **11**, 1–22 (2011)
12. Niizumi, H., Thorsen, T.: PlayStation Network Platform Detailed (2006). https://www.gamespot.com/articles/playstation-network-platform-detailed/1100-6145981/
13. Hamari, J.: Framework for designing and evaluating game achievements. In: Proc. DiGRA 2011 Conference: Think Design Play, p. 20 (2011)
14. Kumar, R., Verma, B.K., Sunder Rastogi, S.: Social Popularity based SVD++ Recommender System. Int. J. Comput. Appl. **87**, 33–37 (2014)
15. Hug, N.: Surprise, a Python library for recommender systems (2017). http://surpriselib.com
16. Tang, J., Hu, X., Liu, H.: Social recommendation: a review. Soc. Netw. Anal. Min. (2013)

17. Konstas, I., Stathopoulos, V., Jose, J.M.: On Social networks and collaborative recommendation. In: Proceedings of the 32nd International ACM SIGIR Conference on Research and Development in Information Retrieval (2009)
18. Bonifati, A., Ciucanu, R., Lemay, A.: Learning path queries on graph databases. In: 18th International Conference on Extending Database Technology (EDBT) (2015)
19. Barceló, P., Libkin, L., Lin, A.W., Wood, P.T.: Expressive languages for path queries over graph-structured data. ACM Trans. Datab. Syst. **37**, 31:1–31:46 (2012)
20. Arenas, M., Diaz, G.I., Kostylev, E. V.: Reverse engineering SPARQL queries. In: Proceedings of the 25th International Conference on World Wide Web – WWW 2016 (2016)
21. Angles, R., Arenas, M., Barceló, P., Hogan, A., Reutter, J., Vrgoč, D.: Foundations of modern query languages for graph databases. ACM Comput. Surv. **50**, 68 (2017)
22. Maddison, C.J., Mnih, A., Teh, Y.W.: The concrete distribution: a continuous relaxation of discrete random variables. arXiv Preprint arXiv1611.00712 (2016)
23. Kingma, D.P., Welling, M.: Auto-encoding variational Bayes. arXiv Preprint arXiv1312.6114 (2013)

Human Factors in Software Projects for Complex Industrial Processes

Marja Liinasuo(✉), Toni Lastusilta, Jouni Savolainen, and Timo Kuula

VTT Technical Research Centre of Finland Ltd, Vuorimiehentie 3, P.O. Box 1000, Espoo, Finland
{marja.liinasuo,toni.lastusilta,jouni.savolainen, timo.kuula}@vtt.fi

Abstract. Process industry has a need to increase product quality while also reducing operating costs and environmental footprint. However, a plant comprises of a complex set of dynamic processes, as well as human operators that control the processes. The best improvement is achieved by an extensive intervention of both aspects. In practice, human factor aspects have often a marginal role in carrying out a technical project. This has not been the case in the present, EU-funded project. The systematic inclusion of human factors perspective requires parallel and collaborative proceeding of technical development and human factors. The flexible approach is central as the human matters not only affects technical development but is also involved with the usage context of the new technical solution.

Keywords: Human factors · Process industry · Optimization

1 Introduction

Process industry has a need to increase operating efficiency while reducing environmental footprint and maintaining an adequate product quality. However, a plant comprises of a complex set of dynamic processes, as well as human operators that control the processes. The best improvement is achieved by an extensive intervention of both aspects. Typically, large processes are split into sub-processes, which are controlled separately. In some cases, there exists a higher level of automation monitoring the whole plant, but in others this is not the case. Especially, if the sub-processes are heavily connected, a plant-wide monitoring and control system is especially useful. Therefore, to start with, a technical plant-wide monitoring and control system is many times the best way to improve the performance of a plant. Hence, the result of this should be that the whole operation of the plant is optimized using the present processes as effectively as possible.

In an EU funded project COCOP (Coordinating Optimisation of Complex Industrial Processes) [1], the target is to develop a plant-wide optimization system for process industry. The new system is to be model-based and predictive. The core concept is to decompose a large optimization problem into smaller sub-process

W. Karwowski and T. Ahram (Eds.): IHSI 2019, AISC 903, pp. 517–523, 2019.
https://doi.org/10.1007/978-3-030-11051-2_78

optimization problems and to coordinate these for a plant-wide optimized result. The optimization system will be piloted in two specific plants: a steel factory and copper smelter.

Such a system is a novelty and thus, in addition to the development of the optimization system, the human share of the work should be taken into account.

Technology-driven projects usually have their own proven methodology or procedures to follow, resulting in a good solution (if the development is performed by experts). Likewise, human centric (human factors) related research has its own methods, which enable the finding of relevant and qualitative data that is important in producing a system, which is meaningful and easy to use for its users. Furthermore, if possible, human factors is involved with other factors affecting the usage of the system, such as work processes, which may be affected with the new tool.

An issue is how to ensure collaboration between the technical and human factors experts in a technical development process. Such collaboration is needed in technical projects, which aim to consider also human related matters in the development of a technological solution. Such an aim is not trivial. It has been shown that, for instance, without human factors approach, a development project may become more costly, the solution may have low user-friendliness, and end users can have insufficient knowledge of safe usage and potential risks of the new tool [2].

In this paper, we will describe how human factors approach has been utilized in the essentially technical COCOP project, and we will discuss the possibilities of different types of software projects, combined with the characteristics of an efficient human factors approach.

2 Human Factors in COCOP Software Project

One of the human factor prerequisites for a technical project with an efficient share of human factors is fulfilled in this project. That prerequisite is that in the beginning of the project, the human factors experts have the task to understand the user perspective and the usage context of the new technological solution, i.e. in this case the new system. Human factors experts could then become acquainted with the main qualities of the work, important to take into account in project decisions, including the technical ones.

Firstly, the performed user interviews in the beginning of the project enabled the acquisition of information relevant for the identification of core matters in the future usage of the new system. Such a matter is, for instance, core qualities of work, as they may affect both the qualities of the new system as well as the context in which the new systems will be used. Among other things, it was found out that in one plant, the work of operators is mainly problem solving. This information can be used in the development process by concentrating on the identification of operator problems. Finding solutions for operators' problems may be relevant for the development of the optimization system. Furthermore, later in the project, this can be used in supporting user acceptance by emphasizing how the new system supports operator problem solving.

Secondly, an entity of its own is the characterization of the qualities that the users of the new system find important. Interviews related to this were also conducted in the beginning of the project. When including such qualities in the new system, the system may even perform better, based on, for instance, tacit knowledge users have about their work. Additionally, when users can contribute to the development of the new system, it will be better accepted.

Furthermore, some important questions were answered already in the beginning of the project. For instance, the role of the future user is not self-evident. It was decided that the user will have the choice of accepting or rejecting the suggestion made by the optimization system. When the role is set, interviews allow the probing of the meaning of this role to the operator. We found that the sense of autonomy is important for operators in one plant. This is an important piece of information as the new system is to guide process control towards the “general good” (plant-wide optimization) instead of “local good” (sub-process optimization), which is presently the target to operators. If the system only suggests and does not make decisions on behalf of the factory operators, users controlling the process are key personnel in the success of the new system. If the system is not used, the sophistication of the solution has no effect on improving the production process.

Solid framework is needed to ensure all relevant factors are taken into account. So far, this is performed by dividing the human factors related matters into two main categories, the ones between the user and the new system, and the ones between users and the usage context. The matters to be done in collaboration with the technical experts belong to the first category, such as user interface qualities and functionalities preferred by users. The matters, which require discussion with the target company, belong to the latter one, such as anticipated changes in the proceeding of work, and the need for training related to plant-wide processes and the new system.

Presently, the development of the plant-wide optimization system is ongoing. During the development process of the new system, participatory design is planned to take place in order to ensure that the new system will be easy to use and will be visually and functionally embedded with the presently used system.

Later on, when the new system is almost finalized, it will be reasonable to test, or rather, validate the new system with users. This way the final system will be refined to suit in the usage of real life.

3 Flexibility in the Human Factors Approach

In order to succeed in collaboration with the different parties of the project and to talk the language of various stakeholders of the companies to which the new system will be implemented, human factors should not withdraw from discussions behind its own concepts and approaches.

In this project, this principle is followed. We have been using the concept of Key Performance Indicator (KPI). It is used, e.g., for marketing and for measuring business benefits, but in our project it has been used to measure the successfulness of the usage of the new system. To be precise, one KPIs is the frequency of the acceptance of the suggestions provided by the new optimization system.

Moreover, the important matters to be taken into account are expressed in the form of “human factors requirements”. This is in accordance with the proceeding of technical development process. As the format of these requirements is the same as the one used by the technical experts, the human factors related requirements become more commonly understood in the project. Especially the features in the user interface, important for usability, are good to express in the same format as other, more technical qualities. Hence, these requirements are taken into account in software development like any other requirement.

Still another factor, perhaps not planned but as a serendipity, is that in human factors meetings, also some representatives of a more technical approach participate. This ensures that aspects important for the technical development are taken into account in human factors' share of the project and, last but not least, the language used by human factors is to a large extent the one used in the technical meetings as well.

4 Discussion

The realization of human factors has been successful in the project, based on the good project plan and cooperative interdisciplinary project members.

Possibilities for improvement exist as well. The definition of requirements has been rather cumbersome and we do not yet know how well the human factors requirements can be taken into account in the technical development process. The part of human factors matters not used by technical project partners could have also been described, in a more simple way, without the format of requirements but rather by describing in the project meetings the plans, and in deliverables the results, with plain words.

Furthermore, the interaction between technical and human factors experts could be more frequent, regarding the way important matters are discussed about in the project. Presently, some information has been acquired through informal discussions, as part of the workrooms of technical experts are physically located near the ones of some human factors experts. This setting is quite vulnerable to changes in seating and personnel.

4.1 Development Process in a Traditional Software Project

Traditional software project management can be characterized by long-term and detailed project plans, detailed requirements specifications, and no design work before all the requirements, analysis, and design documents are completed [3]. Furthermore, they are based on cost- and risk-control, as well as, predictability [3]. The most famous of this type of development process is the waterfall model. In its fundamental form, it includes the steps of gathering requirements, designing, implementation, verification and maintenance. The first step is sometimes referred to as requirements engineering in which requirements are gathered, defined, documented and maintained. A requirement is, based on Brennan [4], as follows:

1. A condition or capability needed by a stakeholder to solve a problem or achieve an objective.
2. A condition or capability that must be met or possessed by a solution or solution component to satisfy a contract, standard, specification, or other formally imposed documents.
3. A documented representation of a condition or capability as in (1) or (2).

When gathering requirements, various approaches can be utilized, e.g. questionnaires, following people around/observation, models, use cases/scenarios/user stories, document analysis, reverse engineering and tools (e.g. user stories and requirements management) [5].

In the waterfall model, one proceeds to the next step only when the preceding step is completed, which makes the model quite rigid, and which has induced variations of the model. In fact, the waterfall model has been so widely criticized that altogether alternative approaches, for example agile software methods, are widely used, especially in the software engineering domain.

In a development process like this, human factors is easily left out or it may only have a marginal role, as there is no explicit role for human factors. The easiest way to proceed for human factors in this line is to focus on user interface related requirements, related to usability; matters beyond that are more difficult to realize through the requirements. Interestingly, the human factors share in COCOP project resembles traditional software project, as the aim has been to define all future tasks in the form of requirements.

4.2 Development Process in an Agile Software Project

The agile software development refers to a set of methods and practices, aiming at flexible and rapid development process with specific principles. Agile manifesto for software development has the following four key points:

1. Individuals and interactions over processes and tools;
2. Working software over comprehensive documentation;
3. Customer collaboration over contract negotiation;
4. Responding to change over following a plan [6].

Hence, the agile method is based on teamwork and bringing value to the customer.

Several tools can be used to achieve project goals and in the following, some commonly used are described [3]. Requirements are recorded in use cases and user stories are continuously improved. The product backlog is a list of items to be implemented. The release plan is a simple calendar schedule showing when each fixed-length iteration starts and ends. At a minimum, the release plan will have a number of iterations that implements features from the backlog and a release iteration at the end of the project.

Human factors has an easier entry to agile software project as in agile development process, the approach is open and frequent collaboration takes place. Especially the usage of other than requirements in guiding the development facilitates the inclusion of human factors approach. For instance, the usage of use cases or scenarios in frequent meetings provide a way to affect development in a form understandable to all stakeholders, irrespective of their educational or experiential background.

However, human factors approach is not as flexible as a pure agile software development process. The methodology used in human factors is often cumbersome. For instance, in order to acquire understanding about work and task qualities of the future users of the new system, interviews are often used to gain deeper insight. They take time to perform, then they need to be transcribed and analyzed, and eventually, the results are to be concluded. Thus, quick turns in the development process are hard to take into account unless there are abundantly resources available for this work. On the other hand, the role of human factors in the context of new system development is to provide human or usage centric information to the project and such information is probably of such nature that it does not become obsolete during the development process, even if changes in the course of development would take place. Last but not least, human factors experts have a different overall perspective and may therefore more easily identify some issue or matter that needs to be tackled. This supports the choice of an agile development process over a traditional one.

5 Conclusions

In the development of a technical system, the flexible and effective inclusion of human factors perspective is challenging but possible. It requires a development process shared in the whole project group. Preferably, there is a technical representative taking part in human factors work to ensure smooth cooperation and collaboration. Human factors part should be defined clearly, so that all stakeholders understand the meaning and importance of it. Furthermore, time should be reserved for the technical team, end users and other relevant stakeholders to perform work initiated by human factors.

The scope of human factors is rather large, as it aims not only in affecting technical solution but also in the understanding of user and usage context. Agile methods, contrasting to traditional development methods, seem promising as they appear to provide methodology common to the whole development process and entry points to human factors approach. Especially flexibility in the proceeding of the project and resources for human factors in several points through the project seem important aspects for human factors, coinciding the principles of agile development.

The project is ongoing and the present ideas are not yet fully tested. Furthermore, the way to express human factors input to the technical development process is here left open; possibly, a combination of several types of means are needed. Future work shows how well these points truly serve the inclusion of human factors approach to the development process as a whole. Furthermore, it will suggest ways to transfer the human factors knowledge in the best possible ways to the technical development group.

Acknowledgments. This work was supported by the COCOP-project. This project has received funding from the European Union's Horizon 2020 research and innovation programme under grant agreement No. 723661. This paper reflects only the author's views and the Commission is not responsible for any use that may be made of the information contained therein.

References

1. EU project COCOP. https://www.cocop-spire.eu/
2. Saetren, G.B., Hogenboom, S., Laumann, K.: A study of a technological development process: human factors - the forgotten factors? Cogn. Tech. Work **18**, 595–611 (2016)

3. Cline, A.: Agile Development in the Real World. Apress, New York City (2015)
4. Brennan, K.A.: Guide to the Business Analysis Boxy of Knowledge (BABOK Guide). http://www.iiba.org/babok-guide.aspx
5. Koelsch, G.: Requirements Writing for System Engineering. Apress, Berkeley (2016)
6. Beck, K., et al.: Manifesto for Agile Software Development (2001)

Research on Aesthetic Cognition Characteristics of Product Form: Case Study of Purple Clay Teapot

Ming Li(✉), Jie Zhang, and Yiping Hou

East China University of Science and Technology, Shanghai 200237, China
1013372753@qq.com, {lzlglm01,lzlglm00}@163.com

Abstract. For analyzing the relationship between the product form and the emotional demand of consumers objectively, the aesthetic cognition characteristics of product form was presented based on the evaluation model of aesthetics measure. According to the principle of formal aesthetic, principle of visual perception and gestalt psychology, 10 aesthetic measure indexes is established. Then the cluster analysis and the structural equation model applied to reduce data dimension and test the rationality of cognition logic for target product. Through compare the linear regression analysis with the neural network model, the aesthetic cognition can be confirmed linear and nonlinear characteristic. Taking the teapot as a research case, the result showed that aesthetic cognition of teapot should be nonlinear characteristic and the structure of aesthetic cognition seems like logsig-tansig transfer functions type. The aesthetic prediction prototype system could provide effective aids for product.

Keywords: Product form · Aesthetic cognition · Nonlinear characteristic Evaluation model · Neural network

1 Introduction

With the continuous improvement of society, people have been in great satisfaction with their material life. The emotional and personalized needs of the product have surpassed the function. The significant changes in consumption patterns have directly aroused companies and designers to rethink the function of products, so producing products that meet the needs of consumers at the spiritual level is one of the most concerned issues for enterprises.

The modern aesthetic studies have focused on metaphysical philosophical discussions. Gustav Fechner who is the father of modern aesthetics and proposed psychological aesthetics by experimental method, and the aesthetics should be transformed from philosophical aesthetics to scientific aesthetics in <Vorschule der Aesthetik> by Fechner [1]. Many scholars have begun to engage in scientific aesthetic research and have derived a number of genres, such as computational aesthetics, psychoanalytic aesthetics, gestalt psychology aesthetics, and experimental aesthetics. Birkhoff who is the founder of computational aesthetics proposed a mathematical model of aesthetic degree in 1933, and the "Aesthetic measure" was expressed as the ratio of order to

W. Karwowski and T. Ahram (Eds.): IHSI 2019, AISC 903, pp. 524–530, 2019.
https://doi.org/10.1007/978-3-030-11051-2_79

complexity, i.e. M = O/C [2]. Some scholars use this method to carry out research on form beauty, like Ngo and Teo established a relevant formula for calculating the layout elements in the interface design, which realizes the quantification of the interface layout features [3]. Hsiao and Chou used the fuzzy information entropy method to construct the aesthetic cognitive model of web design which based on the principle of gestalt psychology [4]. At present, the information processing of aesthetic pleasure is complex which from the explicit of visual perception to the implicit of aesthetic cognition, and most of the aesthetic evaluation of product forms are subjective based on expert experience and questionnaires to determine weights. In this paper, we tried to explore the black box mechanism in the process of aesthetic cognition and discussed the cognitive characteristic between product form and aesthetics, and ten aesthetic indexes were measured and four representative indexes were selected by cluster analysis and structural equation model, then we used linear and nonlinear models to compare which through linear regression analysis and the neural network model, respectively.

2 Aesthetic Cognition

The aesthetic is to know, perceive and judge the world in terms of the perceptual and the rational, subjectivity and objectivity, which also is a special form of human understanding of the world. No matter aesthetic is objective or subjective existence, which make people happy is beauty. The aesthetic cognition which human based on their experience and professional knowledge is a subjective psychological activity and also subject to objective factors. Designers usually apply aesthetic laws and rules which consider and design the interior and exterior of the product to improve the product form beauty.

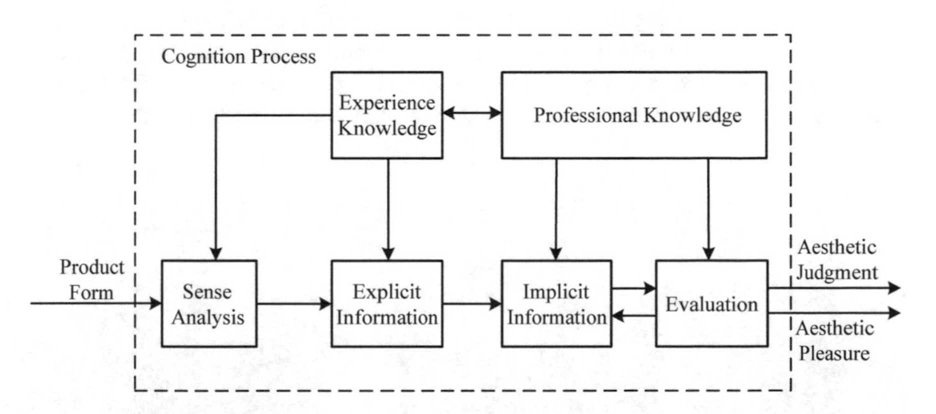

Fig. 1. Aesthetic cognition process

The cognition is a series of psychological processes in which cognitive information is analyzed, processed, stored and perceived by the sensory system, and then the conscious state is used to recognize practical problems (Fig. 1). The perception and cognition of beauty are closely related to the form of products and the cognitive psychology of human beings. The purpose of studying the aesthetic cognition mechanism of products is to determine the relationship between cognitive psychology and aesthetics and help users have a relatively clear understanding of the aesthetic feeling of products in an unstable system (Fig. 1), so that we can provide relevant guidance for designers to express the aesthetic feeling.

3 Product Form Aesthetic Index System

3.1 Aesthetic Measures

Product form is the appearance of designer's design idea, and it is also a concrete embodiment of product's practical function and aesthetic value. All the design ideas are ultimately materialized into form, only through the product form can people perceive and consciousness of product potential functions and values. Therefore, the designer could design good products only they accurately grasp the product form. The aesthetic is the unity of objective regularity and social purpose. The aesthetic feeling is implicit and is a product of social history. The aesthetic is synthesis of the attributes of human, material products, spiritual products, and artistic works, it has a variety of features which include balance, proportion, harmony, distinctive, novelty, formal suitability and completeness, form and content consistency, etc. Therefore, the aesthetic index accumulated from various concrete forms can be used to objectively evaluate the aesthetic of the product.

3.2 Aesthetic Measures

The formal beauty law is the experience summary and abstract of formal in the process of creating beauty, some of its usage include balance and equilibrium, contrast and blend, change and unity, cadence and rhythm, etc. In this study, constructing a coordinate system for calculating the aesthetic index from a certain perspective of the product and the midpoint of the bottom of product is used as the center of the coordinate system (Fig. 2). Ten indexes of product form aesthetic were calculated in Table 1 which based on research on aesthetics degree evaluation method of product form [5].

Fig. 2. Aesthetic measure coordinate (degree of balance)

4 Case Study

4.1 Selection Sample

The purple clay teapot is a unique hand-made clay craft which began in the Zhengde period of the Ming Dynasty in China. It is so popular that artistic and practical combination perfect and culture of zen Buddhism with tea, so we selected teapot as example which been choose from masters and workshops in china. 20 representative teapots (Fig. 3) which remodeled by grayscale process have been scored through interviewed 31 students who have design background and analyzed result of data in Table 1.

Table 1. Aesthetic survey results

Sample	Aesthetic	Sample	Aesthetic	Sample	Aesthetic	Sample	Aesthetic
1	0.8391	6	0.8469	11	0.6750	16	0.9094
2	0.7609	7	0.6750	12	0.8938	17	0.8078
3	0.7141	8	0.7922	13	0.7375	18	0.5813
4	0.6828	9	0.7141	14	0.6359	19	0.7688
5	0.5578	10	0.7297	15	0.6359	20	0.8625

4.2 Determination Representative Indexes

According to the form of the teapot, 10 indexes of aesthetic were measured including degree of balance, degree of equilibrium, degree of unity, degree of coordinate, degree of deviation, degree of economy, degree of homogeneity, degree of symmetry, degree of proportion and degree of order. The K-means cluster analysis was used by SPSS which all errors below 0.08, finally 4 representative indexes of aesthetic is degree of order, degree of symmetry, degree of deviation and degree of proportion in Table 2.

Table 2. Aesthetic degree indexes with cluster analysis

No.	Aesthetic index	Class	Distance	No.	Aesthetic index	Class	Distance
1	Degree of balance	2	0.376	6	Degree of coordinate	2	0.343
2	Degree of symmetry	2	0.170	7	Degree of economy	3	0.202
3	Degree of proportion	3	0.202	8	Degree of order	4	0.000
4	Degree of equilibrium	2	0.355	9	Degree of homogeneity	2	0.356
5	Degree of unity	2	0.207	10	Degree of deviation	1	0.000

The structural equation model was applied to verity the four representative indexes. Then a hypothetical model of aesthetic cognition which based on visual perception principle and logical judgment was constructed by AMOS, and the logical relationship network consists of ten indicators and aesthetics. This model's $\chi 2$ and RMSEA value are used as the validation parameters. $\chi 2$ has the advantage of judging the fit of the structural equation model, and the smaller the value, the better the model. RMSEA value are not affected by sample size and model complexity, and the hypothetical model is better when it less than 0.06, in addition, r1, r2 and e1 are error terms. The hypothetical model which $\chi 2$ and RMSEA value is 39.857 and 0.06 has a good result that 4 representational indexes is reasonable and reliable (Fig. 4).

4.3 Quantitative Mapping Relationship Between Product Form and Aesthetic

Through construction linear regression equation and BP neural network model of aesthetic evaluation, respectively, we can predict the results of two aesthetic cognitive model. The independent variables X of linear regression equation are degree of order, degree of symmetry, degree of deviation and degree of proportion, and the dependent

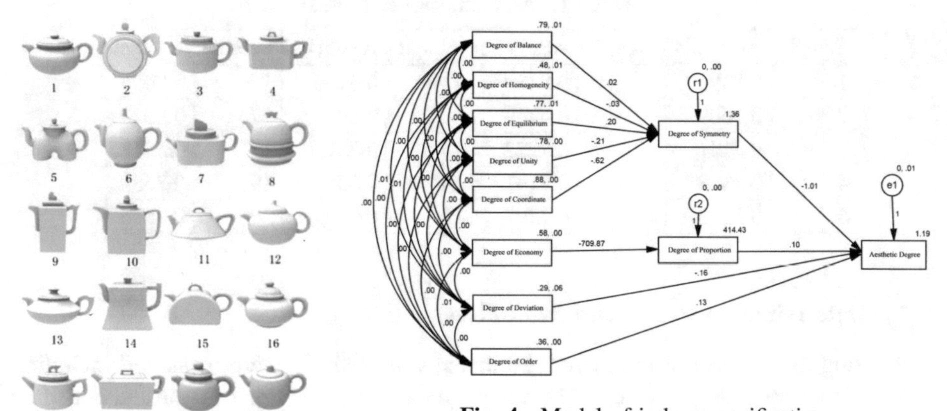

Fig. 3. Samples of purple clay teapot

Fig. 4. Model of indexes verification

variable Y of linear regression equation is aesthetic feeling. Analysis result shown that R^2 value is 0.316 less than 0.6 and Sig value is 0.196 greater than 0.05, The fitting results of the linear regression equation are poor and was analyzed by SPSS, and function was given by

$$M = 0.133 \times DP - 0.846 \times DS + 0.229 \times DO - 0.154 \times DD + 1.304 \qquad (1)$$

The BP Neural network is essentially a nonlinear model structure and commonly used to construct complex relationships between input variables and output variables, it has many advantages for predictive solutions of nonlinear models, such as it is relatively simple to build a model, it does not require prior knowledge and rules for solving problems. The input parameters are degree of order, degree of symmetry, degree of deviation and degree of proportion, and the output parameter is aesthetic feeling, and tansig(), logsig() and purelin() are respectively used for transfer functions in hidden layer and output layer, a total of 9 transfer combination functions (Fig. 5). In addition, this network was set up 10 nodes in hidden layer, 400 maximum learning, 10^{-4} convergence error target and trained by trainlm function in MATLAB toolbox.

Based on the linear and nonlinear prediction of aesthetic cognition model in Table 3, two designed samples were tested (Fig. 6). The linear prediction of aesthetic evaluation which is 0.8327 and 0.6915 have poor precision, besides R^2 and Sig value out of fit. on the other hand, the transfer functions of hidden-output layer include logsig-purelin, log sig-tansig, purelin-logsig and tansig-tansig have high fit 99.743%, 99.851%, 99.804% and 99.79%, respectively, and the remaining transfer functions for network training is less than 75%. So the prototype system (Fig. 7) evaluation of nonlinear which trained by logsig-tansig transfer functions is 0.6850 and 0.6414. The error of prototype linear/nonlinear model is 0.15 and 0.01. The result showed that aesthetic cognition more like nonlinear and model could provide an effective aid to the design of product form.

Table 3. Verification of aesthetic degree evaluation

No.	Symmetry	Proportion	Order	Deviation	Linear/nonlinear prediction	Aesthetic
1	0.771	0.707	0.404	0.036	0.833/0.685	0.691
2	0.827	0.500	0.333	0.361	0.692/0.641	0.637

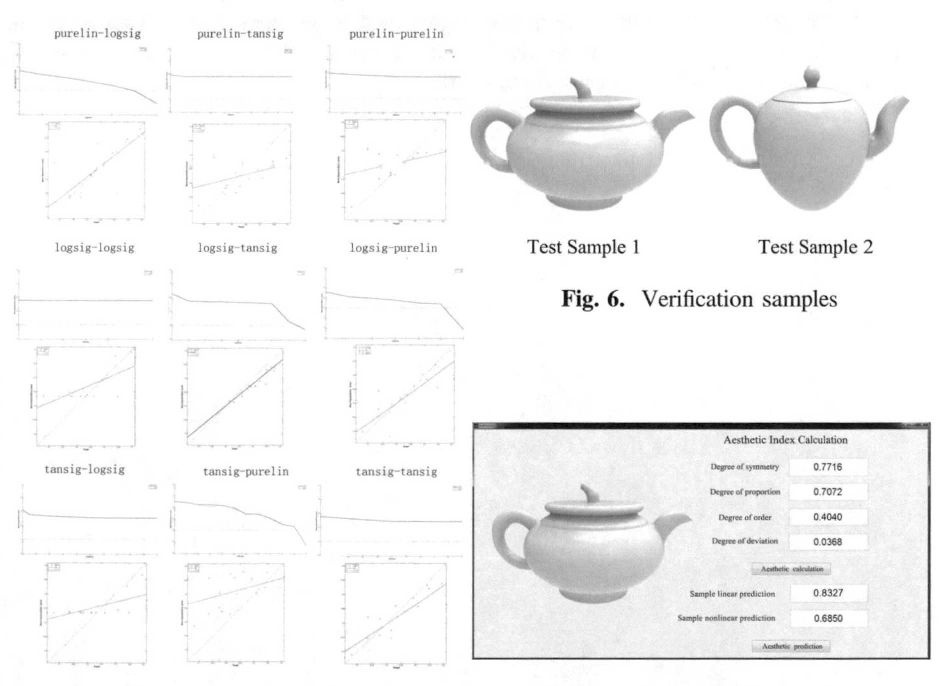

Fig. 6. Verification samples

Fig. 5. 9 transfer functions

Fig. 7. Aesthetic prototype system

5 Conclusion

In this paper, we have presented two prototype system of aesthetic cognition prediction and comparative analysis of linear and nonlinear aesthetic cognitive characteristics. Besides, the research methods include the cluster analysis, the structural equation model, the linear regression equation and BP neural network were applied to reduce data dimension, test cognitive logic, build linear and nonlinear aesthetic cognitive models. In the future, it will have a lot of work to enrich the knowledge from formal, technical, functional, artistic, ecological aesthetic.

References

1. Fechner, G.T.: Vorschule der aesthetik. Breitkopf & Härtel (1876)
2. Birkhoff, G.D.: Aesthetic Measure. Cambridge University Press, Cambrige (1933)
3. Ngo, D.C.L., Teo, L.S., Byrne, J.G.: Modelling interface aesthetics. Inf. Sci. **152**, 25–46 (2003)
4. Hsiao, S.W., Chou, J.R.: A Gestalt-like perceptual measure for home page design using a fuzzy entropy approach. Int. J. Hum.-Comput. Stud. **64**(2), 137–156 (2006)
5. Li, M., Zhang, J.: Research on aesthetics degree evaluation method of product form. In: International Conference on Human Systems Engineering and Design: Future Trends and Applications, pp. 68–75. Springer, Cham (2018)

Humans and Artificial Systems Complexity

Model-Based Multi-modal Human-System Interaction

Daniela Elisabeth Ströckl[1(✉)] and Heinrich C. Mayr[2]

[1] Institute for Applied Research on Ageing, Carinthia University of Applied Sciences, 9020 Klagenfurt, Austria
d.stroeckl@fh-kaernten.at
[2] Alpen-Adria-Universität Klagenfurt, 9020 Klagenfurt, Austria
heinrich.mayr@aau.at

Abstract. The paper presents a model-based approach to the abstract specification of multi-modal interfaces of assistive systems within the domain of Active and Assisted Living (AAL). We introduce an appropriate metamodel and a related domain specific modeling language that allow for a generic definition of multi-modal, situation dependent interactions. The language has been implemented using the metamodeling framework ADOxx®.

Keywords: Human-system interaction · Model centered architecture Conceptual modeling · Multi-modal interface · AAL · Smart home

1 Introduction

Within the last years, voice based interaction with technical systems has experienced a tremendous improvement. Examples are car infotainment systems or voice controlled assistants like Amazon Alexa. Although such speech-based assistive devices enhance the human-system interaction substantially, there are settings in which other media or at least combinations of other media and voice control are necessary for a trouble-free interaction.

This paper addresses one of such settings, namely the use of assistive systems by elder persons living in smart homes [1]. Aging comes with multimorbidity, i.e., combinations of locomotion or visual impairments, loss of hearing, speaking problems and psychological impairments like learning problems and forgetfulness [2]. Purely speech-based communication would not be effective when a person suffers from problems with hearing (e.g., in situations where headphones, hearing aids or high volume are not applicable) and/or problems with processing mentally heard advices.

In such situations, multi-modal interaction using text, pictures, video clips, voice, light effects etc., comes into play. Clearly, multi-modal interaction is nothing new per se. What we propose in this paper, however, is an approach to a generic interface definition based on a domain specific modeling language (DSML). I.e., we provide an

W. Karwowski and T. Ahram (Eds.): IHSI 2019, AISC 903, pp. 533–539, 2019.
https://doi.org/10.1007/978-3-030-11051-2_80

artifact, i.e., a metamodel, and a related representation language that allow for an abstract specification of interaction units by models (instantiated from the metamodel) according to the needs of a given setting.[1]

The paper's organization is as follows: In Sect. 2, we outline the fundaments of domain specific modeling languages and the framework in which our research is embedded. Section 3 introduces the proposed metamodel and DSML. In Sect. 4, we sketch the implementation based on the metamodeling framework ADOxx®[2]. The paper closes with a short outlook and the list of references.

2 Domain Specific Modeling

2.1 Defining a Domain Specific Modeling Language

A DSML serves for modeling specific aspects of a certain domain. Consequently. It comes with only few but tailored concepts [3]. In terms of the 4-level model hierarchy (see, e.g., [4]), a DSML represents a metamodel which is an extension of level M3 and is used to instantiate models on level M1. Much work has been published on evaluating modeling languages or on how to use a DSML, but only few on DSML design for interface modeling.

In our approach, we follow the design process presented in [3]. I.e., we develop a DSML for multimodel interface modeling by applying the steps (1) Preparation (Clarification of Scope and Purpose, Requirements Analysis and Context Analysis), (2) Modeling Language definition, (3) Modeling Process (defining a stepwise procedure of how a particular model may be systematically built using the given DSML), (4) Modeling Tool implementation, and Evaluation.

2.2 Model Centered Architecture

The paradigm of "Model Centered Architecture (MCA)" [5] treats any kind of information managed and/or processed by a part of a digital ecosystem (DES) as an instance of an explicitly specified or implicitly underlying model. The same is true for the processes as well as for the models themselves, the latter being instances of metamodels. Consequently, any DES is seen as a construct consisting of model handlers (consumers and/or producers). MCA thus is a generalization of Model Driven Architecture (MDA) and Model Driven Software Development (MDSD) [6], as well as models@runtime [7]. Like multilevel modeling, MCA advocates, for any system aspect, the use of (possibly recursive) hierarchies of DSMLs. I.e., MCA focuses on models (and their metamodels) in any design and development step up to the running system, so this also applies to system links [8] and interfaces.

---

[1] The work presented here is part of the HBMS (Human Behavior Monitoring and Support) project in the field of Active and Assisted Living (AAL) which has been partly funded by Klaus Tschira Stiftung GmbH, Heidelberg.

[2] https://www.adoxx.org/live/home.

2.3 The Reference Project HBMS

For a proof of concept, we developed our DSML for multimodal interface specification as part of the HBMS (Human Behavior Monitoring and Support) system [9]. HBMS is an ambient assistance system supporting daily life activities developed along the MCA paradigm. It abstracts, aggregates and integrates the observed behavior data into an individual Human Cognitive Model (HCM) which acts as a knowledge base, and assists the supported person via a multimodal interface by retrieving knowledge from his/her HCM. The HBMS input consists of sequences of recognized behavioral actions transmitted (again via model-based specified interfaces) from one or more Human Activity Recognition systems [10]. The "observation engine" (see Fig. 1) of the HBMS system integrates these sequences into HCM as behavioral clusters based on reasoning algorithms that deduce the goal of a sequence of actions. In support mode, the support engine exploits the HCM, a case base of concrete observations, and a potentially existing domain ontology to assist the target person via a multimodal interface, when needed, in taking the appropriate actions for reaching her/his current goal.

Figure 1 shows a simplified architecture of the HBMS system. Activity monitoring and context acquisition is done outside the HBMS boundaries.

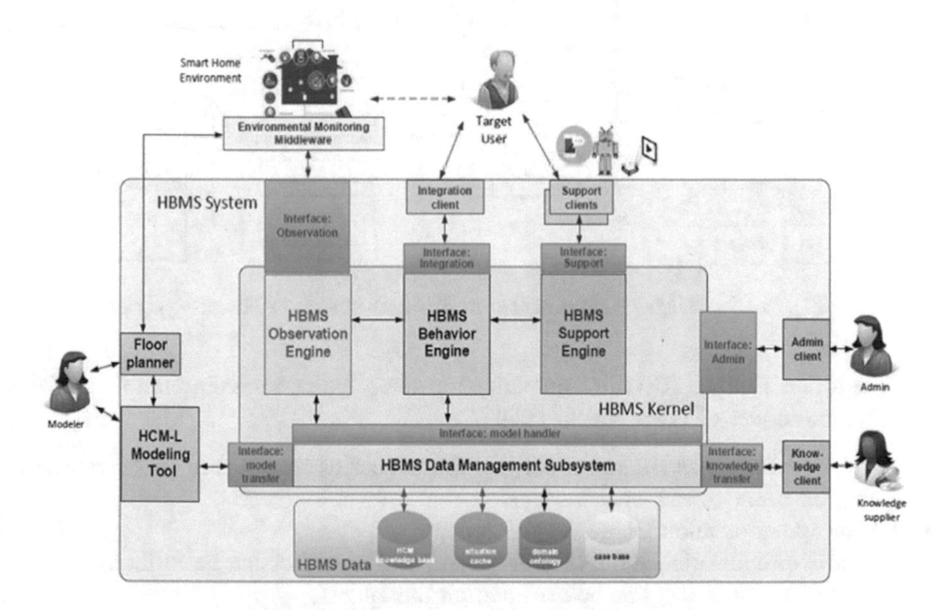

Fig. 1. Simplified HBMS-system architecture [3]

3 A DSML for Multimodal Interface Specification

The core element of the Multimodal Interface Modeling Language (MMI-ML) presented here is the abstract concept "Interface Element" as shown in Fig. 2. It is defined as an aggregation of different, again abstract concepts, namely "Object" and

"Relationship", which in turn are generalizations of the domain specific concepts "Interaction Entity" and "Interaction".

Interactions can occur between two devices (hardware-to-hardware), two interaction entities (software-to-software) and between a device and an interaction entity (hardware-to-software). Devices could be, i.e., monitors, tablets, microphones, beamers or speakers. An IE can be nested, in particular, if it leads to others during the interaction process. In addition, interactions and the interaction mode can be combined to realize multi-modality.

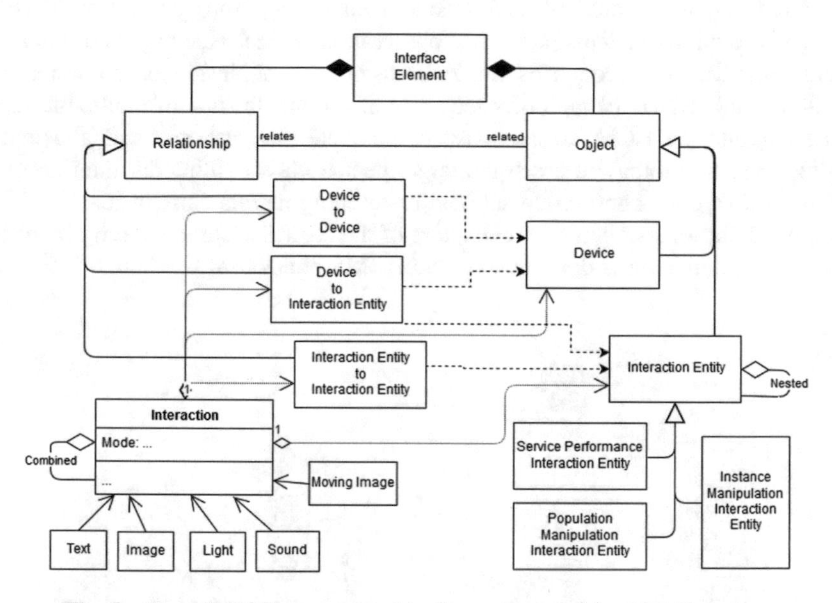

Fig. 2. The MMI-ML represented in the meta-model for HBMS system

Interaction Entities (IE) are summarized to three types, according the idea of the presentation model of Oscar Pastor [11]:

- Service performance IE: serves for defining scenarios like: *a user interacts with a system in order to execute a service.*
- Instance manipulation IE: serves for defining
  - the communication about the manipulation of an object and its attributes w.r.t. a list of services that can be executed on this object,
  - related information to which the user can navigate.
- Population manipulation IE: serves for defining the communication about the manipulation of a collection of objects of any given classes [11].

MMI-ML has been implemented using the Metamodeling Framework ADOxx®: By exploiting the ADOxx® Development Toolkit, we specified the metamodel given in Fig. 1 as an extension of the ADOxx® predefined standard metamodel. For all concepts of our metamodel, we provided a graphical notation (via GraphRep code) as depicted in Table 1, thus defining the language.

Based on that, the ADOxx® Modeling Toolkit provides a modeling platform for MMI-ML, i.e., the interaction classes can be modeled using that platform. Figure 3 shows a very simple example for the interaction classes occurring in an scenario "washing face" as part of a more complex activity "morning routine". The concrete interactions are instances of the elements of that model.

Table 1. Diagram key for the instances of the MMI-ML meta-model

Model Element	**Model Element:** represents the interaction scenario		**Mode – Sound:** Speech input and output
Device	**Device:** examples: boundary microphone, keyboard, speakers, tablet computer		**Mode – Picture:** Output: picture on screen
Service Performance	**Interaction Entity – Service Performance:** Input or output element		**Mode – Locomotion:** Input: gesture control
Instance Manipulation	**Interaction Entity – Instance Manipulation:** An Action according former input		**Mode – Moving Image:** Output: video on screen
Population Manipulation	**Interaction Entity – Population Manipulation:** Searching or filtering data	————	**Relation – Combined Interaction or Nested Interaction Entity**
Interaction	**Interaction:** system-to-system interaction without a mode	....	**Interaction Entity interacts with Interaction Entity**
	**Mode – Text:** Input: text written with input Output: text on screen	– – – – –	**Device interacts with Interaction Entity**
	**Mode – Light:** Output: light beam on object	– · – · –	**Device interacts with Device**

Clearly, the model elements including the relationships as depicted in Fig. 3 have to be specified in detail using attributes, however, due space restrictions we cannot go in details here.

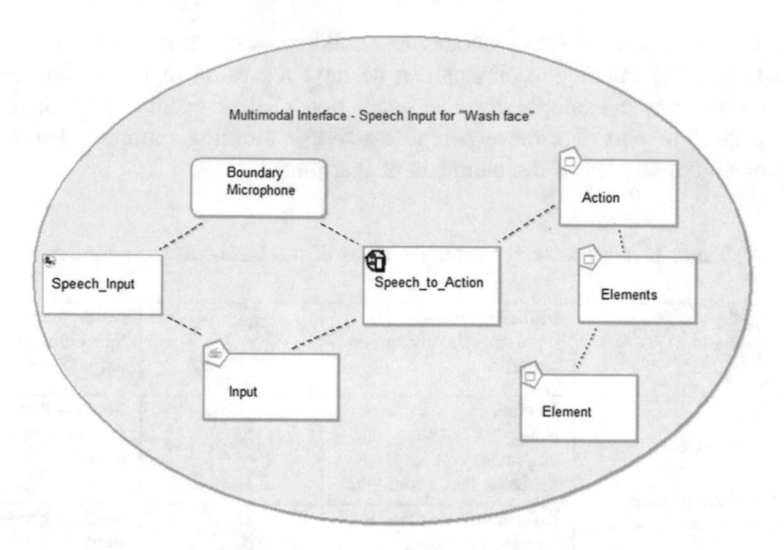

Fig. 3. Speech input interaction for next step to get support during washing the face

4 Outlook

The next development step will be to integrate the MMI-ML approach including the necessary adaptations of the support engine into the HBMS system and test it with people under lab conditions. The model based defined multi-modal interaction specification will then replace the already currently static interaction unit of the system.

References

1. Pal, D., Funilkul, S., Charoenkitkarn, N., Kanthamanon, P.: Internet-of-Things and smart homes for elderly healthcare: an end user perspective. IEEE Access **6**, 10483–10496 (2018)
2. Rubisch, M., et.al.: Sozialministerium, Abt. IV/A/1, Bericht der Bundesregierung über die Lage der Menschen mit Behinderungen in Österreich 2016, Bundesministerium für Arbeit, Soziales und Konsumentenschutz (2016)
3. Mayr, H.C., et al.: HCM-L: domain-specific modeling for active and assisted living. In: Domain-Specific Conceptual Modeling, pp. 527–552. Springer, Heidelberg (2016)
4. Batsakis, S., Petrakis, E.: SOWL: spatio-temporal representation, reasoning and querying over the semantic web. In: Proceedings of 6th International Conference on Semantic Systems. ACM (2010)
5. Mayr, H.C., Michael, J., Ranasinghe, S., Shekhovtsov, V., Steinberger, C.: Model centered architecture. In: Conceptual Modeling Perspectives, pp. 85–104. Springer, Heidelberg (2017)
6. Liddle, S.W.: Model-driven software development. In: Handbook of Conceptual Modeling, pp. 17–54. Springer, Berlin, Heidelberg (2011)
7. Bencomo, N., France, R.B., Cheng, B.H.C., Aßmann, U.: Models@run.time - Foundations, Applications, and Roadmaps. Springer, Heidelberg (2014)

8. Shekhovtsov, V., Ranasinghe, S., Mayr, H.C., Michael, J.: Domain Specific Models as System Links. In: Woo, C., et al. (eds.) Advances in Conceptual Modeling. LNCS, vol. 11158. Springer, Heidelberg (2018)
9. Michael, J., et al.: The HBMS story. Enterp. Model. Inf. Syst. Arch. **13**, 345–370 (2018)
10. Ranasinghe, S., et al.: A review on applications of activity recognition systems with regard to performance and evaluation. Int. J. Distrib. Sens. Netw. **12** (2016)
11. Pastor, Ó., Molina, J.C.: Model-Driven Architecture in Practice: A Software Production Environment Based on Conceptual Modeling. Springer, Berlin, Heidelberg (2007)

A Model Driven Development Approach Using AADL and Code Generation to Develop Modular Distributed Electronic Travel Aid Devices

Florian von Zabiensky(✉), Michael Kreutzer, and Diethelm Bienhaus

Technische Hochschule Mittelhessen, University of Applied Sciences, Wiesenstr. 14, 35390 Gießen, Germany
{Florian.von.Zabiensky,Michael.Kreutzer,Diethelm.Bienhaus}@mni.thm.de

Abstract. Electronic Travel Aids (ETAs) are devices that support people who are blind or visual impaired in tasks as obstacle detection/avoidance or navigation. ETAs consist of sensing, processing and presentation components. Many research projects develop new kinds of ETAs, introducing new presentation or sensing components. An optimal ETA should be a distributed system where each component is a communicating node. In this work we propose an MDD workflow to develop distributed ETAs. This workflow uses AADL to describe the system, its components and their communication. This supports an easy exchange of a component in its system model. So, research projects, for new presentation devices or presentation components, only have to change the node they are working on instead of developing a whole new system. Automatic code generation assures that the newly developed ETA component fits into the system.

Keywords: Electronic travel aids · AADL · ROS 2 · MDD · Development workflow

1 Introduction

Electronic Travel Aids (ETAs) are systems that help people who are blind or visual impaired to navigate in new environments or to get environmental information like obstacle positions and distances. Most of today's ETAs are monolithic systems, hence lacking flexibility, extensibility and interoperability. We are working on a framework to develop a distributed and modular architecture for ETAs, so they can be built with lower effort.

Advantages of this approach are high flexibility based on inter-operable and exchangeable subsystems for specialized purposes. Functionality of the whole system is gained by composition and orchestration of those subsystems. An acoustic display is an example for such a subsystem. It provides information to the user via an acoustic representation. If an ETA has only this type of interaction system integrated, the user is forced to use it. It would be more user-friendly if a choice could be made between

W. Karwowski and T. Ahram (Eds.): IHSI 2019, AISC 903, pp. 540–545, 2019.
https://doi.org/10.1007/978-3-030-11051-2_81

different interaction techniques and devices, depending on the user's personal preferences and needs, even providing a clean interface to other alternatives that will be developed in the future.

The first step to enable exchangeability of such subsystems is to use a unified way for designing the architecture and communication interfaces of the different parts. This enables better cooperation between subsystem developers. By using the same way to describe the architecture and interfaces between subsystems, the integration to a complete ETA is easier and clearer.

In this work we propose a development workflow which uses the Architecture Analysis and Design Language (AADL) to design the subsystems and the ROS 2 Framework to implement it. AADL is a language for designing and analysing architectures in embedded real-time systems. It supports the description of software- as well as hardware-components.

For even more development efficacy and more consistency between architecture and source code we implemented a code generation mechanism from AADL to C++ code. It uses the ROS 2 Framework for communication between subsystems. The Robot Operating System 2 (ROS 2) is a Framework that enables the implementation of a distributed robot system with low effort.

In conclusion we present a model driven development (MDD) approach using AADL and ROS2 to design, analyse and generate code of ETA subsystems.

This paper is structured as follows: First, we will introduce AADL and ROS 2. Next, we present related work and highlight the differences to this work. Finally, we propose an MDD workflow to develop distributed ETAs with AADL and ROS 2.

2 Architecture Analysis and Design Language

AADL [1] is a standardized language to describe the architecture of real-time critical applications. It covers typical software applications as well as embedded systems. Therefor AADL provides a component-based architecture description consisting of software, hardware and abstract components. The software components are data, subprograms, threads and processes, hardware components are processor-, memory-, device- and bus-components. To structure the architecture of a system, AADL provides the definition of systems that may contain other systems as well as software and hardware components. With such methods, AADL allows a system-architect to describe a whole (embedded) software system within the deployment to hardware.

Compared to other modeling languages, AADL has a formalism that prevents different interpretations of a model. This enables the analysis of architecture in several areas like

- Safety, Dependability and Performance analysis [2]
- Schedulability analysis [3]
- Flow latency analysis [4]
- ...

Resulting from the very precise design description, an automatic code generation from an AADL model is possible [5–7].

There are several tools available, open source as well as commercial, using AADL to analyse architectures or generate code. We like to focus on open source tools like OSATE[1] as an IDE for AADL and Ocarina[2] as a model processor supporting model analysis and code generation.

3 Robot Operating System

ROS is an open source framework to develop robot applications. Therefor ROS provides libraries and tools to support the development process. It includes state-of-the-art algorithms e.g. for image processing or self-localization. The ROS community also provides drivers to various hardware components. ROS systems are distributed systems. Each algorithm, hardware driver or control logic is implemented as a node that communicates with the rest of the system. Standard data structures for messages are defined, so each node can be easily exchanged. E.g. to change a self-localization algorithm, only this very node has to be exchanged, if the interface can be kept the same.

The communication between nodes is based on a publish/subscribe mechanism or service calls with service providers and clients. There are also systemwide parameters to parametrize a system instance.

ROS 2 is a new version of ROS that is currently under heavy development. The development of ROS 2 was triggered due to a change in requirements to the robot systems being developed with ROS because of changed use cases.[3] ROS 2 has several advantages for the use in ETAs.

- In ROS 2 there is no need for a central master that organizes the orchestration of the entire system, because of the underlying Data Distribution Service (DDS) standard.
- In ROS 1 an embedded system is attached by implementing a ROS-driver for the system. The plan in ROS 2 is letting embedded systems communicate directly to other ROS nodes. This is also supported using DDS.

However, the big advantage of ROS 1 is the existence of a comprehensive repository of algorithms and drivers. So, in ROS 2 it is possible to access existing nodes by using a ROS 1 – ROS 2 communication bridge.

The direct integration of embedded systems, no need for a central master and the possibility to use existing ROS 1 nodes lead to the use of ROS 2.

[1] http://osate.org.

[2] https://github.com/OpenAADL/ocarina.

[3] For more information: http://design.ros2.org/articles/why_ros2.html.

4 Related Work

MDD is a method to develop a system based on models. A key feature is code generation, which reduces inconsistencies between the source code and the system model during development by requiring the model to be revised instead of the code. In MDD the development is driven by the model of a system and automated in several steps up to create an executable binary. If the system changes over time, fist the model must be adjusted. So, changes take automatically effect in the rest of the development process.

The use of AADL for MDD is no novelty, there are several works on this topic. The IST-ASSERT project [5] is just one to mention. Hugues et al. describe in this work an MDD workflow and toolchain to develop real-time embedded systems for the high integrity domain.

The requirements and restrictions to ETAs are not as hard as they are in space- and aircraft systems. Therefore, we want to focus on ROS as a basic part of ETAs instead of using the Poly-ORB-HI code generation of Ocarina [5]. Using AADL to describe ROS architectures and generate code to support the development process has been published in [6, 8]. In those works, Bardaro et al. developed a mapping from AADL elements to ROS concepts. This enables code generation for ROS 1 architectures.

Since ROS 2 is the underlying framework that we propose to use in ETAs, and several concepts differ from ROS 1, another mapping of AADL elements to ROS 2 concepts is required. Therefore, we wrote a new code generation backend for the existing Ocarina tool.

5 MDD Workflow for Distributed Electronic Travel Aid Systems

ETAs are systems that help people who are blind or visual impaired to get information about their environment. There are several tasks for ETAs like obstacle detection and obstacle avoidance or navigation. The problems of those tasks are similar to those in mobile robotics. There is only one big difference. The data must be presented to a user instead of being directly processed by the robot.

For this purpose, there are view methods like acoustic displays with e.g. speech or auditory icons [9] that can be placed somewhere around the head of the user. Also, virtual directional information may be presented. Another method is using vibrating motors to present such information.

So, several sensing and presenting devices can be part of an ETA and there are also different ways in using this hardware, e.g. speech with or without directional information. Also, there are many ways in processing the given data. Therefore, we want to present a workflow allowing to exchange nodes of an ETA easily. This approach has two big advantages.

1. Research projects on ETAs have the possibility to focus on their main research by replacing parts of a given ETA instead of developing a completely new ETA. This supports quick results. Also, easy and consistent comparison to other systems are

possible. If a project is about the presentation layer and all other parts are the same, then a comparison between presentation layers is unambiguous. Otherwise, the results of a comparison may also result from unknown influences of the rest of the system.

2. Research results can be more easily translated into products. This would reduce the gap between research and market. In [10] Dakopoulos and Bourbakis pointed out such a big gap.

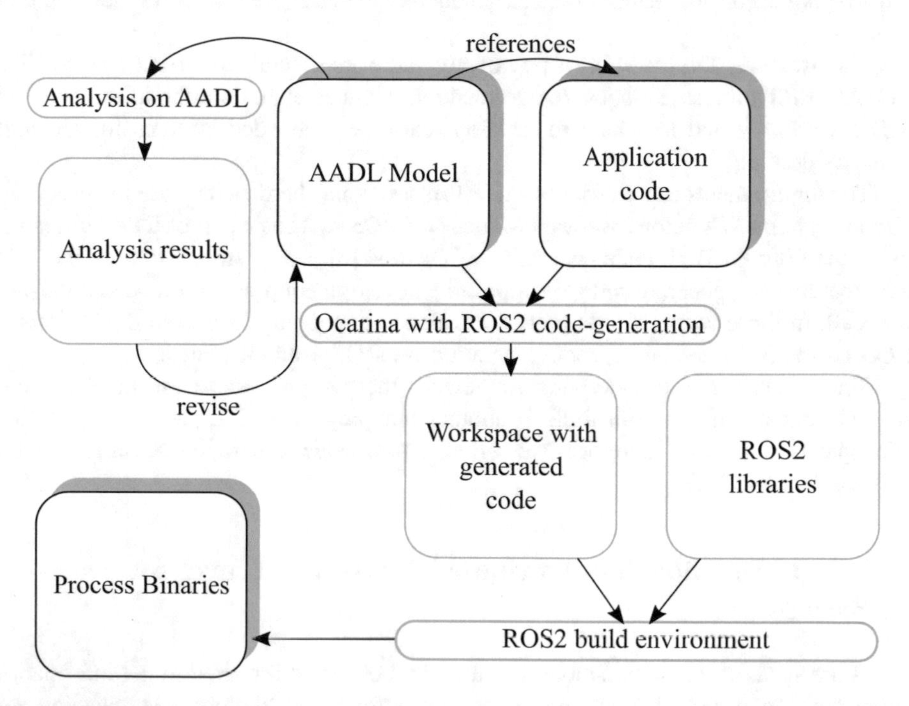

Fig. 1. Illustration of the MDD workflow to create a distributed ETA with AADL and ROS 2

In Fig. 1 we present an MDD workflow that uses AADL to model a system with reference to application code for its behaviour. A system can be a whole ETA system to define the communication between subsystems or the behaviour in several cases. But a system may also be a subsystem like an acoustic display. This subsystem can then readily be used in modelling the complete ETA, integrating the acoustic display.

To use of the advantages of ROS 2 we created a backend for the Ocarina tool which produces C++ code using ROS 2 out of an AADL model. For its behaviour, the model references to C++ code that is inserted into the generated classes. So, the Ocarina tool using our backend can generate a ROS 2 workspace. The executable binaries may then be compiled using the standard development tools of ROS 2.

6 Conclusion and Future Work

In conclusion we presented a workflow using AADL to describe ETAs as connected AADL models of its subcomponents like sensing, processing or presentation soft- and hardware. From this model a ROS 2 workspace can be generated, allowing to compile executable binaries using standard ROS 2 development tools.

The workflow also supports the exchange of ETA-components to integrate new functionality as well as to compare the performance of different subcomponents for purpose of research. This is the first step to reduce the gap between research and market by supporting the transformation process from a research device to a product.

Currently, only nodes for standard computer systems can be generated. To get the best out of this workflow we also have to develop a backend for Ocarina generating firmware for embedded systems from an AADL model.

References

1. AS-2C Architecture Analysis and Design Language: Architecture Analysis & Design Language (AADL). SAE International
2. Bozzano, M., Cimatti, A., Katoen, J.-P., Nguyen, V.Y., Noll, T., Roveri, M.: Safety, dependability and performance analysis of extended AADL models. Comput. J. **54**, 754–775 (2011)
3. Sokolsky, O., Lee, I., Clarke, D.: Schedulability analysis of AADL models. In: Proceedings 20th IEEE International Parallel Distributed Processing Symposium, 8 pp. (2006)
4. Feiler, P., Hansson, J.: Flow Latency Analysis with the Architecture Analysis & Design Language (AADL). Carnegie-Mellon Univ Pittsburgh PA Software Engineering Inst (2007)
5. Hugues, J., Pautet, L., Zalila, B., Dissaux, P., Perrotin, M.: Using AADL to build critical real-time systems: experiments in the IST-ASSERT project. In: 4th European Congress ERTS, Toulouse, Paris (2008)
6. Bardaro, G., Semprebon, A., Matteucci, M.: A use case in model-based robot development using AADL and ROS. In: Proceedings of the 1st International Workshop on Robotics Software Engineering, pp. 9–16. ACM, New York, NY, USA (2018)
7. Hugues, J., Pautet, L., Zalila, B.: From MDD to full industrial process: building distributed real-time embedded systems for the high-integrity domain. In: Kordon, F., Sokolsky, O. (eds.) Composition of embedded systems. Scientific and Industrial Issues, pp. 35–52. Springer Berlin Heidelberg, Berlin, Heidelberg (2008)
8. Bardaro, G., Matteucci, M.: Using AADL to model and develop ROS-based robotic application. In: 2017 First IEEE International Conference on Robotic Computing (IRC), pp. 204–207. IEEE, Taichung, Taiwan (2017)
9. McGookin, D.K., Brewster, S.A.: Understanding concurrent earcons: applying auditory scene analysis principles to concurrent earcon recognition. ACM Trans. Appl. Percept. (TAP) **1**, 130–155 (2004)
10. Dakopoulos, D., Bourbakis, N.G.: Wearable obstacle avoidance electronic travel aids for blind: a survey. IEEE Trans. Syst. Man Cybern. Part C (Appl. Rev.) **40**, 25–35 (2010)

Intelligent Systems in Everyday Work Practices: Integrations and Sociotechnical Calibrations

Christine T. Wolf[(✉)] and Jeanette L. Blomberg

IBM Research, Almaden San Jose, CA, USA
{ctwolf, blomberg}@us.ibm.com

Abstract. A key challenge to the implementation and adoption of intelligent machines in the workplace is their integration with situated work practices and organizational processes. We examine these issues through a qualitative field study in the domain of information technology (IT) services procurement, where highly-skilled IT architects spend considerable effort reading and digesting client RFPs to design technical solutions. Our field study focuses on the design and development of an intelligent tool meant to augment architects' design work. Along with usability issues and curiosity about the tool's underlying intelligent features, architects raised a number of questions about how the tool would be used in relation to other processes and workflows that comprised their design work. Our findings consider how intelligent systems are actors within a sociotechnical system and how new relations emerge through their introduction, raising questions for future intelligent system design and integration.

Keywords: Sociotechnical systems · Human-intelligent systems collaboration Everyday work practices · Workplace transformation · Natural language processing · IT services · Requirements analysis

1 Introduction

While workplace automation has been a topic of concern for several decades, rapid advances in the fields of artificial intelligence (AI) and machine learning (ML) have ushered in a renewed interest in the vision that intelligent machines might fruitfully augment everyday work practices. A key challenge to the implementation and adoption of intelligent machines in the workplace, though, is their integration with situated work practices. Work practices are often complex and dynamic, involving multiple stakeholders and shaped by a number of organizational and domain factors. What's more, individual work practices do not stand alone but instead are part of sociotechnical systems, where people, practices, and technologies are inter-related. Introducing intelligent machines, then, reconfigures not only the immediate work practices they are designed to augment, but also the broader sociotechnical system they are introduced into. These reconfigurations have the potential to be both disruptive and transformative – supporting some work practices, eliminating others, and even still, creating entirely new ones. Such dynamics raise a number of important considerations integral to

W. Karwowski and T. Ahram (Eds.): IHSI 2019, AISC 903, pp. 546–550, 2019.
https://doi.org/10.1007/978-3-030-11051-2_82

effective intelligent machine design. Which activities within a complex sociotechnical workflow are best suited for intelligent assistance? Which activities are best left to workers? What can be eliminated, and what must remain? This cannot be fully discerned a priori but must be continually calibrated throughout the design lifecycle of projects that focus on intelligent machines.

We examine these topics through a qualitative field study in the domain of information technology (IT) services procurement. We report on an early adopter program with IT architects and the feedback they provided on their interactions with various prototypes of an intelligent tool meant to augment the work of requirements definition and solution design. Along with usability issues and uncertainty about the algorithmic underpinnings of the tool's intelligent features, architects raised a number of questions about how the tool would be used in relation to other processes and workflows that comprised their solutioning work. Our findings elaborate on how intelligent systems are actors within a sociotechnical system and how new relations emerge through their introduction, raising questions for future intelligent system design and integration.

2 Field Study Context: IT Infrastructure Services Design

Our study context focuses on the domain of information technology (IT) services procurement, and in particular the design of IT infrastructure architectures. When an organization decides to outsource all or part of their IT infrastructure, they write up their technical requirements in request for proposal (RFP) documentation, which are often comprised of large sets of digital files with unstructured content. Highly skilled IT architects spend considerable effort reading and digesting the content in RFPs and then designing technical solutions to service those requirements. Our field study focuses on the design and development of an intelligent tool meant to augment the work of requirements definition and solution design at a large, global technology services firm. Development of the system follows the Agile software management method, a "continuous delivery" model where initial features of a software application are released, and then iteratively enhanced over time – features are (re)planned and implemented based on cycles of feedback, reflection, and planning. Central to Agile is the active and ongoing involvement of stakeholders throughout the development process, of whom a key constituency are the intended users of the application.

In this paper, we report on insights gathered from feedback provided by the tool's user community, IT architects within the firm. They provided feedback after interacting with various prototypes of the system. In particular, we report on two user feedback programs that ran from October 2017 to July 2018. In total, Phase I included interviews with eight (8) architects, averaging one hour in length. The second program (Phase II) was an "early adopter" program, which included a larger cohort of architects and combined both usability testing (looking for system bugs/defects and whether the system was working as designed) and usefulness (evaluating how well the system aligned with the architects' work practices, evaluating whether the system was fit for purpose). Both the first and second authors carried out Phase II, which ran from January to July 2018 and involved sessions with seventeen (17) individual architects in the early adopter program. These sessions were 1 h long, and included a semi-structured interview portion (discussing the architect's work practices) and then involved real-time use of the tool, where architects were asked to share their screen with the researchers and complete a series of tasks within the system while using the "think aloud" method [1]. In addition to data gathered from these individual sessions, the authors also held focus group sessions with members of the broader early adopter cohort (three (3) focus groups with 8, 10, and 10 architects attending each respectively, for a total of 28 architects) and solicited feedback via email surveys and an online chat forum.

3 Findings

Architects gave feedback on the tool's interface design and overall usability. They also raised a number of questions about the tool's underlying intelligent features, powered by various machine learning capabilities. How does it work? What makes it smart? How do I know its accurate? In addition to these types of questions and feedback about the tool itself, another theme in architects' accounts was the need to understand how the tool – and particularly the various outputs it was capable of producing – were meant to be incorporated into the existing workflow processes for IT architecture design. This arose as a practical concern: "*How do you get that stuff from the tool, which is great, into the documents that people are expected to use today for auditing and record-keeping*?" Angie,[1] an architect wondered. "*I was not sure how to do that, other than some kind of weird manual cut and paste right*?" When new systems are introduced into an existing work practices, workers must develop such hands-on, know-how as they create micro-integrations capable of cohering novel and existing artifacts. In addition to these micro-integrations, workers must also align integration of novel outputs with the larger organizational processes they implicate. "*We've got to think through how does that all fit with what is a fairly industrialized process*," Nathan, an architect, said. "*The minute you've got a process and it's understood by a lot of people to work a particular way, if you want to change it*," he explained, "*somebody would have to articulate the changes and tell people what they can and cannot do and are supposed to do especially based on what they couldn't do before or did do before. That would be the best to me to understand that linkage.*" In Nathan's account, we can see a search for explicit guidance on how the tool was intended to be integrated into robust, established organizational processes – such comments point to integration as a collective effort that brings together the micro-concerns of everyday work practice with the complex logistical workflows that include but also exceed any individual workers' efforts.

Another dimension of artifact integration is partiality – architects were not only concerned about transforming the tool's outputs into other templates or formats, but also what to do with partial outputs. When is the tool uncertain about its analysis or unable to handle something? How will I know what is left unattended to, and what do I do to close that gap? An example of this was in the tool's optimization piece that takes

[1] All names are pseudonyms.

extracted IT requirements and matches them to the company's catalog of services. "*So what this tells me,*" Stacey, one architect wondered aloud as she inspected the interface, "*is that 70% of the solution is covered with standard offerings from the catalog…is the other 30% something that I would have to then go and generate something on the side to get the custom pieces?*" This concern highlights the importance of hybridity in intelligent sociotechnical systems implementations – workers will interact with and utilize the intelligent tool's outputs, but in some cases will importantly need to augment those outputs, transforming them into situated, hybrid variations. Understanding the requisites for such hybridity – and steps that enable its accomplishment – are key components for the successful integration of intelligent systems into everyday work practices.

In addition to their individual interactions with the tool's outputs, architects also highlighted the distributed and collaborative nature of their existing work practices. Their design work was often accomplished through team efforts of several architects, with each architect typically working on a particular piece of the RFP individually, based on their technical expertise (servers, for example, or end-user services). They then bring those sub-designs together for the comprehensive solution design, which is delivered to the client. A key feature architects requested was a nuanced role differentiation inside the tool, which would be particularly useful in handling distributed design work. What this adds to our understanding of intelligent system integration is understanding how intelligent outputs become collaborative artifacts – integration is not solely an issue of individual users and intelligent machines. Rather, existing coordinative practices must reconfigure to accommodate the addition of new sociotechnical actors, like the intelligent system and its outputs.

4 Discussion: Integration and Sociotechnical Calibrations

In this short paper, we have set out some initial findings from our field study, investigating the feedback of workers during the design and development of an intelligent tool meant to support their everyday design work. In doing so, we have outlined a number of concerns raised – from practical questions of micro-integration (do I copy and paste it?) to broader questions of organizational change and collaboration teaming support. Central in these accounts are a concern of how an intelligent tool and its outputs will be integrated into existing organizational practices and systems. While these issues may, arguably, be resolvable by "better user education" (a common refrain in information systems implementations) they provide helpful examples in examining more fully the sociotechnical implications of intelligent tool deployments. Such deployments add to the array of actors in a sociotechnical system. Intelligent tools are new elements, but also too are the outputs they produce – and here, we see workers grasping for guidance on how they are meant to integrate such outputs into their existing work practices and how those work practices should be calibrated or tweaked as a result of these new additions.

Davis et al. [2] draw attention to sociotechnical systems' "design incompletion" – meaning that design in sociotechnical systems is an ever ongoing and unfurling practice, rather than an activity that can be called "done." Sociotechnical systems are continuously made, remade, and transformed through everyday practice. These incompletion dynamics, Davis et al. claim, call for more "predictive" design focus – a perspective that looks at how systems can, should, or might be deployed and to what effect. Such a future-facing perspective provokes consideration of the role of imaginaries in sociotechnical system design – what kinds of intelligent futures are we creating and what kinds do we want to create? To say that sociotechnical systems are never fully "designed" but rather always in the processing of designing also provides a site to more closely consider issues of both agency, visibility, and complexity. Who gets to participate in such design activities, who articulates how things will or should be? Where do such articulations take place? To the extent that sociotechnical decisions get made, where do those decisions take place and how are such decision-making practices made visible to actors across a system? The "garbage can" theory of organizational decision-making is instructive here – which outlines how decision-making processes are not always explicit or rational, but instead a dynamic process whereby technical solutions are always "in search of" problems to solve [3]. Such complexity complicates narratives of linear, progressive organizational-change decisions that are capable of being carefully articulated through strategic communication (like "better user education"). Rather than simply machines that are designed and deployed, intelligent systems are actors that engage a variety of processes and practices within a sociotechnical system. With their introduction, new sociotechnical relations emerge – questions of intelligent system integration, then, require an ongoing interrogation of such emergence and the calibrations it instigates.

Acknowledgements. Thank you to the IT architects who graciously shared their time, energy, and insights during the field study. All opinions are our own and do not represent any institutional endorsement.

References

1. Someren, M.W., Barnard, Y.F., Sandberg, J.A.C.: The Think Aloud Method: A Practical Guide to Modelling Cognitive Processes. Academic Press, London (1994)
2. Davis, M.C., Challenger, R., Jayewardene, D.N.W., Clegg, C.W.: Advancing socio-technical systems thinking: a call for bravery. Appl. Ergon. **45**(2A), 171–180 (2014)
3. Cohen, M.D., March, J.G., Olsen, J.P.: A garbage can model of organizational choice. Adm. Sci. Q. **17**(1), 1–25 (1972)

Exploring the Acceptance of Video-Based Medical Support

Carsten Röcker[(✉)]

Fraunhofer IOSB-INA & Institute Industrial IT (inIT), Ostwestfalen-Lippe University of Applied Sciences, Langenbruch 6, 32657 Lemgo, Germany
carsten.roecker@iosb-ina.fraunhofer.de

Abstract. This paper reports on a study ($N = 471$) exploring the acceptance of video-based home monitoring systems as well as criteria influencing their acceptance. While most participants stated that they would home monitoring solutions under certain conditions, the majority of participants is rather reluctant to use systems that transmit visual and acoustical information to remote medical personnel. Besides age, most user characteristics, which played important roles in technology acceptance research for many years, do not appear to be decisive factors for the acceptance of electronic home-monitoring services.

Keywords: Active assisted living · Electronic homecare · e-health
Video-based monitoring · Technology acceptance · User-centered design
Study

1 Introduction

Today, the design of new homecare solutions is mainly driven by technical considerations of medical professionals and system providers [1]. Even if the value of interdisciplinary collaboration is widely accepted, discipline spanning research is a rare sight, in particular in the field of health technologies and aging [2]. Consequently, developments in this field are often demonstrations of technological possibilities [3] rather than responses to the actual needs of potential users, which is often cited as one of the main reasons for poor adoption behavior [4]. Despite these obvious shortcomings, user integration still does not take place in many companies [5]. Financial constraints and time pressure are predominant reasons for not integrating users in the design process of new technologies [6].

At the same time, the importance of user-centered design approaches was demonstrated in numerous studies. For example, Ziefle and Bay [7] showed that age-sensitive design concepts could significantly reduce age-related handicaps and thereby enable older adults to efficiently operate new technologies. However, user-centered design does not only bring benefits to end users due to better usability of medical products, but is also likely to lead to substantial financial advantages for manufacturers and service providers as the costs of adapting technical concepts and service functionalities are considerably lower in early design stages.

W. Karwowski and T. Ahram (Eds.): IHSI 2019, AISC 903, pp. 551–556, 2019.
https://doi.org/10.1007/978-3-030-11051-2_83

2 Research Questions and Approach

Technology-enhanced homecare services, and in particular video-based monitoring solutions as a low-cost approach for proving remote (emergency) support, became more prevalent in the last years [8, 9]. While first commercial products are available on the market for several years now, there exists very little knowledge about the actual needs and wants of potential users of such systems. Therefore, this paper aims at providing a basic understanding about the acceptance of video-based home monitoring systems as well as criteria that influence the acceptance decision.

As existing research revealed significant influences of personal user factors on the acceptance of different electronic healthcare applications (see, e.g., [10, 11] or [12]), special attention was given to the influences of a broad variety of user characteristics. Consequently, a relatively large number of independent variables was considered (see Fig. 1 for an overview of the different user factors that were included). With respect to their age, the participants were clustered into four groups. The first three groups are based on the technical generations identified by Sackmann and Weymann [13]: the early-technical generation (ETG), the household revolution generation (HRG), and the computer generation (CG). In order to also include younger participants, the classification scheme was extended by a fourth generation, the gameboy generation (GG), previously introduced by Gaul and Ziefle [14].

For data collection, the questionnaire method was chosen in order reach a large number of participants and thereby capture comprehensive feedback from a diverse audience. As the various aspects and technologies explored in this study are rather abstract and might appear futuristic or far-fetched for some participants, the scenario technique was chosen to make the general idea of video-based homecare services more understandable.

3 Results

In total, $N = 461$ participants (56% female) aged 12 to 89 years ($M = 37.9$, $SD = 17.7$) took part in the study. The data were analyzed using different statistical procedures. After briefly describing the general acceptance of video-based monitoring, the influences of the individual user factors on the different acceptance measures were analyzed. In order to do this, the correlations among personal user characteristics and acceptance measures were determined in a first step. Due to the high number of dichotomous variables, Spearman rank analyses were used for correlations. In a second step, the influences of selected factors were explored in more detail using (multivariate) analyses of variance. The significances of the omnibus F-tests were determined based on Pillai's Trace. Post-hoc tests were performed using Tukey's HSD. The significance level was set to 5% for all statistical analyses.

In order to get a more differentiated understanding about the acceptance of the presented monitoring system, three different levels of acceptance were distinguished: unconditional acceptance (*"In general, would you accept the monitoring of your home by medical personnel?"*), conditional acceptance (*"For the benefit of better medical assistance, I would accept a reduction of my privacy."*), and reluctant conditional acceptance (*"The idea of ubiquitous medical assistance depresses me, but if it helps, I would accept it."*). Response for all three levels were measured on a 4-point Likert scale (1 = no, 2 = rather no, 3 = rather yes, 4 = yes). The results show that the acceptance of the presented monitoring solution is rather low ($M = 2.17$, $SD = 0.911$). Nearly 30% of the participants stated that they would not accept video-based monitoring in their home, while another 35% said they would probably not accept it. However, the majority of participants would use it under certain conditions. Around 55% of the participants would (probably) use the described system in favor of better medical care ($M = 2.58$, $SD = 0.841$). Over 80% stated that the idea ubiquitous medical assistance depresses them, but that they would accept such a system if it actually helped ($M = 3.06$, $SD = 0.791$).

In a second step, Spearman rank correlations were calculated to get more detailed insights into the data and to reveal relationships among the individual user characteristics and the three acceptance measures. Figure 1 provides an overview over the correlations between the tested user factors and the three acceptance measures.

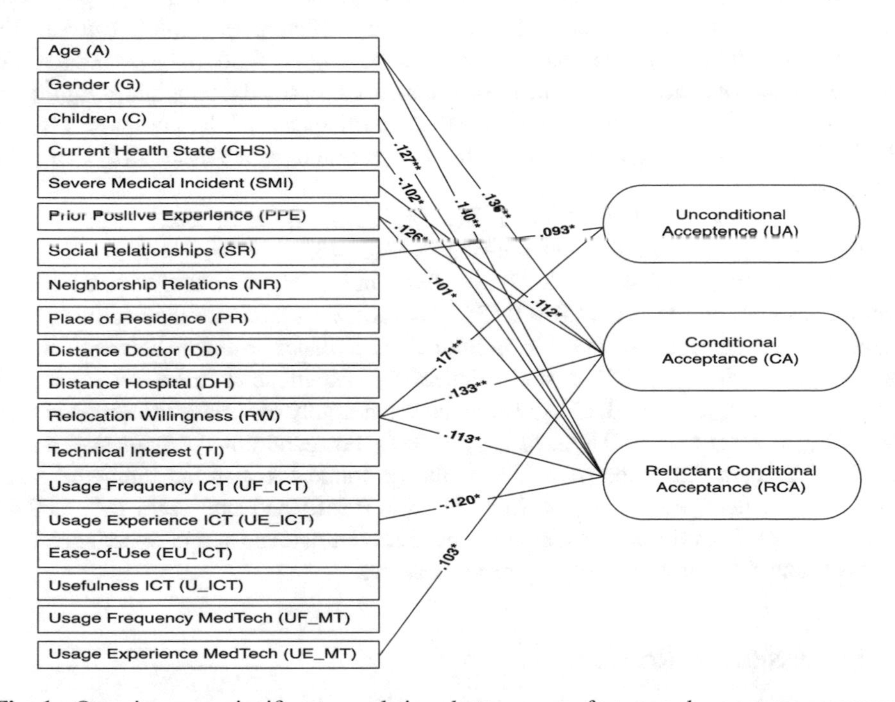

Fig. 1. Overview over significant correlations between user factors and acceptance measures.

As shown above, the individual correlations are rather weak. However, they show important tendencies, which are worth to be explored in more detail. The following paragraphs, therefore, take a closer look at the influences of age, medical history and the experience in using ICT and medical devices on the different acceptance measures as these factors showed significant effects on the acceptance of ICT technologies in related studies (see, e.g., [11, 14] or [7]). A multivariate analysis of variance was conducted with age (A), current health state (CHS), severe medical incident (SMI), prior positive experience (PPE), and the experience of using information and communication (UE_ICT) and medical devices (UE_MT) as factors and unconditional (UC), conditional (CA) and reluctant conditional acceptance (RCA) as dependent variables. The effects of social relationships and relocation willingness were excluded at this point. The results show only significant main effects for age ($F(9, 1053) = 2.30$, $p = .015$, $\eta_P^2 = .019$) and the experience of using medical devices ($F(6, 700) = 2.67$, $p = .015$, $\eta_P^2 = .022$). The interaction effect between both factors was not significant ($F(9, 1053) = 0.470$, p = n.s.).

Univariate analyses also showed that the effect of age on unconditional acceptance ($F(3, 447) = 1.04$, p = n.s.) is not significant, while the effects on conditional ($F(3, 439) = 3.39$, $p = .018$) and reluctant conditional acceptance ($F(3, 435) = 2.89$, $p = .035$) are significant. Post-hoc tests using Tukey's HSD showed that the ratings for conditional acceptance are significantly different ($p = .019$) between the household revolution ($M_{HRG} = 2.76$, $SD_{HRG} = 0.906$) and the gameboy generation ($M_{GG} = 2.48$, $SD_{GG} = 0.763$) as well as marginally significantly different ($p = .052$) between the household revolution and the computer generation ($M_{CG} = 2.50$, $SD_{CG} = 0.840$). For reluctant conditional acceptance, the post hoc analysis revealed a marginally significant difference ($p = .062$) between the responses of participants of the computer ($M_{GG} = 2.99$, $SD_{GG} = 0.737$) and the household revolution generation ($M_{HRG} = 3.20$, $SD_{HRG} = 0.817$).

Univariate analyses showed marginally significant effects of usage experience of medical devices on unconditional acceptance ($F(2, 455) = 2.76$, $p = .064$) and conditional acceptance ($F(2, 447) = 2.77$, $p = .064$), but no significant effect on reluctant conditional acceptance ($F(2, 443) = 0.523$, p = n.s.). Post-hoc tests showed marginally significant differences in the unconditional acceptance ratings between highly ($M_{high} = 3.33$, $SD_{high} = 0.577$) and moderately experienced users ($M_{low} = 2.19$, $SD_{low} = 0.923$, $p_{high/low} = .076$) as well as between highly experienced users and non-users ($M_{no} = 2.12$, $SD_{no} = 0.878$, $p_{high/no} = .057$). For conditional acceptance, a marginally significant difference ($p = .074$) could be found between the ratings of moderately experienced users ($M_{low} = 2.64$, $SD_{low} = 0.858$) and non-users ($M_{no} = 2.45$, $SD_{no} = 0.858$). Significant interaction effects between age and usage experience have not been found for any of the acceptance measures.

4 Discussion of Results

The results show that the majority of participants is rather reluctant to use home monitoring solutions that transmit visual and acoustical information to remote medical staff. Yet, most participants stated that they would accept such systems under certain conditions. It also seems as if age has a relatively strong effect on the acceptance of the described monitoring technologies. Even if age did not correlate significantly with unconditional acceptance, highly significant correlations were found for both conditional and reluctant conditional acceptance. In this context, it is important to highlight that all effects that have been identified are positive, which indicates that the acceptance of medical homecare technologies is higher for older people than for younger ones. Contrary to these results, many technology acceptance studies exploring new medical technologies observed opposite tendencies (see, e.g., [15] and [12]). In literature, negative age effects are often explained with an unfavorable relationship between the time the benefits of technical solutions can be consumed and the effort that is required to acquire the necessary skills [16]. In the scenario used in this study, the usage of the described service required almost no specific technical skills, which reduced the 'learning phase' to a minimum. Hence, the effort necessary to learn how a specific system has to be operated might not have been a crucial factor when assessing the personal usage intention. At the same time, other factors might have come into play. For example, in a later stage of life, physical deficits might be much more ubiquitous than at younger age. Hence, it is likely that older participants have already invested more thoughts into different aging scenarios and resulting options might therefore be judged more realistic. As a consequence, older participants could be more willing to accept medical homecare solutions.

In contrast, most others factors like, e.g., gender, which played an important role in technology acceptance research for many years, do not appear to be decisive factors for the acceptance of electronic home-monitoring services. With regard to gender, the findings are not reflecting the results of the majority of related studies with other types of medical devices. Gender effects were found with regard to the acceptance of medical technology [17, 18] or the usage intention of electronic healthcare devices [15]. However, there are also studies in which no gender differences were found. For example, Wilkowska et al. [11] explored the general attitude towards the usage of assistive medical technology and found neither age nor gender differences.

Similar to gender, all three health-related factors, describing the participants' past and current health condition, seem to have not significant effects on the acceptance of the described home monitoring system.

5 Conclusion

As technology-enhanced home-monitoring services are likely to be used by a diverse group of users, it is important to understand the personal characteristics that contribute to acceptance in order to appropriately address them in the design process of new systems. Therefore, a special focus of this paper was on the analysis of personal user characteristics and individual living situations as well as their influences on acceptance. While age seems to affect the acceptance of the presented video-based monitoring service, most other personal factors included in this study appear to have no considerable effects on acceptance. Yet, it is important to be aware of the fact that acceptance is not only a quite complex, but also dynamic construct. Hence, individual usage barriers might dissolve once users get accustomed to these technologies. While this paper aimed to deliver first insights further research is necessary. This includes in particular long-term studies, but also an extension to other application fields and usage contexts, which have not been covered in this paper.

References

1. Ballegaard, S.A., Hansen, T.R., Kyng, M.: Healthcare in everyday life - designing healthcare services for daily life. In: Proceedings of the ACM Conference on Human Factors in Computing Systems (CHI 2008), pp. 1807–1816. ACM Press, New York (2008)
2. Hennessy, C., Walker, A.: Promoting multi-disciplinary and inter-disciplinary ageing research in the UK. Ageing Soc. **31**(1), 52–69 (2011)
3. Demiris, G., Oliver, D.P., Dickey, G., Skubic, M., Rantz, M.: Findings from a participatory evaluation of a smart home application for older adults. Technol. Health Care **16**(2), 111–118 (2008)
4. Dewsbury, G., Taylor, B., Edge, M.: The process of designing appropriate smart homes: including the user in the design. Scottish Centre for the Environmental Design Research, Robert Gordon University (2001)
5. Bias, R., Mayhew, D.: Cost-Justifying Usability: An Update for the Internet Age. Elsevier Science, New York (2005)
6. Glende, S., Podtschaske, B., Friesdorf, W.: Senior user integration. In: Proceedings of the Second German Congress on Ambient Assisted Living. VDE, Berlin (2009)
7. Ziefle, M., Bay, S.: How older adults meet cognitive complexity: aging effects on the usability of different cellular phones. Behav. Inf. Technol. **24**(5), 375–389 (2005)
8. Röcker, C., Ziefle, M., Holzinger, A.: From computer innovation to human integration: current trends and challenges for pervasive health technologies. In: Holzinger, A., Ziefle, M., Röcker, C. (eds.) Pervasive Health, pp. 1–17. Springer, London (2014)
9. Röcker, C.: Intelligent environments as a promising solution for addressing current demographic changes. Int. J. Innov. Manag. Technol. (IJIMT) **4**(1), 76–79 (2013)
10. Gaul, S., Wilkowska, W., Ziefle, M.: Accounting for user diversity in the acceptance of medical assistive technologies. In: Proceedings of the Third International ICST Conference on Electronic Healthcare for the 21st Century (eHealth 2010) (2010)
11. Wilkowska, W., Gaul, S., Ziefle, M.: A small but significant difference: the role of gender on the acceptance of medical assistive technologies. In: Leitner, G., Hitz, M., Holzinger, A. (eds.) HCI in Work & Learning, Life & Leisure, pp. 82–100. Springer, Heidelberg (2010)
12. Ziefle, M.: Age perspectives on the usefulness on E-health applications. In: Proceedings of the International Conference on Health Care Systems, Ergonomics, and Patient Safety (HEPS 2008), Strasbourg, France (2008)
13. Sackmann, R., Weymann, A.: Die Technisierung des Alltags – Generationen und technische Innovationen. Campus, Frankfurt (1994)
14. Gaul, S., Ziefle, M.: Smart home technologies: insights into generation-specific acceptance motives. In: Holzinger, A., Miesenberger, K. (eds.) HCI and Usability for e-Inclusion, pp. 312–332. Springer, Heidelberg (2009)
15. Arning, K., Ziefle, M.: Different perspectives on technology acceptance: the role of technology type and age. In: Holzinger, A., Miesenberger, K. (eds.) Human-Computer Interaction for eInclusion, pp. 20–41. Springer, Heidelberg (2009)
16. Melenhorst, A.-S., Rogers, W.A., Bouwhuis, D.G.: Older adults' motivated choice for technological innovation: evidence for benefit-driven selectivity. Psychol. Aging **21**(1), 190–195 (2006)
17. Ziefle, M., Schaar, A.K.: Gender differences in acceptance and attitudes towards an invasive medical stent. Electron. J. Health Inform. **6**(2), 1–18 (2011)
18. Alagöz, F., Ziefle, M., Wilkowska, W., Calero Valdez, A.: Openness to accept medical technology – a cultural view. In: Holzinger, A., Simonic, K.-M. (eds.) Human-Computer Interaction, pp. 151–170. Springer, Heidelberg (2011)

Usability Impact of User Perceptions in mHealth—The Case of Ghanaian Migrants

Eric Owusu(✉) and Joyram Chakraborty

Department of Computer and Information Sciences, Towson University, 7800 York Road, Towson, MD 21252, USA
{eowusu, jchakraborty}@towson.edu

Abstract. The use of mHealth applications in health services delivery is widely documented. However, the literature on actual usability studies conducted among migrant consumers in the United States is sparse. Migrant communities in the U.S. have been sidelined when it comes to studies in usability and acceptance of mobile health applications. This pilot study reports the findings of usability of mHealth applications from the perception of Ghanaian migrants, specifically their satisfaction using mobile technology to manage their health challenges.

Keywords: mHealth · Requirements gathering · Usability · User perception

1 Introduction

Health Information Technology (HIT) is increasingly being used by healthcare providers to improve patient care. HIT is defined as "the application of information processing involving both computer hardware and software that deals with storage, retrieval, sharing, and use of health care information, data and knowledge for communication and decision making" [1]. HIT has grown from the use of mobile devices, such as personal digital assistants in the 1990's, to the use of more complicated and real time electronic health (eHealth) interventions such as mobile health applications [2]. Mobile health (mHealth), is defined by the World Health Organization (WHO), as an area of electronic health that provides health services and information via mobile technologies such as mobile phones and PDAs. mHealth applications allow users to be in charge of their healthcare and have access to real time information through a variety of peripheral devices. With more than 1 billion smartphones and 100 million tablets in use today, mHealth applications promises to be an invaluable tool in healthcare management [3]. However, there is limited research on the accessibility and sustainability of mHealth use in migrant populations in the United States [4].

Although there are few examples of mHealth tools that are quite popular in the United States, the use of mHealth tools is variant from community to community and are usually dependent on simplicity of use [5]. Experts in the field have demonstrated that there has been an increase in access to information and communication technology (ICT) and a growing penetration of smartphones [2, 3]. Health disparities have also been established among migrants especially in ethnic minorities [6]. However, despite

W. Karwowski and T. Ahram (Eds.): IHSI 2019, AISC 903, pp. 557–562, 2019.
https://doi.org/10.1007/978-3-030-11051-2_84

the increase in access to ICT and the well documented health disparities among migrants, there is limited information about the actual adoption, usage and attitudes of migrant populations towards mHealth services.

There is a need for further exploratory studies in migrant groups in relation to mHealth and user accessibility and sustenance. Årsand et al. [7], authenticate the need and importance of more studies in mobile health technology and the potential to engage and empower all users to be in control of, and to manage their healthcare. Fleming et al. [8], also shed light on the importance of usability testing within intended users. These findings buttress the point that exploring the usability of mobile health technology among migrant groups will highlight behavioral patterns and provide useful data to improve the design and utilization of mobile health interventions. The specific contribution of this paper provides a preliminary understanding of the user perceptions of the migrant community in using mHealth tools.

2 Background of the Study

The existing and future technologic capabilities of smartphones have the potential to make them an increasingly essential personal health tool. To realize or optimize the potential of mHealth applications, there is a need to explore how these applications are perceived by end users, especially disparate communities, who stand to benefit by having a wealth of health information placed in their palms via their smartphones. The goal is to improve the quality of healthcare and well-being of marginalized groups such as Ghanaian migrants.

There are many divisions of healthcare that can be improved by using mHealth technology. However, there have been barriers such as ease of use, literacy, access to technology and affordability, which have been a challenge in the role that information technology plays in healthcare improvement and inclusion of disparate populations [9]. These barriers that existed have been reduced, and mobile technologies that can be employed in mHealth applications have become affordable, easy to use and widely adopted across socioeconomic status [10]. Mobile technologies have become strategic tools for health education and intervention because more members of the general population now have access to mobile technologies than in the past. Mobile technologies have also become equalizers, in that access to quality information is placed within the reach of people from all walks of life through smartphones and other mobile devices.

Access to mobile phones has been found to run high within all ethnic groups in the United States, with Hispanics at 76%, Whites at 85%, and Blacks at 79% [9]. This high dispersal of mobile phones among diverse populations makes it a promising tool for patient engagement and healthcare management through mHealth applications. It is estimated that currently there are approximately six billion mobile phone users, and half of the global population use smartphones [11].

According to Tate et al. [12], 12% of American smartphone users have at least one health application to access health information. This is very promising in light of the potential benefits of smartphones, but the question still remains, 'how many migrants such as Ghanaians, are utilizing this technology'? The Health Information Technology for Economic and Clinical Health (HITECH) Act offers a platform that serves as a policy model for mHealth technologies by putting in place incentive structures that promote the use of technology in healthcare. One of the core objectives of HITECH is to provide secure communication and other objectives aimed to improve healthcare, especially in disparately impacted populations, which includes migrant populations.

In the United states, scientific research has long established the presence of healthcare disparities within racial and ethnic minorities. The US Department of Health and Human Services released a report in 1984 which stated that, "while the overall health of the nation showed significant progress, major disparities existed in the burden of death and illness experienced by blacks and other minority Americans as compared with the nation's population as a whole" [13]. A report released by the Institute of Medicine (IOM) in 2003 confirmed the existence of significant racial and ethnic disparities within the United States even among individuals with access to care [14]. There is no consensus about what constitutes a health disparity, but the cause of the disparities is thought to be related to sociocultural, behavioral, economic, environmental, biologic and societal factors [13], and groups identified as underserved and affected by health disparities within the United States include among others, African Americans, racial and ethnic minorities, people with English as a second language, and immigrants [15]. All these criteria describe the characteristics found in Ghanaian migrants in the United States, thus the specific interest in this group of people.

3 Purpose of Study

Understanding usability and user perception of mHealth is key in promoting self-healthcare interventions and improving user experience in mobile technologies [16]. This can be used by user interface designers as they design for the global market with cognition of all ethnic minorities. Literature establishes that there are opportunities for mobile applications to address healthcare needs relative to intervention [2]. The purpose of this study is to investigate if usability affects user perception of mHealth tools using Ghanaian migrants as a case study. The study strives to address the following question: *Can usability impact migrant user perceptions of m-Health?*

To address this research question, an experimental design will be conducted in three phases. The first phase will investigate user perception on mHealth technology usage. The second face will be to design, develop and validate an mHealth application based on user requirements. The final phase will be to enhance the application usage and adoption through incorporation of findings from iterative usability testing within the target population.

4 Methodology

A pilot study was conducted to investigate how Ghanaian migrants perceive mobile technology usage for health interventions. The study was carried out within a 6-week period. A random sampling approach was used in selecting participants comprising of migrants in one specific geographical location. Four locations in Maryland were chosen. Participants were chosen by word of mouth or recommendations. All participants agreed to be interviewed on a one on one basis at a convenient location of their choice. The facilities used for the interviews ranged from shopping malls to participant's residence. Each interview lasted 10–15 min. Data was collected from 27 Ghanaians comprising of 13 males and 14 females. All participants were interviewed face to face. The selected age groups that participated were within the ranges of age 25 to 44, 45 to 64 and 65 and above. At each interview, the interviewer explained the purpose of the study to the participants and assured them of confidentiality. A questionnaire was given to the participants to follow through the interviewer's questions, which allowed for consistency of responses relevant for the study. Data was captured directly by marking or writing responses on the questionnaire for the different question types. The data obtained from the interview was cleaned, analyzed and relevant patterns recorded.

Two focus group discussions were conducted to get an idea of community perception of mHealth. Each group comprised 4 participants. The selection of participants was also done by word of mouth and recommendation. Participants were selected from ages 30s, 40s and 50s to get a fair representation of all groups. During the discussions, the facilitator guided the groups using the questionnaire, and captured data by writing responses on the form. To clean the data, all completed forms were thoroughly examined to ensure that all data was accurate and complete.

5 Results

Findings were as follows; 41% (11) of participants were within the age group 25–44 years, 44% (12) were within the age group 45–64 years, and 15% (4) were within the age group 64 and above. All 27 participants used smartphones. Participants used smartphones for these purposes: banking, communication, socializing, referencing, information gathering, and medical and health education. 93% (25) indicated that they use their smartphones to search for medical related information from the web, while 7% (2) did not. 81% (23) of participants indicated they were very familiar with health conditions like diabetes, 15% (4) were not well familiar with diabetes. 85% (22 out of 26) indicated that they would alter their behavior based on medical related information gained from the web, 15% (4 out of 26) indicated they would not, one participant did not respond. 93% (25) indicated that using smartphone to monitor their health would benefit them, 7% (2) responded they will not benefit. 67% (18) indicated that they would be interested in paying for an app to monitor their health, 33% (9) said no or were unsure. With regards to what specific features they would like in an mHealth app to achieve their health goals, 100% of respondents said that they would like mHealth apps developed specifically with the Ghanaian diet and lifestyle in mind.

One finding from the focus group discussion was that participants were more likely to download and use an mHealth app, if it was recommended by their primary care provider specifically for health condition they may have. One other finding was that participants were inclined to use mHealth apps if it met their specific needs.

6 Discussion

The results obtained indicated that 100% of participants, which included the age group 64 and above, used smartphones. This is very interesting and promising, because the general perception is that the older generation (64 and above) only use their phones for general communication purposes and hardly embrace new technology. The usage among this age group indicates that mHealth approaches can be targeted not only towards the younger generation but also towards this vital age group. This is important because 10.9 million US adults aged 65 years and above are affected by chronic diseases such as diabetes. This number is projected to increase to 26.7 million by 2050. Therefore, effective targeted interventions are needed to address the growing burden and older adults [17].

96% (26) of participants indicated that they use their smartphones to search for medical related information from the web, but only 67% indicated that they would be interested in paying for an app to monitor their health. This highlights the need to consider cost in developing mobile health applications for such targeted groups. The cost of mHealth has been noted to have an effect on reaching underserved populations with mHealth initiatives, which are potentially of high benefit due to the high level of mobile phone ownership in these groups [18]. From the focus group discussions participants discussed reliability and privacy of their personal health data to be an issue of high concern to them. This validates the security and integrity concern that has been addressed by Luxton et al. [2].

From the interviews and discussions, one key factor that was highlighted was that respondents were highly interested in a tailored mHealth app that takes into cognizance Ghanaian preferences in terms of ease of use, interface design, and accessibility. Understanding what motivates consumers from different cultures is important for positioning brands in different markets [19]. Such implicit cultural values need to be considered when designing apps intended to reach end users from all cultural backgrounds [20].

7 Conclusion

Ghanaians living in the United States would use an mHealth application if it was developed with an empirical understanding of the Ghanaian user. The data obtained from this study indicates promising evidence to harness the increasing presence and use of smartphones to deliver mHealth services to migrant communities. For an impactful change in the area of mHealth service delivery and usability, further research is recommended in this line of study to provide key insights into achieving usable applications for migrant users.

References

1. Thompson, T.G., Brailer, D.J.: The decade of health information technology: delivering consumer-centric and information-rich health care. US Department of Health and Human Services, Washington, DC (2004)

2. Luxton, D.D., McCann, R.A., Bush, N.E., Mishkind, M.C., Reger, G.M.: mHealth for mental health: integrating smartphone technology in behavioral healthcare. Prof. Psychol. Res. Pract. **42**, 505–512 (2011)
3. Martínez-Pérez, B., de la Torre-Díez, I., López-Coronado, M.: Mobile health applications for the most prevalent conditions by the world health organization: review and analysis. J. Med. Internet Res. **15**, e120 (2013)
4. Srinivasan, S., O'Fallon, L.R., Dearry, A.: Creating healthy communities, healthy homes, healthy people: initiating a research agenda on the built environment and public health. Am. J. Public Health **93**, 1446–1450 (2003)
5. Sama, P.R., Eapen, Z.J., Weinfurt, K.P., Shah, B.R., Schulman, K.A.: An Evaluation of Mobile Health Application Tools. JMIR mHealth and uHealth **2**, e19 (2014)
6. Benz, J.K., Espinosa, O., Welsh, V., Fontes, A.: Awareness of racial and ethnic health disparities has improved only modestly over a decade. Health Aff. **30**, 1860–1867 (2011)
7. Årsand, E., et al.: Mobile health applications to assist patients with diabetes: lessons learned and design implications. J. Diab. Sci. Technol. **6**, 1197–1206 (2012)
8. Fleming, J.B., Hill, Y.N., Burns, M.N.: Usability of a culturally informed mHealth intervention for symptoms of anxiety and depression: feedback from young sexual minority men. JMIR Hum. Factors **4**, e22 (2017)
9. Martin, T.: Assessing mHealth: opportunities and barriers to patient engagement. J. Health Care Poor Underserved **23**, 935–941 (2012)
10. Klasnja, P., Pratt, W.: Healthcare in the pocket: mapping the space of mobile-phone health interventions. J. Biomed. Inform. **45**, 184–198 (2012)
11. Dalkou, M., Nikopoulou, V.-A., Panagopoulou, E.: Why mHealth interventions are the new trend in health psychology? Effectiveness, applicability and critical points. Eur. Health Psychol. **17**, 129–136 (2015)
12. Tate, E.B., et al.: mHealth approaches to child obesity prevention: successes, unique challenges, and next directions. Transl. Behav. Med. **3**, 406–415 (2013)
13. Gibbons, M.C.: A historical overview of health disparities and the potential of eHealth solutions. J. Med. Internet Res. **7**, e50 (2005)
14. Bach, P.B.: Book review unequal treatment: confronting racial and ethnic disparities in health care. In: Smedley, B.D., Stith, A.Y., Nelson, A.R. (eds.) 764 p. National Academies Press, Washington, D.C. (2003). $79.95. 0-309-08532-2. New Engl. J. Med. **349**, 1296–1297 (2003)
15. Montague, E., Perchonok, J.: Health and wellness technology use by historically underserved health consumers: systematic review. J. Med. Internet Res. **14**, e78 (2012)
16. Azhar, F.A.B., Dhillon, J.S.: A systematic review of factors influencing the effective use of mHealth apps for self-care. In: 2016 3rd International Conference on Computer and Information Sciences (ICCOINS). IEEE (2016)
17. Caspersen, C.J., Thomas, G.D., Boseman, L.A., Beckles, G.L.A., Albright, A.L.: Aging, diabetes, and the public health system in the United States. Am. J. Public Health **102**, 1482–1497 (2012)
18. Whittaker, R.: Issues in mHealth: findings from key informant interviews. J. Med. Internet Res. **14**, e129 (2012)
19. De Mooij, M., Hofstede, G.: Cross-cultural consumer behavior: a review of research findings. J. Int. Consum. Mark. **23**, 181–192 (2011)
20. Leidner, D.E., Kayworth, T.: A review of culture in information systems research: toward a theory of information technology culture conflict. MIS Q. **30**, 357–399 (2006)

Making HSI More Intelligent: Human Systems Exploration Versus Experiment for the Integration of Humans and Artificial Cognitive Systems

Frank Flemisch[1,2](✉), Marcel C. A. Baltzer[1], Shadan Sadeghian[1], Ronald Meyer[2], Daniel López Hernández[1], and Ralph Baier[2]

[1] Fraunhofer FKIE, Fraunhoferstraße 20, 53343 Wachtberg/Bonn, Germany
{frank.flemisch, marcel.baltzer, shadan.sadeghian.borojeni, daniel.lopez.hernandez}@fkie.fraunhofer.de
[2] Institut Für Arbeitswissenschaft (IAW), RWTH Aachen, Bergdriesch 27, 52062 Aachen, Germany
{f.flemisch, r.meyer, r.baier}@iaw.rwth-aachen.de

Abstract. When it comes to integrating humans, technical systems and organizations, the interplay of constructive and critical methods and tools is crucial for making the process and product of HSI really intelligent. This becomes even more important, when HSI is applied to the development of artificial cognitive systems, automation and autonomous systems e.g. in aircraft, ships, cars, intelligent factories or cyber defense systems. Experiments as a set of methods to test hypothesizes are well formulated and tested for decades. In contrast to experiments, concept and methods to systematically explore new human machine systems are relatively new and have to be systematized. Human Systems Exploration is a concept of connected activities and methods to systematically invent, conceptualize and test design and use spaces of human machine systems/socio-cyber-physical systems.

Keywords: HSI · Automation · Cognitive systems · Autonomous systems Participatory design · Human system exploration

1 Introduction and Overview: The Need for a More Intelligent Human Systems Integration

The proper testing of complex systems has been the focus of research for many decades. An example for this is the method set of experiments, e.g. in physics or in experimental psychology, which are already quite sophisticated and structured. In contrast to experiments, the proper construction and systems engineering of technical systems is not that crisply defined. At least there are approaches to structure and describe these kind of processes for many years. Increasingly, this systemic thinking is also applied to socio-cyber-physical systems. As part of this development towards an established HSI, the early stage of system development is now getting into the focus of research and development. With increasing technical, social and organizational options,

W. Karwowski and T. Ahram (Eds.): IHSI 2019, AISC 903, pp. 563–569, 2019.
https://doi.org/10.1007/978-3-030-11051-2_85

more heterogeneous tasks and users, and less time, especially the exploration of new design options of human machine systems demand to be systematically investigated, defined and tested.

This contribution addresses these demands. It is based on a longer concept paper on exploroscopes in the military domain [1], which has to be updated and focused on explorations in civil as well as in military applications.

2 A Brief Summary of HSI Challenges: Balancing Tension Fields

Besides the challenge of integrating different – or sometimes even diverging – views, a design group's major challenge is to balance the tension fields of requirements, ideas and methodologies on the one hand, and money, time, quality and quantity on the other hand. Balance can be described as a guiding motif for a combination of methods or as a single method for solving conflicts and dilemmas of opposing perspectives or conflicting requirements in the exploration process. In order to reach a balanced human-systems design subjective perspectives, e.g. of the user, have to be combined with objective perspectives, e.g. of a system evaluation. To accomplish this, qualitative techniques can be combined with quantitative techniques.

Professionals working in the ergonomics or human factors domain often experience a lack of time and capacities to find creative, innovative solutions, which are able to fulfill challenging demands on time to production deadline, budgets and operational usability. The concepts of exploration aim at efficient innovation and set a framework of methodologies and principles to meet the ambitious demand of balanced HSI. The very early examples of explorations introduced in the presentation to this paper demonstrate the usefulness of the exploration framework as a methodological framework. Based on these fruitful experiences, there is a high potential to investigate the human factors of explorations even further.

3 Exploration as a Dedicated Phase in Human Systems Integration

In general, to explore means

a. to investigate, study, or analyze,
b. to become familiar with by testing or experimenting, e.g."explore new cuisines",
c. to travel over (new territory) for adventure or discovery,
d. to examine especially for diagnostic purposes (Merriam-Webster online 2013).

In the context of design and development of human-machine-systems, the term "Human Systems Exploration" or "(design) exploration (of human machine systems)" stands for a temporally and thematically connected series of activities, techniques and tools to invent, design, prototype and assess the effects of different options of human-machine systems. The concept of exploration is on a similar ontological level as the concept of experiment. While an experiment uses methods, tools and techniques optimized to test hypotheses and to control the probability of error, e.g. with the help of statistics, explorations try to optimize the way from initial questions via ideas to an understanding of the different options, selection of options and to one or more working system designs. "System" in the context of this paper is understood as socio-technical system of humans and technical subsystem (human-machine system). Explorations can also be applied to socio-cyber-physical systems.

The concept of exploration has connections to methods from the early stages of the designing process in the mechanical engineering domain. Intuitive methods like brainstorming, the Gallery Methode, the Delphi Method and synectics have proven their worth [2] and have clearly exploratory aspects. In contrast to those stated methods exploration shifts the focus and combines the perspectives of developers, the users and other stakeholders.

What is explored in design explorations? Helpful metaphors here are "design space" and "use space". Like physical space can be thought of in at least 3 dimensions (plus time), the design space can be thought to be made up of design dimensions, i.e. all aspects or qualities of an artifact that can make a difference in design. Examples for simple design dimensions are size, color or haptic quality. Design dimensions can also be more complex, like assistance and automation levels in highly automated air-, ground- or maritime vehicles, or connectivity in a computer network. Design dimensions span the design space. During the exploration the design space can be limited by functional, cost or human factors aspects (Fig. 1), but can be explored in width and depth virtually unlimited.

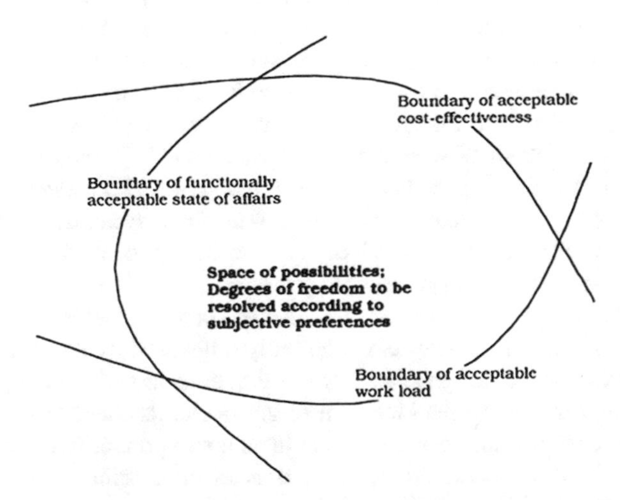

Fig. 1. Boundaries in the design space, (adapted from Rasmussen et al. [3])

The design space consists of all the design dimensions as well as their combination (see Fig. 1). To structure what Rasmussen calls "space of possibilities" it is crucial to give the many subjective decisions in the exploration process an objective structure. Figure 2 shows a mind-net map of the design space, a special form of mind map that helps to structure the many dimensions of a design space in a way that the goal of the whole exercise, the design variant as integrated combination of dimensions, are in the center. This mind-net map can also be used as a creativity tool to explore new combinations that nobody has thought of before.

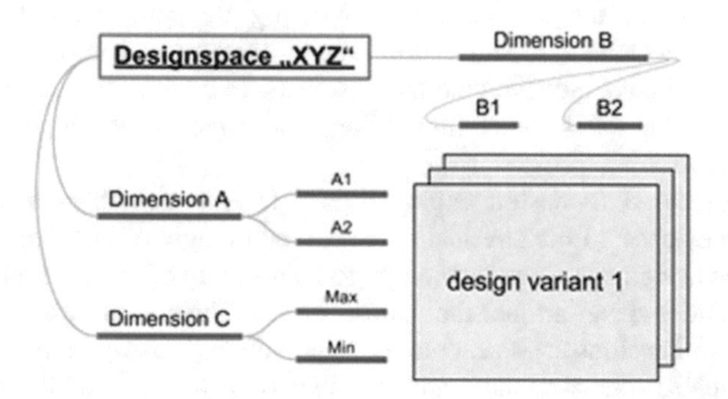

Fig. 2. Mind-Net map of design space (adapted from Flemisch et al. 2008 [4])

Exploration is not only about exploring the design space, but also the use space. Use space is the combination of all possible instances of use with all possible combinations of design dimensions. To structure the use space, it can make sense to group certain parts of the use space into use situations or their abstract sisters use cases, which can be easily understood by users, designers and developers alike.

The challenge here is that the situation in which design and development happens can be vastly different from the situations in which the actual use happens, which in analogy to Normans "gulf of execution and evaluation" [5] could be called the fundamental "gulf of time and space between design and use". To make things even more challenging, it is very clear that in most human-machine systems, the number of use situations will be far too high for all of them to be assessed in appropriate design situations. Vast spaces to explore, easy to get lost!

Not only the design and use space but also the potential users of a system can be explored, since the design of a system is not only influenced by the theoretical degrees of freedom in design (the design space) or the situations in which the system might be used (use space), but also by the kind of user group and its characteristics. Even if the members of the design team or other stakeholders are important in the exploration, the characteristics of a design team are certainly a factor influencing the design – a design often depends on the kind of user and his or her experience, expectations, mental models and motivation. It makes a difference if the user is a highly skilled and trained formula one driver or if the user is just a normal driver of a passenger car. One method to explore and structure the variety of users is to use "Personas" (see e.g. [6]), where virtual persons are invented and even re-enacted by members of the design team. The most important way to explore the "space" of users is to involve a realistic but high variety of real users into the exploration.

Now after a basic introduction to explorations, design and use spaces, let us take a deeper look into the human factors that "fuel" an exploration.

4 Human Factors and HSI of Explorations: Towards a Convergence and Crispness of Mental Models and Implementation

Especially important for explorations is the psychology of the design and evaluation teams who come together with constructive and critical approaches. Proposed methods and techniques for explorations take into account human strengths and limitations on creativity, sharing mental models and dealing with complex decisions in the process of designing and understanding working systems and their effects.

At the beginning of investigating the design space the complexity that the design team and other relevant stakeholders are confronted with could be huge. For design teams the complexity in terms of the amount and variety of design options, the combination of design options and use cases as well as the challenges in learning new procedures and methods might be a source of perceived uncertainty. When uncertainty is overwhelming, individuals tend to encounter resistance against the whole activity. The feeling of high demands and stress due to the complexity also leads to cognitive biases and the tendency of so called "premature cognitive closure" [7]. The feeling of uncertainty that individuals might feel exploring an unknown space or complex problems is likely to be triggered by typical and inherent characteristics of non-routine design situations. Examples for those characteristics are the multiple design options or the amount of relevant design factors with multiple interdependencies and their changes over the time as well as the lack of a design routine. Other examples of dimensions of complexity are the fuzziness of the design goal or preferred use case or the unknown end state to be reached for the final design prototype These tension fields between fuzziness – sharpness and certainty – uncertainty can be quite critical. People with important roles in design and development process, like engineers and managers are mostly trained in pursuing solutions with a high certainty rather than allowing for metaphors or fuzzy concepts. Consequently, facing situations with uncertainty and fuzziness puts them out of their "comfort zone", might make them uncomfortable in the beginning and need some extra care in the process (think of Fig. 3).

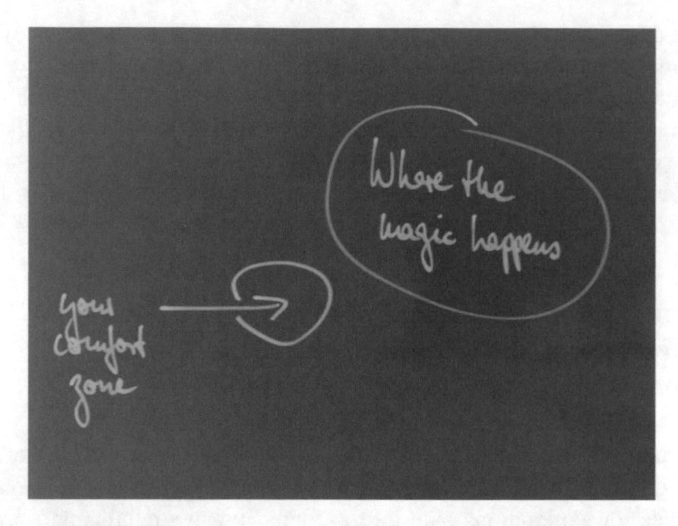

Fig. 3. "Getting out the comfort zone" (http://chrisredmond100.blogspot.de/2012/09/easy-exit-from-your-comfort-zone.html)

Comparable to an auto-focus in cameras, their cognition tries to sharpen the thinking process and, based on their fallacy of competence, they might be tempted to fall back into routine behavior, which is mandatory later on in the development process but sometimes too early and counter-productive in the design process [8]. Besides the potential for uncomfortable affective states, explorative behavior also offers the experience of an increase in intrinsic motivation and the perception of flow when positive feedback and self-control is experienced. If this reinforcing loop is established, the more new spaces are explored the higher the variation of experience and the lower the feeling of uncertainty will be experienced [9].

This perception of flow can indeed fuel explorations and can be an important driving force for exploring new ideas. Another major ingredient for the fuel is certainly curiosity, which is "the desire to know" (Merriam-Webster Online) and ambition, a desire to achieve a particular end even if it seems impossible at a first glance. The so called "Gestaltungsmotiv" [10] is a valuable resource for the design motivation and should be stimulated by the exploroscope methodology. In this respect, design explorations are not so different from explorations of physical space like the Lewis & Clark expeditions or the moon landing, they are new, investigative and promising in the case of success.

The balance of negative and positive dynamics in the exploration process is quite crucial: Positive dynamics has to be stimulated and hedged in order to activate creativity and divergent thinking in the design team, to get and keep things going. Negative dynamics have to be taken serious and can nevertheless be used to shape the exploration process.

This underlines the important role of the HSI experts to open the design space at the beginning of the process and encourage design teams to use the motivation of curiosity and ambition and cope with high levels of uncertainty. This is to be realized in a semi-structured process when at the beginning some feed-forward on a first structure for the space to be explored is provided with the provision of prototypical use cases or scenarios. Continuous feed-back or even feed-forward on the working results should be provided and cognitive challenges and competences in the meaning of knowledge, experience and operational background of the design team members has to be balanced.

5 Outlook: Towards a Systematic Exploration and Description of Human Systems Exploration

This paper could only be a first sketch of the concept of Human Systems Exploration. In a couple of applications at NASA, DLR, RWTH Aachen University and Fraunhofer FKIE, explorations with this explicit theme have already shown good results and led to well working human machine systems. Laboratories specialized for Human Systems Exploration, so called exploroscopes, have been built and used in human systems integration e.g. of human automation systems. What is needed now is a systematic exploration of the concept, leading to a more concise description of the design space, use space and effect space of explorations and exploroscopes. A wide future of balanced socio-cyber-physical systems is right in front of us, let's get ready to explore!

References

1. Flemisch, F., et al.: Towards a balanced human systems integration beyond time and space: exploroscopes for a structured exploration of human–machine design spaces. In: HFM-231 Symposium on Beyond Time and Space, NATO-STO HFM Panel, Orlando (2013)
2. Pahl, G., Beitz, W., Feldhusen, J., Grote, K.-H. (eds.): Engineering Design: A Systematic Approach. Springer, London (2007)
3. Rasmussen, J., Pejtersen, A.M., Goodstein, L.P.: Cognitive Systems Engineering. Wiley Series in Systems Engineering and Management, New York (1995)
4. Flemisch, F., Schindler, J., Kelsch, J., Schieben, A., Damböck, D.: Some bridging methods towards a balanced design of human-machine systems, applied to highly automated vehicles. In: Presented at Applied Ergonomics International Conference, Las Vegas, USA (2008)
5. Norman, D.A., Draper, S.W. (eds.): User Centered System Design: New Perspectives on Human-Computer Interaction. Lawrence Erlbaum Associates, Hillsdale (1986)
6. Pruitt, J., Adlin, T.: The Persona Lifecycle. Keeping People in Mind Through Product Design. Morgan Kaufmann, San Francisco (2006)
7. McDermott, R.: Decision making under uncertainty. In: Proceedings of a Workshop on Deterring CyberAttacks: Informing Strategies and Developing Options for U.S. Policy (2010). http://www.nap.edu/catalog/12997.html
8. Badke-Schaub, P., Daalhuizen, J., Roozenburg, N.: The towards a designer-centred methodology: descriptive considerations and prescriptive reflections. In: Birkenhofer, H. (ed.) Future of Design Methodology, pp. 181–197. Springer, Stuttgart (2011)
9. Dörner, D.: Emotion und Handeln. In: Badke-Schaub, P., Hofinger, G., Lauche, K. (eds.) Human Factors Psychologie Sicheren Handelns in Risikobranchen. Springer, Stuttgart (2008)
10. Hossiep, R., Paschen, M.: Das Bochumer Inventar zur berufsbezogenen Persönlichkeits: beschreibung. Testmanual. Hogrefe-Verlag, Göttingen (1998)

Design Methodologies for Human-Artificial Systems Design: An Automotive AR-HUD Design Case Study

Cuiqiong Cheng[1], Fang You[1(✉)], Preben Hansen[2], and Jianmin Wang[1]

[1] School of Arts and Media, Tongji University, Shanghai, China
chengcuiq@qq.com, {youfang,wangjianmin}@tongji.edu.cn
[2] Department of Computer and Systems Sciences, Stockholm University, Stockholm, Sweden
preben@dsv.su.se

Abstract. With the development of society, technical products become more complex and intelligent. Interaction design of intelligent products is meeting new challenges. It requires practitioners to wade into the whole product development process and grasp basic interdisciplinary knowledge. This paper focuses on design methodologies for Human-Artificial Systems design with interdisciplinary knowledge. In the Interaction Design Method Framework (IDMF) we proposed, design runs through the whole product development process to help practitioners propose emerging technology-based designs. The IDMF framework is composed of 102 design methods and six phases. We deconstructed each phase of the IDMF according to three dimensions, design, information, and business. This paper offers a practical guide to using the IDMF framework by providing a high-level summary of the automotive AR-HUD design. It presented how the IDMF framework help practitioners effectively guide the design team and propose emerging technology-based designs.

Keywords: Design method · Interaction design · Human-Artificial systems design · Connected and automated vehicles · Augmented reality Head-Up display

1 Introduction

Interaction design is the practice of designing interactive digital products, environments, systems, and services. It could design behaviors, animations, and sounds as well as shapes with its specific methods and practices [1, 2]. In the transition stage of intelligent products, technology is constantly evolving and intelligent products are becoming more complex. Interaction design plays an important role in reducing the complexity of intelligent products and improving user acceptance of them. With the emphasis on interaction design, academia has continuously developed interaction design methods and approaches [3]. Publications related to design methods are also popular in the industry.

W. Karwowski and T. Ahram (Eds.): IHSI 2019, AISC 903, pp. 570–575, 2019.
https://doi.org/10.1007/978-3-030-11051-2_86

To make successful intelligent products, enterprises tend to recruit practitioners from design, business, and information technology. To propose emerging technology-based designs and build design influence, design practitioners should grasp interdisciplinary knowledge and not just be a decision recipient.

In this paper, we presented a systematic and comprehensive framework for Interaction Design Method Framework (IDMF) which aimed to be utilized as a practical design tool with groups of team members entailing. The IDMF design process develops consists of six different phases and involves interdisciplinary knowledge.

2 Related Work

Interaction design plays a major role in the user experience of products, services or applications [4]. In the software engineering discipline, researchers have proposed different life-cycle models: waterfall model [5], the spiral model [6]. In the design discipline, Zimmerman et al. [7] proposed the six components of design process; Norman and Draper [8] brought User-Centred Design methods; Cooper et al. [1] built Goal-Directed Design methods. Furthermore, many companies began to set up their design process. IDEO Human-Centred Design Process has three main phases [9].

However, most design processes are just as part of the whole product development process. Missing designers at the early process will have a degree of its further negative influence on the whole process. Kwiatkowska et al. [10] indicated the gap between design and business, such as the lack of trust and partnership. Design methods for facilitating collaboration and communication with stakeholders (business stakeholders and developers) are needed [11]. Design participants should master interdisciplinary knowledge. Therefore, this paper attempts to put forward the design process from a macro point of view, to elaborate product design systematically. In the IDMF framework we proposed, design runs through the whole product development process to propose new creative designs based on emerging technologies to help users solve problems. The teamwork using IDMF framework can easily cooperate among design, information technology, and business.

3 The Interaction Design Method Framework (IDMF)

The IDMF framework is composed of 102 different design methods distributed to six phases. It involves interdisciplinary knowledge: design, information, and business. The IDMF framework is a comprehensive and systematic design method model and aimed to be utilized as a practical design tool with groups of team members entailing:

1. enhance the understanding and the methods used in Human-Artificial Systems design project;
2. develop and expand single units of methodologies or processes within the IDMF for future use, and
3. the methods established through experience and hands-on from real commercial projects can be used to develop better design methodologies for teaching purposes.

3.1 Six-Phase Model of the IDMF

The six phases of the IDMF are as follows (see Fig. 1):

Fig. 1. Six-phase model of IDMF

1. Market Research and Design Research. Analyze commercial markets, existing designs and application fields of emerging technologies to insight market need, investigate and understand the industrial situation.
2. User Research. Conduct a series of surveys and analysis on users to understand their needs, motivations and scenarios. Moreover, we could summarize user behavior patterns, figure out how intelligent products could meet their core needs.
3. Business Model and Concept Design. Build a business model to effectively deliver value to users. Define with the stakeholders what service we should provide to users. Create innovative concept design based on the business model, user scenarios, and emerging technologies.
4. Information Architecture and Design Implementation. Organize information to build an information architecture that can be understood clearly by the user. Then, design task flow based on the understanding of user behavior and draw the information layout and feedback mechanism through prototypes.
5. Design Evaluation and User Testing. To refine our prototypes, we get feedback from users and stakeholders. A series of evaluation frameworks and test methods we can use to find usability issues and then iterate the design.
6. System Development and Operation Tracking. Deliver design instruction to programmers and actively communicate with them. Track operation data to analyze user behavior and make changes for the next iteration.

The six-phased modules of the IDMF framework are designed to help design practitioners learn design thinking, understand the design process. The six phases of the IDMF framework are neither isolated nor linear, but interrelated, forming a spiral and iterative process of development. In practical work, practitioners don't have to follow the IDMF phase by phase but can see it as a small loop in the iterative process.

3.2 Three Perspectives of IDMF

A successful product is desirable, viable, and buildable. It must balance business and technology concerns with user concerns [1]. Design, business and information thinking are both important perspectives of successful product development. These three perspectives are represented as:

- Design: focuses on user model and product design. What user goals are, what do they need and how to meet them.
- Business: focuses on the business model and business plan. How to sell our service to target users, how to achieve profit and sustain a business.
- Information: focuses on content, data and information technology. What core technologies will be used and estimate technical feasibility.

It provides a structure for designers to grasp basic interdisciplinary knowledge and explore Human-Artificial Systems design challenges. In each phase of the IDMF, there are different phases from Design, Business and Information perspectives.

4 An Automotive AR-HUD Design Case Study

This section is to describe how to use the IDMF in Human-Artificial Systems design project at the anonymous auto company G.

Project Description. This project is to design an Augmented Reality Head-up Display (AR-HUD) for connected and automated vehicles (CAVs) with Advanced Driver Assistant System(ADAS) to improve driving safety and driving experience. This was a brand new design project. We cooperated with a business manager, a project manager and several developers from company G. Thanks to the IDMF, we designed the project process together and propose emerging technology-based designs.

Project Implementation Process (see Fig. 2).

Fig. 2. Company G's AR-HUD design process according to the 6-phase IDMF model (Fig. 1)

To make our study more manageable, we focus on Loop 1 of the project process.

Phase 1. Market Research and Design Research. The design methods we chose in this phase are Literature Reviews from Information perspective, Competitor Analysis from Business perspective and Critical Incident Technique from Design perspective. We understand CAVs' emerging technologies by Literature Review. Using Competitor

Analysis, we assessed the strengths and weaknesses of competitors and identify our opportunities. To collect problems driver encountered, we asked drivers to focus on Critical Incidents during driving. Then, we preliminary determined business objectives, product functions, and target users.

Phase 2. User Research. The design methods we chose in this phase are Questionnaire, Deeply Interview and Natural Observation from Information perspective, Storyboard and Scenario Design from Design perspective. Through Questionnaire on safe driving, we got rough factors that affect safe driving, users' demand and driving behavior. We concluded that the driving environment in China is complicated and drivers are easily distracted. To better understand drivers' need for safe driving, we interviewed nine drivers and sort out typical driving scenarios and difficulties. We learn drivers' real behavior and reaction by Natural Observation. After that, we designed a storyboard described what information or functions users expected. It was used to support our insight during meetings with company G. By using Scenario Design, we created possible scenarios in detail to make the story more lively.

Phase 3. Business Model and Concept Design. The design methods we chose in this phase are Brand Positioning from Business perspective, Affinity Diagram from Information perspective and Blueprint from Design perspective. We recognized that safety is the most crucial to users so that our product to "safety" for the brand positioning. We switched drivers' needs into notes and diagrammed according to their affinities. These came to four main information needs: traffic information, environmental information, vehicle status, driver status. Together with company G, we draw a blueprint to map the entire driving experience to find opportunity points.

Phase 4. Information Architecture and Design Implementation. The design methods we chose in this phase are Organization System Design from Information perspective, Layout Design, Paper Prototyping, Interface Style Design and Interface Component Design from Design perspective. A clear Organization System could help drivers find needed information immediately. We divided information into three categories: Core information is about safety, such as ADAS; Secondary accessibility information is personalized, such as navigation; Random information is real-time information about the road and surroundings based on networking and big data. To reduce driver's workload, we divided the Layout of HUD into three fixed blocks: warning area, normal area, and auxiliary area. Paper Prototyping allowed us to simulate the flow of information switching quickly. For testing design in a real situation, we designed several sets of Interface Style and Interface Component.

Phase 5. Design Evaluation and User Testing. The design methods we chose in this phase are Cognitive Walkthrough and Usability Testing from Design perspective. Driver's attention is valuable during driving, especially for novice drivers. We used Cognitive Walkthrough to identify how easy it is for novice drivers to obtain information with the AR-HUD. In addition, we conducted a Usability Testing on a driving simulator to figure out usability issues. Although most of the participants completed the tasks successfully, there were several things we need to improve in the next loop.

Phase 6. System Development and Operation Tracking. Since the design of Loop 1 needs to improve, we did not enter Phases 6. We just into Phase 4 of Loop 2.

From the automotive AR-HUD case study, the usability and feasibility to apply IDMF framework to a Human-Artificial Systems design had been verified.

5 Conclusion

This paper presented a systematic and comprehensive Interaction Design Method Framework (IDMF) for Human-Artificial Systems. In order to facilitate designers work in close collaboration with other stakeholders and propose emerging technology-based designs, we deconstructed the IDMF according to the three dimensions, design, information and business, which offer basic interdisciplinary knowledge. We offer a practical guide to using the IDMF framework by providing a high-level summary of the automotive AR-HUD design.

Acknowledgements. This work was supported by The National Key Research and Development Program of China (No. 2018YFB1004903) and Shanghai Automotive Industry Science and Technology Development Fundation (No. 1717).

References

1. Cooper, A., Reimann, R., Cronin, D., Noessel, C.: About Face: the Essentials of Interaction Design. Wiley, Indianapolis (2014)
2. Moggridge, B., Atkinson, B.: Designing Interactions. The MIT press, Cambridge (2006)
3. Rogers, Y.: New theoretical approaches for human-computer interaction. Annu. Rev. Inf. Sci. **38**, 87–143 (2004)
4. Battarbee, K., Koskinen, I.: Co-experience: user experience as interaction. CoDesign **1**(1), 5–18 (2005)
5. Royce, W.W.: Managing the development of large software systems: concepts and techniques. In: 9th International Conference on Software Engineering, pp. 328–338. IEEE Press, Los Alamitos (1987)
6. Boehm, B.W.: A spiral model of software development and enhancement. Computer **21**(5), 61–72 (1998)
7. Zimmerman, J., Forlizzi, J., Evenson, S.: Taxonomy for extracting design knowledge from research conducted during design cases. In: Proceedings of Futureground (2004)
8. Norman, D.A., Draper, S.W.: User Centered System Design: New Perspectives on Human-Computer Interaction. L. Erlbaum Associates, Hillsdale (1986)
9. IDEO.ORG. http://www.designkit.org/human-centered-design
10. Kwiatkowska, J., Szóstek, A., Lamas, D.: Design and business gaps: from literature to practice. In: 2014 Mulitmedia, Interaction, Design and Innovation International Conference on Multimedia, Interaction, Design and Innovation, pp. 1–7. ACM, New York (2014)
11. Hansen, P., Järvelin, K.: Collaborative information retrieval in an information-intensive domain. Inf. Process. Manag. **41**, 1101–1119 (2005)

Study on Movement Characteristics of Fingers During Hand Grabbing Process

Zhelin Li[1,2], Zunfu Wang[1], Yongyi Zhu[1], and Lijun Jiang[1,2](✉)

[1] School of Design, South China University of Technology, Guangzhou, China
{zhelinli,ljjiang}@scut.edu.cn,
{350554889,845218740}@qq.com
[2] Human-Computer Interaction Design Engineering Technology Research Center of Guangdong, Guangzhou, Guangdong, China

Abstract. Grabbing objects is the daily behavior of the human hand. The joint angle and correlation during the grasping process are the main movement characteristics of the finger. In this study, the OptiTrack motion capture system was used to collect the data. The analysis shows that: (1) In the three grasping modes, except thumb, the angle of the middle joint on the same finger changes the most, the proximal joint is the second, and the distal joint is the smallest. (2) The movement of the joint of the thumb has the lowest correlation with other four fingers. The conclusions of this study can be used on human-computer interaction research, rehabilitation analysis, prosthetic design and bionic robot development.

Keywords: Finger · Grasping · Joint · Correlation · Angle

1 Introduction

The hand is one of the most important parts of human body and the main tool for human interaction with the objective world. With the continuous development of research on bionic manipulators and prostheses, the dynamics and kinematics of the hand have always been the focus of research on grasping behavior.

There have been many literatures on the grabbing behavior of the human hand. Yamaguchi [1] used electric angle meters to measure the flexion change of each joint of the fingers. Santello et al. [2] used the motion capture system to collect the bending angles of the 15 joint points of the hand (3 per finger), and found that the movement of the thumb and index finger accounted for 80% of the total range of motion. Eduardo [3] and others used the correlation coefficient to calculate the correlation of the angle change between each two fingers, and defined the correlation coefficient r of the two fingers greater than 0.7, which has a strong correlation, and establishes the function of the relationship between the two fingers through the linear regression equation, this function was used to guide the control of the data glove to the robot. Liu [4] and others analyzed the independence of each finger and joint by collecting the changes of finger joint and wrist angle under different grasping postures. Qiaofei [5] used CyberGlove 2 data gloves to collect the angles of the joints of the fingers, analyzed the correlation between the fingers, and analyzed the synergy of the finger grip posture.

W. Karwowski and T. Ahram (Eds.): IHSI 2019, AISC 903, pp. 576–581, 2019.
https://doi.org/10.1007/978-3-030-11051-2_87

This article will explore the effects of hand grabbing posture and the size of the object being grasped on the angle change and motion correlation between the joints of the fingers.

2 Experiment and Analysis Methods

Bullock [6] et al. pointed out that the five types of grab motions, Medium wrap, Precision disk, Lateral pinch, Tripod, and Lateral Tripod, accounted for 80% of the total crawl time. Referring to the classification method of grasping action in the literature, this paper selects the three kinds of grasping movements: Medium wrap, Precision disk and Tripod, according to the problem of shade from other fingers in the pre-experiment.

The literature [7] proposes that in the grasp of daily life, 55% of the objects have a size of at least one dimension greater than 15 cm, and such objects cannot be grasped through their longest axis. 92% of the objects weigh less than 500 g, and 96% of the grab positions are less than 7 cm wide. The minimum number of dimensions of the object grabbed was 94% of the total number of crawls. It can be seen that the smallest dimension of grasping may be the default and most trustworthy way. In this paper, three kinds of shape objects satisfying the grasping method are set for the three kinds of grasping method studied.

Related hand grab research uses data gloves and angle meters as devices for collecting hand joint data [3–5]. This invasive measurement method affects the movement of the finger and causes disturbance to the grasping action. Miyata et al. [8] used a motion capture system to collect the position of hand-attached reflective balls to analyze the structural characteristics of the human hand. Therefore, this paper uses the OptiTrack motion capture system for non-contact hand motion capture.

In this paper, three kinds of shape objects suitable for the grasping method are set for the three kinds of grasping methods studied. Each object has three kinds of sizes. The captured object is printed by a 3d printer, as shown in Fig. 1 (Medium wrap grabs a cylinder with a height of 10 cm and a diameter of 3 cm, 4 cm, and 5 cm respectively. Precision disk grabs a sphere with a diameter of 3 cm, 4 cm, and 5 cm respectively. Tripod grabs a cylinder with a height of 1 cm and a diameter of 3 cm, 5 cm, and 7 cm, respectively).

In this paper, the OptiTrack motion capture system is used for non-contact hand motion acquisition, as shown in Fig. 2(a). The OptiTrack system includes six ReFlex13 cameras (sampling frequency is 120fps, accuracy is 0.240 mm). As shown in Fig. 2(b), there are 24 reflective hollow glass balls (with a diameter of 6 mm and the ball is filled with reflective powder,a wristband with 4 reflective balls on the wrist).

A total of 11 right-handed male youths with an average age of 22 years (±4.15) were selected as participants in the experiment. Each of the tested hands was not injured or had other movement disorders, and the hands were marked with signs. Two points *A* and *B* at a distance of 20 cm on the table are used as item placement positions. The grasping process is as follows: the subject first straightens the arm and fully opens the hand; then begins to grasp the object at point *A* and raise it to the upper position for about 5 s; put the object directly back to point B, and then open the hand completely. Each grab gesture is repeated twice to complete the experiment. The process is shown in Fig. 3(a) to (d).

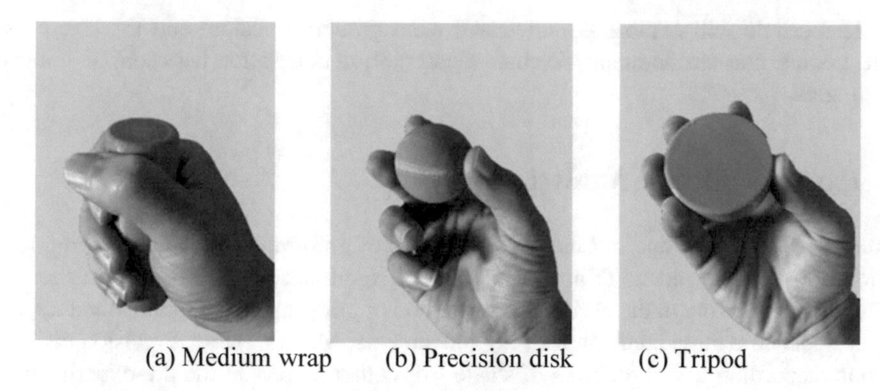

(a) Medium wrap (b) Precision disk (c) Tripod

Fig. 1. The objects and grasp type used

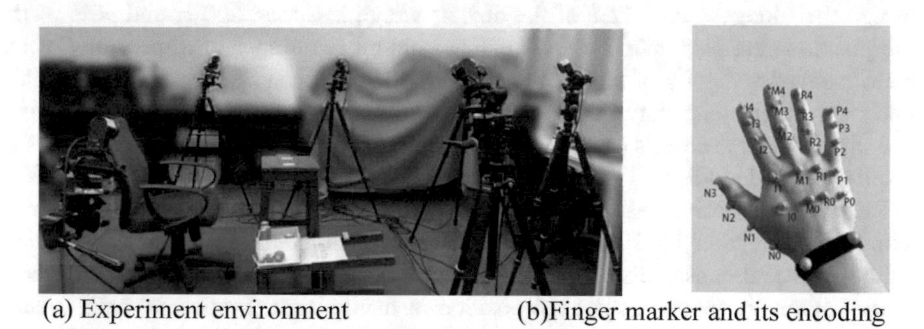

(a) Experiment environment (b)Finger marker and its encoding

Fig. 2. The experiment environment and device

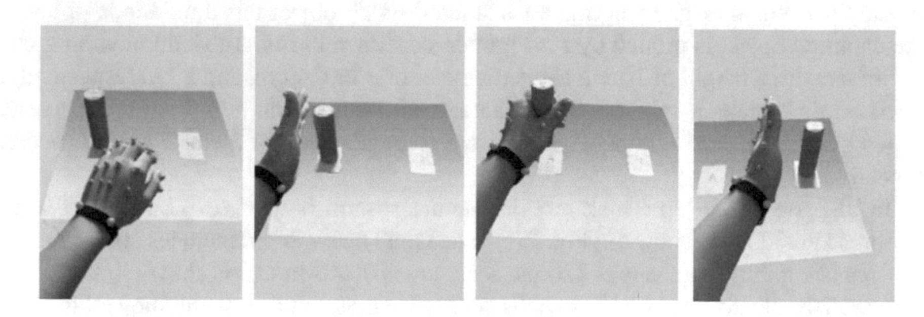

(a) Preparation (b) Starting posture (c) Grab A position item (d)Placed in position B

Fig. 3. Experiment procedure

By using the motion capture system, the angel of each joint at any time can be calculated, the angle data of each joint point in a certain grab process is sorted by time to form a set of data. Assuming that the two sets of data formed by P_A and P_B are A and B, respectively A_i and $\bar{A}$ are the values and the mean of the i-th element in A, respectively (B is also the case), and the correlation coefficient r between the two sets of data can be calculated by the following formula.

$$\gamma^{-} = \frac{\sum_{i=1}^{n}(A_i - \bar{A})(B_i - \bar{B})}{\sqrt{\sum_{i=1}^{n}(A_i - \bar{A})^2}\sqrt{\cdot\sum_{i=1}^{n}(B_i - \bar{B})^2}} \tag{1}$$

3 Result

In the above experiment, in each grabbing action every subject grabs three different sizes of objects separately, and each grabbing action of each subject is repeated once. A total of 198 ($11 \times 3 \times 3 \times 2$) valid data were obtained, and the coordinate information of each marker was extracted by OptiTrack supporting data processing software Motive (version 1.9). Each marker of the finger is defined in the manner as shown in Fig. 2(b), and the calculation is performed by Formula 1 to obtain the angle of each joint point in the grasping motion and the correlation coefficient between each two joint points.

The range of the angle of the finger joint during the grasping process can represent the range of motion of the finger joint in the process. Therefore, the range of the joint angles of all the grabbing processes is averaged, and the finger joints at the time of grasping can be obtained, as shown in Fig. 4.

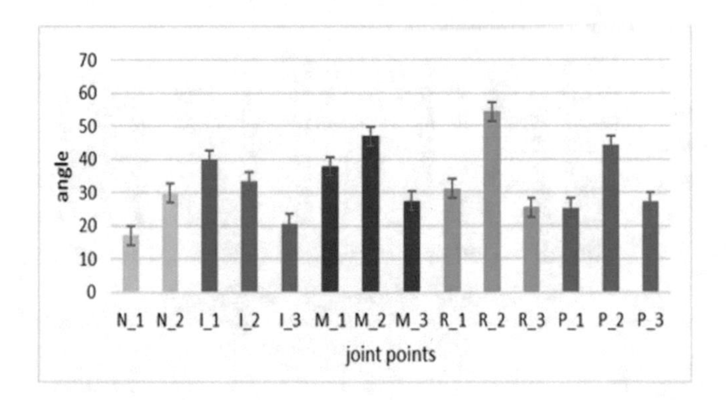

Fig. 4. Range of each finger joint

According to Fig. 4, in the selected grasping mode, the second joint of the middle finger, the ring finger and the little finger has the largest range of motion; the proximal joint of the thumb has the smallest angle of motion; For the remaining four fingers except the thumb, the distal joint has the smallest range of motion on the same finger, and the middle or proximal joint has the largest range of motion.

In order to investigate the influence of the size of the grasped object on the correlation between the joints of the fingers, the correlations of the same object in the same grasping posture are averaged together, and Fig. 5 is obtained.

In Fig. 5, the abscissa represents every two joints, the ordinate represents its correlation, and the medium wrap_1 represents the entire gripping process when the grasped object in the medium wrap mode is the first size, and the others are also the same. The three graphs (a), (b), and (c) are the correlations between the fingers and the remaining fingers after the joint correlation is averaged by fingers. As can be seen from the above figure, when the captured objects are in three different sizes, the correlation between the joint points is not much different. Throughout the grasping process, the correlation trends of joint points under different sizes are consistent. It shows that under the same grasping mode, the size of the grasped object has little effect on the correlation between the joints of the fingers.

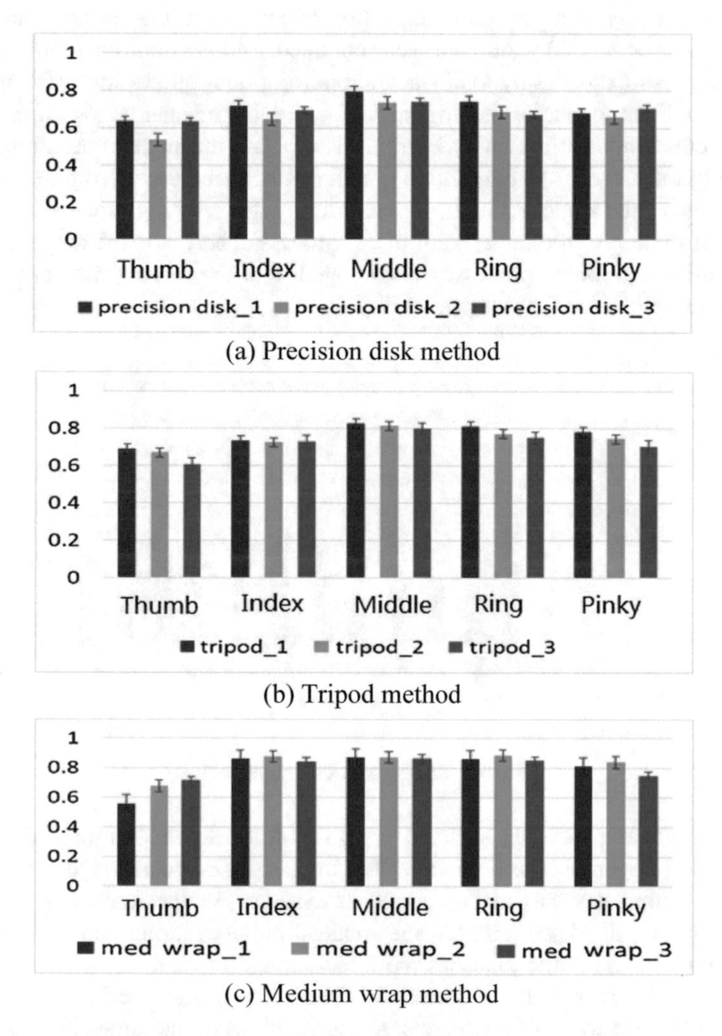

(a) Precision disk method

(b) Tripod method

(c) Medium wrap method

Fig. 5. Finger correlation between the fingers in each grab mode

4 Discussion and Future Work

Due to the structural characteristics of the human hand, the middle joint and the proximal joint on the same finger have a larger range of motion and the distal joint is the smallest. Regarding the contrast of the angle changes between different fingers, this also has a lot to do with the way of grasping. In different capture modes, the participation degree of each finger is not the same as its function.

When the objects of different sizes are grasped in the same posture, the initial state of each finger is completely consistent, and the posture at the time of stable grasping is also highly similar, except that the bending angles of the fingers are different. But in this process, the process of changing each finger is basically similar. Therefore, under the same grasping mode, the size of the grasped object has little effect on the correlation between the joints of the fingers.

Research on the characteristics of finger movement in the grasping movement requires a large number of data in different grabbing states for comparative analysis. Subsequent more experimental data will be collected to enhance the persuasiveness of the data. At the same time, different inter-group and intra-group variables are set to explore the effects of different gripping attitudes on finger angle changes, angular velocity changes, fingertip motion trajectories and changes.

Acknowledgements. This research is supported by the Fundamental Research Funds for the Central Universities 2017ZX013, and the Specialized Science Research Fund from Guangzhou Science Technology and Innovation Commission 201607010308.

References

1. Kamakura Moments: Finger kinematics. Phys. Med. Rehabil. **31**, 122–125 (1991)
2. Santello, M., Flanders, M., Soechting, J.F.: Postural hand synergies for tool use. Neuroscience **18**(23), 10105–10115 (1998)
3. Eduardo, G.N., Faanha, F., Francisco, A.P., Davi, J., Barreto, C.M.: Modeling by correlation of the human hand for application in the robotic hand. In: 2017 IEEE 6th Global Conference on Consumer Electronics (GCCE 2017) (2017)
4. Liu, Y., Jiang, L., Yang, D., Liu, Y., Zhao, J., Liu, H.: Analysis on the joint independence of hand and wrist. In: AIM, pp. 31–37 (2016)
5. Qiaofei, Z.: Analysis of the characteristics of human hand movement and mechanical realization. Huazhong University of Science and Technology, Wuhan (2015)
6. Bullock, I.M., Zheng, J.Z., De La Rosa, S., Guertler, C., Dollar, A.M.: Grasp frequency and usage in daily household and machine shop tasks. IEEE Trans. Haptics **6**(3), 296–308 (2013)
7. Feix, T., Bullock, I.M., Dollar, A.M.: Analysis of human grasping behavior: object characteristics and grasp type. IEEE Trans. Haptics **7**(4), 430–441 (2014)
8. Miyata, N., Kouchi, M., Kurih, T., Mochimaru, M.: Modeling of human hand link structure from optical motion capture data. In: Proceedings of 2004 IEEElRSJ International Conference on Intelligent Robots and Systems, pp. 2129–2135 (2004)

Research on Game Incentive Strategy Design of Highly Automated Driving Takeover System

Zhelin Li[1,2], Shanxiao Jiang[1], Yu Zhang[1], Lijun Jiang[1,2](✉), Xiaohua Li[1], and Zhiyong Xiong[1,2]

[1] School of Design, South China University of Technology, Guangzhou, China
{zhelinli, zhangyu, ljjiang, xili, zyxiong}@scut.edu.cn, mykakaqq@163.com

[2] Human-Computer Interaction Design Engineering Technology Research Center of Guangdong, Guangzhou, Guangdong, China

Abstract. *Objective* The objective of this paper is to study the effect of multilevel game incentive strategy on the takeover performance of highly automated driving takeover systems. *Method* According to the present research results of the automatic takeover system, the game incentive strategy of the automated driving takcover system is designed. Under the simulated automated driving system environment, the usability of the strategy is demonstrated through experiments, and the test data is analyzed. *Result* The application of game incentive strategies has improved both user takeover performance and user experience. *Conclusion* Design of game incentive strategy and interface can provide the basis for the human-machine interaction strategy design of the takeover system.

Keywords: Automated driving · Takeover system · Game incentive strategy

1 Introduction

With the gradual deepening and gradual application of automated driving technology research, safety issues in autonomous driving are receiving more and more attention. In the case of automatic driving at the L3 and L4 levels [1], the driver's automatic driving system cannot make autonomous decision, and the driver needs to take over the driving right again. People in the process of taking over have a significant impact on driving safety. Therefore, it is necessary to rationally design an automatic driving takeover man-machine interaction strategy to achieve safe and reliable driving experience switching with good user experience.

Louw et al. [2] believe that the design of the takeover system should plan better ways for drivers to participate more in the driver's mission, paying attention to vehicle status and road traffic scenarios, in order to maintain a good situational awareness and it is necessary to plan a better warning mode being easily accessible by the driver. Situational awareness explains how the driver manages the relationship between long-term goals of driving (reaching the destination) and short-term goals (avoiding accidents) [3]. Horswill et al. [4] clearly stated that situational awareness plays an important role in detecting drivers' potentially dangerous events in the traffic environment.

W. Karwowski and T. Ahram (Eds.): IHSI 2019, AISC 903, pp. 582–588, 2019.
https://doi.org/10.1007/978-3-030-11051-2_88

Among them, the game incentive mechanism has attracted the attention of many scholars. Schroeter et al. [5] applied gamification techniques (such as glory rewards, point rewards, privilege awards, extended ability rewards, etc.) to reduce the risky driving behavior of young men. The experimental results show that the driver can classify the preceding vehicles on the windshield display using AR technology to reduce the driver's impulse to overtake. Steinberger et al. [6] explored the gamification of driving missions. They develop and evaluate driving gamification applications. The results show that gamification conditions increase driver engagement and reduce driving speed. These studies provide a reference for introducing gamification methods in the automatic driving takeover system.

This study explores an interactive design strategy that maintains a balance between safety and experience. Some game elements are introduced into the autopilot takeover system to increase the fun of the takeover process.

2 Design of Game Incentive Strategy

2.1 Design of Game Incentive Strategy

The function of the game incentive strategy is mainly used to improve the driver's takeover motivation and increase the driving load, thus improving the situational awareness and satisfy the high-level user experience requirements, which can meet the user's emotional experience needs, social experience needs and personality. experience needs. The structure of the game incentive strategy includes three categories.

Takeover Behavior Scoring Mechanism The takeover behavior scoring mechanism is divided into five information elements, including the current takeover status, the current takeover time, the current takeover score, the comprehensive takeover score, and the number of consecutive successful takeovers.

Windscreen Social Mechanism The windscreen social mechanism refers to simple interaction with other cars, drivers, environments, etc. on the car window directly through gestures or touching screens. The icons and titles in the social module can be obtained by raising the comprehensive takeover score, or by unlocking the achievement when completing the incentive task.

Mission Achievement Mechanism Mission achievement mechanism is divided into "Today's Mission" and "Task Achievement". "Today's Mission" includes "completed tasks" and "unfinished tasks." The purpose of these tasks is to encourage drivers to use automated driving system and windscreen social mechanism. By completing the tasks drivers can improve the overall score. "Task Achievements" also includes "Achieved Achievements" and "Unfulfilled Achievements". Achieving the achievements can unlock icons and titles in the social module.

2.2 Interface Design of Game Incentive Strategy

We add game elements to the interface design of the game motivation strategy to match the functional positioning of the strategy. The relevant interface information of the game incentive strategy does not belong to the warning information. In order to avoid excessive occupation of the driver's attention resources, no enhanced edge processing is performed. Figure 1 is the interface design of the takeover behavior scoring mechanism.

Fig. 1. Interface design of the takeover behavior scoring mechanism

3 Experimental Study

3.1 Experimental Hypothesis

This experiment contains one experimental hypotheses: In the simulated automated driving takeover system, applying the game incentive strategy can improve the takeover performance.

3.2 Experimental Subject

The experiment recruited 20 participants who had a driver's license and had basic driving knowledge and operational skills. The 20 subjects were randomly divided into Group A and Group B. Group A was a group without game incentive strategy, and Group B was the game incentive strategy group, with 10 people in each group. The driving experience and proficiency of each group of participants in the group were similar, and the proportion of men and women in each group was close to eliminate the influence of driving experience, proficiency and gender factors on the experimental results.

3.3 Experimental Environment

The experiment uses a self-developed simulation driving system. The hardware includes: a high-performance computer and a 46-inch LCD screen, a Logitech G27 steering wheel, a brake and throttle device, and a DXRacer e-sports chair. The software platform is developed by Unity3D, which can realize the main functions of manual driving, automatic driving and driving control. It can also simultaneously record the number of collisions, takeover time, takeover score and other functions related to testing related data. The simulated driving system is shown in Fig. 2(a). The automatic road section design in this test is based on the research results of Körber et al. [7], and is set to three types: low traffic density section, medium traffic density section and high traffic density section. The traffic density mainly include road environment, road shape, vehicle traffic density, pedestrian traffic density, traffic behavior, number of intersections, weather, etc. Figure 2(b) shows the high traffic density section for example.

(a) Virtual driving test experimental environment (b) High traffic density section

Fig. 2. Virtual driving test experimental environment

4 Result Analysis

4.1 Data Analysis

Data Analysis of Successful Takeovers The number of successful takeovers in the two groups is shown in Table 1.

Table 1. Number of successful takeovers in each group (times)

No.	1	2	3	4	5	6	7	8	9	10	Average
Group A	11	12	10	11	11	12	9	10	11	12	10.9
Group B	12	12	11	12	11	12	12	12	12	11	11.7

Anomalous data test was performed on the above successful take-over data, and the result showed that there was no abnormal value.

An independent sample T test was performed on the number of successful takeovers of Group A and Group B, given a significant level α of 0.05. The significance probability (Sig.) was 0.034, less than 0.05. Within the 95% confidence interval, group B was significantly lower than group A, demonstrating that the takeover behavior scoring mechanism improved the number of successful takeovers.

Data Analysis of Average Takeover Time The average takeover time of the two groups of control groups is shown in Table 2.

Table 2. Average takeover time of each group(s)

No.	1	2	3	4	5	6	7	8	9	10	Average
Group A	2.43	2.14	2.92	2.61	2.42	2.25	3.17	2.83	2.33	2.14	2.524
Group B	2.19	2.33	2.40	2.12	2.42	2.36	1.92	2.25	1.83	2.58	2.240

Anomalous data test was performed on the above average takeover time, and the result showed that there was no abnormal value.

An independent sample T test was performed on the average takeover time data of Group A and Group B, given a significant level α of 0.05. The significance probability (Sig.) was 0.046, less than 0.05. Within the 95% confidence interval, group B was significantly lower than group A, demonstrating that the takeover behavior scoring mechanism improved the average takeover time.

Data Analysis of Secondary Task Performance The secondary task performance of the two groups are shown in Table 3.

Table 3. Secondary task performance (points)

No.	1	2	3	4	5	6	7	8	9	10	Average
Group A	45	32	68	56	31	46	74	27	38	23	44
Group B	39	37	44	20	75	46	34	28	47	80	45

Anomalous data test was performed on the above secondary task performance, and the result showed that there was no abnormal value.

An independent sample T test was performed on the secondary task performance of Group A and Group B, given a significant level α of 0.05. The significance probability (Sig.) was 0.904, more than 0.05. Within the 95% confidence interval, there was no significant difference between the two groups, demonstrating that the takeover behavior scoring mechanism didn't improve the performance of secondary task.

Data Analysis of Successful Hits on DRT Target The number of successful hits on DRT target are shown in Table 4.

Table 4. The number of successful hits on DRT target (times)

No.	1	2	3	4	5	6	7	8	9	10	Average
Group A	1	1	0	2	0	3	0	1	0	0	0.8
Group B	2	0	1	6	1	3	1	4	1	3	2.2

Anomalous data test was performed on the above data of successful hits on DRT target, and the result showed that there was no abnormal value.

An independent sample T test was performed on the data of successful hits on DRT target of Group A and Group B, given a significant level α of 0.05. The significance probability (Sig.) was 0.048, less than 0.05. Within the 95% confidence interval, group B was significantly lower than group A, demonstrating that the takeover behavior scoring mechanism improved the number of successful hits on DRT target.

Data Analysis of Time of Sight Shifting The time of sight shifting in two groups are shown in Table 5.

Table 5. Time of sight shifting in two groups (s)

No.	1	2	3	4	5	6	7	8	9	10	Average
Group A	11.22	12.46	11.56	10.58	9.84	10.23	9.42	10.63	11.21	12.16	10.931
Group B	12.46	13.24	12.22	13.58	10.77	11.39	12.21	15.32	10.40	11.45	12.304

Anomalous data test was performed on the above data of time of sight shifting, and the result showed that there was no abnormal value.

An independent sample T test was performed on the data of successful hits on DRT target of Group A and Group B, given a significant level α of 0.05. The significance probability (Sig.) was 0.024, less than 0.05. Within the 95% confidence interval, group B was significantly lower than group A, demonstrating that the takeover behavior scoring mechanism improved the time of sight shifting.

4.2 Summary of Analysis Conclusions

The above experimental results show that the game incentive strategy will improve the takeover performance as well as be beneficial to the number of successful hits on DRT target and the increase of time of sight shifting. Therefore, the experimental hypothesis is established.

Through the analysis of the results of the questionnaire, the group participants with game incentives have higher takeover motives, so the number of heads up in the secondary task state increases, and the increase in the number of heads raises the situational awareness of the driver's non-takeover time, so that the driver always maintains certain situational awareness prevents excessive immersion in other recreational activities. At the same time, the increase in the number of head-ups proves that the game incentive mechanism effectively increases the driving load of the subject, which in turn reduces the driving load sudden increase at the takeover time, so that the situational awareness recovers faster, and the takeover performance is improved.

5 Conclusion

In this paper, the game incentive strategy is designed in the automated driving takeover system. By adding game elements, the takeover system is "gamified", converting the "need to take over" into "want to take over", and the scoring mechanism to improve the takeover performance is verified by experiments. In the evaluation of the five design elements in the takeover behavior scoring mechanism, in the automatic driving mode, the driver pays more attention to the current takeover status and the number of consecutive successful takeovers, and the current takeover score is second. Therefore, these elements can be focused on in the design process. The next step will be to verify the effectiveness of the interconnected social mechanism and the mission achievement mechanism.

Acknowledgements. This research is supported by the Fundamental Research Funds for the Central Universities 2017ZX013, and the Specialized Science Research Fund from Guangzhou Science Technology and Innovation Commission 201607010308.

References

1. Marinik, A., Bishop, R., Fitchett, V., et al.: Human factors evaluation of level 2 and level 3 automated driving concepts: concepts of operation. J. Hum. Mach. Syst. (2014)
2. Louw, T., Merat, N., Jamson, H.: Engaging with highly automated driving: to be or not to be in the loop? In: International Driving Symposium on Human Factors in Driver Assessment, Training and Vehicle Design (2015)
3. Cunningham, M., Regan, M.A.: Autonomous vehicles: human factors issues and future research. In: Australasian College of Road Safety Conference (2015)
4. Horswill, M.S., Mckenna, F.P.: Drivers' hazard perception ability: situation awareness on the road. J. Cogn. Approach Situat. Aware. Theory Appl. 155–175 (2004)
5. Schroeter, R., Oxtoby, J., Johnson, D.: AR and gamification concepts to reduce driver boredom and risk taking behaviors. In: International Conference on Automotive User Interfaces & Interactive Vehicular Applications. ACM, (2014)
6. Steinberger, F., Schroeter, R., Foth, M., et al.: Designing gamified applications that make safe driving more engaging. In: CHI Conference (2017)
7. Körber, M., Radlmayr, J., Bengler, K.: Bayesian highest density intervals of take-over times for highly automated driving in different traffic densities. J. **60**(1), 2009–2013 (2016)

A Study of User Experience in Knowledge-Based QA Chatbot Design

Rongjia Liu and Zhanxun Dong(✉)

School of Design, Shanghai Jiao Tong University, No. 800 Dongchuan Road, Minhang District, Shanghai, China
mcrk@outlook.com, dongzx@sjtu.edu.cn

Abstract. A chatbot is a computer program which can interact with humans through natural language. In this paper, we introduced Jiao Xiao Tong, an interactive question answering chatbot designed in assisting students and faculties in Shanghai Jiao Tong University to accomplish various campus activities such as setting remainders, answering school trivia questions, route navigation and etc. A two-part experiment was conducted to evaluate two aspects of chatbot user experience. The results show an 6.7% increase in user satisfaction rate using the chatbot in route navigation tasks and interests towards a more interactive approach. The study also suggests that users expect action cues when interacting with a chatbot. In this paper, we described a solution of knowledge-based QA chatbot design in a university context and discussed what level of optimizations was achieved and what remains to be done. This pilot study provides opportunities for further improvement on chatbots design and application.

Keywords: User experience · Chatbot · Question answering · KBQA

1 Introduction

With the evolution of artificial intelligence and virtual assistant, the use of chatbots are gradually arising in fields such as customer support, education, and entertainment. Most conversational agents or chatbots interact with users in text or speech and may offer different features or functions depending on their conversation platform. Our work in this paper focuses on evaluating the user experience design of a chatbot.

In this paper, we first introduced Jiao Xiao Tong, an interactive question answering chatbot designed in assisting students and faculties in Shanghai Jiao Tong University to accomplish various campus activities such as setting remainders, answering school trivia questions, route navigation and etc.

A two-part experiment was conducted among 30 students in Shanghai Jiao Tong University. The first is to compare the route navigation feature in Jiao Xiao Tong with a common map app, participants were asked to complete the same route-finding tasks using the two different methods. The second is to evaluate the effect of chatbot response time and prompts on user satisfaction rate using two beta versions of the chatbot, one with prompts and loading animation and one without. We then analyzes

W. Karwowski and T. Ahram (Eds.): IHSI 2019, AISC 903, pp. 589–593, 2019.
https://doi.org/10.1007/978-3-030-11051-2_89

the experiment results based on both the content of interaction and a detailed user survey, and finally we discussed future research directions in this area.

2 Related Work

Question answering (QA) systems can be seen as information retrieval systems which aim at responding to natural language queries by returning answers rather than lists of documents [1]. QA Systems can be generally categorized into factoids, restricted domain, text-based, knowledge-based and so on [2]. Previous work [3] has shown that providing a QA system with a dialogue interface would encourage the submission of multiple related questions. In recent years, the research and commercial use of chatbots is constantly increasing, tech companies like Google and Microsoft has developed bot frameworks to make chatbots more accessible by common developers and consumers, and it has much potential in a university context. Preliminary work has been done in exploring the possibilities of using chatbots as an undergraduate advisor in student information desk [4]. Users tend to make natural assumptions about the abilities and intelligence of a chatbot [5]. To build such complex and capable agents, more efforts need to be put into gathering and processing unstructured data, and the design of a natural user experience.

3 System Design

3.1 System Architecture

The Jiao Xiao Tong chatbot is designed for university students and faculties with intentions of it working as a campus assistant, helping users to get help and information in their daily campus life through natural language interactions in both text and speech.

The three main components of the chatbot are the data module, the bot module and the web module. The database is in the form of quadruple sets, including the entity, intent and relations in a natural sentence. The bot module utilizes Microsoft Bot Services and is in the form of a website which helps it to be accessible to more potential users.

The complete system architecture is shown in Fig. 1.

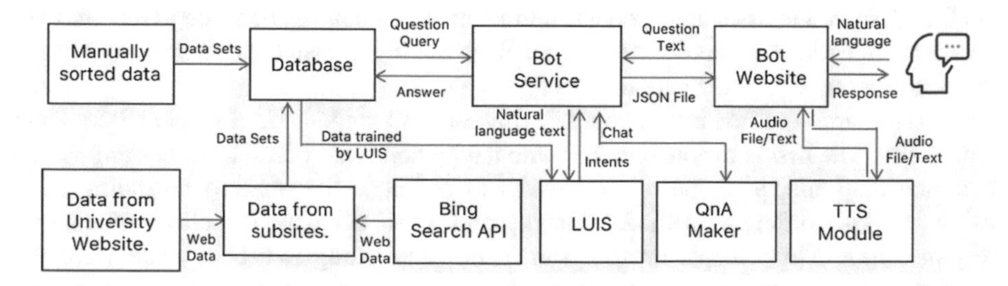

Fig. 1. System architecture of the Jiao Xiao Tong chatbot.

3.2 Chatbot Experience Design

In designing the user experience of the chatbot, we want to introduce as much multimedia approaches as possible as to increase the interactive factor, considering most chatbots in the market only provides a simple text response (Fig. 2).

a) Jiao Xiao Tong Website

b) Jiao Xiao Tong Chatbot Interface

Fig. 2. Jiao Xiao Tong website and chatbot interface.

The bot itself utilizes a series of cloud technologies provided by Microsoft Azure, including the Bot Service, Web Service, Azure Virtual Machine and Azure Cosmos Database. And it uses Microsoft Cognitive Services to provide speech-to-text and custom speaker recognition functions. It also provides multimedia responses such as image, audio, selections, etc.

4 Experiment and Discussion

4.1 Procedure

A two-part experiment was conducted among 30 students (17 female, 13 male) in Shanghai Jiao Tong University. The key objective of the experiment is to answer two questions in designing the user experience of a knowledge-based QA chatbot – one, whether users can successfully complete basic tasks with a chatbot, and two, to what extent do users expect the chatbot to offer response.

In part 1 of the experiment, after a brief introduction, participants were asked to first freely explore and interact with the chatbot for 5 min and then complete a route navigation task first with the chatbot, then with a common navigation mobile app – Baidu Map. With the chatbot, users can either text their destination or send a short recording of their voice command, the chatbot will then prompt the user to state their current location. Upon receiving a valid answer, the chatbot will respond with an interactive map showing a suggested route. Figure 3 shows the interfaces of both methods. The map API used in Jiao Xiao Tong is provided by Baidu Map so the search results should be the same.

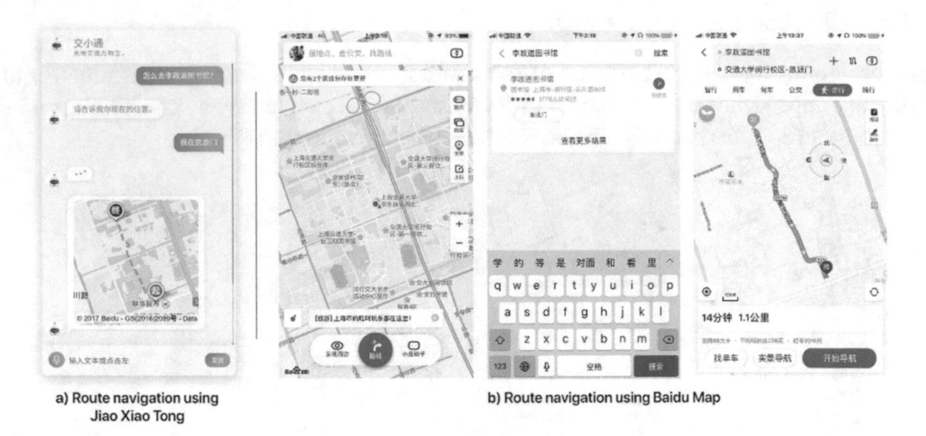

Fig. 3. Route navigation interfaces of Jiao Xiao Tong and Baidu Map.

In part 2 of the experiment, participants were asked to interact freely with two beta versions of the chatbot for 5 min, one with question prompts and loading animations and one without. Then the participants will ask two versions of the chatbot the same list of questions. Figure 4 shows the interface of the two chatbot versions.

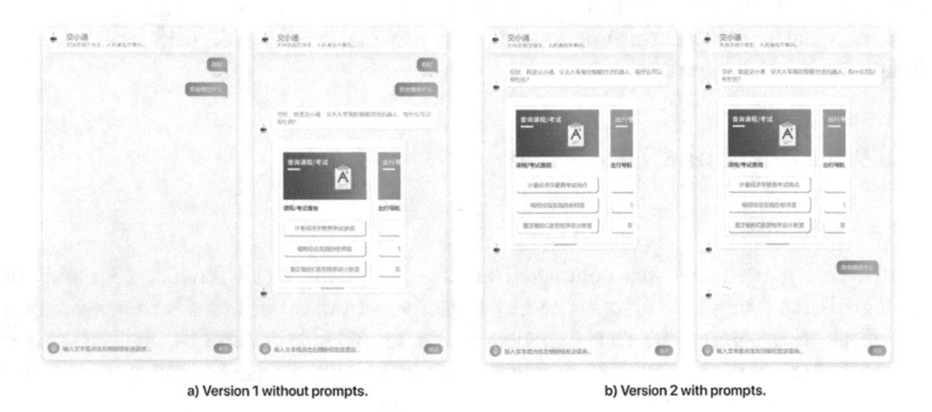

a) Version 1 without prompts.
b) Version 2 with prompts.

Fig. 4. Route navigation interfaces of Jiao Xiao Tong and Baidu Map.

User conversation and activities were logged, and finally all participants were asked to complete a user experience questionnaire.

4.2 Results and Analysis

The analysis is based on users' behavior, conversation logs and a detailed user survey, the results are as shown in Table 1. 93.3% of the participants successfully complete the route-finding task using the chatbot by themselves and the results show an overall 90% of satisfaction rate, 6.7% higher than Baidu Map. Most participants find it interesting to use conversational UI in route navigation tasks. Although Baidu Map shows a higher success rate, for most participants already frequently use similar navigation apps.

However, 93.3% of the participants are willing to continue using chatbots as their navigation assistant and show interests towards a more interactive approach.

Table 1. User satisfaction survey.

Objectives	Jiao Xiao Tong chatbot	Baidu Map
Successfully found route	93.3%	93.3%
Willingness to continue use	93.3%	80%
Satisfaction rate	90%	83.3%

The study also suggests that users expect action prompts when interacting with a chatbot and will lose patience after 4 s when not given prompt response. From our evaluation, we draw optimistic conclusions on the feasibility of interactive knowledge-based QA chatbot in a university context.

5 Conclusion and Future Work

In this paper, we described a solution of knowledge-based QA chatbot design in a university context and discussed what level of optimizations was achieved and what remains to be done. A two-part experiment was conducted based on both the content of user interaction and a detailed user experience survey. The results show 90% of user satisfaction when using Jiao Xiao Tong in route navigation tasks, 6.7% higher than using a common mobile map app, and interests towards a more interactive approach. The study also suggests that users expect action prompts when interacting with a chatbot and will lose patience after 4 s when not given prompt response.

Our future work will focus on studying user behavior when interacting with a chatbot, and conduct a more thorough evaluation of the interface design. This pilot study provides opportunities for further improvement on chatbots design and application.

References

1. Quarteroni, S., Manandhar, S.: A chatbot-based interactive question answering system. Decalog **83** (2007)
2. Mervin, R.: An overview of question answering system. Int. J. Res. Adv. Technol. (IJRATE) **1** (2013)
3. Hobbs, J.R.: From question-answering to information-seeking dialogs (2002)
4. Ghose, S., Barua, J.J.: Toward the implementation of a topic specific dialogue based natural language chatbot as an undergraduate advisor. In: 2013 International Conference on Informatics, Electronics & Vision (ICIEV). IEEE (2013)
5. Xiao, J., Stasko, J., Catrambone, R.: Embodied conversational agents as a UI paradigm: a framework for evaluation. Embodied conversational agents-let's specify and evaluate them (2002)

The Relationship Among the Optical Aspects of Photographic Composition and the Quality, Perception and Interpretation of the Realism in Virtual Images

Marcia Campos[1(✉)], Fabio Campos[2], Marnix Van Gisbergen[3], and Michelle Kovacs[3]

[1] Catholic University of Pernambuco, Recife, Brazil
spot4m@gmail.com
[2] Federal University of Pernambuco, Recife, Brazil
fc2005@gmail.com
[3] Breda University of Applied Sciences, Breda, Netherlands
Gisbergen.M, kovacs.m@buas.nl

Abstract. This work investigates the connection among optical aspects of photographic composition and the quality, perception and interpretation of the level of realism of images. Therefore, to investigate this connection, an experiment was carried out in two steps: The first step consisted of performing analyzes of the optical or photographic contrasts of previously selected images. The second step was the elaboration of a questionnaire with 19 images selected in the first step, aiming to collect data about the perception and opinion of the users. Finally, the objective data from the first step was crossed with the subjective data from the second. The conclusion indicates evidence of the connection or convergence between images that obey the principles of photographic composition that are perceived as having better realism by the users. It is plausible to consider the importance of photo-graphic theory to image design for the users perceive the images as more realistic.

Keywords: Realism · Photography · Perception · Virtual reality
User experience

1 Introduction

Games and other VR applications have been advancing in the implementation of virtual environments with increasing levels of realism. The role of the images and their design principles in this context is evident, affecting the quality of the realism achievable.

Despite the importance of the image, its process of conception and understanding is still ad hoc and intuitive. This is evident with the way the realism comes both in published studies and commercial VR products, where the care with the formal photographic principles of the images is usually absent, despite being deeply connected with human vision and its uptake and perception of reality [1].

W. Karwowski and T. Ahram (Eds.): IHSI 2019, AISC 903, pp. 594–599, 2019.
https://doi.org/10.1007/978-3-030-11051-2_90

This work investigates the connection among optical aspects of photographic composition and the quality, perception and interpretation of the level of realism of images.

Therefore, to investigate this connection, an experiment was carried out in two steps:

The first step consisted of performing analyzes of the optical or photographic contrasts of previously selected images for the elaboration of a questionnaire used in the second step of the experiment. These analyzes were done using a method that makes uses of aspects of photographic composition, via contrast theory [2–5]; this method makes possible to perform analyzes of the quality level of the representation of the optical contrasts of the photographic composition of the images.

In short, this phase consisted on the collection of technical data, ignoring subjective aspects of interpretation or rationalization of these images.

The second step of the experiment was the elaboration of a questionnaire with 19 images selected in the first step, aiming to collect data about the perception and opinion of the users regarding the level of realism of each image. In this stage we obtained a totally subjective data of the perception and interpretation by the users on the level of realism of each image.

Finally, the objective data from the first step of the experiment was crossed with the subjective data from the second, with the aim of verifying the convergence, or divergence, of the analyses of the images using both kind of data.

The conclusion of this study indicates evidence of the connection or convergence between images that obey the principles of photographic composition based on contrast theory with images that are perceived as having better realism by the users. Therefore, it is plausible to consider relevant the importance of photographic theory to image design if one aims to make the users perceive the images as more realistic.

2 Experiment – First Step

This step consisted on the analysis of 19 images, chosen at random from an image bank of 400 images consisting of virtual and real scenes with different levels of compliance with the optical or photographic principles of contrast. The analysis was carried out by a specialist in photography using the Campos method [2, 4], which makes it possible to consider only the light information perceived and analyzed at the technical level of the optical or photographic contrasts of the selected images.

According to the method [2], the optical aspects of photographic composition used in this first experiment were:

- Optical Aspect of Lighting: is the variation of the levels of contrasts between the presence of light (white) and the absence of it (black).
- Optical Aspect of Field Depth: understands the variation of sharpness contrasts of the image.
- Optical Aspect of Movement: corresponds to the variation of contrasts of the motion blur effect of the images captured with movement.

The analysis of the optical aspects mentioned above took place as follows:

1. Mapping of the image area with the perfect focus, once there will be the main object, which is reference for observation of the continuous contrast of the optical aspects of the scene [2–5].
2. Analysis of the optical representation quality of light propagation and mapping of the contrast level of the image illumination planning (always from the reference of the main object).
3. Analysis of the quality of the optical representation of the depth of field aspect and mapping of the contrast level of the image sharpness (always from the reference of the main object).
4. Analysis of the quality of optical representation of the motion aspect and mapping of the contrast level of the motion effect of the image (always from the reference of the main object).

The response model used a 7-points Likert scale, ranging from "poor optical quality" (−−−) to "excellent optical quality" (+++), with a "neutral option" "absence of representation of the optical aspects" (0).

In Table 1 are grouped the evaluation results of the 19 images from the first stage of the experiment, on the quality technical of visual representation of the optical aspects of photographic composition via its levels of contrasts

Table 1. Results from the first step of the experiment

Image	Quality of the optical aspects
Image 1	Absence (0)
Image 2	Excellent (+++)
Image 3	Good (+)
Image 4	Excellent (+++)
Image 5	Excellent (+++)
Image 6	Too Bad (− −)
Image 7	Excellent (+++)
Image 8	Too Bad (− −)
Image 9	Good (+)
Image 10	Excellent (+++)
Image 11	Absence (0)
Image 12	Excellent (+++)
Image 13	Excellent (+++)
Image 14	Excellent (+++)
Image 15	Absence (0)
Image 16	Good (+)
Image 17	Excellent (+++)
Image 18	Excellent (+++)
Image 19	Excellent (+++)

3 Experiment – Second Step

The objective of the second stage was to carry out a questionnaire to collect user perception data about the level of realism of the images used in it. The images were exactly the same as those used in the first step and the questionnaire was answered by 299 people select at random from a set of 4.000 Facebook "friends". Using a worst-case scenario model of infinite population, this sample would result in a Confidence Interval of 7.46% with a Confidence Level of 99%.

On the questionnaire was asked just to answer their opinion about the level of realism of each image using a Likert scale with seven points, ranging from "total realism (+++)" to "no realism (− − −)", and with a neutral point.

In Table 2 are shown the results obtained in the second stage of the experiment, in percentage of the respondents for each of the images realism categories.

Table 2. Results from the second step of the experiment

Image	NR (− − −)	WR (− −)	MR (−)	Neutral (0)	MR (+)	SR (+ +)	TR (+ + +)
Image 1	**50.7%**	**34.2%**	6.0%	6.0%	2.3%	0.3%	0.0%
Image 2	3.4%	11.8%	**21.9%**	5.7%	**30.6%**	**23.9%**	2.7%
Image 3	0.3%	23.9%	**13.4%**	5.4%	**32.6%**	**38.9%**	6.7%
Image 4	0.0%	0.0%	1.7%	1.0%	**14.4%**	**51.7%**	**32.1%**
Image 5	0.0%	1.0%	0.0%	1.3%	2.3%	**11.4%**	**83.9%**
Image 6	5.0%	**19.8%**	**28.5%**	4.4%	**32.9%**	7.7%	1.7%
Image 7	0.0%	0.7%	0.7%	2.7%	4.7%	**32.9%**	**58.1%**
Image 8	3.0%	**30%**	**27.6%**	4.7%	**27.6%**	6.4%	0.7%
Image 9	2.0%	**12.4%**	**26.8%**	5.7%	**27.9%**	**23.5%**	1.7%
Image 10	0.3%	0.7%	0.7%	2.0%	5.7%	**16.8%**	**73.8%**
Image 11	**15.9%**	**46.6%**	**24%**	6.4%	**10.8%**	1.0%	0.3%
Image 12	4.4%	2.7%	3.0%	3.0%	**13.8%**	**20.8%**	**52.3%**
Image 13	2.3%	5.0%	10.4%	2.0%	**24.5%**	**46.6%**	9.1%
Image 14	0.0%	0.3%	0.7%	1.0%	1.3%	**15.1%**	**81.5%**
Image 15	**50.3%**	**35.6%**	7.0%	4.0%	2.0%	0.7%	0.0%
Image 16	**12.1%**	**34.9%**	**32.2%**	5.4%	**12.4%**	2.7%	0.3%
Image 17	3.4%	6.1%	**12.5%**	3.4%	**24.7%**	**39.9%**	**10.1%**
Image 18	6.4%	5.4%	7.4%	4.7%	**19.5%**	**37.2%**	**19.5%**
Image 19	0.0%	0.0%	0.7%	1.0%	1.7%	7.7%	**88.9%**

4 Comparison of the Results from the Two Steps

The objective of the comparison of the results obtained in the two stages of the experiment carried out in this study, is the verification of the possible influence of the optical aspects of photographic composition, through the representation of its levels of contrasts [2, 3], in the user experience regarding its perception and interpretation of the level of realism in the images with real and virtual environments.

Discussion of the results:

- Observation 1: images 1, 11 and 15 were classified by the specialist as having "absence of representation of the optical aspects"; 87.5% or more of the users classified each of those images on the negative side of the Likert scale regarding perceived realism.
- Observation 2: images 6 and 8 were classified by the specialist as having "too bad optical quality" by the specialist; 53.3% or more of the users classified each of those images on the negative side of the perceived realism.
- Observation 3: images 3, 9 and 16 were classified by the specialist as having "good representation of the optical principles", from those images, images 3 and 9 received 53.1% or more responses on the positive side of the scale of perceived realism, and image 16 received 79.2% of the responses on the negative side of the perceived realism
- Observation 4: images 2, 4, 5, 7, 10, 12, 13, 14, 17, 18 and 19 were classified on the first step as having "excellent principles of optical quality", from this group we have one image (Image 2) receiving 57.2% of the responses on the positive side, and the other 10 images receiving 74.7% of more on the positive side (with 4 of them receiving 95.7% or more.

Scale and hypothesis tests show that observations 2 and 3 don't have statistical significance, therefore, only observations 1 and 4 will be considered from now on.

It should be noted as a hint for further studies that in most (4 out of 5 images) of the images of observations 2 and 3 there were a human figure, denoting a possibility of consideration of testing of this variable in further studies.

Therefore, observations 1 and 4 points to the correlation between the observance of the optical principles of photography, regarding the levels of contrasts, and the perceived realism by the user.

5 Conclusions

The results obtained in the experiment carried out in this work suggests the close relationship between the quality of the image via the representation of the contrasts of the optical aspects of the photographic composition, with the user experience, in particular, the realism perceived by the users.

To support that assumption the two testing conditions which revealed statistical significance (observations 1 and 4), that is the images with absence of compliance and images with strong compliance with the optical principles of photographic contrasts, exhibited contrary and therefore convergent evaluations of their realism by users.

Images with absence of the optical photographic principles are bad evaluated in which regards their realism, and, on the other hand, images with good compliance to those principles are well evaluated for their realism.

For sure, further studies must be carried out to isolate other variables that could potentially influence this phenomenon. Nonetheless, this study supports the importance of the aspects of photographic composition through its contrasts, in the problematic of realism perception around the RV devices, to enrich the understanding of the relation of the user experience with the images, to deepen the knowledge of the professionals of the RV industry on the image itself and its main optical aspects and its impact on the level of realism.

References

1. Heeft, M.: Virtual Realisms. NHTV Breda University of Applied Science. Master Media Innovation (2017)
2. Campos, M.M.M.M.: A Fotografia na Concepção da Imagem dos Games. Universidade Federal de Pernambuco tese Ph.D. (2014) http://www.ufpe.br/sib/
3. Birren, F.: The Elements of Color: A Treatise on the Color System of Jo-hannes Itten Based on His Book the Art of Color. Van Nostrand Reinhold, New York (1970)
4. Campos, M.M.M.M., Teixeira, C., Carvalho, B.: Aspectos Fotográficos na Construção de Marcas: estratégias de aproximação com o consumidor. In: II Congresso Internacional de Marcas/Branding: conexões e experiências 2015 – Lajeado – RS. Anais do II Congresso de Marcas/Branding: conexões e experiências, 2015 – Lajeado: Ed da Univates, p. 277 (2016)
5. Freeman, M.: The Image: Collins Photography Workshop Series. William Collins, London (1988)

Sustainable Competitive Advantages in the Industrial Service Business

Aappo Kontu[1(✉)], Jussi Kantola[1], Hannu Vanharanta[1], and Kaisa Kontu[2]

[1] School of Technology and Innovations, University of Vaasa (UVA), Vaasa, Finland
aappo.kontu@gmail.com
[2] Department of Built Environment, Aalto University, Espoo, Finland

Abstract. The industrial service business in Finland has developed swiftly during the last two decades, mainly because of outsourcing, market openings and European Union (EU) regulations. Throughout this time, transformations in the energy, telecom and forestry industries have created a new industrial services sector. This article analysis the changes and structuring of Finnish industrial service companies, their customers and business networks. In the research methodology, the following methods were used to find out the changes and transformations in the industrial service sector: financial analysis desk studies, direct questionnaires and interviews with customers and service companies. Based on research results, a new competitive advantage framework was developed. The framework contains four main conceptual processes: (1) Critical Competence Resource Planning Model, (2) Profitable Growth Model, (3) Market Analysis and Customer Proximity Model and (4) Service Business Development Model. In this paper, we report in detail regarding our conceptualised Critical Resource Plan.

Keywords: Competitive advantage · Enablers · Service science
Service economy · Industrial services business

1 Introduction

The industrial service business in Finland has developed and grown remarkably during the last two decades. The main reasons for these transformations have been market-opening trends, based in part on Finland joining the European Union (EU). In particular, electricity market law [1] has had a strong influence on the transformation, requiring the monopoly network business to separate from other business units in the energy utilities. The old business model, in which all functions – production, distribution, operation, maintenance and construction – were operated as internal services, was no longer efficient enough due to different business drivers.

Based on the transformation, the industrial services sector was born and developed. Most service companies operate in a multi-customer market environment and develop their services to meet market needs [2, 3]. In the service business, the key drivers are flexibility (both personnel and work tool resources), an efficient and mobile workforce,

W. Karwowski and T. Ahram (Eds.): IHSI 2019, AISC 903, pp. 600–606, 2019.
https://doi.org/10.1007/978-3-030-11051-2_91

customer proximity and a light balance sheet [4]. Moreover, margins (EBITDA) are low (3%–10%) with rather limited investments. These reasons are why service businesses need different business models, plans and management to asset-based businesses.

For the most part, these newly founded service companies were originally outsourced from electrical and/or telecom utilities at the start of the industry's transformation. After this service business foundation phase, from, a very active consolidation phase saw numerous mergers and acquisitions – a very fast growth phase in the whole service industry. In addition, internationalisation also occurred.

Today, the Finnish industrial service sector's total turnover is a few billion euros, and there are more than 10,000 employees. Many private or municipally owned service providers have been founded as well as novel service companies with new service models and products. The ownership has also diversified.

2 Problem Formulation and Research Objectives

This service industry transformation has dramatically changed the structures and competence needs of these companies, both in asset owner utilities and service providers. Nevertheless, the transformation of the industrial service sector has not received theoretical and academic research. This research has examined this transformation regarding both service companies and their customers (asset owners, outsourcers) in selected industrial electrical and telecom network services in Finland.

The problem formulation covers many qualitative and quantitative questions, as follows: What were the original reasons and objectives for this transformation, and have the targets been achieved? Have the targets changed. What has happened to the competence requirements? In this research, the following subjects were also studied: (a) competitive advantages, as viewed in strategy plans, (b) critical competence and resource requirements, (c) service providers' differentiation plans and actions against their competitors, (d) new service models and product development plans and resources, (e) the role of the authorities in the transformation, (f) influences on ownership changes, as well as (h) digitalisation, or plans in business development.

This research covered questions and subjects concerning the future objectives and the main goal was to produce a plan to develop the framework and tools to create a sustainable competitive advantage in the industrial service business. In this paper we report the answers to the above questions and give detailed findings concerning the Critical Competences and Resource plan using VRIO modelling [5].

3 Research Strategy and Methodology

3.1 General approach

Substantial research has been published on the subjects of business competitiveness and a sustainable competitive advantage. Here, we have taken three definitions of *competitiveness* that provide the framework for this research:

- For a firm, *competitiveness* is the ability to produce the right goods and services of the right quality, at the right price, at the right time. It means meeting customers' needs more efficiently and more effectively than other firms [6].
- Competitiveness is constantly changing its features. Therefore, a presently competitive firm may not be competitive in five years' time. The best description for *competitiveness* could be a firm's ability to get customers to choose just the company's products instead of competing products [7].
- A competitive advantage exists if the profitability sustained is greater than rivals' and if there is an understanding and knowledge of whether the advantage comes from, for example higher prices, lower costs or a combination of both [8].

3.2 Research Design and methods

Sustainable Competitive Advantage Analysis Methods

In the literature, many business models and methods have been described to analyse the 'sustainable competitive advantage' of companies or businesses: external environment models (macro environment, PESTE [9], five competitive forces [10]), internal environment models (Value chain analysis [11], BCG matrix [12], VRIO resources [5]) and combination external/internal models (SWOT analysis [13]) and company performance measurement (accounting [14, 15]).

The main targets of the methods selected are to obtain an understanding and knowledge of how to create a sustainable competitive advantage by differentiation or cost advantage. We have selected the following methodology and tools to analyse competitive advantage: (a) financial analysis: Simple accounting measurement, growth and profit (EBITDA) rates; (b) Operational analysis by value chain analysis (supported by a SWOT and financial analysis) and VRIO analysis, based on the results of a value chain and financial analysis to define critical competences and resources.

Data Collection

Financial and operative data and other needed information of researched companies were collected by using publicly available data (financials and annual reports etc.). Beside these analysing methods, quantitative and qualitative data of both customer and service companies as well as the service industry were asked by three different questionnaire forms and in-depth interviews of. The number questions were 20 in the customer questionnaire, 13 in the industrial service business questionnaire and 24 in the service company questionnaire.

4 Research Results

4.1 Customer Survey

Customer Survey included 15 electrical, three telecom network companies/utilities and three individuals. The number of respondents were 25, covering more than 70% of the total market [16]. The results show that the service outsourcing created remarkable, immediate efficiency improvements/cost cuts in five to 10 years in 20% to 50% of all companies. The market is expected to continue working, with prices still dropping. Network companies are very satisfied with outsourcing. After five to ten years of the experience of the outsourcing the satisfaction rate improved on the average from 2.5/5 to 4/5. The service market has been created and developed. The most important evaluation criteria for service providers are price (80%–90%), quality, competence, safety and reliability, whereas being local and Finnish, solvency and language are of minor importance. In the future, outsourced service packages will grow, and new business models will be developed (e.g. networking, alliance). Authorities' roles have met and experienced positively and they have been in key roles and drivers in this business transformation.

4.2 Business Survey

Business Survey included 18 industrial service companies and four individuals [16]. The result shows that the service sector believes the future involves growth, bigger service packages, networking, digitalisation, consolidation, internationalisation and tough competition, all of which continues with low margins. Through differentiation, innovative services, cost efficiency and customer proximity, it is possible to create sustainable competitive advantages. Critical success enablers are continuous business development, engaging management/personnel and profitability, but not growth. By taking care of critical competences, customer surveys and work safety, it is possible to retain sustainable competence. Service companies that are part of the energy group clearly harm the service industry market. Most of the recommended service company owners are management, private equity (PE) and public owners but municipalities are less favourable.

4.3 Company Survey

The company survey included questionnaires to 19 service companies (18 respondents), comprising electrical, telecom, district heating, industrial and ICT services. They are more than 70% of the industry [16–18]. The results show that the surveyed service companies' growth has stopped (see Fig. 1). Moreover, their profitability (EBITDA) has dropped (see Fig. 2), despite their total dramatic market growth (30%–50%) and service companies' efficiency improvements (2%–3%/year, totalling 10%–30% over 10 years). Market service prices have dropped continuously and new competitors have appeared.

All companies use a strategy process, but targets have not been achieved. All companies have growth targets. SWOT, unit costs, customer surveys and developing competences are the main tools for exploring critical success factors. However, value chains, the BCG matrix and the VRIO model have not been used. Very low, marginal investments have been made in business and service development. Monthly profit reviews are in; however, new service thinking is not a high priority. When developing competitiveness, it is critical to take care of key competences, profitability, customer proximity and new services. Nevertheless, these are not high priority today.

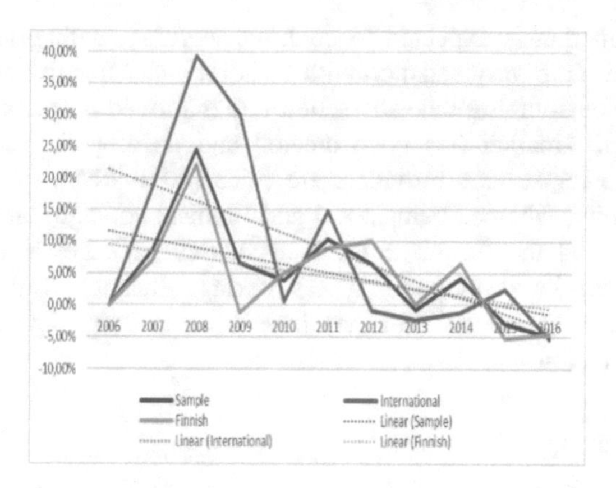

Fig. 1. Service companies' (10 pcs.) growth rate

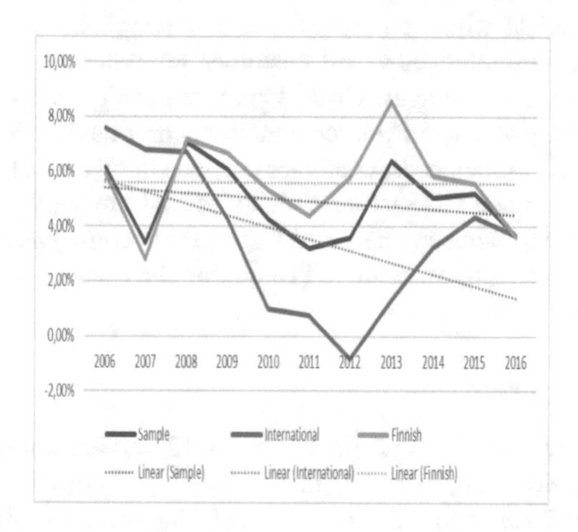

Fig. 2. Service companies' (10 pcs.) profit (EBITDA) rates

5 Discussions and Conclusions

5.1 How to Achieve a Sustainable Competitive Advantage in Industrial Service Companies

This study's results show that service companies need added development process concepts to achieve a sustainable competitive advantage. The process framework is separated into the four subprocesses described in Fig. 3. The biggest handicaps and development potentials in the 20 service companies studied are: a systematic Critical Competence Resource plan, a Customer Proximity plan jointly with customer and Service Development/Differentiation plans and investments.

Fig. 3. Sustainable competitive advantage process chart

5.2 Critical Competence Resource Plan

The VRIO tool has been used to identify Critical Competence resources. See Fig. 4, VRIO Competitive Resource test framework. In the first step, VRIO resources were identified through a value chain analysis supported by SWOT analysis and financial analysis by breaking down critical business Through the use of the VRIO test for all selected competences, they create sustained competitiveness or are at temporary or competitive disadvantages.

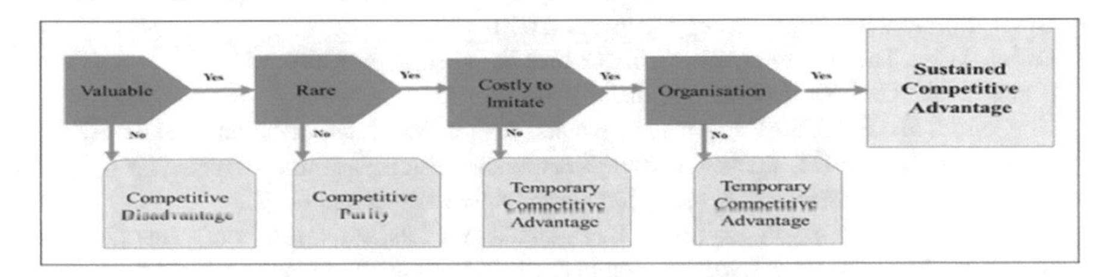

Fig. 4. VRIO competitive resource test framework

In the second step, the company needs to build tools and means to exploit these resources based on its strategy, management, motivation and reward systems. In the third step, these VRIO resources, which lower costs and/or differentiate our products or services, have to protect and sustain motivation. New ideas should be created to make products and services more difficult and costlier to imitate. In the fourth step, companies should constantly review their VRIO resources and capabilities in the whole organisation – the required competences and resources are changing over time.

5.3 Discussions and Future Research

Industrial service companies in Finland have not found sustainable competitive advantages in their businesses. Despite a growing service market, customers' purchasing power is continuously very strong, and the studied service companies have not found profitable growth. New players have taken over market growth, and the studied service companies have not developed new service products and models and or differentiated services. In this survey, we made four proposal process concepts for industrial service companies regarding how to create sustainable competitiveness. Our plan is now to develop and build work tools and systems (ICT) to take over and manage these four sustainable competitive advantage processes – the Critical Competence Resource plan, Profitable Growth plan, Market Analysis/Customer Proximity plan and Service Development plan.

References

1. Electricity market law 386/1995, Finland
2. Järventausta, P.: Ostopalvelujen käyttö verkkoliiketoiminnassa. Technical University of Tampere, Finland (2009)
3. Makkonen, H., Olkkonen, R., Partanen, J., Tahvanainen, K.: Palvelusuhteiden ja verkostojen johtaminen jakeluverkkotoiminnassa, PAVE-hankkeen loppuraportti. Technical University of Lappeenranta, Finland (2012)
4. Kontu, A.: Palvelumarkkinoiden kipupisteet. Network seminar Finnish Energy (2012)
5. Barney, J.: VIRO Framework in strategic management and competitive advantage (2010)
6. Edmonds, T.: Regional competitiveness & the role of the knowledge economy (2000)
7. Feurer, R., Chaharbaghi, K.: Defining competitiveness: a holistic approach (1994)
8. Porter, M.E. Competitive advantage: creating and sustaining superior performance (1985)
9. Grant, R.: Contemporary strategy analysis (2010)
10. Porter, M.E.: The five competitive forces that shape strategy (2008)
11. Porter, M.E.: Competitive strategy (1980)
12. Henderson, B.: BCG/Boston – matrix – stars, cash cows, dogs, question marks (1970)
13. Humphrey, A.: SWOT analysis – strengths/weaknesses/opportunities/threats (1960)
14. Barney, J.: Firm resources and sustained competitive advantage (1991)
15. Ikäheimo, S., Laitinen, E., Laitinen, T., Puttonen, V.: Laskentatoimi ja rahoitus (2011)
16. Kontu, A., Seppälä, R., Kantola, J., Vanharanta, H.: Sustainable Competitive Advantages in Industrial Service Business – Finnish Energy (2018)
17. Balance Consulting/Valor Partners Oy financial reports, annual reports, other public data of surveyed companies
18. Seppälä, R.: pro gradu thesis. The performance analysis of an infrastructure service industry. Vaasa University, Finland (2018)

Electronic Systems Designed to Guide Visually Impaired People in Public Areas: Importance of Environmental Audio Description

Eliete Mariani[1(✉)] and Jaldomir da Silva Filho[2]

[1] The São Paulo Metropolitan Company, Operation Management, Rua Vergueiro, 1200, 01504000 São Paulo, SP, Brazil
eliete.mariani@gmail.com
[2] Faculty of Architecture and Urbanism, University of São Paulo, Rua do Lago 876, 05508080 São Paulo, SP, Brazil
jaldomir@usp.br

Abstract. This research had as main objective to contribute with the design of assistive technologies to assist the locomotion of visually impaired people. We sought to investigate aspects of audio description in portable electronic devices with synthesized voice for the guidance in public environments, to in' public and mass transports. Previous research has shown that the use of electronic equipment with a specific format of audio description, with environmental information, may be of great importance for the autonomous mobility of visually impaired people. In this study, after bibliographic explorations, practical observations were performed in controlled and public environments in Sao Paulo city, using qualitative research techniques with visually impaired volunteers. The analysis of the experiments has demonstrated the importance of studying the phrases to be transmitted, with the purpose of correctly informing about the environment and the route.

Keywords: Visually impaired · Spatial awareness · Audio navigation Metro

1 Introduction

The people with disabilities comprise a portion of the population for whom forms of urban planning have a direct impact on their quality of life, with visual impairment being the most comprehensive among disabilities, affecting more than 35 million people in Brazil [1, 2]. In the last decades, Brazilian norms and laws to promote accessibility have trying to create conditions for people with disabilities to have a more active social and economic life using the precepts of universal design in accessibility projects for public spaces and means of transport. However, there is still a shortage of concrete and centralized information that guides architecture designers and product developers when it comes to accessibility [3, 4].

Some electronic technologies have been developed seeking universal design, among them the current audio description forms of TV programs, cinemas and theaters, as well as the development of synthesized voice technologies so that people with visual

W. Karwowski and T. Ahram (Eds.): IHSI 2019, AISC 903, pp. 607–612, 2019.
https://doi.org/10.1007/978-3-030-11051-2_92

impairments can make practical use of smartphones. For the concepts of audio description to also benefit orientation and mobility, it is necessary to carry the concept to an interface model that has as a principle to guide people with visual impairment, promoting their autonomy with security in public spaces [5–8].

This article aims to demonstrate the importance of readapting the environmental audio description to aspects of orientation and mobility, bringing principles for audio navigation design that can be applied in projects of portable electronic synthesized voice interfaces to guide people with visual impairment, thus seeking an increase in the autonomy of these people for their wayfinding in public environments and mass transportation.

2 Method

This research was preceded by surveys reported by Mariani [3], Silva Filho and Mariani [9], and was performed according to reports by Silva Filho [4], Mariani and Giacaglia [10]. An important part of the conclusion of these surveys refers to the verbal commands informed to the user, concluding that the information verbally passed in audio form should use references known and/or likely to be easily found in the environment. The research reported here was accompanied by the authors of this article, which here demonstrate the behavior of the user before the information "spoken" by the devices.

The qualitative research was carried out under the influence of Angrosino [11]. which recommends this type of research when it is interesting to have access to experiences in its natural context, and Flick [12], for whom qualitative research is indicated to better interpret and identify the user's needs, which cannot always be found in literatures.

The volunteers were invited based on the different personal characteristics, to verify if this would cause differences in the use of audio navigation. In total, seven volunteers with different physical characteristics and personal experiences were identified through semi structured or semi standardized interviews [12, 13]. Among the characteristics of each volunteer, the most important ones for the research were the knowledge of orientation and mobility techniques and familiarity with the use of smartphones, besides the type of visual incapacity. Participated in the survey related users in Table 1.

Table 1. Characteristics and experiences of volunteers.

Volunteer	Type of disability	Have orientation and mobility training	Guide dog user	Walks with autonomy	Smartphone user
RB	Acquired blindness	–	–	x	x
IC	Congenital blindness	x	–	–	x
EA	Congenital blindness	x	–	x	x
SE	Acquired blindness	x	x	x	x
RC	Congenital blindness	x	–	x	–
MR	Acquired low vision	–	–	x	x
DA	Congenital blindness	x	–	x	–

The volunteers received a free and informed consent term, guaranteeing their anonymity and previous knowledge of the research objectives [13].

Thus, data were obtained through participant observation [12] and [13] with visually impaired volunteers guided by portable electronic device in a route with predefined origin and destination, but whose route could be of voluntary choice.

The portable electronic device used was the NavGATe [9], an integrated system in which users carry information receivers transmitted to them from devices installed in the environment. The system provides the exact location instructions for user to follow. In this experiment, the NavGATe was programmed to guide the users from the disembarkation of the trains in the Vergueiro station until its exit to the Cultural Center of São Paulo.

Content analysis was done by reading the transcriptions of spoken interviews, testimonials and observation protocols. This material is summarized after treatment and classification, to guide the selection of the relevant information in the recorded transcriptions, as indicated by Gerhardt and Silveira [14].

3 Development

During the experiment inside the subway station, it was observed that some volunteers made changes of direction a little earlier than necessary, causing them to find walls or take wrong directions, performing the actions before the correct moment. For example, at some point when the "left" information was passed, the volunteers turned to the left immediately, but it would still require a further step or two to reach the right path to the left. These actions are observed with volunteers "RB", "IC" and "RC" in some locations of Vergueiro Station. The tactile floor did not prove to be a facilitator in this case.

At other times, the volunteer initiated an action before even hearing the complete information, passed through the audio of the device. The volunteers "EA", "RC", "MR" and "DA" anticipated the instructions, changing direction before the end of the sentence, in some places of the station. Like the first example, these volunteers reported that the doubt occurred because of the density of information needed to describe this location by audio, sharing the attention of the walking. This turned out to show that people tend to follow incomplete parts of the instructions as they get them, instead of listening to the full instruction before acting.

The observation of the practical experiments also allowed the understanding that it is desirable that the system be designed in a way that allows customization or differentiation between novice and advanced user [15].

4 Discussion

The experiments enabled the understanding that information by audio navigation is like visual information about environments. However, were observed perceptual differences of the volunteers related to the sequence with which these elements are perceived, which may be like the gestalt concepts, but in the sphere of auditory perception, since the cognitive abilities of vision tended to repeat themselves in auditory perception, with peculiarities inherent in the two different ways of perceiving the world.

The diversity of situations that can occur along the walk in public places is very great. The examples presented below illustrate some of the conceptual studies that must be carried out to define the messages spoken through a portable electronic de-vice to guide the visually impaired. Figure 1 shows one example course to be per-formed when the person wants to walk from point A to point B.

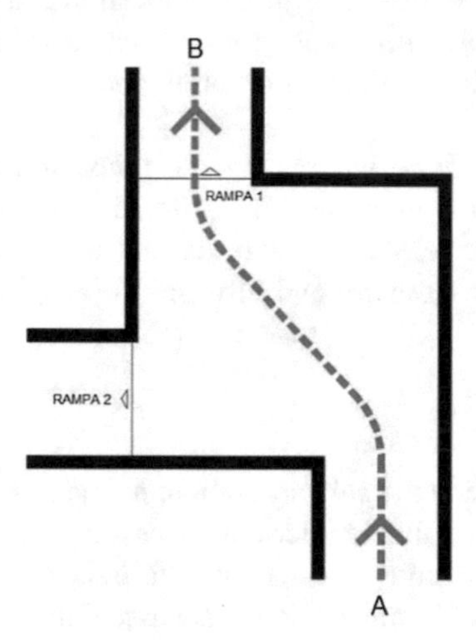

Fig. 1. Desired A-B course.

Using a hypothetical navigation system to guide one visually impaired person through the A-B course, when this person leaves the A point, the system talks the phrase "ramp to the left" (Fig. 2-1). However, since the person has no other reference than the sentence dictated by the system, that person may confuse the instruction and adopt a direction different from the one desired (Fig. 2-2).

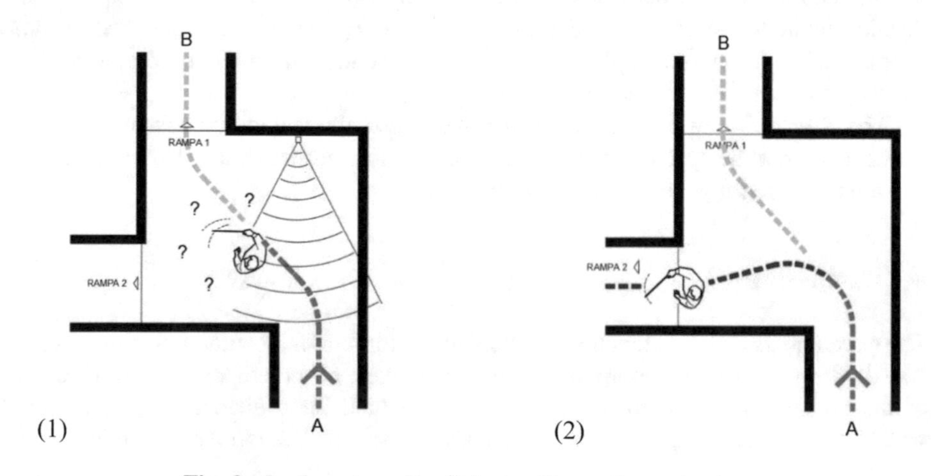

Fig. 2. Inadequate or insufficient guidance for navigation.

5 Results

The diversity of situations that can occur along the public places walk is very great. Through the research, it was observed that memory is the most required cognitive process in audio description interfaces, especially if it is used to navigate public environments. The fact that the visually impaired person is moving, occupying the attention process, causes his short-term memory to be occupied with the tasks of recognizing the elements of the environment. For this reason, the definition of spoken messages by means of a portable electronic device must be carefully elaborated by those who carry out their design.

The environmental elements to be informed, the content of this information, the extent and timing of each information, the moment in which the information will be started and the moment the individual will understand it must be studied. It should also be noted if the elements of the information will not be conflicting or dependent on other information, especially information that has not yet been made available to the user.

The verbalizations cannot conflict with any of the preexisting resources about the orientations that they indicate. In this way, a survey should be made of the characteristics of possible directions to be taken in the environment, places of interest, reference points, and interferences of environmental objects. Next to the elaboration of the verbalizations is the transmission time of the message, so also the importance of a logical sequence for its presentation, considering a hierarchy of presentation according to the user's interest and relating it all to the average speed of the walking of the people.

6 Conclusion

The audio description for navigation of people with visual impairment in public environments is little known and exploited, but of paramount importance for these users to walking. This research demonstrates the need to study precise forms of environmental audio description for the orientation and mobility of visually impaired people in public places where there is usually a high density of visual information. The simple sound reproduction of existing visual communication may not be enough for orientation, so, the designer needs to develop an information model that transcribes the environment without overloading the user's memory. The theoretical precepts demonstrated by the study of the cognitive processes and wayfinding that the visual information translated by the audio description allows the understanding of the environment, remaining the cognitive processing to people follow the desired direction.

References

1. Instituto Brasileiro de Geografia e Estatística – IBGE: Censo Demográfico 2010. Características Gerais da População, Religião e Pessoas com Deficiência. Compilation 2012. Brasil, Rio de Janeiro (2012). Retrieved 2014-06-13, from ftp://ftp.ibge.gov.

br/Censos/Censo_ Demografico_2010/ Caracteristi cas_Gerais_Religiao_Deficiencia/ caracteristi cas_religiao_ deficiencia.pdf
2. World Health Organization – WHO: Visual Impairment and Blindness. Fact Sheet N°282. [Internet - Updated August 2014]. Retrieved 2015-04-15, from http://www.who.int/mediacentre/factsheets/fs282/en/
3. Mariani, E.: Delineamento de Sistemas Eletrônicos para Guiar Pessoas com Deficiência Visual em Redes de Metrô (Guidelines for Electronic Systems Designed for Aiding the Visually Impaired People in Metro Networks). Master's Dissertation, Faculdade de Arquitetura e Urbanismo, University of São Paulo, São Paulo (2016). Retrieved 2017-02-28, from http://www.teses.usp.br/teses/disponiveis/16/16132/tde-02092016-151522/
4. Silva Filho, J.: Princípios para o design de audionavegação em ambientes públicos para pessoas com deficiência visual. Dissertação de Mestrado, Faculdade de Arquitetura e Urbanismo, Universidade de São Paulo, São Paulo (2017). https://doi.org/10.11606/d.16.2018.tde-26062017-115225. Retrieved 2018-09-27, de www.teses.usp.br
5. Rogers, Y., Sharp, H., Preece, J.: Interaction Design: Beyond Human–computer Interaction, 3rd edn. Wiley Higher Education (2013)
6. Hersh, M.A., Johnson, M.A. (Org.): Assistive Technology for Visually Impaired and Blind People. Springer, London (2008)
7. Virtanen, A., Koskinen, S.: NOPPA - Navigation and Guidance System for the Visually Impaired. VTT Industrial Systems. Fin-33101 Tampere, Finland (2014)
8. Yatani, K., Banovic N., Truong, K.N.: SpaceSense: Representing Geographical Information to Visually Impaired People Using Spatial Tactile Feedback. University of Toronto, Department of Computer Science. Microsoft Research Asia (2012)
9. Silva Filho, J., Mariani, E.: Research About Usability and Intuitiveness of the Synthesized Voice in Electronic Navigation and Guidance Unit for Visually Impaired People. 1º Congresso Brasileiro de Pesquisa & Desenvolvimento em Tecnologia Assistiva – CBTA, pp. 50–58. Curitiba, Brasil (2016)
10. Mariani, E., Giacaglia, M. E.: Guidelines for Electronic Systems Designed for Aiding the Visually Impaired People in Metro Networks, In International Conference on Applied Human Factors and Ergonomics 2017 Jul 17, pp. 1010–1021. Springer, Cham (2017)
11. Angrosino, M.: Doing Ethnographic and Observational Research. SAGE Publications (2007)
12. Flick, U.: An Introduction to Qualitative Research, 4th edn. SAGE Publications (2009)
13. Creswell, J.W.: Research Design: Qualitative, Quantitative and Mixed Methods Approaches. SAGE Publications (2014)
14. Gerhardt, T.E., Silveira, D.T. (Org.): Métodos de Pesquisa. Editora da UFRGS, Porto Alegre (2009)
15. Booch, G., Maksimchuck, R.A., Engle, M.W., Young, B.J., Conallen, J., Houston, K.A.: Object Oriented Design with Applications, 3rd edn. Pearson Education, Inc., Boston (2007)

Design Pattern as a Practical Tool for Designing Adaptive Interactions Connecting Human and Social Robots

Ke Ma and Jing Cao(✉)

College of Design and Innovation, Tongji University, Shanghai, China
suta@tongji.edu.cn, caojing.china@163.com

Abstract. We demonstrated the design pattern as a practical design method utilized for designing adaptive human-robot interaction for social robots. Our research distilled a pattern library to instruct designers and practitioners how to construct compound interaction patterns with interaction blocks. We also devised an iterative framework to investigate the implementation of interaction patterns in the workflow of intelligent robot systems. The authors developed a slew of social robots applying design pattern in adaptive HRI design and prototyping to showcase the advantages and outcomes.

Keywords: Adaptive Human-robot interaction · Design pattern
Social robots

1 Introduction

Social robots are increasingly emerging to the public regarding the technical advances in artificial intelligence and human-robot interaction (HRI). The research community has provided insights that social robots are supposed to survive in social circumstances by cultivating adaptive interactions with humans [1]. However, adaptive interactions require robots to perform sensible actions and reactions to humans towards complex scenarios. Issues are confusing designers for social robots to design suitable actions and responses for each scene and unexpected situations. It is therefore desirable to utilize practical design method to overcome the complexity. This research strives to address the challenge by applying design pattern as a tool to facilitate adaptive HRI design which empowers designers to simulate possible interaction patterns for complex social conditions.

2 Design Pattern Library

HRI researchers investigated to apply the design pattern to construct compound interactions as solutions for varying social scenarios [2, 3]. Design pattern is feasible to facilitate design exploration and prototyping for HRI. First, we defined a pattern library distilled from human-human interactions, human-robot interactions, and human-animal interactions to enable the adaptive selection and combination of interaction units. Considering different relationships between humans and robots, we classified the

W. Karwowski and T. Ahram (Eds.): IHSI 2019, AISC 903, pp. 613–617, 2019.
https://doi.org/10.1007/978-3-030-11051-2_93

patterns into three models: (1) human-generate-robot-react, (2) robot-generate-human-react, and (3) human-robot collaboration. The model differs from the intention of generators and corresponding expressions of reactors. For each model, we respectively defined interaction blocks (Fig. 1) as the essential constitutes which can be hierarchically integrated into more complex interactions. It means that the pattern library gives rise to basic units of interaction between two counterparts, and designers can select proper bricks to build a compound sequence of interactions according to conditional statements. The interaction block follows the "action-reaction" mode of two counterparts and the intermediated manipulations including modalities and interfaces which form an iterative chain between human and robot. For instance, the robot generates an inquiry to human via dialogue and human answers it through non-verbal gestures or additional social cues.

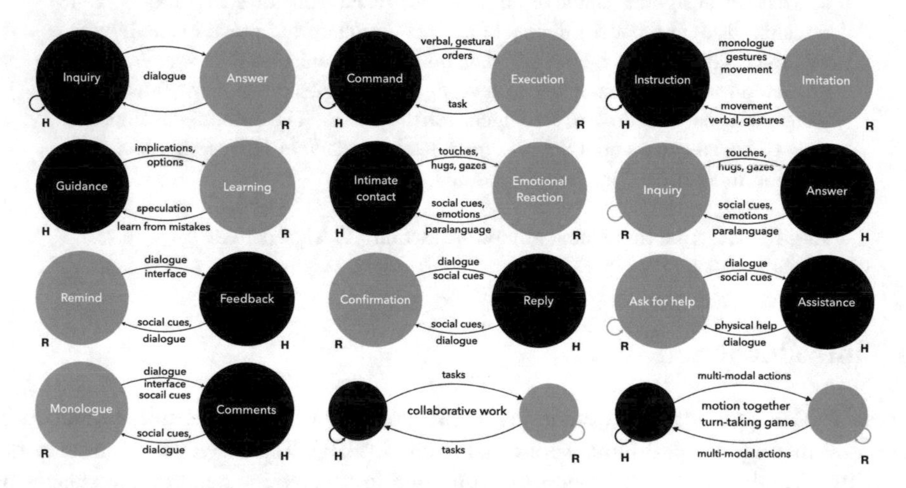

Fig. 1. The interaction blocks for the pattern library.

2.1 Human-Generate-Robot-React

In this mode, human users initiate an intentional request to robots and robots set out to execute specific actions to meet the goal.

- inquiry-answer: users propose an inquiry such as a question or confirmation for robots to answer. This unit commonly presents in the dialogue.
- command-execution: the request is an explicit command via verbal orders or non-verbal gestures. Robots could plan multi-modal actions to accomplish the task such as delivering an object to a designated place.
- instruction-imitation: users are willing to instruct robots to learn a specific task like a movement or a sentence. The intuitive response is to imitate and repeat the learning process and result.
- guidance-learning: it differs from direct instruction. Users would guide the robot to learn and respond through providing implications or options. Robots should speculate to find the answer and rapidly learn lessons from mistakes.

- intimate contact-emotional reaction: users are inclined to perform intimate contacts like touches, hugs, gazes with the robot regarding it as a lovely child or pet. In this situation, the robots react with emotional expressions.

2.2 Robot-Generate-Human-React

The robot is serving as the initiator to communicate with humans to acquire necessary information or feedback.

- inquiry-answer: the robot generates an inquiry for a specific need. The answer from users would help for decision-making and adaptive learning.
- remind-feedback: the robot reminds users to start or stop the task according to the user's command or built-in computational prediction.
- confirmation-reply: when the robot is not confident about the decision, a confirmation to acquire the agreement will help. Users can reply with verbal and non-verbal cues like a nod for approval.
- ask for help-assistance: the robot would be wise to ask for help as confronting problematic physical limitations. For instance, the robot can ask the user to help carry it onto the table.
- monologue-comments: when the robot is introducing its functions or descriptive information, users can wait and provide comments.

2.3 Human-Robot Collaboration

Human users and the robot are jointly collaborating to accomplish the assignment, movement, and game. This pattern is more likely to create physical interactions for company and entertainment.

- collaborative work on the same task: some specific assignment requires the robot and human to cooperate in finishing appropriate jobs.
- in motion together: it is a natural behavioral pattern that the robot moves with users as a follower or partner to show the close relationship.
- turn-taking in the game context: playing turn-taking game is a continuous activity, which demands simultaneous collaborations.

3 Iterative Framework

Next, we consider the issue of incorporating interaction patterns devised by the pattern library into the workflow of robot systems for the prototyping of adaptive HRI. Our research first reviewed the technical architecture from perception to action of intelligent robot systems (Fig. 2) to help designers understand how robots recognize input from humans and environments, afterward decide to execute reactions.

We proposed the iterative framework (Fig. 3) to look into the implementation of interaction patterns in the robot system. We classified the workflow into perception, representation, decision, and reaction in which any iteration follows the procedures and advised sub-steps. Taking the "human-generate-robot-react" model as an instance, we describe the essential modules below. First, whenever a user generates a request to the robot, the internal system begins identifying the user's presence and understanding verbal and non-verbal communications with the detected user input. With the recognized information, the next module represents the context, extracts the main tasks, and analyzes the underlying reasons to provide evidence for the decision-making. Importantly, the analytics determines the interaction pattern generated by human users, for instance, a command. The robot system will afterward designate the appropriate executions according to command-execution block in decision module. Finally, reactions are estimated by selecting modalities, generating multi-modal expressions leveraging the perceived metadata and transferred knowledge, and regulating temporal orders. An iteration ends up with the responses from the robot to the user.

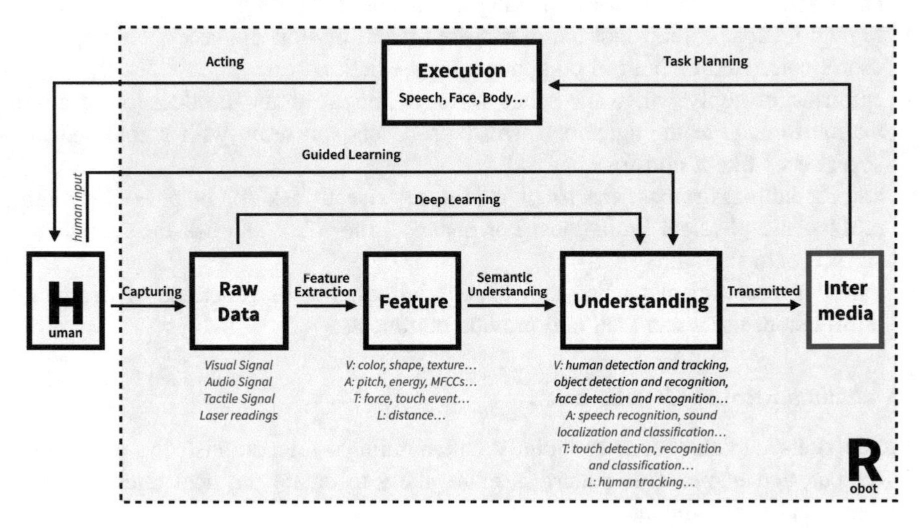

Fig. 2. The technical architecture of social robot systems from intelligent perception to action between humans and robots.

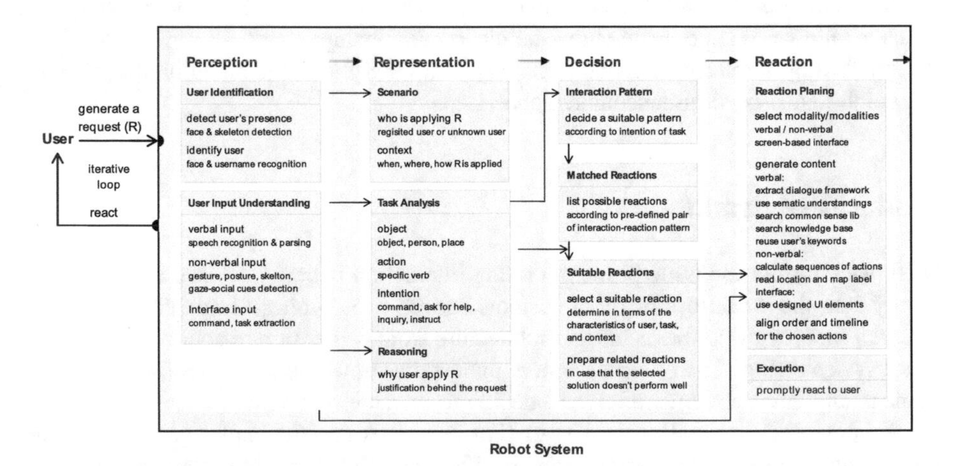

Fig. 3. The iterative framework of robot systems consisting of perception, representation, decision, and reaction. The workflow incorporates interaction patterns into the implementation of HRI and social robot systems.

4 Case

We developed two social robot prototypes (Neko and Chirp) that employ design pattern as a method for adaptive HRI design and iterative framework for computational prototyping of HRI. The two cases demonstrated the empirical evidence by constructing interaction patterns to solve the complex social interactions in context between humans and social robots.

Neko is a personal robot for taking care of plants. He handles advanced computer vision capabilities to identify users and objects for manipulating intimate interactions. He also cultivates an agile watering mechanism including humid measurement, watering inquiry, water absorption, and transmission. Neko acts as an emotional partner for people like both plants and robots. Chirp is a personal robot that can inspire children to promote skills and enthusiasm for playing pianos. We designed diverse musical games at different levels for children to learn music knowledge and practice playing techniques. Children and Chirp are cultivating an intimate relationship like either a tutor or partner.

We presented an online collection[1] of intelligent robot systems which utilized design pattern as a tool in the process of interaction design and prototyping implementation. Readers can access more details about the design and tangible interactions.

5 Conclusions

Consequently, this research presents to use design pattern as a practical tool for social robot designers to design and implement adaptive HRI in a systematic way. Pattern library and the iterative framework can facilitate designers to decrease the complexity and ambiguity of HRI design issues.

Acknowledgements. This research is supported by Shanghai Summit Discipline in Design (DB18116) and research project of Intelligent medical service design strategy for elderly cognitive behavior promotion (1400219033).

References

1. Breazeal, C.: Social interactions in HRI: the robot view. IEEE Trans. Syst. Man Cybern. Part C (Appl. Rev.) **34**(2), 181–186 (2004)
2. Kahn, P.H., Freier, N.G., Kanda, T., Ishiguro, H., Ruckert, J.H., Severson, R.L., Kane, S.K.: Design patterns for sociality in human-robot interaction. In: Proceedings of the 3rd ACM/IEEE International Conference on Human Robot Interaction, pp. 97–104. ACM (2008)
3. Sauppé, A., Mutlu, B.: Design patterns for exploring and prototyping human-robot interactions. In: Proceedings of the SIGCHI Conference on Human Factors in Computing Systems, pp. 1439–1448. ACM (2014)

[1] http://www.makeinteractions.com/hri.html.

Smart Materials and Inclusive Human Systems

Approach to a Design Guideline Regarding the Interaction of Shape Memory Alloys and Fused Deposition Modeling

Felix Oberhofer(✉), Andrea Hein, Daniel Holder, and Thomas Maier

Institute for Engineering Design and Industrial Design, Research and Teaching Department Industrial Design Engineering, University of Stuttgart, Pfaffenwaldring 9, 70569 Stuttgart, Germany
{Felix.Oberhofer, Andrea.Hein, Daniel.Holder, Thomas.Maier}@iktd.uni-stuttgart.de

Abstract. The combination of Shape Memory Alloys (SMAs) and Fused Deposition Modeling (FDM) is promising regarding individual and innovative products with controllable properties. The focus is on the intelligent integration of SMAs in FDM-manufactured parts or products. The objective is to establish a basis for a guideline regarding the interaction of SMAs and FDM. Therefore, important parameters of the combination of SMAs and FDM were derived using a basic classification of integration, which was identified, described and clustered. Using those findings, prototypes comprising flexible structures were designed, manufactured and analyzed in order to evaluate the established parameters and gather practical experiences.

Keywords: Design methods · Additive manufacturing · Fused deposition modeling · Smart materials · Shape memory alloys · User-centered design

1 Introduction

In the field of product development, a considerable number of technical innovations are due to new materials. Promising are above all Smart Materials (SM) as they generate innovative products through controllable properties and thereby realize an adaptable design. Especially SMAs are interesting regarding user-centered design [1]. To extend their use and application, research in the field of connection, integration and design methods is necessary. Considering the possibilities in product design by using SM and pending design questions, there is a strong link to the advantages of Additive Manufacturing (AM), for example functional integration, lightweight structures and design freedom. FDM is particularly interesting due to its process-related flexibility, robustness and integration of additional structures and components. However, there is still a considerable challenge regarding the design process [2]. A review of the literature has shown that studies on the combination of SMA and FDM have been conducted using one specific application or type of integration without a systematic investigation of a generally valid approach. The objective of the present study is to derive important

W. Karwowski and T. Ahram (Eds.): IHSI 2019, AISC 903, pp. 621–627, 2019.
https://doi.org/10.1007/978-3-030-11051-2_94

parameters of the combination of SMAs and FDM and to establish a basis for a guideline regarding the interaction of SMAs and FDM.

2 Shape Memory Alloys and Fused Deposition Modeling

Regarding material and process properties as well as areas of application, the combination of SMAs and FDM is a promising approach, whereby user-centered, individualized, adaptive and structurally complex components are particularly interesting.

SMAs are able to convert into a previously imprinted shape using thermal or electrical activation. They can be strained up to 8 % and are still able to return to their previous shape. This effect relies on the phase transformation from low temperature martensite to high temperature austenite. As forward and reverse transformation occur at different temperatures, there is a hysteresis between austenite start to austenite finish and martensite start to martensite finish temperature. The shape strongly depends on the thermomechanical treatment [3]. There exist three shape memory effects (SME) [4]. (i) To initiate the one-way SME an external force is needed to deform the shape within the martensitic phase. Upon heating, the material reconverts to its original shape. When cooled again no shape change emerges. (ii) The two-way SME incorporates a high and a low temperature shape. To undergo the effect the material must be trained. (iii) Pseudoelasticity (superelasticity) emerges due to stress induced martensitic phase transitions. In a certain temperature range, the transitions from austenite into martensite occur if a critical stress is reached. The component is deformed at high temperature. This effect therefore requires no temperature change [4].

One of the most common AM technologies is FDM [5]. Due to the melt layering process, it has several advantages over other AM technologies. Typical materials used are polymers, such as acrylonitrile butadiene styrene (ABS), polylactides (PLA) and flexible materials like thermoplastic elastomers (TPE) [6]. It is possible to print different materials at the same time, which allows producing multi-material topology optimized structures [7]. This is one major factor to design flexible structures. Another advantage is the possibility to create complex internal features and to consolidate a multi-part assembly into a single printable part to increase functionality and improve performance [5]. Thus, assemblies with movable parts, like hinges or joints, can be produced directly. In comparison to other AM technologies, FDM offers the opportunity to insert parts such as other materials, electronics or SMA wires at a defined point by interrupting and later continuing the building process [7]. This extends the possibility of function integration beyond the structural design.

In order to benefit from the advantages of the FDM process with regard to SMA, efforts in the area of design are necessary to synthesize functions, shapes, structures and materials of a construction with the AM process capabilities.

3 Method Regarding the Interaction of SMA and FDM

The main advantage of the combination of SMA and FDM is the possibility to generate flexible, moving structures, which adapt to changing external conditions. Nevertheless, the connection and integration of SMAs in structural systems is still a great challenge. Therefore, a basic classification of the integration was identified.

An essential aspect to fulfill the required task of a moving structure is the flexibility of structural components. Therefore, the design of those structures was analyzed. An extensive literature review showed flexible structures in the area of FDM and SMA. Analyzing those structures, we defined two design possibilities for the FDM structure: (i) using the elastic deformation of the material or (ii) generating a flexible connection (kinematic pair) through at least two moving components (non-contact). Table 1 shows selected and important examples of those structures. In general, it is possible to divide the detected structures in four categories: structures using flexible materials, geometric adaptions, combinations of the previous and joints.

Table 1. Classification of flexible structures and derived examples.

	Categories	Examples
Material	Flexible material	Soft morphing hand [8]: SMA & SSC (smart soft composite)
		Bending actuator [9]: SMA & PDMS (polydimethylsiloxane)
		Soft robot [10]: SMA & TPE (thermoplastic elastomers)
	Geometry	Textile-based structures [11]: ABS
		3D printing of living hinges [12]
	Combination	Prosthetic finger [13]: TPE with elliptic geometry & wire
Kinematic pairs	Joints	3D-Print joints [7]
		Plastic joints [14]

To combine SMA and FDM (Fig. 1) there are two general methods: (i) the direct (inserting SMAs during the printing process) and (ii) the indirect method (inserting SMAs after the printing process). Printing special materials like SM with a 3D printer wherein the objects change shape post-production is called 4D printing. This method will not be considered further due to the used SMA wire. Feeding SMA while printing has the advantage, that it is possible to optimally position the SMA within the product. If the feeding is performed automated, the print process can continue, otherwise, it has to be interrupted and disadvantages such as material adhesion and warping have to be considered. The indirect method is subdivided into two aspects, which mainly differ regarding the way of assembling. It is possible to print the finished product and later add the SMA or to use the printed piece as a mold and recast the SMA. In this paper, we considered feeding and assembling as a promising approach.

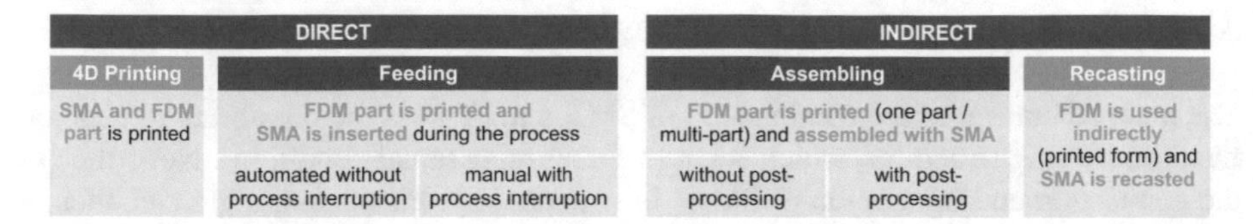

Fig. 1. Possible combinations of SMA and FDM.

Based on the general procedure for combining SMA and FDM, relevant parameters for the construction were derived and summarized in a matrix (Fig. 2) to set up the basis for a design guideline. The matrix includes parameters regarding FDM and SMAs, focusing on parameters required for flexible structures. The parameters were compared to reveal high influences, which were analyzed and described within the matrix, considering both: the dependences between FDM and SMA parameters as well as the effects of adapting parameters within the sections of FDM or SMAs.

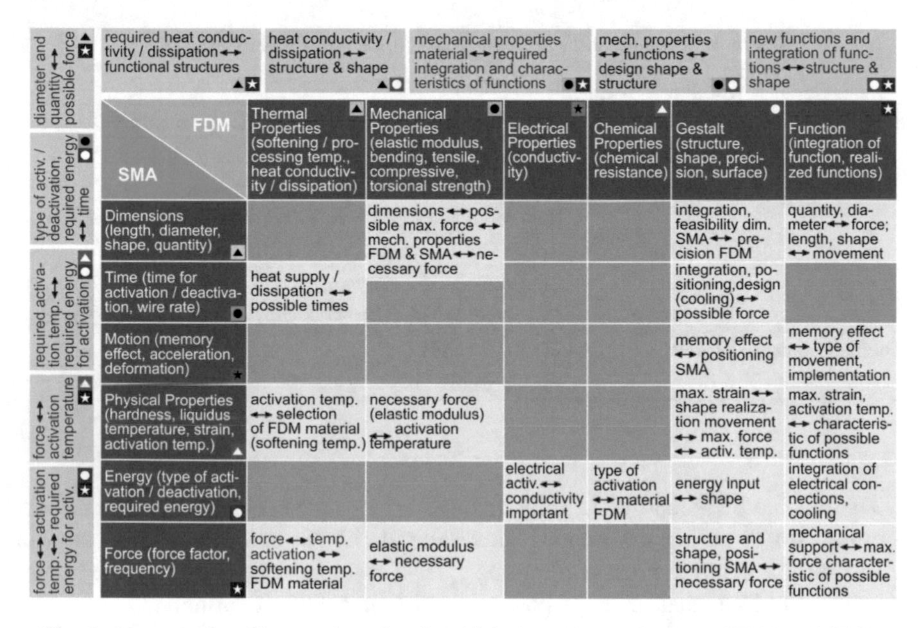

Fig. 2. Parameters of integration showing high dependences between FDM and SMA.

The matrix provides a guideline to design systems that integrate SMAs in additive manufactured components. An essential part are the derived parameters and their dependencies. Hence, it possible to estimate the effects on the system if a parameter is changed. The examination has shown that mainly physical properties (activation temperature), energy (type of activation/deactivation), force, gestalt (structure & shape) and function display multiple interactions and dependencies. Therefore, they have to be considered in particular. The further development of the approach covers the initial design of the product, the manufacturing process as well as the final assembling.

4 Experimental Study

Using those findings, prototypes comprising flexible structures were designed, manufactured and analyzed in order to evaluate the established parameters and gather practical experiences. Therefore, different basic design samples with focus on bending using the classification of flexible structures were developed (Fig. 3).

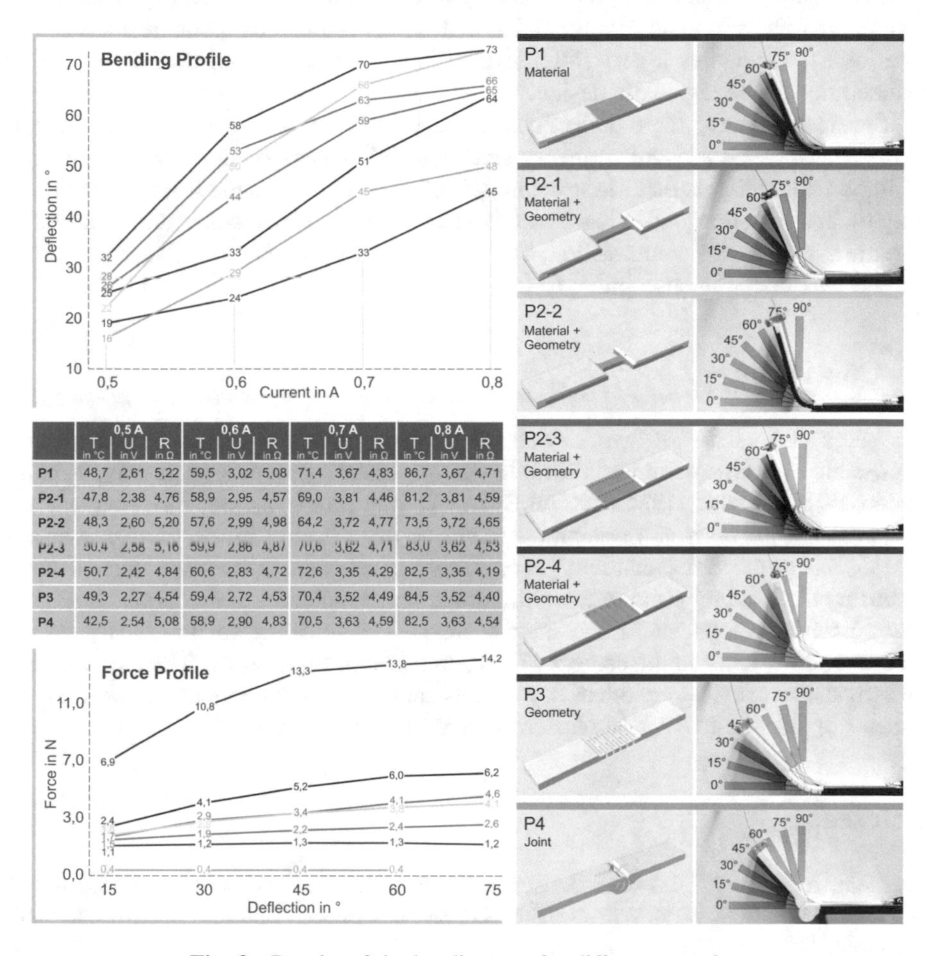

	0,5 A			0,6 A			0,7 A			0,8 A		
	T in °C	U in V	R in Ω	T in °C	U in V	R in Ω	T in °C	U in V	R in Ω	T in °C	U in V	R in Ω
P1	48,7	2,61	5,22	59,5	3,02	5,08	71,4	3,67	4,83	86,7	3,67	4,71
P2-1	47,8	2,38	4,76	58,9	2,95	4,57	69,0	3,81	4,46	81,2	3,81	4,59
P2-2	48,3	2,60	5,20	57,6	2,99	4,98	64,2	3,72	4,77	73,5	3,72	4,65
P2-3	50,4	2,58	5,16	59,9	2,86	4,87	70,6	3,62	4,71	83,0	3,62	4,53
P2-4	50,7	2,42	4,84	60,6	2,83	4,72	72,6	3,35	4,29	82,5	3,35	4,19
P3	49,3	2,27	4,54	59,4	2,72	4,53	70,4	3,52	4,49	84,5	3,52	4,40
P4	42,5	2,54	5,08	58,9	2,90	4,83	70,5	3,63	4,59	82,5	3,63	4,54

Fig. 3. Results of the bending test for different samples.

Sample P1 and the samples P2 consist of two different parts. The rigid part is made of ABS (Ultimaker), which has a high Vicat softening temperature of 97 °C (ISO 306). The flexible section consists of TPE (Verbatim Primalloy) with a Shore hardness of A85. The samples P3 and P4 are completely made of ABS.

The SMA wire (Flexinol, austenite finish 70 °C, dynalloy) was inserted after the printing process and placed in a brass tube, to ensure a low-friction movement and to avoid damage to the FDM part (heat and force). The SMA wire has a diameter of

0.25 mm and a length of 170 mm, with a clamp length within the sample of 108 mm. The achieved contraction was 5 %.

The realizable deflection of the different samples illustrates the influence of the samples' gestalt and mechanical properties. P2-2 and P2-3 nearly have the same force characteristics but the values regarding temperature and voltage and the resulting deflection differ considerably. This is due to the geometrical structure of the flexible part. Comparing P1 and P2-3, the necessary force to bend was reduced by adapting the geometry. This leads to an equally higher deflection at similar temperature and voltage. The joint (P4) requires a very little, leveled force, but only small deflections can be realized due to the kinematic design.

So, in summary, high deflections are achieved by a low bending resistance, implemented by an adapted geometry and/or flexible materials. In terms of geometry, an important finding is that the structure should prevent the sample from shortening in longitudinal direction under load, as this leads to a lower deflection due to an unfavorable use of the wire contraction. The geometry of P3 is due to missing support in longitudinal direction susceptible to this behavior.

5 Conclusion

Our study shows that FDM works well for generating passive but flexible structures. It is possible to design complex, movable and customized products or prototypes combining FDM, flexible structures and SMA. This is promising regarding innovative, user-centered products and complex systems as the adaption to the user (individualization) and to the task (adaptivity) is realizable. The results of the experimental study regarding the bending prototypes contribute to the completion of the matrix in Fig. 2. The further development of the matrix provides a guideline to holistically design products and systems in the context of industrial design engineering. Therefore, further investigations considering other movements and their combinations as well as the transfer to practice-oriented applications have to be conducted.

References

1. Hein, A., Holder, D., Maier, J., Maier, T.: Potential Analysis of Smart Materials and Methodical Approach developing Adaptive Designs using Shape Memory Alloys. Norddesign (2018)
2. Oberhofer, F., Janny, B., Maier, T., Wulle, F., Verl, A.: Chancen und Grenzen der Additiven Fertigung bei der Gestaltung physischer Mensch-Maschine-Schnittstellen. In: 63. Frühjahrskongress der Gesellschaft für Arbeitswissenschaft. GfA Press, Dortmund (2017)
3. Janocha, H.: Adaptronics and Smart Structures. Springer, Heidelberg (2007)
4. Langbein, S., Czechowicz, A.: Konstruktionspraxis Formgedächtnistechnik. Springer Fachmedien, Wiesbaden (2013)
5. Thompson, M.K., Moroni, G., Vaneker, T., Fadel, G., Campbell, I., Gibson, I.: Design for additive manufacturing: trends, opportunities, considerations, and constraints. CIRP Ann. **64**, 737–760 (2016)

6. Wulle, F., Coupek, D., Schäffner, F., Verl, A., Oberhofer, F., Maier, T.: Workpiece and machine design in additive manufacturing for multi-axis fused deposition modeling. Procedia CIRP **60**, 229–234 (2017)
7. Breuninger, J., Becker, R., Wolf, A., Rommel, S., Verl, A.: Generative Fertigung mit Kunststoffen. Springer, Berlin, Heidelberg (2013)
8. Kim, H., Han, M., Song, S., Ahn, S.: Soft morphing hand driven by SMA tendon wire. Compos. Part B Eng. **105**, 138–148 (2016)
9. Wang, A., Rodrigue, H., Kim, H., Han, M., Shn, S.: Soft composite hinge actuator and application to compliant robotic gripper. Compos. Part B Eng. **98**, 397–405 (2016)
10. Umedachi, T., Vikas, V., Trimmer, B.A.: Highly deformable 3-D printed soft robot generating inching and crawling locomotions with variable friction legs. In: IEEE/RSJ International Conference on Intelligent Robots and Systems (IROS). Tokyo, Japan (2013)
11. Gürcüm, B.H., Börklü, H.R., Sezer, K., Eren, O.: Implementing 3D printed structures as the newest textile form. J. Fashion Technol. Text. Eng. (2018)
12. Cahoon, S.: Living Hinge: Design Guidelines and Material Selection. http://www.matterhackers.com/news/living-hinge:–design-guidelines-and-material-selection
13. Mutlu, R., Alici, G., in het Panhuis, M., Spinks, G.: Effect of flexure hinge type on A 3D printed fully compliant prosthetic finger. In: IEEE International Conference on Advanced Intelligent Mechatronics (AIM). Busan, Korea (2015)
14. Schulz, S., Schlattmann, J., Rosenthal, S.: Konstruktionsrichtlinien für die funktionsgerechte Gestaltung additiv gefertigter Kunststoffgelenke. SSP (2017)

Approaches to Digital Manufacturing: Designing Through Materials

Caterina Dastoli[1(✉)], Patrizia Bolzan[1], Massimo Bianchini[1], Barbara Del Curto[2], and Stefano Maffei[1]

[1] Department of Design, Politecnico Di Milano, Milan, Italy
{caterina.dastoli,patrizia.bolzan,massimo.bianchini, stefano.maffei}@polimi.it

[2] Department of Chemistry, Materials and Chemical Engineering "Giulio Natta", Politecnico Di Milano, Milan, Italy
barbara.delcurto@polimi.it

Abstract. Until a few years ago designing through materials has always referred to fully consolidated manufacturing technologies, despite the increasing development and democratization of digital manufacturing. This article reports on an investigation into the second edition of the experimental crash course "Design, digital fabrication and 3D printing". The educational experience was held at Polifactory, Fab Lab and makerspace of Politecnico di Milano, and has focused on the role that additive manufacturing plays in concept development when engineering thermoplastics are given as a design constraint. Therefore, design students experienced materials and digital technology issues by working in team and practicing the design creative process. The results indicate that additive manufacturing plays a competitive role over the design process when engineering thermoplastic materials are used. In addition, students' motivation and capability in managing both technical and design thinking skills over a project development confirmed the effectiveness of the teaching method adopted.

Keywords: 3D printing · FDM technology · Direct digital manufacturing
Engineering thermoplastics · Design course · Design teaching

1 3D Printing for Direct Digital Manufacturing

Additive Manufacturing (AM) is expected to revolutionize production in many fields of the industrial scenario. The technology related peculiarities are well known and reported by several authors [1]. Advantages in comparison to other technologies mainly refers to the possibility to economically build custom products in small quantities, the ability to easily share designs and outsource manufacturing, and the speed and ease of designing and modifying products [2].

However, 3D printing current limitations such as higher costs for large production runs, lower precision and reduced choice for materials, colors and surface finishes, make the technology seen primarily as a tool for Direct Prototyping (DP) [3] when product design and development are concerned.

W. Karwowski and T. Ahram (Eds.): IHSI 2019, AISC 903, pp. 628–633, 2019.
https://doi.org/10.1007/978-3-030-11051-2_95

Indeed, if 3D printers were previously considered as several tools configured only for experts, now we are witnessing an increasing democratization of technological practices [4]. Widely available production labs, low cost and fabrication tools, and open and shared knowledge about production processes, give design students the opportunity to identify digital technologies as design tools, thus broadening creative possibilities and expanding design methodological approaches. Digital technologies enable designers to push boundaries, to work through ideas, factors and issues by assisting design, mock-ups and physically making pieces that would be difficult even impossible to envisage otherwise [5].

Authors such as Berman [3] report the competitive role of AM when used for printing finished goods. Indeed, advances in printing materials have enabled 3D printers to produce objects that can compete with those manufactured by traditional technologies. This is for instance the case of the streamlined fabrication of medical items, such as dental implants and prosthetics [6].

The here presented study aims to highlight that materials with specific properties, such as the engineering thermoplastics, give to AM the possibility of being recognized as a "direct digital manufacturing" (DDM) technology when they are used for product development, where DDM is defined as the direct production of finished goods from a rapid prototyping device [7]. Indeed, designing through them not only allows designers to have less limitations in aesthetic languages expression, but also and mostly make AM identifiable as a technology able to produce finished and functional products in a short period of time.

2 Methodology: When Theory Meets Practice

Teachings in materials science have always been a fundamental part of the Design School of Politecnico di Milano curricula. However, the rapid evolution of the industrial scenario requires more and more figures capable to merge technical and design thinking skills for product innovation [8]. For this purpose, the Design School offers courses which follow the project-based learning approach (PBL) [9, 10]. Students are required to put into practice what they have learnt from the theory, thus experiencing specific materials issues by working in team and practicing the design creative process in concepts generation.

Being well aware of digital fabrication technologies potential [11], "Polifactory", the Fab Lab and makerspace of Politecnico di Milano organized several "Design, digital fabrication and 3D printing" crash courses for design students following the PBL approach. While the first edition aimed to make students aware about additive and subtractive manufacturing technological features and possibilities [12], the second one focused on the analysis of the role that additive manufacturing (AM), and in particular the Fused Deposition Modeling (FDM) technology, plays on the design process when engineering thermoplastics are given as a design constraint. Starting from the theoretical explanation of morphological, mechanical and physical properties of engineering thermoplastic materials such as the Polyester added with carbon fibers (Ca-PET) and the Acrylate Styrene Acrylonitrile (ASA), the aim of the course was the realization of 3D printed and finished products characterized by functionality given from materials.

Students also felt out a questionnaire divided into two main sections. The first was related to the description of the design process adopted when materials and digital technologies drive the final product requirements. The other required them some suggestions for improvements to be implemented in future educational experiences.

3 Engineering Thermoplastics for 3D Printed and Finished Goods

Among conventional materials for FDM technology, ABS is one of the most used when a good performance for 3D printed products is required. Indeed, it shows improved mechanical properties compared to the traditionally used PLA. Nevertheless, ABS printed products still lack of high-performance characteristics when compete with ones obtained by consolidated manufacturing technologies, thus being acknowledged as prototypes. Indeed, in injection molding, the polymer is gelled or melted before being compressed into the mold where it solidifies. In 3D printing, the material is only deposited onto the previously laid layer. This lack of pressure coupled with the low contact area between the various layer, results in a lack of mechanical properties of the final printed object [13].

Materials knowledge allows experts to produce high-performance materials, namely engineering thermoplastics, which address FDM manufacturing process characteristics, by maintaining high mechanical properties over the entire printing cycle [14].

The company Threedfilaments produces several materials for 3D printing, including the Acrylate Styrene Acrylonitrile (ASA) and a particular type of polyester added with carbon fibers (Ca-PET). ASA is characterized by excellent UV resistance and high toughness. These mechanical features are the reason why the material is mainly applied for outdoor applications, such as profiles, elements installed on roofs, parabolas, industrial bodywork parts, and other applications that require high weather and impact resistance. Ca-PET is a composite material where the polymeric matrix, which already presents a good mechanical resistance, is boosted in its thermal and mechanical properties as it is filled with carbon fibers. Moreover, PET shows high resistance to chemicals, does not delaminate over the printing and is quite hygroscopic.

The "Design, digital fabrication and 3D printing" crash course then focused on the application of this kind of materials, in order to spread awareness among students on high-performance materials available for enhanced final printed outputs.

Thus, an expert from the company Threedfilaments has been invited during the first day of the course for showcase to students the above described engineering thermoplastic materials properties, in order to make them aware about the product development possibilities throughout the concept development. In addition, students were provided by several 3D printing technical tips that need to be known when peculiar materials are used. Aim of the first day of the course was to provide students with tools and information useful for the brief to be satisfied over the second day: in order to put into practice technical teachings, students was asked to design and present a product that enhanced the properties of the chosen material. The project had to be developed by working in team. Teams who decided to use ASA had to design a product for outdoor applications, in particular for gardening or outdoor urban sports. Those who chose Ca-PET had to design heat-resistant products, with particular attention to heat sources in kitchen or in lighting.

The students were then required to prepare a presentation and fill out an anonymous questionnaire, giving feedbacks on the effectiveness of the PBL approach adopted by the course.

4 Achieved Results

Students designed, modeled, verified and finally presented projects developed by using FDM technology and thermoplastic engineering materials. Figure 1 shows four achieved results, two designed for the outdoor environment by using ASA (Fig. 1a, b), the others for heat-resistance application by using Ca-PET (Fig. 1c, d).

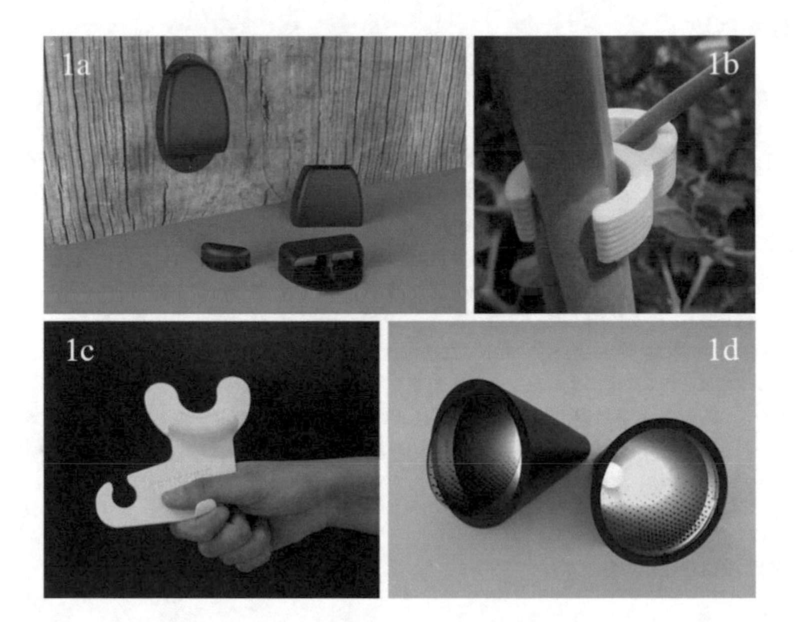

Fig. 1. Four selected finished products developed over the crash course. Batcove (*1a*) and Plang (*1b*) have been achieved by using ASA, while Trinchino (*1c*) and Cove (*1d*) by using Ca-PET.

The first project, "Batcove" (Fig. 1a) is a batbox for domestic use. Designers, F. Cavaliere, M. Ceruti and P. Ciarfuglia, decided to develop a compact and light-weight version of the cave, unlike the existing alternatives available on the market. The ASA made it resistant to UV rays and weatherproof.

The second project is "Plang" (Fig. 1b), designed by M. Assad, E. Belà, G. Tagnin and R. Vangal. Plang is an accessory made of ASA that aims to simplify the process of supporting crop plants during the period of growth. One part clamps to the support pile, while the other holds the stem of the plant without any ligature required, thus avoiding damages to the plant.

Figure 1c shows "Trinchino", a tool that helps people to hold a plate and a glass simultaneously while eating and standing up, as it usually happens during an aperitif. In this project, designers A. Kayalibay, A. Baratto and J. Nalin chose to use Ca-PET, thus enhancing the thermal resistance of the product while it is in contact with warm objects.

The last project is "Cone" (Fig. 1d), designed by F. Inzani, D. Pisanello, A. Torrone and E. Ukolova. The lamp emphasizes the heat resistance of the Ca-PET material. The external shape is characterized by small dimensions, thus making the bulb very close to the inner surfaces of the object.

Achieved results and questionnaires collected answers showed that the course was extremely formative and appreciated. However, the total duration of the course seemed quite short if an in depth development of the design concept is required.

5 Discussion

Quality of knowledge and development of adequate skills are particularly required to manage with the processes of scientific and technological changes in a continuous evolving reality. The ability of facing up problems seems to be dependent more and more on the transversal attitude of the applied knowledge [15]. Indeed, students have been allowed to experience and verify theory over the design process, succeeding in achieving required results in a short period of time.

Additional time for project development would have allow them to take advantage of more verification moments, such as a prolonged interface with the 3D printer or the possibility to review the design process with teachers. Thereby, the complexity of final results was not particularly high. However, what is interesting to note is that students were provided of knowledge relating engineering thermoplastics and AM potentialities for DDM, that has been successfully experienced by involving them within the described theoretical-practical approach.

6 Conclusions

This work has shown the crash course structure, methodology and the team-works achieved results. From the analysis of final outputs, it was concluded that AM plays a competitive role over the design process when engineering thermoplastic materials are used. Indeed, designing through them not only allows designers to have less limitations in aesthetic languages expression, but also and mostly make AM identifiable as a technology able to produce finished and functional products in a short period of time. Moreover, questionnaires confirmed the effectiveness of the PBL approach, showing how it enhances students' motivation and capability in managing both technical and design thinking skills over a project development.

Acknowledgements. This research was supported by Polifactory, Fab Lab and Makerspace of Politecnico di Milano.

References

1. Mueller, B.: Additive manufacturing technologies—rapid prototyping to direct digital manufacturing. Assembly Autom. **32**(2), (2012)
2. Wohlers, T.: Wohlers report 2016. Wohlers Associates, (2016)
3. Berman, B.: 3-D printing: The new industrial revolution. Bus. Horiz. **55**(2), 155–162 (2012)
4. Rognoli, V., Bianchini, M., Maffei, S., Karana, E.: DIY materials. Mater. Des. **86**, 692–702, dic. (2015)
5. Shillito, A.M.: Digital crafts: industrial technologies for applied artists and designer makers. Bloomsbury, (2013)
6. Ventola, C.L.: Medical applications for 3D printing: current and projected uses. Pharma. Ther. **39**(10), 704, (2014)
7. Bak, D.: Rapid prototyping or rapid production? 3D printing processes move industry towards the latter. Assembly Autom. **23**(4), 340–345, (2003)
8. Findeli, A.: Rethinking design education for the 21st century: theoretical, methodological, and ethical discussion. Des. Issues **17**(1),5–17, gen. (2001)
9. Piselli, A., Dastoli, C., Santi, R., Curto, B.D.: Design tools in materials teaching: bridging the gap between theoretical knowledge and professional practice. In: Proceedings of the 20th International Conference on Engineering and Product Design Education (E&PDE 2018), p. 6. Dyson School of Engineering, Imperial College, London, 6th–7th Sept 2018
10. Carulli, M., Bordegoni, M., Bianchini, M.: A novel educational model based on "knowing how to do" paradigm implemented in an academic makerspace. ID&A Interact. Des. Architect., p. 23, (2017)
11. Edgar, J., Tint, S.: Additive manufacturing technologies: 3D printing, rapid prototyping, and direct digital manufacturing, 2nd edn., Johnson Matthey Technology Review, **59**(3), 193–198, lug. (2015)
12. Dastoli, C., Bolzan, P., Bianchini, M., Curto, B.D., Maffei, S.: Design, digital fabrication & 3d printing: a crash course for design students. In: 10th International Conference on Education and New Learning Technologies, Palma, Spain, pp. 4774–4781, (2018)
13. Valkenburg, R., Dekkers, C., Sluijs, J.: On the making of things: taking a risk with 3D printing. In: Proceedings of the 4th Participatory Innovation Conference 2015, p. 512, (2015)
14. Ashby, M.F., Shercliff, H. Cebon, D.: Materials: engineering, science, processing and design. Butterworth-Heinemann, (2013)
15. Palmieri, S.: Design at work. In: Professional workshops of the Faculty of Design 2004. Edizioni POLI.design, (2004)

Bio-smart Materials for Product Design Innovation: Going Through Qualities and Applications

Marinella Ferrara[1](✉), Carla Langella[2], and Sabrina Lucibello[3]

[1] Politecnico di Milano, Diparimento di Design, Milan, Italy
marinella.ferrara@polimi.it
[2] Università degli Studi della Campania "Luigi Vanvitelli", Naples, Italy
[3] Sapienza Università di Roma, Rome, Italy

Abstract. Based on the principles of emerging *bio-smart materials*, in the framework of design-driven material innovation approach and cross-disciplinary research practices, the aim of this paper is to make clear how these materials are creating new opportunities to realize answers to the complex needs of contemporary society, while defining bio-smart materials main qualities, and questioning the implications on research design practices. We present a review of case studies of bio-smart materials applications in order to demonstrate that their diffusion is underway, especially in some application sector characterized by more complexity and a higher propensity for the use of biotechnological innovations. The examples described, are framed in three specific areas: biomedical design, sports design, and design for the environment that are particularly interested of the development and application of novel materials for both their performances and sustainability.

Keywords: Bio-smart materials · Design research practices · Biomedical design · Sport design · Design for environment

1 Introduction

Materials for design are changing and Design Culture is consequently changing. In the approaching bio-technological era, the most adequate artifacts to *survive* the emerging conditions of extreme complexity and mutability will be those able to express a kind of smartness as close as possible to the *intelligence in nature* [1]. The concept of bio-smartness associated with materials and material systems, therefore, assumes increasingly hybrid connotations between the synthetic world and the biological world that lead to the expansion of the traditional concept of smartness beyond the digital computational content. Up to date, this is one of the most desirable and promising research vision expanding from the sciences of life to the design, as it put right the old dichotomy between nature and artifice. Nature is an infinitely variable and dense model of stimuli for the project. In the bio-smart vision, through cybernetic, bionics and biomimetics, Nature is increasingly mixed with the new green technologies, breaking the patterns imposed by physical and sensorial limits to give rise to hybrid and

W. Karwowski and T. Ahram (Eds.): IHSI 2019, AISC 903, pp. 634–640, 2019.
https://doi.org/10.1007/978-3-030-11051-2_96

intelligent objects able even to live their own almost-autonomous-life. As in an inverse process, the complexity of contemporary technologies - and in particular the interaction between biomimetics, robotics, and neuroscience - influences the design project, causing it to replicate, invade and disseminate nature, to reproduce the apparent simplicity through extremely sophisticated techniques [2].

Based on a previous study and definition of *bio-smart materials* [3], this paper opens a scenario of possible cross-disciplinary practices in the framework of design-driven material innovation approach. Four are the main principles of bio-smart intelligence: i. be sustainable: biodegradable and intelligent release; use multi-functional design; ii. be adaptable and responsive: use feedback; self-renewable materials; resilient through variation, redundancy, and decentralization; iii. evolve to survive: integrate the unexpected; iv. to develop: self-organize; alive and growing.

In this paper, we will question how bio-smart materials, responding to the aforementioned principles, are already changing our daily reality by introducing products innovation.

2 Method

In order to make clear the big potentiality of bio-smart materials, we present a review of worldwide case-studies of bio-smart materials applications at different levels of development - from a first experiment of a concept to a product prototype ready to be assessed for market - in order to demonstrate that their diffusion is underprocess, especially in some sector of applications characterized by complexity and a higher propensity for the use of biotechnological innovations.

Presenting the following selection of case studies, allows us to clarify the peculiarities of smart materials, clearly distinguishing them from biomaterials and biomimetic materials. While bio-smart materials can be biomaterials, not all these last can be identified with the bio-smart category. Many bio-smart materials are biomimetic, that is, they draw inspiration from principles and logics observed by biologists in living systems, linked to forms of smartness such as the ability to react to external inputs, adapt to disputes or integrate multiple functions into one detail to achieve synergistic performance. Not all biomimetic materials show characters of smartness from Nature, so not all are bio-smart. Similarly, bio-smart materials have a natural origin or integrate biological components such as microorganisms, but it is not correct to declare that a bio-smart material must necessarily be of natural origin since even synthetic materials can have a similar intelligence content. Then, we have chosen to showcase studies in which the materials have a biological-like intelligence, or they yield, the intelligence of nature integrating it with artificial intelligence systems.

The selection is framed in three specific areas: biomedical design, sports design, and design for the environment, particularly interested in the development and application of novel materials for both their sustainability and high performance. The analysis of the following examples will allow us first consideration of their distinctive qualities.

3 Bio-smart Materials and Biomedical Design

The evolution of healthcare intervention model is taking advantage of the progress in biotechnologies and materials, stimulating research in the design of products for home use, and wearable devices. The healthcare sector, therefore, requires a multidisciplinary approach linked to the design know-how. Bio-smart materials are particularly adequate to be applied in the biomedical field, in which they have a high potential for disruptive innovation impacting on people's quality of life. We will present below, some of the most exciting researches in this sector, to demonstrate the great opportunity offered by bio-materials to the improvement of people wellness.

The project *Anura* is a materials and processes system to produce an instant beneficial patch [4] (Fig. 1). The designer Giuliana Califano developed it in 2018, during the Master D.RE.A.M. Academy – Design, Research, Advanced Manufacturing, at Città della Scienza di Napoli, in collaboration with the CNR Institute IPCB, and Hybrid Design Lab. The instant soft patches are composed of a matrix of biological hydrogels including active ingredients. Realized through 3D printing techniques patches are biocompatible, biodegradable and bio-active at the interface between body and device, through flows and exchanges of the beneficial principles, thanks to the chemical structure and the transdermal microstructure. They are adaptable, responsive and could be deposed on a specific and delimitated area of the skin with a desired shape like a tattoo, in case of injury, inflammation or abrasion. Tattoo patches can be used for therapeutic applications related to joint trauma. In this case, the transdermal patch includes active pain-relieving and anti-inflammatory ingredients that reduce healing times. At the same time, the part covered by the tattoo, thanks to the thickness and the mechanical characteristics of the material, is also more mechanically protected, if compared to the use of a cream or a simple plaster as it absorbs shocks at the traumatized area.

The biosensor can wirelessly transmit data about the body's electrical activities to a smartphone, like a microsystem developed in 2017 by a joint research team that includes DGIST from the Northwestern University, and scientist from the USA, Korea, China and Singapore [5]. This soft electronics microsystem contains a variety of sensors, connected by a unique network of tiny wire coils, all placed in a soft silicone material to protect its components and to make an easily attached pad to the skin. The field of biosensors is very promising from a huge number of applications from monitoring to implantable sensing.

4 Bio-smart Materials and Sports Design

Sports design is a growing application field of design, engineering, and also biology, driven by the global sports market, benefit the multi-billion dollar sports industry. Sports design actually focus on improving the performance and safety of athletes. It includes the design of sports equipment but also the design of sport clothing that can increase athletic or practitioners performance.

A research group from the University of New South Wales, in Australia, coordinated by Melissa L. Knothe Tate [6] has developed an intelligent biomimetic fabric useful for clothing, with a particular structure that ensure high elasticity and impact resistance, like bones (Fig. 1). The research project inspired by the *periosteal*, a membrane that covers the external surface of most human and animal bones, characterized of a greater ability to withstand strong impacts. In addition to the functional characteristics of the fibers, the weaving methods observed under the microscope and modeled in 3D were also transferred. Starting from these models the researchers have made a further evolution by expanding the weaving patterns similar to those natural ones suitable for processing with an innovative digital jacquard frame. Prototype samples were then made which, subjected to mechanical tests, showed mechanical properties similar to the *periosteum* tissue. Among the potential applications of the fabric there are skiers sportswear and clothing for race car drivers able to protect some parts of the body from strong impacts.

Living materials as microbial cultures integrated with electronic circuits are about to dramatically change the very essence of product design, performing sport and daily life workout activity, monitor biochemical vitals during workout. The *MIT Lab* project for a highly innovative shoes concept in partnership with athletic sportswear company *Puma*, and developed in collaboration with Biorealize, applies the new available technology *Deep Learning Insoles* (Fig. 1) in the next pair of performing sport shoes to prevents fatigue and improves athletes' performance [7]. Deep Learning Insoles are silicone based disposable inlays containing microbial cultures, able to monitor biochemical vitals that normally change during running or workout. Since the very early stage of dissemination, also in terms of marketing, just as reported by the launching campaign, the role of bacteria was made quite clear and loud stating that "Microbial layer is composed of mini cavities that are filled with bacteria and media that are specialized in sensing different compounds present in sweat". Bacteria then respond to what they sense with specific chemicals causing a pH and a conductivity change in the sole itself, which gets recorded by a network of electrical circuits, connected to microcontrollers positioned in the third layer. Invisible living organisms are about then to dramatically change the workout and endurance routine and such a new step in bridging science and design is being broadly communicated also to potential mass consumers.

5 Bio-smart Materials and Design for the Environment

Design for the environment plays an essential role in innovation based on sustainability, establishing scenarios for the new bio-smart materials and giving value and meaning to these technological advances. In this area, microbiology and biotechnology are showing a great promise in solving a plethora of problems filling the gap of knowledge in the field of environmental sustainability and covering the different technologies available to sustain the environment. One of the actions in which bio-smart materials can prove particularly useful is detection pollutant and substances harmful to human health, such as particles or radiation, present in air and water.

In this direction, a team of researchers from the City College of New York is working. The team is led by the chemical engineer Teresa Bandosz, who has developed smart fabrics capable of adsorbing, degrading and detecting substances that fall within the most dangerous and developed class of Chemical Warfare Agents (CWAs), the Organophosphate (OP) Nerve Agents. The fabric is made of a natural cotton filaments on which a coating of a heterogeneous porosity nanocomposite (MOFgCNox) is obtained by impregnation obtained from the combination of oxidized graphite carbon nitride (g-C3N4-ox) and Cu-BTC MOF. The fabric gradually changes color when it detects nerve gas and, following a photocatalytic effect, oxidizes it into non-toxic compounds [8]. The photocatalytic oxidation effect is very similar to chemical processes that occur in some plants through photosynthesis and which make plants natural scavengers. Based on this material research, the designer Nikolas Gregory Bentel applied this technology to a T-shirt, an object that can be used every day and is visible from any direction. In 2016, he released *Aerochromic*, a new line of shirts in three designs that change colors according to pollution levels in the air [9]. Chemical salts turn carbon monoxide into carbon dioxide, and that oxidation process transforms the color of the shirts (Fig. 1).

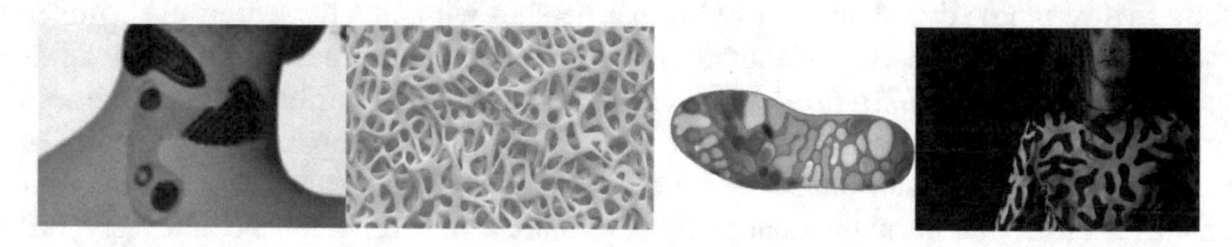

Fig. 1. From right to left: G. Califano, *Anura*, 2018; M.L. Knothe Tate, intelligent biomimetic fabric like bones, 2017; MIT lab and Puma with Biorealize, *Deep Learning Insoles*, 2018; N. G. Bentel, *Aerochromic*, 2016.

6 Conclusion

Which are the main qualities of bio-smart materials while envisioning next design products, and questioning the implications of product design practices, cultures based on material design, and finally the impact on the user perspective?

The principles listed in the introduction were decoded into the material peculiarity used as a criterion for selecting the case study described in the paper. These do not necessarily have to be all in a material to ensure the belonging to the bio-smart materials category. What is important is the presence of an intelligent "behavior" in a biological sense. Some of the peculiarity observed in bio-smart materials are: biological origin (vegetable or animal); biodegradability, reactivity; self-adaptability; ability to process inputs to output variable outputs according to internal or external conditions; renewability; functional redundancy able to respond to unexpected events; self-organization capability; ability to evolve, grow or develop; incorporation of living biological matter.

While the principles propose design logics useful to develop new bio-smart materials (tools for materials designers), the peculiarities could be used as a filter, a sort of checklist, and tools for a critical design study of the bio-smart materials and interpret this in new product qualities. This list emerges not only from the translation of principles but also from a critical case studies analysis that helped to identify qualities that can be considered tangible, observable, recognizable and correlated to a new user experience.

The *Anura* case study demonstrates how materials with biodegradability and biocompatibility characteristics can be rendered bio-active, therefore implemented with intelligence that makes them capable of interacting with the human body. The case study of sports shoes developed by MIT and Puma presents a promising field of bio-smart materials based on incorporating smartness through embedding living matter such as bacteria, microorganisms, and cells. The DGIST case demonstrates how the material of petrochemical origin, like silicon, can acquire multiple sensitive capacities that place it in the bio-smart category. Material *periosteum* tissue inspired illustrate how the in-depth observation of nature details conducted with advanced instruments, like sophisticated microscopes, allows changing not only the structure but also the intelligent behavior of biological materials.

This contribution shows how the design proposes an important opportunity to associate the intelligence of these materials with applications in products that interpret this intelligence to meet the new needs of contemporary living, therefore through equally intelligent products, which can also refer to the principles on which the adaptability and fitting strategies of biological systems are based.

Novel bio-smart materials and cross-disciplinary research are creating a huge opportunity for design practices and user experiences. Due to their biological-like peculiarities and smartness, bio-smart materials are particularly promising to give right answers to the complex needs of contemporary society, stimulating disruptive innovations.

References

1. Langella, C.: Hybrid design. Progettare tra tecnologia e natura. Milan Franco Angeli (2007)
2. Lucibello, S., La Rocca, F.: Innovazione e Utopia nel design italiano. Rome, Rdesignpress (2014)
3. Lucibello, S., Ferrara, M., Langella, C., Cecchini, C., Carullo, R.: Bio-smart materials: the binomial of the future. In: Karwowski, W., Ahram, T. (eds) Intelligent Human Systems Integration, pp. 745–750. Proceedings of the 1st International Conference on Intelligent Human Systems Integration (IHSI 2018): Integrating People and Intelligent Systems, January 7–9, 2018, Dubai, United Arab Emirates. Springer, Cham (2018)
4. Jang, K.I., Chung, H.U., Xu, S., Lee, K., et al. (2017). Self-assembled three dimensional network designs for soft electronics. Nat. Commun. **8**, 15894. https://doi.org/10.1038/ncomms15894
5. Califano, G.: "Anura": ornamental and transdermal patches printed in 3D, Digicult (2018). https://digicult.it/design/anura-ornamental-and-transdermal-patches-printed-in-3d/
6. Jang, K., et al.: Self-assembled three dimensional network designs for soft electronics. Nat. Commun. **8**, 15894 (2017). https://doi.org/10.1038/ncomms15894
7. Ng, J.L., Knothe, L.E., Whan, R.M., Knothe, U., Tate, M.L.K.: Scale-up of nature's tissue weaving algorithms to engineer advanced functional materials. Sci. Rep. **7**, 40396 (2017)

8. Giannakoudakis, D.A., Hu, Y., Florent, M., Bandosz, T.J.: Smart textiles of MOF/gC 3 N 4 nanospheres for the rapid detection/detoxification of chemical warfare agents. Nanoscale Horiz. **2**(6), 356–364 (2017)
9. Bentel, N.G.: Aerochromic (2016). http://www.nikolasbentelstudio.com/aerochromics

Adaptive Structures and Systems: Interaction of Application, Passive and Active Structure

Andrea Hein(✉), Daniel Holder, and Thomas Maier

Institute for Engineering Design and Industrial Design, Research and Teaching Department Industrial Design Engineering, University of Stuttgart, Pfaffenwaldring 9, Stuttgart 70569, Germany
{Andrea.Hein,Daniel.Holder,Thomas.Maier}@iktd.uni-stuttgart.de

Abstract. In the field of product development, the application of high-tech materials enables new concepts. Smart materials generate innovative products through controllable properties. An increasing number of research studies and first successful experiences in industrial use support the importance. Promising are above all shape memory alloys (SMAs), which are investigated in the following, regarding the area of the product development process. Special focus is on necessary constraints and parameters of application, passive and active structure. This leads to the detection of possible applications within adaptive designs and a first approach to a method for the design and integration of SMAs.

Keywords: Adaptive structures and systems · Smart materials Shape memory alloys · User centered design · Design methods

1 Introduction

Smart materials, especially SMAs are applied as multifunctional elements in adaptive structures and systems (ASS). Hein [1] provides a definition of these structures and systems. The property of SMAs of remembering the original form after deformation leads to the development of lighter, more efficient and innovative products. For example, the use of SMAs in automotive context could allow for weight reduction, aerodynamic improvement, and increased user comfort and ergonomics. Despite their importance in research, actual use has not increased to the same extent. Reasons are high development costs and a lack of development methods. To extend their use, application-specific integration, and development and design methods are necessary.

To facilitate a classification three main fields [1] of applications for ASS were identified, categorized and proved by examples in research, design and development. This study [2] and the resulting correlation matrix have shown that SMAs are already greatly researched and applicated in the field of technology, but there is still great potential regarding the fields of ergonomics and design. Promising are, above all, areas of application that combine the three main fields ergonomics, technology and design. It is thereby possible to exploit the full potential smart materials offer.

W. Karwowski and T. Ahram (Eds.): IHSI 2019, AISC 903, pp. 641–647, 2019.
https://doi.org/10.1007/978-3-030-11051-2_97

2 Primary Parameters for Designing ASS

In order to detect new fields of application, evaluate possible implementations and derive a design method, it is necessary to establish objective criteria. Adaptive structures and systems contain a multifunctional element (smart material) to adapt to changing external conditions, henceforth referred to as active structure. Furthermore, the supporting passive structure establishes a link to the application and the user. As adaptive structures and systems contain application, passive and active structure, constraints and parameters to connect those elements are important. Those parameters were methodically derived using feature guidelines [3] (Fig. 1).

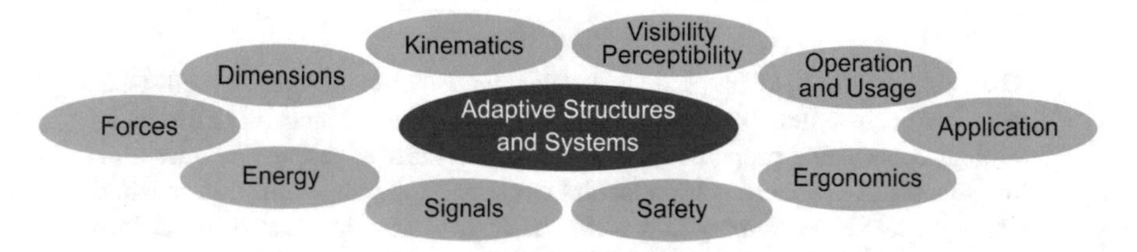

Fig. 1. Constraints for adaptive structures and systems using feature guidelines.

The constraints were subdivided into precise parameters regarding application, passive and active structure (Table 1). It was shown that the primary parameters are similar, but the description and subcategorization vary regarding application, passive and active structure. The parameter energy serves as an example to explain the differences. Within the active structure, the temperature to activate the change within the material's properties is decisive. Regarding the passive structure, the temperature at which the material permanently deforms is significant. As to the application, the operating temperature is important.

Table 1. Primary parameters regarding application, active and passive structure.

Parameter \ ASS	Active Structure	Passive Structure	Application
Dimensions	length, diameter, shape, quantity	general dimensions, shape	installation space, assembly, quantity
Time	time for activation, time for deactivation, wire rate	delay (movement of active structure), lifespan	required time activation and deactivation, lifespan
Motion	type of movement, acceleration, deformation, orientation, memory effect	type of movement, acceleration, stiffness, deformation	required acceleration, required movement
Physical Properties	density, liquidus temperature, plasticity, strain, hardness		-
Energy	temp. activation, type of activation / deactivation	ultimate temperature material	operation temperature, tolerable temperature
Forces	force factor (tension) and frequency (cycles)		required force frequency and force transmission
Operation & Usage	adjustment	designing for user and application	type, frequency, user, location
Visibility & Perceptibility	-	shape, interface	perception, interface

As the three elements presented above are to be designed as an efficient, well-functioning system, it is important to compare the different parameters. The direct comparison of pairs and a modification of the house of quality demonstrated critical parameters (Table 2). It was defined, that the parameters should be reduced by one-third to identify the most critical parameters. The implementation of these methods showed the most critical parameters ($\geq 37\%$), which were marked grey. Thereby it was demonstrated, that the parameters of the active structure are mostly represented within the critical parameters. Therefore, they are most affected if parameters within the application or the passive structure are changed and vice versa.

Table 2. Interferences of application, active and passive structure (0: noncritical, +: critical, ++: highly critical).

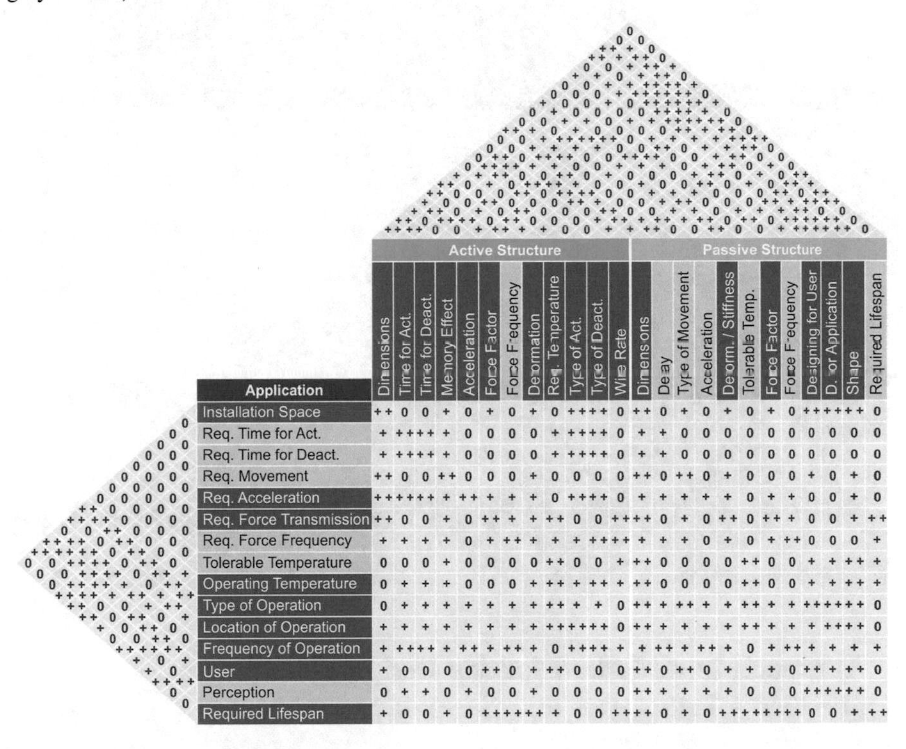

Application	Active Structure												Passive Structure											
	Dimensions	Time for Act.	Time for Deact.	Memory Effect	Acceleration	Force Factor	Force Frequency	Deformation	Req. Temperature	Type of Act.	Type of Deact.	Wire Rate	Dimensions	Decay	Type of Movement	Acceleration	Deform. / Stiffness	Tolerable Temp.	Force Factor	Force Frequency	Designing for User	D. for Application	Shape	Required Lifespan
Installation Space	++	0	0	+	0	+	0	+	0	++	++	0	++	0	+	0	+	0	+	0	++	++	++	0
Req. Time for Act.	+	++	++	+	0	0	0	0	+	++	++	0	+	+	0	0	0	0	0	0	0	0	0	0
Req. Time for Deact.	+	++	++	+	0	0	0	0	+	++	++	0	+	+	0	0	0	0	0	0	0	0	0	0
Req. Movement	++	0	0	++	0	0	0	+	0	0	0	0	++	0	++	0	+	0	0	0	+	0	+	0
Req. Acceleration	++	++	++	+	++	+	+	+	0	++	++	0	+	+	+	+	+	0	+	+	0	0	+	0
Req. Force Transmission	++	0	0	+	0	++	+	+	++	0	0	++	++	0	+	0	++	0	++	+	0	0	+	++
Req. Force Frequency	+	+	+	+	0	+	++	+	+	+	++	++	+	+	+	0	+	0	+	++	0	0	0	+
Tolerable Temperature	0	0	0	+	0	0	0	0	++	0	0	+	++	0	0	0	0	++	0	0	+	+	++	+
Operating Temperature	0	+	+	+	0	0	0	+	++	+	++	+	++	0	0	0	0	++	0	0	+	+	++	+
Type of Operation	0	+	+	+	+	+	+	+	++	+	+	0	++	+	++	+	+	++	+	+	++	++	++	0
Location of Operation	+	+	+	+	+	+	+	+	++	++	++	0	++	+	+	+	+	++	+	+	+	++	++	0
Frequency of Operation	+	++	++	+	++	+	++	+	0	++	++	+	+	++	+	++	+	0	+	++	+	+	+	+
User	+	0	0	0	0	++	0	+	++	0	0	0	++	0	++	0	+	+	+	0	++	+	++	0
Perception	0	+	+	0	+	0	0	+	0	0	0	0	++	+	+	+	+	0	0	0	++	++	++	0
Required Lifespan	+	0	0	+	0	++	++	++	+	0	0	++	++	0	+	0	++	++	++	++	0	0	+	++

3 Method to Derive Secondary Parameters for Designing ASS

Using these critical parameters, it is possible to analyze potential applications regarding a non- or less-critical implementation. In order to enable a significant and clear analysis, parameters were summarized. Table 3 combines interesting applications, based on the correlation matrix [2] with the critical parameters of Table 2 in order to detect applications, which qualify for operational prototypes. Thereby, nine critical examples were detected. Those examples are predestined to serve as first prototypes. By designing critical prototypes, it is possible to derive further important parameters (secondary parameters) and to reveal additional dependencies within application, active and passive structure. Analyzing a potential implementation, five different prototypes regarding their critical parameters were selected (marked grey) to derive general design parameters.

Table 3. Interferences of application, active and passive structure based on representative, selected examples (0: noncritical, +: critical, ++: highly critical).

Automotive	Ergonomics	Technology	Design	Dimensions	Installation Space	Force	Activation / Deactivation	Acceleration	Temperature	Operating	Deformation	Shape	Memory Effect	User Application	Lifespan	
Window Darkening	x	x	x	0	++	++	+	+	+	0	++	+	++	+	+	14
Seat Functions	x	x	x	+	+	0	++	+	+	0	+	++	++	+	0	12
Exterior Door Handle	x	x	x	+	++	++	+	+	++	++	++	++	+	++	++	20
Adjustable Interfaces	x	x	x	+	++	+	+	+	++	+	+	++	+	++	0	21
Steering Interface	x	x	x	+	++	0	+	+	++	+	+	++	+	++	0	14
Interior Air Louvre	x	x	x	0	+	+	+	0	+	0	++	+	0	+	0	8
Mirror Adjustment	x	x	x	0	0	++	+	0	0	0	++	0	0	0	0	5
Aerodynamic Elements		x	x	0	+	++	+	+	+	0	++	++	++	+	+	14
Control System		x		+	+	0	++	+	+	0	0	+	0	0	0	7
Tools																
Haptic Feedback	x	x	x	+	++	+	+	+	++	+	+	++	+	++	0	15
Adaptive Handle	x	x	x	+	++	++	+	+	++	++	++	++	++	++	++	21
Adjustable Interfaces	x	x	x	+	++	+	+	+	++	+	+	++	+	++	0	21
Wearables																
Information Transfer	x	x	x	+	+	0	+	+	++	++	+	+	+	+	0	12
Haptic Feedback Jewelery	x	x	x	++	++	0	+	+	++	++	+	++	+	+	0	15

4 Approach to a Method for the Design and Integration of SMAs - Prototypes and Secondary Parameters

To establish a generally valid method it is essential to evaluate applications containing critical constraints. Analyzing critical applications will lead to further detailed and important parameters as well as prevent failures within complex applications. The method will be derived using two main parts: primary and secondary parameters. As described, primary parameters are set up through research on application, passive and active structure. Secondary parameters will be established designing and testing specific prototypes of adaptive structures and systems.

After defining primary parameters and deriving applications, it is necessary to transfer the knowledge to the prototypes. Thereby it is possible to confirm the primary parameters and to derive secondary parameters from the implementation. For designing and manufacturing prototypes, additive manufacturing, especially fused deposition modeling (FDM) is predestined. Among others, it is possible to generate fast and inexpensive different concepts, design complex geometries, model cavities for the integration of SMAs, customize products individually and test and scrutinize them further before mass production occurs, print multiple materials at the same time or complete systems, as well as interrupt the manufacturing process anytime.

Adaptive structures and systems are characterized by performing a movement to adapt for example to different external conditions. Therefore, the movement of the active structure (SMA) must be transmitted to the movement of the passive structure to achieve the desired motion of the application. There are two general design possibilities for the passive structure to realize the motion (Fig. 2): using the elastic deformation of the material (low elastic modulus, geometry and structure adapted to workload) or generating a flexible connection (kinematic pair) through at least two moving components (degrees of freedom, non-contact).

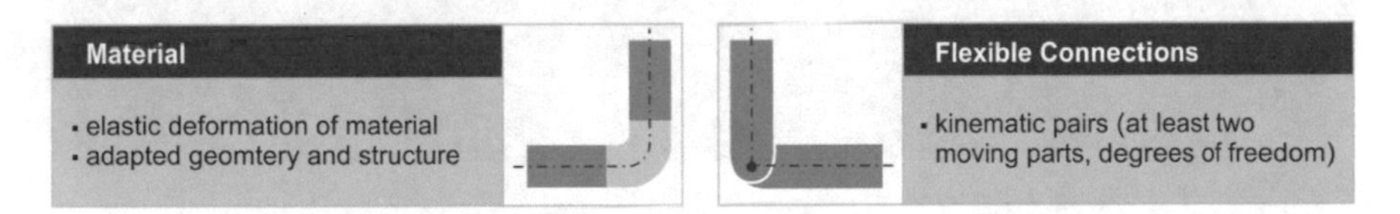

Fig. 2. General design possibilities for the passive structure to realize the motion.

In order to design prototypes, it is important to perform preinvestigations regarding possible types of deformation and movement. In general, there are two main movements: translation and rotation. These are achieved through three main deformations: tension/compression, bending and torsion. Table 4 contains the types of deformation and the position of the SMA wire. Depicted is a possible example to position the SMA wire to achieve the desired deformation (wire movement: contraction). These general movements and deformations can later be applied to the different prototypes.

Table 4. Examples of the general types of deformation.

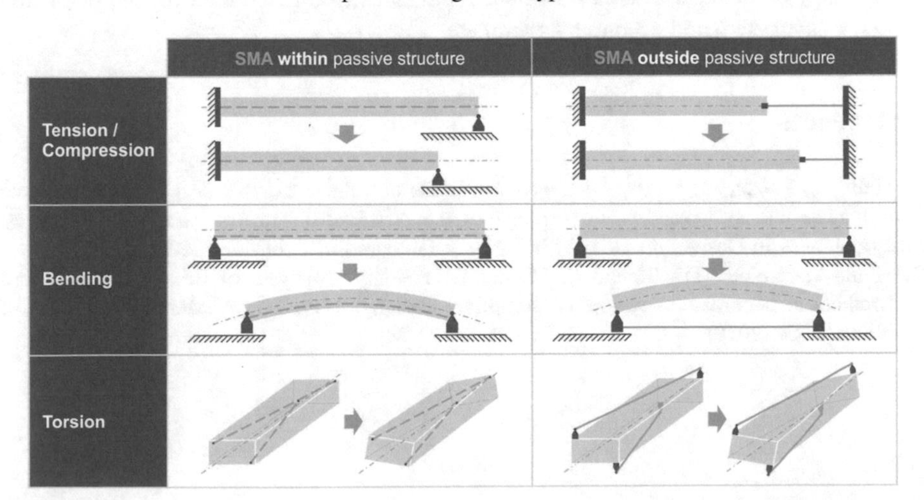

As an approach to a method for designing and integrating SMAs the design processes of the three main parts application, passive and active structure were located within the product development process (Fig. 3). As the parameters of the application specify the required demands, the problem analysis starts with a list of requirements focusing on the application. This list is transferred to the active and passive structure. The conceptual and embodiment design comprise calculation and design of active and passive structure. While finishing the detail design the application containing the parameters operation and usage as well as visibility and perceptibility gains importance.

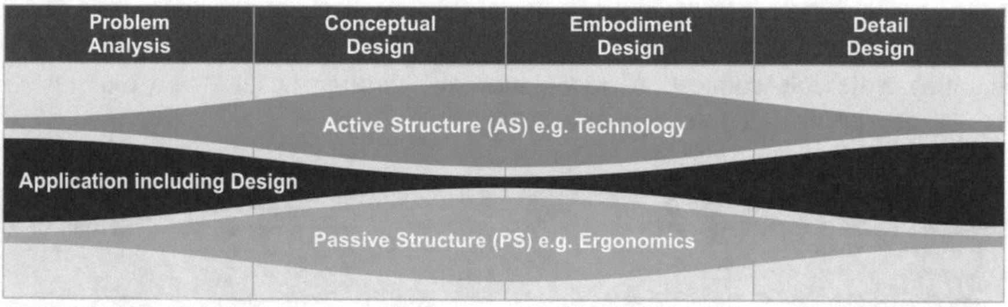

Fig. 3. Locating application, passive and active structure within the product development process according to VDI 2221 [4].

5 Conclusion and Outlook

An approach to a method was derived using two main parts: primary and secondary parameters. Primary parameters were set up through research on application, passive and active structure. Secondary parameters are established designing and testing specific prototypes of adaptive structures and systems. Findings of this research contribute to guidelines to design adaptive structures and systems methodically. The detected important applications will be implemented as prototypes to derive further (secondary) parameters. Based on this a method will be derived using the evaluated primary parameters and secondary parameters.

References

1. Hein, A., Patzer, E., Maier, T.: Improving HMIs of vehicle exterior design using adaptive structures and systems. In: Proceedings of the AHFE 2017 International Conference on Ergonomics in Design, pp. 261–273. Springer International Publishing AG (2017)
2. Hein, A., Holder, D., Maier, J., Maier, T.: Potential Analysis of Smart Materials and Methodical Approach developing Adaptive Designs using Shape Memory Alloys. In: NordDesign (2018)

3. Feldhusen, J., et al.: Vorgehen bei einzelnen Schritten des Produktentstehungsprozesses. In: Feldhusen, J., Grote, K.-H. (eds.) Pahl/Beitz Konstruktionslehre. Springer, Berlin (2013)
4. VDI 2221: Methodik zum Entwickeln und Konstruieren technischer Systeme und Produkte. Beuth Verlag GmbH, Berlin (1993)

A Material Database Framework to Support the Design of Shape-Changing Products

Marius Hölter[1], Agnese Piselli[2(✉)], Sara Colombo[3], and Barbara Del Curto[2]

[1] Design Department, Politecnico di Milano, Milan, Italy
[2] Chemistry Materials and Chemical Engineering Department "Giulio Natta", Making Materials LAB, Politecnico di Milano, Milan, Italy
agnese.piselli@polimi.it
[3] MIT Design Lab, Massachusetts Institute of Technology, Cambridge, USA

Abstract. New classes of smart materials are emerging, revolutionizing the way we design and interact with products. Their dynamic properties are changing our perception and understanding of what a material is in itself (a system), and especially what it is able to do (its performance). The use of smart materials generates new opportunities in the creation of future forms of interaction, promoting the concepts of material move, material turn, material lens, and dynamic products. This study aims to provide design students and industrial designers with information and inspiration on the topic of Shape-Changing Material Systems (SCMSs) by developing a framework for an explorative database. The integrated digital tool, implemented with a set of 25 case studies, fosters the further development of these materials, and opens up new opportunities of application in multisensory dynamic products.

Keywords: Shape-Changing Material Systems (SCMSs) · Smart materials Materials database · Human factors · Product design

1 Introduction

The study of materials and their technical, sensorial and manufacturing properties is an essential element in the education of a product designer [1]. Materials and manufacturing processes are at the core of any physical artefact, and not only do they contribute to its function, but they also have aesthetic and emotional values which allow the designer to shape the character of a product [2, 3]. Having an up-to-date knowledge about advanced materials and manufacturing technologies is of greatest importance for industrial designers to not miss out on new opportunities that might present a solution to their next design problem [4].

In this panorama, new classes of materials are emerging (e.g., smart materials, bio-based materials, DIY materials) [5], revolutionizing the way we design experiences and interact with products [6, 7]. Among them, smart materials are recognized more and more as systems, characterized by digital–physical substrates [8], rather than a simple material class [9]. Their dynamic properties are changing our perception and understanding about what a material is in itself (a system), and especially what it is able to do

W. Karwowski and T. Ahram (Eds.): IHSI 2019, AISC 903, pp. 648–654, 2019.
https://doi.org/10.1007/978-3-030-11051-2_98

(its performance). Materials are becoming something that is 'alive', and so will the future products that incorporate them.

The use of smart materials generates new opportunities in creating future forms of interaction [10], promoting the concepts of material move [11], material turn [12], material lens [13], and dynamic products [14]. Within the research area of smart material composites and Tangible User Interfaces (TUIs), one type of interaction seems to be of particular interest for industrial designers: dynamic, physical shape change [15]. Shape-Changing Material Systems (SCMSs) are becoming of increasing interest [16], as they have promising product applications. Shape-Changing Material Systems (SCMSs) identify a wide range of material concepts as shape-changing composites, dynamic, digital and programmable materials (e.g., programmable carbon-fiber based composite [17]) that are characterized by a shape transformation.

2 Problem Statement and Research Aim

Smart materials have the unique ability to respond to stimuli and adapt to the environment. Through their unconventional behavior, they offer novel possibilities for designers, especially when it comes to the design of interactions and experiences. To take advantage of smart material behaviors, it becomes fundamental to have knowledge of their properties (technical, sensorial, etc.) and their current applications [6]. To provide basic knowledge and encourage the use of "traditional" smart materials in product design, informative tools have been developed [6, 14, 18]. Despite this, there is still no evidence of a systematic, easily implementable and digital tool to serve this purpose. For this reason, the need of guiding designers through a structured smart material selection practice emerged, with the aim to increase the integration of smart material thinking in their design practice.

This research presents a new framework of an explorative and implementable database on Shape-Changing Material Systems to support design students and industrial designers in developing a deep understanding of what constitutes them (i), their sensorial and technical properties (ii) and the enabling technologies and fabrication processes that can be used to manufacture them (iii). The database represents a digital tool that aims to provide designers with information and inspiration on the topic of SCMSs, fostering the further development of these materials, and opening up new opportunities for their application in multisensory products.

3 Materials and Methods

To develop the new database framework, a literature review on the most established tools for material selection and exploration was conducted. Based on their structure, language and approach, five different databases were compared: design-based (Material ConneXion®; MaterialDistrict), engineering-based (MatWeb; MatBase), and integrated material databases (Granta Design CES EduPack). The Products, Materials and Processes Database (PMPDb), developed in 2016 by Figuerola [19], has been analyzed as it represents the first approach towards equally engaging designers and engineers.

These databases were selected not only to identify the material properties and parameters that are interesting when choosing materials, but also to understand how to correlate the transformation phenomena of SCMSs to their technical properties and the constituents that enable the shape-change behavior.

Moreover, 40 projects, experiments and concept studies on the field of dynamic shape change in physical products and materials were analyzed. This research provided a deep investigation of the technologies and tools used for the fabrication of SCMSs, their application areas, key features and main functional principles.

Based on these reviews and on the framework of the most comprehensive commercial material database analyzed (CES Selector), a set of properties and parameters was selected to characterize and describe SCMSs.

- *General Information* [Study/Concept name; Year of publication/development; Research area; Reference source; Development status] is important to get a better understanding of the current SCMS feasibility and status of development, assessing also potential risks (technical failures, development costs, etc).
- *Application* [Potential markets; Scale] helps designers to find inspiration in a specific area of research or application. The "Scale" is related to the size of the system as well as the magnitude of shape change that can be achieved.
- *Material Information* [Input Stimulus; Intrinsic Transformation; Material Composition; Active and Passive Material Units]. When working with SCMSs, different types of Input Stimuli (light, temperature/heat, moisture, electric field/potential, magnetic field, chemical, pressure) can be used for the materials classification [6, 15, 20, 21]. Intrinsic Transformation describes the material's first response to the input stimulus: in material-based soft body transformations, the structural elements function as actuators themselves [22]. The physical structure and the active/passive constituents of a system (Material Units) [15] play a key role in enabling the physical shape change of a SCMS. To better explain this concept, Textuators case study (Table 1, n. 23) is based on the use of an electroactive polymer (EAP) mix that was used to coat the passive material units, and requires to be immersed in electrolyte to activate the shape changing behaviour.

Table 1. List of 25 Shape-Changing Material Systems case studies

	Project name	Source	Status of development
1	4DPrinting: Multi-Material Shape-Change	Tibbits (2014)	Experiment
2	aeroMorph	Ou et al. (2016)	Prototype
3	Artificial Muscles from Fishing Line	Haines et al. (2014)	Experiment
4	bioLogic	Yao et al. (2015)	Prototype
5	Biomimetic 4D Printing	Gladman et al. (2017)	Experiment
6	Exoskin	Tome (2015)	Prototype
7	Ferromagnetic Soft Materials	Kim et al. (2018)	Experiment
8	Fluid-Driven Origami-Inspired Muscels	Li et al. (2017)	Prototype

(*continued*)

Table 1. (*continued*)

	Project name	Source	Status of development
9	Granular Jamming	Jiang et al. (2013)	Prototype
10	HygroSkin - Metrosensitive Pavilion	Krieg (2014)	Prototype
11	jamSheets	Ou et al. (2014)	Prototype
12	Lilies	Dana Zelig (2017)	Experiment
13	Multidirectional Muscles from Nylon	Mirvakili and Hunter (2014)	Experiment
14	Multimaterial 4D Printing	Ge et al. (2016)	Experiment
15	PneUI	Yao et al. (2013)	Prototype
16	Programmable Carbon Fiber	Papadopoulou et al. (2015)	In Development
17	Programmable Knitting	Scott (2015)	Experiment
18	Programmable Wood	Correa et al. (2015)	Prototype
19	Shutters	Coelho and Maes (2009)	Prototype
20	Smart Granular Materials	Dierichs et al. (2017)	Experiment
21	Soft Materials for Soft Actuators	Miriyev et al. (2017)	Prototype
22	Stress Ball	Schramm (2016)	Prototype
23	Textuators	Maziz et al. (2017)	Experiment
24	Thermorph	An et al. (2018)	Prototype
25	uniMorph	Heibeck et al. (2015)	Prototype

– *Dynamic Effects* [Structure; Volume; Geometry; Orientation; Surface Texture; Softness/Hardness; Flexibility/Stiffness] comprises the different types of shape-change related outputs that can result from a transformation and that can be perceived by the user. These parameters, linked to sensorial properties, have been derived from literature review and databases analysis [18, 22, 23].
– *Performance Properties* [Programmability; Directionality; Self-Recovery; Transformation Speed; Deformation Strength; Power Requirement; Functional Environment] describe the inherent requirements and technical capabilities that characterize SCMSs. They are mainly anchored in the results of the literature review and are specific for SCMSs. The integration of such properties in the new database represents an original contribution of this research.
– *Fabrication* [Process; Tools and Technology].
– *Additional Notes.*

4 Results and Discussion

Using the set of parameters described above, the framework of material database has been built. Its framework is inspired by the structural setup of the CES EduPack PMPDb: the proposed database is centered on SCMSs Concept data-table linked to four further data-tables, as described in Fig. 1.

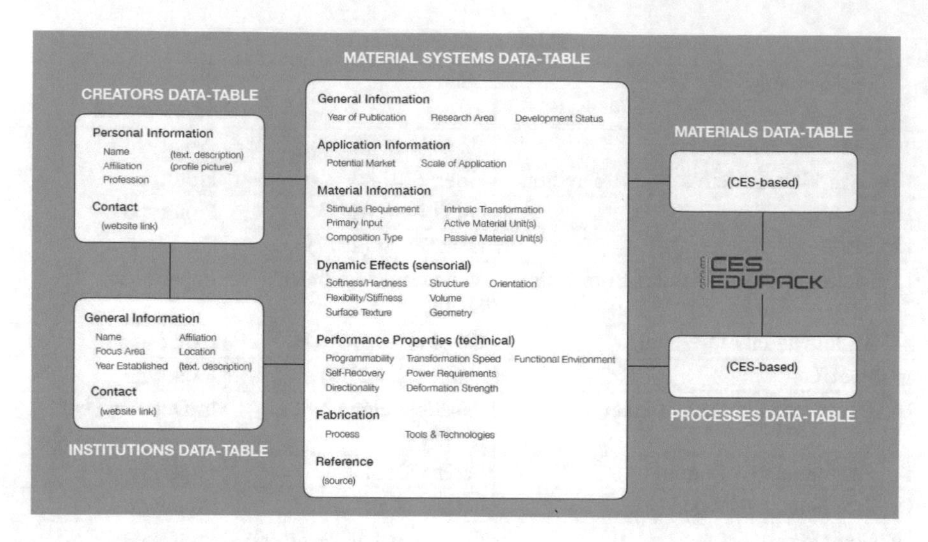

Fig. 1. SCMSs database framework and links to the five data-tables

A set of 25 SCMSs case studies (Table 1), selected among the 40 previously analysed and representative of the variety of behaviors for dynamic shape change in physical objects, was implemented in the database (Fig. 2).

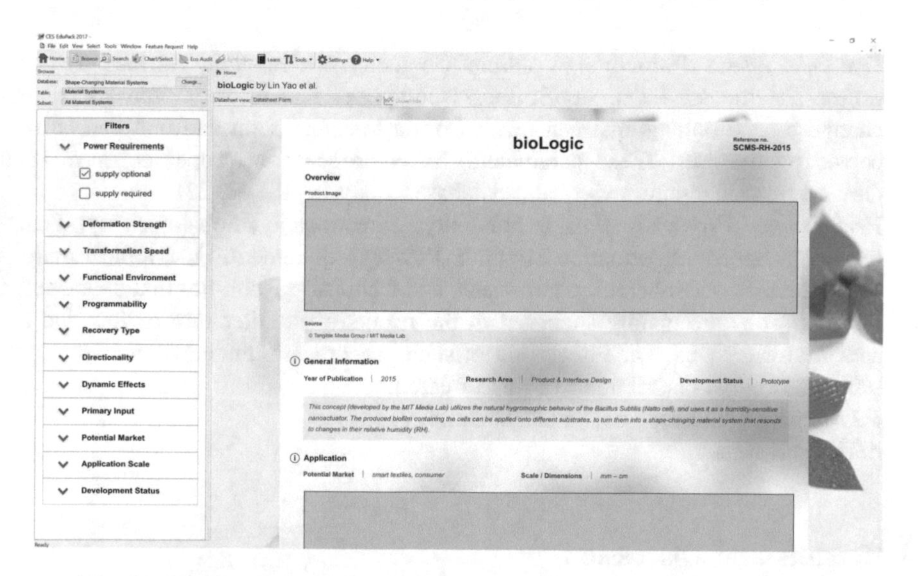

Fig. 2. SCMS record showing the overview and general information sections

5 Conclusion and Further Developments

The main aim of this work was to develop an integrated and informative tool to educate designers on Shape-Changing Material Systems (SCMSs) properties, providing also their classification. The following points illustrate the research main findings:

- Literature research on traditional material databases has been performed to guide the new database structure development.
- 40 case studies have been studied to describe a number of properties and parameters able to characterize SCMSs.
- The framework for a digital SCMSs database consists of five data-tables.
- 25 SCMSs case studies, representative of various behaviors of dynamic shape change in physical objects, are implemented in the first database prototype.

Further investigation will be done to:

- Validate the SCMSs database, conducting user tests with a group of design students, design professionals, and material scientists.
- Expand the database with other stimuli-responsive materials and systems (e.g., thermoresponsive SCMSs, etc.).
- Implement an open multidisciplinary platform, where experts with different backgrounds contribute in updating on new material advancements.

References

1. Ashby, M., Johnson, K.: Materials and Design: The Art and Science of Material Selection in Product Design. Butterworth-Heinemann, Oxford (2002)
2. Karana, E., et al.: Emerging material experiences. Mater. Des. **90**, 1248–1250 (2016)
3. Piselli, A.: Material selection in the professional appliances industry. In: Polimi Design PhD_018. Franco Angeli, Milano, pp. 19–41 (2018)
4. Dastoli, C., Bolzan, P., Bianchini, M., Del Curto, B., Maffei, S.: Design, digital fabrication & 3D printing: a crash course for design students. In: EDULEARN18, pp. 4774–4781 (2018)
5. Veelaert, L., Bois, E., Ragaert, K.: Design from recycling. In: EKSIG 2017, pp. 129–143 (2017)
6. Lefebvre, E., et al.: Smart materials: development of new sensory experiences through stimuli responsive materials. In: Proceedings of the 5th STS Conference, p. 10 (2014)
7. Colombo, S.: Merging Digital and Physical: Tangible Interactions. SpringerBriefs in Applied Sciences and Technology, pp. 21–30 (2016)
8. Wiberg, M.: Interaction, new materials & computing - beyond the disappearing computer, towards material interactions. Mater. Des. **90**, 1200–1206 (2014)
9. Parisi, S., et al.: Mapping ICS materials: interactive, connected, and smart materials. In: Proceedings of the 1st IHSI Conference, pp. 739–744 (2018)
10. Piselli, A., Garbagnoli, P., Cavarretta, G., Del Curto, B.: The shape of light: an interactive approach to smart materials. In: Proceedings of the 20th ICED Conference, vol 9, pp. 219–228 (2015)

11. Fernaeus, Y., Sundström, P.: The material move how materials matter in interaction design research. In: Proceedings of the Designing Interactive Systems Conference (DIS), p. 486 (2012)
12. Wiberg, M., Robles, E.: Computational compositions: aesthetics, materials, and interaction design. Int. J. Des. **4**(2), 65–76 (2010)
13. Jung, H., Wiltse, H., Wiberg, M., Stolterman, E.: Metaphors, materialities, and affordances: hybrid morphologies in the design of interactive artifacts. Des. Stud. **53**, 24–46 (2017)
14. Colombo, S.: Sensory experiences. Informing, engaging and persuading through dynamic products. Ph.D. Thesis, Politecnico Di Milano (2014)
15. Yao, L.: Shape changing composite material design for interactions. Ph.D. Thesis, Massachusetts Institute of Technology (2017)
16. Tibbits, S.: Active Matter. MIT Press, Boston (2017)
17. Self-Assembly Lab MIT, https://selfassemblylab.mit.edu/programmable-materials
18. Bergamaschi, S., Lefebvre, E., Colombo, S., Del Curto, B., Rampino, L.: Material and immaterial: new product experience. Int. J. Des. Objects **10**(1), 11–22 (2016)
19. Figuerola, M., Lai, Q., Ashby, M.: The CES EduPack Products, Materials and Processes Database - White Paper (2016)
20. Bengisu, M., Ferrara, M.: Materials that Move. SpringerBriefs in Applied Sciences and Technology, pp. 5–38 (2018)
21. Addington, M., Schodek, D.L.: Smart Materials and New Technologies. For the Architecture and Design Professions. Elsevier, Amsterdam (2005)
22. Coelho, M., Zigelbaum, J.: Shape-changing interfaces. Pers. Ubiquitous Comput. **15**, 161–173 (2011)
23. Piselli, A., et al.: Development and evaluation of a methodology to integrate technical and sensorial properties in materials selection. Mater. Des. **153**, 259–272 (2018)

Materials for Design. An Experience of Symbolic/Communicative Characterization

Stefania Camplone(✉), Ivo Spitilli, and Giuseppe Di Bucchianico

University of Chieti-Pescara, Viale Pindaro, 42, 65127 Pescara, Italy
stefania.camplone@unich.it

Abstract. If the sensory properties of materials can for a large part assume aspects of objectivity, the same can not be said of their symbolic and communicative properties. But the two aspects are intimately related. In fact, if already from the studies conducted in the Bauhaus school by Itten and Albers research has focused on the sensory mechanisms of knowledge of things, actually we can consider that the actions of observing, listening, touching and smelling, and especially the passage from the sensorial dimension to the perceptive one, are also influenced by the personal experience of each person, reporting sometimes unexpected, unpredictable and inexplicable emotional effects, linking even very distant areas in space and time, thus representing a further classification of the materials. Also human diversity, under a physical, sensory, but above all social and cultural point of view, takes on a significant weight on the symbolic and emotional, and therefore aesthetic and expressive, dimensions of the materials. The knowledge of this "personal" dimension of materials (even in part largely attributable to classification) would allow the designer to foresee the probable "effects" on those individuals that will interact with them. It is however difficult to built a taxonomy system of adjectives capable of clearly "describe" the symbolic-communicative characteristics of the materials, effective for new and conscious attributions of "meaning". The purpose of this study is therefore to verify this possibility, using a sufficiently large sample of users, thus opening the door for a new taxonomic system of materials. This is a first result, towards a "open" classification system, which is useful both to define a symbolic/communicative characterization of materials (and consequently of products), and to relate their sensorial characteristics with their aesthetic and communicative "meanings".

Keywords: Materials for design · Human diversity · Sensory Emotions · Taxonomy system

1 Introduction

The research on materials, in the last decades, has mainly investigated the technological and engineering characteristics, differentiating them in the performances and in the responses to the different solicitations. These are objective classifications that can be used to predict effects or make decisions during the design phase in relation to functional or use requirements. However, these aspects do not deepen those data that pertain to the experience of perception which, due to their often subjective and personal nature,

W. Karwowski and T. Ahram (Eds.): IHSI 2019, AISC 903, pp. 655–660, 2019.
https://doi.org/10.1007/978-3-030-11051-2_99

are hardly censored. And yet, more and more often, product design researches and uses materials especially for the message and for the sensations that they are able, even unconsciously, to evoke and suggest. This happens because every time you use an object, you trigger relationships with the user that depend on his personal history and that are so powerful as to influence the pleasantness of the whole experience with it. It is therefore necessary to try to understand more deeply the nature of the materials that the psychoanalyst, anthropologist and philosopher Jung defines as "a concentration of living forces, both bodily and psychic": also helped by the most recent discoveries of neuroscience, it is possible to describe it also in its qualitative aspects. A greater awareness of these aspects, in fact, would help the designer in the choice of materials and shapes, and would represent a very effective tool for verification of satisfaction both of the engineering needs of a product, and those including the communication and symbolic issues.

The starting point is the response of the sensory organs with which each individual interacts with the world: sight, smell, touch, taste and hearing. In the first phase of using a new product, attention is focused on its technological and superficial performance. At the same time, however, memories and emotions are activated that depend both on personal history and on the context of use: this secondary reaction is difficult to be read, also because it is generally different in each person. However, one can try to find a common emotional character that is explained by different experiences: it is a strong "message", which one can think that everyone receives if he interacts with the same material.

This contribution presents the results of a research experience whose purpose was to relate and to associate in a more objective way the materials to the symbols, the forms and the sensations that they evoke, organizing them in a database of symbolic and communicative characteristics of materials.

2 Objective

The research, which was based on the model of interpretation and the symbolic relationship already developed by the Bauhaus master Itten [1], set the goal of identifying the sensory properties and the meanings that are conveyed by the materials. It is an interpretation, which relates forms and materials, and that very often is considered purely subjective. In particular, the aim of the research was to overcome this subjective and personal sensorial and perceptive "interpretation" of forms and materials, in order to identify and to define their symbolic and communicative characteristics that can be shared by a significant sample of people, up to define an original and useful database for the project.

3 Method

Overcoming the subjectivity of personal experience in the interaction with a material, in order to define an objective classification of the symbolic aspects, required to find the right classification method that could deal with the profound complexity of the system of our senses. In a context in which new materials are developed every day, it is necessary to know the sensorial qualities in order to be able to design and interpret them. This was done by letting a structured sample of individuals choose a set of materials that would involve one of their five senses; subsequently they were asked to grasp their characteristics first on a superficial and phenomenological level, and then deeper and metaphorically, leaving in any case them free to note their personal attributes and effects.

For this purpose, the research has been articulated in some key moments: the first phase was that of the project of the "observation protocol" and of the "detection formats", as well as the identification of the sample of individuals to be observed in their interaction with a specific selection of materials; the second phase concerned the gathering of information, subjecting such individuals to an interactive direct experience with the aforementioned materials and leaving them free to note the emotions and the symbolic effects that arose from touching, seeing or listening them. Subsequently, the information was elaborated, trying to group the materials into families and their expressive characters in keywords, in order to define a database of objective relationships.

3.1 Identification of the Sample and Preparation of the Observation Formats

It was decided to define a sufficiently large sample of individuals. To this end, a sample of 300 people aged between 19 and 25 was identified, both male and female, with a high school diploma, interested in design and applied arts and therefore capable of critical and practical elaboration. At the same time, an observation protocol was defined and four different formats/tables were prepared to be compiled during the observation.

3.2 Observation

The sample of individuals was asked to fill out the formats individually, within a few days.

The first exercise, related to the first format/table, entitled "expressive-sensory experiences 1", concerns the research of 9 samples of materials (referred to 3 sensory aspects, visual-tactile-acoustic, on a 3-level gradation each), directly in their own contexts of daily life, to be included in the first table/format. For each sensory aspect, the format also suggests pairs of adjectives referring to opposite qualities of the materials (e.g. opaque-transparent, hard-soft, reflecting-absorbing, etc.), which each person must choose when filling the format.

The second exercise, related to the second format/table, entitled "expressive-sensorial experiences 2" provides for an in-depth analysis of 5 of the 9 previously selected samples. For each of them it is necessary to freely associate the material with a form (geometric or free, open or closed, etc.: es: circle, curve, polygon, etc.), with some nouns (e.g.: elegance, luxury, tradition, modernity) and with a specific design product.

The third exercise, relating to the third format/table, entitled "abacus of materials", asks for the decomposition of one of the five products previously identified through its components and respective materials, hypothesizing also the type of their production process.

The fourth exercise, relating to the fourth format/table, entitled "sensory and emotional properties", reports the results of the three previous exercises on a summary table which attribute a specific expressive and sensory characteristic (or a combination of them) for each product and component (made with a specific material).

3.3 Comparison with the Parameter of "Human Diversity"

In this type of research, human diversity, in physical and sensory terms, but above all social and cultural, takes on a significant weight on the symbolic and emotional dimension, and therefore aesthetic and expressive, of materials. For this reason it would be necessary to deal with this parameter, in order to highlight any discrepancies in the interaction with the materials. It may indeed happen that the same material or finish refers to different meanings in different cultural contexts. For example, a black material can be associated with a feeling of extreme seriousness simply because in Western culture black color is associated with death, to which, instead, in other oriental cultural contexts, it is associated to white color.

Likewise, differences attributable to different sensory capacities among individuals, also caused by possible different levels of disability (acoustic, visual, tactile or cognitive), could influence the results of a research that investigates precisely the sensorial and expressive aspects of materials.

In this research, therefore, the parameter of human diversity was only partially taken into account: it did not enter into the construction of the formats/tables to be submitted to individuals, but at the same time it was tried not to invalidate the results of the investigation, making sure that the sample of selected individuals came essentially from a homogeneous cultural context and with uniform sensory capacities.

4 Results

The exercises conducted during the observation phase have therefore allowed us to identify specific relationships between materials, forms, senses and symbolic aspects contained in the materials of products and of their individual components. The interpretation of the data, also synthesized through specific graphic schemes, highlights these associative tendencies, up to the possibility of defining a specific "expressive" vocabulary for each family of materials.

In particular, the research included the gathered information in a summary table, in order to be able to interpret them: the sampled materials were in fact over 400. The table was constructed by inserting the results of the first exercise in the first column: the materials chosen by the individuals and the respective families of belonging (ceramics, stone and glass, composites, woods, plastics, metals, fabrics and natural fibers). In the second column it was indicated which of the three senses (sight, touch, smell) was predominantly stimulated. In the third column the three words associated with the material were reported directly by the individuals in the second exercise. In the fourth column the "form" that is inspired by the material has been transcribed. In this column the terms referring to the same concept have also been grouped, to highlight the objectivity of the indicated expressive character. so, for example, wood, which most users have associated with the words “simple”, “economic”, “common”, has been associated with the tag of “economic”. The same synthesis operation was carried out in the last column for the forms associated with the material.

Below are the reflections derived from a first interpretation of the relationships between materials, senses, forms and symbolic/expressive characters.

4.1 Relationships Between Materials and Stimulated Senses

The materials most commonly experienced by individuals have been chosen (woods, plastics and fabrics or derived from processing of natural fibers), highlighting that innovative materials determine greater difficulties in analysis and interpretation.

In the experience with natural materials and fabrics touch was the most stimulated sense, followed by stones and ceramics and plastics. This indicates that the materials that are felt as more natural induce more than others to a tactile approach.

The sight is stressed above all by the finishing of materials and by parameters such as gloss and reflection: 26% of the cases have suggested sight as prevalent sense for plastics while 24% for metals.

Hearing, on the other hand, is stimulated above all by metals that are able, more than any other family, to resonate and produce sound effects. It is interesting to note that even plastics are considered capable of stimulating this sense.

4.2 Relations Between Materials and Shapes

For what concerns the association between shapes and materials, plastics and fabrics suggest circular and rounded shapes, while the ceramic and the woods are associated mostly to geometric and regular shapes, such as rectangles and squares. This leads us to think that some materials are perceived as not workable or deformable and therefore capable of taking only certain defined forms, while others are associated to the possibility of being bent and worked. In particular, it is noted that the fabrics and the papers are mostly associated with open and curved shapes while the plastics are seen in their simplicity and are related to closed and linear shapes.

4.3 Relationship Between Materials and Expressive-Symbolic Terms

A further aspect investigated concerns the associations between the materials and their expressive and symbolic aspects: through the terms associated with the materials by the individuals in the second exercise, the construction of an expressive vocabulary was started which describes the character of each family of materials.

For stones, ceramic materials and glass, the most common terms are related to their mechanical performance (strength, strength, fragility), to their classicism, to the idea of exclusivity, luxury and minimal language. For glass, the idea of lightness and transparency emerges among other features.

Composite materials are described as innovative, and are characterized by the possibility of production processes generating technological and hyper efficient products.

The wood family instead tends to be associated with nature and natural products. Indirectly, this also leads to associating them with wellness and luxury, although they are often defined as light, simple and cheap. It is interesting to note how the woods also suggest resistance and elasticity.

Metals, on the other hand, recall the concepts of strength and workability. They represent the idea of a minimal and contemporary language, with an industrial character. There has also been a tendency to interpret them as cold and aseptic materials; some of them also refer to a retro and luxurious taste.

Plastics are considered modern and contemporary, versatile and adaptable to any context. However, this feature is often combined with the concept of economy, which in turn is associated with low value. It is interesting to note that sometimes plastics are also associated with the concept of eco-sustainability.

The fabrics and natural fibers, finally, communicate softness and adaptability. They are considered natural and therefore far from the artificial world. This associates them with a traditional language, even if poor, but warm and comfortable. The skins convey luxury and elegance.

5 Conclusions

The research started a process of possible association between materials, forms and expressive characters, in an attempt to overcome the simple technical-engineering characteristics. The first results have shown that, while making simplification, it is possible to associate a specific personality to each family of materials.

The construction of a database that manages the sensorial, communicative and symbolic aspects of the materials is therefore useful for governing with consistency and awareness the expressive and communicative aspects of the materials, especially in the meta-design phase of the development of a product concept.

Furthermore, it is desirable that the research can be extended to different cultural contexts, as well as to groups of individuals with different cognitive and sensorial capacities, in order to start a comparison between different cultures and sensitivities with respect to the materials.

Acknowledgments. This paper presents the results of a research experience developed at the Department of Architecture of the University "G. D'Annunzio" of Chieti-Pescara (Italy). In particular, this contribution was written by Stefania Camplone (paragraphs 1, 2, 3) Ivo Spitilli (paragraph 4), Giuseppe Di Bucchianico (Abstract and paragraph 5).

Reference

1. Itten, J.: Kunst der Farbe. Otto Mayer, Ravensburg (1961)

Design for ICS Materials: A Tentative Methodology for Interactive, Connected, and Smart Materials Applied to Yacht Design

Stefano Parisi(✉), Arianna Bionda, Andrea Ratti, and Valentina Rognoli

Design Department, Politecnico di Milano, Via Durando 38a, 20158 Milan, Italy
{stefano.parisi,arianna.bionda,andrea.ratti, valentina.rognoli}@polimi.it

Abstract. The domain of materials for design is changing under the influence of an increasingly technological advancement, which brings miniaturization of technology and material augmentation with the use of sensors, actuators, and microprocessors. Examples of new hybrid material systems with dynamic and computational qualities are increasingly emerging. These are called ICS Materials, an acronym that stands for Interactive, Connected, and Smart. While laboratories and designers around the world are experimenting with these new advanced materials, there is the need to forecast their potentials in the design space and to reflect on their future application critically. This paper drafts the main theoretical foundations and depicts the workshop 'NautICS Materials' – ICS Materials for the Nautical sector – by its objectives, structure, methodology, tools, and results, in order to present a model to transfer to other sectors or to scale up in larger experimental and applied actions for the integration of smart materials in the design space.

Keywords: ICS Materials · Materials experience · Yacht design
Design tools

1 Introduction

New materials with dynamic and interactive qualities, able to sense, process, and materialize data, are emerging under the influence of an increasingly technological advancement, that brings miniaturization of technology and material augmentation. These are investigated by the basic research project 'ICS Materials' [1–4], an acronym that stands for Interactive, Connected, and Smart Materials. Such materials are at their experimental and prototypical stage without a clear destination of use. In the past and present days, the practice of design has always facilitated the development and integration of novel materials [5]. However, the design practice lacks a methodology to approach these materials. The purposes of the research are: (i) to develop methods and tools for design practitioners and students to understand, conceptualize, and design (with) them; (ii) to forecast their potentials in the design space; (iii) and to reflect on their future application critically. This paper is a step in this direction, by proposing a

W. Karwowski and T. Ahram (Eds.): IHSI 2019, AISC 903, pp. 661–666, 2019.
https://doi.org/10.1007/978-3-030-11051-2_100

tentative methodology to design for ICS Materials and describing its application and results in a Yacht Design educational workshop.

The definition of ICS Materials is here proposed as *Hybrid Material Systems*, i.e., material-based systems with different degrees of complexity combining inactive materials, smart material components, and embedded sensing, computing, and actuating technologies [6–8]. They perform shape-shifting, light-emitting, and color-changing behaviors. The seamless combination of elements into a material system might enable less intrusive and more inclusive experiences, a more immediate and responsive interaction, and sustainable integration of technologies into everyday practice. The research is positioned in the intersection of design, new materials, and interaction. We assume a 'behavioristic view' of Interaction [9], which underpins a broad meaning of the term, by also considering other means of interaction different from digital and computational and adopting an inclusive approach [10]. Thus, we consider a broad range of materials empowered by computational, mechanical, chemical, and biological components as sensors and actuators.

ICS Materials arise as potential enablers of meaningful dynamic and interactive materials experiences [11] as tangible interfaces for a diversity of applications, from interactive architecture to smart fashion, from autonomous vehicles to smart and conversational objects. Among the diverse potential areas of use of ICS Materials, Nautical transportation demonstrates to be one of the most suitable and competitive sectors of integration. With new technologies and cutting-edge materials, the yachting industry is evolving rapidly to meet the needs of modern yacht owners and is growing both on sales volume and on boat size. The sector, indeed, is continuously growing, and international market researches [12–14] clearly show a moving towards the large yacht industry with the rising demand for yacht charter and water-based luxury experiences: between 2010 and 2017, the up 60 m market segment, so-called 'megayacht', has grown by an average of 11%, confirming the theory that the high-end sector appears to be more resilient to any crisis in international markets [15]. This type of boat is part of what is called 'luxury design' where a project is highly influenced by the client personality and aesthetic and where all the phenomena are revealed with special characters [16]. Despite the traditional preservative nature of this industry, as yachts are evolving into superyachts and megayachts, the design projects are moving away from the past ergonomic-based use of space and are experimenting new features redefined by the technology.

Looking with the lens of experience design we can identify the following key elements as yacht design trends: (i) *Experience the sea*. Yacht design projects are involving 'soft' features for higher sensory expression enhancing the sense of communion with the sea and a real continuity between indoors and outdoors; (ii) *Innovative layouts*. The General Arrangement is moving away from traditional structures with divided interiors and smaller outdoor spaces. The structural constraints are becoming less, leading to more interesting ways of designing spaces and combining areas with organic structures and pop-up spaces explored in several yacht design concept; (iii) *Focus on health & wellness*. Owners are looking to carry their balanced lifestyle into the world of yachting. The spa experience is accelerating: gyms, cryotherapy chambers, salt inhalation rooms, hot yoga studios are just a few of the ideas on the drawing board.

2 The NautICS Materials Workshop

In a Yacht Design project, ICS Materials may be enablers of meaningful and unique experiences, considering new practices of interaction between the yacht, the sea, and the human behavior. On this theme, a 3-day educational workshop named 'NautICS Materials' was organized and run by the authors, involving 28 students from different backgrounds, with the goal to foresee and ideate future scenarios in the yachting sector, by conceptualizing new ICS materials and applying them in Future Yacht design concepts. A specific set of activities framed in a tentative design methodology – namely 'Design for ICS Materials' – with their supporting tools has been developed: Yachting Scenario Boards, ICS Materials Cards, and a Concept Canvas.

The workshop objective was twofold: (i) to exploit the potential of ICS Material in creating new design concepts of functional and aesthetic elements embedded in the interior layout of the vessel; (ii) to experiment and test a tentative methodology to design for ICS Materials.

2.1 Workshop Methodology

'NautICS Materials' is a three-day workshop program designed as a first step in approaching ICS Material for a yacht design project, organized and run by the authors. This workshop was designed to be adopted as a training program of a design firm or a shipyard new yacht department, as involving participants with a different background – architects, product designers, and engineers – who had no knowledge of ICS material and materials experience. To achieve its objectives, the workshop had three features. First, the participants were divided into 5 multidisciplinary groups of at least 5 members to reflect a common yacht design studio. Second, the work time period was divided into sections for each task, to give a rhythm to the design activity, verify time and tasks and meet efficiency. Finally, a personalized toolkit was given to each team to drive the different design phases. The toolkit contained Yachting Scenarios Boards, ICS Materials Cards, and a Concept Canvas specially designed for the workshop (to know more, go to http://icsmaterials.polimi.it/).

The workshop was organized in the following seven sections conducted in 8 h per day, with a 1 h lunch break:

- *Introduction and preparation* [2 h]: presentation of the macro-trend of the yachting sector and ICS Materials research by the tutors; individuation of the main workshop drivers; division of the participant into groups and toolkit distribution.
- *Exploration* [2 h]: opening of the toolkit and getting familiar with trends, ICS materials, and Yachting scenarios. Answer the question "what does the future hold for superyacht design?" to start thinking how the next level of yachting would be like with new smart materials. Tools: Yachting Scenario Boards and ICS Materials Cards.
- *Definition* [2 h]: to narrow the area of intervention, selection of a part of a yacht journey (sailing, mooring/at anchor) or a part of the day and definition of an on-board space in which develop the concept project of the new material system. Tools: Yachting Scenario Boards and ICS Materials Cards.

- *Conceptualization* [4 h]: ideation of new material system concepts and visualization of yacht experiences enhanced by ICS materials, through sketches, moodboards, storyboards, and textual notes. Tools: Yachting Scenario Boards, ICS Materials Cards, and Concept Canvas.
- *Integration* [4 h]: integration of the material system concept ideas into feasible design proposals. Tools: Concept Canvas.
- *Design* [8 h]: development of Yacht concepts, by using conventional design and representation tools and techniques, i.e. drawing and rendering by hand and software.
- *Delivering* [2 h]: exhibit presentation of the final work to the other teams and open roundtable discussion. Outcomes: 3 poster A2 landscape format for each group.

2.2 Tools

Yachting Scenario Boards. Based on the first experiences of ICS_Materials research project, 'Mapping ICS Materials Workshop 2017' [3] and 'ICS4YD Workshop 2017' [2], five new scenarios for the yachting industry were finalized for this workshop: *The warty jellyfish mood*, *Moisture poetry, Wave of good noise, Thermo-taste*, and *Dynamic equilibrium* [2]. Each scenario was presented through an inspirational A4 board providing a moodboard, an envisioning storytelling, and different keywords regarding the sensorial, emotional, interpretive, and performative layers of the proposed on-board experiences. This tool was used mainly in the *Exploration* phase.

ICS Materials Cards. A deck of 48 cards was designed with the purpose to provide unskilled participants all the elements to understand ICS Materials and build concepts with them, by gaining an understanding of what they are, how they are made, how they work, and how they appear, and identifying major inputs and outputs. Each card depicted an example of ICS Material, with pictures and information, i.e. name of the project, name of the author, a short text describing how it functions and performs, and a graphical schematic representation showing its components, inputs and outputs. To do that each example was deconstructed into its constituting components. The examples selected for the cards encompassed materials, surfaces, and material-based objects and systems used in many applications, with different behaviors and different technological readiness levels. In the *Exploration* stage, the participants were asked to read the content of the cards, cluster them and select the most promising examples according to their scenario.

Concept Canvas. A Concept Canvas in A2 size was designed to be used mainly in the *Conceptualization* phase. The purpose of the canvas was to illustrate and guide the participants through the novel design methodology to conceptualize a new ICS Material. The canvas was divided into three sections, namely (i) *material system building*, (ii) *material system sketching/picturing*, and (iii) *material system description*. The first section provided an empty schematic graphical representation of a material system with blank spaces to complete with the names of components, input, and output. This recalled the same design used in the cards. The purpose was to use the scheme to build a novel material system by getting inspirations from the cards and combining the elements in a new design. The second section provided a blank box where participants could start materializing the first concept idea with sketching, collages of pictures, or mixed techniques. The third section asked to outline the concept with textual technical description ('how it works'), performative description ('what it does'), sensory and experiential description ('how it feels, looks, and sounds'), basing on the Materials Experience framework [17]. This last section aimed to reflect upon the performances and experiences enabled and implied by the concept. Even if we suggested to follow the steps sequentially, the three activities might be carried out parallelly with an iterative approach, as each section inform the others.

3 Results and Discussion

As a result, students conceptualized novel material systems, through the recombination of depicted components. Plus, they purposefully integrated them into design concepts of functional and aesthetic elements embedded in the interior layout of the vessel: (i) *the underwater breathing nest* reinterprets the yacht interior as a living creature able to react to the human presence and heat creating comfortable areas through a shape-shifting smart textiles covering the interior surfaces; (ii) *dynamic flow* materializes the wave sound frequencies in an interior waterfall thanks to external sound sensors; (iii) *heckquilibrium* shows through light the effect of wind and water forces on a sailing boat, covering the interior with movable plywood panels covered by light-emitting smart textiles and optic fibers responding to pressure sensors; (iv) *the floating forest* uses inboard moisture to create a futuristic biosphere providing on-board water and light through a hybrid material system; and (v) *glowrious* reimages the relationship between the on-board natural and artificial light transforming the yacht hull into a luminescent night illusion system, by embedding photo-luminescent pigments into a smart glass controlled through Arduino.

As tangible interfaces, they materialize external and imperceptible environmental data, so that human could experience them through augmented expressions. Changing their characteristics on external stimuli, ICS Materials influence the aesthetics and perception of spaces, encouraging sensory experience while sailing or mooring. Taking inspiration from another industrial sector, the workshop design concepts implemented a new generation of material for composite structure, exterior and interior design and sails with dynamic, augmented, and proactive properties.

The 'NautICS Materials' workshop confirmed the effectiveness of the tentative methodology in achieving the objectives. All the participants, with no previous knowledge on ICS Materials and materials experience, were able to conceptualize novel material systems with different degrees of complexity combining inactive materials and smart material components. The toolkit proved its potential in guiding the design phases from the material understanding, to the new materials conceptualization, and their integration into yacht design concepts. The cards overcame the limitations caused by the lack of physical samples of the actual materials. However, future development of the methodology may integrate material samples and prototyping. Furthermore, the workshop proved the potential of ICS Material to influence the yacht spaces perception enhancing the onboard experience.

The paper drafted the main theoretical foundations about the research 'ICS Materials' and depicted the workshop 'NautICS Materials' by its objectives, structure, methodology, tools, and results, in order to present a model to transfer to other sectors or to scale up in larger experimental and applied actions – not only in education, but also in practice with industrial partners – for the integration of smart materials and technologies in the design space.

Acknowledgments. This work is part of 'ICS_Materials', a basic research project carried out at Design Department of Politecnico di Milano and funded by the University Basic Research Funding FARB 2015. We thank the colleagues taking part to the research: Camilo Ayala Garcia, Venanzio Arquilla, Mauro Attilio Cecconello, Marinella Ferrara, Venere Ferraro, and Davide Spallazzo. The workshop was organised at MYD, Master in Yacht Design of Politecnico di Milano. The research is partially funded by Fondazione F.lli Confalonieri (Ph.D. student grant).

References

1. Ferrara, M., Rognoli, V., Arquilla, V., Parisi, S.: Interactive, connected, smart materials: ICS materiality. In: IHSI, Advances in Intelligent Systems and Computing, vol. 722 (2018)
2. Bionda, A., Ratti, A.: Exploring scenarios for ICS materials in the Yacht Design framework. In: IHSI, Advances in Intelligent Systems and Computing, vol. 722 (2018)
3. Parisi, S., Spallazzo, D., Ferraro, V., Ferrara, M., Ceconello, M.A., Ayala-Garcia, C., Rognoli V.: Mapping ICS materials: interactive, connected, and smart materials. In: IHSI, Advances in Intelligent Systems and Computing, vol. 722 (2018)
4. Parisi, S., Rognoli, V., Spallazzo, D., Petrelli, D.: ICS materials. Towards a re-interpretation of material qualities through interactive, connected, and smart materials. In: DRS18 (2018)
5. Ashby, M., Johnson, K.: Materials and design: the art and science of material selection in product design. Butterworth-Heinemann, Oxford (2002)
6. Vallgårda, A., Sokoler, T.: A material strategy: exploring material properties of computers. Int. J. Des. **4**(3), 1–14 (2010)
7. Razzaque, M.A., Dobson, S., Delaney, K.: Augmented materials: spatially embodied sensor networks. Int. J. Commun. Netw. Distrib. Syst. **11**(4), 453–477 (2013)
8. Barati, B., Giaccardi, E., Karana, E.: The making of performativity in designing [with] smart material composites. In: CHI'18 (2018)
9. Saffer, D.: Designing for Interaction: Creating Innovative Applications and Devices. New Riders Pub., Berkeley (2009)
10. Buchanan, R.: Design research and the new learning. Des. Issues **17**, 3–23 (2001)
11. Karana, E., Pedgley, O., Rognoli, V.: On materials experience. Des. Issues **31** (2015)
12. Deloitte: Boating market monitor. Market insight of the international recreational boating industry (2018)
13. Boat International: Global order Book. The Global Picture (2018)
14. Global Industry Analysts Inc.: Yacht Industry. In A global strategic business report (2017)
15. Campolongo, M.: House and Yacht: the aesthetics of the interior as a link between different sectors. Des. J. **20**(sup1) (2017)
16. Celaschi, F., Cappellieri, A., Vasile, A.: Lusso Versus Design. FrancoAngeli, Milano (2005)
17. Giaccardi, E., Karana, E.: Foundations of materials experience: an approach for HCI. In: CHI'15 (2015)

Augmented Materials for Tangible Interfaces: Experimenting with Young Designers. Outcomes and Analysis

Marinella Ferrara(✉) and Anna Cecilia Russo

Design Department, Politecnico di Milano, Milan, Italy
{marinella.ferrara,anna.russo}@polimi.it

Abstract. This paper deals with the design of emerging, *augmented, ICS materials* in the framework of a *Design-Driven Material Innovation* approach, presenting a selection of concepts of Tangible Interfaces, developed during a workshop held at *Politecnico di Milano*. In the attempt to include digital technologies in product and interior design, enhancing human experience and new perspectives for next interactive products, young designers were asked to explore the new possibilities offered by these novel materials in an *Experiencing Prototyping* and *Smart Aesthetics* framework. The core of the paper focuses on the outcomes and the analysis of this experience between research and educational activity. The last part refers to the ongoing shift within Design Schools, where the educational mission is increasingly promoting workshops and partnerships with design oriented companies. The novel technological landscape implies indeed this sort of collaborations, as Design Schools can effectively assist companies in making evidence based decisions.

Keywords: Augmented materials · *ICS* materials · Material interfaces
Smart aesthetics · Material design teaching · Product design · Interaction design

1 Introduction

Since the early nineties it was already clear that a new methodology was needed to successfully introduce emerging technologies within design educational environments [1]. Hence structuring more innovative approaches, especially today, has become an essential didactic requirement. During the last decade, product and interaction designers, as well as *HCI* experts, have drawn the attention to tangible interfaces, embedding interactive technologies in products, surfaces and high tactile textured traditional materials with an emphasis on the look and feel of materials themselves. In line with these guidelines, design teachers and instructors are increasingly dealing with applying effective methodological approaches, likely to let their students include digital technologies in product and interior design, enhancing human experience and new perspectives for next interactive products. In this paper we will then analyze the outcomes of a didactic experience in relation to the two days' workshop *Exploring Interactive Material Interfaces*, held at the Design School of *Politecnico di Milano* in April 2017. Inspired by current design practices, aiming to foster creative engagement

W. Karwowski and T. Ahram (Eds.): IHSI 2019, AISC 903, pp. 667–672, 2019.
https://doi.org/10.1007/978-3-030-11051-2_101

through *Tinkering*, speculative play instinct [2], and cross-disciplinary creativity, we asked design students to explore the new possibilities offered by interactive, *augmented*, also named *hybrid* [3] and *ICS* [4], *materials*, in the framework of a *Design-Driven Material Innovation* approach. Pedagogical paths recalling the importance of focusing on experimental and inductive reasoning, based on direct experience, were already part of Dewey's research [5], that encouraged a pragmatic method while structuring even a philosophical discourse centered on creative activities in general. While exploring the potential of novel materials, students were indeed stimulated to enhance their tactile and sensory reaction to inspire a new level of perception. Such a perspective roots back to Bruno Munari's lesson, who used to warmly encourage a sensorial relationship with objects of everyday use, but also to Richard Shusterman's *Full Body Thinking Approach* [6], highly appreciated especially in the field of Interaction Design studies. Several outstanding Labs, such as the *Self-Assembly Lab* or the *XLab*, both at *MIT,* are now applying multisensory and cross-field approaches to the implementation of their revolutionary technologies, letting science and hi-tech be nourished by emotional responses and soft-qualities insights. In such a perspective we fostered a multidisciplinary attitude, creating an osmotic exchange also with humanities, helping to structure a more critical and speculative thinking.

2 An Experience Between Research and Teaching Through *Smart Aesthetics* and *Experiencing Prototyping*

The outcomes of the research and teaching experience analyzed throughout this paper refer then to a workshop addressed to twenty-four second year's design students. The educational insights specifically focused on stimulating these young designers-to-be to develop prototypes of material interfaces to be used in a domestic environment, making them reactive and interactive. All the new implementations, related to the embedding of emerging technologies in surfaces and objects of everyday use, have so far generated a new category of objects, likely to be defined as almost-living ones [7] as capable to somehow interact with human beings, determining interesting connections and emotional responses in terms of what can be defined, as a sub branch of the applied aesthetics domain, *Smart Aesthetics* [7]. Multisensory stimulation and involvement are now having a big part in the whole *Design Thinking* approach and providing students with inspirational settings and a basic technical know-how is then starting to play an essential role in the way future designers get trained. Consequently, we have gradually experienced a shift from a *Technology-Centered* perspective, once essentially anchored on a digital base, to a *Behaviorist* approach, mainly centered on a *Human-System Interaction* scale, implying a true dialogic intercourse, now even between human beings and even sophisticated artificial intelligences. Design students, though extremely fascinated by emerging technologies and intelligent systems, are often intimidated by the technical expertise required to embed electronics, sensors and any sort of highly technical equipment in their concepts. So we thought the best way to break the ice and let their creativity flow was to let them directly experience the potential of technical components in a framework of *Randomized Experiment,* just like science practitioners and more recently also makers and start uppers do to eliminate bias and better achieve the greatest reliability and statistical validity, and *Experiencing Prototyping* [8]. *Experiencing Prototyping* has been often used to facilitate different activities during the design process, including understanding, exploiting and communicating the experiential aspects of designing and using not-yet exiting artifacts. Such an approach allows then each student to seize and get the sense of real experiences themselves, unlocking new creative and implementation potential. This methodological approach was also based on the remarks made by Don Norman to the world of Design Education, on how poorly trained design students were to successfully meet the contemporary demand [9], stressing also on how designers need to pay more attention to social and behavioral aspects, while getting a better know-how in terms of science and technology, without anyway turning themselves into scientists or even engineers [9]. Consequently, even the delivery of science and technology insights to design students has to meet contemporary standards, adapting to the level reached today in terms of sophistication and breakthroughs, instead of referring to what only learnt at school in previous years [10]. Based on these premises, we will now present a selection of four concepts analyzing, according to a *HCD Human Centered Design* and *UCD User Centered Design* perspective, the possible implications and the level of involvement potentially generated.

3 From *Ready-at-Hand* Technology and *Tinkering* to Concepts Delivery and Outcomes

Before presenting the selected concepts, we would like to detail how young designers were introduced to electronic components, sensors or even *Arduino* microprocessors. First of all, a briefing was provided to clarify the definition of *Augmented Material* in order for them to understand such an emerging tangible quality in relation to objects, that hybridizes the material dimension with the digital one, disclosing as a whole system able to enhance a highly sensorial and a new experiential involvement. Subsequently, the idea of *Augmented Material* was transferred to the concept. Out of ten groups of students, consisting of two or in some cases three members, here follow four concepts that positively implemented what outlined during the briefing. Starting from an object of everyday use, but with a strong tactile appealing, like a sponge, two of the young designers chose to install within four switches all connected to a speaker, able to reproduce the different sounds related to each of the switchers. Beyond generating a playful almost-living-object, they were able to augment a common yellow sponge only by adding basic components to its physical structure, generating a perceptive displacement, while adding a stress-relieving connotation. Such a basic concept disclosed in all its simplicity as an interesting starting point for further developments and future designs. The manipulation of *ready-at-hand* technology, in a *Tinkering* scenario, led then to a natural process of implementation of what just the day before represented an unexplored field, somehow challenging and intimidating. The second concept deals with a more poetic and conceptual design, providing an interesting outcome on the side of augmented textile interfaces. Starting again with a *Tinkering* based session, two students were intrigued by the direct observation and manipulation of a 12 RGB led ring that inspired them to create a design offering a highly sensorial experience, thanks just to a piece of fabric embedded with an *Arduino* microprocessor. Connecting indeed a 12 RGB led ring to an *Arduino* the two students were able to revisit the traditional design of a clock, while rethinking the communicational code relating a human being to a watch or a clock. They used a different color for the led lights, setting the color red for the hours and the blue for the minutes. When the two rays combined, the concept automatically released a purple light. And just to provide a stronger involvement on the user's side, they manually embroidered the Roman numeral, from zero to twelve, on the fabric. As often suggested while implementing technology to objects of common use, especially for those with a strong culturally rooted design, referring to metaphors helps enhance a *UCD User Centered Design* approach [11]. Expressing an idea in terms of another, which is basically what metaphors do, both through the language as well as the graphic code, represents a way to shorten the distance, instead of generating a skeptical attitude. So that, metaphors, but also archetypal shapes, while awakening the evocative moments humans tend to go through when approaching novel products or objects [12], do work as useful artifices to stem any prejudice eventually linked to the implementation of emerging technology on objects of everyday use. If well balanced with a multisensory stimulation they can then produce new *Smart User Experiences* and full-body involvements (Fig. 1).

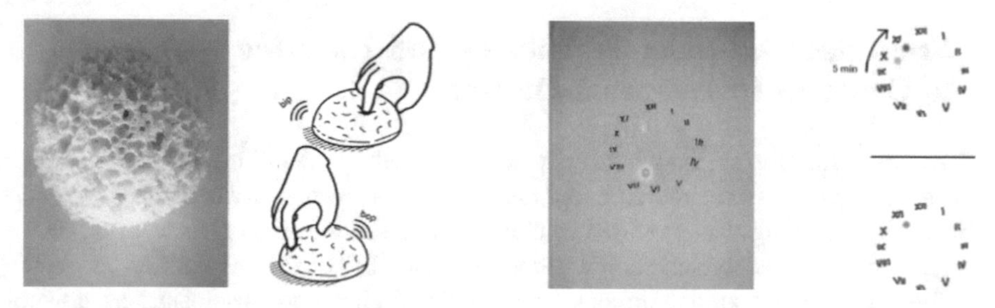

Fig. 1. Left to right: Concept by Laura Casella and Chen Hao Yang; Concept by Chiara Bay and Francesca Zuccheri.

A third concept was inspired by the basic idea of a flower blooming in the presence of light. It aimed to bring the chosen polymeric material to life, letting the petals bloom, once enlightened by a light source. A photo resistor captures light and activates a servomotor that makes the petals move, while switching on the led lights positioned on the corolla. Again, thanks to *Arduino* and a basic circuit, the concept produced interesting outcomes, succeeding in meeting the desired sensorial involvement. The more basic the idea is, with the embedding of electronics just where not expected, the deeper the *Smart User Experience* can be. Reproducing all the magic of a blooming flower through a simple structure in an artificial context delivered a strong payoff to the students themselves, encouraging their research and inspiring for future scenarios. Finally, the fourth concept consists in an interactive panel especially conceived for chromotherapy. The students involved installed between two thin layers of a squared white stretch cotton fabric an RGB recognition sensor connected to an *Arduino* card. Once activated, the light shades produced a sort of "wow effect" in contrast with the minimalism of the design. The communicational code associated to colors, combined with the electronic components, generated an interesting prototype of an interactive material interface. Another starting point then for future implementations, in a *Somaesthetics* framework, where human body discloses as the tools of tools [13], the medium of our being, perception and action in relation to the physical world (Fig. 2).

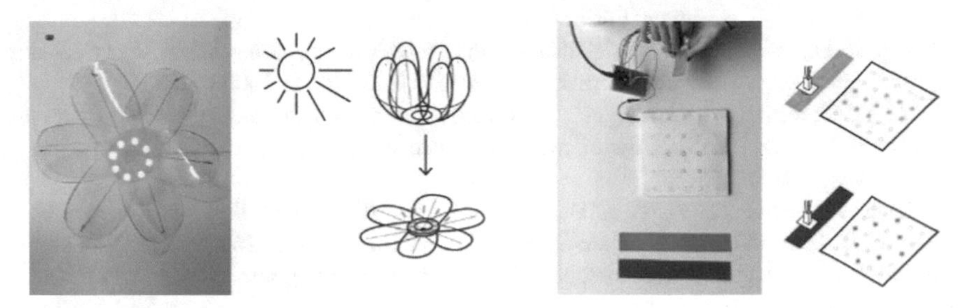

Fig. 2. Left to right: Concept by Camilla Costa and Alice Mingotti; Concept by Lucia Gualdi and Francesca Inzani.

4 Conclusion

In a time when questioning empathy in relation to machines, or discussing the ability of arguing by artificial intelligence [14] disclose as relevant issues within the scientific community, providing young designers with basic insights on emerging technologies and training them in running experiments has turned into an essential requirement of any course/workshop offering an overview on interactive interfaces and augmented materials. Giving then an overview on the several implications on the side of applied Aesthetics and Humanities provides digital natives with effective intangible tools to develop their own critical and speculative attitude towards the transitioning stage we are currently going through. When everything seems programmable and likely to turn into a reactive and interactive machine or interface, being aware of how humans deal with emotions in a framework of body and mind consciousness [6] discloses indeed as a priority. Paradoxically, the more sophisticated technology becomes, the deeper the emotional involvement grows, and especially in terms of multisensory/*Full Body Thinking*. Making students aware of such a next scenario may also help design companies to increase their interest in emerging technology, embedding interfaces and electronics in general in the design of their products. These are indeed the main guidelines usually followed at the School of Design at *Politecnico di Milano*, where design courses and workshops are always run in partnership with design oriented companies. All this has so far generated a relevant shift also in terms of methodology. Nowadays the general attitude is indeed that of assisting companies in making evidence based decisions, rather than just sticking to their briefing, increasing the evidence available to new products and businesses to make the best choices with disposal resources.

Acknowledgments. Many thanks to Luca Metta and Gabriele Lorusso for tutoring the young designers during the workshop.

References

1. Collins, A.: Toward a design science of education. In: Scanlon, E., O'Shea, T. (eds.) New Directions in Educational Technology. NATO ASI Series (Series F: Computer and Systems Sciences), vol. 96. Springer, Berlin (1992)
2. Gross, M.D., Yi-Luen-Do, E.: Educating the New Makers. Cross Disciplinary Creativity (2009). https://www.mitpressjournals.org/doi/pdf/10.1162/leon.2009.42.3.210
3. Lim, Y., Colombo, S., Casalegno, F.: ReActive: exploring hybrid interactive materials in craftsmanship. In: Proceedings of the Conference on Design and Semantics of Form and Movement. IntechOpen
4. Ferrara, M., Rognoli, V., Arquilla, V., Parisi, S.: ICS materiality. In: Karwowski, W., Ahram, T. (eds) Intelligent Human Systems Integration, pp. 763–769. Proceedings of the 1st International Conference on Intelligent Human Systems Integration (IHSI 2018). Springer, Cham (2018)
5. Dewey, J.: Art as Experience. Balch & Company, New York, Minton (1934)
6. Shusterman, R.: Contents. In: Thinking Through the Body: Essays in Somaesthetics, pp. VII–VIII. Cambridge University Press, Cambridge (2012)
7. Russo, A.C., Ferrara, M.: Smart solutions, "smart aesthetics"? Des. J. **20**(sup1), S342–S353 (2017). https://doi.org/10.1080/14606925.2017.1352872
8. Buchenau, M., Suri, J.F.: Experience prototyping. In: Proceedings of the 3rd Conference on Designing Interactive Systems: Processes, Practices, Methods, and Techniques, pp. 424–433. ACM, New York (2000)
9. Norman, D.: Why design education must change. In: Core 77 (2010). https://www.core77.com/posts/17993/Why-Design-Education-Must-Change
10. Ferrara, M., Ceppi, G. (eds.): Ideas and the Matter. ListLab, Trento (2017)
11. Ferrara, M., Russo, A.C.: Next smart design: inclusion, emotions, interaction in the concept of baby soothing, caring and monitoring smart solutions. In: Karwowski, W., Ahram, T. (eds) Intelligent Human Systems Integration. IHSI 2018. Springer, Cham (2018)
12. Russo, A.C.: The emotional side of smartness: intelligent materials and everyday aesthetics. In: Karwowski, W., Ahram, T. (eds) Intelligent Human Systems Integration. IHSI 2018. Advances in Intelligent Systems and Computing, vol. 722. Springer, Cham (2018)
13. Shusterman, R.: Interaction design foundation, www.interaction-design.org (2013)
14. van Eemeren, F.H., Garssen, B., Krabbe, E.C.W., Henkemans, A.F.S., Verheij, B., Wagemans, J.H.M.: Argumentation and artificial intelligence. In: Handbook of Argumentation Theory. Springer, Dordrecht (2014)

Analysis of the Mental Workloads Applied to Press Operators During the Reuse and Recycling of Materials

Hebert R. Silva(✉)

Industrial Engineering Course, Federal University of Uberlandia, Ituiutaba, Brazil
hebert@ufu.br

Abstract. In Brazil, cooperatives formed for the reuse and recycling of materials and employing several workers are currently excluded from the formal labor market. In the cooperative environment, press operators are essential. They work under various time pressures, physical loads, stresses, and tensions because their work is a vital point in the production process. Despite its central importance for the processing of various types of raw materials, the press that is used to make recyclable materials poses a significant accident risk. The objective of this study is to evaluate, from the ergonomics point of view, the mental workload, tasks, and activities of press operators in a recycling cooperative in the city of Ituiutaba. Many observations and interviews were carried out, videos were recorded, and the Task/Activity Description Form and the NASA-Task Load Index was used to record and analyze the data from the press operators.

Keywords: Solidarity economy · Sustainability · Mental workload
NASA-TLX method

1 Introduction

Selective collection involves the collection of wastes that have been previously separated according to their constitution or composition, such as waste with similar characteristics, selected by the generator and made available for separate collections. According to the Brazilian National Policy on Solid Waste [1], selective waste collection implementation is a municipal responsibility and selective collection goals must be included in the municipal solid waste integrated management plan. The recycling process differs for each type of waste. When different types of solid waste are mixed, recycling becomes more expensive or even unfeasible, because it may be difficult to separate the wastes according to their constitution or composition. For this reason, the National Solid Waste Policy has recommended that selective waste collection in Brazilian municipalities should at least separate dry recyclable waste from wet waste. Dry recyclable waste consists mainly of metals (such as steel and aluminum), paper, cardboard, milk cartons, different types of plastic, and glass. Wet waste is non-recyclable and includes toilet and cleaning waste. This policy incentivizes the creation

W. Karwowski and T. Ahram (Eds.): IHSI 2019, AISC 903, pp. 673–678, 2019.
https://doi.org/10.1007/978-3-030-11051-2_102

of recycling cooperatives to allocate workers aiming to develop work, income generation, and social inclusion [2].

However, according to [3], the economic sector in which the recycling cooperatives operate has become the target of negative analyses regarding work organizations, health, ergonomics, and safety, with low efficiency and productivity indicators. Thus, a critical organization level in a cooperative directly influences the achievement of its final results.

2 Methods

The objective of this study is to analyze press operator work using the techniques and tools present in an Ergonomic Work Analysis (EWA), such as the Task and Activity Description Form and the NASA Task Load Index (NASA-TLX). We aimed to identify relevant points that need to be modified and propose solutions related to ergonomics and work safety. Figure 1 shows the work station of a press operator during an evaluation.

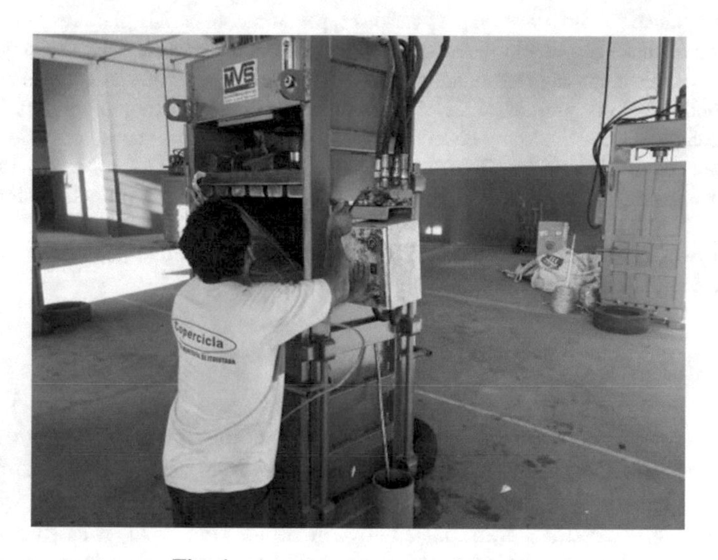

Fig. 1. A press operator work station.

The NASA-TLX method is used for cognitive evaluations and was created by Hart and Staveland (1988) in [4]. NASA-TLX assesses mental workload across six dimensions: Mental Demand, which is the amount of mental and perceptual activity required by the task (such as reasoning, calculating, thinking, seeking, deciding, etc.); Physical Demand, which is the amount of physical activity required to perform the task (such as pushing, lifting, pulling, loading, etc.); Time Demand, which is the degree of time pressure that is felt and calculated as the ratio between time needed and time available; Satisfaction/Performance, which refers to the level of satisfaction that an employee feels with respect to his performance at work; Effort, which is the amount of physical and mental effort that the employee requires to reach the required level of performance; and Frustration Level, which is how irritated, stressed, dissatisfied, and insecure the worker feels while performing his activities.

The objective of this exploratory research is to better understand the press operator's work, which will improve our understanding of the study's hypothesis. Thus, this investigation involved: (a) bibliographic research; (b) interviews with press operators; (c) the observation of cases that contributed to the understanding of press operator work, and (d) the application of ergonomic tools, such as the Task/Activity Description Form and the NASA-TLX.

3 Task/Activity Description Form

The task/activity description form was used, along with photos and videos, to describe the activity performed by the cooperative worker and the amount of time used to accomplish each activity cycle. The activities performed by the press operators are presented in Table 1.

Table 1. Task/Activity Description Form for press operators.

Photograph	Operation name	Operation description	Time
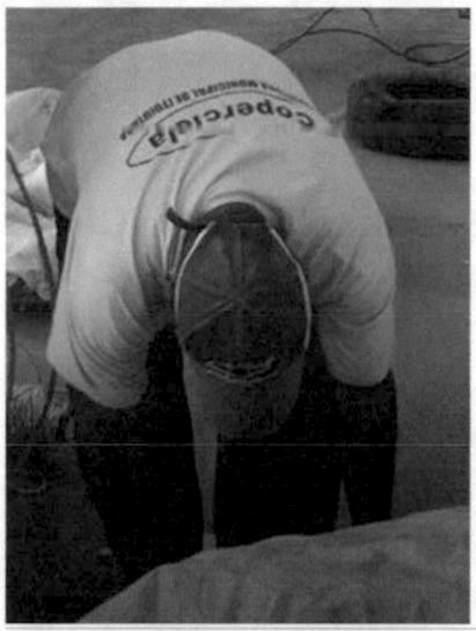	Pick up material to put into the press	The employee performs trunk flexion, cervical flexion, shoulder abduction and adduction, elbow extension, finger flexion and extension	4 s
	Put material into the press	The employee performs cervical and trunk rotation, shoulder elevation with right internal rotation, left elbow flexion, right elbow flexion and extension with the left fist slightly extended and the right fist in pronosupination with palmar flexion and extension	3 s

(*continued*)

Table 1. (*continued*)

Photograph	Operation name	Operation description	Time
	Finishing the placement of material into the press	At this moment, the employee performs ankle flexion, generating a trunk flexion force to hold the material, left and right shoulder elevation in neutral position, right elbow flexion, left elbow extension, right wrist in ulnar flexion, extended and with fingers in adduction, and left wrist in extension	20 s
	Start to operate the press	At this moment, the employee performs a trunk rotation with shoulder elevation, elbow flexion and extension, with wrists in neutral position and palmar flexion, and finger extension	2 s
	Operate the press	At this moment, the cervical spine rotates, with the right shoulder in a neutral position and the left shoulder elevated, adducted, and in internal rotation. The right elbow is slightly flexed and the left elbow flexed. The right fist is in pronation, first four fingers extended, and thumb in digit pressure. The left wrist is in pronation and flexed with palmar flexion	20 s

4 Mental Load Evaluation with NASA-TLX

The cognitive evaluation that is used in the NASA-TLX method analyzes mental workload by interrogating its dimensions and its subjective importance (weight). The questionnaire was used to assess the mental load of the activity, considering its priorities, and generates individual and global results.

The data was collected directly from the information source. The data collection process started with individual natural observation, as the observer was inserted in the working environment. The study proposal was presented, making the operator aware that his answers would be used for academic purposes and kept confidential. This approach is necessary, because the individual approach in face-to-face surveys for each socioeconomic group must be cautious, since it is necessary to adapt to the environments, customs, and culture, to make the research subject comfortable. Table 2 presents the collection results in rates and weight.

The NASA-TLX method was applied to three press operators, and the results are presented in Table 2 and Fig. 2.

Table 2. Representation of the NASA-TLX rates, weight, and indices for press operators.

Factor	Weight	Rate	Weight × rate	Individual loading index
Mental demand	1	20	20	1
Physical demand	5	90	450	30
Time demand	3	30	90	6
Satisfaction/performance	0	20	0	0
Effort	2	30	60	4
Frustration	4	60	240	16
Perceived workload - global index				57

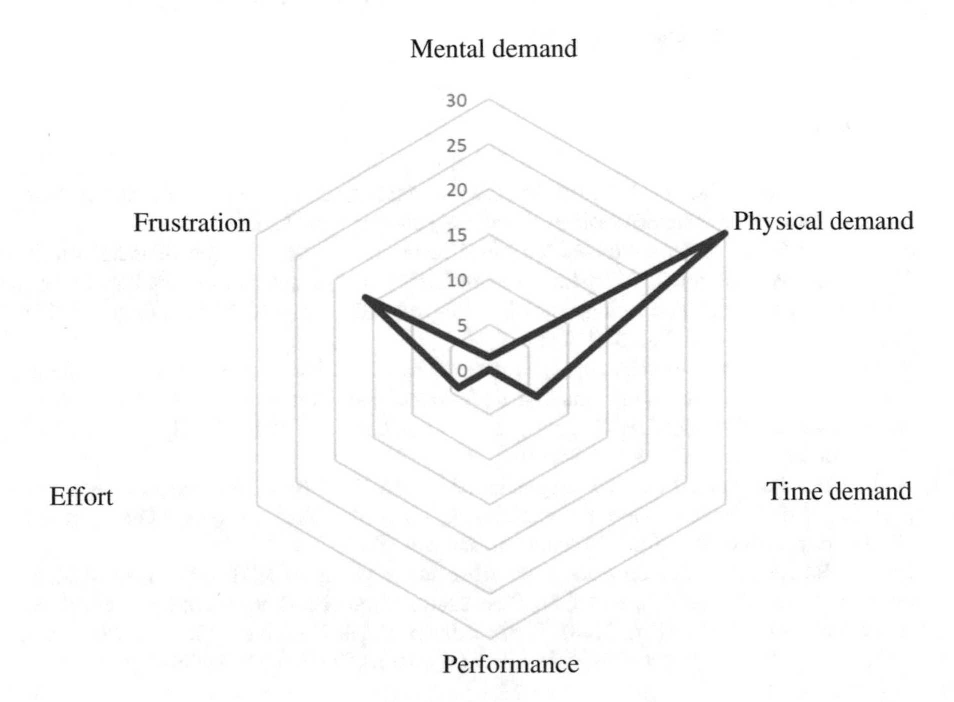

Fig. 2. Workload rate in the general category.

Table 2 and Fig. 2 show that physical demand and frustration level are high compared to the other factors, followed by temporal demand and effort/commitment levels. Based on the overall results, the weighted rate generated a value of 57%, which indicates acceptable mental workloads. However, a systematic control of these conditions is necessary. In addition, the NASA-TLX method confirms the high physical demand indices that are perceived in the Task/Activity Description Form, which are constantly present in reuse and recycling cooperatives, as observed in [2, 3, 5].

5 Conclusions

The main objective of this study was to evaluate the work of press operators, describe their tasks and activities, and measure their work/mental load. We hypothesized that a mental load could be associated with high physical demands during their work operations.

Although the mental load is not a significant factor, the results of this study are similar to those obtained in [5], in which several critical postures are adopted when handling materials. These findings demonstrate the need to propose ergonomic solutions to eliminate heavy tasks, trunk and neck flexion, and to reduce the static posture of the lower limbs. Such situations are constantly present in the tasks/activities that are implemented during the reuse and recycling of materials and frequently occur during the daily work of press operators.

Additional investigations should be conducted into the work and condition of press operators, because there have not been enough subjective mental workload evaluations in the area of material reuse and recycling.

References

1. Ministry of the Environment. Brazil. http://www.mma.gov.br/cidades-sustentaveis/residuos-solidos/catadores-de-materiais-reciclaveis/reciclagem-e-reaproveitamento
2. Silva, H.R.: Study of the ergonomics applied to the reuse and recycling of materials. In: Karwowski, W., Ahram, T. (eds) Advances in Intelligent Systems and Computing. 1st edn. Springer International Publishing, Dubai, 2018, vol. 722, pp. 770–775. Springer (2018). https://doi.org/10.1007/978-3-319-73888-8_119
3. Silva, H.R.: Labor ergonomic analysis applied to a brazilian solid materials recycling cooperative. In: Advances in Intelligent Systems and Computing. 1st edn. Springer International Publishing, Los Angeles, 2018, vol. 605, pp. 191–200. Springer (2018). https://doi.org/10.1007/978-3-319-94000-7_13
4. Hart, S.G., Staveland, L.E.: Development of NASA-TLX (task load index): results of empirical and theoretical research. In: Hancock, P.A., Meschkati, N. (eds) Human Mental Workload, pp. 138–183. North-Holland, Amsterdam (1988)
5. Silva, H.R.: Analysis of ergonomics in the reuse and recycling of solid materials in Brazilian cooperatives. In: Bagnara, S., et al. (eds) Proceedings of the 20th Congress of the International Ergonomics Association (IEA 2018). Springer International Publishing, Florence, 2019, vol. VIII, pp. 960–969. Springer (2019). https://doi.org/10.1007/978-3-319-96068-5_104

Cross-Fertilization to Innovate the Guitar Design

Antonio Marano(✉)

University of Chieti-Pescara, Viale Pindaro, 42, 65127 Pescara, Italy
a.marano@unich.it

Abstract. The paper suggests to reflect and discuss on cross-fertilization and, especially, on innovated materials, electronics and 3d printing technologies applied in the guitar design. The making of stringent instrument concerns different levels of design and production system. Regarding the electric guitar, the cross-fertilization's dynamics concern methods and tools in terms of product and process innovation. The purpose of this paper is to review my recent research conducted with the help of companies, luthiers, musicians, engineers and designers, in order to address the complexity of cross-fertilization for the innovation of the guitar design. The findings of this study suggest new research opportunities linked to the concepts of physical and psychological nature, such as, for instance, usefulness, aesthetics, understandable, environmentally, user friendly.

Keywords: Cross-fertilization · Guitar design · Design process
New materials

1 Introduction

In guitar design, cross-fertilization takes on a strategic role to allow the transfer of knowledge and practices developed in other fields of research to new and possible areas of application and design.

This requires the analysis of the boundary disciplines that, through tools, methods and solutions offer more interesting possibilities to trigger dynamic innovation's processes. On this basis, the exploratory phase of research made it possible to identify and highlight the areas of experimentation and application in which these dynamics of innovation, transfer of knowledge and good practices were more evident. The choice of best practice cases was aimed at bringing out the themes of innovation that can be traced back to the design processes. Thus, the processes of transfer to guitar design were observed, coming mainly from the fields of new materials [1], electronics, 3D printing [2], and from the disciplines of ergonomics applied to industrial products [3], interactivity, modeling and design for the sustainability [4].

In the field of electric guitar design, the cross-fertilization dynamics concern the adoption of methodologies of design process in terms of product and process innovation. At the product level, the musical instrument evolves through the use of new materials with high qualities of lightness, strength and sound, or high sustainability (carbon fiber, titanium, polymer, aluminium, bamboo), and for the new possibilities of

W. Karwowski and T. Ahram (Eds.): IHSI 2019, AISC 903, pp. 679–683, 2019.
https://doi.org/10.1007/978-3-030-11051-2_103

interaction offered by electronics (wireless midi controller, air guitar, smart connection interfaces, internet, video, digital audio, synth controller, touchscreen, touch sensitive etc.). Compared to process innovation, on the other hand, cross-fertilization concerns the use of specific three-dimensional modeling software and the development of 3d printing technologies for the creation of highly personalized [5] musical instruments and morphological complexity.

2 Research Method

The hypothesis underlying the research is the existence of a rapid process of technological innovation and evolution of language in guitar design. In fact, nowadays, in the art of lutherie, referring in general to the various families of stringed instruments and guitars in particular [6], there are new design expressions that combine the forms of the past with the best technologies of the present, the skills of luthier and the experiences of the craft shops [7] with the innovative processes of research, design, prototyping and industrial production, conducted by companies, design studios and designers, often in collaboration with luthiers, musicians and experts in the technologies of materials, electronics, acoustics and computer engineering.

On this hypothesis, a research line based on research questions arises: how does the guitar design evolve due to the phenomenon of cross-fertilization? What are the main levels of innovation and the dynamics of cross-fertilization? How does usability, functionality and aesthetics of the musical instrument change?

The research methodology deals with the combined approach between bibliographic research, the Internet, case studies, interviews, direct empirical research carried out in collaboration with luthiers, designers, but also with rock, fusion and jazz guitarists. The research program is divided into three phases: preliminary phase of an exploratory nature; phase of understanding and evaluation of the levels of innovation highlighted; critical reflection phase for the argumentative and comparative formalization of the results.

The research is based, both on the investigation and collection of data, and on the comparative study of the applications of cross-fertilization in a large number of case studies solved at the level of concept or production. In a first phase, the research identified about 500 contacts between companies, luthiers and designers, in order to select about 120 contacts as the international expression of about 200 case studies related to product and process innovation.

3 Results

The interpretative phase of the research has allowed to highlight how cross-fertilization takes place on different levels related to process innovation and product innovation.

In relation to the first level, the cross-fertilization dynamics mainly concern the adoption of anomalous design process methodologies and procedures for the lutherie sector and, on the contrary, widespread in other areas of industrial design.

In particular, the use of specific three-dimensional modeling software and the development of 3D printing technologies for rapid prototyping and the creation of objects through additive manufacturing, favor new design processes increasingly oriented to the search for complex and customized forms of the musical instruments. Designers, engineers and small companies, with the collaboration of musicians and luthiers, experiment with new languages and advanced production techniques to make guitars, but also very light and transparent electric violins, up to the collections of small wind musical instruments, as well as the prototypes of wind instruments or the concept of a piano.

In the case of Monad studio by Eric Goldemberg and Veronica Zalcberg, the potentials of 3d modeling software and 3d printing processes are exploited to create the Multi system, a collection of five musical instruments inspired by nature. The innovative process of digital design allows to experiment and elaborate a complex morphology of musical instruments that, appropriately controlled through the optimal balance between morphological solution, material and internal honeycomb structure and separate channels, allows to obtain very interesting resonances and strong character of sound identity and, at the same time, guarantees an ergonomic efficiency and an optimal posture of the musician. In the case of the bodies of electric guitars produced by the Swedish company ODD, founded by Olaf Diegel, the 3d printing processes are used to create complex skeletal structures. The whole process is based on 3D digital design and modeling and uses the Solidworks CAD program and the Selective Laser Sintering (SLS) printing technology used by 3d Systems. Questto Nó design studio exploits the flexibility of digital design process to experiment and develop a new formal research on the musical instrument, through digital modeling and 3D printing software, and to ensure the necessary functionality and musical efficiency. Scott Summit of Summitid industrial design studio explores the new opportunities offered by 3D printing to experiment with new forms of interaction between the body, human needs and technology. 3Dvarius, designed by Laurent Bernadac, French engineer and violinist, is an electric violin realized with 3d printing technology and the Stereolithography technique (SLA). The design of the violin is transparent, light and is characterized by the optimization of weight distribution and stability of the center of gravity that facilitate the movements of the violinist.

In terms of product innovation, the transfer processes concern the adoption of technological solutions developed in other sectors, such as the case of new materials and the case of the interaction between lutherie, microelectronics and smart connection software. This is how many luthiers-designers assert themselves, who design, experiment and create guitars characterized by contemporary languages and forms, advanced and environmentally friendly technological solutions and materials, details of great beauty and functional precision. On the one hand, the languages of minimalism, of deconstructivism, of high-tech, up to the organic forms conceived in phytomorphic analogy and to ethical models, evolve. Above all, the musical instrument evolves for the use of new materials with high qualities of lightness, strength and sound, or high sustainability (carbon fiber, titanium, polymer, aluminum, bamboo). Especially, in the travel guitar sector among the guitars that require to reconcile the sound of a full size guitar and the characteristics of compactness, lightness and resistance appropriate to the journey, carbon fiber instruments are spreading on the market.

Composite acoustics develops different models of acoustic guitars in high quality aesthetic carbon and technological efficiency, including The Cargo travel. Among the main features: 22.75″ length of the scale with 14 keys; carbon fiber bridge for maximum sound transfer; rigid structure of the handle/body for greater stability; available with pickups; polymer keyboard reinforced with 21 keys in stainless steel; ultra-precise tuners; gold for decentralized sound with integrated superior reinforcement technology.

The steel-string Rider guitar produced by Blackbird guitars is made of carbon fiber using Cad design technology and automated production. It has an elongated shape with a total length of 35.5″ and is very light (3 lbs). Even Alpaca guitar is very light and durable. It is 32.25″ length, has a scale length of 24.75″ and weighs 3 lbs. and is made of carbon fiber with bio-derived resins. The designer Chris Duncan develops in 3D Cad/Cam this headless travel guitar with a large circular resonance hole placed at the top and in a decentralized position, a pocket made at the end of the body to integrate the tuning mechanics, and a curvilinear profile to improve the comfort of the support on the user's leg. As well as the body of the X7 series produced by Emerald guitar is made of carbon fiber in the back and side bands, while the top is in Padauk wood. A hybrid solution, recognizable by the original curvilinear cut of the resonance opening, which makes it particularly light (3.85 lbs) and resistant and therefore suitable for travel.

In the smart guitar the sound effects and the musical execution modalities of an electric guitar are extended, integrating the synth controller electronics, the wireless, the touch-sensitive keyboards, the touch screen, the Multi-touch LEDs, the electronic devices and the User-friendly apps that encourage the user to study and learn the guitar, even for fun. The typologies evolve with the Midi portable guitar, Midi attached and sensing surfaces controller versions, up to the wearable Air guitar that exploit Motion technology. The case of the Sensus electric guitar is emblematic. Created by Michele Benincaso and developed by a team of experts in various fields, from acoustic physics [8] to wireless networks, to computer engineering, Sensus uses unprecedented features and excellent usability levels, which extend and improve the interactivity and the expressive and communicative potential of the guitarist. It allows, in fact, to add to the musical performances an infinite number of modulations and sound effects (Digital Audio Workstation) without the use of pedals and amplifiers. It is able to create loops, receive, process and share music through the internet connection.

4 Conclusions

In guitar design, cross-fertilization takes on a strategic role to allow the transfer of knowledge and practices developed in other areas of research to new and possible areas of application and design. The research shows that in the field of electric guitar design, the cross-fertilization dynamics concern the adoption of methodologies and methods of design process in terms of product innovation (use of new materials and electronics) and process (3D modeling software and 3D printing technologies).

This line of research and contemporary experimentation allows a competitive development of the market on the basis of technological innovation and strategic choices guided by design conducted by new and evolved figures of luthiers with a solid project training. At the same time, the results obtained open up new possible directions of research that address and restore the complexity linked to the phenomenon of cross-fertilization for innovation in guitar design. The new research questions could be: what are the physical and psychological advantages and disadvantages for the guitarist involved in experimentation and musical research? How the concepts of usefulness, aesthetics, understandable, environmentally, user friendly change in relation to the guitarist?

References

1. Ferrara, M., Lucibello, S.: Design Follows Materials. Alinea editrice, Firenze (2009)
2. Shillito, A.M.: Digital Crafts. Industrial Technologies for Applied Artists and Designer Makers (2013)
3. Bandini Buti, L.: Ergonomia olistica. Il progetto per la variabilità umana. Franco Angeli, Milano (2008)
4. Vezzoli, C.: Design per la sostenibilità ambientale. Progettare il ciclo di vita dei prodotti. Zanichelli, Bologna (2016)
5. Pine, B.J.: Mass Customization. Harward Bussiness School Press (1993)
6. Cassiani Ingoni, M.: Chitarre. Elementi di liuteria. Casa Musicale Eco (2008)
7. Sennet, R.: L'uomo artigiano. Feltrinelli, Milano (2008)
8. Cingolani, S., Spagnolo, R. (cura di): Acustica musicale e architettonica. Città studi edizioni, Torino (2012)

Impact of Emerging Technologies for Sustainable Fashion, Textile and Design

Muhammad Faisal Waheed(✉) and Ahmad Mukhtar Khalid

University of Sialkot, 1 Km, Main Daska Road, Sialkot, Pakistan
{Faisal.waheed,am.khalid}@uskt.edu.pk

Abstract. The impact of Emerging technologies for sustainability of NANO materials and Artificial Intelligence in the area of Fashion, Textiles and Design is explored in this paper. This century is an era of new technologies viz., Biotechnology, Nanotechnology, Artificial Intelligence. These emerging technologies have introduced the world an entirely new canvas of work and have contributed immensely in diverse fields such as clothing, war ware, novel materials and entirely a new line of biomaterials in textiles like, highly tensile, unique surface structure, self cleansing fabrics, dye ability, flame retardant fabrics, Ultra Violet protection, anti static, anti bacterial, soil resistance, wrinkle resistance, stain repellant, Antimicrobial, fire retardant, water repellent, durability 3-D technology, and bulletproof fabrics, etc. All these technologies have revolutionized the textile industry, which is basic component of fashion and design industry. In present times, indeed a new era in all these industries is going to be imparting a sustainable progress and development.

Keywords: NANO materials · Artificial intelligence · Fashion Textile · Design · Creativity · Sustainable · Insect resistant

1 Introduction

The "Sustainable" factor in the domain of Fashion, Textile and Design has further strengthen by the advancements of Emerging Technologies (ET) like NANO materials, and Artificial Intelligence (AI) [1, 2]. The contribution of these unique technologies has an immense impact on our daily life. The involvement of "ET" has changed the usual ways of design, the products, and materials have transformed the manufacturing industry into absolute new paradigm. In the area of clothing, which is directly involved with textiles and design, has ultimately converted into centre of attraction [3]. The importance of clothing is something, which is awfully close to human body, and it has extremely positive and negative effect of its wearer.

In this era when the advancements of technology has created the enormous collision in the area of textiles, which is somehow at the verge of disaster and is in vulnerable state. It is the well-known fact that Textile Industry status is one of the most Polluting industries in present time [4]. Design in apparel clothing is the most important factor, which is directly related to textiles and fabrics, hence the fabric production is the main root cause of destabilizing and deforestation, which have

W. Karwowski and T. Ahram (Eds.): IHSI 2019, AISC 903, pp. 684–689, 2019.
https://doi.org/10.1007/978-3-030-11051-2_104

already affected environment, health and rapid climate change, heading towards the global warming [3, 5].

If we look into the past, brands and designers were in a phase of improvement in design and production. They were far away from the innovation in terms of sustainability criteria. Therefore, with the help of ET the methods of production have revolutionized, creating enormous impact on society, health and environment [6]. It is the history now, when the sustainable clothing was shapeless and of dull color. Now, one can see a range of bright and vibrant colors, and perfectly fitted outfits with the tag of eco-friendly clothing.

Emerging Technologies like Biotechnology, Nanotechnology, and Artificial Intelligence, has created awareness in consumers about eco-friendly environment [1, 2]. Researchers who are constantly working and developing great technologies are helping Textile and Fashion Designers to create their products, which help to become socially and environmentally acceptable. These technologies have changed their status from textiles to eco-friendly textiles and from designers to ethical designers and have adding the sense of responsibility and sustainability for the love of Mother Nature [7].

To understand the impact of emerging technologies for the sustainable Fashion, Textile and Design the following developments in the ET will be explained in detail.

2 Materials and Methods

Making composite materials with the help of Biomaterials and NANO sized particles or fibers have revolutionized the textile manufacturing industry. It has changed the overall impact of producing materials keeping in mind the style as well as environment. By implementations of these methods, these novel materials have introduced us entirely new and different form of qualities and properties, which we had never seen and experienced before. Now they have been used almost all kind of clothing [6].

In result of implementing of these innovative materials, the new design strategy and production methods have been introduced. As a whole, these advancements have developed a new paradigm in the area of textile, Fashion and design, and most importantly, they have become user friendly, eco-friendly, biodegradable and helping to create the positive impact of environment [5].

All the areas of textiles have changed their properties by implementing the Nanotechnology and Biotechnology have developed the direction of clothing to intelligent, multifunctional, and higher in performance, which are now the need of time. By Implementation of these technologies have immensely changed the properties from fiber to fabric, which are somehow helpful for the fashion designers. In past there were limitations in design in terms of materials like, drape, stiffness, rigidness, behavior, Print, Shrinkage, Color bleed, cutting, Strength, absorbency etc. These limitations of materials have developed the constraints in creativity for the fashion designers. Now the things have completely opposite as advancements of these ET have given boost to creativity. Designers are coming up with more bold and creative ideas, keeping the sustainability in mind.

Clothing is the wearable art and for this, designers have the excellent support and new and advanced tools in shape of materials to meet the diversified consumer demand.

3 Biotechnology Techniques

Biotechnology is based on DNA technology heading to Enzymes synthesis to save resources like, energy, time, and most importantly water. This advanced technology has directed the manufacturing industry towards the new horizon, where the possibilities of success and productivity are endless. In current times, Biotechnology is playing a key role to save this planet and creating it more sustainable and safe for the future generations.

Biotechnology in textiles deals in innovative and advance technologies, where it applies on composed structure of textile fibers, which are designed to use in specific design industry. This is an updated and performance based technology, in result of it many high tech novel fabrics have been developed, which contains the high performance properties, such as water and dirt repellant, shock proof, lightweight, temperature regulating etc. [8].

Biotechnology is currently a driving force in design industry, although it has been used in multiple domains such as, textiles, medicine, agriculture, fashion and design. In textiles it is mainly integrates of natural and synthetic materials [9]. It has developed the enormous development of multiple properties in one material, which is somehow beneficial for the designers in many ways like, apparel, home fashion, luxury automobiles and climate based materials, which are normally used in outdoor areas. Biotechnology is playing a vital role in terms of innovations like.

3.1 Self-cleansing Surface

Self-cleansing fabric has huge impact upon fashion and design industry for the outer looks; these fabrics repel the dirt and can be cleaned easily. They are not soiled whatsoever [7].

3.2 Naturally Colored Cotton

Who can visualize the naturally colored cotton better than designers? One of the marvelous innovations of Biotechnology is the production of naturally colored cotton through genetic engineering, though color range is limited. However, in future, it can be so interesting to see fields of cotton with primary, secondary and tertiary colors, and world would be far better place, without dyes and pigments, which are so injurious to human health, as well as devastating the environment [10].

3.3 Animal Fiber

To get the valuable wool for the outerwear, Biotechnology vaccines are there which when injected into sheep, after specific time breaks appear and wool fiber can be pulled off. This procedure will take half the time of labor for shearing the sheep [10]. Another important breakthrough is the scorpion goat, which produces wool fit to endure extremely high temperature, and used in making astronauts space suits.

4 Nanotechnology Techniques

Nanotechnology is an industrial revolution, in which the properties of materials will drastically change when they are reduced to NANO scale. To treat textiles with the coating of NANO materials is the NANO technology, which improves the properties of the material and making it more durable [11]. It is great news for the designers that when you see the NANO particles through proper equipment, it changes its color at this NANO level. Innovations in NANO technology have changed the commercial aspect of all, related to design, textile and Fashion business.

As the concepts and nature of products have been changed in term of its properties, this effect is very drastic and more effective. One can foresee the economical boom because of this advance innovation. About design and Fashion designers, NANO technology advancements have created endless possibilities for design and practicality of product [12]. By NANO technology techniques textile sector functionalities, have changed as the innovative fabrics with high performance properties came to exist. Stain repellant, Water repellant, UV protected, Anti static, Wrinkle free, Anti bacterial, Fire retardant, Bio degradable, Bulletproof and defense clothing to name a few [1, 2, 7].

4.1 Bullet Proof and Defense Clothing

Earliest innovation of NANO materials are the bulletproof vest and clothing. Nanotubes fiber has developed this innovative product, which is seventeen times tougher, then the KEVLAR and lighter. For the design and fashion industry this innovative materials is as appealing as designers can think of many other wearable products with such material.

4.2 Water Repellant

Water repellant fabrics in terms of design and fashion are unique in terms of those countries where rain and humid climate prevails a lot. This innovation is great as water remains on the surface of the fabric and does not penetrate [7].

5 Genetic Engineering Techniques

In Genetic Engineering techniques, the major development is the Colored Florescent Silk category of material, which is somehow look very fashionable, and design oriented. As the material, it is so much attractive and appealing to design. The technique which researchers have used, that they have inserted glowing proteins which are the taken from Corals and Jellyfish, into the silk worm Genome. In result of this genetic engineering innovation is the properties of the material is more or like same as silk but it become slightly weaker after processing [13]. In the area of Fashion and design this whole process of genetically engineered silk is very fascinating and it will open up new directions of taking and using these materials in a more creative and sustainable way.

Another latest and more sustainable genetic engineering material is "SYNTHETIC SPIDER SILK". Though it is synthetic but it is more Biodegradable, because it mainly contains the cellulose, Silica, and Water [14]. The properties of this material have astonished the researchers as well as designers. The behavior of this material is mind blowing as it is stronger than steel and tougher than Kevlar, but interestingly more flexible than the Lycra and spandex yarn. This futuristic material has great possibilities to design in multi disciplinary domains such as apparel, sportswear, automobiles, bulletproof clothing, and sporting goods etc.

6 Artificial Intelligence

In this advanced technology world, Artificial intelligence, (AI) is playing very important role in the area of Fashion and Design industry. In the creative design process, the presence of Artificial Intelligence technology is the need of an hour. Most of the retail industry and brand are doing their sales online, they have to keep themselves up to the mark and quick response to buyers is so much important that it can affect the sales either way. With the passage of time, the time span of the online product is getting shorter and shorter. However, manufacturing and supply chain process is more or less the same in which the Artificial intelligence is also taking part to resolve these matters.

In the presence of Artificial intelligence the role of designers have become more critical, as they will have to equip themselves with the next generation's tools and technology. AI tools in terms of Fashion and Design are so important in this technological world. As it is difficult for the designers to look up multiple season collections and figure out the data about what was hot selling what was not. AI has come up with this solution, as there is a complete database for all the previous collections and incredible amount of information is available with the click of the button.

In this case traditional process to design is still the same as to do research, collecting fabrics, making prototypes etc. but to catch the latest technology development, designers will be able to learn about new tools, which are definitely improving the design process day by day. Designer's role will remain the same, as they will have much to share as knowledge available with them at just a click away [15, 16].

7 Conclusions

The impact of Emerging Technologies and sustainability in the area of Design, Fashion and Textiles is inevitable in the coming times. These technologies have been developed for multi-disciplinary purposes aiming at betterment of the humankind and to save the planet earth. Awareness about health, environment and sustainability is in demand. Researchers and designers are working on these areas with more efficient manner with the help of these advanced and technical tools. These designed domains are skilled base, and skill never gets old. One has to accept and equipped with advanced technologies and technical tools to work in this environment.

Acknowledgements. Authors gratefully appreciate the encouragement given by the Dr. Ijaz A. Qureshi the VC, University of Sialkot. We appreciate the fruitful discussion with Dr. Javed Anjum Sheikh. We are also thankful to Mr. Faisal Manzoor, Chairman Board of Governors, and Mr. Muhammad Rehan Younas, the Executive Director for facilitating this work.

References

1. Carp, B.: Profiles of 21 Taiwanese Performance Textile Companies Performance Apparel Markets **1**(57), 83–113 (2017)
2. Carp, B.: Profiles of 21 Innovative Taiwanese Textile Companies Textile Outlook International **216**(183), 113–143 (2016)
3. Sustainability- The Future of Fashion. www.fibre2fashion.com
4. Fashionable Fabrics Leading to Deforestation. www.fibre2fashion.com
5. Transforming the Fashion Industry One Bio Based Fabric at a Time. www.fibre2fashion.com
6. de Araunjo, M.: Fabre science understanding how it works and speculations on its future. In: Fandgueiro, R., Rana, S. (eds.) Natural Fibre Advances in Science and Technology Towards Industrial applications, vol. 12, pp. 3–17. RILEM Books, Springer, Dordrecht (2016)
7. Application of Nanotechnology in Textiles Industry. http://textilelearner.blogspot.com
8. Innovative Textiles Made Possible by Biotechnology. https://www.gesundheitsindustrie-bw.de
9. Bio Textile|Application of Biotechnology in Textiles|Importance of Bio Technology in Textile Processing. http://textilelearner.blogspot.com
10. Bio Tech Textiles - Future Trend. https://www.fibre2fashion.com
11. Nanotechnology in Textiles – Manufacturing and Applications. https://www.textilemates.com
12. Nanotechnology in Textiles. www.azonano.com
13. Genetically Modified Fashion. www.the-scientist.com
14. New Artificial Spider Silk: Stronger than steel and 98% water. www.smithsonianmag.com
15. AI is changing the future of Fashion Design. www.businessinsider.com
16. AI in Fashion Design – New Era of Creativity? https://wtvox.com/

How to Choose One Sustainable Design Method Over Another: A Consumer-Product Optimizing Prototype

Shaoping Guan(✉), Rui Cao, and Lu Shen

School of Design, South China University of Technology, Guangzhou 51006, China
shpguan@scut.edu.cn, {cao.r, 201721050606}@mail.scut.edu.cn

Abstract. This paper developed a prototype aiding designers to optimize sustainability of durable consumer-products. The theory of Emotionally Durable Design indicates emotional attachment between users and products should be strengthened to lengthen psychological life span of products. Yet attachment degree depends more on subjective evaluation of users, who tend to have different degree on different products. In the present study, a questionnaire was used to interview participants about their attachment degree to products. Then the products were investigated about their duration of indefective functioning. The prototype is a two-dimensional coordinate model based on Attachment Degree and Duration of Indefective Functioning. It can be used as a reference to choose among numerous sustainable design methods. For example, products with low Attachment Degree and long Duration of Indefective Functioning need to be redesigned with methods of reducing usage of non-environmentally friendly materials, recycling or building resilience into relationships between users and products.

Keywords: Sustainable design · Product attachment · Emotionally durable design

1 Introduction

Sustainable design covers a wide range of ideas and scopes, and it has been proposed and guided to practice for many years. While sustainable design has made some achievements, it has proven to be limited. On the one hand, environmental problems caused by industrial products continue to deteriorate. On the other hand, the scope of application of sustainable design is still limited [1].

Consumerism is a hindrance to achieving sustainable development. In some cases, consumers get rid of durable products that still function properly at the time of disposal [2], because they look old fashioned, because new products on the market offer more possibilities, and so on [3].

One possible strategy to slow down product life cycle is by increasing the attachment people experience towards the product the use and own [4]. Consumer-product attachment implies the existence of an emotional tie between a person and an

W. Karwowski and T. Ahram (Eds.): IHSI 2019, AISC 903, pp. 690–695, 2019.
https://doi.org/10.1007/978-3-030-11051-2_105

object. When a person becomes attached to an object, he or she is more likely to handle the object with care, repair it when it breaks down, and postpone its replacement as long as possible [3].

In the present study, we used a questionnaire to investigate the degree of consumer-product attachment that people feel to six durable products they typically own: a computer, a sofa, a desk, a mobile phone, a kitchen knife and a washing machine. The Duration of Indefective Functioning is obtained by interviewing the manufacturer about their durability testing.

1.1 Emotionally Durable Design

When examining sustainable design from a technical perspective, it is scientific and rigorous [1]. What is the reason for its little effect? Jonathan Chapman, founder of the theory of emotional perseverance, argues that consumerism is the interpretation of this problem. Consumerism, the driving force behind the modern operation of capitalism, has prompted people to continue to have a desire for products, and they are constantly tired. So people are phasing out many existing products that are still functioning well, and this cycle is accelerating now.

Research on emotion has focused more on evaluation of design, whereas social sciences are being adapted to the early, 'fuzzy front end' of design in order to understand people and products more holistically [5]. The theory of Emotional Durable Design indicates that emotional attachment between users and products should be strengthened to lengthen the psychological life span of products. The attachment is personally given by users to a product through satisfaction and involvement produced by their intimate experiences [6]. Looking closely at the products around us, we can find that different products play different emotional roles in our lives. In our opinion, products that belong to any emotional role have their specific value.

1.2 Consumer-Product Attachment

The degree of consumer-product attachment is defined as the strength of the emotional bond a consumer experiences with a durable product. In recent studies about product attachment, there are three constructs related to attachment: irreplaceability, self-extension and indispensability. However, only irreplaceability and self-extension has a strong correlation with attachment [1]. Whereupon in this study, only irreplaceability and self-extension were used as two other measures of attachment.

In our opinion, the attachment degree depends more on subjective evaluation of users, who tend to have different degree on different products. We hypothesized that people of similar age and income would have similar attachment performance to products.

1.3 Duration of Indefective Functioning

Environmental and social benefits of the wider global community compared with the desires of the individual are not strong enough to motivate a different lifestyle [7]. Many objects that were once bought for their functional, hedonic or psychosocial benefits are eventually discarded. In some cases, consumers get rid of durable products that still function properly at the time of disposal, because they look old-fashioned, because they are no longer compatible with other products, because new products on the market offer more possibilities, and so on. When the product is discarded more depends on the user's subjective judgment. Differing from the length of time before the product is discarded, we define Duration of Indefective Functioning as the length of time before the product fails for normal use for the first time. This length is likely to be longer than the length of time before a product being discarded.

2 Method

2.1 Respondents

This study selected 70s and 80s generations of people who accounted for 26% of the Chinese population [8]. The 70s and 80s generations has high disposable income and rich experience in product use. We selected the middle-income group to conduct the questionnaire study.

A questionnaire was sent to 150 members of a consumer panel based on a random sample of the local community. 125 usable questionnaire were returned in time, a response rate of 83.3%. Of the 125 respondents, 71 (56.8%) were females. Ages ranged from 28 to 43 years, with an average of 37.

2.2 Questionnaires

We chose six durable consumer products: a computer, a sofa, a desk, a mobile phone, a kitchen knife and a washing machine. They fall into four product categories: home appliances, consumer electronics, furniture and kitchen utensils. Among them, computers and mobile phones are consumer electronics, and sofas and desks belong to furniture. We developed a questionnaire based on five-point Likert scales. Respondents indicated to what extent they agree with six statements regarding their relationship with their products.

2.3 Interview

Respondents also fill in the brand and model of the product they use. We contacted the producer via email to interview them about the durability test results for these products. Many producers don't provide the length of time. For example, mobile phone battery manufacturers provide the maximum charging times. After investigating the frequency of use by consumers, all durability test results are converted into durations.

3 Results and Discussion

According statistical analysis on SPSS, there are both significant positive correlations between attachment and irreplaceability, attachment and self-extension. Meanwhile, the statistical data shows that the attachment degree of the same product presents a significant centralized distribution ($p = 0.000$), which proves our hypothesis that people of 70s and 80s generations with similar income would have similar attachment performance to products.

3.1 Mean Degree of Attachment

According to the correlation of irreplaceability and self-extension to attachment in correlation analysis, three selected weight coefficients are: attachment (50%), irreplaceability (30%) and self-extension (20%). The calculated Mean Degree of Attachment is shown in Fig. 1.

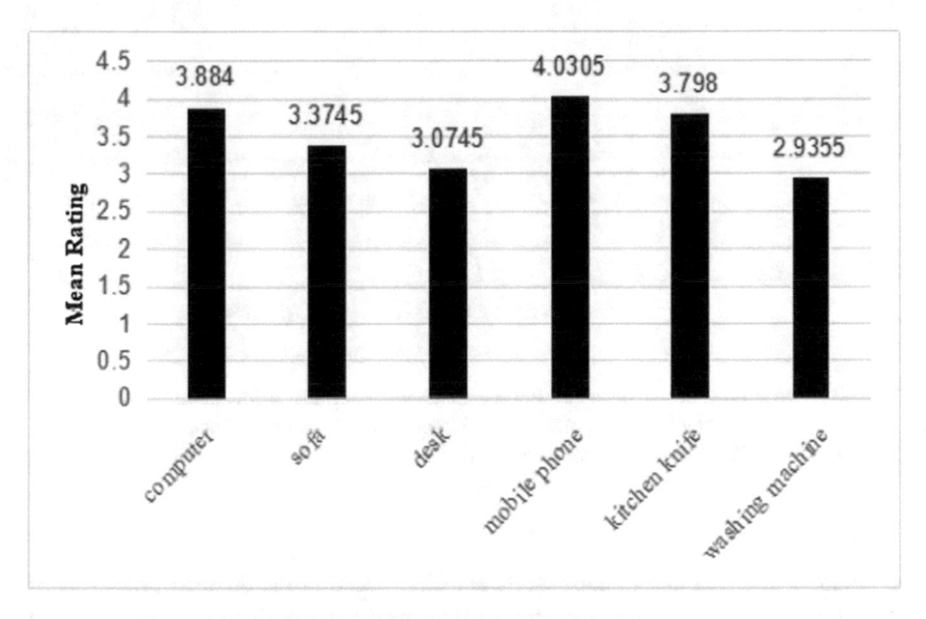

Fig. 1. Mean degree of attachment

3.2 Mean Duration of Indefective Functioning

The Mean Duration of Indefective Functioning (MDIF) calculated by the durability test results is shown in Table 1.

Table 1. Mean duration of indefective functioning

Product	Computer	Sofa	Desk	Mobile phone	Kitchen knife	Washing machine
MDIF (years)	4.5	11	21	3.6	6.7	8.5

3.3 The Prototype

Combining the two sets of data: Mean Degree of Attachment and Mean Duration of Indefective Functioning, we draw the two-dimensional coordinate model in Fig. 2.

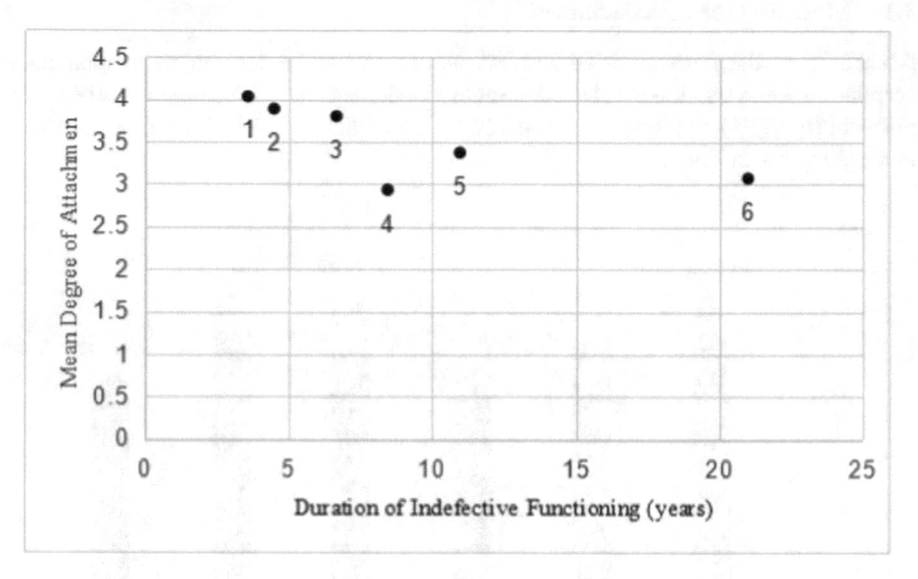

Fig. 2. The prototype based on mean degree of attachment and mean duration of indefective functioning. (The products corresponding to the number: 1. Mobile Phone, 2. Computer, 3. Kitchen Knife, 4. Washing Machine, 5. Sofa, 6. Desk)

3.4 Discussion

The results show that users have significant differences in the degree of attachment to different products, and in the same category of products, the degree of attachment will also show significant differences. We do not want to discuss the influence factors behind product attachment, there are already many studies in this area. What interests us is the mismatch between attachment degree and duration of Indefective Functioning. Since there is no ideal matching degree between the two, our analysis is proceeded in the comparison of the products.

Computer and Mobile Phone. As shown in Fig. 2, the mobile phone has the highest degree of attachment among all products, followed by the computer. They are typical products whose attachment degree are high and duration of indefective functioning are short. From the perspective of user experience, discarding products that we attached to is a negative emotional experience; from the perspective of sustainable design, the rapid retirement of electronic products is also a huge waste of resources. Due to the high degree of attachment to the product, users will even retain the retired electronic products instead of making them to be recycled, which is another waste of resources. For electronic products such as mobile phones and computers, our sustainable design recommendation is to moderately lengthen their physical life span. For example, use more ageing resistance batteries, screens and other accessories in the product. In order to cope with the high upgrading rate in the electronics industry, we recommend the use of replaceable electronic components to enable upgrades on existing equipment.

Kitchen Knife. The kitchen knife has a close attachment degree to the computer and has a longer Duration of Indefective Functioning, so the matching of the kitchen knife is reasonable in our evaluation.

Washing Machine. The washing machine has the lowest degree of attachment, but has a relatively long Duration of Indefective Functioning. We conclude that users may only retain products because of the functionality of the product. For this type of product, we recommend using more environmentally friendly materials in their design. As can be seen from the existing laundry service and public washer machine service, perhaps servitization and sharing is the trend of products with low attachment degree.

Sofa and Desk. The situation of sofa and desk that are both furniture products is similar: both have long Duration of Indefective Functioning, but have slightly lower attachment degree. For this kind of product, we recommend building resilience into relationships between users and products to increase the attachment degree. One of the approach is to "let the product age slowly and gracefully". Ageing surfaces can be designed to add character to objects, giving them a history, a sense of age and a story.

4 Contribution

The significance of this research is to propose a new way of thinking about sustainable design, and develop the theory of emotionally durable design to a certain extent. The prototype can be used as a reference for designers to choose among numerous sustainable design methods in the early phases of product design.

References

1. Jonathan, C.: Emotionally Durable Design: Objects, Experiences and Empathy. Earthscan, Sterling (2005)
2. Debell, M., Dardis, R.: Extending product life: Technology isn't the only issue. J. Adv. Consum. Res. **6**(1), 381–385 (1979)
3. Schifferstein, H.N.J., Zwartkruis-Pelgrim, E.P.H.: Consumer-product attachment: measurement and design implications. J. Int. J. Des. **2**(3), 1–13 (2008)
4. Page, T.: Product attachment and replacement: implications for sustainable design. J. Sust. Des. **1**(3), 256–282 (2014)
5. Battarbee, K., Koskinen, I.: Co-experience: Product experience as social interaction. J. Prod. Exp. **1**(2), 461–476 (2008)
6. Chic-Hsiang, K.: The association of product metaphors with emotionally durable design. In: 6th IIAI International Congress on Advanced Applied Informatics, pp. 58–63. IEEE Press, New York (2017)
7. Bhamra, T., Lilley, D., Tang, T.: Design for Sustainable Behaviour: Using Products to Change Consumer Behaviour. J. Des. J. **14**(4), 427–445 (2011)
8. China Residents Consumption Development Report. http://www.ndrc.gov.cn/

Human-Autonomy Teaming

Interaction Concept for Mixed-Initiative Mission Planning on Multiple Delegation Levels in Multi-UCAV Fighter Missions

Felix Heilemann(✉) and Axel Schulte

Universität der Bundeswehr München, Werner-Heisenberg-Weg 39, 85577 Neubiberg, Germany
{Felix.Heilemann,Axel.Schulte}@unibw.de

Abstract. The management of several UCAVs from board a manned combat aircraft is a challenging problem. While fully automatic planners lead to complacency and low workload, manual planning leads to a much too heavy workload for the pilot. In this article, we present a mixed-initiative planning concept, consisting of a delegation and an assistance part. The delegation part of the concept enables the pilot to command tasks to the UCAVs on a team-, task- and parameter-based level, which are taken as planning constraints. The assistance part supports the pilot in the plan creation process and can proactively counteract possible failures. An initial validation of the implemented concept is presented.

Keywords: Human-autonomy teaming · Mixed-initiative planner

1 Introduction

The mission planning of multiple vehicles in dynamic mission scenarios by a single pilot is a highly relevant field of research [1, 2]. A major issue here is the risk for excessive pilot mental workload and loss of situation awareness. While highly automatic planners are feasible to solve such multi-vehicle planning problems in reasonable processing time, they increase the risk for automation-induced errors such as the loss of situational awareness, complacency, or opacity [3]. In order to address these problems, our mixed-initiative planner (MIP) enables the pilot to assign tasks on different delegation levels and supports in the planning process [4]. This article presents an interaction concept for the task delegation and assistance. Chapter 2 gives a brief overview of other MIP systems. Chapter 3 introduces the different participants of the proposed system, their relationship and the different delegation levels. Chapter 4 shows the task assignment to the Unmanned Combat Aerial Vehicles (UCAV) and Chap. 5 introduces the different assistance features of the system. Chapter 6 provides a brief evaluation of the concept and possible future modifications.

W. Karwowski and T. Ahram (Eds.): IHSI 2019, AISC 903, pp. 699–705, 2019.
https://doi.org/10.1007/978-3-030-11051-2_106

2 Background

In a previous investigation, we studied the management of the manned-unmanned team on a very high level of delegation. This study indicated that a constant high level of automation temporarily led to complacency effects of the pilots and was lacking in adaptability to balance the pilot's activity and work demands over the course of the mission. The experiments further expressed the desire to be able to assign specific tasks to the UCAVs during mission execution, especially in less demanding situations [5]. In the helicopter domain, the general applicability of the mixed-initiative concept to the multi-UAV MUM-T domain has already been proven in previous studies. There, a prototype was integrated into a helicopter mission simulator and evaluated with German helicopter pilots. Results showed the advantages of the concept such as reduced workload and increased plan awareness and performance [6, 7]. Other work on this topic is conducted at the Air Force Research Laboratory. In their approach, the operator is assisted in the play based guidance of the unmanned system by an intelligent agent [2].

3 Work System

The concept presented here, is based on the work system described in [4], which comprises the pilot, the MIP and the unmanned team members, as well as the own aircraft. The pilot and the MIP work together in a hierarchical as well as in a heterarchical relationship. The MIP additionally is in a hierarchical relationship to the unmanned team members. From this work system design, it follows that the guidance of the individual team members must always be carried out through the MIP, as it is the only participant in a hierarchical relationship with them. The MIP enables the pilot to delegate both, individual tasks, e.g. the reconnaissance of a building, and a complex set of cooperative tasks, e.g. a coordinated target attack, to the team. The MIP further enables the pilot to assign these tasks on the following delegation level:

- Team-based: On this delegation level a task is automatically assigned to the team members.
- Task-based: Here an individual task is assigned to a single team member (e.g. the reconnaissance of an area) at a specific position in the plan.
- Parameter-based: At this delegation level the individual parameters of the tasks can be directly accessed and adapted as required.

The hierarchical task delegation is described in Chap. 4. In addition to the hierarchical relationship, the pilot can be supported in the planning process by the MIP. This support is performed with the heterarchical relationship and described in Chap. 5.

4 Interaction Concept Delegation

This chapter deals with the hierarchical relationship between the pilot and the MIP. In this context, it is explained how to create tasks for individual targets, how to assign them to the team members at the different delegation levels, and how these tasks can be parameterized. This interaction takes place by use of a multifunctional display in our experimental fighter aircraft cockpit simulator, shown in Fig. 1. On the left and right side of the display the pilot can select individual pop-up pages (a). In this example, the task delegation page presented later is selected (b). In the center at the bottom of the tactical map, we see our own manned combat aircraft (c) and the unmanned team members (d), located in the blue circle. The red circles on the map represent the enemy air defense (e) and the yellow symbols indicate primary and secondary targets (f).

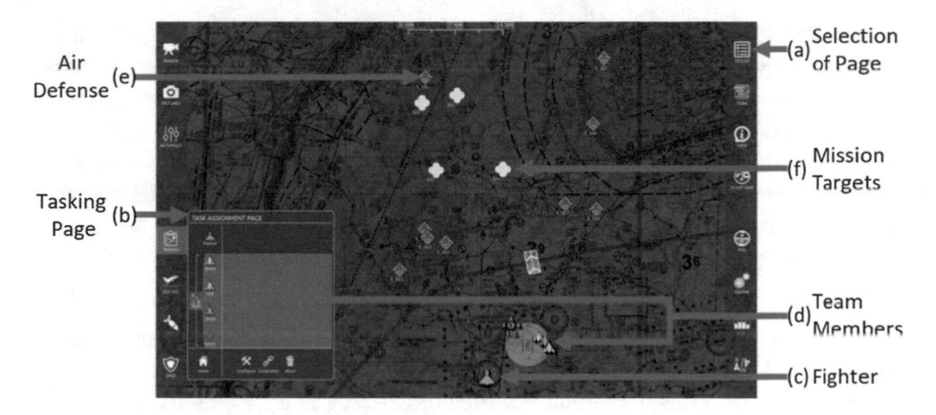

Fig. 1. Tactical map with task assignment page

4.1 Target and Task Selection

The selection of a target is performed by finger pointing on the tactical map. Around the selected target a radial context menu with target specific tasks opens. For an enemy air defense station, the pilot is offered the possibility to suppress the air defense (SEAD), shown in Fig. 2. When selecting a primary attack target, e.g. building B01, the pilot can choose between an individual sensor task and a cooperative attack task, compare Fig. 3. If the pilot selects the attack task, a second level in the radial context menu opens. In this second level, the sub-tasks belonging to the cooperative task are displayed. The pilot can then decide between the cooperative attack task and the corresponding subtasks. This allows the pilot to either delegate all subtasks himself or to delegate the cooperative task to the mission planner. For the attack of a primary ground target these are the subtasks reconnaissance (RECCE), laser designation (LASER), weapon engagement (WPN) and battle damage assessment (BDA), compare Fig. 4.

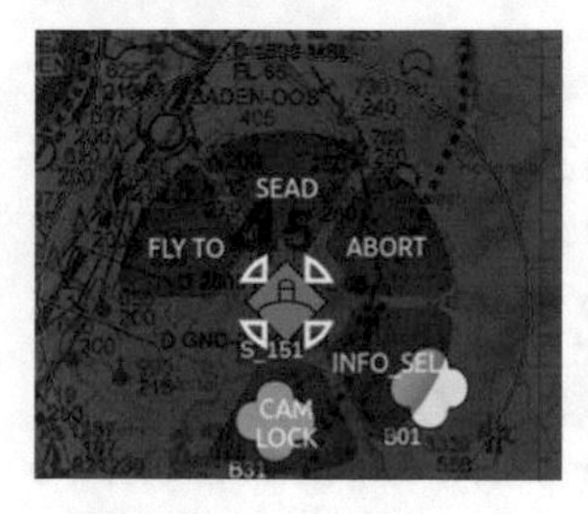

Fig. 2. Selection of SEAD target

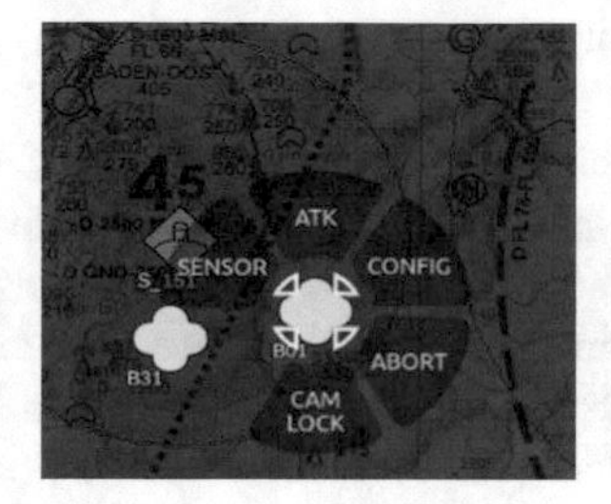

Fig. 3. Selection of a building

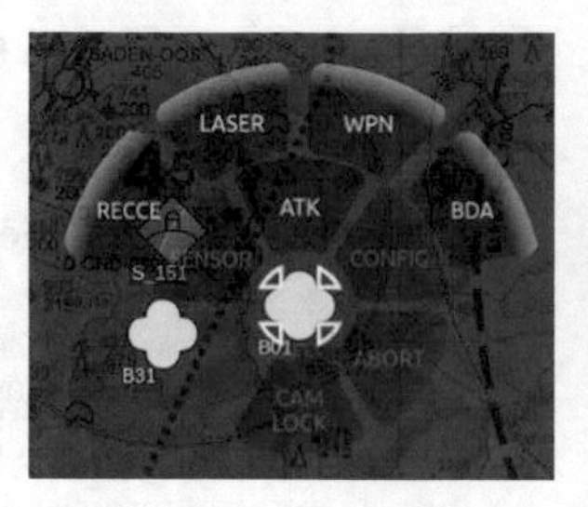

Fig. 4. Cooperative tasks for an attack

4.2 Task Delegation

After the selection of a task, the pilot can allocate it to the team members using the task assignment page. This page offers different delegation options depending whether a cooperative or individual task is selected. The possible delegation positions are marked with the symbol. As shown in Fig. 5, cooperative tasks can only be delegated to the team by selecting the team (Fig. 5a) button. If the pilot selects the team delegation the cooperative task is decomposed into the subtasks and these tasks are assigned to the best suited team members and coordinated. Figure 6 shows the result of a team-based delegated cooperative attack task on target B01. The assignment of an individual task can take place at the following delegation levels:

- Team delegation: The MIP assigns the task to the best team member and inserts the task at the best position (Fig. 6a).
- Individual delegation: The pilot specifies the UCAV and assigns the task:
  - Floated: The MIP determines the best position for the task in the UCAVs plan (Fig. 6b).
  - Fixed: The pilot specifies the exact position in the UCAVs plan where the task shall be inserted (Fig. 6c).

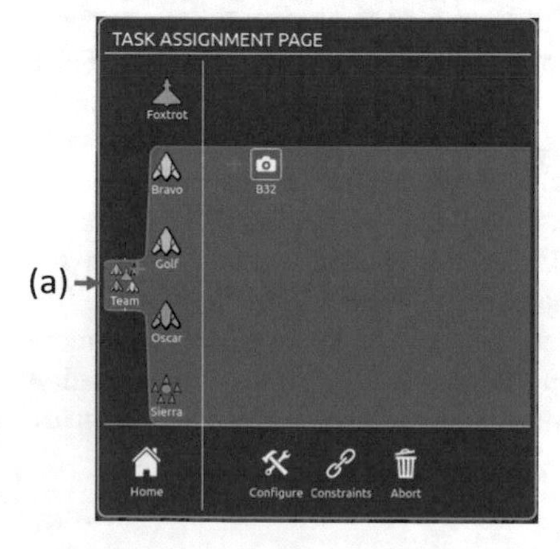

Fig. 5. Assignment possibilities for cooperative tasks

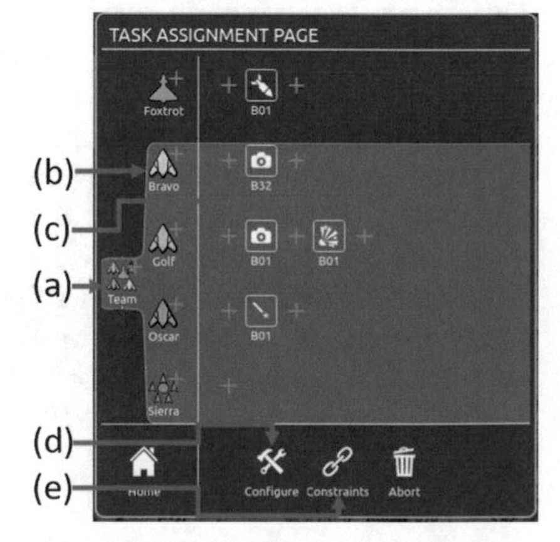

Fig. 6. Assignment positions for elementary tasks

During the assignment of a task, it is automatically filled with default parameters from the task model [4].

4.3 Task Parametrization

After delegation, the pilot can further adapt the parameters of the assigned tasks. For this the pilot selects the task which is to be parameterized and either selects if the parameters (Fig. 6d) or constraints (Fig. 6e) are modified. Depending on the task, the configuration parameters differ. Figure 7 shows the modifiable parameters for a reconnaissance task. In this case, the pilot has the possibility to modify the offset of the pattern (Fig. 7a), as well as the corresponding reconnaissance pattern (b), duration (c) and sensor technology (d). The adjustment of the constraints is shown in Fig. 8. The pilot can either remove existing constraints (marked blue) or add further constraints.

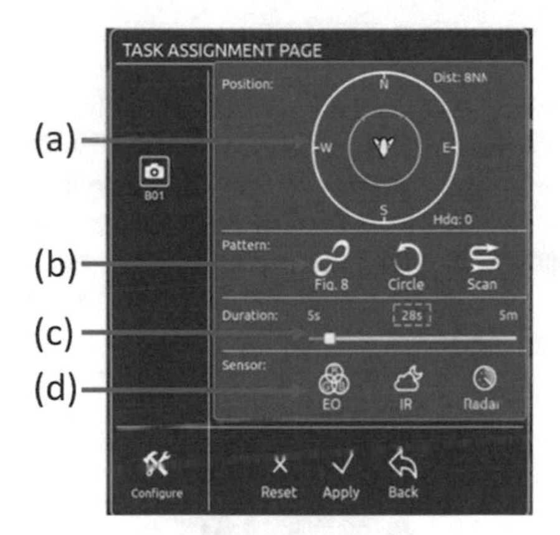

Fig. 7. Parameter configuration

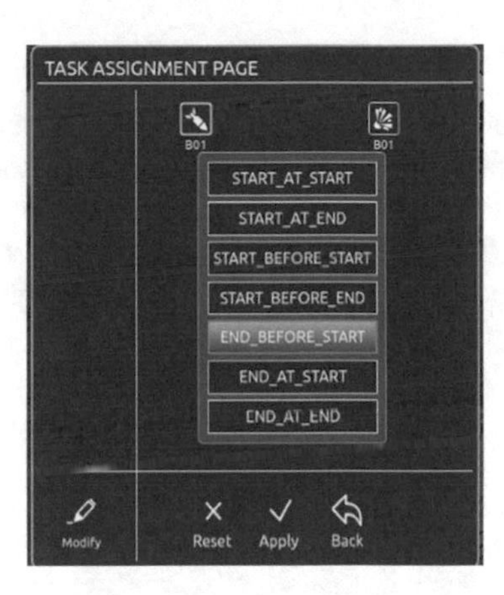

Fig. 8. Constraint modification

5 Interaction Concept Assistance

This chapter deals with the heterarchical relationship between the pilot and the planner agent. The assistance of the mission planner ranges from indications of missing goals, over the identification of planning conflicts, to suggestions for improvements for sub-optimal plans.

5.1 Missing Tasks

In case the pilot has forgotten to assign a task to the team, which is necessary for the successful mission execution, it is possible to direct the attention of the pilot to this task at different levels, depending on the workload. In situations with low workload and when the pilot has enough time to assign the task, only a simple hint is given (Fig. 9a). In situations with increased workload or time-critical situations, the task is simplified for the pilot by allowing the pilot to directly assign the task to the UCAV in the popup dialog (Fig. 9b). In situations with very high demands and/or very high urgency, the mission planner can automatically delegate the task to the team (Fig. 9c). The pilot, however, can still revise this decision at any time.

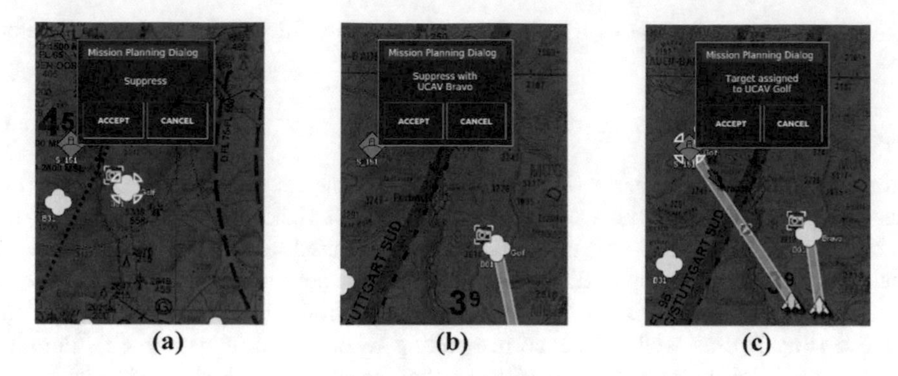

(a) (b) (c)

Fig. 9. Different levels of assistance for missing tasks

5.2 Improvements and Conflicts

When the mission planner identifies a better plan, it can present the necessary steps for improvement to the pilot. Figure 10 shows such a plan improvement resulting in a 15 s faster plan. Furthermore, planning conflicts resulting from an incorrect task assignment of the pilot can be resolved and presented to the pilot. In Fig. 11 such an incorrect allocation, where the battle damage assessment is assigned before the attack is displayed.

Fig. 10. Plan improvement

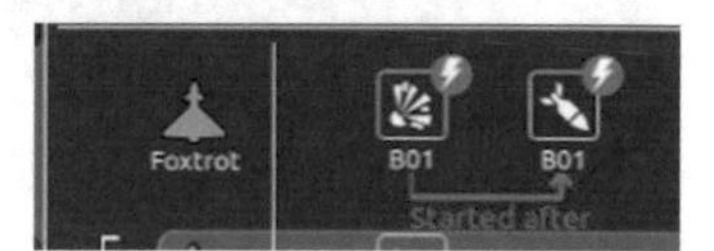

Fig. 11. Planning conflict

6 Evaluation and Conclusion

In this article, we presented an interaction concept for the mixed-initiative planning of several UCAVs from aboard a manned fighter aircraft. In this context, we addressed the different delegation possibilities of the hierarchical mode, as well as the assistance possibilities of the heterarchical mode offered by the system. The presented interaction concept was experimentally investigated with two pilots of the German Air Force and 12 aerospace students. All test subjects were able to successfully complete the missions. The team- and the individual-based assignment of tasks were positively evaluated by the test persons. In the case of fixed task assignment, however, the participants expressed the desire that the tasks could be added directly to a time line and parameterized there. The interaction concept will be adapted to the understandings extracted from this experiment.

References

1. Chen, J.Y.C., Barnes, M.J.: Human–agent teaming for multirobot control: a review of human factors issues. IEEE Trans. Human-Mach. Syst. **44**(1), 13–29 (2014)
2. Behymer, K., Rothwell, C., Ruff, H., Patzek, M., Calhoun, G., Draper, M., Douglass, S., Kingston, D., Lange, D.: Initial evaluation of the intelligent multi-U × V planner with adaptive collaborative/control technologies (IMPACT) (2017)
3. Wiener, E.L., Curry, R.E.: Flight-deck automation: promises and problems. Ergonomics **23** (10), 995–1011 (1980)
4. Felix Heilemann, A.S., Schmitt, F.: Mixed-Initiative Mission Planning of Multiple UCAVs from Aboard a Single Seat Fighter Aircraft. SciTech (2019)
5. Gangl, S., Lettl, B., Schulte, A.: Management of multiple unmanned combat aerial vehicles from a single-seat fighter cockpit in manned-unmanned fighter missions. In: AIAA Infotech@Aerospace (I@A) Conference (2013)
6. Schmitt, F., Schulte,, A.: Experimental evaluation of a scalable mixed-initiative planning associate for future military helicopter missions. Lecture Notes Computer Science (including Subser. Lect. Notes Artif. Intell. Lect. Notes Bioinformatics)
7. Schmitt, F., Roth, G., Barber, D., Chen, J., Schulte, A.: Experimental Validation of Pilot Situation Awareness Enhancement Through Transparency Design of a Scalable Mixed-Initiative Mission Planner, vol. 722 (2018)

Development of an Autonomous Manager for Dyadic Human-Machine Teams in an Applied Multitasking Surveillance Environment

Mary E. Frame[1](✉), Alan S. Boydstun[1], Anna M. Maresca[1], and Jennifer S. Lopez[2]

[1] Wright State Research Institute, Wright State University, 4035 Colonel Glenn Hwy, Beavercreek, OH 45431, USA
{mary.frame,alan.boydstun,anna.maresca}@wright.edu
[2] Air Force Research Laboratory, 1864 4th St, Wright-Patterson AFB, Dayton, OH 45433, USA
jennifer.lopez.11@us.af.mil

Abstract. Automation is crucial in increasingly many workplaces. Though automation is often associated with job replacement, humans and machines have divergent proficiencies. Thus, human-machine teaming is generally favored over replacement. Within applied surveillance environments, automation is leveraged for cognitively intensive tasks. To maintain optimal performance within a dyadic human-machine team, we developed an Autonomous Manager (AM) that dynamically redistributes tasks between human and machine. Participants performed four simultaneous image identification tasks while paired with a simulated autonomous partner. Our AM was responsible for monitoring team performance and redistributing tasks when performance fell sub-threshold. We manipulated the refresh rate of the images, affording us the opportunity to measure improvement under multiple conditions.

Keywords: Task Delegation Systems · Human-Machine Teaming Adaptive Automation

1 Introduction

Many working environments have experienced a shift from a model of one person working on a single task, to a single person multitasking between cognitively demanding tasks. Increasingly demanding working environments have required the use of automation to facilitate efficiency and alleviate fatigue for overworked humans. Effectively leveraging technology can improve performance and reduce or moderate workload [1]. However, insertion of technology into working environments alone is not necessarily effective at improving performance due to differences of proficiencies between humans and machines. Rather than replacing humans, automation and human workers often must collaborate and communicate as teams [4]. Human-Machine Teams

W. Karwowski and T. Ahram (Eds.): IHSI 2019, AISC 903, pp. 706–711, 2019.
https://doi.org/10.1007/978-3-030-11051-2_107

(HMTs) can vary on the degree of control over tasking managed by each teammate. For the human, this ranges from a passive supervisory role, to an action-vetoing role, to full control over tasks [1]. Sophisticated HMTs are increasingly leveraged in surveillance environments, where automation has been incorporated into daily mission activities alongside humans. There is still a need to co-manage these human and machine teammates with their differing skill sets under dynamic working conditions that can effect overall performance.

Identifying events of interest in static or dynamic surveillance footage is crucial and can have profound influence on decision-making and the respective consequences of those decisions. Modern surveillance environments, such as building security, often require a single human to monitor multiple feeds of visual data simultaneously. As workload demands increase for humans, workload must be delegated between humans and their automation. When humans are overworked, their performance can suffer, making interventions from automation necessary, even if that automation is inferior to a human on a particular task in isolation. It would be impossible for a human manager to effectively reallocate tasks in real-time for every dyadic human-automation team as it would be impossible to ascertain a consistent measure of current workload and performance, let alone quickly and efficiently parse tasks. To solve this managerial dilemma, we developed an Autonomous Manager (AM) to track performance and subjective assessment of workload in real-time to appropriately and dynamically reallocate tasks between humans and machines and maintain an acceptable level of performance.

Previous research on HMTs has demonstrated a great deal of benefit from implementation of adaptive, rather than static, automation [3]. Automation that adapts based on the human teammate's physical state or behavior directly benefits humans: well-balanced workload, better situation awareness, increased trust in automation, and overall higher performance [2]. When automation is incorporated without proper adaptability, this can lead to higher workload, increased confusion, and a loss of situation awareness [6]. Adaptability may be achieved and overall system performance enhanced by continuously monitoring human performance or physiological metrics, such as electroencephalography (EEG) [5]. Based on the resulting real-time estimates of human workload, automation can be utilized when it is needed most and can be reallocated to other tasks once the human is capable of adequately balancing the tasks at hand [3].

2 Autonomous Manager Decision Logic

To emulate the priorities in surveillance working environments, the AM's parsing decision logic prioritized maintaining performance over moderating workload. When the HMT performed above threshold (85% team performance for this task), the AM assessed subjective workload as reported by subjects in the NASA-TLX. If workload was above a maximum acceptable threshold, the automated partner would assume control over the lowest performing task. Conversely, if workload was excessively low, indicating subject under-work, the lowest performing task by the automated partner was shifted to human control. When HMT performance was below threshold, the AM determined which task had the lowest performance and switched from the current operating human or machine to the other. There was also a simple memory parameter, where script would reallocate the tasks to the previous iteration in the simulation if the previous iteration yielded a substantially (at least 10%) higher score.

3 Method

Participants were tasked with observing images within four simultaneously occurring tasks based on whether certain key elements were present in the picture (see Fig. 1) For this experiment, the target elements were: (1) blue vehicles, and (2) groups of 4 or more people gathered together. Images were similar to over-head surveillance photographs (see top right corner of Fig. 1 for an example with both targets).

The task environment was generated in a customized Javascript program. Participants completed a short practice block controlling all four tasks. This served to familiarize participants with the tasks and provide baseline performance information. The refresh timing was 10 seconds for each image, refreshing concurrently. Following the practice block, participants completed two 25-minute test blocks with either all four images refreshing simultaneously (Synchronous) or refreshing at different rates (Asynchronous). For both blocks, mean refresh time for each image was 8 seconds. Participants would click a button on the interface indicating "Present" if a target was contained in the image and "Absent" if not. Red gears superimposed over an image and the graying out of the response buttons served to indicate when the automated teammate was controlling a given task. On both blocks, the initial configuration was set for participants to perform all four tasks. After each minute, subjective workload was measured using pertinent questions from the NASA-TLX embedded within the task including the degree of mental workload, time pressure, and effort exerted. Workload and performance scores were used to compute the task distribution for the next minute of the task.

4 Results

Due to a coding error with one of the two conditions, for some participants, the final two minutes were omitted from the data. We first tested what we classified as the most essential function of the AM: to balance human and automation task distribution such that the HMT's performance would be greater than either of the two operators working in isolation. Automation performance was uniformly distributed between 50% and 100%. The automation's average baseline for performing all tasks was 75%. The human baseline for our sample was ($M = 68.62\%$, $SD = 11.60\%$). Figure 2 illustrates that for both refresh rate conditions, team performance was sustained above both human and automation baseline performance for nearly every time sample. The first time point was roughly equivalent to the human baseline performance since the task initiated with the human participant performing all tasks. The superiority of the HMT is readily apparent as soon as the second minute of the task.

Comparisons were made between the short baseline prior to the task and each of the two tasking conditions as a sanity check to ensure that human performance and efficiency with the AM's parsing was superior to performance of the human performing all tasks alone without adaptive redistribution. Refresh timing was slightly different between the practice and full blocks; in the practice block participants had more time to evaluate images and respond to four simultaneous tasks. Despite this additional time to respond, baseline performance was significantly lower than human accuracy in both the Asynchronous (Mean Difference = 9.36%), $t(14) = 3.46$, $p < .01$, and the Synchronous (Mean Difference = 9.49%), $t(14) = 3.31$, $p < .01$, blocks. Participants were also able to respond significantly faster in both the Asynchronous, (Mean Difference = 3325ms), $t(14) = 15.36$, $p < .001$, and Synchronous, (Mean Difference = 2239ms), $t(14) = 8.25$, $p < .001$, blocks, compared to the baseline. Human performance and speed improved resulting from effective integration with an automated partner, yielding superior performance to the human performing alone. As a secondary check, human accuracy was binned within trial (excluding baseline) into five temporal bins. For both Synchronous, $F(4,340) = 3.36$, $p = .01$, and Asynchronous blocks, $F(4,338) = 3.99$, $p < .01$, the first temporal bin yielded a significantly lower performance score than all 4 later time bins.

Accuracy was compared as a function of image refresh rate condition. There were no significant differences in accuracy of the whole human-machine team or human accuracy on the tasks performed. Although a priori we hypothesized that differences should occur as a function of difficulty, the lack of difference demonstrates that the AM was effectively parsing tasks. This conclusion is bolstered by significant differences in the average number of tasks being completed by a human in each condition. Human participants managed significantly more tasks over time in the Synchronous condition (M = 2.98, SD = 0.40), than in the Asynchronous condition (M = 2.69, SD = 0.32), $t(14) = 3.47$, $p < .01$. This is indicative that in the Asynchronous condition it may indeed have been more difficult for humans to manage all four tasks. Interestingly, despite differences in the number of tasks being simultaneously managed by participants, there were no differences in perceived workload between the two conditions. There was a robust difference in response times between conditions, with significantly longer response times in the Synchronous condition (Mean Difference = 1075ms), $t(14) = 9.88$, $p < .001$, than in the Asynchronous condition.

5 Discussion

The AM successfully improved human and HMT performance above baseline levels in both conditions by effectively redistributing tasks between the human participant and the automation, further reinforcing results from previous studies on the beneficial impact of adaptive automation [3] [5] [6]. Taken together, participants' performances on the subtasks they controlled were higher than the overall system performance due to the high stochasticity of the automation's performance (uniformly distributed between 50-100%). The AM was robust to the chaotic automated partner, which represented a worst-case scenario of an impossible to predict teammate. This can realistically occur when an automated tool is initially employed, especially for machine learning algorithms needing large amounts of training data. These realistic scenarios underscore the importance of the human member of HMTs as underscored by Parasuraman and colleagues [4]. The AM maintained overall performance of the human-machine team even with unpredictable automation. Adaptive management of highly chaotic automation has been relatively unexplored in the literature on adaptive automation.

Manipulating the refresh rate between the tasks afforded an excellent opportunity to test the AM's parsing effectiveness when there is variability in performance based on task difficulty. There were significant differences between refresh synchronization conditions regarding how many tasks subjects could perform simultaneously. Subjects were able to manage more tasks when all four images refreshed together. Despite the differences in difficulty, the AM led to performance maintenance by effectively balancing distribution of tasks, regardless of the optimal number of tasks for the human.

These results provide an excellent demonstration of the value of adaptive parsing and redistribution of tasking for human machine teams that is full of possibilities for expansion into future efforts. One concurrent ongoing study relies on a narrower proficiency of the automation to emulate an expert (uniformly distributed between 85-100%) or novice (uniformly distributed between 70-85%). In the present study, situations where system accuracy was lower than the human's performance reflect incidences where the automation's proficiency was randomly selected as low for that iteration. Real-world implementations of automation vary in terms of reliability and thus both chaotic and reliable automation are worthy of testing. Another concurrent study focuses on displaying performance feedback since this may have an impact on situation awareness similar to varying levels of automation [1]. Future efforts will determine the effect of longer baselines, visual performance feedback, learning effects, and task familiarization on automated task distribution using the AM.

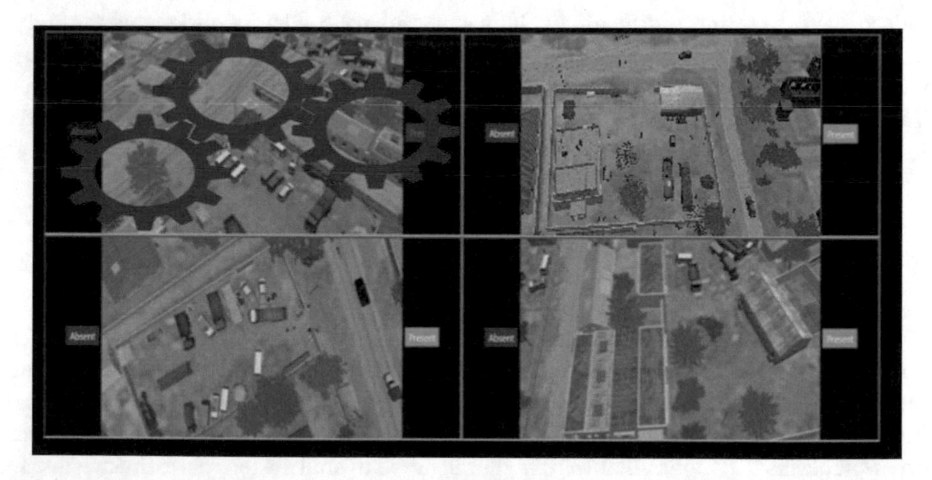

Fig. 1. Snapshot of the tasking environment. Images were generated in-house using Meta-VR. The automation is controlling the task in the top left corner as indicated by red gears superimposed over the image. The task in the top right corner provides an example of both targets: blue vehicles and groups of four or more individuals.

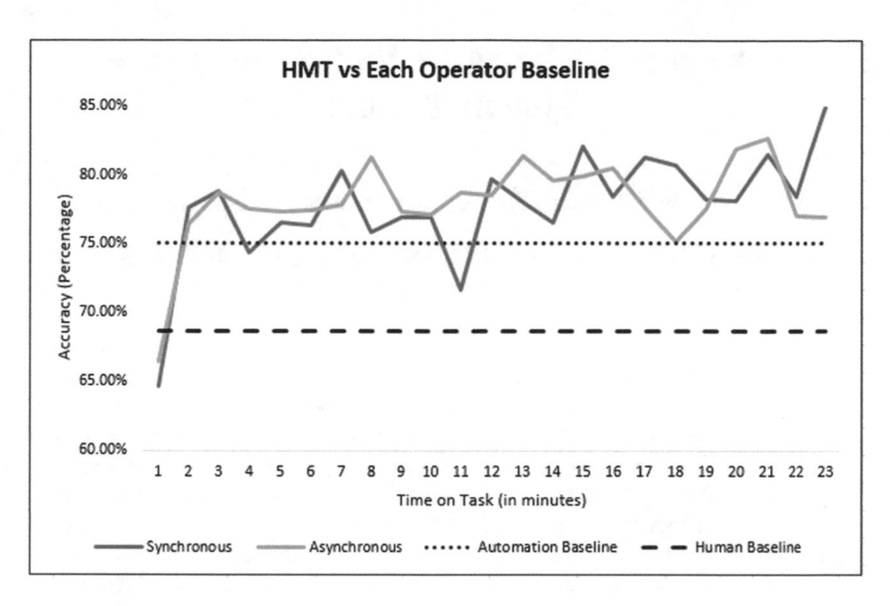

Fig. 2. Accuracy of the HMT over time in both refresh rate conditions compared to baseline of human and automation working in isolation. The results show a sustained performance improvement over both baselines after the first minute (once teaming begins), illustrating the value of the AM to overall task performance.

References

1. Kaber, D.B., Endsley, M.R.: The effects of level of automation and adaptive automation on human performance, situation awareness and workload in a dynamic control task. Theoretical Issues in Ergonomics Science. **5**, 113–153 (2004)
2. Miller, C.A., Parasuraman, R.: Designing for flexible interaction between humans and automation: Delegation interfaces for supervisory control. Human Factors. **49**, 57–75 (2007)
3. Parasuraman, R., Mouloua, M., Molloy, R., Hilburn, B.: Adaptive function allocation reduces performance cost of static automation. In: 7th International Symposium on Aviation Psychology, pp.37–42. (1993)
4. Parasuraman, R., Wickens, C.D.: Humans: Still vital after all these years of automation. Human Factors. **50**, 511–520 (2008)
5. Prinzel, L.J., Freeman, F.G., Scerbo, M.W., Mikulka, P.J., Pope, A.T.: Effects of a psychophysiological system for adaptive automation on performance, workload, and the event-related potential P300 component. Human Factors. **45**, 601–614 (2003)
6. Young, M.S., Stanton, N.A.: Automotive automation: Investigating the impact on drivers' mental workload. International journal on cognitive ergonomics. **1**, 325–336 (1997)

Crossing the Uncanny Valley of Human-System Teaming

Nathan Schurr(✉), Adam Fouse, Jared Freeman, and Daniel Serfaty

Aptima, Inc., 12 Gill Street Suite 1400, Woburn, MA 01801, USA
{nschurr, afouse, freeman, serfaty}@aptima.com

Abstract. Increasingly advanced systems are becoming pervasive in society. They are taking on roles and have abilities that necessitate their teaming with humans. However, the authors argue that we will soon reach a point where the increased embedding of artificially intelligent systems into collaborative teams results in reduced performance. We believe that there is an upcoming "uncanny valley of collaboration" that must be traversed on the way to natural truly productive hybrid teams. In order to avoid potential obstacles to hybrid team progress and safety, we argue that this challenge should be clearly identified and addressed across the larger human system integration community. In this paper, we describe the problem, propose potential solutions, and provide some examples where we are making strides toward bridging this valley.

Keywords: Human-Machine Teaming · Human-System Integration
Context Modeling · Hybrid Team Design · Symbiotic Teams · Uncanny Valley

1 Introduction

In 1960, J.C.R. Licklider wrote: "Man-computer symbiosis is an expected development in cooperative interaction between men and electronic computers. It will involve very close coupling between the human and the electronic members of the partnership." Almost 60 years later, we are far from natural and tight collaboration with our technological systems.

We propose that much like with visual realism, there is an "uncanny valley of collaboration" that needs to be crossed for systems to be accepted and productive as teammates. If we veer into this uncanny valley without crossing it, we will create systems that appear collaborative but that fail to meet users' expectations, frustrate users in their interactions, and degrade the quality of their work. These symbiotic and truly collaborative systems will understand and anticipate teammate expectations and needs. They will even go so far as communicating their own collaboration limitations.

In this paper, we will define this uncanny valley and argue that recent work provides a foundation—the anchors and piers—on which to build a bridge across it. In addition, we present an innovative approach to enable these Human-System teams based on current research programs that we lead.

The foundations for designing symbiotic systems lie in recent models of tasks and teams (drawing on research about human cognition and skills), communication between team members (from the field of language analysis), and coordination (in the

W. Karwowski and T. Ahram (Eds.): IHSI 2019, AISC 903, pp. 712–718, 2019.
https://doi.org/10.1007/978-3-030-11051-2_108

field of task/resource allocation and mixed initiative systems). In sum, recent research addresses tasks that the teams perform, their skills, and goals.

Several innovations will enable symbiotic, human-machine, hybrid teaming. These are: (1) enhanced communication modalities, (2) reasoning about goals and rewards and (3) algorithms for designing and adapting team structure and process. The new communication modalities (ranging from gestures to implicit communications) are exciting and provide much more natural ways to collaborate. The reasoning about goals and rewards are enabled by machine learning inference algorithms that identify what matter and why without overburdening the human teammates. And the adaptation of teams is enabled by more agile and aware team members coupled with innovative approaches to best select team modifications.

In this paper, we will frame the issue, challenges and highlight key enabling capabilities to bridge the gap. In addition, we will call out projects that are relevant to this human-machine teaming domain. Then, we will present three key innovations and how advances in them will help to bridge across the uncanny valley of collaboration between humans and systems. We believe that that pushing toward this will finally help realize Licklider's original vision of "man-computer symbiosis."

1.1 Human System Collaboration

We would like to describe briefly here what is meant by human-system collaboration and the hybrid teams that are enabled by it. As systems become more complex in reasoning and advanced in capabilities, they gradually are taking on more responsibility across a variety of domains. Human-system interaction has grown to encompass not just using as a system or tool, but to include multimodal human-system communication, human-system work delegation of tasks, and even joint human-system performance of tasks. What began as merely human-system interaction will increasingly need to evolve into human-system collaboration. This human-system collaboration will begin to have the same costs (coordination overhead) and benefits (improved performance and robustness) of human teams. In addition, these systems will be treated and thought of more as peers.

We argue that these "hybrid teams" of human and system will be enabled by this human-system collaboration and will gradually become the dominant form of working and living in the future. Later, in Sect. 3 below, we will explain some key examples of technologies that we are developing toward this goal.

2 Uncanny Valley of Collaboration

2.1 Origins

The original uncanny valley concept was identified by the robotics professor Masahiro Mori in 1970 [6]. Mori hypothesized that as the appearance of a robot is made more human, some observers' emotional response to the robot becomes increasingly positive and empathetic, until it reaches a point beyond which the response quickly becomes strong revulsion. This phenomenon has been observed in both robotics and computer graphics [9].

Much like in areas of physical appearance and physical movement, we argue that in the areas of teaming and collaboration actions, there will begin to be a similar phenomenon that we are calling the "uncanny valley of collaboration." As shown below in Fig. 1, we propose that as we increase the collaborative ability of systems, we will eventually hit a point at which the hybrid team's performance will drop dramatically and hybrid team members will have a similar, deeply negative reaction to working together.

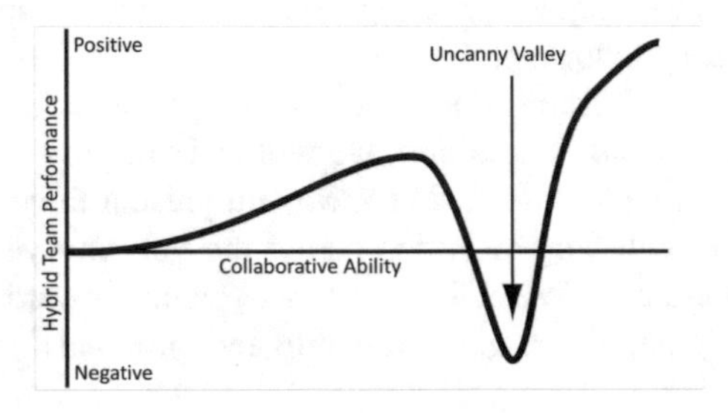

Fig. 1. Proposed uncanny valley of collaboration.

2.2 Implications and Challenges

Given the existence of this upcoming "uncanny valley of collaboration," there are several related issues and challenges that the human-system integration community should begin to discuss and address. We hope that this paper helps to spur such discussions. For example, as increased collaborative abilities are implemented in systems, hybrid team participants can experience intense frustrations or increased workload when dealing with a system that was near-capable of full collaboration but falls short. Even more importantly, this can lead to decreased performance and even hybrid team safety issues as they are deployed to critical domains. Hybrid team frustrations, miscoordination and failures will pose longer term obstacles to advances in system complexity, applications and hybrid teaming itself.

This will likely result in lost progress, lost opportunities and lost applications of hybrid human-system teams. This could result in a situation where people are more questioning and doubting of integrating with systems because of prior experiences and misguided expectations. In addition, it could result in reduced funding and public perception of systems and the related field of artificial intelligence.

3 Bridging the Valley

3.1 Addressing this Uncanny Valley

Understanding that this "uncanny valley of collaboration" exists will be beneficial if only to better prepare and better design systems. As in the real world, when approaching a deep valley, you can choose to steer clear of it or attempt to cross a bridge above the gap if you would like to avoid the steep drop off (see Fig. 1). To steer clear of it in this case would lead the community working on human system collaboration to potentially avoid inserting increasing collaborative abilities into their systems or also focus more efforts and resources on clearly conveying to human counterparts the collaboration limitations of these systems. This is akin to the practice in robotics and graphics, of intentionally portraying more cartoon-like or non-human character to alter or even lower expectations. We realize that this "steering clear" may be a perfectly viable and especially desirable near-term solution. However, to help cross the uncanny valley as fast as possible and minimizing the undesirable negative performance, we propose that in addition to making more capable system, we must also work toward bridging technologies.

We argue these bridging technologies enable productive progress toward symbiotic, hybrid team collaboration. Below are two examples of recent work that where we are aiming at these bridging technologies. The first is that of capturing the full context of the hybrid team which enables avoiding performance pitfalls but adapting to this context. Toward this, we have built a Context Platform applied it to a growing number of domains. The second example aims to simulate and design better hybrid team structures and tasks. With the Adaptive Teams project, the eventual aim would be to build a hybrid team that is robust to inevitable changes in the world and drops in performance.

3.2 Context Platform

To better understand not only system performance or human performance but the context of the overall hybrid teams and their goals, Aptima has been developing the Context Platform [8]. Using the representations architecture described below, the Context Platform has been applied to domains such as assisting civil affairs personnel [7], cyber defenders [2], and information analysts [1]. We believe that measuring this context can not only can this make the system more capable but also help identify upcoming collaboration breakdowns (such as that in the "uncanny valley of collaboration").

Figure 2 aims to express how broadly we are modeling context in this situation. At the core of the Context Platform is a conceptual model that embodies our inclusive definition of context in an abstract form. Our approach is instantiated for a particular application by creating a specific model of context that captures the key concepts and relationships for a domain. Entity attributes and relationships are tracked over time, so at any moment our picture of context consists of the current state of entities and relationships, as well as all historical values.

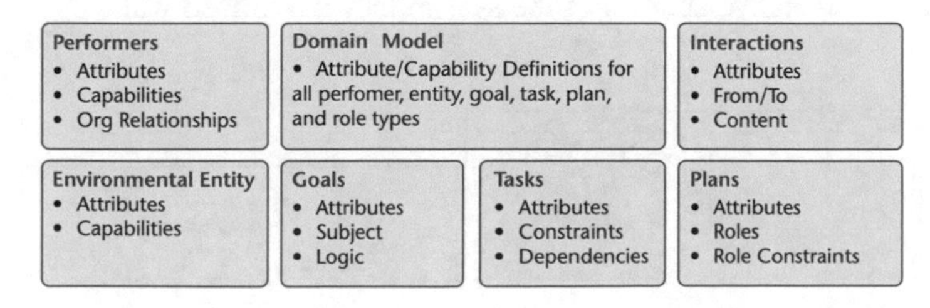

Fig. 2. Model for context for hybrid teams.

The core technology underlying Context Platform is a graph-centric multi-model knowledge fusion, storage, and analysis platform. It enables context-aware inference in applications by modeling domain data and capturing user interactions with that data to extract generalized context elements in a graphical data structure. The Context Platform has been engineered using a scalable, microservice-based architecture to provide flexible data processing and enrichment pipelines for streaming and batch ingest of structured and unstructured data (as shown in Fig. 3).

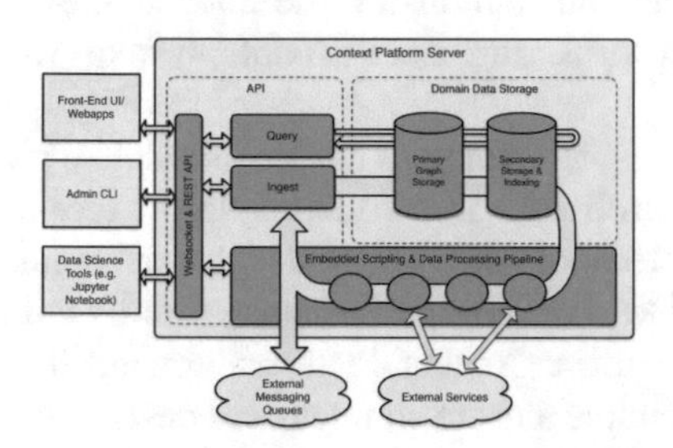

Fig. 3. Context platform architecture.

Given the above representation and architecture, the Context Platform has served as the framework enabling both adaptation to user goals [1] and novel multimodal interactions [3]. Successfully utilizing multiple modalities of interaction requires modeling and adapting to the context of the human-system relationship and the content of the communication.

3.3 Adaptive Teams

To explore the bridging technology of Adaptive Teams we have been exploring how to best design and optimize hybrid teams [5]. Figure 4 shows a potential team and how they can be adapted in both structure and in role. The proposed adaptations are based on a generalizable mathematical abstraction for the design of adaptive human-machine teams, based on application of active inference theory and the free energy principle to hybrid teams of human and machine actors. The free energy principle tries to explain how self-organizing systems, such as the brain, restrict themselves to a small number of attracting states to avoid disorder in uncertain environments and minimize surprise [4]. This surprise is a way to frame the hybrid team member's reaction to the "uncanny valley of collaboration." The team member is surprised by very capable but ultimately detrimental collaborative abilities of the system they are teamed with.

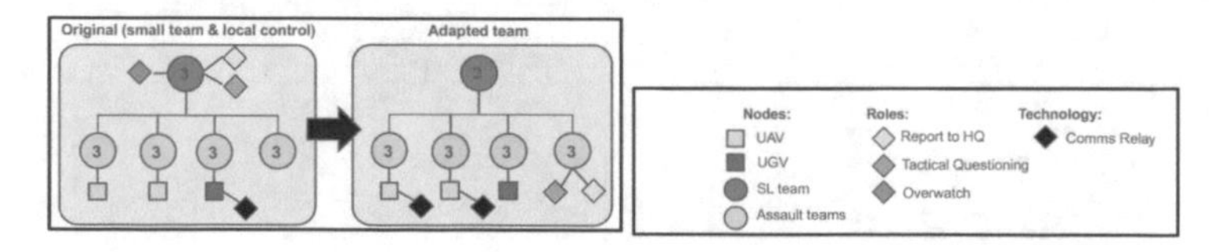

Fig. 4. Team structure adaptation for infantry squad for 15-person squad.

To test the design, structure, dynamics and performance of such hybrid teams we have developed a mechanism for simulating the performance of teams, which we call the Simulation of Task Assignment for Teams to Evaluate Structure (STATES). STATES takes as input team structures and mission scenarios. The core of STATES consists of probabilistic generation of task assignment vectors based on the ranges, probabilities, and locations defined in the mission. Through multiple simulations, it generates instantiated event/task networks based on the probabilities defined by the mission and simulates task assignment and performance to generate metrics of the teams' processes and outcomes. STATES could be eventually used to anticipate and mitigate potential drops in hybrid team performance.

4 Discussion and Future Work

In this paper we have presented an upcoming challenge: as systems become more capable and integrated, there will inevitably be encountered an "uncanny valley of collaboration." Recognizing this valley exists will enable system designers to plan for it, to intentionally steer clear of it or begin working on bridging technologies to cross it. Our aim is to begin a discussion about these critical bridging technologies.

We propose three key areas to enabling symbiotic, hybrid teams: (1) enhanced communication modalities, (2) reasoning about goals and rewards and (3) algorithms for designing and adapting team structure and process. We later introduced recent technologies: Context Platform is enabling and measuring area (1) supporting technology for area (2) and Adaptive Teams is addressing area (3).

In the future, we hope to further define and quantify the width and depths of this "uncanny valley of collaboration." We also plan to refine and explore additional bridging technologies. We hope that doing so will minimize the inevitable setbacks to the field of hybrid team collaboration.

References

1. Fouse, A., Mullins, R.S., Ganberg, G., Weiss, C.: The Evolution of User Experiences and Interfaces for Delivering Context-Aware Recommendations to Information Analysts. In: Ahram T., Falcão C. (eds) Advances in Usability and User Experience. AHFE 2017. Advances in Intelligent Systems and Computing, vol 607. Springer. (2017)
2. Fouse, A., Mullins, R.S., & Ziemkiewicz, C.: A Framework for Context-Aware Visualization in Cyber Defense. Proc. of IEEE Symposium on Visualization for Cyber Security (VizSec) (2016)
3. Fouse, A., Weiss, C., Mullins, R.S., Hanna, C., Nargi, B., & Keefe, D.F.: Multimodal Interactions in Multi-Display Semi-Immersive Environments. In 2018 IEEE Conference on Cognitive and Computational Aspects of Situation Management, Proceedings of. Boston, MA. (2018)
4. Friston, K.: The free-energy principle: a unified brain theory? Nature Reviews Neuroscience **11**(2), 127–138 (2010)

5. Levchuk, G., Pattipati, K., Fouse, A., Serfaty, D., & McCormack, R.: Active learning and structure adaptation in teams of heterogeneous agents: designing organizations of the future. Proc. SPIE 10653, Next-Generation Analyst VI, 1065305 (21 May 2018)
6. Mori, M., MacDorman, K.F., Kageki, N.: The Uncanny Valley [From the Field]. Proc. IEEE Robotics & Automation Magazine **19**(2), 98–100 (2012)
7. Mullins, R., Fouse, A., Mccormack, R., & Lovell Pfautz, S.: A Context-aware Decision Support Tool for Assessing and Mitigating Drivers of Civil Instability. Proc. of the 6th International Conference on Applied Human Factors and Ergonomics. (2015)
8. Pfautz, S.L., Ganberg, G., Fouse, A., Schurr, N.: A general context-aware framework for improved human-system interactions. AI Magazine **36**(2), 42–49 (2015)
9. Plantec, P.: "Crossing the Great Uncanny Valley—Animation World Network". URL: https://www.awn.com/vfxworld/crossing-great-uncanny-valley (December 19, 2007)

Hierarchical Planning Guided by Genetic Algorithms for Multiple HAPS in a Time-Varying Environment

Jane Jean Kiam[1](✉), Valerie Hehtke[1], Eva Besada-Portas[2], and Axel Schulte[1]

[1] Institute of Flight Systems, University of the Bundeswehr, Werner-Heisenberg-Weg 39, 85579 Munich, Neubiberg, Germany
{jane.kiam, valerie.hehtke, axel.schulte}@unibw.de
[2] Departamento de Arquitectura de Computadores y Automatica, Universidad Complutense de Madrid, Plaza de Ciencias 1, Ciudad Universitaria, 28040 Madrid, Spain
ebesada@ucm.es

Abstract. A hierarchical task planning structure is favorable for its capability to accommodate constraints at different abstraction levels and also for the similarity of its planning approach as a human. This structure is adopted for the task planning for multiple HAPS. However, the combinatorial search problem grows with the presence of multiple agents. This work proposes a method to guide the decomposition of the tasks down the hierarchy with genetic algorithm in order to find quality plans within limited time.

Keywords: Multiple agents · Hierarchical planning · HAPS · Genetic algorithm

1 Introduction

Light-weight ($\sim$75 kg), solar-powered High-Altitude Long-Endurance (HALE) platforms cruising slowly ($\sim$28 m/s) in the stratosphere, are a promising alternative to fixed-orbit satellites [1]. Also known as High-Altitude Pseudo-Satellites (HAPS) (see Fig. 1), although sensitive to even moderate weather conditions, HAPS are ready to be deployed.[1] It is reasonable to optimize mission reward and operation safety by increasing autonomy in flight planning. This is especially critical when multiple HAPS are deployed to increase coverage of mission areas. Increasing autonomy is also lucrative to make the operation more economically viable.

A hierarchical planning approach that consists of decomposing a task of higher abstraction into subtasks following a known recipe of task execution, is suitable for solving the mission planning problem for HAPS, since constraints raised by airspace regulations, mission requirements, aircraft limitations etc. can be represented at different abstraction levels, or rather space resolutions. Being intuitive for human cognition,

[1] https://www.airbus.com/newsroom/press-releases/en/2018/07/Airbus-opens-first-serial-production-facility-for-Zephyr-High-Altitude-Pseudo-Satellites.html

W. Karwowski and T. Ahram (Eds.): IHSI 2019, AISC 903, pp. 719–724, 2019.
https://doi.org/10.1007/978-3-030-11051-2_109

hierarchical planning methods are used in many human-in-the-loop applications [2, 3]. Since the operation of HAPS will be monitored by human operators, it is essential that the planning model and constraints, as well as the plans are "explainable".

Fig. 1. The launching of a HAPS-platform (Zephyr 7 © Airbus DS GmbH)

Although convenient, hierarchical planning in existing works mainly decides the sequence of task execution, but not when to execute [2, 3]. The current works also cannot cope with large combinatorial problems where a substantial number of task decompositions are possible, which is the case when multiple HAPS are involved [4–6].

In this article, we first state the multiple-HAPS task planning problem. Then, we recapitulate the hierarchical planning structure for HAPS and describe the extension to multiple-HAPS. Subsequently, we report on our work to guide the planner to explore the most promising decompositions using Genetic Algorithm (GA), a flexible method to solve for Multiple-Objective Optimization Problems (MOOP), which is our case due to the various constraints to meet, e.g. mission requirements, airspace constraints, weather conditions etc. The GA was implemented, and some preliminary results are shown. Finally, an example visualization of the solution plans will be provided to demonstrate the effortless comprehension for the human operator.

2 Problem Statement

HAPS operate in the stratosphere at an altitude of about 18 km, where the airspace is relatively calm. A well-defined airspace is necessary [7] so that the number of vehicles operating at that flight level can be scaled up without conflict. A typical realistic

Table 1. Constraints resulting from airspace regulations and mission requirements

Abstraction level	Constraints
MA	• An MA is accessible only within certain time windows as according to airspace availability • At any instant, only one HAPS is allowed to be in an MA • If the HAPS flies into an MA, it must perform the monitoring tasks there, i.e. MAs should not be used as corridors
LOI	• Minimum revisit frequency within a period of time • Minimum revisit time • Minimum image coverage of the ground to be rewarded

continuous surveillance mission is shown in Fig. 4d, in which several Locations Of Interest (LOIs) marked in green are landmarks to be monitored. The Mission Areas (MAs) in blue encompassing LOIs of the same client denote the allocated airspace for carrying out the task. Waiting Areas (WAs) in yellow polygons are airspace in which the HAPS can loiter freely. Mission elements are connected by the dedicated Corridors (C).

In our work, we focus on the offline planning for landmark-monitoring using an on-board electro-optical (EO) camera; therefore, the clouds between the operating altitude and ground can affect the mission efficiency. The mission planning problem is subject to inhomogeneous constraints concerning elements in the scenario at different abstraction levels, or rather spatial resolution, as listed in Table 1. The main objective is to optimize the rewards the HAPS team receive, given that a mission is considered successful if the minimum image coverage of the ground is fulfilled.

3 Hierarchical Planning Structure

In a hierarchical planner, a method is applied to decompose each non-primitive task into subtasks [4]. [8] describes the hierarchical planner for the time-dependent task planning of a single HAPS. The duration of each primitive task is computed linearly, and the duration of non-primitive tasks equals the sum of the durations of the primitive tasks. In this article, we only consider total-ordered tasks.

Figure 2 depicts the structure of the hierarchical task planner for two HAPS. Each HAPS has its own task network, represented by a different color. The task networks are however interdependent so that constraints described in Table 1 are checked in order to ensure non-conflicting concurrency. Each HAPS can also have independent start and goal conditions. The decomposition of a high-level task is done in a top-down-forward manner, i.e. monitoring MA1 with the blue HAPS in Fig. 2 is first decomposed into a monitoring sequence of LOIs from the nearest to the farthest (LOI12, LOI11), which will be further decomposed into a sequence of Points-Of-Interest (POI13, POI14, POI11, POI17, POI15). The sequence of POIs defines the flight pattern; with a linear estimation of travel time between two consecutive POIs, the duration of each higher-level task can also be estimated. The decomposition stops when the planning temporal horizon is exceeded.

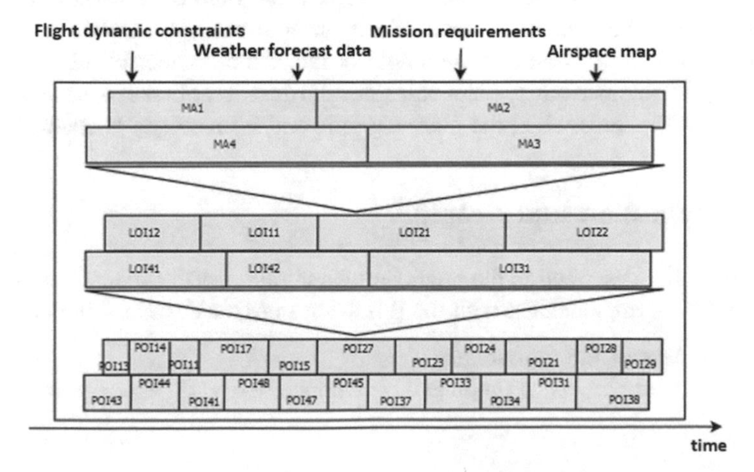

Fig. 2. Hierarchical decomposition of tasks of two task networks (blue and gray)

3.1 Search for Promising Decompositions Guided by Genetic Algorithm

The multitude of decomposition methods at the MA-level when multiple HAPS are involved results in a classical combinatorial search problem that is considered NP-hard. Therefore, evaluating all possible solution plans within allocated planning time might be impossible. We intend to use Genetic Algorithm (GA) to guide the planner at the MA-level to explore first the most promising decompositions. Another advantage of GA is that feasible solutions are always available even before the optimal solution is obtained.

3.2 Encoding of the Task Allocation with Multiple-String Chromosomes

In GA, often the chromosome encoding is done using a single string; however, for planning problems involving multiple agents, a multiple-string representation can be used [9]. We use here multiple-string chromosomes for the encoding, with each string representing the sequence of genes or rather mission elements to visit for a HAPS, as shown in Fig. 3a.

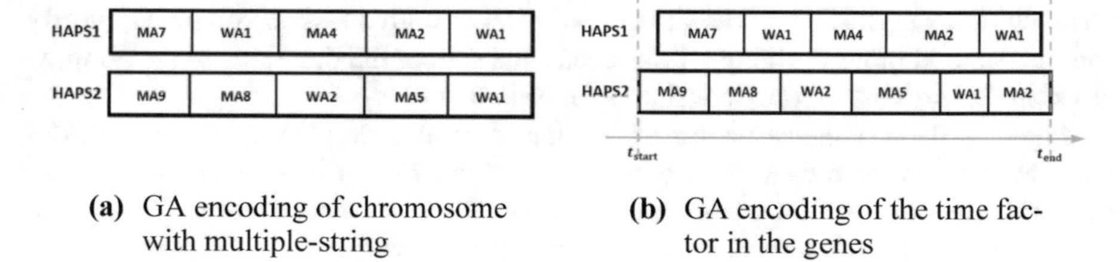

(a) GA encoding of chromosome with multiple-string

(b) GA encoding of the time factor in the genes

Fig. 3. GA encoding of chromosome for two HAPS

As mentioned in the previous section, the underlying planning problem is time-dependent, since the reward function depends on the time-varying cloud coverage forecast and the mission requirements are subject to time constraints. It is hence important to also encode the time factor in the genes of each chromosome. In order to accommodate the time dependency in the task planning, the encoding of the chromosomes must consider the duration of each task. Figure 3b shows the adapted encoding method we use. Each gene of the chromosome is assigned a length that represents the duration. Note that the length or rather the number of mission elements in each string is not necessarily identical, since the duration of each gene can vary, and the decomposition stops when the planning temporal horizon t_{end} is reached.

4 Implementation and Results

With the encoding described in the previous subsection, after the generation of an initial population, GA is implemented with the following steps, as described in detail by [10]:

1. evaluate fitness of the individuals;
2. select the fittest parents to produce a new population via a single-point crossover;

3. introduce mutation with a small probability into each new individual in order to avoid repetitions.

When evaluating the fitness of an individual, if a gene in the chromosome does not fulfill the constraints described in Table 1, the gene becomes the cut-off point for the cumulative fitness value. The above steps are repeated until convergence is reached or if planning time is exceeded.

The method is integrated in the offline mission planner to extend its planning capability for multiple HAPS. Historical cloud coverage forecast data extracted from COSMO-DE [11] on the 20th April 2018 were used for the test. Figure 4a summarizes the cloud coverage over time of the day for each MA. Since the GA is governed by some randomness, the results are not reproduceable. We tested the GA with 20 random runs and in average (the solid blue line in Fig. 4b), the maximum reward of each generation increases with each iteration, and eventually converges.

As explained earlier, a hierarchical planning structure has some advantages with regard to plan explainability. When planning for long duration, the tasks involved and the flight routes can be confusing and difficult to digest by an operator if the whole plan is displayed at once. Figure 4c shows the task plan over a day. With a hierarchical planning structure, the partial route for each element of each abstraction level can be visualized as desired by the operator. In Fig. 4d, the partial routes of the selected elements in Fig. 4c (frame highlighted in red and orange colors respectively) are shown.

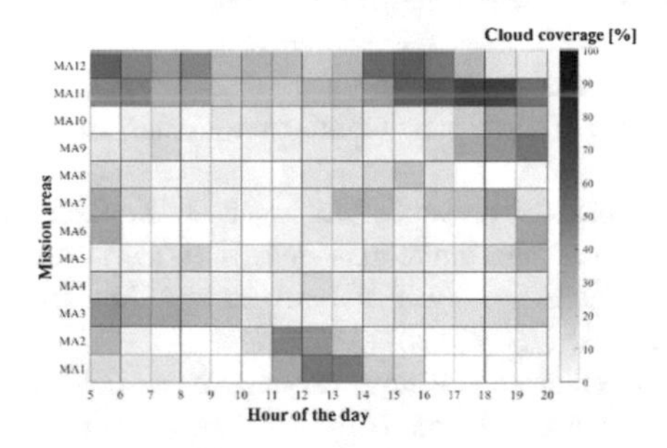

a. Cloud coverage over time of the day

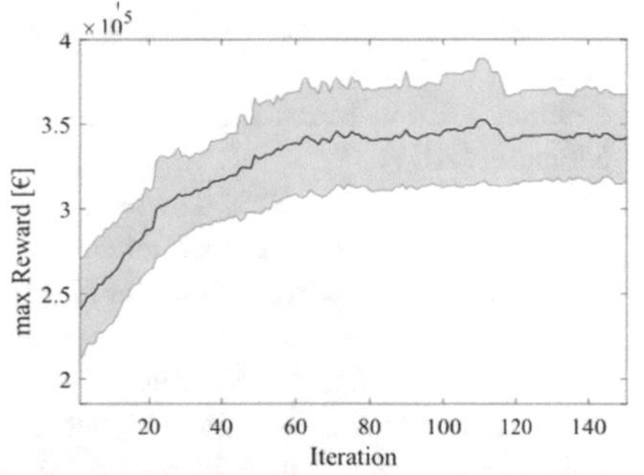

b. Increasing maximum reward with each iteration

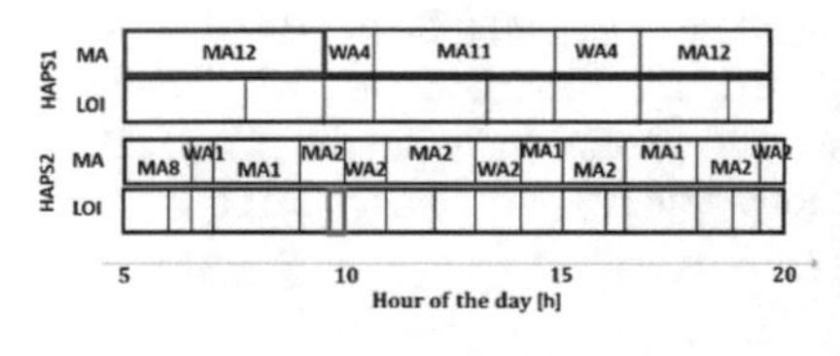

c. Task plans

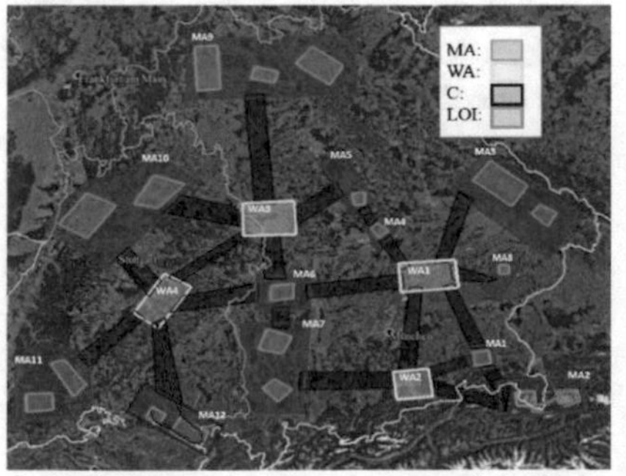

d. Mission scenario and partial flight paths

Fig. 4. Test with historical weather data

5 Conclusion

In this article, a hierarchical task planning structure for multiple HAPS is proposed and the search for the most promising decompositions is made efficient by using the genetic algorithm to guide the search. In this work, we adapted the GA to consider the time-dependency of the genes in the chromosomes. For future works, benchmarking with different parameters can be performed to ensure early convergence to optimality.

References

1. Müller, R., Kiam, J.J., Mothes, F.: Multiphysical simulation of a semi-autonomous solar powered high altitude pseudo-satellite. In IEEE Aerospace Conference, Montana (2018)
2. Benton, J., Smith, D., Kaneshige, J., Keely, L., Stucky, T.: CHAP-E: A plan execution assistant for pilots. In: 28th International Conference on Automated Planning and Scheduling (ICAPS), Delft, The Netherlands (2018)
3. Höller, D., Bercher, P., Behnke, G., Biundo, S.: A generic method to guide HTN progression search with classical heuristics. In: 28th International Conference on Automated Planning and Scheduling (ICAPS), Delft, The Netherlands (2018)
4. Nau, D., Au, T.C., Ilghami, O., Kuter, U., Murdock, J.W., Wu, D., Yaman, F.: SHOP2-An HTN planning system. J. Artif. Intell. Res. (JAIR) **2003**(20), 379–404 (2003)
5. Chen, K., Xu, J., Reiff-Marganiec, S.: Markov-HTN planning approach to enhance flexibility of automatic web service composition. IEEE International Conference on Web Services, Los Angeles, California (2009)
6. Kiam, J.J., Schulte, A.: Multilateral quality mission planning for solar-powered long-endurance UAV. In IEEE Aerospace Conference, Yellowstone Conference Center, Big Sky, Montana (2017)
7. Johnson, M., Jung, J., Rios, J., Mercer, J., Homola, J., Prevot, T., Mulfinger, D., Kopardekar, P.: Flight test evaluation of an unmanned aircraft system traffic management (UTM) concept for multiple beyond-visual-line-of-sight operations. In Twelfth USA/Europe Air Traffic Management Research and Development Seminar (ATM) (2017)
8. Kiam, J.J., Schulte, A.: Multilateral mission planning in a time-varying vector field with dynamic constraints. Man, and Cybernetics, Miyazaki, Japan, IEEE Systems (2018)
9. Shima, T., Rasmussen, S.J., Sparks, A.G., Passino, K.M.: Multiple task assignments for cooperating uninhabited aerial vehicles using genetic algorithms. Comput. Oper. Res. **33** (11), 3252–3269 (2006)
10. Mitchell, M.: An Introduction to Genetic Algorithms. The MIT Press (1999)
11. Baldauf, M., Seifert, A., Förstner, J., Majewski, D., Raschendorfer, M., Reinhardt, T.: Operational convective-scale numerical weather prediction with the COSMO model: description and sensitivities. Mon. Weather Rev. **139**(12), 3887–3905 (2011)

Can Intelligent Agent Improve Human-Machine Team Performance Under Cyberattacks?

Wen Ding[1], Sonwoo Kim[2], Daniel Xu[1], and Inki Kim[1(✉)]

[1] School of Engineering and Applied Science, Engineering Systems and Environment, University of Virginia, Charlottesville, VA 22904, USA
{wd3kf, dlx8dr, ik3r}@virginia.edu
[2] Computer Science, School of Engineering and Applied Science, University of Virginia, Charlottesville, VA 22904, USA
sak2km@virginia.edu

Abstract. This paper presents a preliminary work of a simulation study for the evaluation of human-machine team (HMT) performance with assistance of a sentinel system under cyberattacks. Sentinel system is an intelligent agent of cyberattack-detection kit, whose user interface (UI) was designed and implemented as a proof-of-concept in the simulation. The goal for this sentinel system is to improve the pilots' Situation Awareness (SA) of the system vulnerability and resiliency, such that the HMT performance can be improved against disruptive events. Based on the literature of cyberattacks on unmanned aerial vehicle (UAV) control systems, realistic mission operation and cyberattack scenarios were identified and implemented on a simulated UAV ground control station (GCS). In the follow-up experiment, up to twenty-four Air Force pilots will be instructed to supervise the UAVs under specific mission scenarios, and to respond for system recovery solutions generated on Sentinel UI after detection of cyberattacks. Understanding the interactive behaviors of the pilot and Sentinel under mission contexts and cyberattacks is expected to help improve HMT performance.

Keywords: Sentinel system · Human machine team · Cyberattacks simulation system · Performance modeling

1 Introduction

With increasing tactic values of UAVs in military operations, there are growing concerns about the vulnerability of the vehicle and their control systems against malicious enemy attacks. The concerns have seen incidents, such as the suspected hijacking of a U.S. military RQ-170 UAV [1], and the detection of a keylogging virus among the UAV fleet at the Creech Air Force Base [2]. An increasing complexity of autonomous technologies in the UAV control also aggravate the system vulnerability [3]. In this regard, this study proposes to enhance system resiliency by exploiting collaborative operations of human-machine team (HMT). Due to the nature of cyberattacks being adaptive and rare events [4], there are significant challenges in testing and evaluating

W. Karwowski and T. Ahram (Eds.): IHSI 2019, AISC 903, pp. 725–730, 2019.
https://doi.org/10.1007/978-3-030-11051-2_110

HMT behaviors, which in turn prevents from developing more advanced recovery mechanisms from simple repetitive experiments. Effective post-cyberattack recovery depends not only on the cyber capabilities of systems, but also on the sound judgment and timely responses of human operators in the mission context [5]. To examine HMT performance empirically, a high-fidelity supervisory control platform that carries battlefield mission contexts and post-cyberattack situations is necessary. Besides, the knowledge and skills of the human operator needs be evaluated in an experimental setting, in order to obtain repeatable and representative findings.

To that end, this study intends to reproduce mission operations and cyberattacks on an unmanned aerial-vehicle (UAV) control system. In practice, such systems are supervised by the team of operators at the ground control stations (GCSs). In the current simulation, the single pilot supervises semi-autonomous UAV control operations under given mission goals and constraints, when malicious cyberattacks to the GCS occurs at a random point in time. Upon such unexpected cyberattacks, even if these response solutions are generated by sentinel system, selecting the most appropriate options for system recovery is well beyond machine intelligence [6]. Typically, artificial agents are not trained against disruptive situations, and little data exists for setting mission priorities or out-of-the-box strategies. Thus, the machine needs to incorporate human insights that have derived from the situation awareness (SA) of ongoing mission and cyberattacks. In this regard, a sentinel system is expected to accelerate man-machine collaboration in tactic and strategic battlefield decision-making. For this study, a sentinel system was designed, and the impact of the sentinel system on HMT performance will be evaluated in the follow-up experiment.

2 Related Works

2.1 Cyberattacks on Unmanned Aerial Systems

Increasing concerns about the UAVs' vulnerability to cyberattacks have prompted a significant body of research [7–9], where virtually every component of an unmanned aerial system (UAS) is examined for vulnerabilities under hypothetical attacks and countermeasures being considered for each one of these components. All these cyberattacks, real and imagined, fall under the three categories of implementation [3], hardware attack (the attack is preinstalled on the UAS), wireless attack (the attack is carried via the UAS wireless channels), and sensor spoofing (false data is passed to the UAS sensors). Although the values and needs for intelligent agent for security intervention has been discussed for a while [10], its effects on emergency responses of human-machine team remains unknown.

2.2 Human-Machine Team Collaboration

In emergency, human intervention to the system can be crucial to effective recovery thanks to the flexibility of SA and judgment under uncertainty of situations [11]. Especially, under unprecedented cyberattacks with no prior information, effective cooperation for early responses and adaptive strategies is vital to minimizing the damages to the precious military assets [12].

To better manage a human-machine teamwork, an independent system of systems can play an important role. An intelligent agent of attack detection and response instructions has potential for mediating between human and machine. A sentinel system can serve as an artificial cognitive unit [10], to assist in human decision under a variety of battlefield missions and cyberattack situations.

For instance, sentinel systems can further help interpret operational information and contextual situations based on the computational inferences of sensor data, and formulate rule-based strategies for human operators. Although the pilots with a higher level of situation awareness (SA) can better formulate strategies themselves suited to the specific mission contexts, a team SA through cooperation with the machine is promising for enhanced situational judgment and decision making [13]. The current study hypothesizes that the assistance of Sentinel can augment human SA, and contributes to improved resiliency under cyber-attacks.

3 Construction of Human-in-the-Loop Simulation

3.1 Simulation System Architecture

Ardupilot, an autopilot suite with various drone properties is used to construct the simulation system [14]. Meanwhile, the DroneKit-SITL architecture is used to simulate the aerial-vehicle dynamics without reliance on the actual hardware [15]. Under this architecture, several vehicle telemetry information such as altitude, battery level, and flight plans, can be accessed and manipulated by using python scripts for mimicking normal and abnormal system/vehicle behaviors. This script-based manipulation was a main mechanism to simulate cyber-attacks. For the vehicle types in simulation, the group-two, fixed-wing UAVs (typically 20–50 pounds) were chosen.

3.2 Cyberattack Scenario and Mission Operation

The known type of cyberattacks and their categories [3] in Sect. 2.1 were referred to in selecting cyberattack scenarios for the simulation development. The selection criteria considered similarity to the reported attack cases, as well as the ease of implementation. The specific cyberattack was a Trojan horse virus, which would modify the waypoints stored aboard the UAV to prevent it from overflying undesired areas. This particular type of cyberattack was introduced in [5]. The main cyberattack scenario was that the vehicle waypoints for one of the swarm-UAV operations abruptly changed by the adversary as an initial attack. The system interface integrated itself with the UGCS (SPH Engineering Inc., Canada), a visual UAV control software that allows the pilot to control multiple UAVs.

Depending on experiment settings, cyberattacks are triggered randomly after satisfying certain parameter requirements, such as location, battery level, flight-time of the vehicle. Once the initial attack occurs and pilots are warned by sentinel system, the pilot could choose to add new waypoints manually to the current vehicle mission, to load an alternate preset flight plan from the system, or to recover the original flight plan. Based on system damage type, the sentinel system will inform a list of recovery responses with or without information about recover quality and time, as an instruction for the pilot responses. Pilots can choose to conduct one or more recovery responses in a sequence. In this process, it is highly likely that pilots will face tradeoffs between feasible responses, and that they are forced to exploit their own SA of the ongoing situation.

4 Sentinel Design

In general, the pilots are only able to gain limited access to the status of UAV on disruptive events, because its control is mainly based on graphical user interface (GUI). The resultant low-level SA can become a critical problem under cyberattacks, when the pilot may have trouble formulating effective recovery strategies. To accelerate man-machine collaboration for battlefield decision-making, a novel sentinel UI panel is designed to send security messages that reflect vehicle status and attack type, as well as guide the pilots into possible response solutions. The primary purpose of the sentinel UI design is to increase the SA level for pilot, which can hypothetically improve HMT performance under cyberattack scenarios.

4.1 Rationale of Sentinel UI Design

Upon cyberattacks, the sentinel system would detect them and alert their immediate impact on the system as illustrated in Fig. 1. For the pilot, it takes time to be activated with this warning together with other possible signs of system anomalies. In order to better comprehend the current situation, the pilot may examine the system status and available resources. Then, a set of recovery or reconfiguration options will be considered both by human and machine. Throughout this short period of alert, observation, and response selection, the sentinel UI can help in a number ways to improve the pilots' SA of immediate cyberattacks and better decision-responses.

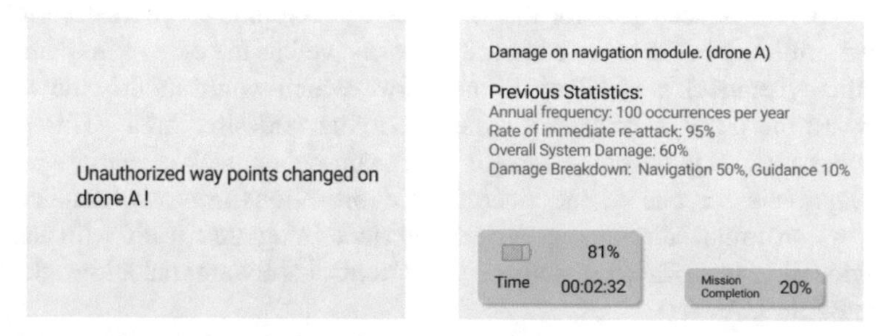

Fig. 1. The sentinel warning on a detection of cyberattack (Left Panel). The sentinel diagnosis of system damage, together with a history report of similar attacks (Right Panel).

As is likely with deceptive manipulation of the control display at the GCS, the pilot may become sluggish in recognizing the attacked status. To prompt the pilot's recognition, the sentinel system needs to execute an automated reconfiguration of the display. If available, presenting a history report of similar attacks on the sentinel can help project enemy doctrine. For fear that other telemetry information on the GCS interface might be also disrupted, presenting key status parameters (e.g., battery level, time of flight, or mission progress) on the sentinel UI can help avoid additional data verification processes and make a more confident decision for the pilot. As the pilots handle high-security intelligent information while operating time-critical missions, they are often faced with an urge to make out-of-the-box, highly-uncertain strategic decisions, such as abandoning UAVs (if the mission completion level is low), or aborting the current mission. Furthermore, under multiple-drone swarm operations, the pilots can become quickly overloaded with the urge to simultaneously monitor and reconfigure multiple vehicles.

4.2 Recovery Response

After the sentinel interaction with the pilot on initial attacks, a list of available response solutions may be presented to the pilots. The pilots could choose one or a combination of multiple responses, depending on the time constraints, mission context, system status, security concerns, or other factors of reasoning. From the cognitive standpoint, this choice involves the engagement of multiple SA information, reasoning and problem solving under the tradeoffs among mission criteria and other dilemmatic situations. Table 1 illustrates a list of possible response solutions against malicious disruption of waypoints.

Table 1. A list of resiliency solutions against Waypoint attacks

System damage	Possible resiliency options
Damage on navigation module	- Replace the navigation module
	- As an added precaution, replace the internal UAV communications to a backup communication channel between the navigation and flight control sub-system
	- Reconfigure the control display to permit presentation of the updated trajectories
Damage on guidance module	- Recover waypoints (reset the navigation software and upload the original flight plan)
	- Use alternate waypoints (reset the navigation software and upload an alternate flight plan)
	- Switch to a manual guidance
	- Reconfigure the control display to permit presentation of the updated trajectories (An operator may want to check the actual waypoints that the suspicious vehicle has been following.)
Others	- Force land both UAV(s) - Recover the UAV that the Sentinel indicated as attacked - Apply the diverse redundancy solution for both UAV(s) - Call in a new UAV from home
	- Continue mission while ignoring Sentinel alert
	- Abort mission and go home with all UAVs

5 Future Research

The current development of simulation system and sentinel UI design will be followed by experimental scenarios and plan. The follow-up experiment will test and evaluate the HMT performance for trained Air Force pilots.

References

1. Downed US drone: How Iran caught the "beast." https://www.csmonitor.com/World/Middle-East/2011/1209/Downed-US-drone-How-Iran-caught-the-beast
2. Virus infects program that controls U.S. drones, https://www.cnn.com/2011/10/10/us/drone-program-virus/index.html
3. Kim, A., Wampler, B., Goppert, J., Hwang, I., Aldridge, H.: Cyber attack vulnerabilities analysis for unmanned aerial vehicles. In *Infotech@aerosp 2012*, pp. 1–30 (2012)
4. Statistique, L. De., Sabatier, U.P.: Rare event simulation. Probab. Eng. Informational Sci. **20**, 45–66 (2006)
5. Jajodia, S., Liu, P., Swarup, V., Wang, C.: Cyber situational awareness. Springer, US (2009)
6. Dutt, V., Ahn, Y., Gonzalez, C.: Cyber situation awareness?.: Modeling the security analyst in a cyber-attack scenario through instance-based learning. IFIP Annu. Conf. Data Appl. Secur. Priv. **6818**, 280–292 (2011)
7. Krishna, C.G.L., Murphy, R.R.: A review on cybersecurity vulnerabilities for unmanned aerial vehicles. 2017 IEEEInt. Symp. Safety, Secur. Rescue Robot. 194–199 (2017)
8. Yag, E.: A study on cyber-security of autonomous and unmanned vehicles. J. Def. Model. Simul. **12**, 369–381 (2015)
9. Rivera, E., Baykov, R., Gu, G.: A study on unmanned vehicles and cyber security. Texas. pp. 1–3 (2014)
10. Onken, R.: Cognitive cooperation for the sake of the human-machine team effectiveness. In: UNIVERSITAET DER BUNDESWEHR MUENCHEN NEUBIBERG (GERMANY FR). pp. 7–9 (2003)
11. Millot, P.: A common work space for a mutual enrichment of human-machine cooperation and team-situation awareness. In: IFAC HMS. pp. 387–394 (2013)
12. Carver, B.Y.L.I.Z., Turoff, M.: The human and computer as a team in emergency management information systems, ACM. **50**, 33–38
13. Salmon, M., Salmon, P.M., Stanton, N.A., Walker, G.H., Baber, C., Daniel, P., Mcmaster, R., Young, M.S., Salmon, P.M., Stanton, N.A., Walker, G.H., Baber, C., Daniel, P.: What really is going on?? Review of situation awareness models for individuals and teams. Theor. Issues Ergon. Sci. 297–323 (2008)
14. Ardupilot, http://ardupilot.org/ardupilot/
15. Dronekit, http://dronekit.io/

Full-Mission Human-in-the-Loop Experiments to Evaluate an Automatic Activity Determination System for Adaptive Automation

Fabian Honecker(✉) and Axel Schulte

Institute of Flight Systems (IFS), Bundeswehr University Munich, Werner-Heisenberg-Weg 39, 85577 Neubiberg, Germany
{fabian.honecker, axel.schulte}@unibw.de

Abstract. To support helicopter pilots, a task-centered workload-adaptive associate system has been developed. This article briefly describes the concept and concentrates on the implementation and experimental validation of an automated system to determine pilot activity and estimate mental workload.

Keywords: Adaptive automation · Dempster-Shafer theory · Human mental state estimation · Manned-unmanned teaming · MUM-T

1 Theoretical Background and Concept

Introduction. The management of multiple unmanned aerial vehicles (UAVs) from the cockpit of a manned helicopter (manned-unmanned teaming, MUM-T) is one approach to achieve new operational capabilities for future military helicopter missions. The drawback of this approach is the increased task demand placed on the pilots, especially the commander. Particular cognitive tasks such as mission management, UAV-guidance and reconnaissance can lead to very demanding situations, where the mental workload (MWL) of the pilots may exceed a tolerable level.

In the past, effort has been made to reduce MWL through automation. It is well known, that automating everything is not the ideal solution since it leads to classical human factors problems. A better way focuses on the entire work system consisting of human operators with their tools and use to adaptive automation [10, 11] to assist the operators. Therefore, we developed a workload-adaptive associate system for MUM-T helicopter missions. Hence, we address three basic questions: How can a workload adaptive system be realized? How can MWL be operationalized, i.e. described in a technical system? How can the activity of a pilot be detected automatically and MWL estimated?

Concept. As the work system, we consider a classical helicopter cockpit, where pilots control the aircraft assisted by a workload-adaptive associate system. This approach requires the operationalization of MWL to qualify an associate system for adaptive automation. Most of the operationalizations of MWL mentioned in the literature do not fit to the engineering problem to construct an associate system because they do not account for the task context. Therefore, we are using a context-rich operationalization

W. Karwowski and T. Ahram (Eds.): IHSI 2019, AISC 903, pp. 731–737, 2019.
https://doi.org/10.1007/978-3-030-11051-2_111

of MWL and a task-centered approach according to [13, 5], where MWL is described by the tasks the operators have to do (plan), the actual activity of the pilots (a combination of the currently-executed tasks), the associated mental resource demands of the tasks in the activity and the behavior of the pilots (the way in which the tasks are performed). As key element, a hierarchical operator task model, containing all tasks from mission tasks down to small interactions, is used to describe the working domain. Based on this task model, a mission plan is generated using mixed-initiative mission planning [12]. During the mission, the interactions of the operators with the technical system are monitored and the current activity is determined in real-time [6]. Based on the current activity, mental resource demands can be estimated. By using the mission plan, the adaptive associate system can project the mental state of the operator into the future and determine intervention triggers, e.g. workload peaks, neglected tasks or critical events [1]. To close the adaptation loop, the associate system uses behavior rules for a beneficial human-machine relationship [9] to create and apply intervention strategies like attention guiding, task simplification or task adoption [1]. The following sections describing the implementation and experiments concentrate on pilot activity determination (AD) and MWL estimation.

2 Implementation

Determination of pilot activity. We are using an indirect evidential reasoning approach to determine the actual activity of a pilot, since the execution of many tasks, especially cognitive tasks, cannot be directly observed. Therefore, by observing the interactions of the operators with measurement equipment including gaze tracking, one or multiple evidences are collected that support or reject the execution of a certain task. After collecting evidences, these evidences are combined by using a simplified version of Dempster's rule of combination [3, 14]. In our theory, every evidence is stored in the operator task model and described as a normalized belief triplet $\boldsymbol{Q}(X) = (p, q, r)$ according to [2]. Here, p, q and r are values in the range from 0 to 1. $p(X)$ describes the belief, which supports the evidence for task X in the task, $q(X)$ the doubt, which rejects the task and $r(X)$ the ignorance. The normalization is given by $p+q+r=1$. Two evidences can be combined by using the following rule of combination [6]:

$$\boldsymbol{Q}_1(X) \oplus \boldsymbol{Q}_2(X) := \boldsymbol{Q}_{12}(X) = \begin{pmatrix} p_{12}(X) \\ q_{12}(X) \\ r_{12}(X) \end{pmatrix} \tag{1}$$

$$p_{12}(X) = \left| \frac{p_1 p_2 + p_1 r_2 + r_1 p_2}{1-(p_1 q_2 + q_1 p_2)} q_{12}(X) = \frac{q_1 q_2 + q_1 r_2 + r_1 q_2}{1-(p_1 q_2 + q_1 p_2)} \right.$$

Combining multiple evidences is achieved by applying this rule iteratively. Finally, tasks with high belief and low doubt values are selected as the activity.

Estimating mental resource demands. To describe mental demands, we are using the 8 resource channels of multiple resources theory (MRT) [15]. Every resource component d_r

can take values in the range of 0 to 1. A demand of 1 indicates, that given task execution, there is no free capacity left on this resource. To model resource demands, one task model is stored for every task in the task model. For multitasking situations, where the operator conducts N tasks in parallel, we sum up all resource demands of every task t by extending the VACP model [8] leading the total demand for resource r:

$$d_r = \sum_{t=1}^{N} d_r^t \tag{2}$$

If one of the individual resources demands exceeds a value of 1, we assume an overload on this mental resource. By using this theory, a resource adaptive associate system can be designed [7].

Our motivation is to additionally calculate an overall scalar workload value based on these resource demands on a computer. Our workload metric based on resource demands is motivated by [16], where workload is stated as the demand of mental resources d related to the free mental capacity r. According to the MRT [15], mental resources interfere with each other and lead to resource conflicts c. We assume that conflicts reduce the free capacity. Putting this into a formula leads to:

$$W = \frac{\text{demand}}{\text{capacity}} = \frac{d}{r} = \frac{d}{1-c} \tag{3}$$

Since this formula is only a rough translation of the theoretical description, we need further assumptions, especially to account for multiple tasks and multiple resources. We postulate the following empirical formula, to calculate a scalar workload value:

$$W = \max_r \left\{ \sum_{t=1}^{N} \frac{d_r^t}{1 - c_r^t} \right\} = \max_r \left\{ \sum_{t=1}^{N} \frac{d_r^t}{1 - \sum_{\substack{k=1 \\ k \neq t}}^{N} \sum_{j=1}^{8} K_{rj} d_j^k} \right\} \tag{4}$$

Here, d_r^t is the demand for task t and mental resource r. The denominator quantifies the free resource capacity of resource r and task t, where all resource conflicts of task t with other tasks are summed up by using Wickens' conflict matrix K_{rj}^W [15]. The conflict matrix tells us the magnitude, how individual resources interfere with each other. Since this matrix is not normalized and tends to overestimate workload in our setup, we are using a normalization factor based on the averaging the row-sums to rescale the matrix components: $K_{rj} = \frac{1}{4.175} K_{rj}^W$. By using another sum over all N tasks, we can calculate a workload measure for each individual resource given the currently-executed tasks. This measure is similar to the VACP approach, but in contrast to the sum described above, it accounts for resource conflicts. At this point, we need further assumptions how to combine these values into a single one. Since our motivation is to detect overload rather than fatigue, we suggest calculating the overall workload value as the maximum of every resource r. Mental overload is indicated if this workload measure exceeds a value of 1.

3 Experiments and Results

Experimental design. To evaluate the methods described above for pilot activity determination (AD) and workload estimation, we conducted human-in-the-loop-experiments with four crews consisting of two experienced helicopter pilots (3900 average flying hours and 650 h of military operations). Each crew conducted 6 full-scale helicopter missions and an additional traffic pattern for every pilot in a simulator.

Subjective ratings. In the first step, subjective ratings for 7 different pilots have been conducted. One pilot was flying a traffic pattern observed by the AD system. For evaluation purposes, the AD system reported in real-time the pilot's activity (currently-executed tasks) using speech synthesis during the flight. After the experiment, the pilot had to rate the system performance. All pilots accepted the scenario but stated, that the speech output was very disturbing. Completeness of the task model was subjectively evaluated at $80 \pm 16\%$, correctness of the activity determination was subjectively evaluated at $87 \pm 11\%$, and on a scale from +3 (very good) to -3 (very bad), the overall performance was subjectively evaluated at 1.2 ± 1.7. All errors are given as standard deviations.

Data acquisition and analysis. For deeper insights, we asked the pilots for their activity, mental resource demands and MWL in selected 20 to 60-s use cases. Instead of using questionnaires, we developed an interactive tool (*TaskKladde*) for this tedious job. Software ergonomic features, like an interruptible video playback, a dynamic task tree, and demand vector mapping have been included. We asked the pilots, to split the use cases into situations of parallel task activity (task situations) by selecting tasks from the task model. In the second step, the pilots were asked to rate their mental resource demands and MWL for every task situation. Data from AD and resource estimation was sampled with a rate of 2 Hz. Data analysis was also done with the *TaskKladde* program to compare the AD with human ratings for selected use cases in the missions. Signal data for resource demands and scalar MWL values were averaged over every task situation and the standard deviation together with Gaussian error propagation was used to describe statistical errors. In total, from the accumulation of the various use cases, 17 min of the 19 h of mission execution time were analyzed (Fig. 1).

Task situation complexity. To examine task situation complexity, the number of parallel performed tasks were counted. Differences between the pilot flying and the commander could be found. In the evaluated 17 min, the pilots reported 169 task situations for the pilot flying and 77 for the commander. In the same time span, the AD system detected 329 situations for the pilot flying and 214 for the commander. For the pilot flying, 27% (40%) of all situations were single-tasking and 85% (54%) multi-tasking (values for the AD system are given in brackets). For the commander 57% (36%) single-tasking and 38% (44%) multi-tasking. The difference between the two pilots reflects the persisting manual control task of manually flying the helicopter.

Dynamic properties and granularity. For dynamic task properties, 359 (548) tasks in typical use cases for the pilot flying and 110 (297) for the commander have been considered and characteristic numbers like average task duration or task state switches

Fig. 1. *TaskKladde* program for data acquisition.

(TSS [4]) calculated. The average duration of a task was $4.3 \pm 0.8s$ $(2.0 \pm 0.9s)$ for the pilot flying and $8.1 \pm 1.3s$ $(2.3 \pm 0.6s)$ for the commander. For the pilot flying, 20.1 ± 5.0 (34.6 ± 0.8) task set switches per minute (TSS) could be observed and 13.4 ± 3.8 (29.3 ± 6.6) for the commander. The large difference in the average and TSS values between human data and automatic is due to the fine granular pilot task model and short sample times of the automated activity determination system. Task granularity can be adjusted by filtering signal data, e.g. with a first order low-pass filter. While the unfiltered activity of the pilot contains many small tasks of short duration, the filtered data only reflects the long enduring mission tasks. The signal curves for resource demands, resource conflicts, and mental workload of the unfiltered data are very noisy, whereas the filtered curves are very smooth.

Automatic task analysis for whole missions. We conducted an automatic mission analysis with the prototype to illustrate the capability and power of an automatic pilot activity determination system for future task analysis. We found, that task situation complexity was similar in all missions but showed differences due to the manual flying in the roles between the pilot flying and the helicopter commander.

Closed-loop adaptive assistance. In particular use cases during the experiments could be shown, how adaptive assistance based on automatic activity determination and workload estimation can be achieved. Different intervention strategies like attention guiding, task simplification and task adoption triggered by workload peaks or neglected tasks could be observed [1]. In one particular scenario, the helicopter pilots were performing a difficult takeoff inside a dangerous area. Because such a takeoff is a demanding use case, their workload was high. To transition from takeoff to transit flight, multiple system setting modifications must be performed. The associate system was aware of the required tasks and recognized the high workload situation of the pilot flying. To support the pilot and simplify the task situation, the transit configuration of the helicopter was set automatically and the pilots were notified of the action.

4 Conclusions

In this experiment, it was shown that the method for real-time activity determination based on evidential reasoning and workload estimation has been implemented in a helicopter simulator. Although, the model shows large errors on small time scales (0.5 s), both, task granularity and MWL signal curves converge against the human expectations if time filters are applied. The main open questions are, how to acquire model parameters from domain experts, what is the best level of granularity in the task model and what time constant best fits the prediction of MWL. Furthermore, the impact of ignorance on the decision of an associate system has to be examined in detail.

References

1. Brand, Y., Schulte, A.: Design and evaluation of a workload-adaptive associate system for cockpit crews. In: Harris, D. (ed.) Engineering Psychology and Cognitive Ergonomics, pp. 3–18. Springer International Publishing, Cham (2018)
2. Dempster, A.P.: The Dempster-Shafer calculus for statisticians. Int. J. Approx. Reason. **48** (2), 365–377 (2008)
3. Dempster, A.P.: Upper and lower probabilities induced by a multivalued mapping. Ann. Math. Stat. **38**(2), 325–339 (1967)
4. Grootjen, M., et al.: Task-based interpretation of operator state information for adaptive support. Found. Augment. Cogn. section **2**, 236–242 (2006)
5. Honecker, F. et al.: A Task-centered Approach for Workload-adaptive Pilot Associate Systems. In: Schwarz, M. and Harfmann, J. (eds.) In: Proceedings of the 32rd Conference of the European Association for Aviation Psychology, Cascais, Portugal (EAAP), Groningen, NL, pp. 485–507 (2017)
6. Honecker, F., Schulte, A.: Automated online determination of pilot activity under uncertainty by using evidential reasoning. In: Harris, D. (ed.) Engineering Psychology and Cognitive Ergonomics: Cognition and Design, pp. 231–250. Springer International Publishing, Cham (2017)
7. Maiwald, F., Schulte, A.: Enhancing military helicopter pilot assistant systems through resource adaptive dialogue management. In: Vidulich, M.A. et al. (eds.) Advances in Aviation Psychology. Ashgate Studies in Human Factors and Flight Operations. Ashgate Publishing, Ltd., Farnham, England, pp. 177–196 (2014)
8. McCracken, J.H., Aldrich, T.B.: Analyses of selected LHX mission functions: implications for operator workload and system automation goals. U.S. Army Research Institute, Fort Rucker, AL (1984)
9. Onken, R., Schulte, A.: System-ergonomic design of cognitive automation: dual-mode cognitive design of vehicle guidance and control work systems. Springer, Berlin Heidelberg (2010)
10. Parasuraman, R., et al.: Theory And Design of Adaptive Automation in Aviation Systems. Naval Air Warfare Center, Aircraft Division, Warminster, PA (1992)
11. Scerbo, M.W.: Adaptive automation. In: Parasuraman, R., Rizzo, M. (eds.) Neuroergonomics. The Brain at Work. pp. 239–252 Oxford University Press, New York (2007)
12. Schmitt, F., Schulte, A.: A scalable mixed-initiative planner for multi-vehicle missions. In: International Conference on Engineering Psychology and Cognitive Ergonomics at HCI-International, Las Vegas, Nevada, USA, 15–20 July 2018

13. Schulte, A. et al.: Human-system interaction analysis for military pilot activity and mental workload determination. In: IEEE International Conference on Systems, Man, and Cybernetics, SMC 2015, Kowloon Tong, Hong Kong, pp. 1375–1380 (2015)
14. Shafer, G.: A mathematical theory of evidence. Princeton University Press, Princeton and London (1976)
15. Wickens, C.D.: Multiple resources and performance prediction. Theor. Issues Ergon. Sci. **3** (2), 159–177 (2002)
16. Young, M.S. et al.: State of science: mental workload in ergonomics. Ergonomics. **58**(1), 1–17 (2015)

Developing a Context Framework to Support Appropriate Trust and Implementation of Automation

Sabrina Moran[1(✉)], Heather Oonk[1], Petra Alfred[1], John Gwynne[1], and Martin Eilders[2]

[1] Pacific Science and Engineering, 9180 Brown Deer Road, San Diego, CA 92121, USA
{SabrinaMoran, HeatherOonk, PetraAlfred, JohnGwynne}@Pacific-science.com

[2] Air Force Research Laboratory, Munitions Directorate, Eglin Air Force Base, Valparaiso, FL 32542, USA
Martin.Eilders@us.af.mil

Abstract. Despite the widespread application of automated systems, use of automation can be problematic if inappropriate trust is placed in the systems. To be most effective, automated systems should be designed to promote appropriate trust from the individuals who interact with them. We propose that a key contributor to achieving appropriate trust in automation is an understanding of the context in which the human-automation team functions. Context encompasses many factors related to the automated system, the human operator, the operational environment, and the missions being performed. This paper describes the development of a conceptual framework that links human-automation team tasks to impactful context factors to promote shared awareness within the team and appropriate trust and usage of the automation. Development of the context framework was based on a review of the trust in automation scientific literature, supplemented by interviews with subject-matter experts (SMEs) in the unmanned system domain.

Keywords: Human factors · Systems engineering · Automation Trust · Context · Trust in automation · Human-Automation teaming

1 Introduction

Automated systems have become ubiquitous in virtually every aspect of contemporary life, including healthcare, transportation, defense, commerce, and manufacturing. Automation is used to accomplish tasks that are repetitive, difficult, or dangerous for humans to perform. To promote effective use, these systems should be designed to ensure they will be appropriately trusted by the individuals who interact with them. This paper provides a brief overview of trust in automation literature and describes the development of a framework of trust-related context factors. This framework considers the automated system type and mission type and allows specific human-automation tasks to be linked to their relevant context factors. Understanding the context factors

W. Karwowski and T. Ahram (Eds.): IHSI 2019, AISC 903, pp. 738–743, 2019.
https://doi.org/10.1007/978-3-030-11051-2_112

that impact various human-automation teams and tasks can be helpful to individuals designing and integrating automation to ensure its implementation promotes shared awareness and appropriate system use.

1.1 Background

Operators often underuse or misuse automation due to an inappropriate level of trust in the system. As social creatures, humans respond to technology socially, resulting in our reliance on automation being greatly impacted by our trust in it [4, 6]. The complexity of the situations in which automation is often implemented keeps most operators from completely understanding the processes underlying the automated system; therefore, trust in the system is impacted by numerous internal and external factors [6]. The development and maintenance of shared awareness becomes more difficult as the size of the human-automation team increases. This problem will likely be exacerbated as more complex team compositions become operational, such as the proposed cross-domain collaboration between unmanned underwater vehicles (UUVs), unmanned surface vehicles (USVs), and unmanned aerial vehicles (UAVs) as part of the U.S. Air Force's Future Vision for remotely piloted aircrafts [3].

Another significant challenge faced by humans interacting with or managing automated systems is that many situation awareness and decision support tools exist in legacy systems that do not completely support users' needs. This shortcoming is often due to the designs of these tools failing to consider the performance capabilities and limitations of their human users. For example, Pacific Science & Engineering (PSE) recently examined issues related to human performance in autonomous systems and found them to be persistent across diverse platforms (e.g., air, ground, marine) [2].

1.2 Purpose

A key contributor to mis-calibrated trust is a lack of understanding of the context, as it relates to the operating environment, type of autonomous system, characteristics of human operators, and the type of mission being performed. Typically, human operators have access to *some* context of use, but this information is often incomplete, ambiguous, and/or presented across disparate systems – requiring users to integrate the information to form a comprehensive operational context and estimate its impact on system performance. Further, current context-sharing is typically unidirectional (i.e., contextual information is not shared with the autonomous system).

The goal of this effort was to create a conceptual framework that could be applied to a variety of human-automation teams to ensure all necessary information is shared between the human and the automated team members. Human-automation teams perform more effectively as system transparency increases and the human team member(s) become more aware of the context impacting their task and the automated system. Moreover, automation's ability to detect the influence of context factors on human operators can be used to adaptively modify the task allocations between the human-automation team members, thereby enhancing overall team performance.

2 Context Framework Development

Development of the context framework began with information gathering. Initial efforts were focused on literary research and knowledge engineering with subject matter experts (SMEs) to establish a substantial knowledge base to support all subsequent tasks. Once the content had been gathered, it was mapped to a context framework to allow for the subject matter to be filtered based on system type, mission type, and context factor.

2.1 Literature Review

Development of the context framework was based on an extensive review of trust in automation literature, including scientific articles and recent news stories. This effort identified properties of human-automation team environments that contribute to mis-calibrated trust. Through the examination of more than 30 recent news stories, it was found that mishaps and issues in human-automation teams stem from a lack of shared understanding and inappropriate trust within the teams.

The scientific literature review included over 500 research articles spanning across eight system types (e.g., UAVs, robots). From this literature, we were able to collect 20+ context factors, 40+ metrics, 14+ algorithms, and 10+ user interface (UI) elements. Additionally, numerous essential trust-related attributes were identified that an effective human-automation team should possess, including:

- ***Context-awareness***: Ability to "understand" situational contexts and adapt to those contexts [7].
- ***Change-awareness***: Ability to detect and interpret changes that occur in operational contexts to maintain understanding of an operational situation.
- ***Support for trust***: Trust in automation literature describes trust in terms of performance, purpose, and process [5, 6].
- ***Trust and shared understanding support for all team members***: Automated system(s) and human(s) must support shared understanding and trust.
- ***Generalizability/feasibility***: Solutions should be generalizable and adaptable to fit across various operational environments.

2.2 Knowledge Engineering

Semi-structured interviews were completed with SMEs from the unmanned system domain to amplify our application-based knowledge, validate trust-related factors, and identify additional factors. Emphasis was placed on the human-automation team context (i.e., environment, tasks, missions, system, and human attributes) and necessities for promoting well-calibrated trust within teams. The main topics discussed in the interviews included: operator roles, unmanned systems, context, and trust.

Thirty-four scenarios emerged from the interviews that describe specific incidents and/or routine situations in which trust in automation impacts human and/or system performance. These scenarios formed the basis for operational use case development. Use cases provided insights into the types of information the operator and system need to know about specific contexts and how trust and system performance is impacted by this awareness.

2.3 Content Refinement

An extensive number of context factors were found throughout the research process. Five inclusion criteria for the context factors were validated in SME interviews and then applied to the comprehensive list (see Fig. 1).

- ***Impact***: Effect of the context factor (or lack of factor) on trust.
- ***Ease of access***: Difficulty of accessing the contextual information.
- ***Commonality***: Prevalence of the factor across scenarios and domains.
- ***Degree of convergence with the research literature***: Prevalence of the context factor in literature.
- ***Gaps in the "state of the art"***: Determination of the degree to which similar context is available in other systems, based on a review of existing technologies.

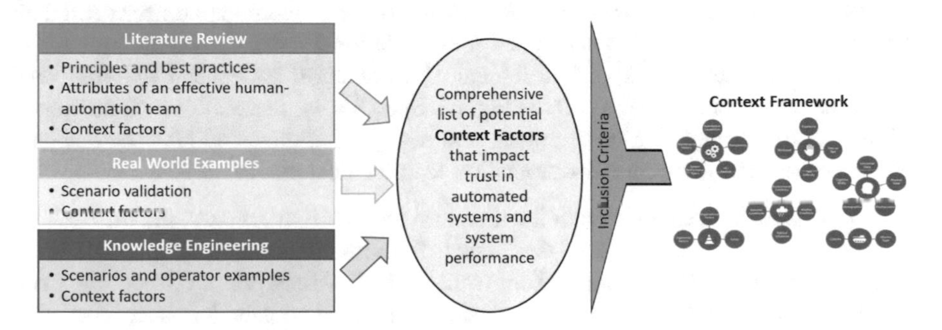

Fig. 1. The content gathering and refinement processes for the context framework.

The application of the inclusion criteria resulted in a final context framework, that comprises 23 context factors organized into 6 categories:

- ***Technical***: Related to performance of technical systems.
- ***Human task-related***: Related to the human's ability to complete the task.
- ***Global:*** Conditions external to the human-automation team.
- ***Human internal***: Related to factors intrinsic to the human operator.
- ***Human environment***: Related to the environment the human operates in.
- ***Mission:*** Related to missions performed by the human-automation team.

Once the list of context factors had been consolidated, the factors were further mapped based on their inter-relationships. Understanding the relationships between context factors and how a change in one can impact another can further increase operational awareness. For example, in Fig. 2, *Safety* is selected (shown in gold), and it has been found to be related to *Weather Conditions* and *Mission Type* (shown in blue).

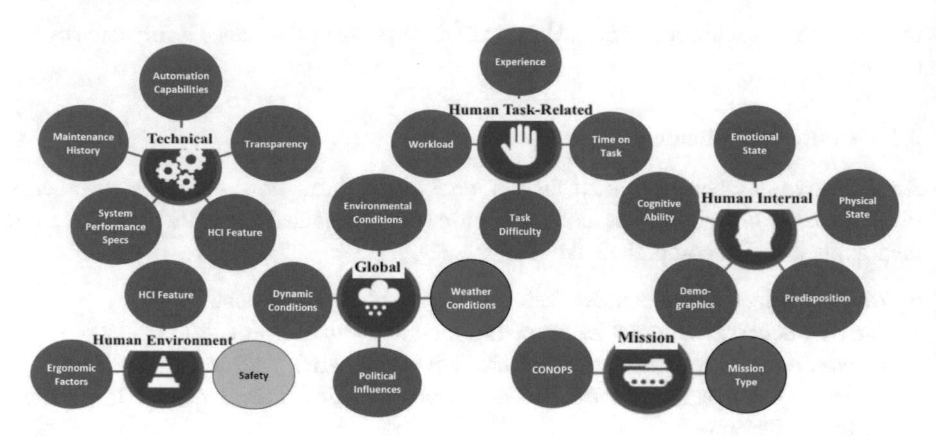

Fig. 2. The context framework with context factors grouped into categories and an example interaction between the factors depicted by color.

2.4 Content Mapping

Once sufficient information was gathered relating to human-automation teams and their operational environments, it was mapped into a relational database containing tools to help researchers, developers, and designers of automation understand the vital relationship between context and trust (see Fig. 3). Specifically, these tools include literary findings, metrics, algorithm descriptions, and design recommendations. The database was created around three human-automation team characteristics:

- ***Context factors***: Extracted from the literature and SME interviews, the trust-related context factors are the basis of the context framework (see Fig. 2).
- ***Mission components***: Adapted from a theoretical framework developed by Beer, Reith, Tran, and Cook [1]. The mission components include: *Area, Assistive, Construction*, *Monitor*, *Resource*, *Target/Inspect*, and *Transit*.
- ***Unmanned systems***: The systems included in the database are: *Robots, Decision Aids, Driverless Cars, Unmanned Aerial Vehicles (UAVs), Unmanned Ground Vehicles (UGVs), Unmanned Space Vehicles, Unmanned Surface Vehicles (USVs), and Unmanned Underwater Vehicles (UUVs).*

Each tool in the database was linked to the relevant context factors, unmanned system type(s), and mission component(s) to ensure the database could provide valid, tailored content to future users.

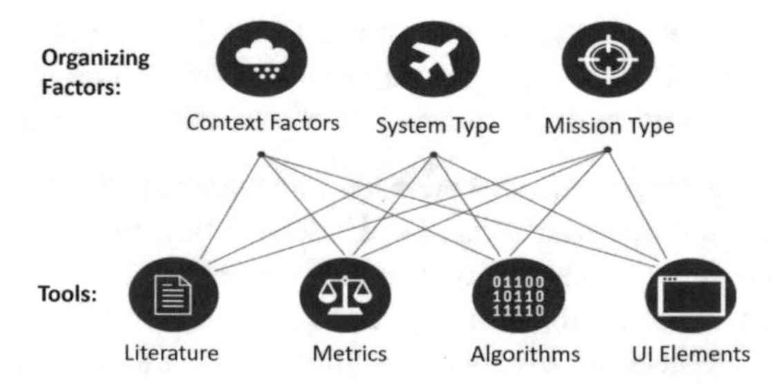

Fig. 3. The structural mapping underlying the trust in automation database.

3 Final Product

The outcomes of the literary research, knowledge engineering, and formation of the human-automation context framework serve as the foundation for the Operator's Context and Trust in Automation Visualization (OCTAV) toolkit. This toolkit was created to give individuals designing, developing, and interacting with automation information about the effect of context on human-automation tasks and access to pertinent references, metrics, algorithm descriptions, and UI elements. With this mapped database, users can specify relevant context factors, system types, and mission components when querying the database to ensure they get pertinent results.

The OCTAV toolkit is web-based and includes a front-end UI that provides education on potential issues with trust in automation, background knowledge, and training for toolkit use. OCTAV will serve as an innovative means for providing tailored information to researchers, system designers, and program managers to support appropriate trust and promote transparency in human-automation systems. This toolkit provides "building blocks" and decision support to aid in the design, creation, and implementation of human-system interfaces to improve human-automation team interactions and performance across numerous application domains.

References

1. Beer, R.D., Reith, C.A., Tran, R., Cook, M.: Framework for multi-human multi-robot interaction: impact of operational context and team configuration on interaction task demands. In: AAAI 2017 Spring Symposium on Computational Context: Why It's Important. Stanford, California (2017)
2. Cook, M.B., Smallman, H.S.: Human-centered command and control of future autonomous systems. In: Proceedings of the 18th International Command and Control Research and Technology Symposium (ICCRTS). Alexandria, Virginia (2013)
3. Eggers, J.W.: A future vision for remotely piloted aircrafts: leveraging interoperability and networked operations. In: Proceedings of the 18th International Command and Control Research and Technology Symposium (ICCRTS). Alexandria, Virginia (2013)
4. Hoff, K.A., Bashir, M.: Trust in automation integrating empirical evidence on factors that influence trust. Hum. Factors **57**, 407–434 (2005)
5. Lee, J., Moray, N.: Trust, control, strategies, and allocation of function in human machine systems. Ergonomics **35**, 1243–1270 (1992)
6. Lee, J., See, K.A.: Trust in automation: designing for appropriate reliance. Hum. Factors **46**, 50–80 (2004)
7. Mouran, T.P., Dourish, P.: Introduction to this special issue on context-aware computing. Human-Comput. Interact. **16**, 87–95 (2001)

An Exploratory Analysis of Physiological Data Aiming to Support an Assistant System for Helicopter Crews

Matthew Masters(✉), Diana Donath, and Axel Schulte

Universität der Bundeswehr München, Neubiberg, Germany
{matthew.masters,diana.donath,axel.schulte}@unibw.de

Abstract. Utilizing a helicopter simulator developed within the Institute of Flight Systems at the University of the Armed Forces Munich, this work investigates how human resource theory and real-time physiological monitoring might support an adaptive assistant system intended to provide mission-relevant support to a helicopter crew during simulated mission scenarios. This investigation is conducted through an analysis of a series of simulated missions flown by subjects of varying experience with the simulator. Across-subject analysis highlights the significant variability of subject physiological responses and perceived workload. Additionally, correlations between various biological signals and assessed and perceived workload are identified. Within-subject analysis illustrates the temporal characteristics of various biological signals in this environment and reveals evidences suggesting future modeling of perceived workload though biological signals and a task-based workload assessment are promising.

Keywords: Adaptive automation · Physiological monitoring · Assistant systems · Human-autonomy-teaming · Manned-unmanned teaming

1 Introduction

It is commonly accepted that successful human-autonomy teaming depends on sustained and effective coordination between people and technology [1]. We suggest that this coordination can be supported through a shared and common understanding of the operating environment including the task objective, system capabilities, system resources, and external influences. This work aims to strengthen the shared understanding of the human system within such human-autonomy teams and investigate how this increased understanding might improve the effectiveness of the team. Specifically, we investigate the potential impact of supplementing an existing adaptive assistant system within a helicopter simulator (described in [2]) with "pilot state" information in the form of physiological data including: heart rate, heart rate variability, electrodermal activity (EDA), EDA variability, blink rate and saccade rate.

Previous work on the adaptive assistant system developed within the aforementioned helicopter simulator inferred an aspect of "pilot state" through the prediction of task execution and the subsequent assignment of resource demands to currently-

W. Karwowski and T. Ahram (Eds.): IHSI 2019, AISC 903, pp. 744–750, 2019.
https://doi.org/10.1007/978-3-030-11051-2_113

executed tasks [3]. The magnitude and interaction of these resource demands are subsequently used to predict the current workload level of the pilots [4]. By now including physiological measurements, we hope to individualize this workload estimate and in general come to a more complete understanding of the state of the human system. With this more complete understanding, it is inferred that assistance can then be more effectively provided as the assistant system can more informatively adapt to the current state of the pilot.

Through the following analysis of simulated helicopter missions, various findings are illustrated including: (1) the relationship between operator assess workload (as currently implemented), operator perceived workload, and physiological markers of the operator; (2) The variability of physiological phenomena among pilots highlighting the need to provide individualized assistance; (3) the impact of various mission events on resource utilization, physiological measures, and operator perceived workload.

2 Methods

2.1 Experiment Design

Although the helicopter simulator supports dual pilot manning, for this experiment, the cockpit was manned by a single subject. This was done to maximize the demands placed on the subject and to ensure consistency of task load across the subjects. Six male subjects ages 28-35 participated in this experiment. Each flew the same simulated mission with the objective of retrieving and transporting wounded soldiers to safety in a hostile environment. In addition to other flight requirements, it was required that all routes be reconned by one of three UAVs within the pilot's control prior to helicopter travel. The subjects were responsible for all aspects of mission execution including: planning for accomplishment of the mission objective, flying, completing checklists (before/after takeoff and entering/leaving helicopter-operating-areas), conducting regular radio communications (before/after takeoff, entering/leaving helicopter-operating-areas, and at other times as required), and UAV management. Every 30 s, a short, high pitch tone prompted the subjects for a personal assessment of their overall workload level. Subjects responded on a scale from 1 to 10 with 1 being "relaxed," 5 "focused," and 10 "overwhelmed."

2.2 Data Collection and Signal Processing

Physiological Data. Single-channel Electrocardiography (ECG) and Electrodermal activity (EDA) data were collected at 200 Hz using the BIOPAC MP160 data acquisition system. Heart rate (HR) was calculated using a 20 s rolling mean of the instantaneous HR which is calculated by taking the inverse of the inter-beat-interval (IBI). HR variability (HR-var) was calculated by taking the standard deviation of normal RR-intervals (outliers removed) over two a minute window. EDA variability (EDA-var) was calculated by down-sampling the EDA signal to 4 Hz, taking its derivative, and then computing the standard deviation of the signal over a one-minute window. Baseline ECG and EDA data were collected for 4 min prior to mission execution.

Blink rate (blinks/minute) and saccade rate (saccades/minute) were calculated according to blink and saccade events as detected by the Smart Eye Pro eye-tracking system.

Workload. For instances in which the subject did not provide their current personal evaluation of workload (Personal WL), the next valid estimate was used assumed. The signal was then interpolated between samples to obtain a continuous estimate.

The system's evaluation of the subject's workload (System WL) is made by evaluating the resource requirements of the currently executed tasks. The process by which this is performed has been thoroughly described in previous publications [4]. This high frequency signal was filtered using a 2nd order Savitzky-Golay filter with a window length of 1 min.

3 Across-Subject Analysis

Table 1 provides a summary of mission statistics for all subjects over the entire duration of their respective missions. Errors are given in standard deviations from the mean. The standard score or "z-score" of HR, HR-var, and EDA-var were calculated using the respective signal's baseline measurements. It is noted that each subject had an elevated mean HR and five of the six had depressed mean HR-var from baseline.

Table 1. Summary of subject population and mission statistics for each.

Sub. num	Sim exp. (hrs)	Mission duration (min)	Personal WL	System WL	HR (bpm)	HR Z-score
1	<2	31.7	5.8 ± 1.6	5.6 ± 2.0	66.9 ± 3.9	1.4 ± 1.3
2	<2	43.3	5.7 ± 1.3	5.4 ± 2.6	76.7 ± 2.8	1.3 ± 0.9
3	~10	37.5	7.8 ± 2.3	4.9 ± 2.3	85.2 ± 4.7	1.5 ± 1.1
4	~4	35.2	7.7 ± 2.6	3.3 ± 2.3	94.2 ± 6.8	2.1 ± 1.3
5	~50	32.7	5.8 ± 1.8	5.9 ± 2.1	79.1 ± 3.3	1.6 ± 1.4
6	~100	31.2	6.0 ± 1.8	6.9 ± 2.2	103.9 ± 10.5	2.3 ± 1.4

Sub. num	HR var	HR var Z-score	EDA var	EDA var Z-score	Blink rate	Saccade rate
1	0.054 ± 0.018	−4.2 ± 2.3	0.0012 ± 0.001	−1.2 ± 5.0	3.1 ± 1.5	135.3 ± 27.2
2	0.054 ± 0.01	−1.8 ± 1.1	0.0018 ± 0.00077	0.4 ± 1.8	4.4 ± 2.8	77.2 ± 16.3
3	0.044 ± 0.006	−4.3 ± 1.2	0.0016 ± 0.00083	−0.9 ± 1.1	5.2 ± 2.9	112.1 ± 16.3
4	0.054 ± 0.014	−2.6 ± 1.9	0.0079 ± 0.0018	−0.0 ± 1.5	7.9 ± 3.5	99.1 ± 19.5
5	0.052 ± 0.01	0.6 ± 1.2	0.0074 ± 0.0026	3.0 ± 3.3	2.0 ± 1.8	87.6 ± 34.4
6	0.05 ± 0.02	−1.8 ± 2.6	0.0025 ± 0.00089	−0.4 ± 2.0	6.1 ± 3.0	111.9 ± 15.5

A correlation matrix between individual signals averaged across all subjects is provided in Fig. 1. Although not particularly strong correlations, it is noted that multiple correlations were anticipated while others were unexpected. For example, it is observed that a positive correlation exists between Personal WL and System WL and HR. It was surprising however to note the strong negative correlation that exists between HR-var and System WL does not exist between HR-var and Personal WL. The lack of correlation between EDA-var and Personal WL seems to support the assertion made by others that EDA is not directly connected with stress [5].Other correlations (and lack thereof) are seen in the figure.

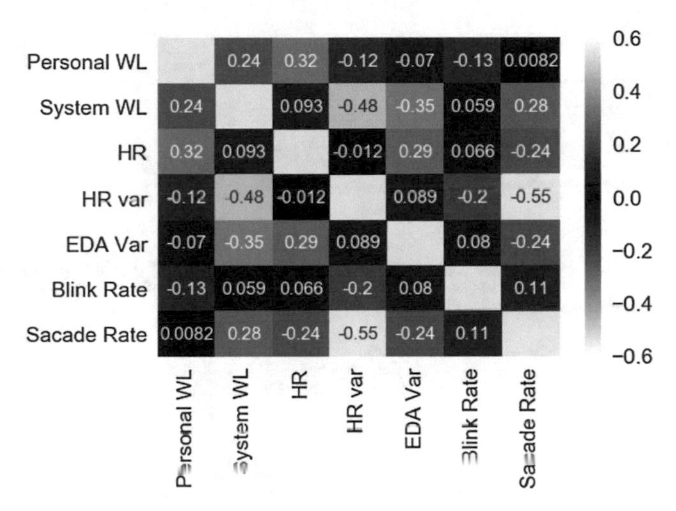

Fig. 1. Correlation matrix between individual signals averaged across all subjects where the values are the Pearson correlation coefficients (PCCs) between the two intersecting signals.

4 Within-Subject Analysis

An important perspective into the relationship between the many signals is gained by viewing each signal over the duration of the mission. Figure 2 provides one such illustration. The subjects' elevated HR and depressed HR-var are clearly represented and suggest his elevated mental stress level [6, 7]. Less clearly interpreted are the fluctuations seen among the EDA-var, blink rate, and saccade rate.

Individual EDA events (peaks) were regularly observed after provoking situations such as mission updates, realization of pilot errors, and following system malfunctions. These observations suggest the potential of augmenting an assistant system with this signal to gain insight into the situational awareness of the pilot.

Initial attempts to perform linear, polynomial, and support vector machine (SVM) regression to train a model to predict Personal WL from the remaining data were unsuccessful. To visualize and more fully understand the problem, Principal Component Analysis (PCA) was used to reduce the dimensionality of the dataset for visualization. Figure 3 illustrates the resulting 2-dimensional data colored by

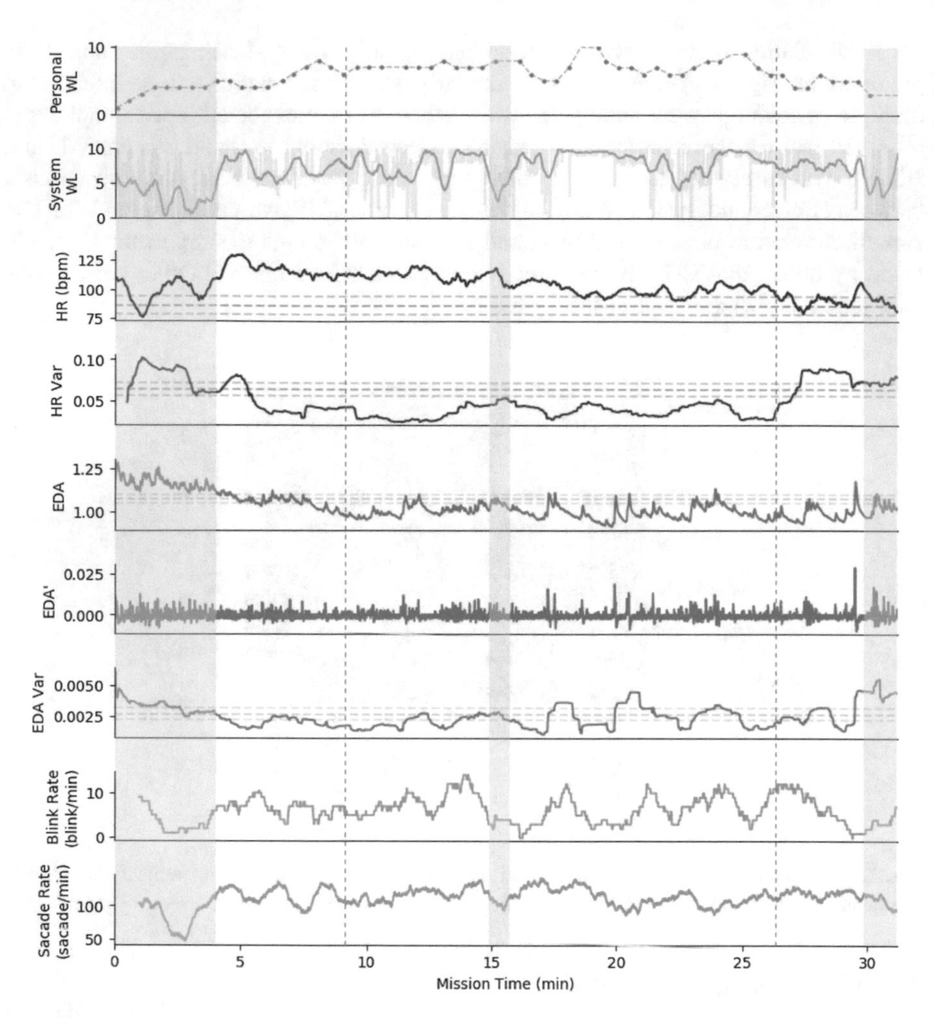

Fig. 2. A variety of signals plotted over the duration of the complete mission for subject 6. Shaded periods denote times in which the helicopter was on the ground. Vertical dashed lines show times in which the helicopter entered (left) and exited (right) areas of potential enemy activity. Both the unfiltered and filtered System WL signals are shown in the same axes. The horizontal dashed lines in multiple axes represent the baseline mean and one standard deviation above and below this mean.

Personal WL. The observable slight separability suggests future regression or classification of the subjects' workload level may be possible yet also explains why linear regression was not successful in modeling Personal WL.

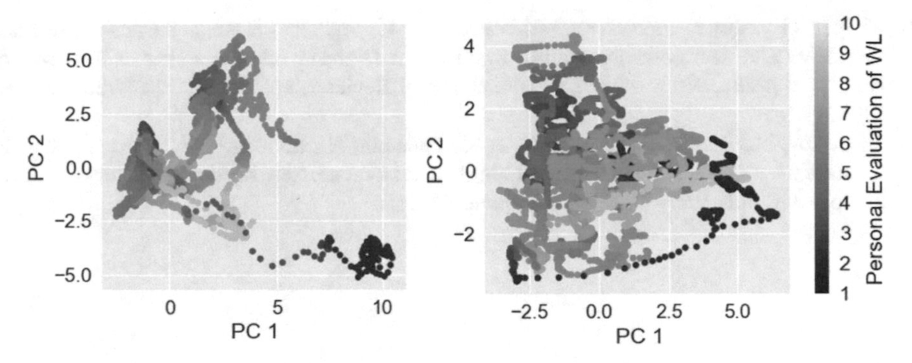

Fig. 3. Dimensionality reduction using PCA to visualize the separability of the dataset for subject 1 (left) and subject 5 (right). The color scale is set by the personal evaluation of workload.

5 Discussion and Conclusion

Physiological data from multiple subjects collected during simulated helicopter missions were analyzed. Across-subject analysis illustrated a weak correlation between HR and individually-perceived workload suggesting possible incorporation of this signal in a workload-adaptive assistant system. This analysis additionally highlighted the significant variability of subject physiological responses and perceived workload reinforcing the need to have a pilot-adapted assistant system. Within-subject analysis illustrated the temporal characteristics of various biological signals in this environment and revealed evidences suggesting future modeling of perceived workload though biological signals and a task-based workload assessment may be possible. Continuing research will attempt to determine the optimal utilization of these signals in a real-time pilot-adapted assistant system.

References

1. Christoffersen, K., Woods, D.D.: How to make automated systems team players. In: Advances in Human Performance and Cognitive Engineering Research, vol. 2, pp. 1–12. Elsevier (2002)
2. Brand, Y., Schulte, A.: Design and evaluation of a workload-adaptive associate system for cockpit crews. In: Engineering Psychology and Cognitive Ergonomics, pp. 3–18. Springer (2018)
3. Honecker, F., Schulte, A.: Automated online determination of pilot activity under uncertainty by using evidential reasoning. In: Engineering Psychology and Cognitive Ergonomics: Cognition and Design, pp. 231–250. Springer (2017)
4. Honecker F., Brand, Y., Schulte, A.: A task-centered approach for workload-adaptive pilot associate systems. In: Proceedings of the 32nd Conference of the European Association for Aviation Psychology, pp. 485–507. Cascais (2016)
5. Bakker, J., Pechenizkiy, M., Sidorova, N.: What's your current stress level? Detection of stress patterns from GSR sensor data. In: 11th IEEE International Conference on Data Mining Workshops. IEEE (2011)

6. Thayer, J.F., Ahs, F., Fredrikson, M., Sollers, J.J., Wager, T.D.: A meta-analysis of heart rate variability and neuroimaging studies: Implications for heart rate variability as a marker of stress and health. Neuroscience & Biobehavioral Reviews, vol. 36, pp. 747–756. Elsevier (2012)
7. Hjortskov, N., Rissen, D., Blangsted, A.K., Fallentin, N., Lundberg, U., Sogaard, K.: The effect of mental stress on heart rate variability and blood pressure during computer work. Eur. J. Appl. Physiol., **92**(1–2), 84–89. Springer (2004)

Does the Type of Visualization Influence the Mode of Cognitive Control in a Dynamic System?

Christine Chauvin[1](✉), Farida Said[2], and Sabine Langlois[3]

[1] Lab-STICC (UMR CNRS 6285), Université Bretagne Sud, Bd. Flandres Dunkerque 17, 56100 Lorient, France
christine.chauvin@univ-ubs.fr

[2] LMBA (UMR CNRS 6205), Université Bretagne Sud, Bd. Flandres Dunkerque 17, 56100 Lorient, France
farida.said@univ-ubs.fr

[3] Renault, Research Department, IRT System X. Technocentre, 1 Avenue du Golf, 78084 Guyancourt Cedex, France
Sabine.langlois@renault.fr

Abstract. This study investigates the influence of a visualization system on the mode of cognitive control adopted in a navigation task. It relies on a driving simulator experiment. It uses data clustering methods, which help identify three patterns of in-vehicle data corresponding to three different modes of control. The study shows that Augmented Reality HUD supports drivers, since it contributes to avoiding the adoption of a scrambled mode of control. However, it promotes an opportunistic mode of control characterised by a higher but limited anticipation.

Keywords: Cognitive control modes · Operator assistance · Visualization system · Augmented reality · Data clustering

1 Introduction

When interacting with a system in a dynamic environment, humans use internal models (a model of the system as well as a model of the dynamics of the environment) and external data. Cognitive control refers to functions that allow information processing and behaviour to vary adaptively rather than remaining rigid and inflexible and to rely in some cases, more on internal data and in other cases, more on external. Within the framework of cognitive system engineering, Hollnagel [1] identified four modes of cognitive control, from the most reactive (determined by the occurrence of external data and characterized by a short processing time) to the most proactive (relying on internal data processing and characterized by a longer processing time). The control mode changes from "scrambled" to "opportunistic" when there is an increase in the predictability of the situation or in the available time, and from "opportunistic" to "tactical" or "strategic" if the predictability of the situation and/or the available time continues to improve. It is assumed that cognition reliability is low for both scrambled and opportunistic modes, whereas it is assumed to be relatively high in tactical and

W. Karwowski and T. Ahram (Eds.): IHSI 2019, AISC 903, pp. 751–757, 2019.
https://doi.org/10.1007/978-3-030-11051-2_114

strategic modes [2]. One major question is whether a man-machine interface can influence the adoption of a specific control mode.

This paper investigates the influence of visualization on the cognitive control mode used by drivers in a navigation task. Navigation in an unknown road environment represents a costly cognitive activity for drivers. Numerous studies have shown the difficulties they encounter in planning and following routes [3]. Different navigation systems have been designed to facilitate this task. Amongst them are Augmented Reality-based Head Up-Displays which are expected to facilitate drivers' perception of useful information and to therefore increase the predictability of a complex situation [4], [5]. This paper aims at assessing the value of an Augmented Reality Head Up Display (AR HUD) in comparison to a classical windshield HUD (W-HUD).

2 Method

2.1 Participants and Equipment

A driving simulator experiment was carried out with 32 participants. It was conducted with the simulator of the Institute for Technological Research System X: the Dr SIHMI (Driving Simulator for Human Machine Interaction studies) platform. This static simulator consists of an automobile cockpit (including a steering wheel, the acceleration and brake pedals, a gear lever, an AR HUD or a classical W-HUD, and a standard GPS), and a 180° curved projection screen giving a simulation of different driving situations with the SCANeR simulation software.

Figure 1a gives a view of the W-HUD, whereas Fig. 1b presents the AR HUD. The classical W-HUD displayed a map indicating the route to follow, reproducing the system used on BMW Series 7 vehicles. With a surface area of 9*15°, the display unit of the AR HUD projects virtual information in the driving environment. In this case, arrows indicated the direction to follow; it appeared 400 m before the place where the manoeuvre is possible. The AR surface of the unit was 8*15°. In both conditions, additional information appeared on a fixed section at the bottom of the windscreen: the speed limit, vehicle speed, next change of direction (shown through a "turn by turn" icon) and the remaining distance before reaching it, and the direction sign.

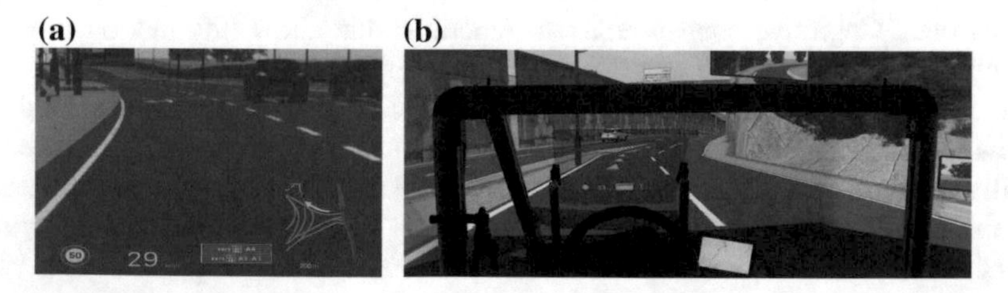

Fig. 1. a. W-HUD and direction sign. b. AR HUD and GPS Waze

2.2 Experiment Design and Procedure

The experiment sessions lasted 1 h30. Participants were first invited to read an information document and then sign consent forms. They were also introduced to the simulator, interfaces, and the test (5 min). Then, a learning phase consisted of driving with and without augmented reality. Afterwards, participants were required to drive in two complex geographical areas. In both areas, half the participants drove with the AR HUD and the other half with the W-HUD. Each participant tested both interfaces. The running order was counterbalanced. In case of failure, the driving phase was interrupted, and participants had to start over again. As participants could replay this phase up to three times (should they fail the first and second driving test), the duration of this phase varied between 20 and 30 min. A clarifying interview was held after each experimental phase. Finally, participants were asked to complete a questionnaire relating to their preferences and their profile (of about 10 min).

2.3 Data Collection

Assessment of interfaces is based upon objective data (e.g. vehicle speed, acceleration, actions on the brake pedal, the acceleration pedal and the steering wheel, Time To Collision or TTC, lane-change duration) and subjective data (such as a comfort score) which was collected during a clarifying interview.

2.4 Hypothesis

As in previous studies [6], we expected to identify behavior showing the drivers' anticipatory capacity and revealing three modes of cognitive control. The scrambled mode should be associated to rough actions on the pedals and steering wheel, short following distances and few mirror checks. Conversely, a tactical mode should lead to smooth actions, adequate following distances and frequent mirror checks. In the opportunistic control mode, drivers' actions are determined by the salient features of the situation; this control mode should lead to reduced speed and few mirror checks.

2.5 Analysis

A two-step multivariate segmentation analysis was carried out, where clustering techniques are used for both variables and participants. In the first step, subsets of strongly correlated variables were detected, and the ones that provide the same kind of information were placed into the same group. The resulting groups reveal the main dimensionalities of the data, and a representative variable was determined for each of them. These representatives were used in the second step to segment the participants via classical clustering algorithms. Finally, the interpretation of the clusters of participants was made by means of univariate tests and Classification and regression trees (CART).

3 Results

This paper presents preliminary results concerning one task (changing lane in an X-shaped interchange). Only 3 out 32 participants had to redo this task due to a failure. For these participants, only the data collected during the second run were analyzed.

Objective variables that measure vehicle's performance (speed, actions on the pedals or the steering wheel…) were used to segment the participants into groups whereas variables related to the interaction with other vehicles (TTC, rearview mirror consultation…) along with subjective variables were used to interpret the resulting clusters.

Figure 2 shows the dendrogram for the clustering of vehicle variables. It appears from the examination of the bootstrapped mean-adjusted rand index that the three-partition solution is the most stable.

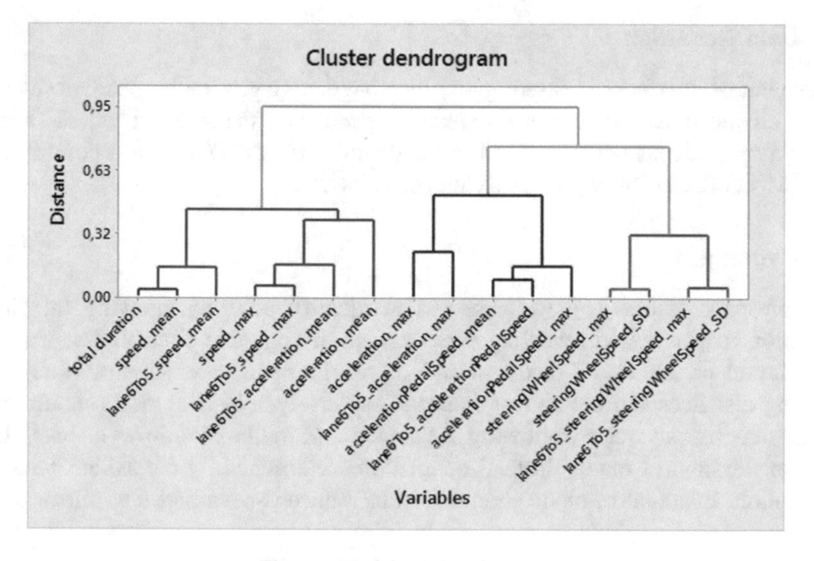

Fig. 2. Variables dendrogram

Examination of the composition of the clusters of variables and their factor loadings (squared correlations between variables and the first principal component of their parent group), suggests that Cluster 1 (in blue) refers to the speed of the vehicles, Cluster 2 (in red) to the smoothness of the actions upon the acceleration pedal and Cluster 3 (in green) to the smoothness of the actions upon the steering wheel.

In each cluster, the variable with the greatest factor loading is chosen as the representative of the group: maximal speed for Cluster 1, maximal acceleration for Cluster 2 and standard deviation of the actions upon the steering wheel for Cluster 3. The subsequent clustering of participants in groups is based on these variables and it is performed via the hierarchical agglomerative criterion of Ward reinforced with kmeans partitioning.

Figure 3 displays the resulting clustering of participants: the first group (in green, N = 10) is composed of participants who didn't experience Augmented Reality or did not make use of it, the second group (in red, N = 6) is mainly composed of participants who experienced AR HUD, and the third group (in black, N = 15) is well balanced between AR HUD and W-HUD.

Univariate tests and CART algorithm were used to describe the main discriminating variables of the clusters of participants. It resulted in the comfort score reported by the participants in the third group being significantly higher than in the two others, $F(2, 30) = 5.14$, $p = 0.013$.

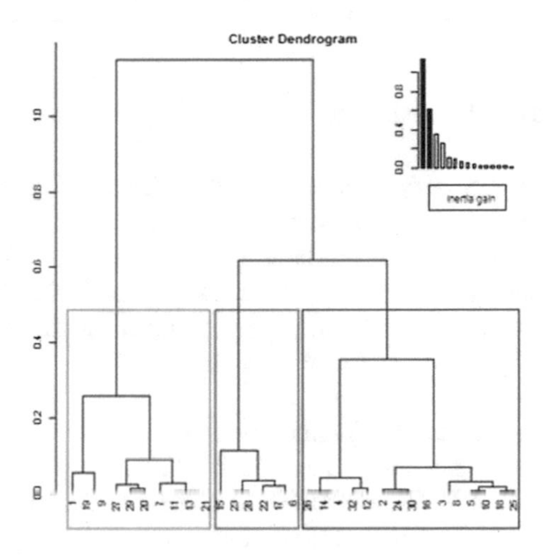

Fig. 3. Clusters of participants from the Ward dendrogram

In the first group, control actions are rougher with more variability than in the other groups (cf. Table 1). Furthermore, the task is performed more quickly and the TTC between the ego vehicle and the front left car is significantly lower.

Table 1. Main discriminating variables of Class 1

Class 1	v.test	Mean in class	Overall mean	sd in class	Overall sd	p.value
SteeringWheelSpeed_SD (rd.s^{-1})	4.49	0.28	0.21	0.04	0.06	<0.001
SteeringWheelSpeed_max (rd.s^{-1})	4.01	2.06	1.48	0.53	0.55	<0.01
Acceleration_max (m. s^{-2})	2.54	4.37	3.72	0.74	0.96	0.01

The second group is characterized by a lower maximal speed, more time to achieve the task, smoother longitudinal and lateral actions with more reduced variability (cf. Table 2) and, furthermore, by a later consultation of the rearview mirrors.

Table 2. Main discriminating variables of Class 2

Class 2	v.test	Mean in class	Overall mean	sd in class	Overall sd	p value
Total_duration (s)	3.74	16.74	13.81	2.72	2.10	<0.01
Speed_max (km/h)	−4.18	41.65	53.10	4.76	7.35	<0.01
Acceleration_max (m s^{-2})	−2.86	2.70	3.73	0.52	0.96	<0.01
SteeringWheelSpeed_max (rd s^{-1})	−2.42	0.99	1.48	0.14	0.55	0.01
SteeringWheelSpeed_SD (rd s^{-1})	−2.92	0.15	0.21	0.03	0.06	<0.01

4 Discussion and Conclusion

Clustering methods help identify three behavioural classes. With a significantly higher score of comfort, one class is distinguished from the others. It is composed of participants who were provided with either the AR-HUD or the W-HUD. It is worth noting that a large majority of them reported, during clarifying interviews, that they mainly used the color of the direction sign displayed on the bottom of HUD and matched it with the sign posts rather than using AR (blue arrows) or the W-HUD (map).

In the other classes, most participants reported a negative feeling. One of them is mostly composed of participants who used the W-HUD. They made abrupt actions and TTC was lower than the TTC observed in the other classes. These features are reminiscent of a scrambled mode. One class is mostly composed of participants who used the AR-HUD. Compared with other participants, they take longer to make the lane change, and they look at the mirrors later. They reduce their speed, in order to give themselves enough time to react. These features are reminiscent of an opportunistic mode of control.

In this study, benefits of AR HUD in terms of driver assistance prove therefore to be mixed. It supports drivers, since it contributes to avoiding the adoption of a scrambled mode of control. However, it promotes an opportunistic mode of control characterised by a higher, but limited anticipation. Referring to the model presented by Ross and Burnett [3], one may assume that AR HUD supports the stage of the navigation task consisting of pinpointing the location of the next maneuver within the road environment and identifying which direction to travel at the maneuver. It does not support, however, the former stage of "preview" consisting in building an early mental picture of the maneuver.

References

1. Hollnagel, E.: Human reliability analysis – context and control. Academic Press, London (1993)
2. Hollnagel, E.: Simplification of complexity: the use of simulation to analyse the reliability of cognition. In: Reliability and Safety Assessment of Dynamic Process Systems, pp. 166–178. Springer, Heidelberg (1994)
3. Ross, T., Burnett, G.: Evaluating the human–machine interface to vehicle navigation systems as an example of ubiquitous computing. Int. J. Hum. Comput. Stud. **55**, 661–674 (2001)
4. Medenica, Z., Kun, A.L., Paek, T., Palinko, O.: Augmented reality vs. street views: a driving simulator study comparing two emerging navigation aids. In: Proceedings of the 13th International Conference on Human Computer Interaction with Mobile Devices and Services, pp. 265–274. ACM (2011)
5. Jose, R., Lee, G.A., Billinghurst, M.: A comparative study of simulated augmented reality displays for vehicle navigation. In: Proceedings of the 28th Australian Conference on Computer-Human Interaction, pp. 40–48. ACM (2016)
6. Eriksson, A., Stanton, N.A.: Driving performance after self-regulated control transitions in highly automated vehicles. Hum. Factors **59**, 1233–1248 (2017)

Using AI-Planning to Solve a Kinodynamic Path Planning Problem and Its Application for HAPS

Jane Jean Kiam[1(✉)], Axel Schulte[1], and Enrico Scala[2]

[1] Institute of Flight Systems, University of the Bundeswehr, Werner-Heisenberg-Weg 39, Neubiberg, 85579 Munich, Germany
{jane.kiam, axel.schulte}@unibw.de
[2] Fondazione Bruno Kessler, Via Sommarive, 18, Povo, 38123 Trento, Italy
escala@fbk.eu

Abstract. This work emphasizes on the ability of a domain-independent AI-planner to solve a kinodynamic path planning problem by recapitulating the encoding in the PDDL+ modelling language and by showing the easy extension for multiple HAPS. The advantage of the approach is highlighted with the concept of an implementation framework that incorporates tools to validate the problem model and to explain the plans to the operator. Some flight path plans are illustrated as well as the validation of plans are described.

Keywords: Kinodynamic · Path planning · AI planning · HAPS
Wind field · Explainable AI

1 Introduction

Path planning for autonomous vehicles with dynamic constraints in a realistic environment is often a kinodynamic planning problem, i.e. the state space is subject to obstacle-based constraints and first-order differential constraints. Existing Rapidly exploring Random Trees (RRT)-based motion planners, kinematic A*, etc., are common approaches for solving this class of path planning problem. While being efficient, these planners are usually tailored for "experts" who have considerable knowledge in search algorithm and programming to understand the behavior and decision of the planners.

Following the Explainable AI (XAI) program launched recently by DARPA which focuses on explainable machine learning, a series of activities in Explainable AI Planning (XAIP) started to take place. Being explainable is important so that in a system where Human-Autonomy Teaming (HAT) is required, human trusts, interacts and understands the AI system [1]. Fox et al. argued in [1] that AI Planning is favored in being explainable, for it is based on models. Some pioneer works have started to venture into this new field to explain AI planners to the end users, e.g. model-checking to verify if the goal conditions are reachable [2], a validator [3] to explain why the determined plan is better, filter violation techniques to explain why replanning is necessary.

W. Karwowski and T. Ahram (Eds.): IHSI 2019, AISC 903, pp. 758–764, 2019.
https://doi.org/10.1007/978-3-030-11051-2_115

In order to benefit from the techniques put forth by the AI planning community, whether for XAIP or for their domain-independent planning capability, a planning problem must first be modelled in a formalism understood by the AI planners. One of the widely-used formalisms is the Problem Domain Definition Language (PDDL), which supports, since PDDL+ [4], temporal and numeric planning. In this article, we demonstrate the analogy between the PDDL+ formalism and the modelling of a control-based motion planning using a rather challenging path planning domain, in which a High-Altitude Pseudo-Satellite (HAPS) flies in a time-varying wind field with dynamic obstacles.

We first recapitulate the modelling of a kinodynamic planning problem and its convertibility into the PDDL+ formalism. We also show a few example scenarios for the use of an AI-planner as a flight path planner with an emphasis on the extension for multi-HAPS flight path planning. A concept of how XAIP can be incorporated into the planning framework is proposed. At last, we illustrate some results of the validation tests in a realistic simulated environment.

2 Kinodynamic Path Planning Problem of a HAPS

2.1 Kinodynamic Planning Problem

As considered in LaValle and Kuffner [5], a path planning problem is referred to as "kinodynamic planning problem" if the state space is subject to obstacle-based global constraints, and has first-order differential constraints, i.e. each state s of the state space $\mathbb{S}$ of the planning problem is defined by $s = (x, \dot{x})$, where x is a vector of geometrical configurations of the vehicle. If U denotes the control space, the state transition will be governed by

$$\dot{x} = f(x, u), u \in U. \tag{1}$$

A common approach to solve the kinodynamic problem is to use the RRT [5]. The feasibility of the plan is ensured via the addition of new vertices that are connected to the tree by a control parameter u and via a collision check.

2.2 Using an AI Hybrid Planner to Solve a Kinodynamic Planning Problem

Flying in the stratosphere at an altitude of about 18 km, and being completely solar-powered, HAPS can operate with extreme long endurance and is a promising alternative to fixed-orbit satellites. HAPS fly at an equivalent airspeed of 9 m/s and are equipped with weak electric motors, causing the motion to be very restrained and strongly influenced by wind. It is hence essential to consider during flight path planning the maneuverability and the wind field to ensure feasibility at its best. Furthermore, hazardous weather zones to avoid such as Cumulonimbus clouds and turbulences, can no longer be considered static obstacles due to the low speed of the HAPS [6].

A main progress made in AI planning in the past two decades is focused on domain-independent planners, i.e. planners that are able to solve a planning problem without knowing the problem beforehand. Such planners understand the problem via a modelling language, with the most popular being PDDL. The main strength of PDDL is to capture state transformations of an object precedented by a ? caused by an action. Starting from PDDL+, numeric and temporal planning problems with automated processes can be well encoded too [4, 7].

The feasibility of the planned path for HAPS can be ensured via the consideration of the first-order movement constraints, the wind effect on the motions, as well as the obstacles in the environment. Below are a few example analogies of the analytical expressions in classical control-based motion planning and the representation in PDDL+ (See Table 1).

Table 1. Encoding a control-based motion planning problem in PDDL+

Action to increase turn rate $\dot{\chi}$

Let $A_{\dot{\chi}} = \{-|\dot{\chi}_{max}|, -|\dot{\chi}_{max}| + \Delta\dot{\chi}, \ldots, |\dot{\chi}_{max}| - \Delta\dot{\chi}, |\dot{\chi}_{max}|\}$ be the action space

Encoding in PDDL+

```
(:action increase_turn_rate
  :parameters (?uav -uav)
  :precondition (and    (< (turn_rate ?uav) (- (max_turn_rate ?uav)
                           (delta_turn_rate ?uav)))))
  :effect (and (increase (turn_rate ?uav) (delta_turn_rate ?uav)) ))
```

Process to update latitude ϕ

$\phi(t+1) = \phi(t) + \dot{\phi}(t+1)\Delta t$, where $\dot{\phi} = \frac{(v_{wind,N} + v^{*}_{TAS} \cos\gamma \cos\chi)}{R+h}$

Encoding in PDDL+

```
(:process update_latitude
 :parameters (?uav -uav)
 :precondition ()
 :effect (and (increase (phi ?uav)
                    (* #t (/ (+ (* (v ?uav) (* (cos (pitch ?uav))
                                     (cos (heading ?uav))))
                         (north_wind ?uav))(+ R (h ?uav)))))))
```

3 Implementation Framework for HAPS

3.1 Applicable Scenarios for HAPS

Figure 1a shows a typical surveillance mission to monitor ground activities, in which several Locations of Interest (LOIs) marked in green are landmarks to be monitored periodically. The Mission Areas (MAs) in blue encompassing LOIs of the same client denote the allocated airspace to carry out the task. The HAPS can loiter freely at night or when idling in Waiting Areas (WAs) represented by yellow polygons. Dedicated Corridors (C) connect mission elements. After performing a higher level strategic planning to decide for the order of LOIs to visit, a tactical planning step involving four-dimensional flight path planning from a start to a goal position is required [8].

Another possible deployment of HAPS is shown in Fig. 1b; several HAPS are contracted to monitor continuously using an electro-optical (EO) camera over a wide catastrophic area marked in green with the populated areas marked in blue during a long-duration search and rescue mission. The red polygons represent dynamic weather critical zones to be avoided, e.g. Cumulonimbus clouds. The mission success rate depends on the cloud coverage between the flight level and the ground.

In both missions, a time-stamped path is critical, since mission requirements are time-dependent and the environment is dynamic. In the tactical planning problem needed in the mission scenario depicted in Fig. 1, in which a start state (position and time) and a goal position are given, the goal condition can be formulated with a tolerated position error of epsilon (see Fig. 2a). For the case where multiple HAPS are involved, an additional global constraint must be included (see Fig. 2b) in the problem domain definition to avoid collisions between any two platforms. The goal is reached if the desired points of interest (POIs) are cleared (Fig. 2c).

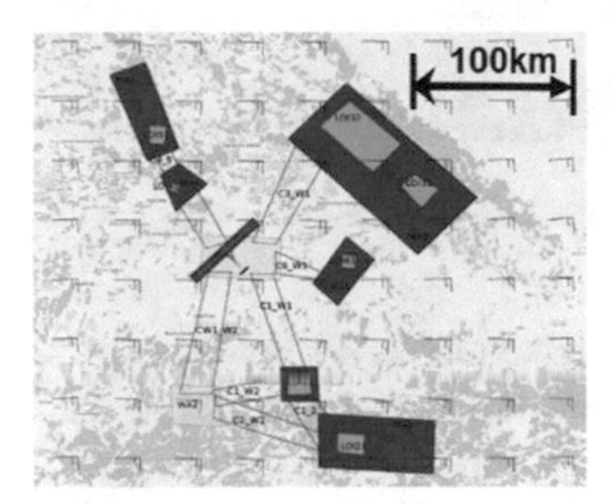

a. Repetitive monitoring of multiple locations

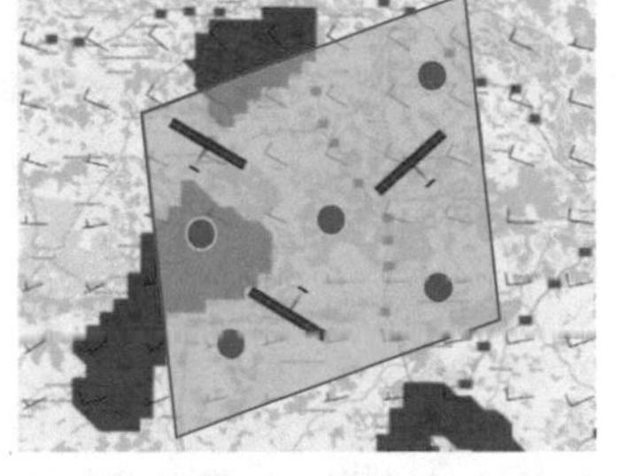

b. Continuous monitoring of a wide area in urgency

Fig. 1. Mission scenarios

```
(:goal
 (and  (>= (longitude haps) (- (goal_longitude) (epsilon)))
       (>= (latitude haps) (- (goal_latitude) (epsilon)))
       (>= (latitude haps) goal_latitude-epsilon)
       (<= (latitude haps) goal_latitude-epsilon) ) )
```

a. Goal condition to reach a goal position expressed in PDDL+

```
(:constraint avoid_traffic_collision
:parameters (?uav1 -uav ?uav2 - uav)
:condition (or  (not (different ?uav1 ?uav2))
                    (> (^ (+ (^ (- (latitude ?uav1) (latitude ?uav2)) 2)
                          (^ (- (longitude ?uav1) (longitude ?uav2)) 2)) 0.5) min_distance) ))
```

b. Global constraint to avoid collisions by imposing a minimum distance

```
(:goal
  (and (cleared poi_1)
       (cleared poi_2))
```

c. Goal condition to clear multiple POIs

Fig. 2. Expression of different constraints and goal conditions for different mission scenarios

4 Implementation Framework for HAPS

A domain-independent AI-planner is meant to be used as a black-box for realistic hybrid planning (see Fig. 3). A domain file is created to model the planning problem, e.g. flight kinematic model, constraints etc. Once created, the domain file remains unchanged. To adapt the parameters of the planning problem, e.g. the number of HAPS involved, the goal positions, the start positions, wind field, position of critical zones etc., a planning instance file is created.

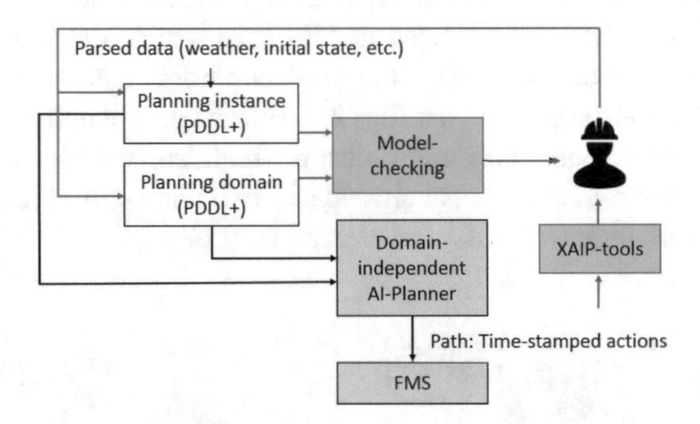

Fig. 3. Implementation framework of an AI-planner

Using the AI-planner comes with several advantages. The physical problem can be formally represented in a more comprehensive way in PDDL+ scripts, as described in the previous sections. The planner can be used off-the-shelf and heuristic algorithms can be exploited blindly. Furthermore, since domain-independent AI-planner are usually made publicly available, they are extensively tested and therefore less prone to error. The blue connectors in Fig. 3 illustrate how the use of AI-planners can also benefit from the XAIP. Some ongoing works have been done on model-checking [2, 9] to verify if a planning problem is solvable or not, and in the case of insolvability, the cause of deadlocks must be explained. XAIP tools to explain for example why a plan is better than others [3, 10] can help the operator to trust and understand the behavior of the planner. Plans will not be blindly executed, making it possible for the operator to intervene reasonably.

5 Test and Validation

The planned flight paths were successfully validated using a six degrees of freedom (6-DoF) HAPS simulator coupled with a four-dimensional flight controller [11] in a realistically simulated physical environment using historical weather data. We observed that the equivalent airspeed remains stable around the optimal airspeed ($\sim$9 m/s), i.e. following the planned path is more energy efficient and operationally safer.

Figure 4a depicts the flight path planning with two given goal positions for two HAPS with known start position and time. On top of the numeric flight path planning while maintaining a minimum distance between HAPS as described in Fig. 2b, the AI-planner also schedules to match the HAPS to the goals.

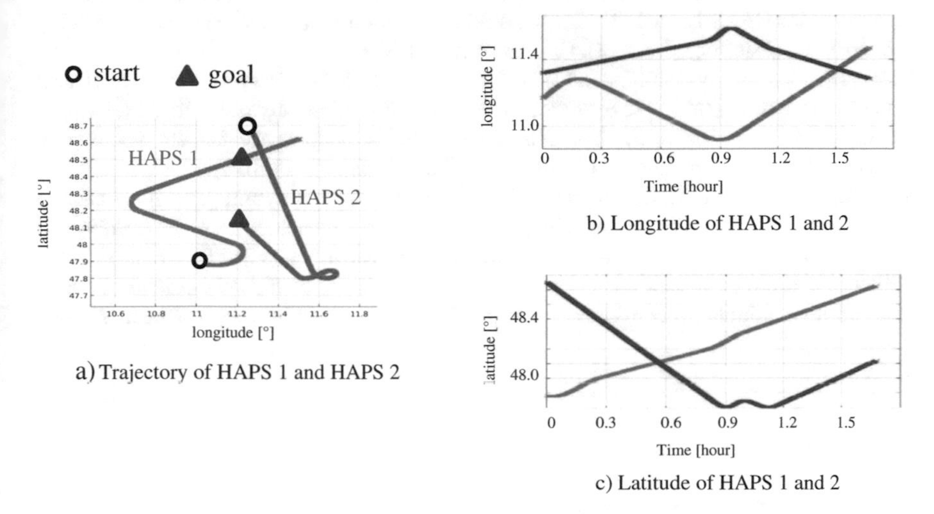

Fig. 4. Centralized flight path planning for two HAPS

6 Conclusion

This paper demonstrated and validated the use of a hybrid AI-planner to solve a kinodynamic path planning problem by using HAPS as a concrete example application. It is also shown here how the work can be extended to plan flight paths for multiple HAPS. A concept for the inclusion of XAIP tools is provided to emphasize on the advantages of using a model-based AI-planning approach. Left for future is to incorporate model-checking and XAIP tools when these are mature enough to cope with our planning problem.

References

1. Fox, M., Long, D., Magazzeni, D.: Explainable Planning. International Joint Conference on Artificial Intelligence (IJCAI) Workshop on Explainable AI, Melbourne (2017)
2. Bogomolov, S., Magazzeni, D., Minopoli, S., Wehrle, M.: PDDL+ planning with hybrid automata: Foundations of translating must behavior. In: Proceedings of International Conference on Automated Planning and Scheduling (ICAPS) (2015)
3. Fox, M., Howey, R., Long, D.: Validating plans in the context of processes and exogenous events. In: Proceedings of AAAI Conference on AI. Pittsburgh, Pennsylvania (2005)
4. Long, D., Fox, M.: Modelling mixed discrete-continuous domains for planning. J. Artif. Intell. Res. **27**, 235–297 (2006)
5. LaValle, S.M., Kuffner, J.J.: Randomized kinodynamic planning. Int. J. Rob. Res. **20**(5), 378–400 (2001)

6. Kiam, J.J., Gerdts, M., Schulte, A.: Fast subset path planning/replanning to avoid obstacles with time-varying probabilistic motion patterns. In: 8th European Starter AI Researcher Symposium, The Hague, The Netherlands (2016)
7. Kiam, J.J., Scala, E., Ramirez, M., Schulte, A.: Using a hybrid AI-Planner to plan feasible flight paths for HAPS-Like UAVs. In: ICAPS Proceedings of the 6th Workshop on Planning and Robotics (PlanRob), Delft, The Netherlands (2018)
8. Kiam, J.J., Schulte, A.: multilateral mission planning in a time-varying vector field with dynamic constraints. In: IEEE Systems, Man, and Cybernetics, Miyazaki, Japan (2018)
9. Steinmetz, M., Hoffmann, J.: Towards clause-learning state space search: learning to recognize dead-ends. In Proceedings of AAAI Conference on AI (2016)
10. Seegebarth, B., Müller, F., Schattenberg, B., Biundo, S.: Making hybrid plans more clear to human users - A formal approach for generating sound explanations. In: Proceedings of International Conference on Automated Planning and Scheduling (ICAPS) (2012)
11. Müller, R., Kiam, J.J., Mothes, F.: Multiphysical simulation of a semi-autonomous solar powered high altitude pseudo-satellite. In: IEEE Aerospace Conference, Montana (2018)

Evaluating the Coordination of Agents in Multi-agent Reinforcement Learning

Sean L. Barton[1(✉)], Erin Zaroukian[1], Derrik E. Asher[1], and Nicholas R. Waytowich[2]

[1] Computational & Information Sciences Directorate, U.S. Army Research Laboratory, Adelphi, USA
sean.l.barton.ctr@mail.mil
[2] Human Research & Engineering Directorate, U.S. Army Research Laboratory, Adelphi, USA

Abstract. The present study provides an in-depth analysis of inter-agent coordination through a complete exploration of agent behavioral dimensions. We evaluate the behavioral dimensions in a multi-agent predator-prey pursuit task where predator agent coordination necessarily exists due to a shared goal. We explore two conditions, one that is void of explicit coordination (fixed-strategy), and one that has the potential for explicit coordination (learning agents). This comprehensive evaluation of multi-agent behavioral dimensions provides theoretical evidence for true inter-agent coordination by a learning algorithm and the behavioral dimensions that agents coordinate in a cooperative task.

Keywords: Coordination · Multi-agent · Reinforcement learning
Predator-prey pursuit · Teaming

1 Introduction

Reinforcement learning (RL) is an attractive option for providing adaptive behavior in computational agents because of its theoretical generalizability to complex problem spaces with complicated dynamics or non-linear features [1, 2]. However using deep RL in multi-agent domains is problematic, as the presence of independent actors violates the assumptions of Markov Decision Processes (MDPs) that are fundamental to RL [2]. A number of solutions have arisen over the past 20 years for utilizing RL in multi-agent problem spaces [2], but only recently have these methods been extended to deep RL [3]. One important challenge for using RL in multi-agent contexts is assessing the degree and efficacy of coordination that develops between learning agents that must function as a team.

In recent work, we presented convergent cross mapping [4] (CCM) as a means of measuring the causal relationships between pairs of agents in a predator-prey pursuit task [5]. The results from this previous work indicated that predators utilizing learned policies (i.e., learning agents) achieved substantially higher scores over non-learning agents. However, the CCM evaluation did not support the hypothesis that learning agents achieve greater performance by learning to coordinate the actions selected by

W. Karwowski and T. Ahram (Eds.): IHSI 2019, AISC 903, pp. 765–770, 2019.
https://doi.org/10.1007/978-3-030-11051-2_116

their policies. This raises questions concerning the extent to which other behavioral dimensions might be coordinated in this cooperative task. Further, the current study provides a means of illuminating explicit coordination between learning agents performing a cooperative task.

2 Methods

Inter-agent coordination was evaluated in a cooperative task in which two predator agents attempted to capture a prey agent as many times as possible within a given time limit. Predator agents were identical in their capabilities[1]. Agent positions were randomized at the start of each episode. Agent actions were continuous accelerations in the X and Y dimensions, truncated by an upper bound. The simulation environment was made available through the OpenAI Gym network [6] and was developed for the multi-agent deep RL problem space [3].

The current study analyzed behavior for two types of predators: fixed-strategy that minimized their distance to the prey (independent of each other's actions and state), and learning predators that followed learned policies which maximized the reward gained by capturing the prey. The policies of learning predators were represented by deep neural networks and trained via a multi-agent deep deterministic policy gradient (MADDPG) [3]. During training, learning predators received a fixed reward when either made contact with the prey. In both conditions (fixed-strategy and learning), the prey was a learning agent that was trained via MADDPG. The prey received a negative reward when captured, and learned policies that minimized the absolute value of this negative reward. Each episode ran for 25 time steps, and agents were trained for 100,000 episodes.

Inter-agent coordination was measured with a modified version of CCM called *multi-spatial convergent cross mapping* [7] (MCCM), which permits calculation of CCM using short time-series repeated across spatial or temporal observations. This method is ideal for simulation studies where the only limit on repeated observations (i.e. testing episodes) is computational power and wall-time. In order to provide an in-depth analysis of the dimensions of learned coordination between predators, we computed the MCCM score between all dimensions of an agent's behaviors that could be observed by another agent. In this task, this was limited to the 2-dimensional position (X, Y), velocity $(\dot{X}, \dot{Y})$, and acceleration $(\ddot{X}, \ddot{Y})$.

As discussed in previous work [5], predators' behaviors are trivially coordinated due to the identical influence of a shared goal (both pursuing the same prey). To measure only the predators' intentional coordination, we modified one predator's behavior at test-time such that its actions were replaced by the mechanics of a double-pendulum. The modified predator followed a chaotic but smooth behavioral trajectory that was independent of prey behavior (removed the shared goal). Using CCM to evaluate the causal influence that the modified predator has on the unmodified predator

[1] A prey agent was capable of 33% greater acceleration and 25% greater maximum velocity than the predator agents.

allows us to measure the intentional coordination resulting from predators working together.

In order to understand how stable the learned coordination was between predators, we generated an average causal influence score that was derived from 50 independent simulations. For each simulation, MCCM was evaluated for each pair of behavioral dimensions obtained from 50 test episodes. MCCM was tested for significance to determine if a true causal influence was detected [7]. If significant, the MCCM value of that causal influence was taken as the causal strength. If no significant causation was detected, the causal strength was set to zero. All causal strength values were then averaged over all 50 simulations to provide a composite measure for each paired behavioral dimension.

3 Results

Figures 1, 2, and 3 present dimension matrices that show averaged causal influence strength for all possible combinations of the 6 behavioral dimensions. For each matrix, cells indicate the average causal strength that the X-axis dimension exerted on the Y-axis dimension. For Figs. 1 and 2, the causal strength is bounded between 0 (no causal dependency) and 1 (complete causal dependency).

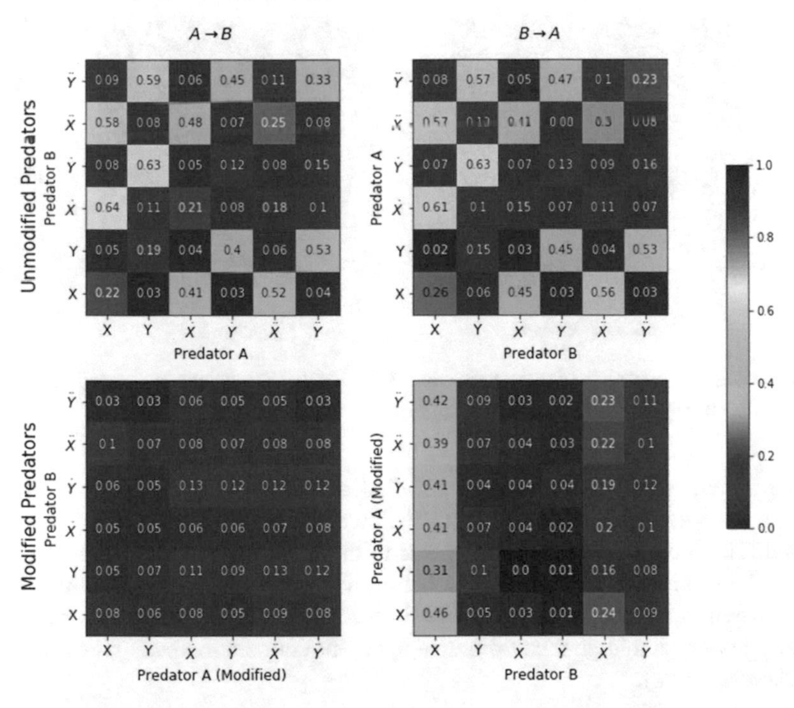

Fig. 1. Averaged causal strength between behavioral dimensions of fixed-strategy predators. X-axis dimensions causally influence Y-axis dimensions. Rows show Unmodified (top) and modified (bottom) predator causal influences, while the columns show the direction of causation (Left: A → B, Right: B → A).

Figure 1 shows the causal influence between fixed-strategy predators at the end of training[2]. Fixed-strategy agents have a shared goal, but are incapable of explicit coordination. Thus, any measured causal dependency is an artifact of the shared goal. The top row of Fig. 1 shows that unmodified predators (with a fixed strategy) appear to have a high degree of bidirectional causal influence. Because the fixed-strategy predators are incapable of influencing each other's actions, we can be certain that this is an artifact resulting from their shared goal.

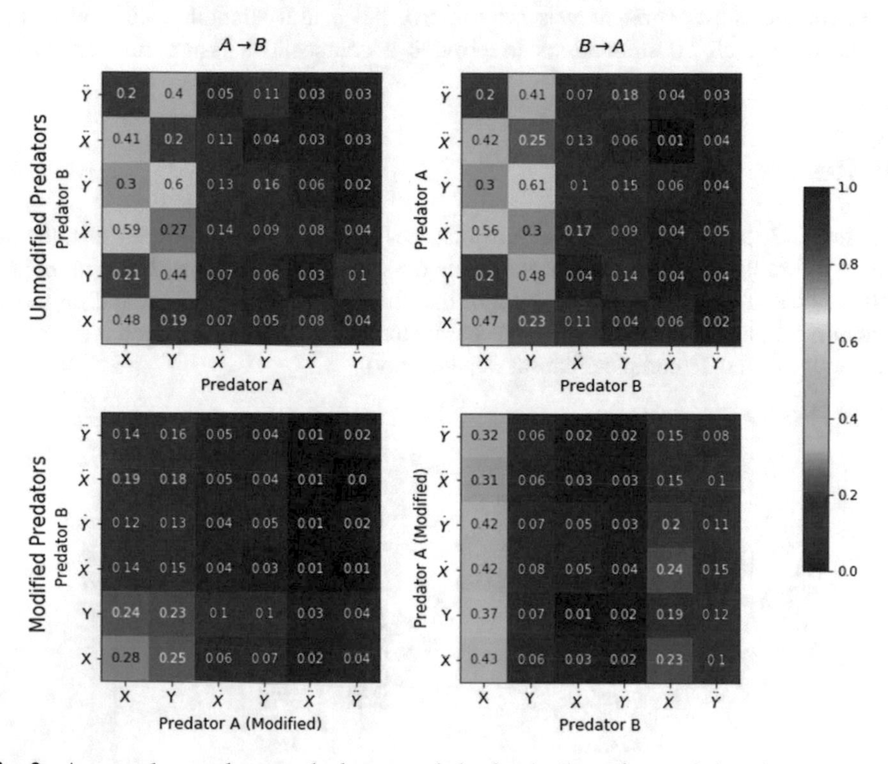

Fig. 2. Averaged causal strength between behavioral dimensions of learning predators. Organized as in Fig. 1.

When one predator's behavior is modified and the MCCM re-run (bottom row of Fig. 1), we see no influence of the modified predator on the fixed-strategy predator, as expected. The unmodified predator appears to influence the modified predator, but this is most likely an artifact of collisions between the two predators resulting from our choice of pendular mechanics as the basis for modified predator behavior. Figure 1 illustrates how a shared goal can produce apparent cooperation even when none exists (as is the case here).

[2] In this case, training only impacted prey behavior, as fixed-strategy predators did not change their pursuit behavior through the learning process.

Figure 2 shows the same analysis as Figure 1, but applied to the learning predator condition. Here, each predator appears primarily influenced by its partner's X and Y position. The bottom row of Fig. 2 shows the remaining causal influences when one predator is modified. If a predator has learned a policy that takes into account the behavior of is partner, then there should be an impact of the partner's behavior on its own even when that partner is modified to no longer pursue the prey. The bottom row of Fig. 2 shows that the majority of the causal influence has been removed. However, there appears to be a weak influence of the modified agent's position on the position of its partner.

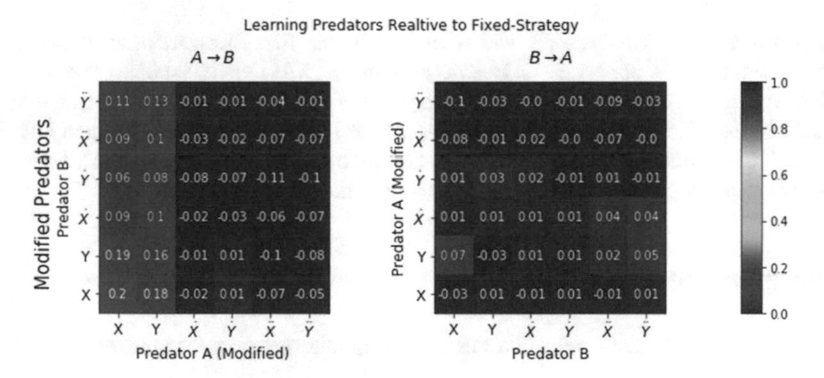

Fig. 3. Subtraction of modified fixed-predator condition from modified learning predator condition. Remaining influence shows true coordination between agents.

We can better isolate this potential effect by comparing it to a baseline. The fixed-strategy predators cannot cooperate and so provide an ideal benchmark against which to evaluate learning predators. Figure 3 shows the causal influence between modified and unmodified learning predators after subtracting any apparent causal influence detected between modified and unmodified fixed-strategy predators. This removes spurious effects that arise from our design choices (such as the pendular behavior modified predator) and leave only the true causal influence between learning predators (Fig. 3 left side).

We can see clearly in Fig. 3 that even when its partner is modified and no longer pursuing the prey, a learning predator continues to modulate its position based on the position of its partner. This is clear in the weak but persistent causal influence exerted by the modified predator on the unmodified predator. By contrast, the unmodified predator is not influenced by the learning predator at all.

4 Discussion

This study expands on the findings discussed in [5]. We show here how a shared goal can produce apparent cooperation when none exists. Further, we illustrate how MCCM can be used to detect true causal relationships in the behaviors of computational agents, and show that for the predator-prey pursuit task learning agents primarily modify their behavior based on the position of their partners, rather than the actions.

One possible explanation for this is that agent actions in this task (i.e., acceleration) converge on a "bang-bang" control scheme, were agents rapidly oscillate between minimum and maximum acceleration values to control velocity and position. Such actions are likely noisier than the resulting behavior (a change in position) and thus from a learning agent's perspective the position of a partner is far more informative about how to make a behavioral adjustment. An important take away from this work is the efficacy of MCCM as a measure of inter-agent coordination as it pertains to behaviors in cooperative tasks.

Acknowledgements. This research was sponsored by the Army Research Laboratory and was accomplished under Cooperative Agreement Number W911NF-18-2-0058. The views and conclusions contained in this document are those of the authors and should not be interpreted as representing the official policies, either expressed or implied, of the Army Research Laboratory or the U.S. Government. The U.S. Government is authorized to reproduce and distribute reprints for Government purposes notwithstanding any copyright notation herein.

References

1. Mnih, V., et al.: Human-level control through deep reinforcement learning. Nature **518**, 529–533 (2015)
2. Matignon, L., Laurent, G.J., Le Fort-Piat, N.: Independent reinforcement learners in cooperative markov games: a survey regarding coordination problems. Knowl. Eng. Rev. **27**, 1–31 (2012)
3. Lowe, R., Wu, Y., Tamar, A., Harb, J., Pieter Abbeel, O., Mordatch, I.: Multi-agent actor-critic for mixed cooperative-competitive environments. In: Guyon, I., Luxburg, U.V., Bengio, S., Wallach, H., Fergus, R., Vishwanathan, S., Garnett, R. (eds.) Advances in Neural Information Processing Systems 30, pp. 6382–6393. Curran Associates, Inc. (2017)
4. Sugihara, G., May, R., Ye, H., Hsieh, C.-h., Deyle, E., Fogarty, M., Munch, S.: Detecting causality in complex ecosystems. Science, 1227079 (2012)
5. Barton, S.L., Waytowich, N.R., Zaroukian, E., Asher, D.E.: Measuring collaborative emergent behavior in multi-agent reinforcement learning. In: 1st International Conference on Human Systems Engineering and Design. IHSED; Springer
6. Brockman, G., et al.: OpenAI Gym. arXiv:1606.01540 [cs] (2016)
7. Clark, A.T., et al.: Spatial convergent cross mapping to detect causal relationships from short time series. Ecology. **96**, 1174–1181 (2015)

Analysis of Force Interaction for Teamwork Assistance by Concern For Others

Genki Sasaki[1](✉) and Hiroshi Igarashi[2]

[1] Department of Electrical and Electronic Engineering, Graduate School of Engineering, Tokyo Denki University, Adachi-ku, Tokyo 120-8551, Japan
g.sasaki@crl.epi.dendai.ac.jp

[2] Department of Electronic Engineering, School of Engineering, Tokyo Denki University, Adachi-ku, Tokyo 120-8551, Japan
h.igarashi@crl.epi.dendai.ac.jp

Abstract. Teamwork assistance and objectively evaluation is a purpose of this research. For that reason, we focus on differences in a solo task and a cooperative task. This method is "Concern For Others: CFO" which is an element of evaluating teamwork in real time. This paper analyzes CFO with force interaction during the cooperative task in order to clarify the strategy of teamwork assistance using CFO in real time. From a result, a teamwork assistance using CFO by force feedback is possible.

Keywords: Teamwork · Cooperative task · Force interaction · AI Prediction

1 Introduction

A cooperative task is an act of mutual cooperation with others, and it is important for human makes to live as a member of a social group. Humans need not only individual skills but also opportune teamwork in a cooperative task, e.g. large-scale construction task, transportation task, and team sports [1].

Many kinds of research have elucidated the teamwork skills of human, and these researches developed a measurement method using a questionnaire of 4 and 5 questions for teamwork evaluation [2, 3]. Since these researches use the questionnaire after their experiment, teamwork cannot be objectively evaluated in real time. Real time evaluation of teamwork is important since it can contribute to the support of the team that responds to the task situation.

Previous research has suggested an objectives evaluations method for teamwork by "Concern For Others: CFO" which is an element of teamwork in a cooperative task [4, 5]. Since CFO is estimated based on the operation input amount, a team is assisted in real time by using CFO. We have been researching using CFO for objective teamwork evaluation and assistance. These researches analyzed the affection of force interaction to both CFO and cooperative performance during the cooperative task [6].

This paper will aim to estimate CFO by force information, and clarify the relationship with CFO and teamwork. In addition, in order to perform force feedback for

W. Karwowski and T. Ahram (Eds.): IHSI 2019, AISC 903, pp. 771–776, 2019.
https://doi.org/10.1007/978-3-030-11051-2_117

teamwork assistance, we will clarify the strategy of teamwork assistance using CFO in real time.

2 Definition and Estimation of CFO

In this research, we focus on differences between solo task and a cooperative task in aim for quantitative evaluation of the team. A change of action an individual is assumed to be caused by the influence of the intervention of the others when the same task in both a solo and a cooperative. Therefore, the change of action due to the intervention of another operator during the cooperative task. "Concern for Others: CFO" was thus defined in this research. The outline of CFO estimation is shown in Fig. 1.

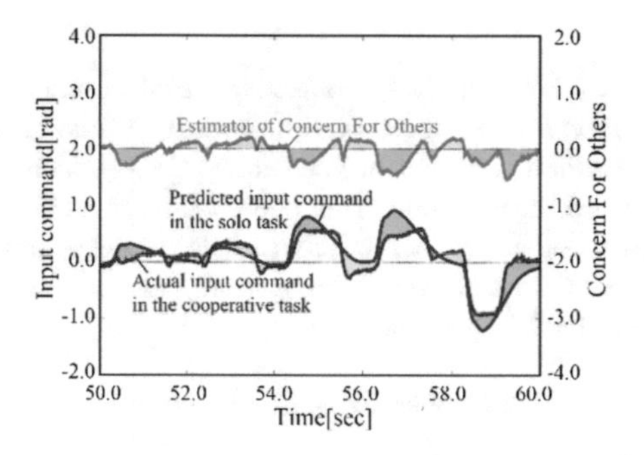

Fig. 1. Estimation of concern for others.

The estimation of CFO is given by:

$$\boldsymbol{\sigma}_i = \boldsymbol{u}_i - \boldsymbol{f}_{NN}(\boldsymbol{X}_i), \tag{1}$$

where $\boldsymbol{\sigma}_i$ is the estimation CFO, i is operator's number, $\boldsymbol{u}_i = \begin{bmatrix} u_{pitch} & u_{roll} \end{bmatrix}^T$ is actual operation command during the cooperative task, $\boldsymbol{f}_{NN}$ is predicted operation command during the solo task, and $\boldsymbol{X}_i \in R^N$ is N variables corresponding to the current environmental information necessary for prediction by the Neural Network (NN).

All of the research that estimate CFO is using operation information to estimation CFO. This operation information is an operating plate angle for the cooperative task, and it is position information. However, force information is important as well as position information is important in human action. Therefore, this research estimate CFO based on position information and force information during the cooperative task as a hybrid. Assuming that CFO using position information is PCFO $\boldsymbol{\sigma}_i^\theta$ and CFO using force information is FCFO $\boldsymbol{\sigma}_i^\tau$, it can be estimated as follows:

$$\boldsymbol{\sigma}_i^\theta = \boldsymbol{\theta}_i - \boldsymbol{\theta}_{NN}(\boldsymbol{X}_i^\theta), \tag{2}$$

$$\boldsymbol{\sigma}_i^\tau = \boldsymbol{\tau}_i - \boldsymbol{\tau}_{NN}(\boldsymbol{X}_i^\tau), \tag{3}$$

where i is operator's number, $\boldsymbol{\theta}_i = \begin{bmatrix} \theta_{pitch} & \theta_{roll} \end{bmatrix}^T$ is actual operation angle during the cooperative task, $\boldsymbol{\theta}_{NN}$ is predicted operation angle during the solo task, $\boldsymbol{\tau}_i = \begin{bmatrix} \tau_{pitch} & \tau_{roll} \end{bmatrix}^T$ is actual operation torque during the cooperative task, $\boldsymbol{\theta}_{NN}$ is predicted operation torque during the solo task, and $\boldsymbol{X}_i^{\theta}$, $\boldsymbol{X}_i^{\tau}$ are similar the (1).

3 Cooperative Task Platform

In the cooperative task in this research, the operator's actions must be quantitatively measured and evaluated. Therefore, we developed the cooperative task platform with force interaction that can participate up to 4 people assuming a general cooperative task.

The operator performs the task in the virtual space displayed on the Head Mounted Display (HMD) using the cooperative task platform shown in Fig. 2(a). In the virtual space of Fig. 2(b), the plate is displayed, and the box and the target are placed on the plate. Since the box moves by the tilt of the plate, the operator can freely control the box using up and down the two bars. The input interface consists of two DC motor, two bars extending from the left and right DC motor, and a microcontroller. The DC motors gives force interaction among operators. The motor control method is applied Disturbance OBserver (DOB) to the motor to realize a robust control system [7]. Furthermore, Reaction Torque OBserver (RTOB) allows estimating input torque of operator [8].

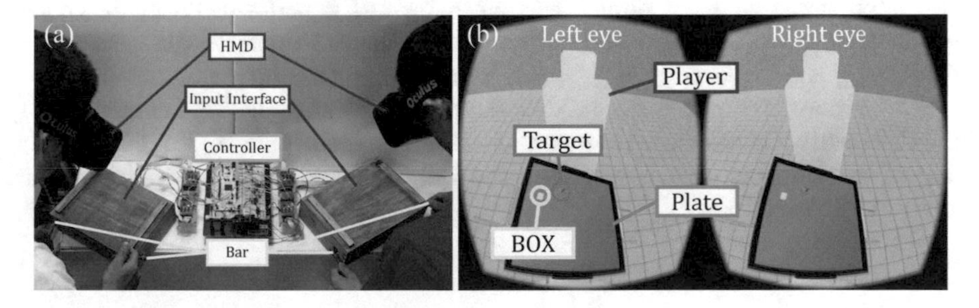

Fig. 2. Cooperative task platform. (a) Overview. (b) Virtual task display on the HMD.

4 Experimental Environment

4.1 Solo Operation Modeling

This research learns operation angle and force of solo task by three layers NN and generate individual solo task model. From the learning result, the solo task model can predict the operation of the solo task in any case. Table 1 shows the setting of the NNs for learning. The time for learning is set as 45 s.

Table 1. NNs settings for solo operation modeling.

Settings	Position	Force
Input layers	20	30
Middle layers	40	55
Output layers	2	2
Learning coefficient	0.005	0.005
Input signals	5 steps of input angle history on the left and right, and 5 steps of coordinate error between the box and the target	5 steps of input angle and torque history on the left and right, and 5 steps of coordinate error between the box and the target

4.2 Plate Control Task

The plate control task is both the solo task and cooperative task. It can participate up to 4 people. The plate control task is performed in the virtual task environment. Since the target moves at random direction and a fixed distance every 2 s, the operator control the box at following the square target. The box moves according to the tilt of the plate, and the operator controls the plate using the input interface.

4.3 Evaluation Plate Control Task

The performance evaluation during the cooperative task is performed from the contact time of the box and the target. The contact time continues to increase while the contact of the target and the box is satisfied, and it measures by the program.

Furthermore, the cooperative task's evaluation requires excluding the influence by individual skill. Therefore, we compare the performance of human group and evaluation of virtual group that is generated the individual model. The Evaluation Plate Control Task (EPCT) is:

$$EPCT = J_H - J_{NN}, \tag{4}$$

where J_H is contact time when a human group performs the plate control task, and J_{NN} is contact time when a virtual group by solo models performs the plate control task.

5 Experiment

This experiment will clarify the force interaction between operators based on PCFO and FCFO during the cooperative task. We will reveal the relation between CFO (PCFO and FCO) and performance, and look for strategy of teamwork by force feedback.

First, the solo operation model is generated by NN. Then, pairs of subjects perform the cooperative task. The cooperative task is performed for 180 s, however, 10 s from the start of the task is excluded from the evaluation section in order to take proficiency for the task of the subject into consideration.

Since CFOs are estimated by each operator, the CFO needs to be evaluated across the group. Therefore, we calculate the summation CFO and subtraction CFO. The summation CFO are: summation PCFO and summation FCFO. The subtraction CFO are: subtraction PCFO and subtraction FCFO.

6 Results

In this experiment, subjects of 10 pairs consisting of 20s 8 people were gathered.

Figure 3 shows the relationship between FCFO of two subjects estimated during cooperative task. Each plot averages the values of FCFO in 2 s until the target moves. Since the evaluation section is 170 s and the group is 10, the graph has a total of 850 plots. The result shows a positive correlation (r = 0.88) between FCFO of the two subjects. Therefore, subject's operation force changes their way of inputting force depending on other subjects. We discuss: subjects receive other operation force by force interaction. Then, the subject's operational intention changes other operational intention to theirs. It occurring as a chain.

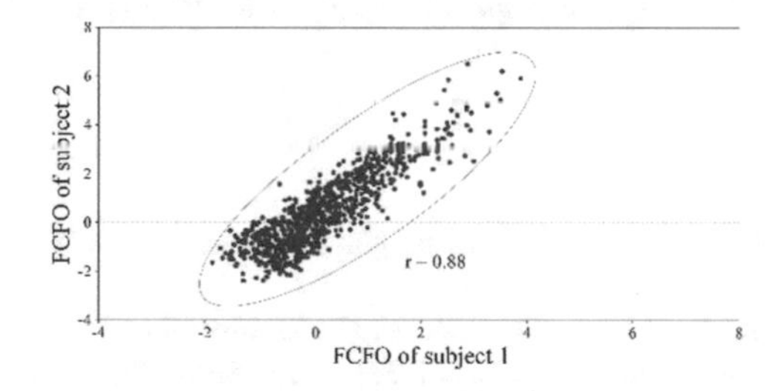

Fig. 3. Correlation of FCFO between each subject.

Figure 4 shows the relationship between CFO (summation and subtraction) and EPCT. Figure 4(a) shows a positive correlation (r = 0.76) between summation PCFO and summation FCFO. In addition, this result suggests the EPCT is decease when

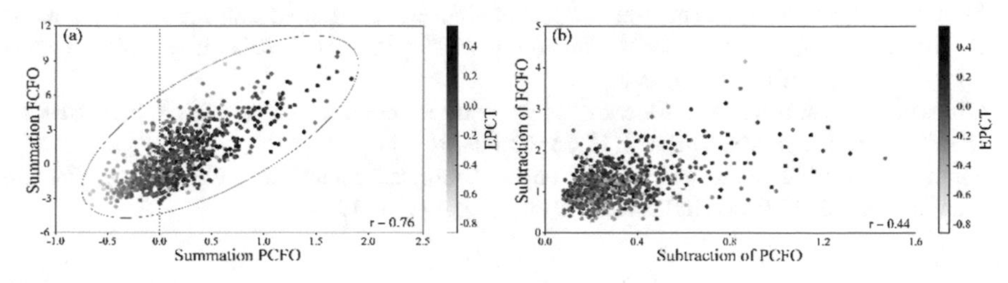

Fig. 4. The maps of EPCT by between PCFO and FCFO. (a) Summation CFO. (b) Subtraction CFO.

negative both summation PCFO and summation FCFO. Figure 4(b) shows a poor positive correlation (r = 0.44) between subtraction PCFO and subtraction FCFO. This result doesn't suggest correlation to the EPCT. Therefore, Summation PCFO and summation FCFO can be teamwork assistance is suggested.

7 Conclusion

This paper proposed FCFO estimation considering force information, in order to experimentally elucidate how the force interaction affects cooperative task. In addition, we clarified the relationship between CFO (PCFO and FCFO) and performance, and suggested the strategy of teamwork assistance by force feedback. As the results, the subject's operational intention changes other operational intention to theirs by force interaction. The result showed the EPCT is decreased by both negative summation PCFO and negative summation FCFO. Therefore, a teamwork assistance using CFO by force feedback is possible.

In the future work, we will construct a force feedback system using the PCFO and FCFO for teamwork assist. Furthermore, we will analyze PCFO and FCFO in more detail and aim for clear teamwork evaluation.

Acknowledgements. This work was supported by JSPS KAKENHI Grant Number 18K11407.

References

1. Feltz, D.L., Lirgg, C.D.: Perceived team and player efficacy in hockey. J. Appl. Psychol. **83**, 557–564 (1998)
2. Sarcevic, A., Marsic, I., Waterhouse, L.J., et al.: Leadership structures in emergency care settings: a study of two trauma centers. Int. J. Med. Inform. **80**(4), 227–238 (2011)
3. Capella, J., Smith, S., Philp, A., et al.: Teamwork training improves the clinical care of trauma patients. J. Surg. Educ. **67**(6), 439–443 (2010)
4. Igarashi, H.: Quantification of concern for others to social skill evaluation. In: The 22nd IEEE International Symposium on Robot and Human Interactive Communication, pp. 559–564. IEEE Press (2013)
5. Nii, S., Igarashi, H.: Estimation of teamwork in cooperative tasks by multiple persons. In: The 2016 RISP International Workshop on Nonlinear Circuits, Communications, and Signal Processing, 8PM1-4-3 (2016)
6. Sasaki, G., Igarashi, H.: Temporal analysis of CFO in cooperative task for teamwork assist. In: IECON 2018 – 44th Annual Conference of the IEEE Industrial Electronics Society, pp. 5487–5492. IEEE Press (2018)
7. Ohnishi, K., Shibata, M., Murakami, T.: Motion control for advanced mechatronics. IEEE/ASME Trans. Mechatron. **1**(1), 55–67 (1996)
8. Murakami, T., Yu, F., Ohnishi, K.: Torque sensorless control in multidegree-of-freedom manipulator. IEEE Trans. Ind. Electron. **40**(2), 259–265 (1993)

Defining Generic Tasks to Guide UAVs in a MUM-T Aerial Combat Environment

Sebastian Lindner(✉), Simon Schwerd, and Axel Schulte

Institute of Flight Systems, Universität der Bundeswehr Munich, 85577 Neubiberg, Germany
{sebastian.lindner, simon.schwerd, axel.schulte}@unibw.de

Abstract. In the fighter jet domain, it is very difficult to delegate tasks to unmanned systems because such a MUM-T scenario contains tasks from different dimensions, e.g. from AI, SEAD, OCA, etc. At the same time, there is a demand for a uniform and transparent operating concept, e.g. using a tactical map. Therefore, a way is looked for to describe the many different tasks with a uniform framework. In the presented approach a task command is split into an environmental object (physical, logical) to which the instruction refers. A military action to be performed on this object and a time parameterization (slack-time).

The object specification corresponds to the classification as it is usual for vectorized GIS data (point, line, area). The military action results from the different types of use (AI, SEAD, etc.) and the domains of action (Intelligence, Effect, etc.).

Keywords: Human-autonomy teaming · Cognitive agent · Task-based control human-system integration · Control pattern · Task formalization

1 Motivation

The concept of task-based UAV guidance have been introduced for different domains, like ground-based control stations [1] and helicopter missions [2]. The concept aims at reducing the human's workload by using cognitive agents in a supervisory control relation. The supervisory relation is one part of the *Dual-Mode Cognitive Automation Concept* [3]. We want to transfer this concept in the domain of military fighter jets. In our understanding, future combined air operation will be a composite of manned and unmanned aerial vehicles (Manned-unmanned-Teaming, MUM-T). Thus, the challenge is to enable a pilot commanding unmanned systems, additionally to his own aircraft. Figure 1 shows a pilot being in supervisory control over an unmanned combat aerial vehicle according the Human-Autonomy Teaming Design Patterns approach introduced in [4].

W. Karwowski and T. Ahram (Eds.): IHSI 2019, AISC 903, pp. 777–782, 2019.
https://doi.org/10.1007/978-3-030-11051-2_118

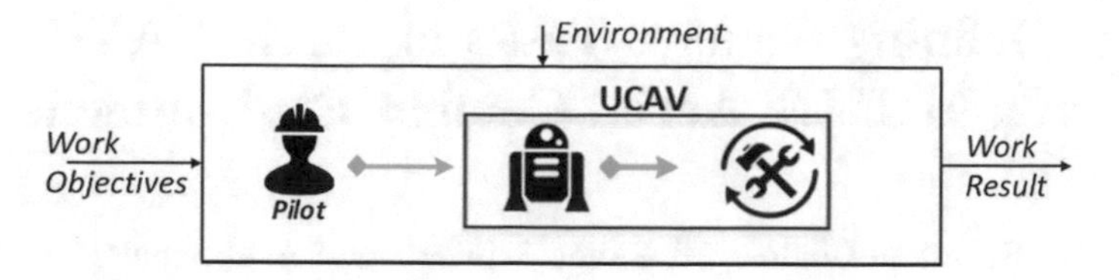

Fig. 1. Supervisory control relation between a pilot and UCAV.

Military fighter jet missions must be performed in permissive but also in highly contested scenarios. Workload is due to the high task load the essential point for the work system design. Introducing high automation may result in a deficit in situational awareness. The challenge is to keep the pilot in the loop without simultaneously overtaxing him. In a first attempt with high automatic functions [5] it was found that the pilot had too few possibilities to intervene – the tasks were formulated too generally. The pilot must not be overwhelmed by the task assignment process on the one hand, but the task must contain enough information on the other hand that the agent can execute the pilot's intention in the desired way. For this reason, it must be analyzed how a generalization of tasks can be performed. Before decomposing a task, we have a look on the domain-specific considerations.

2 Consideration for an Efficient Task Generalization

Application Domain: A combined air operation consists of different force packages, each dealing with a different objective of an air mission. Offensive Counter Air (OCA) deals with the occurrence of enemy fighter jet, Air Interdiction (AI) describes the use of aircraft attacks to destroy tactical ground elements, Suppression of Enemy Air Defense (SEAD) and Electronic Warfare (EW) are both possibilities to attenuate the opponent's air defense systems. A package is a military flight formation consisting mostly of 4 fighter jets. The leader is called flight lead and will be in our understanding always be manned. We want to substitute the subordinates of the flight with unmanned systems so that in dangerous situations as few humans as possible must be on site. Thus, the requirement is to enable each lead to manage his unmanned subordinates, independent of the package and their corresponding objective.

Human-Autonomy Integration: The effectiveness of such a framework relies in the ability to use the strength of both, the human and cognitive agent. The rule of thumb for automation is "as little as possible, as much as necessary". Thus, a user-centered system analysis helps to identify the necessary automation functions [4]. With a higher span of control, the need for automation increases. Thus, the design of the tasking interaction is an essential part in the human-autonomy integration. The use of the tactical map as the tasking interface promises a transparent operating concept. In [6] a tasking interface was designed to deal with a high span of control in MUM-T aerial combat missions. Task assignment can be performed by selecting displayed objects.

Ethical Aspects: Autonomous military operations, including weapon delivery, in which humans are not directly involved in the decision-making process are amongst other open issues ethically highly questionable [7]. Thus, task must be formulized to clearly reflect the pilot's intention with an unambiguous meaning.

Machine Requirements: The cognitive agent must be able to interpret the task and to transfer it in a chain of actions. We want to provide an overview rather than a specific implementation for the agent's design and requirements. In Fig. 2 we present a generic architecture with the task-input and the signal-output of the agent. Another design approach was made in [1].

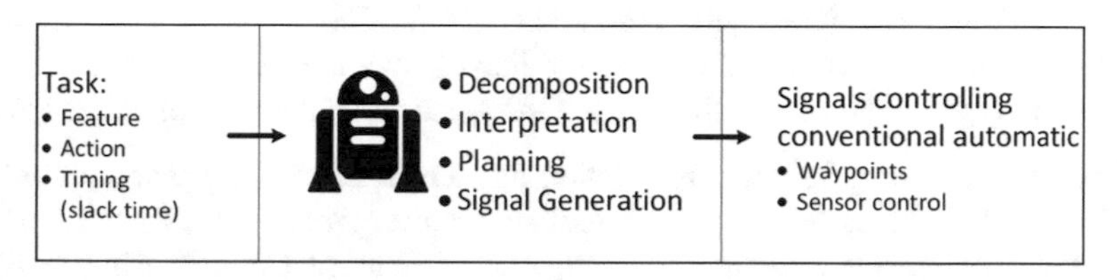

Fig. 2. A cognitive agent dismantling tasks and transforming them into signals.

First, the agent decomposes the tasks into its sub-elements. It selects missing parametrizations based on his situational understanding and tactical interpretation. Nevertheless, the pilot must be able to intervene and adapt the chosen parametrization according to his intention. Optimization and scheduling of the constituents can be realized by a planning module. Execution is performed by transmitting signals to the conventional automatic on board of the UCAVs.

In literature different approaches have been made to accomplish the agent's design challenge. One approach is using (inference based) rule engines [8]. Some use knowledge-based approaches based on ontologies [2]. We want to study the use of self-learning agents based on neural networks. With improvements in reinforcement learning algorithms like the A3C [9] training time of such systems is drastically decreased. However, regardless of the chosen architecture, the inputs of the agent must clearly be defined.

3 Elementary Task Formulation

A MUM-T system in the fighter jet domain has to deal with challenges of various nature. Thus, a general taxonomy needs to be developed to represent different mission objectives. Basically, in military aviation a task can be decomposed to a spatial region (Sect. 3.1), an action (Sect. 3.2) combined with a qualifying information (Sect. 3.3), in our case always a temporal information [10]. In our approach, we use a three steps process to define a task. First, spatial information is expressed through geometrical features. Second, an action is connected to a feature including specifying parameters for execution. Third, the temporal information is derived by a dialog between the manned and the unmanned vehicle. An example would be the surveillance of an airport for 30 min starting at 09:30.

3.1 Features

Every geographic object can in principle be represented by either a point, a line, or an area. This data representation is very common and i.e. used in Geographic Information System (GIS) [11]. Coppin et al. [12] used it to control a swarm of UAVs. We use this geometric reduction approach not only to character physical but also logical entities. Thus, every object and tactical structure in the agent's world view is a feature.

Features

– Points like a building, missile launcher, parking aircraft or even a navigation point
– Moving Points like fighter jets, cars, tankers, AWACS
– Lines like streets, rivers, routes, forward line of own troops
– Areas (Polygon) like marshalling area, CAPs, Airports, radar reconnaissance range

Context-sensitivity: With the application domain in mind we provide a context to each feature. This helps to further differentiate features from each other. The context information implemented so far are Navigation, Offensive Counter Air (OCA), Air Interdiction (AI) and Suppression of Enemy Air Defense (SEAD) and maybe be extended with Electronic Warfare (EW). This structural element gets even more important for complex air missions.

3.2 Actions with Parameters

The pilot's intention is reflected in the chosen action. Since in such a complex human-machine system a multitude of tasks occur, the context of the feature helps to preselect the most important actions. With this reduction of the action space a fast allocation is to be made possible. A military action serves either to gain important information about the target area (*intelligence*) or to achieve an impact (*effect*). Depending on the action, a qualifying action-specific parameter needs to be added. Each of these parameters can automatically be derived by the cognitive agent, based on his situational understanding and the preprogrammed default behavior. Nevertheless, if the pilot has a certain tactic in mind for performing the action based on his experience, he can simply pass it to the agent as a constraint via the qualifying parameters. Let us consider the pilot's objective to find a convoy in enemy territory. With his assumption that the convoy is primarily on roads, he would extract roads in this area as a line feature from the tactical map. He chooses the context as air interdiction. The action to be performed is *SCAN* with the implication that all recognized elements will be reported. Parameters to be influenced are the direction of the reconnaissance, the flight altitude, the desired sensor, how many images are to be taken per minute and so on. If the pilot has only one unmanned subordinate, he is able do this manually. With a higher span of control and a higher task load, the pilot is no longer capable of doing so. Then, he must resort to automation for parametrization.

3.3 Time Scheduling

In military planning, time is the most valuable dimension. Thus, airborne mission success relies on accurate planning and precise execution. Due to the high dynamics of the environment, the mission plan must be continuously adapted to the change. Tasks can be dropped, or previously important tasks can suddenly lose their priority due to other time-sensitive goals. Concerning time, there are two different types of tasks in air operations. Time-critical tasks with the need for exact timing and not time-critical tasks which schedule for execution is flexible within certain boundaries. To be able to differentiate between these tasks, we introduce the *slack time*. In the military term it is comparable to the window of opportunity. Within this slack time, the pilot requests the agent to perform the task. The actual duration of the task is an action parameter and is derived with the action definition. The start time of the task must lie within the slack time, but the agent can freely choose the exact time within these limits. Thus, it results that the time component in our approach consists of the slack time, the estimated start time and the duration of the task. Figure 3 shows the different types of tasks.

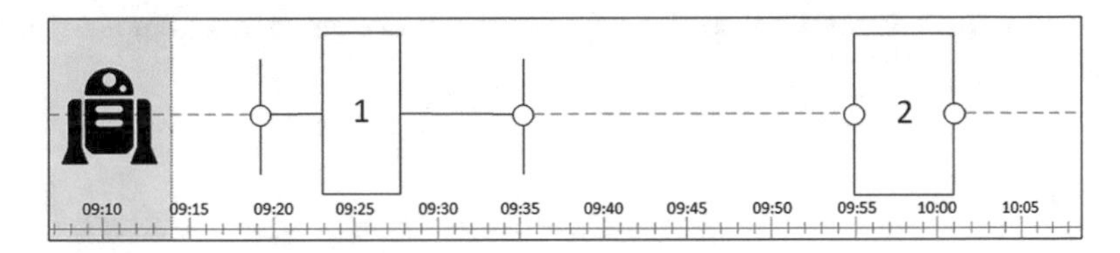

Fig. 3. UCAV time plan with (1) task with slack time and (2) strict task.

4 Task Assignment Interface

With the task formulation in mind we want to have a closer look a realization of the tasking interaction. We know that the effectiveness of framework relies in the ability to use the strength of both, the human and cognitive agent. Thus, automation functions and the pilot need to be in one bidirectional process. We implemented a task assignment process in our fighter jet simulator as shown in Fig. 4.

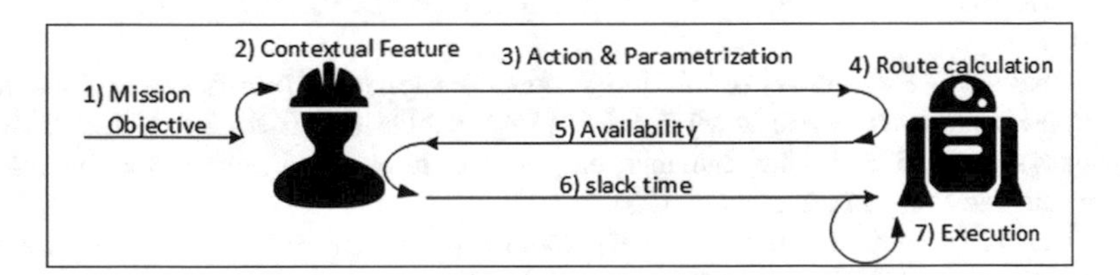

Fig. 4. The sequence of the task assignment shows the bidirectional interaction between the human and the cognitive agent.

The pilot formulates a task, based on his interpretation of the mission objective. The contextual feature with the appropriate action is selected. Depending on the pilot's intent and the task load can specify parameters. This feature-action pair is sent to the cognitive agent. What automation do best are exact calculations, and so the agent delivers accurate statements about his availability. The pilot schedule the task and can optionally add slack time at his discretion.

5 Conclusion

The proposed task formulation helps to transfer the MUM-T approach to the fighter jet domain. With a task standardization development and comparison of cognitive agents is facilitated. Different architectures approaches can be investigated using a generalized input. An interesting aspect is certainly the cooperation between cognitive agents – making the optimization space to a multi-agent problem. With the demand to implement cognitive (rule-based, inference-based, self-learning) algorithms on real systems, the use of a simulator as an intermediate stage is essential to save time, material and financial resources. The Institute of Flight System is building up a generic flight simulator to analyze the requirements for a MUM-T system. An evaluation of the system components will be performed in an experimental campaign with fighter pilots. Operational requirements will be the key factor for designing the future combat air system.

References

1. Clauß, S., Schulte, A. (eds.): Task Delegation in an Agent Supervisory Control Relationship Capability Awareness in a Cognitive Agent. IEEE (2014)
2. Uhrmann, J., Schulte, A.: Concept, design and evaluation of cognitive task-based UAV guidance. J. Adv. Intell. Syst. **5**(1) (2012)
3. Onken, R., Schulte, A.: System-Ergonomic Design of Cognitive Automation: Dual-Mode Cognitive Design of Vehicle Guidance and Control Work Systems. Springer-Verlag, Berlin, Heidelberg (2010)
4. Schulte, A., Donath, D., Lange, D.S. (eds.): Design Patterns for Human-Cognitive Agent Teaming. Springer (2016)
5. Gangl, S., Lettl, B., Schulte, A. (eds.): Management of Multiple Unmanned Combat Aerial Vehicles from a Single-Seat Fighter Cockpit in Manned-Unmanned Fighter Missions. AIAA (2013)
6. Heilemann, F., Schulte, A. (eds.): Interaction Concept for Mixed-Initiative Planning on Multiple Delegation Levels in Multi-UCAV Fighter Missions. IHSI, San Diego (2019)
7. Wagner, M. (ed.): Taking humans out of the loop: Implications for international humanitarian law. J. Law Inf. Sci. (2011)
8. Rudnick, G., Schulte, A. (eds.): Scalable Autonomy Concept for Reconnaissance UAVs on the Basis of an HTN Agent Architecture. IEEE (2016)
9. Mnih, V., et al. (eds.): Asynchronous methods for deep reinforcement learning. In: International Conference on Machine Learning (2016)
10. Myers, L.B., Tijerina, L., Geddie, J.C.: Proposed Military Standard for Task Analysis. Battelle Columbus Div OH (1987)
11. Maliene, V., Grigonis, V., Palevičius, V., Griffiths, S.: Geographic information system: old principles with new capabilities. Urban Des. Int. **16**(1), 1–6 (2011)
12. Coppin, G., Legras, F. (eds.): Controlling Swarms of Unmanned Vehicles through User-Centered Commands. AAAI, Toronto (2012)

Task Boundary Inference via Topic Modeling to Predict Interruption Timings for Human-Machine Teaming

Nia S. Peters[1(✉)], George C. Bradley[2], and Tina Marshall-Bradley[2]

[1] Air Force Research Laboratory, 711th Human Performance Wing, Battlespace Acoustic Branch, Wright-Patterson AFB, OH 45433, USA
nia.peters.1@us.af.mil

[2] Walden University, 100 Washington Ave S, Unit 900, Minneapolis, MN 55401, USA
{george.bradley,tina.marshall-bradley}@mail.waldenu.edu

Abstract. Human-machine teaming aims to meld human cognitive strengths with the unique capabilities of smart machines. An issue within human-machine teaming is a lack of communication skills on the part of the machine such as the inability to know when to *interrupt* human teammates. A proposed solution to this issue is an *intelligent interruption system* that monitors the spoken communication of human teammates and predicts appropriate times to interrupt without disrupting the teaming interaction. The current research expands on a prosody-only task boundary model as an *intelligent interruption system* with a topic-only task boundary model. The topic-only task boundary model outperforms the prosody-only model with a 9.5% increase in the F1 score, but is limited in its ability to process topical data in real-time, a previous benefit of the prosody-only task boundary model.

Keywords: Human machine teaming · Intelligent interruption systems
Topic modeling · Collaborative communication

1 Introduction

Human-machine teaming will enable humans and machines to communicate and share information. The ability of machine teammates to know when to communicate information to human teammates within human-machine teaming has attracted the attention of the U.S. military and commercial industries to support human-robot teams, unmanned aerial vehicle (UAV) operations, systems for spotting anomalies and preventing cyber-attacks, control stations for power and chemical processing plants, air traffic control stations, commercial and military pilots in cockpits, and human-computer technical support teams.

If the objective is for machines to initiate actions on their own, they should know when to automatically share information with or *interrupt* human teammates. An *interruption* is defined as an unanticipated requests for task switching from a person, an object, or an event [1]. A proposed automatic intrusion technique is the development of

W. Karwowski and T. Ahram (Eds.): IHSI 2019, AISC 903, pp. 783–788, 2019.
https://doi.org/10.1007/978-3-030-11051-2_119

an *intelligent interruption system*. Such a system leverages information from the dialogue of human-machine interactions and applies hand-crafted or machine learning techniques to disseminate information at appropriate times with minimal cost of disrupting the overall teaming interaction.

A topic-only task boundary model is proposed as an *intelligent interruption system.* This model leverages the lexical content from the human utterances within a human-machine teaming dialogue, converts the dialogue words into topical features that illustrate the probability a user is speaking about a topic, and classifies the utterance as a task boundary if it precedes the end of a task. The rationale behind the topic-only task boundary model is: (1) task boundaries have been shown to be good indicators of appropriate interruption timings within the literature (2) topic modeling via lexical content may assist in discriminating *end of task* words such as "got it," "copy that," "done," etc. from words associated with the ongoing task.

The proposed topic-only task boundary model outperforms the prosody-only task boundary model [2] with a 9.5% increase in F1 score for human-machine dialogues with more robust dialogues, but is limited in its ability to process the lexical data in real-time which was a previous benefit of the prosody-only task boundary model. The goal of future work is to leverage the real-time processing power of the prosody-only model and the accuracy of the topic-only model to build an effective and efficient *intelligent interruption system* for human-machine teaming interactions.

2 Background

There is literature that recommends appropriate points of interruptibility as boundaries within task execution. A *task boundary* can be defined as a time instance between two moments of task execution. Task boundary modeling has been used to indicate appropriate points of interruptibility [3–6] and shown that deferring delivery of notifications until a boundary is reached can meaningfully reduce costs of interruptions. Conversely, interrupting tasks at random moments can cause users to take up to 30% longer to resume the tasks, commit up to twice the errors, and experience up to twice the negative affect than when interrupted at boundaries [3, 4, 7].

The result of a previously proposed prosody-only task boundary model resulted in a low latency processing system that was integrated into a real-world human-machine teaming interaction and accurately detected task boundaries within a simple human-machine teaming dialogue, but failed to produce accurate task boundary detection within a robust human-machine dialogue [2]. The objective of the current research is to replace the prosody-only task boundary model with a topic-only task boundary model to increase the task boundary detection performance in more complicated and robust human-machine teaming dialogues.

3 Data and Method

Prior to developing an *intelligent interruption system*, it is necessary to simulate a human-machine teaming interaction which in this research is a dual- human, dual-task team. In the simulated dual-human, dual- task interaction, two humans communicate using a push-to- talk communication interface regarding information related to a human-human task. The *intelligent interruption system* "listens" to the dyadic human dialogue, processes dialogue at a push-to-talk level, and makes inferences on when to send information related to an orthogonal human-machine task or the *interruption task*. The entire data collection is described in [2].

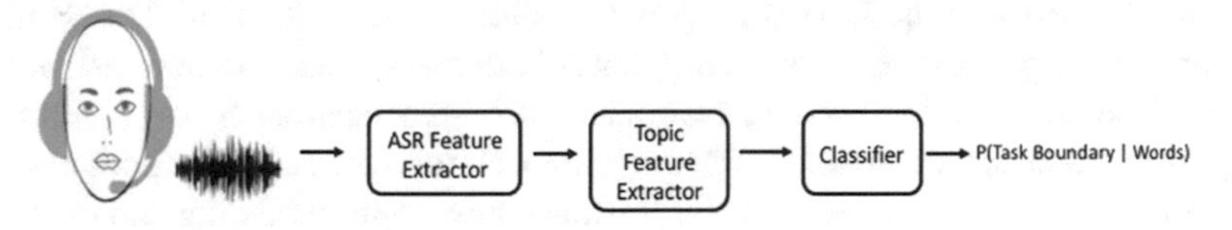

Fig. 1. Topic-only task boundary model

To construct a topic-only task boundary model, the raw audio of the push-to-talk utterances from the human-human task are sent to an automatic speech recognizer (Google Speech API) to convert speech to text. From this text, topic features are extracted using topic modeling via the Latent Dirchlet Allocation (LDA) [8] algorithm. These features are then used by a binary classifier to discriminate utterances that precede a task boundary from those that do not. The entire process is illustrated in Fig. 1. An utterance is labeled as belonging to the task boundary class or non-task boundary class based on whether it precedes a task boundary indicated by the team participating in the human-human task pressing the DONE button. The class distribution N(nonTB,TB) is N(7478,4838) for the Tangram task [2] and N(7089,1569) for the Uncertainty Map task [2], both human-human tasks. The *intelligent interruption system* via the topic-only task boundary model makes interruption decisions based on the detection of a task boundary and sends information related to the Keeping Track task [2], the human-machine task.

The topic-only model is based on the Maximum Entropy (MAXENT) modeling techniques used in [9] which uses maximum entropy lexical classifiers with expectation constraints to specify affinities between words and labels via the software Mallet [10]. The difference between the topic-only and prosody-only [2] task boundary models is not only the input features (prosody vs. lexical/topic), but also the binary classifier, Random Forest [11] vs. MAXENT [9] for prosody and topic respectively. Several binary classification models were evaluated and the presented models produced the best results for their respective feature input and modeling technique.

4 Results and Discussion

The first evaluation of interest are the keywords associated with each output topic from LDA. It was assumed that words associated with confirmation cues (indicative of a task boundary) could be discriminated from other words within the ongoing tasks.

NON-Task Boundary: the, got, and, then, have, triangle, tree, okay, bird, boat, with, sailboat, left, down, upside, palm, bottom, flying, are, corner
Task Boundary: the, and, like, one, then, got, have, with, down, that, guy, first, dog, you, looks, last, triangle, right, person, it's

Fig. 2. Tangram task topic modeling non-task boundary vs. task boundary keywords

Figure 2 illustrates the LDA topic keywords for the Tangram task. The word "got" is highlighted in red because it is a word that is both associated with non-task boundary and task boundary topics because teammates use sentences such as "I *got* a swan, turkey, turtle, and upside down duck" (ongoing task utterance). Teammates also use "I *got* it" (task boundary utterance). This illustrates how topic modeling can capture two different uses of the same word. Since utterances associated with the ongoing task have more topical words such as the shapes the teammates are describing, the topic model does a good job at discriminating task boundaries from non-task boundaries based on topical features via lexical content. As illustrated in Fig. 4, the F1 score for task boundary detection for the Tangram task is 85.9%.

Figure 3 illustrates the LDA outputs topic keywords for the Uncertainty Map task.

NON-Task Boundary: the, house, like, and, that, front, roof, sidewalk, it's, has, one, there's, side, driveway, kind, out, corner, with, there, this
Task Boundary: you, got, have, see, like, that, okay, think, yeah, i'm, one, don't, but, know, it's, what, pictures, three, just, this

Fig. 3. Uncertainty map task topic modeling non-task boundary vs. task boundary keywords

From the keywords, the non-task boundary words seem to be associated with users describing the interface or engaging in the ongoing task and the words in the task boundary class seem to be associated with finalizing the task such as "got it" or "I think I got it" or "okay" as confirmation cues which are highlighted in blue. The task boundary detection performance for the Uncertainty Map task is a F1 score of 71.1%. The topic-only task boundary model F1 score exceeds random performance (F1 score of 50%) by 71.8% for the Tangram task and 40% for the Uncertainty Map task.

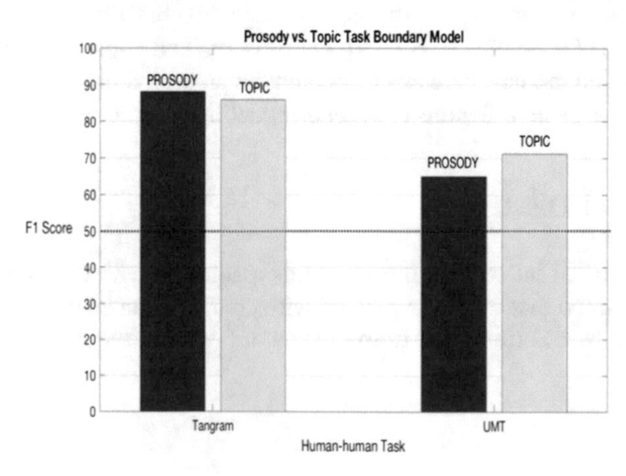

Fig. 4. Performance comparison of the prosody-only and topic-only models for the tangram and uncertainty map tasks

In Fig. 4 the performance of the prosody-only [2] and topic-only task boundary models as proposed as *intelligent interruption systems* are compared. Both the prosody-only and topic-only models perform similarity in inferring task boundaries for the Tangram task. The Tangram task has a very predictable dialogue where teammates alternate between single utterances related to the ongoing task of describing the Tangram shapes and a single utterance reply from the teammate associated with a confirmation cue before moving on to the next Tangram task. Such a predictable dialogue would result in more accurate task boundary detection accuracies since short utterances could be associated with task boundaries and long utterances associated with non-task boundaries. Discriminative prosodic features such as duration or energy could separate these two classes of utterances.

For the Uncertainty Map task which has a much more complex and robust dialogue, the topic-only model outperform the prosody-only model with 9.5% increase in the F1 score as illustrated in Fig. 4. If prosody aids in the discrimination of short and long utterances as in the Tangram task, for the Uncertainty Map task these features may not be as useful because short utterances in this task may not only be confirmation cues as they are in the Tangram task, but also fillers such as "ok," "um-hum" and other shorter sentences inherent to longer and more robust dialogues. The topic model does capture the ongoing task topics and discriminates those words from confirmation cues associated with the end of a task. A drawback of the topic-only model is its intractability in real-time because of the long latency in processing the lexical content prior to putting this information through the model.

5 Conclusion and Future Work

In comparing the topic-only task boundary model as a proposal to an *intelligent interruption system* to the prosody-only task boundary model [2], the topic-only model does a better job at predicting task boundaries for more robust dialogues. The topic modeling task boundary detection performance is like the prosody-only model for the Tangram task but showed improvement in detecting task boundaries for the Uncertainty Map task which has a much more robust dialogue compared to the Tangram task. This is a promising result that indicates a prosody-only model could either be augmented with a topic modeling model or a voting scheme between models could provide some improved performance in the overall output. One drawback of the proposed topic modeling strategy is the reality that these models use lexical information and to extract this information in a real-time system it is necessary to pass the dialogue utterances through an automatic speech recognition system. This introduces latency into the overall system which could slow system performance down and hinder the overall performance of making interruption decisions at optimal times especially since the latency from the topic model results in interruptions that are far from the intended point of interruption. The goal of future work is to leverage the real-time processing power of the prosody-only model and the accuracy of the topic-only model to build an effective and efficient *intelligent interruption system* for human-machine teaming interactions.

References

1. Arroyo, E., Selker, T.: Attention and Intention Goals Can Mediate Disruption in Human-Computer Interaction, pp. 454–470 (2011)
2. Peters, N.: Interruption timing prediction via prosodic task boundary model for human-machine teaming. In: Proceedings of the 2019 Future Information and Communication Conference (2019)
3. Adamczyk, P.D., Bailey, B.P.: If Not Now, When?: The Effects of Interruption at Different Moments Within Task Execution, vol. 6, no. 1, pp. 271–278 (2004)
4. Bailey, B.P., Konstan, J.A.: On the need for attention-aware systems: measuring effects of interruption on task performance, error rate, and affective state. Comput. Hum. Behav. **22**(4), 685–708 (2006)
5. Czerwinski, M., Cutrell, E., Horvitz, E.: Instant messaging and interruption: influence of task type on performance. In: OZCHI Conference Proceedings, SRC, vol. 356, pp. 361–367 (2000)
6. Iqbal, S.T., Bailey, B.P.: Leveraging characteristics of task structure to predict the cost of interruption. In: Proceedings of the SIGCHI Conference on Human Factors Computer Systems - CHI '06, p. 741 (2006)
7. Iqbal, S.T., Bailey, B.P.: Investigating the Effectiveness of Mental Workload as a Predictor of Opportune Moments for Interruption
8. Blei, D., Ng, A., Jordan, M., Bohus, D., Horvitz, E.: Latent dirichlet allocation. J. Mach. Learn. Res. pp nd Learn. to Predict Engagem. with a Spok. Dialog Syst. openworld settings Proc. SIGDIAL, pp. 993–1022 SRC-GoogleScholar FG-0 (2009)
9. Druck, G., Mann, G., McCallum, A.: Learning from labeled features using generalized expectation criteria. In: Proceedings of the 31st Annual International ACM SIGIR Conference on Research and Development in Information Retrieval - SIGIR '08, no. 1, p. 595 (2008)
10. McCallum, A.K.: MALLET: Learning for Language Toolkit (2002)
11. Banerjee, S.: Random Forest Classifier (2016)

Cardiovascular Parameters for Mental Workload Detection of Air Traffic Controllers

Thea Radüntz[1(✉)], Thorsten Mühlhausen[2], Norbert Fürstenau[2], Emilia Cheladze[1], and Beate Meffert[3]

[1] Federal Institute for Occupational Safety and Health, Unit "Mental Health and Cognitive Capacity", Nöldnerstr. 40/42, 10317 Berlin, Germany
{raduentz.thea, cheladze.emilia}@baua.bund.de

[2] German Aerospace Center, Institute of Flight Guidance, Lilienthalplatz 7, 38108 Brunswick, Germany
{thorsten.muehlhausen, norbert.fuerstenau}@dlr.de

[3] Department of Computer Science, Humboldt-Universität zu Berlin, Rudower Chaussee 25, 12489 Berlin, Germany
meffert@informatik.hu-berlin.de

Abstract. In our study, we focused on air traffic controller's working position for arrival management. Our aim was to evaluate cardiovascular parameters regarding their ability to distinguish between conditions with different traffic volumes and between conditions with and without the occurrence of an extraordinary event. Our sample consisted of 21 subjects. During an interactive simulation, we varied the load situations with two independent variables: the traffic volume and the occurrence of a priority flight request. Dependent variables for registering mental workload were cardiovascular parameters, i.e., the heart rate, relative low-frequency and high-frequency band powers, and band-power ratio of the low- and high-frequency bands. Heart rate was the only parameter able to differentiate significantly between simulations with minimal and high air-traffic volume, while the effect of the priority-flight request remained doubtful. No significant interaction between traffic volume and priority request could be identified for any of the cardiovascular parameters.

Keywords: Mental workload · Heart rate · Heart rate variability
Signal processing · Air traffic controllers

1 Introduction

In our digitalized society, inappropriate mental workload has a number of negative consequences on employee's health and the safety of persons. This is especially true in occupations with safety-critical tasks such as air traffic control. Air traffic controllers have to keep engaged and try to maintain their performance even under difficult situations. They experience high cognitive demands and responsibility, both leading to increased mental workload.

W. Karwowski and T. Ahram (Eds.): IHSI 2019, AISC 903, pp. 789–794, 2019.
https://doi.org/10.1007/978-3-030-11051-2_120

In order to register mental workload for understanding and enhancing work conditions, researchers have been studying different measuring methods since decades. In this context, cardiovascular parameters were used as an objective and continuous measure of mental workload. The most prominent cardiovascular parameters were the heart rate that increased during load as well as parameters from the frequency domain, i.e. power values of different frequency-band ranges [1]. Although definitions of the frequency ranges varied across studies [2, 3], most researchers considered the band around the 0.1 Hz as workload relevant [4]. This frequency band reflects blood-pressure regulation mechanisms. Researchers evaluated also the frequency band above 0.1 Hz [5]. This upper frequency band was connected to respiratory activity [3, 6]. In general, power in both frequency bands decreased with increasing load conditions [3, 6, 7] although there were also studies with contradictory results [8, 9]. Furthermore, the ratio between both frequency-band powers was an additional workload-relevant parameter that in general decreased with increasing load [10]. However, it is still unclear which cardiovascular parameter is best suited for assessing mental workload in the field.

Thus, the aim of our study was to evaluate cardiovascular parameters regarding their ability to distinguish between conditions with different load levels. For this, we focused on air traffic controller's working position for arrival management and conducted a study in a simulator. According to [11] variations in mental workload of air traffic controllers were mainly induced by the traffic volume but might also arise by unexpected events. To this end, we formulated the following three research questions:

1. Which cardiovascular parameter is best suited for assessing workload differences that arise from different traffic-volume conditions?
2. Which cardiovascular parameter is best suited for assessing workload differences that arise from extraordinary events?
3. Which cardiovascular parameter is able to capture interaction effects between traffic volume and extraordinary events affecting mental workload?

2 Method

2.1 Procedure and Subjects

During an interactive real-time simulation at the Air Traffic Management and Operations simulator (ATMOS) of the German Aerospace Center (DLR) in Braunschweig, we varied the load situations and registered the pulse signal of the air traffic controllers. Load variation was perform by means of two factors: the traffic volume and the occurrence of a priority-flight request. Traffic volume comprised four levels related to the number of aircraft per hour (ac/h) in the arrival sector, i.e., 25 ac/h, 35 ac/h, 45 ac/h, and 55 ac/h. The second factor was by nature dichotomous: occurrence vs. absence of a priority-flight request because of a sick passenger on board. The combination of both led to eight simulation scenarios with a duration of 20 min or 25 min. The potential priority-flight request occurred after the 10th min.

21 subjects (2 female, 19 male, 22 to 64 years, mean age 38 ± 11) participated in a two-day experiment where they had to complete the above-mentioned eight traffic scenarios in randomized order. Four of the simulation scenarios were presented on the first day, the remaining four were conducted on the second day.

The Federal Institute for Occupational Safety and Health (BAuA) in Berlin was in charge of the project. All of the investigations acquired were approved by the local review board of the BAuA and the experiments were conducted in accordance with the Declaration of Helsinki. All procedures were carried out with the adequate understanding and written consent of the subjects.

2.2 Measurements and Statistical Analysis

For scenarios with a priority-flight request, we considered the pulse signal segment of 3 min starting from the request time point. This was around the 10th min of the simulation. For scenarios without priority-flight request we considered the pulse signal segment from minute 10 to 13.

The pulse signal was windowed with a Hamming function and filtered with a bandpass filter (order 100) between 0.5 and 3.5 Hz. Next, peak detection was performed in order to gain the heart rate and the inter-beat intervals. Artifacts were automatically detected by means of statistical analysis, corrected using linear interpolation of the values at neighboring points, and equidistantly resampled with a time resolution of 0.5 s. Heart rate (HR) was determined in beats per minute in the time domain. For evaluation of heart-rate variability parameters in the frequency domain, we calculated the fast Fourier transform of the inter-beat interval signal that had been previously windowed with a Hann window. Workload-relevant frequency-band powers were computed related to [1] for the low-frequency (LF: 0.04-0.15 Hz) and high-frequency (HF: 0.15-0.4 Hz) bands. In order to calculate the relative values of the power components in proportion to the total power, we divided both absolute power values by the total power of the signal (frequency range between 0.02 and 0.4 Hz). Furthermore, we computed the ratio of the absolute low-frequency and high-frequency band powers (LF/HF). For achieving a normal distribution for the further analysis, we computed the logarithms of the above-mentioned three parameters from the frequency domain.

For answering our research question regarding the most appropriate cardiovascular parameter for assessing air traffic controllers' workload, the effect of the traffic volume, occurrence of an exceptional event such as a priority-flight request, and interaction effect between both, we carried out four analysis of variance (ANOVA). The dependent variable of each was either the HR, logarithm of the relative LF and HF band powers, or logarithm of the LF/HF ratio. For each ANOVA we utilized a repeated-measures design with two within-subject factors (two levels for the priority-flight request factor and four levels for the traffic-volume factor). General differences between the levels were examined and tested with a post hoc test (Bonferroni corrected). For testing the differences between priority-flight and no priority-flight event on each traffic-volume level, we used four t-tests for each parameter and adjusted the values accordingly.

3 Results

Results of the four ANOVAs are summarized in Table 1 and shown in Fig. 1.

The effect of the number of aircraft led to a highly significant increase of controllers' HR. Bonferroni corrected post hoc tests showed significant differences between the simulation scenarios of 25 ac/h and 45 ac/h as well as 55 ac/h. A descriptive evaluation showed that during priority-flight request scenarios the HR means increased gradually with increasing traffic volume.

Table 1. Analysis of cardiovascular parameters across simulation conditions.

	Cardiovascular parameters	F	p	η^2
Traffic volume	log (relLF)	1.362	0.217	0.064
	log (relHF)	2.623	0.059	0.116
	log (LF/HF)	2.565	0.063	0.114
	HR	6.352	**0.001**	0.241
Priority-flight request	log (relLF)	0.642	0.432	0.031
	log (relHF)	0.201	0.659	0.01
	log (LF/HF)	0.305	0.587	0.015
	HR	6.505	**0.019**	0.245
Traffic volume and priority-flight request	log (relLF)	0.832	0.832	0.04
	log (relHF)	0.884	0.454	0.042
	log (LF/HF)	0.894	0.45	0.043
	HR	1.345	0.268	0.063

Note. Values of .001 are actually $p \leq 0.001$.

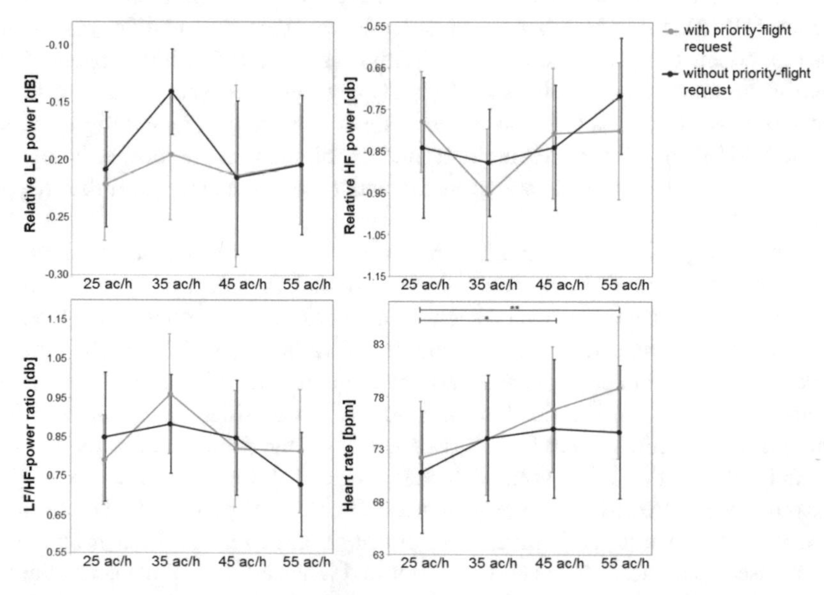

Fig. 1. Means of cardiovascular-parameter values over 21 participants: logarithm of the relative LF (top left) and HF (top right) band powers, logarithm of the LF/HF ratio (bottom left), and HR (bottom right) (Bonferroni corrected post hoc tests: ***: $p \leq 0.001$; **: $0.001 < p \leq 0.01$; *: $0.01 < p$ 0.05; error bars indicate 95% confidence interval).

Although the LF/HF ratio and to some extend also the relative LF band power revealed the expected decrease with increasing load, none of them reached the significance level. The HF band power increased with increasing traffic volume and did not show a significant main effect.

The impact of the priority-flight request became significant for controllers' HR. However, Bonferroni-adjusted t-tests for each traffic-volume level between scenarios with and without priority-flight request were not able to identified a significant difference for none of the levels (at 25 ac/h: $t(20) = -1.180$, $p = 1.008$; at 35 ac/h: $t(20) = 0.055$, $p = 3.827$; at 45 ac/h: $t(20) = -1.200, p = 0.976$; at 55 ac/h : $t(20) = -2.031, p = 0.223$).

None of the parameters from the frequency domain showed a significant main effect for the priority-flight request factor. However, the relative LF band power showed the tendency to decrease between scenarios with and without priority-flight request during the low-traffic scenarios of 25 and 35 ac/h.

No interaction effect could be obtained between the two factors with none of the cardiovascular parameters used.

4 Discussion and Conclusions

HR was the only parameter able to capture significant workload differences connected to traffic volume while the impact of the priority-flight request remained doubtful. A significant interaction effect between both factors was not given. With respect to the traffic volume, significant differences could be identified only between the minimal and the two highest air-traffic volume conditions. We conclude that although HR seems to be a reliable cardiovascular indicator for air traffic controllers' mental workload, it lacks sensitivity regarding neighboring workload levels.

Graphical evaluation of the parameters from the frequency domain revealed a potential for the LF/HF ratio and relative LF band to meet the expected decrease with increasing load. However, none of the cardiovascular parameters from the frequency domain were able to capture significant differences regarding neither the traffic-volume nor the priority-flight request factor. One reason for this could be the small number of participants. A further reason could be the definition of the frequency-band ranges. In our work, we defined the workload-relevant frequency bands in accordance with the frequency-band ranges of [2]. Slight modification of these ranges as proposed by e.g., [6] or [3] could enhance the results. Furthermore, we concentrated on the time slot immediately after the priority-flight request. Evaluation of later time slots could offer a deeper insight in the dynamical changes of workload and exhibit new findings regarding the cardiovascular parameters from the frequency domain.

Finally, the examination of between-subject factors (e.g., age, current job demands) could reveal additional information about controller's mental workload registered by the cardiovascular parameters. However, we currently conclude that HR is the most promising indicator for mental workload registration among the cardiovascular parameters.

Acknowledgements. We would like to thank Kerstin Ruta for her daily operational support, Lea Rabe for conducting the experiments, the numerous pseudo pilots for their contribution during the experiments, and André Tews for his conceptual, technical, and overall support. More information about the project that acquired our data can be found at http://www.baua.de/DE/Aufgaben/Forschung/Forschungsprojekte/f2402.html.

Author Contributions. T.R. initiated the project and was responsible for the overall conception of the investigation. T.R., T.M., and N.F. developed the research design of the study. T.M. was responsible for the implementation of the simulation scenarios and the overall technical support. E.C. conducted the experiments, acquired the data, and provided support for the data analysis with SPSS and graphic editing. The study was supervised by T.R. Data interpretation was performed by T.R. and B.M. The manuscript was written by T.R. Final critical editing was performed by T.M., N.F., and B.M.

References

1. Mulder, L.J.M., De Waard, D., Brookhuis, K.A.: Estimating mental effort using heart rate and heart rate variability. In: Stanton, N., Hedge, A., Brookhuis, K., Salas, E., Hendrick, H. (eds.) Handbook of Human Factors and Ergonomics Methods, pp. 20-1–20-8. CRC Press (2004)
2. Malik, M., Bigger, J.T., Camm, A.J., Kleiger, R.E.; Malliani, A., Moss, A.J., Schwartz, P.J.: Heart rate variability. Standards of measurement, physiological interpretation, and clinical use. Circulation **17**, 354–381 (1996)
3. Jorna, P.G.A.M.: Spectral analysis of heart rate and psychological state: a review of its validity as a workload index. Biol. Psychol. **34**(2), 237–257 (1992)
4. Nickel, P., Nachreiner, F.: Sensitivity and diagnosticity of the 0.1-Hz component of heart rate variability as an indicator of mental workload. Hum. Factors **45**(4), 575–590 (2003)
5. Shaffer, F., Ginsberg, J.P.: An overview of heart rate variability metrics and norms. Front. Public Health **5**, 258 (2017)
6. Mulder, L.J.M.: Measurement and analysis methods of heart rate and respiration for use in applied environments. Biol. Psychol. **34**(2), 205–236 (1992)
7. Veltman, J.A., Gaillard, A.W.K.: Pilot workload evaluated with subjective and physiological measures. In: Brookhuis, K.A., Weikert, C., Moraal, J., De Waard, D. (eds.) Aging and Human Factors. Traffic Research Centre, University of Groningen (1993)
8. Cinaz, B., Arnrich, B., La Marca, R., Tröster, G.: Monitoring of mental workload levels during an everyday life office-work scenario. Pers. Ubiquit. Comput. **17**(2), 229–239 (2013)
9. Nagasawa, T., Hagiwara, H.: Workload induces changes in hemodynamics, respiratory rate and heart rate variability. In: 2016 IEEE 16th International Conference on Bioinformatics and Bioengineering (BIBE), Taichung, pp. 176–181 (2016)
10. Wilson, G.F.: An analysis of mental workload in pilots during flight using multiple psychophysiological measures. Int. J. Aviat. Psychol. **12**, 3–18 (2002)
11. Averty, P., Collet, C., Dittmar, A., Athènes, S., Vernet-Maury, E.: Mental workload in air traffic control: an index constructed from field tests. Aviat. Space Environ. Med. **75**(4), 333–341 (2004)

"I'm Your Personal Co-Driver—How Can I Assist You?" Assessing the Potential of Personal Assistants for Truck Drivers

Jana Fank(✉) and Markus Lienkamp

Institute of Automotive Technology, Technical University Munich,
Boltzmannstr. 15, 85748 Garching, Germany
{fank, lienkamp}@ftm.mw.tum.de

Abstract. User acceptance is widely recognized as a major factor for predicting the intention to use a technical device. As such, personal assistants for truck drivers pose significant design and function challenges. Long, solitary hours and comprehensive interaction with the vehicle open various possibilities for an assistant's service. Including drivers' requirements early in development is hence beneficial. This paper describes the result of an online survey intended to assess truck drivers' attitudes toward personal assistants in the truck cabin. Its authors investigate the potential and the predicted acceptance of personal conversation, virtual, or robotic agents as interaction partners. They furthermore analyze how hedonic and pragmatic attributes affect the intention to use these personal assistants.

Keywords: Personal assistant · Virtual agents · Social robots
Truck drivers · Online survey · Pragmatic attributes · Hedonic attributes

1 Introduction

Current research investigates the effect of personal assistants (PA) in the form of conversational, virtual, or robotic agents in cars [1–3]. Only a few sources investigated the parameters influencing drivers' acceptance of PAs. The literature indicates that user acceptance contributes significantly to a technical device's overall success. Although several approaches describe acceptance factors [4–6], the Technology Acceptance Model (TAM) is probably best known [6]. TAM names "usefulness" and "ease of use" as the main factors affecting the intention to use. Graaf et al. [7] tried to construct an acceptance model specially for social robots in domestic environments. They identify three theoretically based belief structures: attitudinal, normative, and control. The attitudinal belief structure consists of pragmatic and hedonic product aspects [7]. Pragmatic aspects include the already familiar attributes of acceptance research, such as "usefulness" or "ease of use." Hedonic attributes determine factors that affect peoples' intrinsic motivation, meaning voluntary utilization with no apparent reinforcement. This study's authors want to evaluate factors of user acceptance of PAs in voluntary use in truck cabins. The effects of hedonic attributes are therefore measured.

W. Karwowski and T. Ahram (Eds.): IHSI 2019, AISC 903, pp. 795–800, 2019.
https://doi.org/10.1007/978-3-030-11051-2_121

2 Research Question

This contribution aims at assessing the potential of PAs in the form of conversational, virtual, or robotic agents for truck drivers—and furthermore, to investigate the effects of hedonic and pragmatic attributes on PA acceptance. The first research question relates to possible PA functionalities. The second question was posed to investigate factors of PA acceptance for truck drivers.

RQ1: Which functionalities do truck drivers want from a PA?
RQ2: Do hedonic and pragmatic attributes affect truck drivers' intention to use PAs?

3 Methodology

3.1 Participants

A total of 80 subjects replied to the online questionnaire; 71 completed it. Participants took on average 10 min to complete the questionnaire. Thirteen subjects were excluded from the data because they are no longer involved in regular active driving scenarios. This left 58 completed questionnaires to be used in the data analysis. The participants' demographic characteristics in the final sample are displayed in Fig. 1 and compared with the general truck-driving population's demographics taken from [8]. 36.2% of the drivers work in long haul transport, 27.6% in regional transport, and 27.6% in local transport.[1] 8.6% operated in other transport sectors such as construction or driving school (Table 1).

Table 1. Participants' characteristics (n = 58) versus those of German truck drivers [8]

		Sample (in %)	Population (in %)
Gender	Male	93.1%	89.3%
	Female	6.9%	1.7%
Age	<25	0%	2.5%
	25–54	65.5%	69.7%
	55–64	32.8%	25.5%
	>65	1.7%	2.3%

3.2 Questionnaire Construction

A three-part online survey was designed for the analysis. The first part collected demographic data from the participants. The second part determined possible PA functions, which were defined based on a previous conducted interdisciplinary

[1] Local transport: up to 50 km/day; regional transport: 50 to 150 km/day; long haul transport: more than 150 km/day.

workshop. The third part investigated the hedonic and pragmatic attributes pertaining to the intention to use a PA. Constructs for hedonic and pragmatic attributes were selected according to Graaf et al. [7] with some modification to better address PA attributes. A brief definition of PA in the forms of conversational, virtual, and robotic agents was provided in written form to support the participants during the survey. Participants were informed that all data recorded for the study is anonymized.

3.3 Procedure

In August 2018, 153 truck drivers were invited via telephone to voluntarily participate in the online survey. The truck drivers were recruited using a database provided by our institute. Furthermore, a link to the questionnaire was posted on the official Facebook website of the institute's dynamic truck-driver simulator.

3.4 Item Analysis

To verify whether the test items meet their purpose, an item analysis was conducted. The item difficulty shows that the items for "ease of use" were easy to measure psychometrically (Table 2). The other constructs show medium item difficulty. The scientific merit of each item was valued using Cronbach-α for standardized items (Table 2). The reliability of the questionnaire was acceptable ($r > .70$) according to [9]. Only one item for the construct "ease of use" had to be excluded from further analysis, because of an insufficient value ($r = .408$: "I think I can use a PA when there is someone to help me"). The selectivity of every item was calculated to $r\_ix > .30$.

Table 2. Item analysis of the questionnaire

	Construct	Scale range	Item difficulty	α	r_ix
	Functionality (*10 items*)	0 to 4	1.64 to 3.03	.915	0.479 to 0.865
Pragmatic attributes	Usefulness (*2 items*)	0 to 6	3.59 to 3.66	.883	0.800
	Ease of use (*2 items*)	0 to 6	4.03 to 4.43	.762	0.616
	Adaptability (*3 items*)	0 to 6	3.14 to 3.91	.899	0.759 to 0.873
Hedonic attributes	Enjoyment (*3 items*)	0 to 6	3.26 to 3.62	.844	0.552 to 0.831
	Attractiveness (*3 items*)	0 to 6	2.47 to 3.60	.803	0.494 to 0.777
	Companionship (*4 items*)	0 to 6	1.88 to 3.29	.815	0.389 to 0.745
	Animacy (*5 items*)	0 to 4	1.25 to 2.21	.863	0.571 to 0.762
	Intention to use (*3 items*)	0 to 6	3.48 to 4.53	.942	0.823 to 0.906

4 Results

Descriptive statistics were calculated for each of the constructs and items on the questionnaire. The participants were asked which functionality they would like to have for a PA (5-Point Likert-scale range from "strongly disagree" to "strongly agree"). The participants primarily see the provision of information as a suitable task for a PA ("answer questions": 3.03 ± 1.139, "information about surrounding area": 2.84 ± 1.105, Fig. 1). However, they feel that PAs would be helpful to support ancillary activities such as learning a new language (2.69 ± 1.173) or doing sports exercises (2.38 ± 1.24). They also see a personal assistant as a way to connect with friends and family. The driver's ability to have "personal conversations" with a PA is considered to be the least desirable function (1.72 ± 1.412).

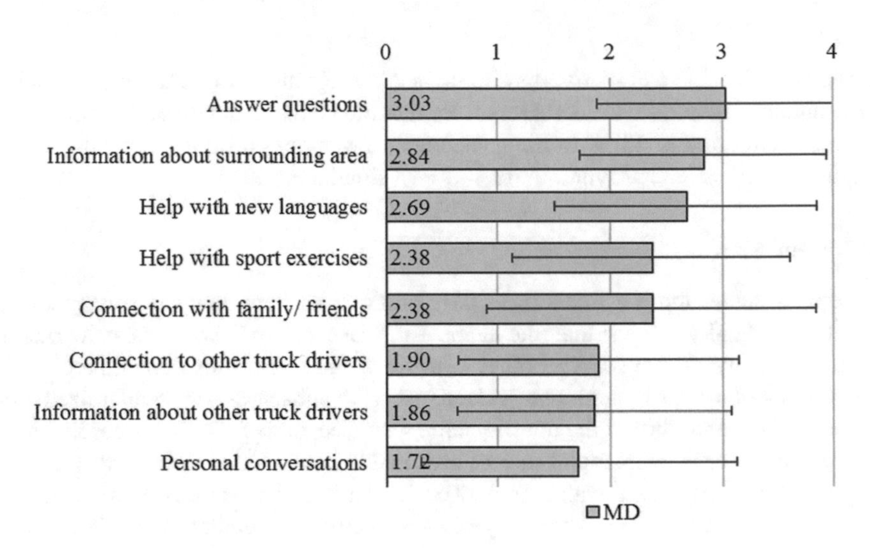

Fig. 1. Evaluation of possible PA functionalities (n = 58)

In the open questions, the participants mentioned "information about free parking spots" or the "goods acceptance times of the customer" from the PA as useful. Furthermore, the participants indicated that they would like to have a PA for "playing games" and assist in "planning pauses and driving times." They would also appreciate a PA that "recognizes if the driver is tired" and starts a conversation to "decrease fatigue." The assistant should also be able to observe surrounding traffic and "indicate traffic rule violations." The participants could imagine trying a PA in the form of a conversational, virtual, or robotic agent (4.53 ± 1.625); however, they are more reluctant to say that they will use it (3.48 ± 1.779) or recommend using it (3.67 ± 1.877). The participants indicated that they are "worried about talking to a machine" or raised "concerns about data misuse." Nevertheless, they would like to "adapt a PA to their wishes," e.g., gender, name, or form.

A multiple regression analysis was conducted to analyze possible correlations between hedonic and pragmatic attitudes affecting intention to use. The regression analyses indicated multicollinearity in the data, as the usefulness score achieved a tolerance value of $TOL < .20$ and the variance-inflations factor was $VIF > 5.00$ [10]. As Graaf et al. [7] stated, usefulness seems to correlate strongly with the other constructs, so the "usefulness" factor was eliminated. The corrected regression [10] indicated that the pragmatic attribute "adaptability" affects participants' intention (Table 3).

Table 3. Regression analysis (*p < .05, **p < .005)

	Construct	Beta	T	p-value	R^2
Pragmatic attributes:	Ease of use	.051	0.782	.438	.917
	Adaptability	.300	3.115	.003**	
Hedonic attributes:	Enjoyment	.466	4.213	.000**	
	Attractiveness	−.205	−1.908	.062	
	Companionship	.338	2.715	.009*	
	Animacy	.053	0.572	.570	

Furthermore, the hedonic attributes "enjoyment" and "companionship" affect a participants' intention to use. These results exhibit an R^2 value of .917, which is satisfactory.

5 Discussion

This study was intended to assess the potential of PAs for truck drivers—especially PAs in the form of conversational, virtual, and robotic agents. The descriptive analysis indicated that truck drivers could imagine using such agents in their truck cabins. They see their main function as being comparable to PAs from other contexts, such as home assistants or smartphone devices. The assistant should primarily provide useful information. However, the desired content differs. Navigation, information on environment, and parking spots are comparable to the context of cars. However, truck drivers need further information about customers or scheduled pauses and driving hours. Furthermore, truck drivers want to use PAs to ameliorate long, monotonous trips—one of truck drivers' main challenges [11]. PAs should therefore assist when learning languages or by playing games, and initiate conversations to prevent drivers from succumbing to fatigue. However, the drivers are reluctant to engage in personal discussions with the PA. This appears to be the upper limit of interaction with PAs in trucks.

The multiple regression analysis indicated that a highly configurable PA produces greater predicted acceptance. The individual statements of the participants underlined this. The regression establishes that a PA should promote a social relationship with truck drivers. How such a relationship can be generated systematically should be investigated in further studies. One possible approach could be to use anthropomorphism to emotionalize the relationship with the PA [12].

The small number of participants limits this study's conclusions. In contrast to Graaf et al. [7] it was not possible to obtain a sufficient number (>100) of subjects to allow for further statistical analysis [9]. It should also be noted that eliminating variables during a regression analysis should always to be done cautiously [10]. It also has to be mentioned that there are more acceptance factors for PAs. Since PAs for truck drivers are currently just beginning to be developed, the participants had no experience with this kind of system. The survey started on the premise that the participants are able to project their thoughts and feelings on the agents presented to them. Although the study cannot ensure that all participants had a comparable agent in mind, they agreed on the benefit of these assistants. Since the study was based on an online survey, a self-selection bias tends to be associated with the results [13].

Acknowledgements. As first author, Jana Fank initiated this article's research idea, and contributed to the study design and data analysis. Markus Lienkamp made an essential contribution to the conception of the research project. He critically revised the paper with regard to important content and gave final approval of the version to be published. Special thanks goes to Ramon Tengel, who contributed to the study's design. The research was conducted with basic research funds from the Institute of Automotive Technology, Technical University of Munich.

References

1. Forster, Y., Naujoks, F., Neukum, A.: Increasing anthropomorphism and trust in automated driving functions by adding speech output. In: 28th IEEE Intelligent Vehicles Symposium (IV), pp. 365–372 (2017)
2. Kraus, J.M., et al.: Human After All: Effects of mere presence and social interaction of a humanoid robot as a co-driver in automated driving. In: 8th International Conference on Automotive User Interfaces and Interactive Vehicular (AutomotiveUI), pp. 129–134 (2016)
3. Häuslschmid, R., von Bülow, M., Pfleging, B., Butz, A.: Supporting trust in autonomous driving. In: 22nd ACM International Conference on Intelligent User Interfaces (IUI), pp. 319–329 (2017)
4. Rothensee, M.: Psychological determinants of the acceptance of future ubiquitous computing applications. Berlin, Humboldt-University, Dissertation. Kovač, Hamburg (2008)
5. Venkatesh, V., Davis, F.D.: A model of the antecedents of perceived ease of use: development and test. Decis. Sci. **27**(3), 451–481 (1996)
6. Davis, F.F.: User acceptance of information technology: system characteristics, user perceptions and behavioral impacts. Int. J. Man Mach. Stud. **38**(3), 475–487 (1993)
7. de Graaf, M.M.A., Allouch, S.B., van Dijk, J.A.G.M.: Why would I use this in my home? A model of domestic social robot acceptance. Hum. Comput. Interact. **16**(1), 1–59 (2017)
8. BAG: Marktbeobachtung Güterverkehr: Auswertung der Arbeitsbedingungen in Güterverkehr und Logistik 2017-I. Available: https://bit.ly/2AaVwNN
9. Blanz, M.: Forschungsmethoden und Statistik für die Soziale Arbeit: Grundlagen und Anwendungen. Kohlhammer, Stuttgart (2015)
10. Urban, J.M.D. Regressionsanalyse: Theorie, Technik und Anwendung, 4th edn. Vs Verlag für Sozialwissenschaften (2011)
11. Evers, C.: Auswirkungen von Belastungen und Stress auf das Verkehrsverhalten von Lkw-Fahrern, Philosophische Fakultät, Rheinische Friedrich-Wilhelms-Universität Bonn, Dissertation, Bonn (2009)
12. Zlotowski, J.A.: Understanding Anthropomorphism in the Interaction Between Users and Robots, HIT Lab NZ, University of Canterbury, Dissertation, Canterbury (2015)
13. Stanton, J.M.: An empirical assessment of data collection using the internet. Pers. Psychol. **51**(3), 709–725 (1998)

Show Me How You Click, and I'll Tell You What You Can: Predicting User Competence and Performance by Mouse Interaction Parameters

Christiane Attig[(✉)], Ester Then, and Josef F. Krems

Chemnitz University of Technology, Cognitive and Engineering Psychology, Wilhelm-Raabe-Str. 43, 09120 Chemnitz, Germany
christiane.attig@psychologie.tu-chemnitz.de

Abstract. Automatically detecting and adapting to user competence is a promising approach for advancing human-technology interaction. With the present work, we demonstrate that perceived user competence and performance can be predicted by easily ascertainable low-level mouse interaction parameters with considerable amounts of explained variance. $N = 71$ users with varying competence interacted with a statistical software while mouse interaction parameters were recorded. Results showed that perceived task competence could best be predicted by clicks per second, maximum mouse velocity, and average duration of pauses > 150 ms ($R^2 = .39$). Perceived system competence could best be predicted by clicks per second, maximum mouse acceleration, and average number of pauses > 150 ms ($R^2 = .28$). Performance could best be predicted by clicks per second, maximum mouse velocity, and average number of pauses > 150 ms ($R^2 = .50$). Results imply that assessing low-level mouse interaction parameters could be a feasible approach for automatic detection of user competence and performance.

Keywords: Adaptive systems · User competence · Mouse interaction

1 Introduction

In a world of growing digitalization and automation, successful interaction with technology is crucial for mastering work (e.g., regarding industry 4.0) and social demands (e.g., regarding use of digital devices in public space). If the interaction with technology fails, that is, if the goal of the interaction is not accomplished, negative affect (e.g., frustration) can arise and user acceptance can be impeded [1]. One promising solution for enhancing ease of use and satisfaction with technical systems is the implementation of adaptive systems, which are able to automatically detect user variables and adapt to the current user. For instance, affect-aware systems recognize users' affective states through analysis of various channels of emotion expression (e.g., facial expressions, physiological data [2]) and can, noticed or unnoticed, adapt to the user (e.g., by adjusting the interface or offering tailored support [3]).

Besides user affect, adaptive systems can also be designed to adapt to the competence level of the user (e.g., [4]). These adaptations hold the potential to increase

W. Karwowski and T. Ahram (Eds.): IHSI 2019, AISC 903, pp. 801–806, 2019.
https://doi.org/10.1007/978-3-030-11051-2_122

performance and support learning processes [5], and are thus particularly suited for digital learning environments. Regardless of the user variable that is adapted to, user monitoring should be as non-intrusive, continuous, and easily implementable as possible. One type of data source which fulfils all three criteria is user input (e.g., mouse interaction, keystrokes). Hence, with the present research, we aim at identifying user competence indicators through low-level mouse interaction parameters in a complex mouse-based computer task.

2 User Competence Detection Based on Mouse Interaction Parameters

When interacting with digital devices, different levels of competence specifity can be distinguished: task competence (i.e., knowledge regarding how to conduct a specific task with a specific device/software) and system competence (i.e., knowledge regarding how to use the specific device/software in general). Moreover, also a more general interaction competence (i.e., computer literacy [6]) can be assumed to support performance.

Examining users of an image editing software, [7] analyzed low-level mouse interaction parameters to predict user task competence. To this end, novice participants repeated two different tasks seven times until they reached a level of skilled use. It was found that skilled use was characterized by a higher mouse acceleration, a smaller amount of clicks on menu buttons, and shorter dwell times.

Besides task competence, [8] also investigated mouse interaction indicators for interaction competence by examining computer novices and IT professionals. All participants repeated a task in an image editing software 15 times to enhance task competence while interaction competence did not increase. The authors found that high task competence was linked to a smaller amount of pauses > 150 ms, a shorter menu navigation time, and a higher mouse velocity. High interaction competence was also linked to a higher mouse velocity and fewer and shorter pauses. The potential of pause measures to distinguish between low and high task competence was later replicated [9].

All three studies [7–9] examined user behavior in laboratory settings with detailed instructions and varied task competence through task repetitions. In contrast, [4] recruited regular system users and observed their free interaction behavior in the laboratory and in situ (i.e., during participants' home use). This study supported the importance of pause measures for identifying competence levels.

In sum, past research has demonstrated the potential of low-level mouse interaction parameters to predict user competence variables. However, natural human-computer interaction has not been examined exhaustively – thus, the question remains if the identified parameters are also able to predict user task and system competence in more natural settings (e.g., during working on tasks without detailed walkthrough-guides). With the present research, we aim at identifying low-level mouse interaction parameters for predicting two types of competence variables (perceived task competence and perceived system competence), and, further, objective performance. To this end, users with varying competence levels worked freely on different tasks with a statistical software.

3 Method

3.1 Participants

To provide a broad range of task and system competence regarding the statistical software SPSS, we recruited students and scientific staff from Chemnitz University of Technology. In sum, $N = 72$ participants took part (one was excluded from the analyses due to linguistic difficulties in understanding instructions; thus, $N = 71$). The average age was $M = 25.07$ ($SD = 4.53$), 79.2% were female. Thirty-five (48.6%) participants were undergraduate students, 31 (43.1%) were graduate students, and six (8.3%) were PostDocs. The majority (63.4%) stated to use SPSS once a year or less, 26.8% once per semester, and 9.9% on a weekly or daily basis.

3.2 Material and Measures

The participants worked on four tasks in SPSS (version 24.0 [10]). For logging mouse interaction parameters, we used Inputlog (version 7.0.0.11 [11]) and the program Pymouse, which was written by the author ET. All instructions and questionnaires were presented in a browser using the survey software Limesurvey [12]. We preprocessed the raw mouse interaction data using Open Office and SPSS and calculated the parameters average and maximum mouse velocity, average and maximum mouse acceleration, clicks per second, and number and average duration of pauses > 150 ms (according to [8]).

Perceived system competence was assessed with two items. The first ("How would you estimate your experience with SPSS?") was answered on a 5-point Likert scale from 1 (*no experience*) to 5 (*thorough experience*). The second ("How competent do you feel when using SPSS in general?") was answered on a 5-point Likert scale from 1 (*not competent at all*) to 5 (*very competent*). Internal consistency was $\alpha = .90$. For assessing participants' perceived task competence, similar questions were presented after each of the four tasks ("How would you estimate your experience with this specific task in SPSS?"; "How competent did you feel during the task completion?"). Answers were provided on the same scales as before. Internal consistency varied between $\alpha = 71$ (Task 4) and .86 (Task 2).

As a performance measure, we assessed the time to completion (TTC). Then, the minimum necessary durations according to the keystroke-level model [13] were subtracted from the respective TTC. In addition, two independent raters evaluated participant's task performance based on the SPSS output file according to predefined criteria as not solved (0), partly solved (1), or completely solved (2). Interrater reliability varied between $\kappa = .84$ (Task 2) and 1.0 (Task 3). Finally, a weighted performance score was calculated by multiplying the TTC with 0.5 (task completely solved), 1 (task partly solved) or 2 (task not solved). Note that lower weighted TTC indicate a better performance.

3.3 Procedure

Before the SPSS tasks, users were asked to rate their perceived SPSS competence. Then, they were instructed to work on four tasks with varying task difficulty in SPSS (1: Pie chart with percentages, 2: Calculating Cronbach's Alpha, 3: Factor analysis, 4: Contrast analysis). For each task, a time limit of five minutes was set. Participants were not allowed to ask the experimenter or search on the internet for help. After each task, participants were asked to rate their perceived task competence (and perceived user state measures, which are not part of this research).

4 Results

For the present analyses, we focused on Task 2, because of the largest range of performance and task competence. For each criterion (task competence, system competence, performance), we calculated various multiple linear regression analyses with three predictors each: (1) clicks per second, (2) one out of four mouse speed parameters (average mouse velocity, maximum mouse velocity, average mouse acceleration, maximum mouse acceleration), and (3) one out of two pause parameters (average number of pauses > 150 ms, average duration of pauses > 150 ms). For each criterion, only the best predictor combinations based on R^2 are reported (see Table 1).

Clicks per second, maximum mouse velocity and average duration of pauses could explain 39% of the variance of perceived task competence. Clicks per second, maximum mouse acceleration, and average number auf pauses > 150 ms could explain 28% of the variance in perceived system competence. Clicks per second, maximum mouse velocity, and average number of pauses could explain 50% of the variance of objective performance. In sum, a higher click count, slower mouse speed (velocity and acceleration), and more, but shorter pauses > 150 ms were linked to higher perceived competence and objective performance.

Table 1. Multiple regression models of best predictors of perceived task competence, perceived system competence, and objective performance.

Criterion	Predictor	β	$p_β$	$R^2_{adj}(R^2)$	p
Perceived task competence	Clicks per second	.36	.003	.36 (.39)	<.001
	Maximum mouse velocity	−.27	.008		
	Average duration of pauses > 150 ms	−.31	.009		
Perceived system competence	Clicks per second	.37	.001	.24 (.28)	<.001
	Maximum mouse acceleration	−.29	.010		
	Average number of pauses > 150 ms	.22	.050		
Weighted time to completion	Clicks per second	−.53	<.001	.48 (.50)	<.001
	Maximum mouse velocity	.20	.028		
	Average number of pauses > 150 ms	−.39	<.001		

5 Discussion

In the present research, we examined the potential of selected low-level mouse interaction parameters for predicting perceived task competence, system competence, and objective performance. Results showed that these non-intrusive and easily ascertainable parameters could explain considerable amounts of variances of the criteria. Hence, for automatically detecting user competence regarding mouse-based computer tasks, the assessment of mouse interaction parameters seems to be a fruitful approach.

The results are largely contradictory to results from past research. Whereas [4, 7, 8] found that lower average mouse velocity/acceleration was linked to lower user competence, we found that it was linked to higher competence (both task and system competence, and also performance). Moreover, we found that higher competence and performance were linked to more clicks per second, which is contradictory to [7]. In addition, more pauses > 150 ms were linked to higher system competence, contradictory to [9]. In contrast, the finding that longer pauses are connected to lower competence [4, 7–9] could be replicated.

The differences between the presented results and results from past research might be due to differences in user competence assessment (subjective ratings vs. expert ratings as in [4] vs. objective assessment through performance measures as in [7, 8]). However, our performance measure (weighted TTC) comprises both the objective time to completion and an expert rating. Thus, at least regarding performance, our results should not be substantially different from past research. Another explanation might be the different type of software, indicating that interaction with different software might result in different mouse parameters indicating user competence. Hence, our study supports the notion that competence indicators might be software-specific. Finally, in the present research, participants did not solve a task multiple times until optimal task solution was ensured and considered skilled use as in [7–9]. Therefore, competence levels might be characterized by other attributes in more natural interaction with free task completion than in highly standardized settings. To summarize, our study yielded promising results for automatic user competence detection through easily implementable mouse interaction parameters.

Acknowledgements. This research was funded by the European Social Fund and the Free State of Saxony under Grant No. 100269974. We want to thank our student assistants Daniel Götz, Katharina Schulzeck and Sabine Wollenberg for their support in data collection and manuscript preparation.

References

1. D'Mello, S.K.: A selective meta-analysis on the relative incidence of discrete affective states during learning with technology. J. Educ. Psychol. **105**, 1082–1099 (2013)
2. Calvo, R.A., D'Mello, S.: Affect detection: an interdisciplinary review of models, methods, and their applications. IEEE T. Aff. Comput. **1**, 18–37 (2010)
3. Lavie, T., Meyer, J.: Benefits and costs of adaptive user interfaces. Int. J. Hum. Comput. Int. **68**, 508–524 (2010)

4. Grossman, T., Fitzmaurice, G.: An investigation of metrics for the in situ detection of software expertise. Hum. Comput. Int. **30**, 64–102 (2015)
5. Magnisalis, I., Demetriadis, S., Karakostas, A.: Adaptive and intelligent systems for collaborative learning support: a review of the field. IEEE T. Learn. Technol. **4**, 5–20 (2011)
6. Sengpiel, M., Jochems, N.: Development of the (adaptive) computer literacy scale (CLS). In: Lindgaard, G., Moore, D. (eds.) The Proceedings of the 19th Triennial Congress of the International Ergonomics Association (2015)
7. Hurst, A., Hudson, S. E., Mankoff, J.: Dynamic detection of novice vs. skilled use without a task model. In: Proceedings of the SIGCHI Conference on Human Factors in Computing Systems (pp. 271–280). ACM, San Jose (2007)
8. Ghazarian, A., Noorhosseini, S.M.: Automatic detection of users' skill levels using high-frequency user interface events. User Model. User-Adap. **20**, 109–146 (2010)
9. Ghazarian, A., Ghazarian, A.: Pauses in man-machine interactions: a clue to users' skill levels and their user interface requirements. Int. J. Cog. Perfor. Supp. **1**, 82–102 (2013)
10. IBM Corp.: IBM SPSS Statistics for Windows (Version 24.0). IBM Corp., Armonk (2016)
11. Leijten, M., Van Waes, L.: Keystroke logging in writing research: using inputlog to analyze writing processes. Writ. Commun. **30**, 358–392 (2013)
12. LimeSurvey GmbH: LimeSurvey—An Open Source Survey Tool. LimeSurvey GmbH, Hamburg (2003)
13. Card, S.K., Moran, T.P., Newell, A.: The Keystroke-level model for user performance time with interactive systems. Commun. ACM **23**, 396–410 (1980)

Applications and Future Trends

Integrating People and Intelligent Systems by Design

Stuart White and Alana Nicastro(✉)

Marine Corps Tactics and Operations Group, United States Marine Corps,
Marine Air Ground Task Force Training Command, Box 788305,
Twentynine Palms, CA 92278-8305, USA
stuart.white@usmc.mil, alana.nicastro@sdsu.edu

Abstract. A profound problem facing the Marine Corps today is its struggle to understand the *complex nature of learning*. The ability to learn is the most essential part of a being a Marine; even though it is the least understood. Integrating people and intelligent systems while expanding the boundaries of current state-of-the-art technologies is an exciting mandate by the United States Marine Corps—yet without partnering with those who can *design* significant learning experiences for our military personnel, we fall short of excellence. Cultivating relationships outside of the Marine Corps, as it turns out, is a strategic asset to drive learning innovation and improve institutional message alignment. Learning is essentially about realizing desired change. The authors of this narrative advocate for dialogue on the design of learning in relationship to maneuver warfare and discuss the need to collaborate with a local research university to understand the critical problems facing the military

Keywords: Learning · Maneuver warfare · Institutional message alignment
Enact intelligence

1 Ability to Learn

The ability to learn is the most essential part of being a Marine; even though it is the least understood. The Marine Corps Tactics and Operations Group (MCTOG)—a Center of Excellence for the Ground Combat Element that prepares Marines for combat —has discovered why it must explore its beliefs about learning and knowledge. Learning, after all, is closely tied to our ability to wage maneuver warfare.

Many of our military learning communities are tasked with creating dynamic instruction based on evolving threats and complex operating environments, However, when confined to the Marine Corps' superficial formalized processes, these learning communities often forgo the total development of the Marine relative to their wartime role. These superficial formalized processes are prevalent in the operational units too. They follow procedural-laced methods to develop training and often fail to achieve the outcomes to execute maneuver warfare in combat.

W. Karwowski and T. Ahram (Eds.): IHSI 2019, AISC 903, pp. 809–813, 2019.
https://doi.org/10.1007/978-3-030-11051-2_123

Teaching and learning extends beyond formal learning communities. The entire Marine Corps is responsible for generating a culture of excellence—marked by intelligence, lethality and tactically gifted Marines.

Yet we fall victim to poorly designed (and poorly assessed) learning. Well-designed learning is a critical requirement; and right now, it is a critical vulnerability. Learning is a social and emotional human experience. The hallmark of learning is to learn from others—it requires empathy, integrative thinking, optimism, collaboration and experimentation. The design of learning can negatively or positively impact morale and/or an end-state. Perhaps one of the most profound problems facing the military today is its struggle to understand the *complex nature of learning*.

The authors of this paper noticed the same limitations inside MCTOG—curricular practices were often trapped in *conventional thinking patterns* by those exposed to formal, military learning environments most of their adult lives. We simply managed training and education to the status quo, rather than invest in *the work of learning* to meet the demands of today's combat environment. MCTOG made a conscious effort to do things differently—beginning with a new mindset that *learning is a platform for experimentation*; suitable for an organization that prides itself on the ability to improvise, adapt and overcome.

We know the value of training and education is mostly psychological—it is an enabling process and a form of empowerment. MCTOG now uses *contemporary thinking practices* to include the art of curriculum design as a process of critical questioning that frames learning and teaching. "Thinking like" a designer is not only a tactic, but an indispensable competency. We must design activities that stimulate original thought, improvisation, dialogue and social negotiation to appropriately tackle the complex problems of today and tomorrow.

Therefore, combat readiness requires more than a checklist mentality. It includes creating a space where Marines can do their best thinking. The design of learning should maximize competencies that are necessary to succeed in combat and use the appropriate learning paradigm (i.e., behaviorism, cognitivism, constructivism) to develop Marines who are *ready for combat* and *ready to learn!*

2 Reinvigoration of Maneuver Warfare

There have been a few recent developments inside the Marine Corps that support MCTOGs aspirations to modernize training and optimize learning. First, there is an initiative to "reinvigorate" maneuver warfare. Maneuver warfare is our warfighting philosophy. Second, a new Marine Corps Doctrinal Publication (MCDP) "on learning" is underway. This new publication will tie-in the Marine Corps values and ethos to its beliefs about intelligence and learning. Third, the institution wants to move away from the factory model of education to reflect the Information Age society in which we live. Fourth, the development of readiness standards may advance. Readiness standards are predetermined clear and worthy intellectual priorities.

The above initiatives are all interconnected. They offer expansive open space for the organization to grasp the *qualitative meaning* of combat proficiency and effectiveness as well as discover the power of institutional message alignment.

Institutional Message Alignment. Institutional message alignment will be a strategic factor in realizing our maneuver warfare philosophy while shaping the culture of the Marine Corps. It brings focus to *what really matters.* Design theorists and practitioners understand the impact that institutional message alignment has on performance.

Based on what we know about combat proficiency and effectiveness, Marines should perform individual and collective tasks within a larger tactical and/or wargame scenario to leverage the *adaptive behavior* needed to solve complex problems. This "adaptation" is a critical element for future Marine Corps forces and the "single most important component" to combat effectiveness.[1] Similar to maneuver warfare, we should not avoid uncertainty and disorder in the design of learning but embrace it as key elements to our success; especially as we focus less on counterinsurgency threats and more on large scale combat operations.

In short, war is the violent clash of human wills and characterized by a growing list of variables (i.e., friction, uncertainty, disorder, complexity). Maneuver warfare is essential when we want to outpace and overtake the adversary. Shattering the cohesion of our opponent is achieved through physical, cognitive and moral constructs.

Warfare is active, multifaceted and network-centric. There are a number of dimensions that define a battlespace—to include land, sea, air, space, cyber, and battle for the mind. Nowadays, the battlefield is open to anyone with access to the Internet. It is perhaps the *human dimension* of war (battle for the mind) that is our greatest fighting power. Boldness, imagination, intelligence and grit are necessary qualities to exercise maneuver warfare.

Maneuver warfare is the total sum of the Marine Corps mission, values, principles and warfighting tenets.

Perhaps the MCDP "on learning" will be the driving force that aligns our institutional messages—a doctrine that conveys the impact of the Information Age mindset, outlines clear and worthy intellectual priorities and unpacks the exercise of maneuver warfare.

New Approaches. New approaches often require that we unlearn what we have learned in the past. As the Marine Corps changes its current learning paradigm from the Industrial Revolution to the Information Age, the following lessons learned from MCTOG may be of help:

(1) a backward design approach safeguards our maneuver warfare philosophy during curricular/alignment activities
(2) a backward design approach safeguards the opportunity for learner enrichment during curricular/alignment activities
(3) similar to learning, many people do not have a fundamental nor conceptual understanding of maneuver warfare
(4) many people do not understand the significant role the Training and Readiness (T&R) enterprise plays in combat proficiency and effectiveness

[1] See DARPA, Measurement of Combat Effectiveness in Marine Corps Infantry Battalions, 1977, p. 13.

(5) a culture characterized by trust, initiative and professionalism requires a fundamental shift to "see" the totality of a warfighter through strengths, achievements and possibilities; inspired, of course, by institutional message alignment
(6) a fresh perspective often promotes new thinking; it provides an opportunity to examine routine practices that may no longer be productive and/or worthwhile.

In preparation of this conference presentation, the authors have discovered how important it is to forecast optimization among human effort and artificial intelligence. Integrating people and intelligent systems suggests a co-constructed learning experience is taking place between human learner and the machine learner. What does that look like? How will Artificial Intelligence (AI) enhance a Marine's ability to think, decide and act while it instantaneously enhances its own cognitive processing?

The Marine Corps should take a closer look at what integrating people and intelligent systems may mean now and in the future. The military is moving full speed to achieve the Secretary of Defense's "25 bloodless battles before the first battle"[2] measure. This directive has increased the importance of simulated training systems for Marine readiness.[3]

While the Marine Corps is focused on technology to enhance *quantitative readiness* (i.e., virtual reality, augmented reality, high-fidelity simulations), the institution may want to take an operational pause and consider the totality of the opportunity. Despite its appearances, this is not a technological challenge, but a *learning design challenge!*
Integrating People and Intelligent Systems. MCTOG has advanced over the years to understand the importance of *understanding by design*[4] and its impact on cognition, self-determined change and the reinvigoration of maneuver warfare. As we continue to design learning experiences that expand complex thought and self-awareness, we also want to learn how AI can play a role in sense-making and the exchange of ideas. Can we create an environment where people and intelligent system(s) dialogue; and the feedback results in deeper learning?

The use of tablets, modeling/simulations and holographic displays are all important tools to enhance learning, but the authors believe the *real integration* between people and intelligent systems are the social and emotional interactions between the human and the system—and their collective ability to solve complex problems and transform information into working knowledge and/or wisdom (i.e., "learning from the gunfight"). Nothing is more important than the investment we make in visualizing our true potential and discussing what that looks like in the context of modernization; therefore, we ask the question, how does the Marine Corps want to be impacted by intelligent systems?

The authors of this paper are currently strengthening their relationships with intellectual partners to help address critical problems facing the military. We know that

---

[2] As mentioned in the 2nd Annual Defense News Conference, 25 bloodless battles: Synthetic training will help prepare for current and future operations, https://www.defensenews.com/smr/defense-news-conference/.

[3] Ibid.

[4] Grant Wiggins and Jay McTighe offer a great perspective on understanding and how it is different than knowledge, see *Understanding by Design, 2nd edition*

excellence in intelligence, lethality and tactics includes embracing theory, research and practice as well as modeling what it means to become a learning laboratory. Cultivating relationships outside of the Marine Corps, as it turns out, is a strategic asset to improve institutional message alignment, help further develop talent and drive necessary innovation—all important to regain and maintain our fighting edge.

3 Operation "Enact Intelligence"

The authors have learned how important it is to intentionally design an environment where people can understand the complex nature of learning and expand the boundaries of current state-of-the-art technologies. Enact intelligence begins by visualizing the ways we want Marines to "act" inside the organization and ends with the realization of that image. Enacting intelligence is a series of deliberate practices to achieve the future state. It involves communicative strategies that are powerful in reshaping institutional behaviours. In essence, enact intelligence is about realizing desired change—and influences every person inside the organization.

As the Marine Corps fully embraces the Information Age, MCTOG will explore a partnership with a local research university and collaborate in a myriad of ways—to include developing original research, applying new theories, creating certificate programs, using high-end virtual reality laboratories, interacting with artificial intelligence, working with software programmers and more. We want to learn how Marine learners and intelligent systems are enacted. Perhaps we will foster an intelligence movement—*enact intelligence!*

To lead the work of intelligence (both human and system), MCTOG is dialoging with a local research university to develop new avenues of learning and technology. The goal is to create an exchange of ideas (for the sake of learning)—and demonstrate *humility* and *responsibility* as we learn and grow—and make sense of the of partnership, critical problems and potential solutions. It requires that we step back from our traditional spaces of work.

MCTOG wants to transform the way it develops its people as a way to improve the quality of learning and research. This transformation requires a shift in our current organizational beliefs and practices. It is also important for us to understand that the current organizational environment was established by traditional military and hierarchical design thinking. We want a blue-print for long-term growth.

The Marine Corps finds itself at a tipping point—train to retain the foundational competencies to succeed today and seek to exploit innovation to adapt for tomorrow. We are faced with an adaptive challenge; especially when it comes to integrating people and intelligent systems. As MCTOG continues to examine the complex nature of learning, we will operate with the belief that no one has the monopoly on adaptive problems. Therefore, we will seek solutions that are informed by original research, dialogue and diverse opinion—while leveraging technology to give us the competitive advantage and learning edge as we prepare for the future of warfare.

From the Simulator to the Road—Realization of an In-Vehicle Interface to Support Fuel-Efficient Eco-Driving

Craig Allison(✉), James Fleming, Xingda Yan, Neville Stanton, and Roberto Lot

Faculty of Engineering and Physical Sciences, University of Southampton, Boldrewood Innovation Campus, Southampton SO16 7QF, UK
{Craig.Allison, J.M.Fleming, X.Yan, N.Stanton, Roberto.Lot}@soton.ac.uk

Abstract. Motivated by the observation that modifying driver behavior can significantly reduce fuel usage and CO_2 emissions, this paper documents the development of a dedicated in-vehicle interface to support eco-driving. This visual interface has been tested in simulator conditions, demonstrating an 8.5% reduction in fuel use, and will soon be deployed on-road. Transitioning from simulator testing to on-road testing presents significant challenges to ensure driver safety and system effectiveness in the presence of changing road conditions and imperfect information about the current driving scenario.

Keywords: Human factors · Interface development · User testing Eco-driving

1 Introduction

The International Energy Agency estimates that transport accounts for 35% of overall global energy use [1], with passenger cars and light duty vehicles alone accounting for 21% of this. Such vehicles are also responsible for 24% of global carbon dioxide (CO_2) emissions. Considered within the context of global warming, it is estimated that 14% of global mean temperature change will be a direct consequence of transportation, with automobiles being the primary contributor [2]. Approaches to addressing the problem of high fuel use and high emission road transportation currently being promoted by governments include the use of alternative transportation schemes such as cycle to work events [3], and the development of alternative, environmentally friendly and fuel efficient vehicular drivetrains, including electric or hybrid vehicles [4]. Although these methods are demonstrating signs of success, the initial high investment cost and in some cases a lack of corresponding infrastructure [5] act as barriers to limit uptake.

One alternative to such schemes, however, is the changing of driver behaviors to improve fuel-efficiency by employing eco-driving techniques [6]. Approximately 5–20% of CO_2 emissions can be eliminated if drivers drove in a more eco-friendly way [6]. Eco-driving is typified by behaviors such as gentle acceleration, prompt gear changes, limiting the engine to approximately 2,500 revolutions per minute (RPM),

W. Karwowski and T. Ahram (Eds.): IHSI 2019, AISC 903, pp. 814–819, 2019.
https://doi.org/10.1007/978-3-030-11051-2_124

anticipating traffic flow to minimize braking, driving below the speed limit, and limiting unnecessary idling, with drivers who exhibit such behaviors recording dramatically reduced fuel use and greenhouse gas emissions [6]. Whilst eco-driving is, from the drivers' perspective, a cost-effective approach to reduce fuel use [7], previous research has demonstrated that some eco-driving behaviors are difficult to maintain long term [8], with the majority of drivers returning to their previous driving style when not under direct observation [9]. One approach to supporting drivers in achieving greater fuel economy is via the use of in-vehicle interfaces [10]. The provision of feedback to driver actions has been essential in supporting eco-driving long-term [10].

However, questions remain regarding how to present this feedback information to drivers and the value that this can have in terms of overall fuel usage. To this end, a visual interface was developed that displays real-time speed recommendations to the driver. This interface forms part of a speed advisory system that attempts to coach the driver into more fuel-efficient behavior by computing the most fuel-efficient speed and acceleration choice given the current driving situation.

2 Interface Display & Driving Simulator

A prototype of the described system was tested within the Southampton University Driving Simulator (SUDS), based at the University of Southampton, UK (Fig. 1). SUDS comprises of a 2015 Land Rover Discovery Sport and a simulated roadway environment that is projected across three forward mounted screens. A fourth projector screen is situated behind the vehicle, and LCD screens are placed on each of the side-view mirrors to simulate the view in the rear-view and side view respectively. Engine noise is simulated using the vehicle's internal audio system. The simulator uses the software package STISIM, which runs on a Windows 7 PC.

Fig. 1. Southampton University Driving Simulator (SUDS)

The speed advisory system calculates the most fuel-efficient speed for the current road environment based upon preceding vehicles, road curvature, road type and current vehicle speed. This speed recommendation is used to update the visual interface, which also receives the updated simulation state from STISIM in order to update the speedometer and RPM counter.

The visual interface was designed and developed in C# using Windows Forms as a graphical library. The resulting application was executed on a Microsoft Surface Pro tablet, which is placed directly behind the steering wheel of the car to replace the physical instrument cluster. The visual design of the interface is shown in Fig. 2 and comprises of a speed and RPM display augmented with a green and yellow 'eco-band' to recommend a near-optimal speed range to the driver. The green-region provides the driver with a range of speeds recommended for fuel-efficient driving based on the current vehicle, road and traffic state. A further yellow region allows some margin for error and has a width chosen to correspond to typical speed variations observed in normal driving [11, 12]. A similar 'eco-speedometer' design was rated highly in perceived usefulness and user acceptance in a previous study [10]. When in use, the interface updates in real-time with the green eco-band stretching from zero speed to the current recommendation. As the situation changes, this speed changes, which has the effect of smoothly interpolating between recommended speed values to gently 'coach' the driver into following the optimal speed profile. Technical details of the speed advisory system can be found in [13], while information regarding the design of the visual interface appears in [14].

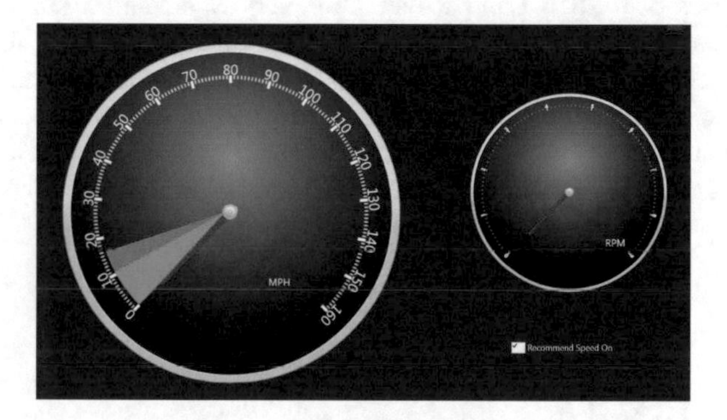

Fig. 2. Visual interface to speed advisory system

2.1 Initial Evaluation

To explore the effectiveness of the interface display, 36 participants (18 male, 18 female) completed a laboratory study within the SUDS laboratory. Participants were required to drive a 13-mile simulated route, based on roads in Southampton, UK. Each participant drove in three test conditions: normal driving, unassisted eco-driving, and driving with use of the speed-advisory system to support their driving. The order in which the different test conditions were completed was fully counterbalanced. A variety of objective data was collected during the simulated drives, including speed, acceleration, throttle position, brake pedal position and headway. This data was processed using detailed vehicle powertrain models to provide an accurate estimate of fuel consumption. In addition, participants were required to complete a variety of subjective measures for each of the three drives including the Dundee Stress State Questionnaire (DSSQ) [15, 16], NASA Task load Index (NASA-TLX) [17] and the System Usability Scale (SUS) [18].

Results demonstrated that whilst participants used the least amount of fuel when asked to drive in a fuel-efficient way, saving approximately 11% in total fuel usage compared to normal driving, this driving style was associated with significantly greater workload. Use of the developed interface in contrast allowed participants to achieve an approximate 8.5% fuel saving, without the corresponding increase in workload, suggesting that drivers would be more able to use the developed system long term.

3 Real Road System Adaption

Following the simulator study, the visual interface was adapted for installation within the University of Southampton Instrumented Vehicle in preparation for on-road testing (Fig. 3). Three key challenges emerged, relating to driver safety, filtering of on-board sensor signals, and initiation of coasting by the driver.

Fig. 3. University of Southampton instrumented vehicle

3.1 Safety

The foremost concern in moving from simulator testing to on-road testing is ensuring that the driver can operate the vehicle safely with the speed advisory system activated. Considered within this context, the interface developed should not be distracting to the driver and, should the system fail, a speedometer should still be available to the driver. To reduce this risk, the interface was modified so that the image is projected on the windscreen as a Head-Up Display (HUD), rather than as a Head-Down Display replacing the instrument cluster. This modification reduces the potential for distraction effects and ensures that the vehicle speedometer remains available to the driver.

3.2 Sensor Signal Filtering

A key technical challenge faced in a real vehicle is inaccuracy of measurements of the road and traffic state, in particular the headway distance to the leading vehicle which is processed to provide the speed recommendation. In the simulator, the position and velocity of all other road users are known precisely. However, in the real world this information must be estimated from vehicle sensors such as a front-mounted radar, the signal from which is noisy and corrupted by spurious measurements from other metallic objects. This necessitated a filtering strategy to provide a reliable measurement of lead vehicle position and velocity, in order that the speed recommendation shown to the driver varies smoothly without any abrupt jumps.

3.3 Initiation of Coasting

One observation from the simulator study is that, although it is effective at reducing fuel consumption overall, the current visual interface is ineffective at promoting 'coasting' behaviors in which the driver reduces speed gradually by releasing the accelerator pedal rather than by braking. Future work carried out prior to the on-road test will consider adding an indicator to the visual interface, potentially combined with an audible notification, to tell the driver when they should release the accelerator pedal to coast to a halt.

4 Conclusions

This paper documented the development of a speed advisory system designed to improve fuel efficiency, motivated by the observation that greater adoption of eco-driving can be achieved using dedicated in-vehicle interfaces. The interface documented within the current paper has been tested in simulator conditions, demonstrating an 8.5% reduction in fuel use, and will soon be deployed on-road. Transitioning from simulator testing to on-road testing presents significant challenges that must be overcome. Whilst on-road testing is an essential step in any interface designed to modify driver behavior, deployment of such technologies in the real-world presents challenges in ensuring safety of the driver, and in ensuring system effectiveness.

Acknowledgements. This work was funded by the UK Engineering and Physical Sciences Research Council (EPSRC) grant EP/N022262/1 "Green Adaptive Control for Future Interconnected Vehicles" (www.g-active.uk).

References

1. International Energy Agency (IEA): Key World Energy Statistics. https://www.iea.org/publications/freepublications/publication/KeyWorld2017.pdf
2. Skeie, R.B., Fuglestvedt, J., Berntsen, T., Lund, M.T., Myhre, G., Rypdal, K.: Global temperature change from the transport sectors: historical development and future scenarios. Atmos. Environ. **43**, 6260–6270 (2009)

3. Rose, G., Marfurt, H.: Travel behaviour change impacts of a major ride to work day event. Trans. Res. Part A Pol. Prac. **41**, 351–364 (2007)
4. Lorf, C., Martínez-Botas, R.F., Howey, D.A., Lytton, L., Cussons, B.: Comparative analysis of the energy consumption and CO2 emissions of 40 electric, plug-in hybrid electric, hybrid electric and internal combustion engine vehicles. Tran. Res. D: Trans. Env. **23**, 12–19 (2013)
5. Yilmaz, M., Krein, P.T.: Review of battery charger topologies, charging power levels, and infrastructure for plug-in electric and hybrid vehicles. IEEE Trans. Power. Elec. **28**, 2151–2169 (2013)
6. Barkenbus, J.N.: Eco-driving: an overlooked climate change initiative. Energy Policy **38**, 762–769 (2010)
7. Birol, F.: CO2 emissions from fuel combustion-highlights. Int. Energy Agency (2011)
8. Delhomme, P., Cristea, M., Paran, F.: Self-reported frequency and perceived difficulty of adopting eco-friendly driving behavior according to gender, age, and environmental concern. Trans. Res. D Trans. Environ. **20**, 55–58 (2013)
9. Lai, W.T.: The effects of eco-driving motivation, knowledge and reward intervention on fuel efficiency. Trans. Res. D Trans. Environ. **34**, 155–160 (2015)
10. Meschtscherjakov, A., Wilfinger, D., Scherndl, T., Tscheligi, M.: Acceptance of future persuasive in-car interfaces towards a more economic driving behaviour. In: Proceedings of the 1st International Conference on Automotive User Interfaces and Interactive Vehicular Applications, pp. 81–88. ACM (2009)
11. Yan, X., Fleming, J., Allison, C., Lot, R.: Portable automobile data acquisition module (ADAM) for naturalistic driving study. In: 15th European Automotive Congress (2017)
12. Fleming, J.M., Allison, C.K., Yan, X., Stanton, N.A., Lot, R.: Adaptive driver modelling in ADAS to improve user acceptance: a study using naturalistic data. Safety Sci. (2018)
13. Fleming, J.M., Yan, X., Allison, C.K., Stanton, N.A., Lot, R.: Driver Modeling and Implementation of a Fuel-saving ADAS. IEEE Conference on Systems, Man and Cybernetics (SMC) (2018)
14. Allison, C.K., Stanton, N.A., Fleming, J.M., Yan, X., Goudarzi, F., Lot, R.: Inception, ideation and implementation; developing interfaces to improve drivers' fuel efficiency. Paper to be presented at Chartered Institute of Ergonomics and Human Factors (CIHEF) Ergonomics & Human Factors, Birmingham, UK (2018)
15. Matthews, G., Joyner, L., Gilliland, K., Campbell, S., Falconer, S., Huggins, J.: Validation of a comprehensive stress state questionnaire: towards a state big three. Pers. Psych. Eur. **7**, 335–350 (1999)
16. Matthews, G., Campbell, S.E., Falconer, S., Joyner, L.A., Huggins, J., Gilliland, K., et al.: Fundamental dimensions of subjective state in performance settings: task engagement, distress, and worry. Emotion **2**(4), 315 (2002)
17. Hart, S.G., Staveland, L.E.: Development of NASA-TLX (task load index): results of empirical and theoretical research. In: Advances in psychology, vol. 52, pp. 139–183. Elsevier (1988)
18. Brooke, J.: SUS: a "quick and dirty usability scale. In: Jordan, P.W., Thomas, B., Weerdmeester, B.A., McClelland, I.L. (eds.) Usability evaluation in industry (189–194). Taylor and Francis, London (1996)

Adaptation of Assistant Based Speech Recognition to New Domains and Its Acceptance by Air Traffic Controllers

Matthias Kleinert[1](✉), Hartmut Helmke[1], Gerald Siol[1], Heiko Ehr[1], Dietrich Klakow[2], Mittul Singh[2], Petr Motlicek[3], Christian Kern[4], Aneta Cerna[5], and Petr Hlousek[5]

[1] Institute of Flight Guidance, German Aerospace Center (DLR), Braunschweig, Germany
{matthias.kleinert,hartmut.helmke,gerald.siol, heiko.ehr}@dlr.de

[2] Spoken Language Systems Group (LSV), Saarland University, Saarbrücken, Germany
{dietrich.klakow,mittul.singh}@lsv.uni-saarland.de

[3] Idiap Research Institute, Martigny, Switzerland
petr.motlicek@idiap.ch

[4] Austro Control, Vienna, Austria
christian.kern@austrocontrol.at

[5] Air Navigation Services of the Czech Republic, Jenec, Czech Republic
{aneta.cerna,petr.hlousek}@ans.cz

Abstract. In air traffic control rooms, paper flight strips are more and more replaced by digital solutions. The digital systems, however, increase the workload for air traffic controllers: For instance, each voice-command must be manually inserted into the system by the controller. Recently the AcListant® project has validated that Assistant Based Speech Recognition (ABSR) can replace the manual inputs by automatically recognized voice commands. Adaptation of ABSR to different environments, however, has shown to be expensive. The Horizon 2020 funded project MALORCA (MAchine Learning Of Speech Recognition Models for Controller Assistance), proposed a more effective adaptation solution integrating a machine learning framework. As a first showcase, ABSR was automatically adapted with radar data and voice recordings for Prague and Vienna. The system reaches command recognition error rates of 0.6% (Prague) resp. 3.2% (Vienna). This paper describes the feedback trials with controllers from Vienna and Prague.

Keywords: Machine learning · Assistant based speech recognition
Automatic speech recognition · Air traffic controller

W. Karwowski and T. Ahram (Eds.): IHSI 2019, AISC 903, pp. 820–826, 2019.
https://doi.org/10.1007/978-3-030-11051-2_125

1 Introduction

Air traffic control (ATC) is a conservative business: Air traffic controller (ATCo) still use paper flight strips in control rooms around the globe. These strips contain vital information about aircraft that are under control of an ATCo in a specified airspace sector (e.g. callsign, type and weight class). For guiding an aircraft through the sector, an ATCo gives instructions (commands) to a pilot via voice communication. These commands are mainly related to speed, direction or flight altitude. Notes about the instructed commands, written on paper flight strips, are only available for the controller herself/himself. If provided in a digital form this information could be valuable for other systems e.g. to provide additional safety functions or for planning purposes. Therefore, many Air Navigation Service Providers (ANSP) already replaced or prepare to replace the traditional paper flight strips with different electronic/digital solutions. These systems, however, require manual digital input and, therefore, increase the controllers' workload.

Recently the AcListant® [1] project has validated that Assistant Based Speech Recognition (ABSR) can offer a solution to reduce manual controller inputs by integrating an assistant system with a speech recognizer [2]. The system analyzes the controller-pilot-communication, automatically extracts the instructed commands and uses them as input for digital flight strip solutions. Since the controller only needs to correct the system in case of a wrongly recognized command, the ATCo has more free cognitive resources for other tasks. For AcListant® ABSR was adapted manually to fit the needs of the Dusseldorf approach area in Germany. In simulation runs with different ATCos the system achieved recognition rates better than 95% and error rates below 2% [3]. However, the whole adaptation process required significant data resources, time and expert knowledge and needs to be repeated every time ABSR will be used at another airport or another ATCo sector.

The Horizon 2020 funded project MALORCA (MAchine Learning Of Speech Recognition Models for Controller Assistance), proposed a cheap and effective solution through a Machine Learning (ML) framework, that takes advantage of the large amounts of radar and voice data being recorded in air traffic control rooms on a daily basis. The developed framework is capable of adapting a generic ABSR system to different airports and controller positions [4]. In order to enable an efficient adaptation process ABSR was divided into several conceptual modules that consist of different building blocks, models and data elements. Only the models need to be adapted to new environments. Prague and Vienna approach area were chosen as first showcases for model adaptation.

In the next section we present related work with respect to speech recognition applications in ATM. Section 3 describes the essential building blocks of an ABSR system, Feedback from the end-users (ATCos) on the completely trained ABSR system was collected during Proof-of-Concept trials in Prague and Vienna [4, 5] and is presented before the conclusions in Sect. 4.

2 Related Work

Large improvements in automatic speech recognition have been reached recently, i.e. also due to the industrial applications such as Google Home or Amazon Echo. A good introduction into the state-of the art of ASR applications in the ATM domain until 2014 is given by Nyuyen and Holone [6]. Recent work has also employed ASR in real applications for pilot read back error detection [7]. Very recently, a speech recognition challenge was released by Airbus to develop ASR for an air traffic control scenario [8], allowing large variety of academy and industry to gain an access to real (i.e. manually transcribed) data and develop new machine learning algorithms in this domain. These proposed algorithms mainly focused on building robust acoustic and language models, combined with some graph (i.e. FST) based algorithms to efficiently deal with command deviations (i.e. especially in case of call-sign detection).

Assistant Based Speech Recognition integrates the information of an assistant system as context information. DLR and Saarland University first used an arrival manager, which analyses the current situation of the airspace to predict possible future commands of the ATCo [9]. This approach has shown to both significantly increase command recognition rate and reduce command recognition error rate in AcListant® [2, 3] and MALORCA [4] project.

3 Conceptual Modules of Assistant Based Speech Recognition

The ABSR systems consists of four main modules, namely, DATA, TEXT, COMMAND and USER module. These modules interact with each other as shown in Fig. 1, in an ATC environment.

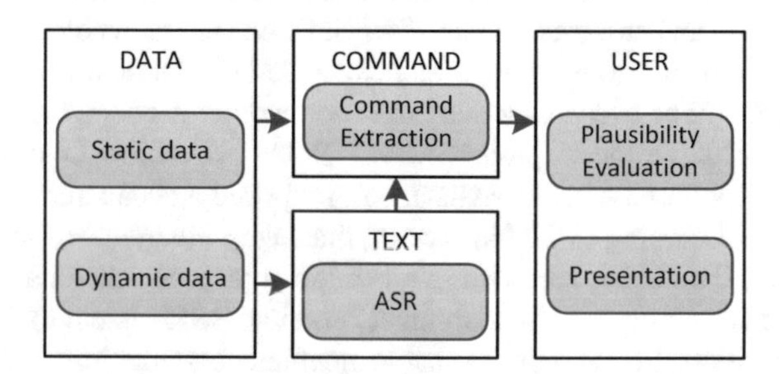

Fig. 1. Conceptual overview of ABSR modules

The DATA module satisfies the airport-specific data requirements of the ABSR system. Static data represents information, which changes rarely like names of waypoints and runways, used frequency values etc. Dynamic data represents information, which is time- or space-dependent information like the ATCos' voice signals, radar data, flight plan information, weather information etc. The TEXT module employs this information to perform Automatic Speech Recognition (ASR) on the ATCos' voice signals. The ASR system converts the voice signal to a sequence of words and in process generating many text hypotheses for the same voice signal.

The COMMAND conceptual module uses information from DATA module and extracts the relevant information from the TEXT module's hypotheses to transform them into ATC command hypotheses. Different hypotheses for a voice signal are generated and hence, the USER module selects a unique output, which is adequately presented to the controller. The USER module utilizes the plausibility values and command hypotheses to perform this task. Further details of these modules, especially, its adaptation to different approach areas are described in [5]. The building blocks are setup as generic part of the system. Data elements and models have to be defined resp. automatically adapted by ML to a specific environment.

4 Proof-of-Concept

The ABSR system was presented to ATCos from Prague and Vienna during the Operational Proof-of-Concept trials of the MALORCA project. The ATCos evaluated the system based on different tasks.

4.1 Operational Trial O1

The operational part O1 for the proof of concept trials performed pairwise comparison of the output of two different ABSR systems. Both systems had to execute command recognitions on recorded voice and radar data sets from Prague and Vienna. Only voice recordings that resulted in different command recognitions for both ABSR systems were taken into account for further analysis. N voice recordings (35 for Vienna and 36 for Prague) were randomly selected from those pre-filtered results.

Experimental Setup. Two ABSR systems were trained: (1) Baseline ABSR, trained only with manually transcribed data and (2) ML (Machine Learning)-Improved ABSR, improved with all available both manually and automatically transcribed data. Each ATCo could replay a voice recording as often as necessary and select between four choices.

1. The output of the ML-improved ABSR system is the better one
2. The output of the Baseline ABSR system is the better one
3. Both outputs are equally good
4. Both outputs are equally bad

The controllers did not know for all presented examples which recognition output is generated by which ABSR system (ML-improved or baseline) and the order of the selectable answers was always randomly generated. The experiment was conducted with four ATCos from Prague and five from Vienna.

Results. The ATCos preferred the results of the ML-improved ABSR system (64%) against the baseline ABSR system (21%). In 15% of the cases they could not decide between one of them.

4.2 Operational Trial O2

The operational part O2 for the proof of concept trials is based on monitoring the output of the ABSR system on a radar screen replaying real radar and audio recordings from the air traffic control rooms of Prague and Vienna.

Experimental Setup. The basic idea of the O2 trials was to give the ATCos an impression on how ABSR would work in a real operation room. The ATCos sat in front of a radar screen. Radar data recorded previously was replayed in real time on the radar screen. Additionally audio recordings that correspond to the radar data were used as input for the ML-improved ABSR system. The voice recordings were injected into the ABSR system and real-time speech recognition was performed. The ATCo listened to the voice recordings, saw the corresponding radar data and the ML-improved ABSR system attempted to recognize the voice recordings in real time. The recognitions were displayed to the ATCo in the radar label. The ATCos task was to monitor the recognitions and to correct the recognitions whenever ABSR fails to deliver the right result. The ATCos could provide feedback both with respect to recognition rate and speed. The scenario lasted approx. 30 min including a 10 min training phase.

Results. The results of the O2 experiments with ATCos from Prague and Vienna are shown in Table 1. Column "corrected by ATCo" indicates that the ATCos did not correct all of the ABSR errors. This happened because the ATCos were not familiar enough with the HMI that was used during the experiments, but all errors were detected by the ATCos and reported to the staff supervising the experiment.

4.3 Debriefing Sessions with ATCos

After the operational Proof of Concept parts O1 (pairwise comparison baseline/ML-improved ABSR) and O2 (monitoring ABSR) a questionnaire was presented to the ATCos. They were asked to answer different questions by assigning a digital value between one and six to each of the questions. "1" means "totally disagree", whereas "6" means "totally agree" (Table 2).

Table 1. Results of operational proof-of-concept O2

Airspace	ATCo	Number of commands	Number of ABSR errors	Corrected by ATCo	Detected by ATCo
Prague	C1	99	9	9	9
	C2	99	9	6	9
	C3	99	9	7	9
	C4	99	9	9	9
Vienna	C1	122	16	16	16
	C2	122	16	16	16
	C3	122	16	16	16
	C4	122	16	15	16
	C5	122	16	16	16
All	9	1006	116	110	116

Table 2. Mean values and standard deviations

	Prague		Vienna		Both	
	μ	σ	μ	σ	μ	σ
I could imagine to work with Speech Recognition support for radar label maintenance	5.8	0.5	5.0	1.0	5.3	0.9
I understood the application of Speech Recognition support for radar label maintenance	6.0	0.0	5.4	0.5	5.7	0.5
Today's support of Speech Recognition was adequate for the presented scenario	4.8	1.3	5.2	0.4	5.0	0.9
Speech Recognition support of MALORCA system would be (already) helpful for my workplace	4.0	1.4	4.4	1.1	4.2	1.2
The application of the MALORCA system would provide an improvement for my work	4.8	1.0	5.2	0.8	5.0	0.9
The number of command corrections was proper with respect to the scenario (traffic density…)	4.3	1.0	5.0	0.7	4.7	0.9
It was easy to do a corrective action and I was able to maintain situational awareness\	5.5	1.0	5.8	0.4	5.7	0.7
ASR will cause safety problems	2.0	0.8	2.5	0.6	2.3	0.7

5 Conclusions

The MALORCA project validated for different controller positions of Prague and Vienna approach area that the automatic adaptation of ABSR is possible by using machine learning technologies. This paper presented the results of the operational Proof of Concept trials from the MALORCA project with ATCos from Prague and Vienna. It showed that ATCos prefer the output of a machine learning adapted ABSR system compared to a basic generic ABSR system. Furthermore the trials with nine different ATCos monitoring recorded real life air traffic from Prague and Vienna showed that recognition rates of machine learning trained system are high enough to reduce controller workload. Even though the ABSR system mostly recognized (88.5%) the given commands, the situation awareness of the ATCos were not negatively affected. They were able to detect all misrecognitions.

The feedback from ATCos after trials showed that they were satisfied with the performance of ABSR and that the system would help in their daily work if available in their operational environment. Currently no system supplier for ANSPs can provide such a technology. The next step must be the integration of ABSR into operational environment.

References

1. The project AcListant® (Active Listening Assistant) http://www.aclistant.de
2. Helmke, H., Ohneiser, O., Mühlhausen, T., Wies, M.: Reducing controller workload with automatic speech recognition. In: IEEE/AIAA 35th Digital Avionics Systems Conference (DASC). Sacramento, California (2016)

3. Helmke, H., Ohneiser, O., Buxbaum, J., Kern, C.: Increasing ATM efficiency with assistant-based speech recognition. In: 12th USA/Europe Air Traffic Management Research and Development Seminar (ATM2017). Seattle, Washington (2017)
4. Kleinert, M., Helmke, H., Siol, G., Ehr, H., Cerna, A., Kern, C., Klakow, D., Motlicek, P. et al.: Semi-supervised Adaptation of Assistant Based Speech Recognition Models for different Approach Areas. In: IEEE/AIAA 37th Digital Avionics Systems Conference (DASC). London, England (2018)
5. Kleinert, M., Helmke, H., Ehr. H., Kern, C., Klakow, D., Motlicek, P., Singh, M., Siol, G.: Building Blocks of Assistant Based Speech Recognition for Air Traffic Management Applications. In: 8th SESAR Innovation Days, Salzburg, 2018, to be published
6. Nguyen, V.N., Holone, H.: Possibilities, challenges and the state of the art of automatic speech recognition in Air Traffic Control. Int. J. Comput. Inf. Eng. **9**(8), 1940–1949 (2015)
7. Chen, S., Kopald, H.D., Chong, R., Wei, Y., Levonian, Z: Read back error detection using automatic speech recognition. In: 12th USA/ Europe Air Traffic Management Research and Development Seminar (ATM2017), Seattle, WA, USA (2017)
8. AIRBUS Air Traffic Control Challenge Workshop: https://www.irit.fr/recherches/SAMOVA/pagechallenge-airbus-atc-workshop.html (2018)
9. Shore, T., Faubel, F., Helmke, H., Klakow, D.: Knowledge-based word lattice rescoring in a dynamic context. In: Interspeech 2012, Sep. 2012, Portland, Oregon (2012)

Ergonomics Index System of Airplane Cockpit Display and Control Resources

Qingyuan Bai(✉), Yang Bai, Xinglong Wang, Xingmei Zhao, and Jin Yu

Shenyang Aircraft Design and Research Institute, No. 40 Tawan St., Shenyang 110035, Liao Ning, China
{baiqingyuan_000, baijordan23, zhaoxingmei9771, 13069871787}@163.com, avic601_wxl@126.com

Abstract. Aiming at cockpit displays and controls system of large commercial airplane, this research discussed characteristics and requirements of ergonomics design of cockpit display and control system. Then, the essential ergonomics factors of display and control system were identified and analyzed based on typical flight task and specific demands. Moreover, the common method of ergonomics research on airplane cockpit were included with the study content of displays and controls. In addition, ergonomics index system was established for evaluation and promotion of airplane cockpit, which could provide theoretical foundation and technical support of interface design of commercial airplane cockpit.

Keywords: Ergonomics evaluation · Index system · Airplane cockpit Display and control

1 Introduction

With the development of modern technology and improvement of industrial manufacture, the requirement of human-machine interaction and ergonomics design of airplane cockpit was increased accordingly. Cockpit was essential for pilot to perform flight task, and its display and control system established direct interaction bridge between pilot and airplane. Therefore, the ergonomics design of airplane cockpit was correlated with the performance of pilot's decision and operation, which had significant influence on flight efficiency and safety. Especially encountered with the technical update of new interaction of displays and controls, airplane cockpit was expected with higher requirement of ergonomics design of information display content and format, as well as control humanity. It meant cockpit ergonomics should be attached with more importance to achieve fully capability of human system integration of pilot and airplane interface. However, the advantages of disadvantages of cockpit design were need verification and validation of reasonable evaluation method and index model, which became popular topic of ergonomics designers and researchers.

W. Karwowski and T. Ahram (Eds.): IHSI 2019, AISC 903, pp. 827–832, 2019.
https://doi.org/10.1007/978-3-030-11051-2_126

Recent studies of ergonomics research on display and control system of airplane cockpit indicated the future tendency of large screen of panoramic display and multimodal interaction [1, 2]. As improvement of aviation electronic technology, the classification and amount of flight information received or produced by airplane subsystem was increasing accordingly that required higher capability of information processing of the whole system. Therefore, the pilots had to be faced with mental overload, which conversely lead to higher requirement of the displays and controls. Moreover, the ergonomics evaluation of airplane cockpit developed many relevant technologies and methods of design assessment, including survey (questionnaire and focus group) and experiment (workload test and psychophysical measurement) [3, 4].

2 Ergonomics Characteristics Analysis

As typical resource of airplane cockpit display and control system, airplane interface played important role of connection with pilot and airplane to perform information and task interaction [5]. As shown in Figs. 1 and 2, there were two types of interface, hardware interface and software interface. The first one included button, wheel, and switch, and the latter one provided visual information interaction of airplane and display realization. And the information interface was mainly contained with visual and auditory interaction, the task interface was heavily relied on physical input device of hand and foot controllers.

Fig. 1. Cockpit of civil aircraft

Fig. 2. Cockpit of military aircraft

2.1 Diversity of Display and Control

Due to incensement of flight information and highly complexity of pilot airplane system, new technologies and applications were adopted in cockpit design to help pilot achieve better performance and ensure flight safety. On the one hand, large screen and head-mounted display became popular component in recent cockpit displays. On the other hand, voice recognition and multi-touch were widely equipped with cockpit controls. Especially the voice recognition could successfully realize a natural way of input method to improve usability and eliminate human error.

2.2 Mental Overload of Pilot

With the development of automation in cockpit system, the pilots were required to process tremendous information and encountered with increment of mental workload, which might cause serials of safety issues related to automation. Such as loss of situation awareness and monitoring inefficiency were difficult for pilot to recover by manual operations. However, under such circumstances, more effective information recognition and interaction methods were necessary, which lead to overload of information observation. Besides, certain flight task required specific manual maneuver and urgent operation within limited moment to complete. Especially during 10 min of landing phase, the pilots need to perform over 100 operations with hundreds of fixation of display interface, which cost only 0.1 to 0.5 s. Therefore, the pilots should keep good match between visual organ and display interface to obtain flight information while they should keep good match between physical organ and control device to ensure coordination with flight safety.

2.3 Variety of Flight Environment

Pilot performance of flight task was significantly affected by various flight environment, which contained both external environment and internal environment. The former one was referred to weather conditions of airplane and the latter one was mainly referred to cockpit environment where glair caused by ambient illuminance had probably severe impact on pilot's recognition and manipulation of airplane.

3 Evaluation Content Analysis

In consideration of physical characteristics of airplane cockpit, the influencing factors of display and control system were consisted of display screens, joystick, throttle, button, foot pedal, and related coordination of control force, device shape, overall layout, and coding [6]. Moreover, those of interaction characteristics were consisted of function definition, interface layout, information content and color coding, maneuver response and format, warning display, voice interaction and touch screen.

3.1 Display

Design of display resources should be fully considered with vision field theory which allowed primary indicators and displays to be located inside the comfort area of pilot's vision. And the display formatting should be designed in reference with relevant standard of human factors guidance to improve pilot's observation and recognition of flight information. Because the displays showed visual information and realized human-machine interaction, its visual characteristics were highly concerned during design and evaluation. Such as contrast ratio, luminance balance, luminance scope, refreshing frequency, resolution and readability were required to involve with ergonomics issues of the displays. And it should be also considered with the display content of flight information, which included of symbol content and allocation, information density and chaos, display format, information cuing and hint, as well as information fusion of hierarchy and priority. In addition, the visual coding of flight information should be also taken into account, especially in color matching, warning display and information salient.

3.2 Control

The movement direction of controls was firstly considered in ergonomics design and evaluation of airplane cockpit and should be in accordance with pilot's input and expected response. Then, the feedback of controls should convey correct physical feelings to the pilots to understand the airplane was under control. In addition, the control lighting should provide appropriate illuminance condition to support clear location and recognition of control operation, which could not cause any interference of external environment view or internal cockpit readability. Besides, layout design of the controls should be followed in reasonable and logical position allocation in consideration of primary function grouping and operation ordering as well as frequency.

4 Ergonomics Index System

As shown in Table 1, the ergonomics design of the cockpit display control system for civil aircraft was selected as the first-layer index of the evaluation index system. The second-layer indices were contained in four categories. The overall layout mainly was referred to the layout of the aircraft cockpit dashboard and console, such as the layout of the power meter device. The display was usually involved with visual or auditory display, such as head-up display, multi-function display and voice talker. Control devices were composed of avionics control equipment and flight control equipment, including buttons, keyboards, knobs, steering column and steering rudder. Besides, alarm warning was another essential index that represented for sound signal frequency, tune and so on. Furthermore, the ergonomic design indices corresponding to each secondary index were the bottom-layer indices.

Table 1. Index system

Top layer	Middle layer	Bottom layer
Ergonomics evaluation	Layout	Visibility of the dashboard
		Accessibility
		Consistency
		Functional grouping
		Corresponding to the supporting equipment
		Clarity of motion position relationship
	Display	Resolution
		Size
		Contrast
		Brightness
		Intuitive clarity of display information forms
		Visibility of information
		Audible of information
		Admissibility of information
		Compatibility of information
	Control	Component accessibility
		Size of control part
		Keeping readability
		Error-proof design
		Handle shape
		Motion range
		Manipulation force
		Sensitivity
		Safety
	Alarm	Clarity warning signal
		Difference between alarm signals of alarm
		Rationality of the number of alarm signals
		The influence of alarm signal on operation
		Mode and category of alarm
		Light position of alarm
		Leakage of alarm

5 Conclusion

This research established three-layer index system of display and control resource of airplane cockpit based on characteristics analysis and evaluation review. And the index system was intended to guide ergonomics evaluation and promotion of human-machine interaction of airplane cockpit.

References

1. Yang, H.: Research on design method of cockpit control and display. China Instrum. (04), 56–58 (2015). (in Chinese)
2. Fu, S., Wang, L.J., Huang, D.: Annual report on theory and method of ergonomics comprehensive evaluation for civil aircraft cockpit. Sci. Technol. Inf. **14**(13), 179–180 (2016). (in Chinese)
3. Yuan, X., Hao, D.J., Liu, H.Y., Jin, Z.F., Dong, D.Y.: Research on evaluation method of man-machine interface for civil aircraft cockpit. Civil Aircr. Des. Res. (01), 17–22 (2017). (in Chinese)
4. Pei, H.J., Ding, Y.Y., Jiang, Y.L., Gao, Z.G., Shi, H.: Research on ergonomics compliance authentication method and evaluation method of civil aircraft cockpit. Civil Aircr. Des. Res. (04), 52–57 (2017). (in Chinese)
5. Ye, K.W., Bao, H.: Application of fuzzy AHP based on entropy weight in aircraft cockpit ergonomics evaluation. Aircr. Des. **37**(04), 54–56 (2017). (in Chinese)
6. Xiao, L.: Mission oriented ergonomics evaluation index system for fighter aircraft cockpit. Mech. Eng. (06), 88–90 (2018). (in Chinese)

Evaluation of Helmet Comfort Based on Flexible Pressure Sensor Matrix

Xiao Chen[1], Cong Zhang[2], Chuang Ma[2], Haixiao Liu[2], Yanling Zheng[2], Yi Jiang[1], Yuanyuan Zu[1], and Jianwei Niu[2(✉)]

[1] Military Institute of Engineering and Technology, Academy of Military Sciences, Beijing 100088, China

[2] School of Mechanical Engineering, University of Science and Technology, Beijing 100083, China

niujw@ustb.edu.cn

Abstract. The helmet comfort has been a focus in infantry equipment fitting evaluation, and has not been solved well for decades. In order to evaluate the stability and comfort of helmets, this paper proposed a novel helmet comfort evaluation method, combined with quantitative analysis and subjective rating. We demonstrated the helmet pressure acquisition system, and we also presented the data analysis and display system. Helmet pressure data between the skins of the wearer and the cushion pads lying in the helmet inner liner was collected by using a pressure headgear, which was composed by a matrix of flexible pressure sensors. The system includes the head pressure sensing matrix composed of more than one hundred flexible pressure sensors, pressure data analysis and display module, and helmet comfort evaluation module. We carefully chose the head model according to the latest anthropometry industrial standards, in which the head dimensions were illustrated in details. We used Geomagic, a popular re-engineering soft package in computer graphics, to construct the head model on which we would put the pressure sensing matrix. The original resistance values of the pressures were transferred by A/D conversion circuit from analog to digital numbers and then output to the data analysis module. In this way, the pressure data of each sensor in the helmet inner liner was collected. We asked several experts to rate the helmet comfort and then used Analytic Hierarchy Process (AHP) to calculate the final comfort score. In the criterion layer of AHP technique, we proposed four criteria, i.e., duration-related metrics, pressure amplitudes, pressure distribution and pressure stability. Finally, the pressure distribution image was showed on the computer by using the Graphic class in C#. Experimental results showed quite good agreement between our evaluation score and the subjective feeling of the subjects. The proposed technique makes the pressure data measurement more intuitive and efficient, and it is quite convenient to analyze and evaluate the factors affecting the ergonomics of the helmet, to optimize the safety and fitting of the helmet rationally. It's also expected to expand this technique in other human body wearable products, such as goggles, glasses, ear-phones and neck brace.

Keywords: Helmet Comfort · Fitting Evaluation · Flexible Sensor Pressure Sensing Matrix

W. Karwowski and T. Ahram (Eds.): IHSI 2019, AISC 903, pp. 833–839, 2019.
https://doi.org/10.1007/978-3-030-11051-2_127

1 Introduction

The modern helmet is used to prevent the shrapnel from penetrating the head, and protect the head from directly colliding with hard and high-energy objects, and it can effectively protect the human head, eyes and face. It's argued the helmet should not affect the human's line of sight and head and neck's movement. It has a wide range of applications in military, medical, sports, entertainment and other industries. Uncomfortable helmets will accelerate the fatigue of the wearer [1] and distract his/her attention, and these seriously affect the efficiency and safety of the wearer. In order to ensure the ergonomic of the helmet, numerous countries have carried out in-depth researches, covering data acquisition, modeling and standardization of three dimensional head, and detection of the physical properties of helmet such as center of mass, and moment of inertia center of mass [2], head geomagic feature description [3], dynamic response of helmet biomechanical system to impact force [4] and other characteristics. US Air Force Operations Center conducted an experiment on the center of mass of the "head-to-flight helmet" system by using the KGR30 mass analyzer produced by Cosmos Electronics (https://cosmoselectron.com/), but the study failed to measure the center of mass of the irregular helmet directly. University of Melbourne established an evaluation device for the dynamic performance of the SPH-4helmet [5], but the device is over complicated and can be easily affected by the physiological factors of the helmet wearer.

Currently, Chinese military helmets were designed according to the head anthropometric dimensions of Chinese soldiers. However, because lacking the standard head model, the commercial helmets on the market in China are usually designed according to the face and head shape of Westerners. In addition, there are few objective evaluation methods for helmet shape, comfort and stability. The purpose of this paper is to develop a helmet testing device and software, to establish a hybrid evaluation model of ergonomic properties of various helmets, and to propose guidelines for optimizing the ergonomic design of current helmets.

2 System Design

The system includes a head matrix, a sensing device, a computer, and a data analysis and display module (see Fig. 1). The system selects the appropriate head models according to the evaluated helmet, then the sensing device transmits the pressure value between the head model and the helmet to the computer, and the data analysis module calculates the comfort index of the helmet, thereby to evaluate whether the helmet is suitable for the anticipated person in sizing.

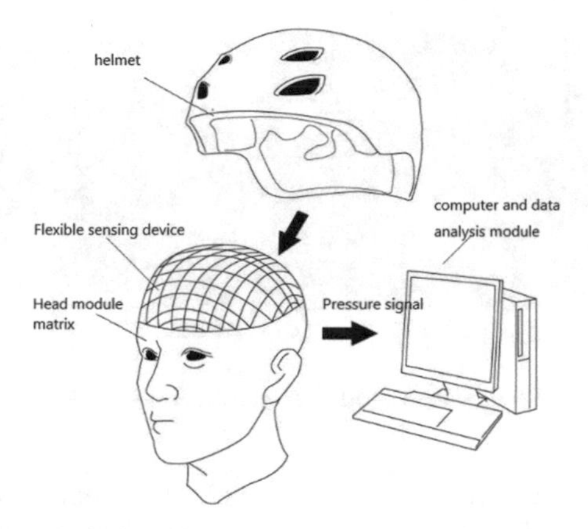

Fig. 1. Overall functional design of the system

We designed a standard head model matrix that was based on the head-size measurements and their relationship, which was described by the national standard GB/T23461-2009 "Three-dimensional size of adult male heads" [6]. We applied reverse modeling technology and used modeling software such as UG (https://www.plm.automation.siemens.com/global/zh/products/nx/) in order to establish a head model close to Chinese head. We established multiple head models to make the conclusion representative of Chinese people. On the external surface of the head model, a flexible sensor network was used to collect the pressure values between the head and the helmet. Through the data acquisition and analysis of the host computer, the system could complete the measurement and conduct the evaluation of helmet fitting.

3 Head Modeling Matrix Design and Data Acquisition Subsystem

For the head modeling process, direct modeling is often used, including the 3D scanning methods [7] and the anatomy-based model construction methods [8]. But the extremely high cost often prevents their wide application, so this paper uses reversing model construction technique. Inspired by the concept of free form deformation (FFD) [9], a powerful digital sculpture tool in Computer Aided Design (CAD), we proposed a novel modeling approach of human head from the unorganized point cloud. As shown in Fig. 2, we constructed a standard head model by stratified laser scanning of point clouds on the human head surface. First, we constructed a standard head from laser scanned 3D unorganized points. Second, we applied FFD to the standard head to perform customization of head size and shape according to the user-specified head dimensions. In this paper, we established head model matrix based on different head types whose circumference is between 52 cm ~ 60 cm, such as Round High, Round Regular, Medium Regular, Medium High, Round Extra High, Super Round High and Super Round Super High. In this way, we obtained forty-four different typical head models finally.

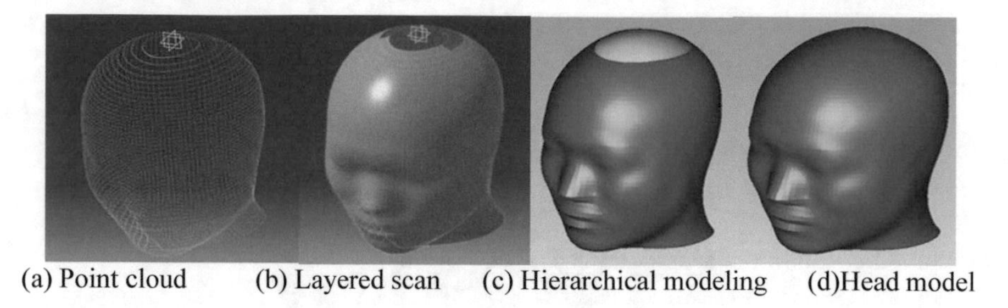

(a) Point cloud (b) Layered scan (c) Hierarchical modeling (d)Head model

Fig. 2. Head mode reconstruction procedure

The data acquisition system (see Fig. 3) includes a bottom silica layer which closely attach to the head surface, and a flexible sensor matrix network on the bottom silicone layer, where one pressure sensor is attached to each intersection of the matrix network. The pressure sensor is welded and integrated into a whole by a coiled wire, and the matrix network is connected to a control module for outputting signals and a power supply module is for supplying power. The control module includes a row of gate switching circuits, transport and amplification pathway circuit, and an A/D conversion circuit. The row gating circuit was used to control the gating of the matrix network row, and the pressure sensors in the gating row sequentially amplify the inductive current values through the op amp channels of each channel, and execute analog to digital conversion by the A/D conversion circuit. The pressure values is converted and outputted to the data analysis module via the computer. A top surface silicone layer was disposed on the bottom layer which the pressure sensor was bonded, the top silicone layer was bonded with the bottom silicone layer, the pressure sensor and the wire into a unitary flexible sensing matrix. Since the sensing device consisted with a silicone layer on the bottom surface and a silicone layer on the top surface, it not only provides a uniform bearing surface for the pressure sensor array, but also plays the role of electrical insulation to ensure stable operation of the circuit in the pressure sensor array.

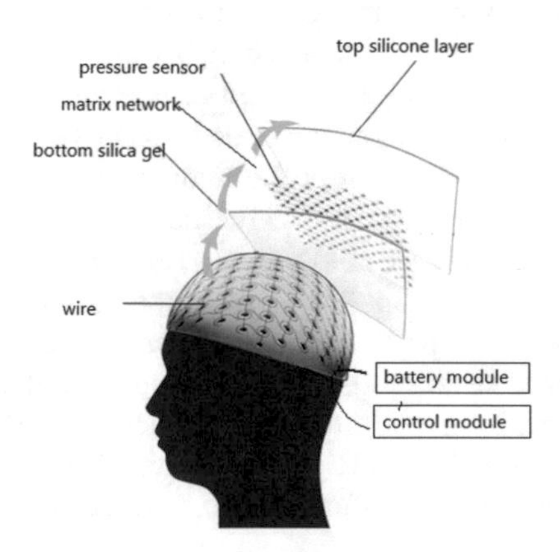

Fig. 3. the schematic of data acquisition system

Once the appropriate head model was selected, the data acquisition system was installed on the selected head model, and the fitting pressure data of the helmet was gathered through the pressure sensing matrix.

4 Data Analysis and Display Subsystem

The data analysis module is composed with a helmet ergonomic analysis unit and a display unit (see Fig. 4). After the sensing device tests the pressure value on the head model, the ergonomic performance analysis unit calculates the relevant helmet fitting evaluation index.

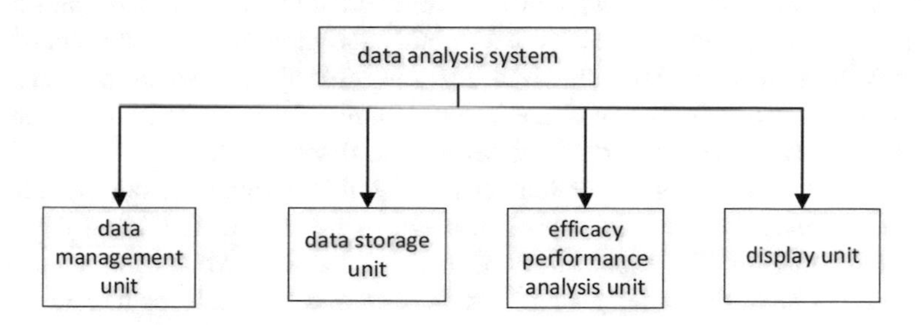

Fig. 4. Schematic diagram of the data analysis system

The display unit includes a cloud map and a bar plot. The user can view the distribution of pressure on the head model from the cloud map. The mean pressure and peak pressure can be displayed via the bar plot. Interpolation was used on the coordinate data and pressure data to make the cloud map smooth, as shown in Fig. 5. The helmet fitting could be seen according to the pressure color on the cloud map.

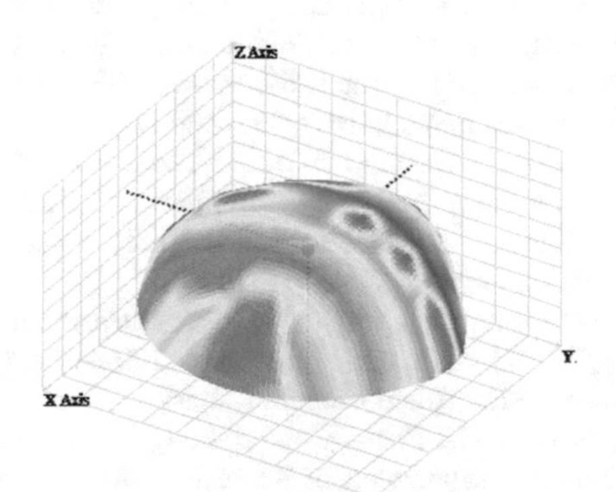

Fig. 5. Pressure cloud displayed by cloud image module

Evaluating helmet whether is suitability by AHP (Analytic Hierarchy Process). The analytic hierarchy process (AHP) is a theory of measurement through pairwise comparisons and relies on the judgements of experts to derive priority scales [10]. Collecting the data of indicator layer, then comparing these data with the standard of criteria level. For example, we compare pressure impulse value with time correlation criterion, then we can give a score by Delphi method. We can evaluate this helmet according to this score.

5 Discussion and Conclusion

The novel helmet comfort evaluation system based on flexible pressure sensor matrix includes three parts: head model matrix, data acquisition module and data analysis module. The head model matrix is aimed to build the more representative head model for Chinese by reverse modeling method. By using this system, when the trial gets the data from the sensors, he/she only need to compare with the recommended standard pressure value, then the comfort of helmet will be determined.

The system is meaningful for helmet comfortable evaluation, but there are still some limitations. In the data transmission between the data analysis and the sensors, wireless transmission was expected more to bring more convenience. In the fugure, shear force between the inner liner of the helmet and the scalp or hair of the head should be taken into consideration.

References

1. Gallagher, H.L., Caldwell, E.: Neck Muscle fatigue resulting from prolonged wear of weighted helmets. Wright-Patterson AFB. Biomechanics Branch, Biosciences and Protection Division, Human Effectiveness Directorate, Air Force Research Laboratory, OH. Report No. AFRL-RH-WP-TR-2008-0096
2. Blake, L.C., Philip, J.O., Stephen, P., et al.: Measuring the mass and center of gravity of helmet systems for underground workers. Int. J. Indus. Ergon. **64**, 23–30 (2018)
3. Robinette, K.M., Whitestone, J.J.: Methods for characterizing the human for the design of helmets. Wright-Patterson Air Force Base **3**, 397–409 (1992)
4. Katherine, M.B., Evan, L.B., Thomas, G.B., et al.: The effect of football helmet facemasks on impact behavior during linear drop tests. J. Biomech. **79**, 227–231 (2018)
5. Vilenius, Asko T.S., Williams, J.F.: Aspects of SPH-4 helmet static and dynamic behaviour. Institution of Engineers, Australia, pp. 543–549 (1996)
6. Institute of Quartermaster Equipment, General Logistics Department of the Chinese People's Liberation Army. GB/T23461-2009 Three-Dimensional Size of Adult Male Heads. Beijing: China National Standardization Administration (2009)
7. Trieb,R., Seidl, A., Hansen G., Pruett C.: 3-d body scanning—systems, methods and applications for automatic interpretation of 3d surface anthropometrical data. In: Human Factors and Ergonomics Society Annual Meeting Proceedings, pp. 844–847 (2000)
8. Kahler K., Haber J., Yamauchi H., Seidel H.P.: Generating animated head models with anatomical structure. In: Proceedings of the ACM SIGGRAPH symposium on computer animation, pp. 55–64 (2000)

9. Sederberg, T.W., Parry, S.R.: Free-form deformation of solid geometric models. SIGGRAPH Comput. Graph. **20**(4), 151–160 (1986)
10. Analytic hierarchy process. https://en.wikipedia.org

Estimation of Mental Workload from Information About Peripheral Vessels by Non-contact Measurement Using Microwave Radar

Satoshi Suzuki[1,2(✉)], Yuta Terazawa[1], Kentaro Kotani[1], and Takafumi Asao[1]

[1] Faculty of Engineering Science, Kansai University, 3-3-35, Yamatecho, Suita, Osaka 564-8680, Japan
ssuzuki@kansai-u.ac.jp
[2] Department of Industrial & Manufacturing Engineering, Pennsylbania Stare University, 310 Leonhard Bldg, University Park, PA 16802, USA

Abstract. In this research, we aimed to estimate changes in mental workload from information on organic changes in peripheral vessels as measured by a non-contact sensing method using microwave radar. Experiments were conducted in a laboratory using two simple tasks, namely, a mental arithmetic task and an auditory task, in reference to the literature. Some previous studies have reported differences in the autonomic reflexes induced by voluntary and involuntary tasks. Moreover, these studies also found two types of physiological reflexes in response to the tasks. Heart rate variability (HRV) was induced by voluntary behavior, while stress was induced by a passive stimulus and showed an association with fluctuation of blood flow due to a change in the resistance of peripheral vessels. In our results, we detected changes in mental stress, undetectable by HRV, by using information about peripheral vessels measured by microwave radar without directly touching the body. We discuss possible application in driver state estimation via a driver–vehicle interface.

Keywords: Mental workload · Non-contact · Physiological measurement Estimation

1 Introduction

To estimate mental workloads from physiological indices, heart rate variability (HRV) is conventionally used to characterize changes in the autonomic nervous system [1]. However, some research has indicated that HRV is not sufficiently sensitive to mental workload, meaning that estimation of mental workload from HRV alone will have limited accuracy. Changes in peripheral vascular resistance as a stress-induced reaction are expected to contribute to the accuracy of estimating mental workload [2, 3]. Therefore, this information about peripheral vascular changes should be taken into consideration.

W. Karwowski and T. Ahram (Eds.): IHSI 2019, AISC 903, pp. 840–845, 2019.
https://doi.org/10.1007/978-3-030-11051-2_128

In recent years, novel non-contact methods of monitoring vital signs using microwaves have been proposed. These non-contact and unrestricted sensing techniques measure motion on an extremely small scale, including movement occurring on the body surface as a result of cardiac activity [4]. This type of technique has not only successfully measured heart rate but has also been used to estimate detailed fundamental information such as changes in HRV [5]. Moreover, these methods can measure cardiac information from peripheral parts of the body, which suggests the possibility of measuring changes in peripheral vascular resistance caused by mental stress.

In this research, we aimed to estimate changes in mental workload from information on organic changes in peripheral vessels measured by a non-contact sensing method using microwave radar.

2 Methods

2.1 Strategies of Sensing Methods and Measurements

Changes in the autonomic nervous system as a stress response in the cardiovascular system serve to control the volume of blood flow sent to the whole body. Since the driving force is based on the pressure produced by the heart, it is necessary to take blood pressure into account. Blood pressure is defined as the product of cardiac output and peripheral vascular resistance. HRV appears mainly as a result of controlling stroke volume. In contrast, peripheral vascular resistance is determined by organic changes in blood vessels. Therefore, it is necessary to capture information on the organic changes of blood vessels if vascular resistance is to be measured.

To estimate these organic changes, it is common to use the characteristics of blood propagation to estimate changes in blood vessels from two parameters: pulse transit time (PTT) or pulse wave velocity (PWV). Pulse volume (PV) is also used as an index of the stiffness of blood vessels during measurement of the pulse wave. For example, PTT is measured by using electrocardiography (ECG) and the pulse wave. The R wave of the ECG indicates the timing when blood is ejected from the heart, and the peak in the signal of the pulse wave at a peripheral part of the body indicates the timing when the pulsatile flow of blood arrives at the peripheral part. The lag between the start of the R wave and the peak of the pulse wave is treated as the PTT. PV is the amplitude of each beat of the pulse wave, and is treated as an indicator of the extent of dilation (or contraction) of the blood vessels. Following this general technique, in this research, microwave sensors were installed near the trunk close to the heart and near a peripheral part of the body, and the changes of these indices were observed.

2.2 Experimental Strategies

On reason why stress reactions are hard to grasp from HRV alone is that the reactions depend on the type of task. Some stress responses from active tasks performed voluntarily, such as lane changes during automobile driving, can be measured by HRV. However, HRV is not observed in passive and involuntary tasks that involve receiving external stimulation unilaterally, such as interruption by another car. In such cases, the changes appear to happen in terms of peripheral vascular resistance [2, 3].

Therefore, we prepared two tasks of different types, following previous studies, namely, a mental arithmetic (voluntary) and an auditory task using sound stimuli with white noise (involuntary). These tasks have been used in many laboratory studies.

In the voluntary task, two single-digit numbers were conveyed by sound to the participants, who mentally added the two numbers and answered orally by providing only one digit of the answer. This was based on the Uchida–Kraepelin test, which has been widely used in research on work performance. The presentation interval was 3 s, the numbers presented were random, and each presented number was different from prior one to avoid duplication.

In the involuntary task, we transmitted white noise from 30 to 90 dB to participants via headphone. The sound level of noise was controlled between 30 and 90 dB with slight changes and repeated fades in and out to avoid activating rapid enhanced activity of the sympathetic nerve system in response to surprising loud sounds.

2.3 Experimental Setting

Five healthy male students were recruited as participants. The experimental tasks (described above) were conducted in the following way. A participant sat in a seat modeled after the driver's seat of an automobile. He was first encouraged to rest for a few minutes, during which control data of NASA Task Load Index (NASA-TLX), ECG, pulse waves, and signals from microwave radar sensors were acquired. After the rest, the tasks were given to the participant. The sequences of tasks were randomized for each participant to avoid order effects. Each task required about 2 min. Including the rest period, each participant spent about 6 min performing the experiments.

A weighted workload (WWL) score was calculated from the NASA-TLX data, which were collected during the rest and after each task. This score was used as the standard for comparison of the results from each task. During the experiments, an ECG (V5 position) and pulse wave were measured from the end of the middle finger of the left hand to serve as reference values for the physiological measurements. At the same time, monitoring by microwave radar sensors installed on the seat was performed. One microwave sensor was installed in the back of the seat, located below the underside of the scapula to information similar to that of an ECG near the heart (below, the trunk radar sensor). The other sensor was installed under the seating face to monitor motion on the left thigh as information of the peripheral part (below, the peripheral radar sensor). The data acquired from all physiological measurement devices and microwave radars were passed through an A/D converter with sampling frequency of 1 kHz and imported to a personal computer with time synchronization. After storing all data on the personal computer, the HRV was calculated from the ECG data, PV was calculated from the pulse wave, and PTT was calculated from the ECG and pulse wave. The values corresponding to HRV, PV, and PTT were also calculated based on the data acquired by the two microwave radar sensors.

3 Results

Figure 1 shows the WWL scores measured by NASA-TLX at rest (left) and during the experiments (right). In the figure, squares indicate the voluntary mental arithmetic task and circles indicate the involuntary auditory task. Although the two distributions differ in small ways, they clearly showed changes in stress caused by both tasks, as seen by comparison with the results from the rest as a control.

Figure 2(a) shows the HRV calculated from ECG data, and Fig. 2(b) shows the HRV as estimated from data acquired by the trunk radar sensor (markers and placement are as in Fig. 1). The HRV calculated from ECG data during the voluntary task (Fig. 2 (a)) shows the same increase as recorded by NASA-TLX relative to rest. However, the HRV calculated from ECG data during the involuntary task does not show an increase compared with the control value. This result is expected from the hypotheses of previous research [2, 3]. Also, the HRV during the voluntary task as estimated from radar data (Fig. 2(b)) shows a slight increase, agreeing with the HRV calculated from ECG data. However, there is very little difference in the HRV estimated from radar data between the rest and voluntary task conditions.

Figure 3(a) shows PV calculated from ECG data, and Fig. 3(b) shows PV estimated from data acquired by the peripheral radar sensor. PV during the voluntary task as calculated from the pulse wave (Fig. 3(a)) is slightly lower than the control value. Similarly, PV during the involuntary task, as calculated from the radar sensor data showed a decrease compared with the control. For PTT, no significant differences were found between rest and the tasks, with no significant difference by method.

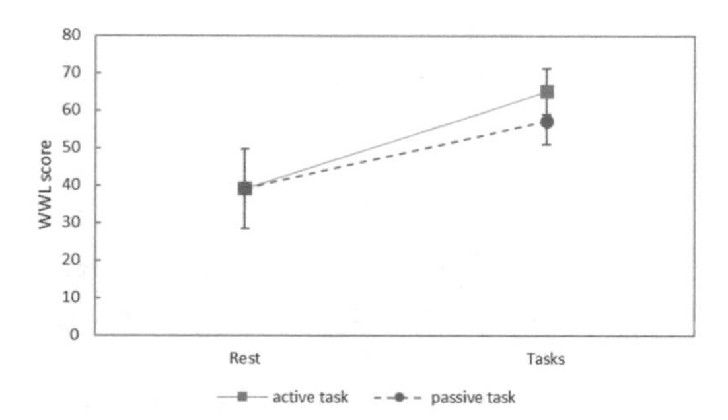

Fig. 1. Results of subjective evaluation by NASA-TLX as standard for comparison with results of objective evaluation using physiological measurements.

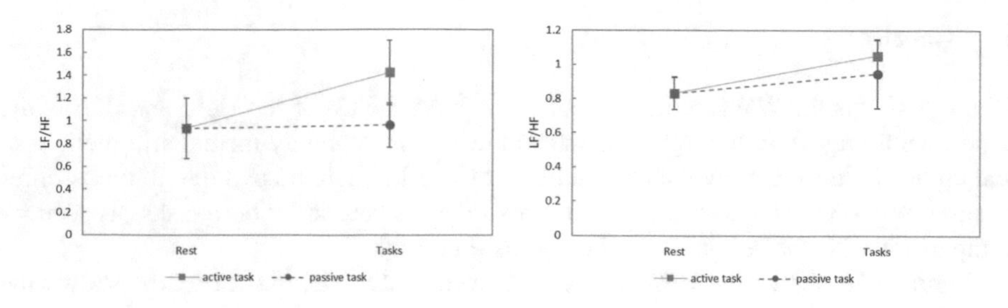

a) HRV calculated from ECG data

b) HRV estimated from data acquired by the trunk radar sensor

Fig. 2. HRV as calculated by two measurement methods.

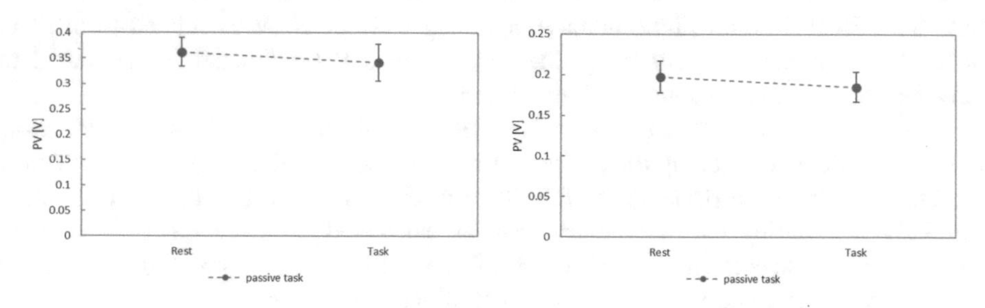

a) PV calculated from pulse wave data

b) PV estimated from data acquired by the peripheral radar sensor

Fig. 3. PV as calculated by two measurement methods.

4 Discussion

The results in this research mostly agree with the findings of previous research [2, 3]. First, we confirmed that mental stress can be estimated using HRV in only some situations. Specifically, the difference in ability to estimate mental stress from HRV depends on whether the task is active or passive. In involuntary tasks, mental stress cannot be estimated from HRV alone, but it may be estimable by using information from peripheral vessels.

According to a previous study [6], task stress increases sympathetic activity via enhanced alpha adrenergic activity, and the constriction of small vessels occurs. The elasticity of arteries is correspondingly decreased, with arteries becoming harder. Along with this, PV is decreased and PTT is shorter. Although the effect of mental stress from involuntary tasks could not be confirmed from changes in PTT sufficiently during our experiments, our results show the possibility of detecting signs mental stress from PV. Moreover, these results indicate the possibility to perform this detection by a non-contact automated sensing technique, and these sensing techniques may open new possibilities.

Currently, we are conducting experiments to monitor mental stress by using this non-contact sensing technique and getting positive results. This sensing technique and estimating method are potential candidates for estimating the mental state of car drivers in the future.

Acknowledgments. This work was supported by the Overseas Research Program of Kansai University.

References

1. Riener, A., et al.: Heart on the road: HRV analysis for monitoring a driver's affective state. In: Proceedings of the First International Conference on Automotive User Interfaces and Interactive Vehicular Applications, pp. 99–106 (2009)
2. Williams, R.B.: Patterns of reactivity and stress. In: Matthews, K.A., Weiss, S.M., Detre, T., et al. (eds.) Handbook of Stress, Reactivity, and Cardiovascular Disease, pp. 109–125. Wiley, New York (1986)
3. Schneiderman, N., McCabe, P.M.: Psychophysiologic strategies in laboratory research. In: Schneiderman, N., Weiss, M., Kaufman, P.G. (eds.) Handbook or Research Methods in Cardiovascular Behavioral Medicine, pp. 349–364. Plenum Press, New York (1989)
4. Suzuki, S., Matsui, T., et al.: An approach to remote monitoring of heart rate variability (HRV) using microwave radar during an input task. J. Physiol. Anthropol. **30**(6), 241–249 (2011)
5. Suzuki, S., Matsui, T., et al.: A novel autonomic activation measurement method for stress monitoring: non-contact measurement of heart rate variability using a compact microwave radar. Med. Biol. Eng. Comput. **46**(7), 709–714 (2008)
6. Allen, M.T., Obrist, P.A., Sherwood, A., Crowell, M.D.: Evaluation of myocardial and peripheral vascular responses during reaction time, mental arithmetic, and cold pressor tasks. Psychophysiology **24**, 648–656 (1987)

Building an Argument for the Use of Science Fiction in HCI Education

Philipp Jordan[1(✉)] and Paula Alexandra Silva[2]

[1] University of Hawai'i at Mānoa, Honolulu, USA
philippj@hawaii.edu
[2] DigiMedia Research Center, University of Aveiro, Aveiro, Portugal
palexa@gmail.com

Abstract. Science fiction literature, comics, cartoons and, in particular, audio-visual materials, such as science fiction movies and shows, can be a valuable addition in Human-computer interaction (HCI) Education. In this paper, we present an overview of research relative to future directions in HCI Education, distinct crossings of science fiction in HCI and Computer Science teaching and the Framework for 21st Century Learning. Next, we provide examples where science fiction can add to the future of HCI Education. In particular, we argue herein first that science fiction, as tangible and intangible cultural artifact, can serve as a trigger for creativity and innovation and thus, support us in exploring the design space. Second, science fiction, as a means to analyze yet-to-come HCI technologies, can assist us in developing an open-minded and reflective dialogue about technological futures, thus creating a singular base for critical thinking and problem solving. Provided that one is cognizant of its potential and limitations, we reason that science fiction can be a meaningful extension of selected aspects of HCI curricula and research.

Keywords: HCI Education · Popular culture in science · Science fiction

1 The Future of HCI Education

In a 2016 summary article, Churchill, Bowser and Preece [6] outline current priorities in Human-computer interaction (HCI) Education. Among others, the article discusses a four-year long initiative [32] by the Special Interest Group on Computer-Human Interaction (SIGCHI), which assessed the needs and requirements of the current and future HCI curriculum subjects.

Using cross-cultural survey data from a broad, international sample, interview studies, results from discussions at CHI workshops and town-hall conferences, the SIGCHI HCI Education project elicited a total of 114 discrete topics in HCI instruction. In addition, the project created a repository of contemporary HCI courses and curricula [33], across a variety of universities and departments.

The survey respondents—a mix of international HCI professors, practitioners and students [32, pages 7–8]—named design research in addition to qualitative evaluation methods, as some of the top priorities in HCI research. The SIGCHI HCI Education project also created an organized and representative overview of relevant HCI classes

W. Karwowski and T. Ahram (Eds.): IHSI 2019, AISC 903, pp. 846–851, 2019.
https://doi.org/10.1007/978-3-030-11051-2_129

and syllabi [33], including (i) Introductory HCI classes, as well as courses on (ii) Design, Values & Ethics or (iii) Artificial Intelligence (AI).

2 Science Fiction, 21st Century Skills and HCI Education

Science Fiction. Science fiction, including Design Fiction[1], in its diverse variations and media forms – written, illustrated, audio-visual or interactive; short story, cartoon or illustrated comic; video clip, show or movie; board or video-game – can be a valuable addition in the future of HCI Education. In Michalsky's essay [19, page 248] from 1979, the author concludes that 'speculative fiction' is:

> […] a tool in coping with the onrushing future. Studying speculative fiction offers the student the opportunity to be more creative in his thinking about the future and thus augment the options for possible tomorrow.

The use of science fiction in educational contexts and classrooms has been both, a subject of debate and research in Computer Science and a diversity of other related Science, Technology, Engineering and Mathematics (STEM) fields [39]. While the thought of a 'science fiction-inspired HCI curriculum and research agenda' [21] is still a pipe dream, noteworthy efforts[2] have been made to integrate science fiction into HCI-relevant classes across universities in the United States (see Table 1); thus, acknowledging the educational potential of science fiction.

Table 1. HCI-relevant classes which use aspects of science fiction.

Course Alpha	Course Title	Institution
CS 190 [37]	Robotics Freshman Seminar	Emory University
CS 201 [20]	AI and Science Fiction	Minnesota State University
CS 463 [10]	Introduction to AI and Science Fiction	University of Kentucky
CS 585 [9]	Science Fiction and Computer Ethics	University of Kentucky
STS 1500 [38]	Science, Technology and Contemporary Issues: Considering the Future through Fiction	University of Kentucky
STS 2500 [38]	Science Fiction and the Future: The Frankenstein Myth in Emerging Biotech	University of Virginia
MAS S64 [3]	Sci Fab: Science Fiction-Inspired Prototyping	Massachusetts Institute of Technology
MAS S65 [4]	Science Fiction to Science Fabrication	Massachusetts Institute of Technology

21st Century Skills. The Partnership for 21st Century Skills (P21) [36] is widely recognized as the crucial and visionary framework in teaching and development of knowledge for the next-generation of students and teachers. In a nutshell, the P21 framework [25] contains of twelve skills, categorized into three larger domains:

[1] Design Fiction is an emerging method in Design Research, see for example [16].

[2] Such as the ASU Center for Science and the Imagination [1] or the UCLA-based Science and Entertainment Exchange [23].

i) Learning and Innovation, ii) Digital Literacy, and iii) Career and Life Skills. In particular, the '4C skills' in the Learning and Innovation domain, are considered as the profound future learning competencies by the National Education Association and former President Obama [24, page 5].

Creativity & Innovation versus Critical Thinking & Problem Solving. In this paper, we argue that two of those four competences – *Creativity & Innovation* and *Critical Thinking & Problem Solving* – can be nurtured by resorting to science fiction in the context of HCI Education.

While both skills, *Creativity & Innovation* and *Critical Thinking & Problem Solving*, are viewed independently in the P21 framework, they are naturally interrelated, with the former usually being the preceding cognitive process to the latter. Lin et al. [15, pages 198–199] provide an example of the co-occurrence of both plus the finding, that science fiction films in the context:

> […] of practical educational activities can stimulate students' imaginations and enhance their ability to design product improvements.

While the first part of above quote highlights how science fiction clearly sparks *Creativity & Innovation*, the remainder underlines how one needs to engage in a critical reflection process—*Critical Thinking & Problem Solving*—in order to be able to effectively achieve product improvements. Thus, according to the P21 framework, Creativity & Innovation can be seen as [26, page 1]:

> the ability to produce and implement new, useful ideas […].

Due to it's strong – in the case of science fiction movie and shows, audio-visual – context, embedded in a rich narrative, science fiction has the potential to allow students and educators to explore the full bandwidth of the design space; from positive (utopian) to negative (dystopian) future visions.

Despite of frequently representing 'unrealistic' and 'technologically implausible' futures, science fiction materials can serve as creativity triggers and be pivotal to uncover new design possibilities. For example, a fictional robot from the Disney movie Big Hero 6 [11], called 'BayMax', has inspired researchers to create (and evaluate) a real-world care robot called 'Puffy' – an innovative companion for children with neurodevelopmental disorder [8].

Critical thinking & Problem Solving can concisely be defined as [27, page 1]:

> strategies we use to think in organized ways to analyze and solve problems.

Again, due to its rich context and narrative, but also as a cultural artifact, science fiction can be useful in outlining ethical concerns, and to spark dialogue, analysis and reflection about yet-to-come interfaces, interactions, devices and technological outcomes. Uncovering potential moral and societal implications can be particularly interesting when the discussion is led by questions on the how, when, whys and why nots a given Human-Technology interaction is acceptable, plausible, and so forth.

By coupling science fiction examples with prompt questions which lead to reflection, we can make our knowledge and concerns explicit and thus, adjust to future design endeavors. A similar strategy has been previously applied in educational and creativity contexts [35].

Science fiction has been used in teaching computer ethics [5] and computer security [14] classes, therefore encouraging alternative viewpoints and extending traditional technical foci in HCI Education. Rogers [28, page 679] refers to science fiction movies and shows as *"culturally current media"*, which can not only illustrate good and bad user interface design, but also support the development of *"design strategies, application and evaluation"* for innovation in HCI curricula.

3 Discussion and Concluding Remarks

Science fiction in HCI research has been discussed prior [12, 13]. The value of fictional visions of the future was introduced as early as 1992 in a CHI panel [18]—about 25 years ago. Science fiction can be used to imagine the tomorrow [17], to inspire HCI research [29] and user interface design [7, 31]. Furthermore, science fiction can showcase future technologies [30] well ahead of time and reportedly stimulated research and development in medical device – [40] or robot-design [34].

However, science fiction is by no means a 'cure-all' for HCI Education. For example, Lin et al. [15] and Barnett et al. [2] found that science fiction movies do have an impact on the understanding and perception of students on scientific mechanisms and concepts—positively or negatively. In addition, Myers and Abd-El-Khalick [22] provide a classroom example, where the assumptions in a science fiction film eventually lead students towards detrimental learning outcomes.

In this paper, we endeavored to highlight intersections of science fiction and HCI Education, in view of two crucial skills of the P21 framework. Although, past research concerning science fiction materials in educational contexts has shown mixed results, we reason that science fiction can be of high-value in classroom settings and the future of HCI Education; when utilized with fore-thought and due care.

Through such a 'conscientious' integration of science fiction in HCI activities and syllabi, we reason that the benefits will outweigh the drawbacks and possibly pave the way toward a science fiction-inspired HCI curriculum.

References

1. Arizona State University: Center for Science and the Imagination (2018). https://csi.asu.edu/
2. Barnett, M., Wagner, H., Gatling, A., Anderson, J., Houle, M., Kafka, A.: The impact of science fiction film on student understanding of science. J. Sci. Educ. Technol. **15**(2), 179–191 (2006)
3. Bonson, J., Novy, D.: MAS.S64 Sci Fab: Science Fiction-Inspired Prototyping (2015). https://scifab.media.mit.edu/syllabus/
4. Brueckner, S., Novy, D.: Syllabus—MAS S65: Science Fiction to Science Fabrication (2013). http://scifi2scifab.media.mit.edu/syllabus-3/
5. Burton, E., Goldsmith, J., Mattei, N.: How to teach computer ethics through science fiction. Commun. ACM **61**(8), 54–64 (2018)

6. Churchill, E.F., Bowser, A., Preece, J.: The future of HCI education: a flexible, global, living curriculum. Interactions **23**(2), 70–73 (2016)
7. Figueiredo, L.S., Gonçalves Maciel Pinheiro, M.G., Vilar Neto, E.X., Teichrieb, V.: An open catalog of hand gestures from Sci-Fi movies. In: Begole, B., Kim, J., Inkpen, K., Woo, W. (eds.) Proceedings of the 33rd Annual ACM Conference Extended Abstracts on Human Factors in Computing Systems—CHI EA '15. pp. 1319–1324. ACM Press, New York, New York, USA (2015)
8. Gelsomini, M., Leonardi, G., Degiorgi, M., Garzotto, F., Penati, S., Silvestri, J., Ramuzat, N., Clasadonte, F.: Puffy-an inflatable mobile interactive companion for children with neurodevelopmental disorders. In: Proceedings of the 2017 CHI Conference Extended Abstracts on Human Factors in Computing Systems, pp. 2599–2606. ACM (2017)
9. Goldsmith, J.: Science fiction and computer ethics (2018). http://www.cs.uky.edu/~goldsmit/sf/syl18.html
10. Goldsmith, J., Mattei, N.: Fiction As an Introduction to Computer Science Research. Trans. Comput. Educ. **14**(1), 4:1–4:14 (Mar 2014)
11. Hall, D., Williams, C.: Big hero, p. 6 (2014)
12. Jordan, P., Auernheimer, B.: The fiction in computer science: a qualitative data analysis of the ACM digital library for traces of star trek. In: Ahram, T., Falc˜ao, C. (eds.) Advances in Usability and User Experience. pp. 508–520. Springer International Publishing, Cham (2018)
13. Jordan, P., Mubin, O., Obaid, M., Silva, P.A.: Exploring the referral and usage of science fiction in HCI literature. In: Marcus, A., Wang, W. (eds.) Design, User Experience, and Usability: Designing Interactions, pp. 19–38. Springer International Publishing, Cham (2018)
14. Kohno, T., Johnson, B.D.: Science fiction prototyping and security education. In: Cortina, T. J., Walker, E.L., King, L.S., Musicant, D.R. (eds.) The 42nd ACM Technical Symposium, p. 9 (2011)
15. Lin, K.Y., Tsai, F.H., Chien, H.M., Chang, L.T.: Effects of a science fiction film on the technological creativity of middle school students. Eurasia J. Math. Sci. Technol. Educ. **9**(2), 191–200 (2013)
16. Lindley, J., Coulton, P.: Pushing the limits of design fiction. In: Kaye, J., Druin, A., Lampe, C., Morris, D., Hourcade, J.P. (eds.) Proceedings of the 2016 CHI Conference on Human Factors in Computing Systems—CHI '16. pp. 4032–4043. ACM Press, New York, New York, USA (2016)
17. Marcus, A.: The Past 100 Years of the Future: Human-Computer Interaction in Science-Fiction Movies and Television (2012)
18. Marcus, A., Norman, D.A., Rucker, R., Sterling, B., Vinge, V.: Sci-fi at CHI. In: Bauersfeld, P., Bennett, J., Lynch, G. (eds.) Proceedings of the SIGCHI conference on Human factors in computing systems—CHI '92. pp. 435–437. ACM Press, New York, New York, USA (1992)
19. Michalsky, W.: Manipulating our futures: the role of science fiction in education: The Clearing House. J. Educ. Strat. Issues Ideas **52**(6), 246–249 (1979)
20. Minnesota State University: Course Descriptions (2014). https:// www.mnsu.edu/supersite/academics/catalogs/undergraduate/2014-2015/computerscience.pdf
21. Mubin, O., Obaid, M., Jordan, P., Alves-Oliveria, P., Eriksson, T., Barendregt, W., Sjolle, D., Fjeld, M., Simoff, S., Billinghurst, M.: Towards an agenda for Sci-Fi inspired HCI research. In: Proceedings of the 13th International Conference on Advances in Computer Entertainment Technology. pp. 10:1–10:6. ACE '16, ACM, New York, NY, USA (2016)

22. Myers, J.Y., Abd-El-Khalick, F.: “A ton of faith in science!” Nature and role of assumptions in, and ideas about, science and epistemology generated upon watching a sci-fi film. J. Res. Sci. Teach. **53**(8), 1143–1171 (2016)
23. National Academy of Sciences: The Science & Entertainment Exchange (2018). http://scienceandentertainmentexchange.org/
24. National Education Association: Preparing 21st Century Students for a Global Society—An Educators Guide to the Four Cs (2016). http://www.nea.org/assets/docs/A-Guide-to-Four-Cs.pdf
25. P21: Framework for 21st Century Learning (2018). http://www.p21.org/our-work/p21-framework
26. P21: What We Know About Creativity (2018). http://www.p21.org/storage/documents/docs/Research/P21_4Cs_Research_Brief_Series_-_Creativity.pdf
27. P21: What We Know About Critical Thinking (2018). http://www.p21.org/storage/documents/docs/Research/P21_4Cs_Research_Brief_Series_-_Critical_Thinking.pdf
28. Rogers, M.L.: Teaching HCI design principles using culturally current media. Proceedings of the Human Factors and Ergonomics Society Annual Meeting **54**(8), 677–680 (2010)
29. Russell, D.M., Yarosh, S.: Can we look to science fiction for innovation in HCI? Interactions **25**(2), 36–40 (2018)
30. Schmitz, M., Endres, C., Butz, A.: A survey of human-computer interaction design in science fiction movies. In: Proceedings of the 2Nd International Conference on INtelligent TEchnologies for Interactive enterTAINment, pp. 7:1–7:10. INTETAIN ’08, ICST (Institute for Computer Sciences, Social-Informatics and Telecommunications Engineering), ICST, Brussels, Belgium, Belgium (2007)
31. Shedroff, N., Noessel, C.: Make It So: Interaction Design Lessons from Science Fiction. Brooklyn N.Y. USA, Rosenfeld Media (2012)
32. SIGCHI: 2011-2014 Education Project (2011 2014). http://prior.sigchi.org/resources/education/2011-education-project-1/report-of-2012-activities/view
33. SIGCHI: HCI syllabi (2011–2014). http://prior.sigchi.org/resources/education/2011-education-project-1/syllabi
34. SIGCHI, A.: Puffy—An Inflatable Mobile Interactive Companion for Children with Neurodevelopmental Disorder (2018). https://www.youtube.com/watch?v=10gCiClVWM0
35. Silva, P.A.: BadIdeas 3.0: A method for creativity and innovation in design. In: Proceedings of the 1st DESIRE Network Conference on Creativity and Innovation in Design, pp. 154–162. DESIRE ’10, Desire Network, Lancaster, UK, UK (2010)
36. Stanley, T.: Authentic Learning: Real-World Experiences That Build 21st-Century Skills. Prufrock Press (2018)
37. Summet, V.: CS 190: Robotics Freshman Seminar (2012). http://www.mathcs.emory.edu/~valerie/courses/fall12/190/syllabus.html
38. University of Virginia: Rosalyn W. Berne (2018). https://engineering.virginia.edu/faculty/rosalyn-w-berne
39. Vrasidas, C., Avraamidou, L., Theodoridou, K., Themistokleous, S., Panaou, P.: Science fiction in education: case studies from classroom implementations. Educ. Media Int. **52**(3), 201–215 (2015)
40. XPRIZE: Qualcomm Tricorder XPRIZE (2018). http://tricorder.xprize.org/

Artificial Intelligence and Human Senses for the Evaluation of Urban Surroundings

Deepank Verma[1(✉)], Arnab Jana[1], and Krithi Ramamritham[2]

[1] Centre for Urban Science and Engineering, Indian Institute of Technology, Bombay, India
{deepank, arnab.jana}@iitb.ac.in

[2] Department of Computer Science and Engineering, Indian Institute of Technology, Bombay, India
krithi@cse.iitb.ac.in

Abstract. Traditional city planning and design tools require major restructuring. Even with the rapid growth in the availability of mobile communication devices, connectivity, data generation, and analysis tools, the idea of the creation of citizen-centric and smart cities has not been fully conceptualized. Individual perception and preferences toward urban spaces play an important role in mental satisfaction and wellbeing. However, the notion has not been studied and experimented along with various planning instruments. This study discusses the recent studies involving Artificial intelligence tools and sensory data collection. This paper further comment on the integrated methodology to collect sensory datasets that will further help in the evaluation of urban surroundings with individual perspectives.

Keywords: Urban perception · Deep learning · Sensory datasets
City planning

1 Introduction

Urban Perception can be described as the study of human preferences and perception of the urban environment. Urban perception studies are essential to city planning and management as they provide scientific data that can be used as evidence to validate land use and site planning by-laws. Further, the methodological approach to measure, judge and evaluate the urban surroundings can be reiterated in different areas of cities or different cities. Such studies help in comparison and benchmarking urban spaces with the aim of retrofitting and redevelopment. The individuals' preferences for surroundings has been extensively studied in a variety of urban and non-urban contexts. The traditional methodology adopted to evaluate the surrounding environment can be summarized in six steps: (a) Identification of survey locations, which included the locations present in the city, forest trails, and parks. (b) Collection of Sensory datasets, such as visual, auditory and olfactory. (c) Selection of properties of the environment that are considered in a particular experiment such as the presence of people, greenery, and buildings. Such properties are also studied as part of environmental attributes [1] and descriptors [2] (d) Selection of survey population. (e) Ratings provided to the

W. Karwowski and T. Ahram (Eds.): IHSI 2019, AISC 903, pp. 852–857, 2019.
https://doi.org/10.1007/978-3-030-11051-2_130

environment scenes by the survey participants in-situ and ex-situ. In the former case, the survey participants rate the surroundings by manually visiting the selected locations. (f) Correlation of perceived emotions with the evaluated environment properties.

Although the discussed approach has been widely experimented, the studies following the methodology have always been questioned for generalizability and applicability in various urban extents [3]. The sense of vision has dominated the appreciation of surroundings, while studies including olfactory and auditory senses have been scarce. Also, such studies have been conducted in small test sites and areas, which do not provide a representative sample for large city extents. Translation of such studies to a broader urban context has not been considered viable due to the requirement of the tools for data collection and analysis, and unavailability of a required number of participants to conduct large-scale surveys. Erstwhile studies have been able to find the relationship between the existence of particular attributes of the environment and the perceived thoughts and emotions in individuals. Several constructs have been postulated such as the presence of natural content induces satisfaction [4], the openness and the presence of people induces a sense of safety [5], and presence of familiar places are preferred than new scenes [6]. However, the validation of such hypotheses has not been successfully conducted in larger domains.

2 Data Collection and Analysis Techniques

In recent years, efficiency in rapid data collection and complex data analysis has been achieved with the help of crowdsourcing, smartphones and Artificial Intelligence algorithms. With the access of ICT to larger masses, the collection of data from online surveys, scraping social media messages, and informed participation has been experimented at larger scales. Similarly, Smartphone devices have become apparent in the general population. The availability of a large array of sensors in smartphone devices has made it the most feasible tool to conduct sensory data collection. The collection of visual datasets through the onboard camera, audio datasets with microphones, and olfactory cues through the mobile-based application can be achieved through these devices. Apart from data collection, the development in the field of computation and data analysis has led to a paradigm shift. Artificial Intelligence techniques have shown near human accuracy in data analysis tasks. Such methods have been able to obtain meaningful results from multimillion samples of datasets. With relevance to urban perception, these methods may help in the extraction of environmental attributes from the urban surroundings, the presence of which can be correlated with perceptual attributes such as safety, liveliness, calmness, and beautifulness.

Machine Learning applications learn the complex structure of raw datasets by extracting patterns [7]. Deep Learning models are the subset of Machine Learning methods which learn the representation of such datasets with multiple levels of abstraction [8]. The Deep Learning models such as Convolutional and Recurrent Neural Networks can be used to identify the presence of visual cues such as trees, roads, people, streetlights, and buildings in street view photographs and the auditory cues such as car honk, chirping birds, and movement of the crowd. With the help of such models, large visual and auditory datasets collected from urban surroundings can be quickly processed and studied in relation to urban perception.

3 Insights from Recent Studies in Urban Perception

Urban surroundings are diverse, apart from the built infrastructure; cross-cultural perspective, traditions, and demography play a crucial role in the perception of the landscapes. This section discusses different tools and techniques followed by recent studies to collect intrinsic properties of sensory realms.

Cloud services such as Google Vision API[1] facilitates the identification of the contents of the image in the form of contextual tags with the help of DL techniques. Hyam [9] used the information provided by the tags in categorizing urban areas into natural and human-made environments. Such studies are helpful in finding relationships between the machine-generated data and the socio-economic prospects of the area. Databases such as Google and Tencent Street view images have become the de facto source for assessment and comparison of visual characteristics of cities and neighborhood. Most of the large-scale scene understanding studies have utilized street view platform and Deep Learning techniques to understand the associated qualities of urban environments. Shen [10] used image segmentation DL technique to divide the GSV images into the amount of greenness, openness, buildings, vehicles, and roads. The extracted details are then stored in the database for quick information retrieval. Place pulse [11] is another such example that utilizes street view scenes in the creation of a crowdsourced dataset.[2] It provides a perception based comparison of street view scenes with the help of attributes such as liveliness, boredom, wealthiness, and safety. This dataset comprises of Google street view scenes of different cities for which the participants from all over the world provide perception-based ratings. Further studies have utilized Place Pulse dataset and deep learning methods to predict the perception of the people in new scenes [12].

Although street view databases provide the most comprehensive visual datasets to conduct large-scale studies, the limiting factors associated with the usage of such datasets are worth noting. The perceived qualities of the place such as liveliness, boredom, and safety highly dependent upon the duration of the day, the presence of people and vehicles on the streets, etc. Street view images are fixed data points in space with no temporal dimension, hence are unsuitable for in-depth analysis. Similarly, crowdsourcing perception based responses such as in place pulse dataset have several drawbacks. The judgments provided by the participants on environmental qualities may depend upon multiple factors such as nationality, culture, and traditions. The degree of familiarity (observer being a resident or a visitor) of the place might also affect the assessment of the observer towards the same. Further light into these aspects of perception-based studies is required in future studies on this topic. Contrary to the use of the Google platform, WebRTC based smartphone application is utilized to collect

---

[1] https://cloud.google.com/vision/.

[2] http://pulse.media.mit.edu/.

street view imagery at different durations [13]. With the help of DL algorithms, such as object detection, segmentation, and classification, the environmental attributes are extracted and plotted in a map to study temporal and spatial variations in the selected area.

The sounds generated by rain, thunder, wind, waves, birds, and animals are studied to have a positive effect on the people [14]. However, urban sounds have been mostly studied concerning noise and its ill effects. While the acoustical parameters such as loudness and sharpness are easier to obtain from recorded sound clips, the identification of the sources of sounds is comparatively complicated. Deep Learning models, especially variants of Recurrent Neural Networks have been used for speech detection, Sound event detection, and sound classification [15]. Although these models have performed significantly over the traditional algorithms on custom sound datasets, they have not been tested in sounds collected from real-life urban scenes. Future studies may conduct extensive sound collection and automate the identification of sound sources along with capturing acoustical features.

Erstwhile studies have been successful in creating soundscape maps and predicting the perception of people in the auditory realm with several algorithms. Traditional auditory perception studies involved sound-walk methods, in which participants evaluate the surroundings based on the sources of sound and the intensity. The recorders and microphones have been used to capture the ambient sounds to simulate the auditory environment artificially. Hong [16] collected sound clips from various locations in the city from which sound pressure and psychoacoustic parameters were obtained. The sound sources such as traffic noises, human, and water sounds were identified in the sound clips collected with the help of participants. The soundscape map is then prepared from the collected data, and spatial regression models were tested. Similarly, 50 sound clips of different urban areas were recorded and evaluated by 100 participants to measure perceptual attributes such as pleasantness, eventfulness, and familiarity in soundscapes [17]. The classification techniques such as Artificial Neural Networks and SVM have been used to predict perceived soundscape quality in urban open spaces [18].

As compared to the number of perception-based studies in the auditory and visual realm, the research in smellscapes assessment has been significantly lesser. The language and cultural barriers have restricted classification of smells at large. The vocabulary and the general terms associated with the particular odors is often misinterpreted in literary texts and dictionaries. These ambiguities in defining and characterizing smells have provided the playground for the researchers to experiment with various means of data collection and representation. Identification of the presence of olfactory cues necessitates subjective evaluation of people's responses. Similar to sound-walks, the researchers have conducted extensive smell-walks to capture discontinuous and episodic smells. Xiao [19] conducted semi-structured interviews from nineteen smell walkers who helped in understanding perceptual qualities of smellscape. Participants' descriptions of the Smellscapes were classified into nine indicators that contribute to smellscape pleasantness. McLean [20] created an artwork "Amsterdam" which included the smell map of the city as experienced by a group of 44 smell walkers in four days. Symbols and colors are selected to describe the smells in the olfactory map thus generated. Smell-walkers provide inputs regarding distinctive odors in the area. Mapping encountered smells helps to visualize the location of particular smells related to spatial settings. Crowdsourcing is one of the emerging dimension in data collection from the masses through a digital medium. Studies have utilized crowdsourcing platforms such as Flickr, Instagram, and Twitter to collect vocabulary to describe sounds and smells [21]. These specific smells and sounds are then tagged to each street segment in the whole city thereby creating sound [22] and smell maps [21] for the city.

4 Conclusions

While such studies have utilized different tools and techniques to collect sensory datasets and analysis, each sensory realm has been viewed in isolation. The individuals' preference to a particular space is dependent on the responses from the multiple human senses. There is a need to bind the discussed approaches to create a simplified but comprehensive toolbox to assess urban surroundings with multimodal sensory analysis. Given the accessibility of the general population to ICT and Smartphone devices, extending the discussed approaches to include public participation in data collection is a viable option to consider. Also, the sensuous character of the surroundings varies with time, a collection of sensory datasets at regular intervals is therefore vital for extensive evaluation. With large computational power, open visual and auditory datasets and tools for analysis, machines can be easily trained to understand the complex datasets to achieve human-like performance in feature extraction and classification tasks.

The urgent demand for better management and planning has become more relevant with the increase in urban population. Utilizing current advance techniques in identification and mapping sensory information as perceptual layers are one direction to move forward. The individuals' perception of urban spaces is essential to focus on human-centric urban planning practices and research than top-bottom urban planning and design approaches. The sense of security, belongingness, happiness revolves around the environment a person dwells, works and recreate. The quantification of the environmental and perceptual attributes of surroundings is, therefore, a building block to understand the inherent complexity in the urban domain.

Acknowledgments. The authors would like to thank the Ministry of Human Resource Development (MHRD), India and Industrial Research and Consultancy Centre (IRCC), IIT Bombay for funding this study under the grant titled Frontier Areas of Science and Technology (FAST), Centre of Excellence in Urban Science and Engineering (grant number 14MHRD005).

References

1. Kasmar, J.: The development of a usable lexicon of environmental descriptors. Environ. Behav. **2**(2), 153–169 (1970)
2. Nasar, J.L.: Perception, cognition, and evaluation of urban places. In: Altman, I., Zube, E.H. (eds.) Public Places and Spaces, pp. 31–56. Springer, US, Boston, MA (1989)

3. Evans, G.W., Smith, C., Pezdek, K.: Cognitive maps and urban form. J. Am. Plan. Assoc. **48** (2), 232–244 (1982)
4. Kaplan, R.: The nature of the view from home: psychological benefits. Environ. Behav. **33** (4), 507–542 (2001)
5. Nasar, J.L.: Environmental correlates of evaluative appraisals of central business district scenes. Landsc. Urban Plan. **14**(C), 117–130 (1987)
6. Herzog, T.R., Kaplan, S., Kaplan, R.: The prediction of preference for familiar urban places. Environ. Behav. **8**(4), 627–645 (1976)
7. Goodfellow, I., Bengio, Y., Courville, A.: Deep Learning. MIT Press (2016)
8. Lecun, Y., Bengio, Y., Hinton, G.: Deep learning. Nature **521**(7553), 436–444 (2015)
9. Hyam, R.: Automated image sampling and classification can be used to explore perceived naturalness of urban spaces. PLoS One **12**(1), e0169357 (2017)
10. Shen, Q., et al.: StreetVizor: visual exploration of human-scale urban forms based on street views. IEEE Trans. Vis. Comput. Graph. **24**(1), 1004–1013 (2018)
11. Salesses, P., Schechtner, K., Hidalgo, C.A.: The collaborative image of the city: mapping the inequality of urban perception. PLoS One **8**(7), e68400 (2013)
12. Liu, L., Wang, H., Wu, C.: A machine learning method for the large-scale evaluation of urban visual environment. Comput. Res. Repos (ArXiv) (2016)
13. Verma, D., Jana, A., Ramamritham, K.: Quantifying urban surroundings using deep learning techniques: a new proposal. Urban Sci. **2**(3), 78 (2018)
14. Yang, M., Kang, J.: Psychoacoustical evaluation of natural and urban sounds in soundscapes. J. Acoust. Soc. Am. **134**(1), 840–851 (2013)
15. Cakir, E., Parascandolo, G., Heittola, T., Huttunen, H., Virtanen, T.: Convolutional recurrent neural networks for polyphonic sound event detection. IEEE/ACM Trans. Audio Speech Lang. Process. **25**(6), 1291–1303 (2017)
16. Hong, J.Y., Jeon, J.Y.: Exploring spatial relationships among soundscape variables in urban areas: a spatial statistical modelling approach. Landsc. Urban Plan. **157**, 352–364 (2017)
17. Axelsson, Ö., Nilsson, M.E., Berglund, B.: A principal components model of soundscape perception. J. Acoust. Soc. Am. **128**(5), 2836–2846 (2010)
18. Yu, L., Kang, J.: Modeling subjective evaluation of soundscape quality in urban open spaces: an artificial neural network approach. J. Acoust. Soc. Am. **126**(3), 1163–1174 (2009)
19. Xiao, J., Tait, M., Kang, J.: A perceptual model of smellscape pleasantness. Cities, 0–1 (2018)
20. McLean, K.: Smellmap: Amsterdam—olfactory art and smell visualization. Leonardo **50**(1), 92–93 (2017)
21. Quercia, D., Schifanella, R., Aiello, L.M., McLean, K.: Smelly maps: the digital life of urban smellscapes. Jacobs 1961 (May 2015)
22. Aiello, L.M., Schifanella, R., Quercia, D., Aletta, F.: Chatty maps: constructing sound maps of urban areas from social media data. R. Soc. Open Sci. **3**(3), 150690 (2016)

Mobile AR Tourist Attraction Guide System Design Based on Image Recognition and User Behavior

Xiaozhou Zhou[1], Zhe Sun[2], Chengqi Xue[1(✉)], Yun Lin[1], and Jing Zhang[1]

[1] School of Mechanical Engineering, Southeast University, Nanjing 211189, China
{zxz,101000270,23015962,zhangjing1026}@seu.edu.cn

[2] Nanjing Research Institute, HUAWEI Technology Co., Ltd., Nanjing 210029, China
sun8023zhe@126.com

Abstract. In this paper, the image recognition technology and augmented reality (AR) technology are combined in the design of the tourist attraction guide system. Through the mobile application scanning the real environment to recognize the scenic spots, and superimposing the virtual scenic spots information in the real scene, it can be more targeted for the users to provide the information of the scenic spots. Aiming at the key technical problems involved in the implementation of the system, a modeling method of AR guide system based on Unity3D and Vuforia is proposed. Based on this, the prototype design and development of the auditorium system of Southeast University, which consists of AR mode and virtual screen display mode, is carried out. This system integrates the scanning and recognition of scenic spots and browsing of navigation information. User testing shows it has a high recognition rate and a comfortable user experience.

Keywords: Augmented reality · Unity3D · Image recognition
User behavior

1 Introduction

With the frequent use of people's mobile phones and the popularity of mobile networks, the emergence of mobile guide systems provides people with access to information on attractions. At present, most of the travel guide APPs identify the current location of the user by positioning, and then push the attraction information on the position. As we know, errors often exist in positioning the current location, so the user could not receive valid information instantly. This has become the most common use problem for similar software. It would be useful to solve this problem by combining the image recognition technology and AR technology in the attraction guide system. Applying AR technology to the guide system, the advantages of immersive, interactive, multimodal display could bring a new experience for the users.

W. Karwowski and T. Ahram (Eds.): IHSI 2019, AISC 903, pp. 858–863, 2019.
https://doi.org/10.1007/978-3-030-11051-2_131

In the research of AR guide system, Yang et al. [1] designed an AR multimodal guide system comprised by two computers, a camera and a head-mounted display. Harvard Innovation Labs developed *PIVOT the World*, which enables historical scenes on real-life scenes on mobile phones [2]. Andrew developed a virtual browsing project *Handheld City* for Toronto Museum, which statically scanning static identification maps on handheld devices to present museum virtual models on phones [3]. Narzt et al. proposed a new mobile guide system *Instar* in 2003, which obtaining location information during moving process and displaying the augmented information on mobile devices [4]. And the mobile AR navigation *HyMoTrack*, developed by Vienna University of Technology, its hybrid tracking system enables centimeter-level position tracking of complex indoor environments and providing route guidance of visualization [5]. *AR CITY*, a guide system with the capable of identifying buildings and displaying the augmented navigation line in the screen of mobile phone was created by Mobile AR Laboratory of New York University in 2015 [6].

As can be seen, to date there has been some research results in guide systems that related to AR and image recognition technologies. Image recognition refers to the technique of analyzing and processing the acquired image to identify the target information contained in the image, usually composed of image capture, image pre-processing, image feature extraction, recognition classification, and result output. AR technology is a human-computer interaction technology that superimposes virtual information in real-time to enhance the expression and enable to interact with people or objects in the real environment. In the mobile AR guide system, the user collects the image information of the scene through the camera of the mobile device and transmits it to the data processing module, and the image would be matched with the identification map in the database to realize image recognition. Then the virtual image information corresponding to the identified map is superimposed displayed to the screen of the mobile device. Thus, the user can obtain the attractions' information in multimodal according to the mobile AR guide system.

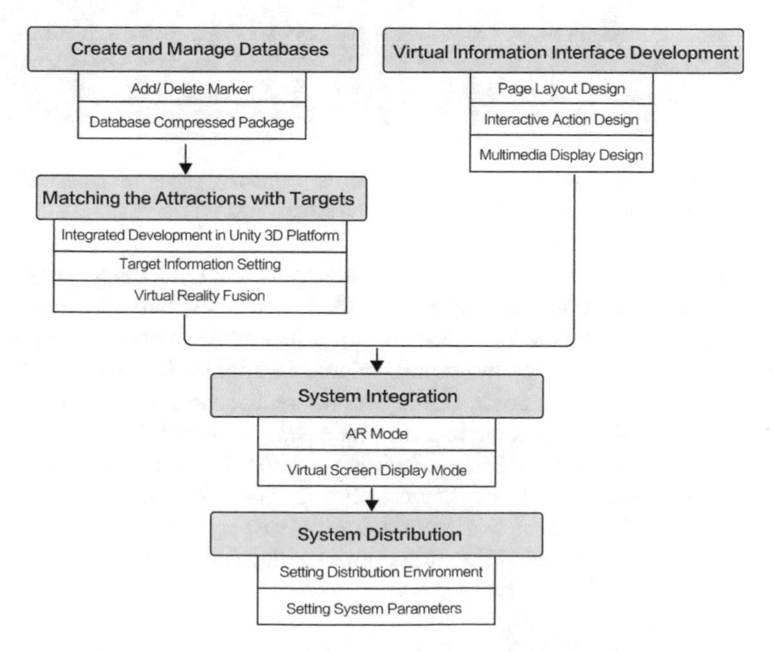

Fig. 1. Mobile AR guide system design process

In this paper, we integrated the registration tracking technology, virtual reality fusion technology and human-computer interaction technology in the mobile guide system. Based on these technologies, we build an AR guide system of the university campus. The system design processing is shown in Fig. 1.

2 Mobile AR Guide System of Campus

Through the prior investigations of the target users, we believe that there are two common scenarios: one is the user identified by scanning sights and attractions to match the image; the other is the user read the multidimensional information related to the attractions. Based on this, the guide system of campus consists of two parts: AR mode and virtual screen display mode. The system allowed users to switch between these two modes. System switches to the virtual screen display mode when the user confirms the recognition in AR mode, and return to the AR mode by performing scanning recognition of the attraction again.

Fig. 2. AR mode in the guide system. From left to right are pages of *recognition object*, *recognition success* and *access to the main menu*, respectively.

In the AR mode, the user can perform attraction matching recognition by scanning the image of the attraction, and the mobile APP accurately provides the attraction information. In this mode, the user needs to raise the mobile phone to align the scene, meanwhile, the background of the mobile phone is bright relatively and the virtual information needs to superimposed display on the real scene captured by the camera. Therefore, the visual information in AR mode was designed in a dynamic, high-contrast pattern, while the increasing effective operating hotspot to support the user's fuzzy input when raising hands, as shown in Fig. 2. The AR recognition part involves image recognition of attractions and information AR display, which can be divided into Vuforia online processing and Unity offline processing. The landmarks in the database were managed through the *Target Manager*. In order to improve the image recognition rate during actual use, we selected multiple images as the identification maps from each perspective that the user used to visit the specific attraction. Then the matchings of scenic spots and attraction information were implemented on the Unity3D platform.

Fig. 3. Virtual screen display in the guide system. Multiple dimensions of information are displayed horizontally, using swipe gestures to switch pages, from left to right are pages of *introduction*, *style features*, *history*, respectively.

When the users view the specific information, they tend to held the mobile phone closer, and the changing light background in AR mode will interfere the user's viewing. Therefore, we have designed another mode to support the user's information browsing. In the virtual screen display mode, the attraction information was classified as multiple dimensions, so the user can browse the classified information conveniently. The development of the virtual information interface is mainly carried out through the UGUI sub-system that comes with Unity. We selected 10 classic buildings for data collection from the Sipailou campus of Southeast University (China), which built in 1920s and has historical significance. The information architecture design, interaction design and visual design were built on these data by then. In the interactive design, according to the operational relevance and logic between the pages, the interaction switch between the pages and the interactive functions of the buttons are mainly achieved by matching to the Scene. The virtual information interfaces of each information module of the guide system were placed in different scenes respectively, and then match to the corresponding scene by swipe gesture or clicking the interactive visual element in the interface, thereby responding to the user's interaction operation, as shown in Fig. 3.

Fig. 4. Real scene test of the application.

3 Conclusions

This paper proposed a modeling method of AR guide system based on Unity3D and Vuforia, targeting the key technical problems involved in the implementation of the system. Through the mobile application scanning the real environment to recognize the scenic spots, and then superimposing the scenic spots information in the real environment, it can be more targeted for the users to provide the information of the scenic spots. What's more, it also can bring new sensory and emotional experiences to users, and change the interaction between people and information. Compared to the QR code recognition method, the experience of scanning the real scenes and displayed in AR mode forms more intuitive and fun.

This mobile guide system has two advantages. Firstly, different display and control method was employed in the AR mode and the virtual screen display mode to provide a more comfortable user experience. Secondly, multiple pictures of the scene taken from all the routes into the scene were defined as identified images to a single target, so the image recognition rate significantly improved in practical use. It runs well on the mobile terminals, and the image recognition accuracy is high, fast and robust. The user's real scene test (as shown in Fig. 4) showed that the mobile guide system has a high recognition rate in actual use, and the AR and screen display mode switching brings a good user experience to the user.

Acknowledgments. The research leading to these results has received funding from The National Natural Science Foundation of China (No. 718710567, 71471037).

References

1. Yang, J., Yang, W., Denecke, M: Smart sight: a tourist assistant system. In: 3rd International Symposium on Wearable Computers, pp. 73–78. IEEE Press, San Francisco (1999)
2. Pivot. http://www.pivottheworld.com
3. Toronto Museum Project. http://www.futurestories.ca/toronto
4. Narzt, W., Pomberger, G., Ferscha, A., et al.: Pervasive information acquisition for mobile AR-navigation systems. In: Proceedings 5th IEEE Workshop on Mobile Computing Systems and Applications. IEEE Press, Monterey (2003)
5. Gerstweiler, G., Vonach, E., Kaufmann, H.: HyMoTrack: a mobile AR navigation system for complex indoor environments. J. Sens. **16**, 1–19 (2015)
6. Augmented City. http://augmentedcitynyu.blogspot.com

Proposal of a Method to Evoke Cross-Cultural Communication by Using Digital Signage

Kimi Ueda[1(✉)], Motoki Urayama[1], Hiroshi Shimoda[1], Hirotake Ishii[1], Rika Mochizuki[2], and Masahiro Watanabe[2]

[1] Kyoto University, Yoshida-Honmachi, Sakyo-ku, Kyoto 606-8501, Japan
{ueda, urayama, shimoda, hirotake}@ei.energy.kyoto-u.ac.jp

[2] Service Evolution Laboratories, Nippon Telegraph and Telephone Corporation, Hikarino-Oka, Yokosuka, Kanagawa 239-0847, Japan
{mochizuki.rika, watanabe.masahiro}@lab.ntt.co.jp

Abstract. A method of evoking cross-cultural communication between Japanese and foreign tourists with using digital signage has been proposed in this study aiming to improve foreigners' sightseeing experience in Japan. The proposed method utilizes a digital signage with interactive design, which resembles some jigsaw puzzle and requires gestures and cooperative work to see complete information, to evoke communications between the signage users. The results of evaluation experiment showed that the proposed method can evoke cross-cultural communication significantly more than conventional slideshow-type information presenting method.

Keywords: Digital signage · Cross-cultural communication · Cooperative work

1 Introduction

1.1 Background and Purpose

Recently, the number of foreign tourists visiting Japan has been increasing [1], and there have been lots of measures implemented to improve their experience in Japan. There is, however, still difference between the amount of information Japanese can get and foreign tourists can get when they travel around Japan, especially most of Japanese can only speak Japanese and there is not enough multilingualized information signs. Cross-cultural communication between Japanese and foreign tourists can be adopted as an alternative way to get information and it may also provide unforgettable experience traveling in Japan.

Japanese government is now promoting the use of digital signage as a part of "Action Plan to Accelerate ICT (Information and Communication Technology)" in anticipation of the 2020 Tokyo Olympic game. This study focuses on the use of digital signage as evoking tool of cross-cultural communication between Japanese and foreign tourists. The purpose of this study is, therefore, to propose a method of evoking cross-

W. Karwowski and T. Ahram (Eds.): IHSI 2019, AISC 903, pp. 864–869, 2019.
https://doi.org/10.1007/978-3-030-11051-2_132

cultural communication between Japanese and foreign touristes by using digital signage.

1.2 Conventional Studies

Many conventional studies showed the possibility of digital signage on evoking communication between signage users to share information in office places [2–6]. Since 21th century started, digital signage has been applied to evoke communication in certain communities such as Opinionizer [7, 8] and CoColage [9], and they showed the possibility of evoking communication between users especially with 'honey-pod effect'. However, these studies focused on certain community consisting of same language speakers or people who have similar interests and cultural backgrounds, which was somewhat easy situation to start communication. On the other hand, this study focuses on the communication between users of different languages or cultural backgrounds.

Although there are also some studies discussed ICT use on supporting cross-cultural communication, such as Open Smart Classroom [10, 11] and Scratch [12], they focused on how to support communication which had already occurred smoothly, not on how to evoke cross-cultural communication.

2 Proposed Method

The proposed method presents some sightseeing information on a digital signage in interactive way to provide some chance of cooperative works between people in front of it in order to evoke their communication. The interactive content was designed like a jigsaw puzzle as shown in Fig. 1, which can be enjoyed by using only gestures and can be understood its way of use by the affordance of its pieces' shape. The puzzle's piece and flame move stick to the head of the users, and when they get closer, the piece fits into the flame, and another piece appears on the user's head again. Users are required to complete a jigsaw puzzle to see complete information.

In order to attract people passing in front of the signage, digital signage with large screen is effective, and pieces stick to users' head are designed to sway around their heads to attract their interest. In addition, the information presented on the signage was chosen among sightseeing information, to supply topics to evoke cross-cultural communication after they completed the jigsaw puzzle and get full information presented on the digital signage.

Fig. 1. Example of contents shown on signage by proposed information presentation method (Cooperative Jigsaw puzzle design)

3 Evaluation Experiment

3.1 Methods and Measurements

An evaluation experiment was conducted from Dec. 23rd 2017 to Jan. 7th 2018. 20 Japanese university students and 20 foreign students participated and they were divided into 10 groups in 4 participants that were unacquainted with each other. Some Japanese sightseeing information (especially Kyoto's souvenirs) had been prepared as contents on digital signage in two information presentation type, (a) slideshow style as a standard method and (b) jigsaw puzzle style as the proposed method. Figure 2 shows the outline of the experiment.

Four evaluation components were set as follows; (1) whether the proposed method can attract users' interest, (2) whether it can evoke cooperative works, (3) whether it can evoke cross-cultural communication and (4) what information exchanged and how they communicated. Based on the evaluation components, participants' behavior observation and questionnaires were conducted. Because of the possibility of some effects of participants' characteristics on the frequency of evoking communication, the Big Five Personality test [13] was also conducted. The sequence of the type of information that each participants experience was counterbalanced.

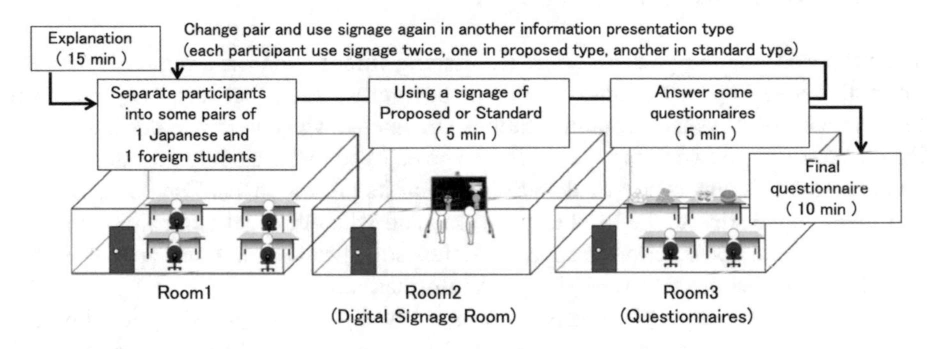

Fig. 2. Outline of experimental procedure.

3.2 Results and Discussion

The component (1) was evaluated by the results of questionnaire "Did Slideshow/Jigsaw-puzzle signage attract your interest?" as shown in Fig. 3. It was showed that the proposed method can attract users interest as much as the standard method. The reason why they were attracted by the signage was different between standard and proposed method. 28 participants out of 40 answered "(when using slideshow-type), the presented information was attractive" and 5 participants answered "(slideshow-type) seems like good for killing time". On the other hand, only 14 participants answered "(when using jigsaw-type), the presented information was attractive" and 4 participants answered "(jigsaw-type) seems like good for killing time", and which was distinctive in the proposed method, 16 participants answered "It was fan to operate the (jigsaw-type) signage by gesture" and 14 participants answered "puzzle's pieces were naturally put into its flame."

About (2), 90% of the pairs did cooperative works when they were in front of the signage of the proposed method, so that it was showed that the method could provide chance of cooperative work. Based on the results of the questionnaire, "Did you understand how to use the jigsaw-type signage?", 28 participants out of 40 understood the usage, 5 participants couldn't understand it and 2 participants didn't answer the question. Thus it was suggested that the proposed signage's comprehensive interface of interactive contents had good effect on evoking cooperative work.

(3) was evaluated by the questionnaires and participants' behavior observation. 12 pairs used the proposed method and started cross-cultural communication, which was significantly more than 4 pairs in the standard method ($p = 9.82 \times 10^{-3}$). Based on the results of the questionnaire "why didn't you start talking with your pair?", 47% in proposed method and 23% in standard method of participants answered "I was worried that we couldn't understand each other because of the language difference", and there was significant difference ($p = 4.39 \times 10^{-3}$). Considering the result that cross-cultural communication was significantly evoked in the proposed method signage, they at first tried to talk with their partner through the presented contents, but after that they faced the difficulty of language difference, then the answer of the worries about language was increased. The proposed method didn't include the language support contents, so it is required to consider how to support them in aspect of language difference in future studies.

About (4), from the result of a questionnaire "please write down what you talked about with your partner as much as you can remember", when using the standard method signage, 4 pairs talked about the sightseeing information on the signage. On the other hand, 12 pairs using the proposed method signage talked about the usage of digital signage itself. In addition, based on the results of participants' behavior observation, 8 pairs spoke Japanese to communicate each other, 2 pairs spoke both Japanese and English, and 6 pairs spoke English. In this experiment, Japanese participants were university students who are thought to be better language-skilled than average, and foreign participants were also students who had stayed in Japan for more than 3 months and were thought to be used to speak some Japanese. Therefore, the results would be different if the experiment was conducted in actual public situation. Some supporting method for language difference are also needed from this aspect.

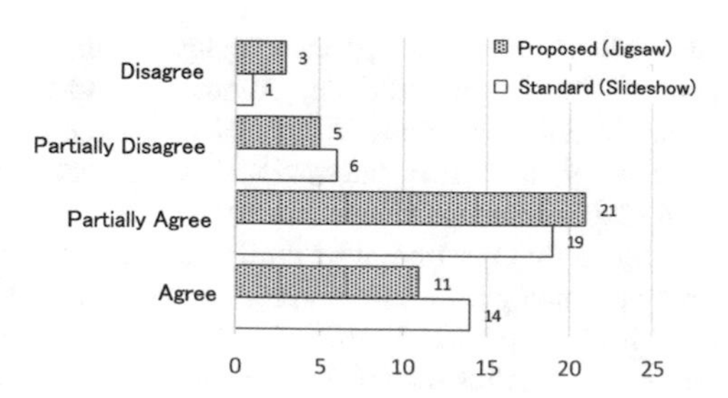

Fig. 3. Example photos of contents shown in signage by Proposed information presentation method (Jigsaw puzzle design)

4 Conclusion

In this study, the authors focused on the effect on evoking communication of digital signage, and the method of evoking cross-cultural communication by using digital signage has been proposed aiming to improve foreign tourists' experience.

The results of evaluation experiment comparing the proposed jigsaw-puzzle method and the standard slideshow method, it was showed that the proposed method can evoke cross-cultural communication significantly more than the standard method by its interactive way of information presentation. However, there are some difficulty to improve users' experience by the proposed method. Significantly more participants were worried about their language difference when using the proposed method signage than the standard method signage, and the information shared in communication evoked by the signage were just the system itself, not the worthwhile sightseeing information.

In future study, some language support method such as real time translation is necessary and also it is required to improve the proposed method to lead the evoked communication to the chance of getting the information about the contents on digital signage, not the signage system itself.

References

1. UNWTO World Tourism Barometer. http://mkt.unwto.org/barometer
2. Mankoff, J., Schilit, B.N.: Supporting knowledge workers beyond the desktop with palplates. In: Proceedings of the SIGGHI Conference on Human Factors in Computing Systems, pp. 550–551 (1997)
3. Houde, S., Bellamy, R., Leahy, L.: In search of design principles for tools and practices to support communication within a learning community. SIGCHI Bull. **30**(2), 113–118 (1998)
4. Sawhney, N., Wheeler, S., Schmandt, C.: Aware community portals: shared information appliances for transitional spaces. Pers. Ubiquitous Comput. **5**(1), 60–70 (2001)
5. Huang, E.M., Mynatt, E.D.: Semi-public displays for small, co-located groups. In: Proceedings of the SIGGHI Conference on Human Factors in Computing Systems, pp. 49–56 (2003)
6. Bardram, J.E., Hansen, T.R., Soegaard, M.: Awaremedia: a shared interactive display supporting social, temporal, and spatial awareness in surgery. In: Proceedings of the 2006 20th Anniversary Conference on Computer Supported Cooperative Work, pp. 109–118 (2006)
7. Rogers, Y., Brignull, H.: Subtle ice-breaking: encouraging socializing and interaction around a large public display. In: Workshop on Public, Community and Situated Displays (2002)
8. Bringnull, H., Rogers, Y.: Enticing people to interact with large public displays in public spaces. In: Proceedings of the IFIP International Conference on Human-Computer Interaction, pp. 17–24 (2003)
9. Farnham, S.D., et al.: Measuring the impact of third place attachment on the adoption of a place-based community technology. In: Proceedings of the SIGGHI Conference on Human Factors in Computing Systems, pp. 2153–2156 (2009)
10. Ishida, T.: Language grid: an infrastructure for intercultural collaboration. In: IEEE/IPSJ Symposium on Applications and the Internet, pp. 96–100 (2006)

11. Suo, Y., Miyata, N., Ishida T., Shi, Y.: Open smart classroom: extensible and scalable smart space using web service technology. In: Advances in Web Based Learning, pp. 428–439 (2008)
12. Resnick, M., et al.: Scratch; programming for all. Commun. ACM **52**(22), 60–67 (2009)
13. Barrick, M.R., Mount, M.K.: The Big Five personality dimensions and job performance. A Meta-Anal. Pers. Psychol. **44**(1), 1–26 (1991)

Designing with Data. Anticipating the Impact of Personal Data Usage on Individuals and Society

Laura Varisco[1(✉)], Sara Colombo[2], and Federico Casalegno[2]

[1] IEX Design Research Lab, Politecnico di Milano, Milan, Italy
laura.varisco@polimi.it
[2] MIT Design Lab, Massachusetts Institute of Technology, Cambridge, USA

Abstract. In a world where digital technologies are becoming human symbionts, in the form of personal companions, wearable devices, and connected environments, digital services can rely more and more on the massive collection of personal data to provide extremely tailored experiences. [1] It becomes necessary to explore the consequences of data harvesting and use during the design of new systems and services, especially before such solutions are fully deployed. In this paper, we propose the Impact Anticipation Method, which collects knowledge on potential issues related to the use of personal data in the design of new physical-digital solutions and we exemplify its application through a use case. The goal is to provide designers of data-rich digital solutions with a tool that analyzes the issues and the perturbations their designs could have on people and society, ultimately supporting the formulation of guidelines to refine the designed solutions.

Keywords: Personal information · Impact Anticipation Method
Ethical impacts

1 Introduction

As pointed out by authors such as Lanier, Mitchell and Greengard [2–4], networked technologies embedded in everyday devices and services raise critical questions in term of opportunities and problems. Approaches such as Value Sensitive Design (VSD) [5] and Responsible Research and Innovation (RRI) [6] consider societal values in different phases of the design process as critical elements to provide social acceptance and ethical acceptability for the technological solution.

While creating innovative solutions that involve the collection, use and management of personal data, designers have to face uncertainties related to the consequences this information might have when the designed solution is released and deployed in the society. Despite the emergence of policies addressing the issue of privacy and personal data usage by private and public organizations, [7] there seems to be little reflection on how the use and misuse of such data can impact people's lives on different levels.

In this paper, we describe the Impact Anticipation Method (IAM), a tool developed to bring the society's point of view into the design process, to encourage and support

W. Karwowski and T. Ahram (Eds.): IHSI 2019, AISC 903, pp. 870–876, 2019.
https://doi.org/10.1007/978-3-030-11051-2_133

ethical discussion among designers on future impacts of their data-intensive solutions. In particular, we considered impacts on four different levels: (i) user's self-perception, (ii) people's behaviors, (iii) interpersonal relationships, and (iv) social and political agency. For instance, customized services can influence the user's idea of self by providing them with virtual mirrors showing what the system 'knows and thinks' of them (e.g. through tailored advertisement, recommended contents such as music playlists or movies). [8, 9] Data about people's preferences, emotions, or habits can affect individuals' actions and behavior [10] as well as how they build social relationships. [1] Ultimately, the analysis of big data often reveals patterns that can lead to the creation of public policies addressing population's needs and desires [3].

The IAM provides designers with insights to raise awareness and stimulate critical thinking during the design phase, by enabling the identification of potential impacts and the generation of guidelines for future design iterations.

2 The Impact Anticipation Method

The Impact Anticipation Method (IAM) consists in (i) creating a database of potential issues (i.e. perceived positive and negative impacts) connected to the intensive use of personal data in digital services and solutions, and (ii) providing designers with tools to support reflection and awareness about ethical implications of their solutions, in different phases of the creative process.

In this paper, we introduce the *Potential Issues Database* and its generation process. We subsequently describe the *Data Impact Tool* and its application to a use case, i.e. the design of the Worker Profile+ concept.

2.1 Phase 1: The Potential Issues Database

The first phase of the IAM consists in generating a database of critical issues and potential impacts of the use of personal data in digital services on society. This knowledge is gathered by looking at the topic through the lens of society and by exploring the actual concerns and enthusiasms of people. In particular, the database is generated through two different activities:

- The analysis of science fiction storytelling artifacts (i.e. movies and TV series)
- The evaluation of ongoing societal discussions represented by current news on research and applications of innovative technologies and opinions of relevant public figures diffused online.

Storytelling Artifacts Analysis. This activity was performed through the investigation of narratives related to future sci-fi scenarios. We used user-generated online contents from the Internet Movies Database (IMDb.com) as data source, to explore both the settings of the narratives and the users' reactions to movies and TV series. Scraping was used to collect the plots for the analysis and to identify the user's ratings for the titles and the recurrence of keywords.

This activity led to the identification of three macro themes collecting emerging fears and hopes about the future:

Living self. This category refers to technological evolution as an opportunity to perfect people, fill the gaps that create differences between them and enhance potentialities of humanity.

Machine control. Science-fiction narratives present machines and robots as a necessary help for humans in their everyday life and describe different ways to interact with and control them.

Alternatives. This cluster illustrates the recreation of worlds in digital environments and the possibility to create different versions of one's own identity

Online News Analysis. The second step of the protocol concerns the collection of samples of contemporary news on emerging and existing digital services, devices and technologies, which involve the use of personal information. A number of issues were identified through the extraction and synthesis of the topics emerged from the analysis of each single news, considering all the four levels the use of personal data may impact on. The identified topics represent specific concerns and expectations of people in relation to the use of their personal data.

ERS Creation and Potential Issues Database. Issues were clustered based on similarity around the three macro themes previously identified. Eight Ethic-oriented Reference Scenarios (ERS) (i.e. clusters of issues) emerged as a result:

Living Self. 1. *Perfect Humanity.* Systems identify patterns or anomalies through the detection and analysis of data about biological characteristics of individuals (such as genetics and biodata).

2. Pervasive Awareness. Pervasive technologies track people's activities supporting the identification of bad or good behaviors and performances.

3. Mnemonic. The capacity of digital data storage fulfills the dream of extending human memory and accessing information anytime and everywhere.

Machine Control. *4. Super Monitor.* Ubiquitous computing, geo-located information and connected cameras allow systems to massive surveil people and their activities, so to increase security and safety, and to predict events.

5 .Human-Behavior Computers. AI, digital prosthetics, and robotics optimize tasks and automate actions by learning from humans how to face problems and perform actions in order to lighten humans' loads.

6. Automation Box. Full automation of activities performed by AI and machines leads to a scenario where machines' decisions remain obscure to human understanding. In extreme automation, users depend on software.

Alternatives. *7. Stargate.* Connections delete distances and the need of physical presence. The contraction of space and time alters interpersonal communication and relationships.

8. Avatar. The creation of alternative worlds using virtual, augmented and mixed realities allows humans not only to be in a different place but also to be different persons, allowing the creation of multiple different digital selves.

ERS represent a synthesis of hopes and fears currently perceived by society on the topic of personal data usage. Together with the list of issues they comprise, they compose the Potential Issues Database [11]. Within each ERS, issues are divided into the four levels the use of personal data may impact on: self-perception, behaviors, relationships, and social agency.

2.2 Phase 2: The Data Impact Tool

The knowledge included in the *Potential Issues Database* can help designers reflect on the consequences of their concepts or solutions on society in different design phases, e.g. to sensitize the designer about potential issues at the very beginning of the process, or to adjust the design after the concept ideation phase, in order to avoid certain issues or to reinforce some beneficial impacts.

In this work, we present the *Data Impact Tool*, a design tool developed to guide the refinement of concepts after the initial ideation phase. The tool consists of five steps that can be followed by designers while developing and detailing solutions relying on extensive use of personal data, in order to anticipate their potential impacts and to adjust their design choices accordingly:

Step 1 – Selection of ERS. The relevant scenarios for the concept are selected.

Step 2 – Issues Selection. Within the identified ERS, the issues relevant to the concept are selected.

Step 3 – Impacts Identification. The selected issues are clustered into a number of critical themes for the concept. They represent potential impacts that the concept can have on individuals and society. The issues composing the critical themes can come from different scenarios and impact levels.

Step 4 – Features assessment. For each critical theme, the concept features that can potentially affect or actualize the identified impacts are outlined.

Step 5 – Guidelines generation. Specific design guidelines in the form of questions are created in order to avoid negative issues and emphasize potential benefits.

3 Validating the Data Impact Tool Through a Use Case

In order to test its validity, the tool was applied to the evaluation of Worker Profile +, a future concept of a database able to collect and analyze personal data coming from workers in the energy industry.

3.1 Use Case: The Worker Profile + Concept

The Worker Profile + is a future concept developed within the Augmented Health and Safety project in collaboration with Eni, a worldwide energy company (www.eni.com). The project was aimed at envisioning how Health and Safety practices in the energy industry could be improved through the use of emerging technologies in the next 5 years.

The Worker Profile + (WP+) is a digital record of the worker's lifetime health statistics. It collects real-time data about the worker's health conditions (heart rate, respiration, hydration, diet, exercise) and exposure to chemical and physical hazards, through both smart wearables and direct inputs from the worker. Selected analytics are accessible to the doctor and the worker via dedicated interfaces. The WP+ analyzes collected data through machine learning algorithms that can identify risks and health issues in advance. It also recognizes real-time acute stress by comparing biometric data coming from wearable sensors with voice frequency analysis performed by the Health

Chatbot, another concept of the solution ecosystem. The WP+ encourages the worker to report activities and to pursue healthy behavior by setting customized goals and by assigning weekly and monthly prizes.

3.2 Assessing WP+ Through the Data Impact Tool

During its development, the WP+ concept raised relevant questions in terms of data usage and management. Therefore, it represents a suitable use case for the application and validation of the Data Impact Tool. The following steps describe its application.

Step 1 - Selection of ERS. The WP+ uses emerging technologies to enhance the potentialities of humanity, promoting health by increasing self-awareness, guiding the worker through alerts and notifications, incentivizing good behavior and predicting health issues. According to these features, we identified the following ERS as the reference scenarios for the concept: *Perfect Humanity, Pervasive Awareness, Mnemonic* and *Super Monitor*.

Step 2 – Issues Selection. Within the identified ERS, we selected the issues relevant to the concept. For instance, (a) Changes in self-knowledge from continuous data/info updates (*Pervasive Awareness*), (b) Alteration of interaction and relationships between employee and employer (*Perfect Humanity*), and (c) Democratized healthcare thanks to automatization and access to information (*Super Monitor*).

Step 3 – Impacts Identification. The selected issues were clustered around 10 critical themes for the concept. Figure 1 summarizes the emerging critical themes for WP+.

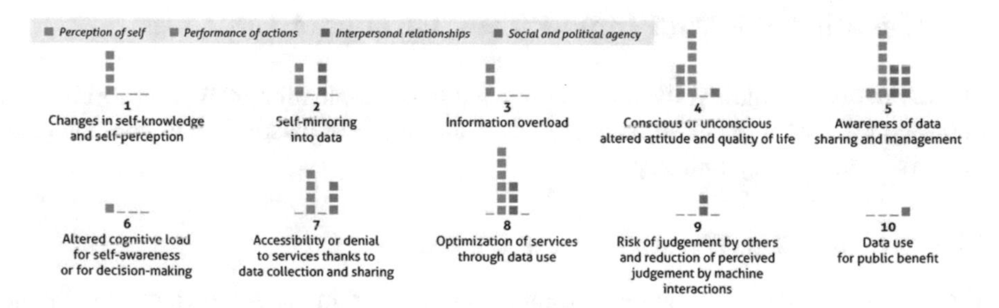

Fig. 1. Identified critical themes. The squares indicate the number of issues related to the theme, their colors represent the impact levels they are related to.

Step 4 – Features Assessment. For each critical theme, the WP+ features that may contribute to the actuation of the identified impacts were outlined. For instance, *Self-mirroring into data* and its consequences on user's identity depend on the following WP+ features: diet and exercise goals that the system sets for the user; prizes received by the user.

Step 5 – Guidelines Generation. After identifying the main potential impacts and the concept features that are connected to them, we generated design guidelines in the form of questions. For instance: *What goals are set by the system? What identity do they want to reinforce? How can the health goals set by the system avoid the* *generation of negative self-perceptions, e.g. making the user feel lazy or unhealthy? How is the user's identity affected by comparisons with other's prizes and achievements?*

Such guidelines, meant to support subsequent iterations and developments of the concept, were defined through discussion around each theme and their connected features, and were affected by the experience the designer intends to create in the interaction with the system (i.e. values, meanings, emotions, etc.).

4 Results and Conclusions

The application of the *Data Impact Tool* to the WP+ concept allowed the designers: (i) to be aware of potential impacts of the designed artifact; (ii) to include the point of view of users and society in the design process; (iii) to formulate guidelines that, together with the identified critical themes, will help to refine the concept's functions and features in future iterations.

The use case application confirmed the validity of the tool and its usefulness in helping designers to make informed design choices by critically reflecting on the ethical impacts of their solution during the design development phase. Some critical elements emerged, concerning the complexity of the tool, the presence of many information layers and clusters, and the number of steps to follow in order to reach the final guidelines. A subsequent and simplified version of the tool will be developed, also supported by visual forms and a step-by-step process explanation. A set of additional tools addressed to different phases of the design process will be developed as part of the Impact Anticipation Method. Finally, the Potential Issues Database is meant to be programmed to be automatically updated, in order to speed up the scraping process and to make it available to non-expert users. This will provide an up-to-date representation of current fears and hopes of people and society concerning the use of personal data.

Acknowledgment. The development of the IAM is part of the doctoral research of the author Laura Varisco that has been funded by TIM S.p.A., Services Innovation Department, Joint Open Lab Digital Life, Milan, Italy. The Worker Profile + concept was developed by the MIT Design Lab in collaboration with the Eni ICT and Health, Safety, Environment and Quality Departments within the MIT Energy Initiative.

References

1. Arslan, P., Casalegno, F., Giusti, L., Ileri, O., Kurt, O.F., Ergüt, S.: Big data as a source for designing services. Web, June 2017
2. Lanier, J.: Who Owns the Future?. Simon and Schuster, New York (2014)
3. Mitchell, W.J.: Me++: The Cyborg Self and the Networked City. Mit Press, Cambridge (2004)
4. Greengard, S.: The Internet of Things. MIT Press, Cambridge (2015)
5. Friedman, B., Bainbridge, W.S.: Value sensitive design (2004
6. Von Schomberg, R.: A vision of responsible innovation. In: Owen, R., Heintz, M., Bessant, J. (eds.) Responsible Innovation, pp. 51–74. Wiley (2013, forthcoming)

7. Directive 95/46/EC (General Data Protection Regulation) Official Journal of the European Union, vol. L119, pp. 1–88, 4 May 2016
8. Lupton, D.: The digitally engaged patient: Self-monitoring and self-care in the digital health era. Soc. Theory Health **11**(3), 256–270 (2013)
9. Li, I., Dey, A.K., Forlizzi, J.: Understanding my data, myself: supporting self-reflection with ubicomp technologies. In: Proceedings of the 13th International Conference on Ubiquitous Computing, pp. 405–414. ACM, September 2011
10. Young, N.: The virtual Self: How Our Digital Lives are Altering the World Around Us. McClelland & Stewart, Toronto (2012)
11. Varisco, L., Pillan, M., Bertolo, M.: Personal digital trails: toward a convenient design of products and services employing digital data. In: 4D Designing Development Developing Design, pp. 188–197 (2017)

Social Consensus: Contribution to Design Methods for AI Agents That Employ Personal Data

Milica Pavlovic[1,3](✉), Francesco Botto[2], Margherita Pillan[1], Carmen Criminisi[3], and Massimo Valla[3]

[1] Interaction & Experience Design Research Lab, Polytechnic University of Milan, 20158 Milan, Italy
{milica.pavlovic,margherita.pillan}@polimi.it
[2] Fondazione Bruno Kessler, 38123 Trento, Italy
fbotto@fbk.eu
[3] Joint Open Lab Digital Life, Services Innovation Department, TIM S.p.A., 20134 Milan, Italy
{carmen.criminisi,massimo.valla}@telecomitalia.it

Abstract. The emerging complex IoT ecosystems, embodied through Artificially Intelligent (AI) Agents on the front-end interaction with the user, rise many new considerations to be taken into account during the design process, among which the use of sensitive personal data. This paper introduces a case study, a concluded project of a system supported by AI algorithms for delivering tailored services to the drivers, including insurance offerings and supporting drivers in practicing safer driving style. We report on a segment of user studies done within this project that relates to the use of personal data, and we discuss the notion of emerged user values within. Accordingly, we observe and propose inclusion of social consensus considerations within the design process and evaluation of the same.

Keywords: Design methods · Human-systems integration · AI agents Personal data · Social consensus

1 Introduction

Advancements in technological capabilities are enabling cross-device interactions and the creation of complex ecosystems of Internet of Things (IoTs), delivering services through personalized Artificially Intelligent (AI) Agents. Such networked systems can produce valuable solutions for both individuals and communities [1, 7]: efficient management of energy, lighting and heating systems; smart transportation; monitoring of physiological parameters for fitness and medical purposes by wearable devices and others, are just some examples of progress produced by the evolution of digital technologies. In this setting, the spreading of devices able to collect and use data gathered from individuals is shaping current socio-technical systems and it induces innovations that are changing everyday scenarios and behaviors [2, 3, 8, 9]. The design of personalized digital services requires knowledge and tools to understand the potential

W. Karwowski and T. Ahram (Eds.): IHSI 2019, AISC 903, pp. 877–883, 2019.
https://doi.org/10.1007/978-3-030-11051-2_134

impact on individuals and communities such services might bring over short and long terms, thus enabling designers with stakeholder teams to make conscious design choices during the design process.

For discussing our viewpoint on design processes in this context, we introduce a case study, project MEMoSa (EIT Digital, Digital Wellbeing Action Line, Activity ID 17160) concluded in December 2017. Project aim is testing and reshaping a design concept for a personalized assistive system for car drivers. The project refers to development of a mobile and cloud service for insurance companies (providing health and car insurance), by supporting drivers in being aware of their physical and psychological status while driving. The system, therefore, is improving road safety and reducing overall insurance costs with benefits for customers and insurance companies. The system (Fig. 1) is based on a combination of a wearable device that tracks driver's physiological conditions, an On-Board Diagnostic (OBD) unit that collects data from the car, a mobile app and cloud components for data integration and analysis. This is the MEMoSa Assistant component (i.e. Agent) that, thanks to artificial intelligence algorithms, provides safety alerts and suggestions to avoid risky situations.

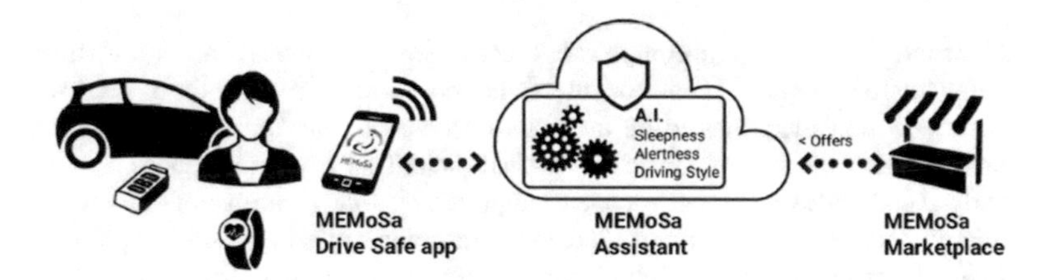

Fig. 1. Elements of the MEMoSa system.

2 Methodology

The design and validation of the MEMoSa AI Agent were driven by user experience values, and they required a collaborative approach oriented towards the alignment [4] of these values with all involved stakeholders, i.e. project partners. The evaluation process with potential users took part in three sequential phases that simultaneously drove the reshaping of the design system (Table 1).

Table 1. Overview of the three user testing phases.

	1st Phase	2nd Phase	3rd Phase
Aim	Validating proposed initial design use case scenarios	Validating the user interface mock-up	Validating prototype in real-environment setting
Method	**Focus groups**	**UI video mock-ups**	**Rea-life testing**
Feedback	Online and offline surveys and open discussions	Online surveys	Surveys during and after the trial period
Participants	n = 39	n = 54	n = 17

Focus Groups [6] (Fig. 2a) lasted 3 hours each and were based on storytelling [5] of proposed design scenarios, and had as an outcome qualitative reflections of targeted users (drivers 20-60 years old) for shaping the initial perceived values of the system. The second phase relied on 5 video mock-ups of the User Interface (Fig. 2b) that presented the application features, and were evaluated through closed as well as open-ended survey questions. Finally, the trial phase (Fig. 2c) was organized through real-life experimentation of the MEMoSa system, providing users with the artefacts and testing version of the mobile application, which took part within a two-week testing period. This paper is reporting on a segment of conducted user testing with the focus on the use of personal data, discussing criticalities and potentialities of personal data in interaction design [8].

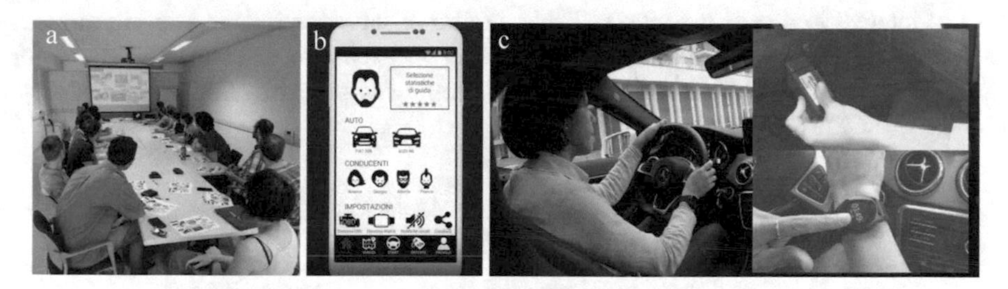

Fig. 2. (a) Participants of a Focus Group, (b) Video representation of the User Interface mock-up, (c) User testing the system prototype in a real-life environment setting.

3 Results & Discussion

3.1 Results on Usage of Personal Data Derived from Three Testing Phases

Focus Groups. For this testing phase, participants were presented with four use case scenarios related to the design system: (1) *On the Spot Insurance* (suggesting micro insurance products in regard to ongoing and emerging situations), (2) *Car Diagnostic* (monitoring car status and being notified about possible dangerous situations), (3) *Safe Driving* (identifying a possible dangerous situation for the driver and risk of an accident), (4) *Entertainment* (offering value added services like entertainment and contextual services to the driver).

On the Spot Insurance scenario was rated with medium interest (1,87 with a max scale of 4). It was observed as positive by 6 participants, who agreed that personalization of the policy, i.e. customized policies, based on the needs of the customer at a certain time are desirable. However, according to 19 participants, the negative point is that the economic side of the negotiation and some conditions may be unclear or not acceptable. Therefore, overall transparency in evaluation and back-end processes are desirable. Even though *Car Diagnostic* scenario was rated with medium-high interest (2,39/4), it has brought up certain considerations with data usage. Namely, few participants were concerned about privacy and treatment of data deriving from the OBD, as such could be spread with subsequent issues. Scenario *Safe Driving* evaluated with medium interest (1,24/4), was according to 8 participants perceived as positive in terms of monitoring psychophysical state and providing feedback to the driver, especially for long travels. Issues of one's safety as well as safety of family members are highlighted as of high interest, however, monitoring of health parameters by wearables is also touching the sensitivity of data privacy, and 7 participants reported that they might find it invasive to give such data to an insurer. They discussed that it brings a sense and feeling of presence of a "big brother". Scenario *Entertainment* did not raise much of questions around the use of personal data, but it was also rated as low by interest (0,66/4).

Before and after the Focus Groups open discussions on presented scenarios, participants were asked to reflect on their willingness and interest to record certain personal data via diverse artefacts in order to receive an elaborated information they might find of interest in a certain context (Table 2). We observed that there is a high interest in providing certain personal data for receiving in exchange elaborated useful information and features. Before and after the storytelling of scenarios the opinions did not change drastically in numbers.

Table 2. Participants' availability to connect wearable devices with the car, while driving, to record physiological and performance data for following scopes.

Scope	Monitor distraction		Measure stress		Receive support	
	Before FG	After FG	Before FG	After FG	Before FG	After FG
Yes	32	32	22	19	33	29
Maybe	6	6	8	9	5	8
No	1	1	9	11	1	2

UI Video Mock-Ups. The validation in this phase leveraged on five online micro-videos (1-2 minutes each) explaining some core functionalities of the MEMoSa mobile application. In terms of use of personal data, the evaluation of this step referred to willingness of sharing data with different entities, such as other drivers using the same car (Table 3), insurance companies (Table 4), and MEMoSa system (Table 5).

Table 3. Participants' expressed willingness to share the following data with drivers with whom they share the same car, implying both the sharing of own data as well as interest in having access to other driver's data.

Type of data	Historic data of individual travels		Ongoing travel in real-time		Driving statistics		Driving behavior	
	Own	Other's	Own	Other's	Own	Other's	Own	Other's
Yes	37	33	39	36	37	37	34	36
No	17	21	15	18	17	17	20	18

Majority is willing to make the data exchange in this situation as there is a certain perceived value to it. However, some stated that driving style and routes, as well as stops and positioning would share only selectively, i.e. only with certain drivers.

Table 4. Participants' expressed willingness to share data with an insurance company.

Type of data	personal data	Age	Driving experience	Planning of trips	Driving style	Ongoing travel	Risky drive estimations
Yes	43	50	48	31	33	23	25
No	11	4	6	23	21	31	29

General personal data, age and driving experience appear to be not questionable when it comes to exchange of data with insurance companies. However, data that relate to real-time situations, ongoing trips and driver's estimations appeared as the opposite.

Table 5. Participants' expressed willingness to share data, derived from the OBD and wearable, with the MEMoSa system for the following scopes.

Scope	Receive support				Monitor well-being status		
Type of data	Speed	Harsh breaks	Harsh acceleration	Anomalies detection	Heart rate	Alertness	Sleep quality
Yes	46	43	40	53	41	38	35
No	8	11	14	1	13	16	19

In overall, there is an interest in having certain exchange of data for receiving support during drive as well as monitoring one's well-being, where the detection of anomalies is seen as the highest value, while alertness and sleep quality did not receive such a high consent.

Real-Life Testing. In this phase, feedback was collected during and after the trial period through two online surveys, distributed respectively after 7 days within the two-week period. This phase was related mostly to mere usability aspects, as the participants had the chance to try out the actual prototype. In regard to the issues of use of personal data, participants confirmed that for majority (82,4%) it was clear the purpose of using data collected from cars, smartwatch and smartphone, and 88,2% of them also stated that it is clear who has access to their personal data. Furthermore, 82,4% observed the balance between the amount of personal data provided and the offers received (both commercial and support/security assistance) as adequate and acceptable. After the real-life testing period, participants were asked to rate the three initial use case scenarios (we did not proceed with the fourth one as it was ranked as low interest), in order to make a comparison with the first testing phase. The values of scenarios in overall raised (*On the Spot Insurance* 2,47/4 as opposed to 1,87 previously rated; *Car Diagnostic* 2,71/4 opposed to 2,39; *Safe Driving* 2,65/4 opposed to 1,24), however privacy, security and transparency in data usage still remained a significant concern.

3.2 Considering A Social Consensus Within the Design Process

For overall testing of the proposed AI supported IoT system we organized three sequential steps that followed the design process, and we reasoned about user values within the following conceptual levels:

1. Usability (physical commodity and comprehension of the system),
2. Desirability (motivations of becoming a user),
3. Acceptability (expanding on a social level).

In the level of usability (2nd & 3rd testing phase), transparency in communication is observed as highly important when it comes to use of personal data. In terms of desirability (1st & 3rd testing phase), participants were willing to share their data, and also getting insight into other people's data, stay informed and receive correspondent services they find as useful in regard to sensitivity of data they shared. Furthermore, they expressed willingness to be familiar with the back-end operations of the system and understand in which way gathered data is being translated into an information, and who has access to such information.

The **results** showed strong presence of considerations from the side of users on data usage as well as sensitivity of certain data, which consequentially translate into user values of the design concept. Privacy and data treatment transparency appear as constant considerations through all three conceptual evaluation levels, especially when it comes to the acceptability (1st, 2nd & 3rd testing phase) of the designed system itself. We emphasize the importance and need of including acceptability on a social level, i.e. achieving social consensus, within the design processes through a critical analysis of the use of personal data in order to support the design of complex personalized services that shape emerging socio-technical systems. Fulfilling considerations that correspond to what is established as a social consensus on a particular issue, i.e. socially acceptable, within the design process would support building trust in connected systems for personalized AI agents. For e.g., for using bio-data extracted from a wearable to provide crafted information for the user, a system has to communicate clearly who are the parties that have access to such data as well, in order to achieve acceptance of use.

4 Conclusion & Future Work

Within the design system of radical innovation in terms of user interaction and experience, such as the one supported by an AI agent, trust plays a significant role. In the conducted case study, we confirmed that such systems that rely on the use of personal sensitive data are desirable, but acceptable under certain conditions. There is a need for a social consensus to be considered and directly employed during the design and evaluation process, in order to target and support the area of user values that deal with data and information exchange. With this notion, we want to provide an initial contribution to design methods that regard evaluation of complex IoT systems embodied through AI agents. As the discussion in this paper is based on one conducted case study, the future work will require application of reflections and concerns derived from this project to be translated and analyzed in other projects of similar nature. Such application could provide comparison and possible solutions to how to address this issue along design processes.

Acknowledgments. This work has been partially funded by TIM S.p.A., Services Innovation Department, Joint Open Lab Digital Life, Milan, Italy.

References

1. Arslan, P., Casalegno, F., Giusti, L., Ileri, O., Kurt, O.F., Ergüt, S.: Big Data as a Source for Designing Services. Web (2017)
2. Colombo, S.: Morals, ethics, and the new design conscience. In: Rampino, L. (eds.) Evolving Perspectives in Product Design: From Mass Production to Social Awareness. Franco-Angeli (2018)
3. Friedman, B., Kahn, P.H., Borning, A., Huldtgren, A.: Value sensitive design and information systems. In: Early Engagement and New Technologies: Opening Up the Laboratory, pp. 55–95. Springer, Dordrecht (2013)
4. Kalbach, J.: Mapping Experiences: A Complete Guide to Creating Value Through Journeys, Blueprints, and Diagrams. O'Reilly Media, Inc. (2016)
5. Kankainen, A., Vaajakallio, K., Kantola, V., Mattelmäki, T.: Storytelling group—a co-design method for service design. Behav. Inf. Technol. **31**(3), 221–230 (2012)
6. Krueger, R.A.: Focus Groups: A Practical Guide for Applied Research. Sage Publications (2014)
7. Mitchell, W.J.: Me++: The Cyborg Self and The Networked City. MIT Press (2004)
8. Pillan, M., Varisco, L., Bertolo, M.: Facing digital dystopias: a discussion about responsibility in the design of smart products. In: Proceedings of the Conference on Design and Semantics of Form and Movement—Sense and Sensitivity, DeSForM 2017. InTech (2017)
9. Taebi, B.: Bridging the gap between social acceptance and ethical acceptability. Risk Anal. **37** (10), 1817–1827 (2017)

Web Search Skill Evaluation from Eye and Mouse Momentum

Takeshi Matsuda[1(✉)], Ryutaro Ushigome[2], Michio Sonoda[3], Masashi Eto[3], Hironobu Satoh[3], Tomohiro Hanada[3], Nobuhiro Kanahama[3], Hiroki Ishikawa[3], Katsumi Ikeda[3], and Daiki Katoh[3]

[1] Department of Information Security, University of Nagasaki, Nagasaki, Japan
tmatsuda@sun.ac.jp
[2] Graduate School of Science and Engineering, Chuo University, Tokyo, Japan
[3] National Institute of Information and Communications Technology, Tokyo, Japan

Abstract. Conventionally, it has been pointed out that the eye movement when using the PC correlates with the movement of the mouse, and those are utilized as additional information on the usability improvement and other observed data. This study had analized the eye movement and the mouse movement data, and had examined the method to estimate the user's skill.

Keywords: Eye tracking · Mouse momentum · Web search skill Skill evaluation

1 Introduction

It had been mentioned that the eye movement and the mouse movement are highly correlated [1]. It is known that the eye movement can be tracked from the light rays that reflect on the eyeballs [2, 3]. The operability of the GUI screen such that a computer is evaluated using the Eq. (1).

$$T = a + b\log_2\left(1 + \frac{d}{w}\right). \tag{1}$$

Here, a and b are constants determined by regression analysis, respectively. The distance d is calculated from the starting point to the center of the target, and the value w is correspond to the size of the target. Equation (1) is called Fitts's law [4]. The ease of finding the target and the operability after finding it will be depending on how the GUI screen is designed. However, there will be users who can find targets quickly regardless of whether the design is good or bad. This is thought to be related to skills, prior knowledge and experience of the users. It is expected that those features of such users will also be reflected in the mouse movement data required for the search operation. Therefore, in this study, in order to examine whether user's work contents and skills can be estimated from the eye and the mouse data, we gave some task to

W. Karwowski and T. Ahram (Eds.): IHSI 2019, AISC 903, pp. 884–888, 2019.
https://doi.org/10.1007/978-3-030-11051-2_135

subjects, and acquired the eye and mouse movement data at the web search. As a result, it was confirmed that the eye and mouse movement has strong correlation as documented in the paper [1]. But, there were also some data that did not show correlation depending on the user's work content. By removing such data, it was confirmed that the eye and the mouse momentum data were almost proportional, and the ratio was different for each subjects.

2 Preliminary

This study will analyze user's Web information retrieval skill from the eye and the mouse movement. The speed to reach to the target information may be depending on the prior knowledge of subjects. But denying the possibility that some skill of information retrieval may exist besides is difficult. In fact, if we ignore whether subjects can understand the target information in detail, there may exist users who find in a short time on the source necessary to understand that information. It is already known that the eye and the mouse movement has strong correlation, but this study confirms the above fact, and also considers the difference between the eye and the mouse movement in the skill and work content of the user. The eye and the mouse movement data was acquired by using Tobbi and UWSC, respectively. Tobbi can colect eye movement data and mouse click information. UWSC is known as the Windows OS automation tool, so it can colect the user's input data.

3 Experiment

This study had acquired the eye and the mouse data of 5 subjects. In the experiment, we gave the theme related to mathematics or automobile according to the prior knowledge of the subjects, and acquired the eye and the mouse movement data.

[Mathematics related theme]:
Explain on how a germ and a stalk are defined or the Fermat's statement of the last theorem.

[Vehicle theme]:
Explain on the high performance model of the C segment of German cars and their strengths and weaknesses or a mirage diamond.

Calibration of the eye and mouse data is required because those data are collected by separate and independent devices. Tobbi to acquire eye movement data can record the coordinate data clicked with mouse, so calibration can be executed by comparing the click data of Tobbi and UWSC. Furthermore, since Tobbi and UWSC differ in the granularity of data acquired in time series, it is necessary to complement of those data in order to make the data acquisition time match. However, in this study, in order to pay attention to the eye and the mouse momentum, instead of complement the data, calculate the eye and the mouse momentum from the coordinate data of each time until the timing when Tobbi and UWSC click data coincide and compute their sum. Specifically, we compute the distance

$$m = \sum_{i=t}^{T} d_i, \quad d_t = \sqrt{(x_{t+1} - x_t)^2 + (y_{t+1} - y_t)^2} \tag{2}$$

between (x_t, y_t) and (x_{t+1}, y_{t+1}). Here, the coordinates (x_t, y_t) is observed at the time t, and the coordinate (x_{t+1}, y_{t+1}) is observed at the next recorded time t + 1. Let m_{UWSC} and m_{Tobii} be a mouse momentum and an eye momentum, respectively.

4 Result

Here, let us show the acquired data $\{(m_{UWSC}, m_{Tobii})\}_{i=1}^{n}$ in this experiment. Amount of the data is one of feature quantity that can be used for the data classification in this case. The fact that the number of observed data is small will correspond to the fact that the man-hour in the work is small. Furthermore, the smaller the value of (m_{UWSC}, m_{Tobii}), for each t, it means that work efficiency is well.

[Experimental result 1]
Figure 1 indicates the data of subjects with high information retrieval skills. The task of this subject was mirage diamonds, and subjects had no prior knowledge of this task. The correlation coefficient between the eye and the mouse momentum of the data in Fig. 1(A) is 0.93, and it can be seen from Fig. 1(A) that they have a strong correlation. The case of Fig. 1(B) is the number of observed data is also small, but it is data which did not reach the correct information. The task of the subject of Fig. 1(B) concerned Fermat's last theorem, and he had no prior knowledge. The correlation coefficient of Fig. 1(B) is 0.60, and when the data at time 255164 is removed, the correlation coefficient becomes to 0.74.

[Experimental result 2]
Figure 2(A) shows the subject's data of the task concerning C segment vehicles, and the he is interested in cars. And, Fig. 2(B) is the same subject as in Fig. 2(A) and is searching for the Fermat's last theorem without prior knowledge. The correlation coefficient of the data in Fig. 2(A) and (B) are 0.72 and 0.59, respectively.

[Experimental result 3]
Factors that weakened the correlation between eye and mouse differed depending on subjects and work contents, but in some subjects there was a tendency that the correlation became weak when the link click operation regardless of the presence (Fig. 3 (A)) or absence (Fig. 3(B)) of prior knowledge.

[Experimental result 4]
Also, in case of the operation of switching tabs of open browsers, It can be seen from Fig. 4 that there is strong correlation between eye and mouse when there is prior knowledge.

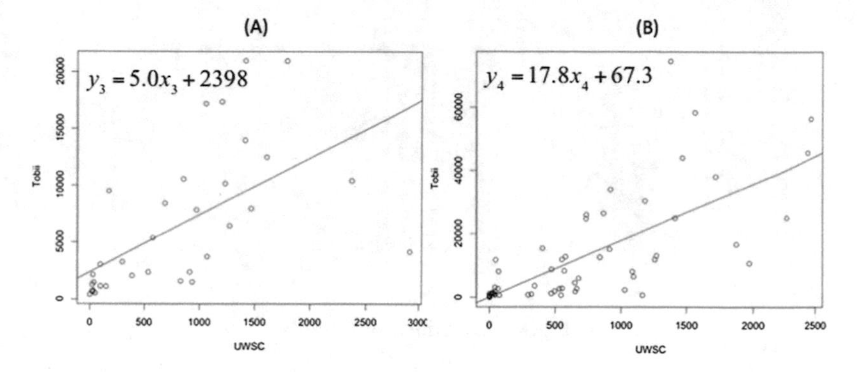

Fig. 1. Experimental result 1

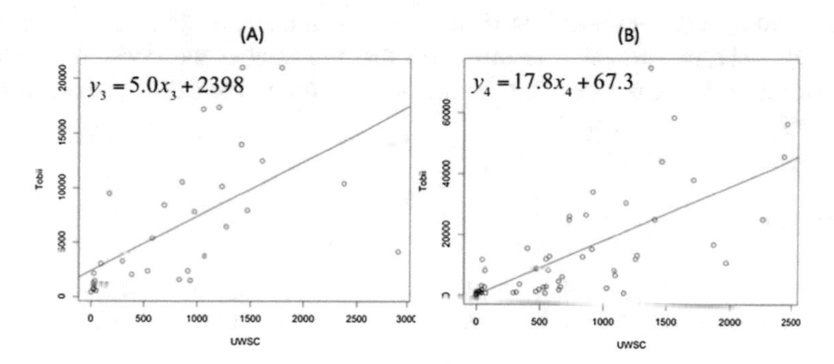

Fig. 2. Experimental result 2

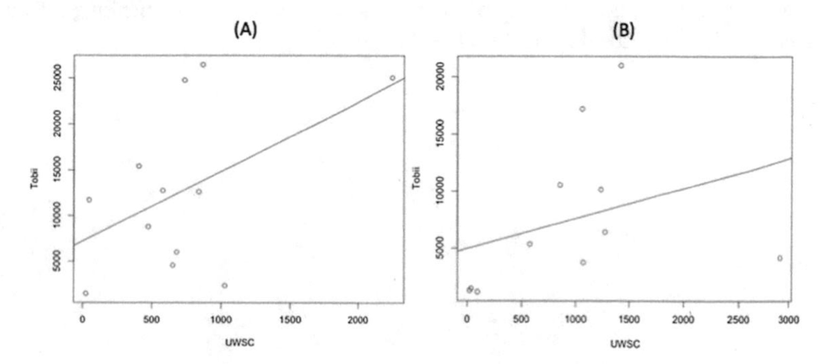

Fig. 3. Experimental result 3

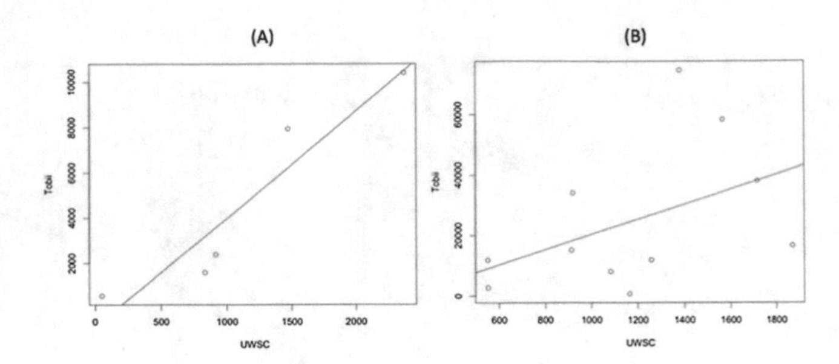

Fig. 4. Experimental result 4

5 Discussion and Summary

As a result of this experiment, the user's skills may be extracted by the slope of the regression line. Here, x-axis and y-axis are a mouse momentum and an eye momentum, respectively. The experimental results show that the smaller the slope, the higher the skill. Our future work is to extract factors that determine the skill by analyzing work contents of subjects in detail.

References

1. Chen, M.C., Anderson, J.R., Sohn, M.H.: What can a mouse cursor tell us more?: correlation of eye/mouse movements on web browsing. In: Proceedings of CHI 2001, pp. 281–282. ACM (2001)
2. Tobii. https://www.tobii.com
3. Punde, P.A., Jadhav, M.E., Manza, R.R.: A study of eye tracking technology and its applications. In: 2017 IEEE International Conference on Intelligent Systems and Information Management (ICISIM), pp. 86–90 (2017)
4. Fitts, P.M.: The information capacity of the human motor system in controlling the amplitude of movement. J. Exp. Psychol. **47**(6), 381–391 (1954)

Evaluation of the Effects of an AC Magnetic Field on Cutaneous Blood Flow Volume by Cold Water Immersion Test

Nur Izyana Faradila Binti Azmi[1(✉)], Hideyuki Okano[2], Hiromi Ishiwatari[3], and Keiichi Watanuki[2,4,5]

[1] Faculty of Engineering, Department of Mechanical Engineering, Saitama University, Saitama, Japan
nur.izyana.f.b.596@ms.saitama-u-ac.jp
[2] Advanced Institute of Innovative Technology, Saitama University, Saitama, Japan
hideyukiokano@aol.com
[3] Soken Medical Co., Ltd., Tokyo, Japan
[4] Graduate School of Science and Engineering, Saitama University, Saitama, Japan
[5] Brain and Body System Science Institute, Saitama University, Saitama, Japan

Abstract. This study focuses on the acute influence of an AC magnetic field (50 Hz, B_{max} 180 mT, for 10 min) on recovery of blood flow after cold water immersion (5 °C water for 1 min) in healthy human subjects. In a randomized, double blind and crossover study design, magnetic field (MF) and sham control (CTL) exposure experiments were carried out. The microcirculation images were recorded and analyzed using a 2D laser speckle flowmetry. The blood flow volume values of the fingers and hand in both MF and CTL groups were significantly reduced immediately after cessation of the immersion and there were significant differences in the recovery rate of finger blood flow between both experiments. The response to cold immersion can be used to detect vascular disorders and the MF-enhanced blood circulation results might prove the physiological role of MF exposure to help eliminating the metabolic waste products in our body.

Keywords: AC magnetic field · Cold water immersion · 2D laser speckle flowmetry · Microcirculation

1 Introduction

Recently, we demonstrated that the blood flow volume in the arteries were significantly increased by acute and local exposure to an AC magnetic field (MF, 50 Hz, B_{max} 180 mT, 15-min duration of exposure) when compared to the sham control (CTL) exposure [1]. However, only a few researchers have described that ice water immersion test can help to determine individual vascular reactivity and marked changes in blood flow when a person immersed an extremity in ice water [2]. Therefore, in this study we

W. Karwowski and T. Ahram (Eds.): IHSI 2019, AISC 903, pp. 889–894, 2019.
https://doi.org/10.1007/978-3-030-11051-2_136

evaluate the effects of AC MF combined with cold stress to determine the blood flow changes in the cold stimulus when exposed to the MF exposure.

2 Methods

2.1 Subjects

Healthy volunteer subjects (5 males, 5 females, age range 21–24 years, heights 158–175 cm, weights 52–65 kg) volunteered to participate in the present study. The study protocol was approved by the Saitama University's institutional review board. Each subject was fully informed about all procedures and voluntarily signed an informed consent document before entering the study. During the study period, subjects did not use any form of physical therapy and did not take any vasoactive medication. Subjects' body temperature, and systolic and diastolic blood pressures were within normal ranges. Subjects were made relax physically and mentally for at least 10 min before the initiation of the procedure. All trials were carried out during daytime (11:00 a.m.–17:00 p.m.) at room temperature (25 ± 0.5 °C) and relative humidity of 50 ± 10%.

2.2 Study Protocol

The study was conducted during this period when the weather conditions were neither too hot nor too cold. This is because the results could be affected due to the cold adaption in winter climate, and increased sensitivity to cold water during summer. The study protocol is shown in Fig. 1. Monitoring of cutaneous microcirculation in the back of the left hand was conducted under hand and forearm exposures to an AC MF (50 Hz, B_{max} 180 mT). The blood flow volume values analyzed from the microcirculation images were compared with two different exposures; MF exposure and sham control (CTL) exposure. In a randomized, double blind and crossover study design, MF and CTL exposure experiments were carried out.

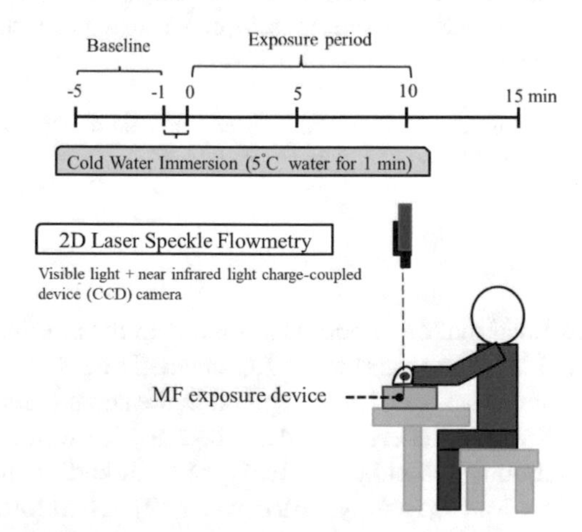

Fig. 1. Study protocol for hand and forearm exposures to an AC MF and monitoring of cutaneous microcirculation in the back of the hand.

Before the experiment started, the blood flow volume was recorded for 5 min for the baseline reading. The conventional method of cold water immersion test was followed in which the left hand up to the wrist was immersed in cold water of 5 °C for 1 min [3] using a waterproof hand covering. The MF or CTL exposure was performed continuously for 10 min immediately after cessation of the immersion during the 15 min experiments. For hand and forearm exposure, the ventral side of the left hand and forearm were positioned on an AC MF exposure device for 20 min to keep the hand and arm motionless as long as possible except for during the immersion.

The microcirculation images were recorded and analyzed at 1-min intervals for 20 min using a 2D laser speckle flowmetry (Omegazone OZ-3, Omegawave, Fuchu, Tokyo, Japan). The MF or CTL exposure was performed continuously for 10 min. For hand and forearm exposures, the hand and forearm were positioned on an MF exposure device for 20 min to keep the arm motionless as long as possible during the clinical trial in each individual as shown in Fig. 2.

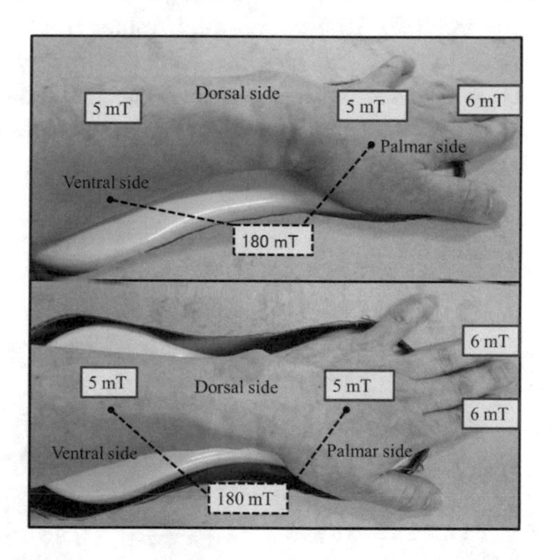

Fig. 2. Hand and forearm exposures using an AC MF exposure device.

The AC MF exposure device (Soken MS, Toride, Ibaraki, Japan) was utilized for research purpose. The spatial distribution of the B_{max} from the surface of the MF exposure device has been reported in detail elsewhere [1].

2.3 2D Laser Speckle Flowmetry

The principle of laser speckle flowmetry has been reported in detail elsewhere [4–6]. It is assumed that the measurement depth for human skin is about 1 mm or less from the surface [6]. The whole sequential 10 images in the back of the left hand for 10 s were captured by the pixel of the CCD camera simultaneously in high resolution mode (1 image/sec with image resolution 750 × 560) at 1-min intervals for 20 min. After recording the images, three measurement points of nail of index finger, nail of ring finger and center of hand were selected and circled using region-of-interest (ROI). The blood flow volume values were analyzed from every ROI using the built-in software in the blood imager.

2.4 Statistical Analysis

Statistical analysis of differences in mean values of MF and CTL groups was made by using the Wilcoxon rank-sum test (between groups) and the Wilcoxon paired signed rank test (within group) for non-parametric data. For all comparisons, a P value less than 0.05 was considered significant.

3 Results and Discussion

Blood flow volume values of ten individuals have been measured one to two times for MF and CTL exposure experiments, and the average values were calculated for each individual. The changing rate (%) from the baseline values of blood flow volume was analyzed because the variability of the baseline values was very large between individuals, which can be associated with physiological fluctuations over time. The response to cold immersion of cutaneous blood flow volume values in the dorsal fingers and hand is shown in Fig. 3.

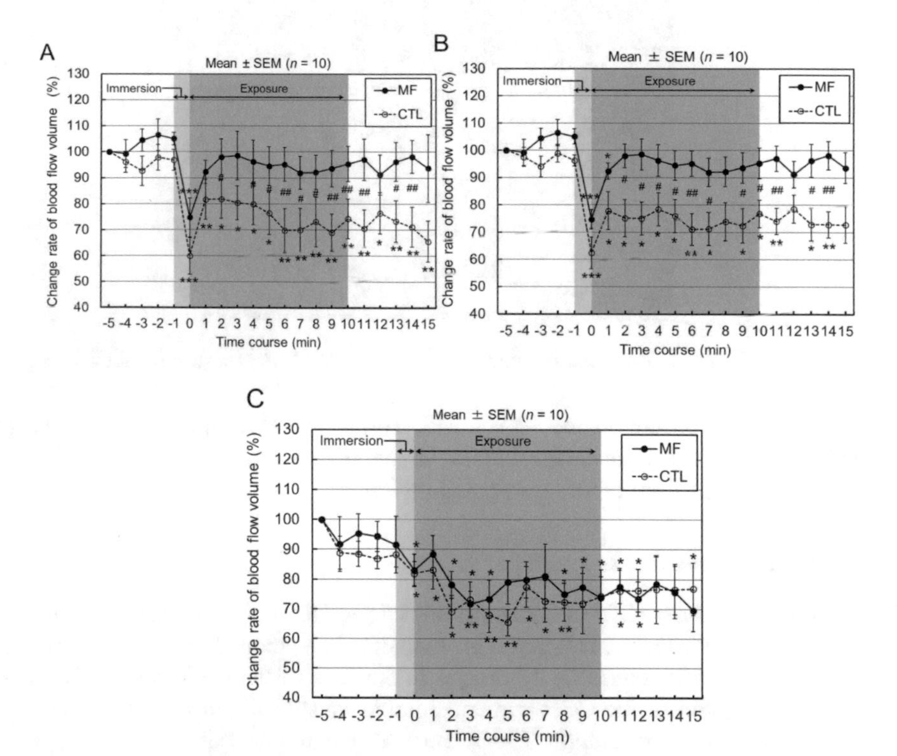

Fig. 3. The time course of the change rate of blood flow volume in magnetic field (MF) and sham control (CTL) groups. A: Nail of ring finger. B: Nail of index finger. C: Central part of hand. Values are expressed as mean ± SEM (n = 10 in each group). $*P < 0.05$, $**P < 0.01$, $***P < 0.001$ compared with the baseline (within group). $\#P < 0.05$, $\#\#P < 0.01$ compared with the CTL (between groups).

Most subjects showed response to the cold immersion, and as a result, the blood flow volume values of the fingers and hand in both MF and CTL groups were significantly reduced immediately after cessation of the immersion. The sudden drop of blood flow volume in the hand represented vascular reactivity to the cold stimulus during the initial 60 s of the experiment. This response is considered as a neurogenic reflex and indicates the pivotal role of autonomic nervous system in regulation of cardiovascular functioning in normal individuals [7].

These changes were mediated by the increase of blood flow after the first 2 min experiment, indicating the hemodynamic adaption to the stimulus and the recovery of blood flow volume back to the normal, underlying the mechanism of autonomic neural pathways. There were significant differences in the recovery rate of finger blood flow volume between MF and CTL groups during and after MF exposure. However, in the case of central part of hand, no significant differences were found neither in MF nor CTL thus this needs further investigation in the future.

Significant rises of blood flow volume were observed during MF exposure compared to the CTL. This is because MF can actively enhance a better blood recovery compared to the normal recovery process. Thus, our findings of the increased blood flow velocity in the previous study could be due to the initiation of enough "eddy current" induced by AC MF exposure [1]. The eddy current plays a crucial role in the activation of parasympathetic nerve and hemodynamic responses via cholinergic pathways together with NO-mediated vasodilation. [1, 8, 9]. Hence, we proved that AC MF could improve blood flow volume in our microcirculation even under the ischemic conditions induced by cold water immersion.

In particular, as a mechanism of its action, it was inferred that NO may be involved. In other words, it is considered that NO decreased in the ischemic state, whereas the MF inhibited the decrease of NO, resulting in suppression of decrease in the rate of blood flow volume. Recent studies have reported that the increase in NO inhibits thrombus formation and increases blood flow [10]. Also, as a clinical trial investigating the action of NO itself, it has been reported the increase of pulmonary blood flow induced by directly inhaling NO [11]. This suggests that the possibility arises that MF may act on NO with such increasing effect of blood flow, and thereby increase NO production, or suppress the decrease of NO production.

In general, blood circulation and microcirculation play a pivotal role on transporting nutrients and growth substances as well as eliminating waste products including pain producing metabolites [12, 13]. Therefore, these results imply that the physiological role of an MF-enhanced blood circulation recovery might help eliminate the metabolic waste products and endogenous pain producing substances inducing muscle stiffness and pain. Moreover, MF could promote wound healing by transporting nutrients and growth substances. Further studies should be needed to investigate the MF-based therapeutic applications and elucidate the underlying mechanisms of MF effects on pain relief and recovery of muscle fatigue.

4 Conclusion

In this study we proved that after the sudden drop of blood flow due to the cold water immersion, MF exposure can induce a higher blood flow volume recovery rate compared to the normal rate in CTL. Therefore, to improve these findings, further research should be done in the near future by also considering the heart rate, blood pressure and respiratory rate of the subjects throughout the experiments.

References

1. Okano, H., Fujimura, A., Ishiwatari, H., Watanuki, K.: The physiological influence of alternating current electromagnetic field exposure on human subjects. In: IEEE International Conference on Systems, Man, and Cybernetics (SMC), pp. 2442–2447 (2017). ISBN 9781538616451
2. Laskar, S., Harada, N.: Different conditions of cold water immersion test for diagnosing hand-arm vibration syndrome. Environ. Health Prev. Med. **10**, 351–359 (2005)
3. Kent, P., Wilkinson, D., Parkin, A., Kester, R.C.: Comparing subjective and objective assessments of the severity of vibration induced white finger. J. Biomed. Eng. **13**, 260–262 (1991)
4. Briers, I.D.: Laser doppler and time-varying speckle: a reconciliation. J. Opt. Soc. Am. A **13**, 345–350 (1996)
5. Forrester, K.R., Stewart, C., Tulip, J., Leonard, C., Bray, R.C.: Comparison of laser speckle and laser doppler perfusion imaging: measurement in human skin and rabbit articular tissue. Med. Biol. Eng. Comput. **40**, 687–697 (2002)
6. Kashima, S.: Spectroscopic measurement of blood volume and its oxygenation in a small volume of tissue using red lasers and differential calculation between two point detections. Opt. Laser Technol. **35**, 485–489 (2003)
7. Garg, S., Kumar, A., Singh, K.D.: Blood pressure response to cold pressor test in the children of hypertensives. Online J. Health Allied Sci. **9**, 7 (2010). ISSN 0972-5997
8. Ravera, S., Bianco, B., Cugnoli, C., Panfoli, I., Calzia, D., Morelli, A., Pepe, I.M.: Sinusoidal ELF magnetic fields affect acetylcholinesterase activity in cerebellum synaptoso mal membrane. Bioelectromagnetics **31**, 270–276 (2010)
9. Patruno, A., Amerio, P., Pesce, M., Vianale, G., Di Luzio, S., Tulli, A., Franceschelli, S., Grilli, A., Muraro, R., Reale, M.: Extremely low frequency electromagnetic fields modulate expression of inducible Nitric Oxide Synthase, Endothelial Nitric Oxide Synthase and Cyclooxygenase-2 in the Human Keratinocyte Cell Line HaCat: potential therapeutic effects in wound healing. Br. J. Dermatol. **162**, 258–266 (2010)
10. Ghimire, K., Altmann, H.M., Straub, A.C., Isenberg, J.S.: Nitric Oxide: what's new to NO? Am. J. Physiol. Cell Physiol. **312**, C254–C262 (2017)
11. Sánchez Crespo, A., Hallberg, J., Lundberg, J.O., Lindahl, S.G., Jacobsson, H., Weitzberg, E., Nyrén, S.: Nasal Nitric Oxide and regulation of human pulmonary blood flow in the upright position. Appl. Physiol. **1985**(108), 181–188 (2010)
12. Plante, G.E.: Vascular response to stress in health and disease. Metabolism **51**, 25–30 (2002)
13. Beard, D.A., Wu, F., Cabrera, M.E., Dash, R.K.: Modeling of cellular metabolism and microcirculatory transport. Microcirculation **15**, 777–793 (2008)

Development of an Individual Joint Controllable Haptic Glove (CRL-Glove) and Apply for CLASS

Kazushige Ashimori(✉) and Hiroshi Igarashi

Department of Electrical and Electronic Engineering,
Graduate School of Engineering, Tokyo Denki University,
Adachi-Ku, Tokyo 120-8551, Japan
{k.ashimori,h.igarashi}@crl.epi.epi.dendai.ac.jp

Abstract. In this paper, we propose a haptic glove: CRL-Glove, it has a wide range of movement for daily tasks while wearing them. Control is possible on six degrees of freedom of the three fingers: index, middle, andthumb. To confirm the possibility of control over the fingers' motion, after performing a free-motion experiment, a mastery assistance task was performed. The mastery assistance task consists of a student and a teacher, both wearing the CRL-Gloves. The student is learning through the teacher's fingers movement. This method is then compared with other education methods: a method that does not use the gloves, a master-slave method and CLASS. The CLASS is a learning assistance method using haptic information exchange between the teacher and the student. Compared to normal education method, the student is learning more intuitively. This paper aims to validate the CLASS method's usefulness compared to the two normal education methods.

Keywords: Haptic glove · Haptics · Learning assist · CLASS
Force interaction · Bilateral · Education · Guitar

1 Introduction

A lot of researches have been focusing on gloves that can reproduce haptics sense for virtual reality or rehabilitation for example. This kind of glove is called haptic glove.

Methods to reproduce haptics sense such as vibration motors method or using a tendon drive system to pull a tied cord are exist but none of them provides feedback for each finger joints' motion [5–7].

A human finger can behave a wide range of the movement with high torque, and it is a most important interface in a human. Although control human finger joints individually are difficult, control of each joint is necessary for haptics recording and reproduction, haptics interaction.

Then this research proposes a type of haptic glove that can control each finger joints' movement by attaching DC motors to an exoskeleton. Furthermore, we applied the glove that is proposed in this research to a complementary learning assist system (CLASS) which method using haptics sense, and we confirm the usability of the glove.

W. Karwowski and T. Ahram (Eds.): IHSI 2019, AISC 903, pp. 895–901, 2019.
https://doi.org/10.1007/978-3-030-11051-2_137

2 Concepts and Design of the CRL-Glove

Figure 1(a) shows the overall view of a haptic glove, and Fig. 1(b) shows the structure of the glove. The CRL-Gloves allow a wide range of movement for virtual reality or rehabilitation use. Control is possible on six degrees of freedom of the three middle fingers: index, middle and ring fingers. Also, as the exoskeleton is placed on the back of the hand, it is possible to perform daily tasks while wearing them.

CRL-Glove controls a human finger motion by using the link mechanism shown in Fig. 1(b), in order to perform force feedback in both directions of hand opening and grasping. Motors are arranged three in each glove fingers, and the links of the glove are fixed to each human finger phalanx with a hook-and-loop fastener. Each link mechanism connecting each node has six degrees of freedom (thus, each glove finger has 18 degrees of freedom). The mount connected at a distal phalanx is fixed to the middle phalanx via links L1 and L2. This fixed end is connected to the motor, and feedback is made to the distal phalanx.

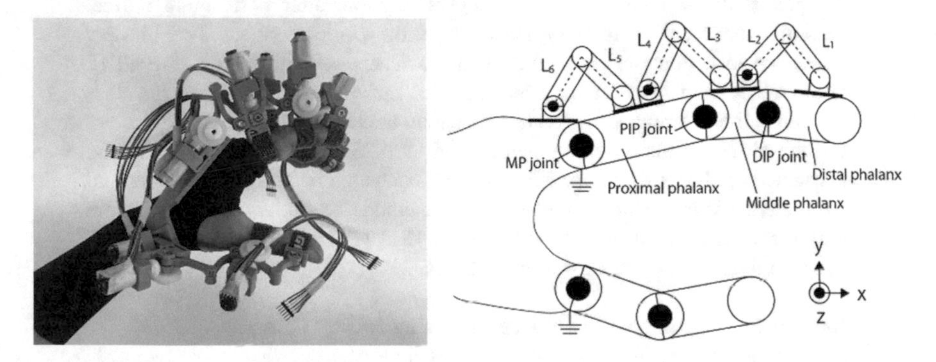

Fig. 1. Proposed method: a haptic glove controllable each joint individually. (a) Overview of the CRL-Glove. (b) Construction of the CRL-Glove.

3 Proposal Method

The purpose of this research is to propose a haptic glove which allows to control human finger joint individually and to apply that glove to CLASS which method proposed us in this paper [1], and to conform to the usability.

The CLASS is a learning assistance method using haptic information exchange between the teacher and the student. Compared to normal education method, the student is learning more intuitively. The CLASS method aims to support teacher's education by using haptics feedback between the student and the teacher. In this research, it is necessary to perform haptics communication between the two gloves in order to perform a learning support experiment using the CLASS. Therefore, in this study, we control the glove by adopting bilateral control which proposed by Iida and Ohnishi [2].

To ensure a wide range of possible movement for the hand wearing the CRL-Glove, the motor controlling the finger joint movement is arranged vertically. The rotation axis of the motor is adapted to the finger joint rotation using bevel gear. However, as the bevel gear greatly increase static friction, reaction force observer or disturbance observer bilateral control which was originally used is no longer effective. Therefore, pressure sensors (FSR: Force Sensing Resisters) are applied to each finger joint. Haptics communication with the CRL-Glove is then performed by measuring pressure change on the fingers. FSR was inserted into the ball of the finger and back of the finger to measure the finger force in both directions of finger opening and grasping.

When wearing the gloves, an initial value of the pressure is not constant and the human finger has individual differences in the thickness. Thus, the calibration was executed each time when wearing the gloves.

Figure 2 shows the control block diagram of the glove in this experiment. In Fig. 2, Km is the position gain, Gm is the gain of the reaction force calculated from the reaction force observer (RFOB) [3] with disturbance observer (DOB) [4], and Sm is the gain of the static friction observer at the pressure Pm by the FSR. In addition, the command value in the motor is θ_M^{ref}, and τ_M^{dis} is a disturbance torque.

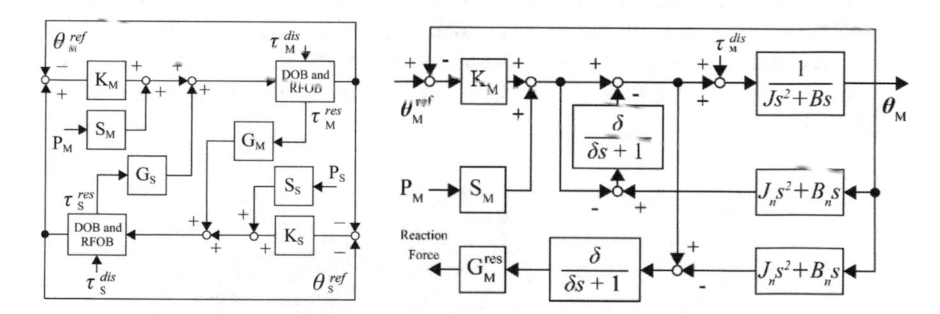

Fig. 2. Control block diagram for the CRL-Glove. (a) Block diagram of bilateral control. (b) Detail of master side block diagram.

4 Experimental Environment

4.1 Task Decision

To confirm the possibility of control over the fingers' motion, after performing a free-motion experiment, a mastery assistance task was performed. The mastery assistance task consists of a student and a teacher, both wearing the CRL-Gloves. The student is learning through the teacher's fingers movement. This method is then compared with other education methods: a method that doesn't use the gloves, a master-slave method and CLASS.

Since CRL-Glove can be controlled independently for each joint, we decided to perform a guitar performance as a mastery task to reveal its usefulness.

In the musical instruments, the guitar needs peculiarly finger motion, since when a player makes a hand motion of the guitar chord; the guitar fingerboard is handled three-dimensionally.

Therefore, a beginner of the guitar playing is difficult to understand mastering adequately with only visual information and auditory information. In addition, as a task, guitar performance has ease of evaluation and ease of adjustment of difficulty.

4.2 Experimental Method

In this experiment, in order to master beginners of guitar performance, we compared the usual mastery support method (without gloves) and the method of using the gloves. From the experiment in a paper [1], we confirmed that feedback gain has an effect on the efficiency of mastery support. Thus, subjects of the experiment were divided into four teams (100%, 70%, 40%, and 0%) by difference of the feedback gain for a teacher comparing with without glove team. In which, a team of 0% feedback team is the same as the master-slave method.

A task of the experiments is to play eight types chord shown in Fig. 3 expressed in TAB score. Learning contents are to play a total of 8 codes with the rhythm of BPM 60. Each chord is four seconds and which is repeated two times, thus the task of one time is 32 s. The output sound of the guitar was recorded using Roland's audio interface: QUAD CAPTURE.

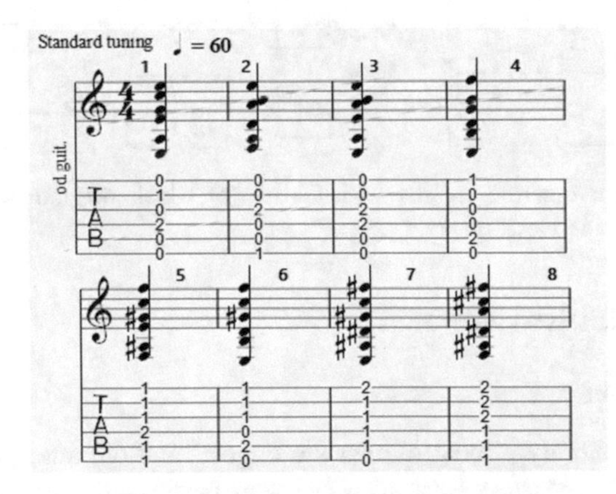

Fig. 3. Experimental task: eight chords of the guitar (TAB score).

As an experimental procedure, we first taught subjects basic information on guitar, let me hear the sound source for 32 s twice, and during that time subjects did self-learning. During this 64 s, the teacher does not teach, and the subject learns using tab staff and sound source only. As soon as self-study finished, we recorded the subject's output sound to the same tempo metronome. Subjects are free to decide the beginning of the recording, subjects perform while watching tab notation. Let this result be the basic ability of the subject, and consider how much it has grown from this value in the evaluation.

Subsequently, subjects who conduct experiments using gloves wear the gloves and teacher educated for 3 min. At this time, the subjects ware not told the effect of the feedback gain for the teacher. After the education, subjects output was recorded, after removed the glove.

Then, the subject who wared the glove ware the glove again with same feedback gain, and teacher educated for 5 min. Finally, the output of the subjects was recorded again.

5 Experimental Results

For the evaluation of the experiment, we decided to compare the recording of the first time with the recording of the second time and the third time, and obtain the growth rate by the correlation coefficient, SN ratio, sustain. The calculation formula used is the same as in [1]. Subjects are totally 12, the twenties, without glove team has 4 subjects, and another team has 2 subjects. Also, the proposed glove allow to control only three fingers, the subjects used the same three fingers when it plays.

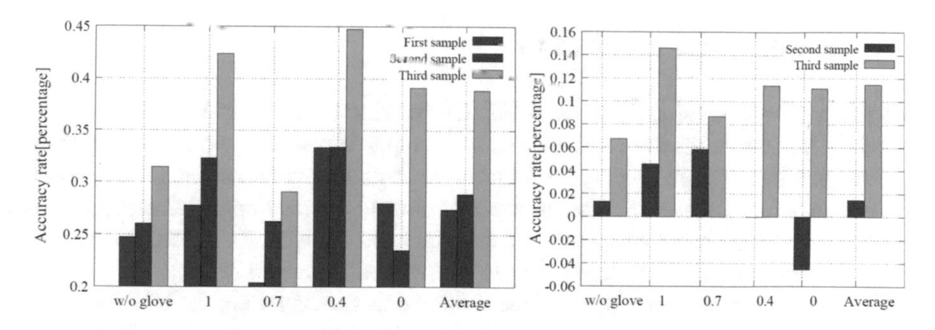

Fig. 4. Experimental results: correlation ratio comparing with teacher's output. (a)Average on eight chords. (b) Growth ratio from first sampling.

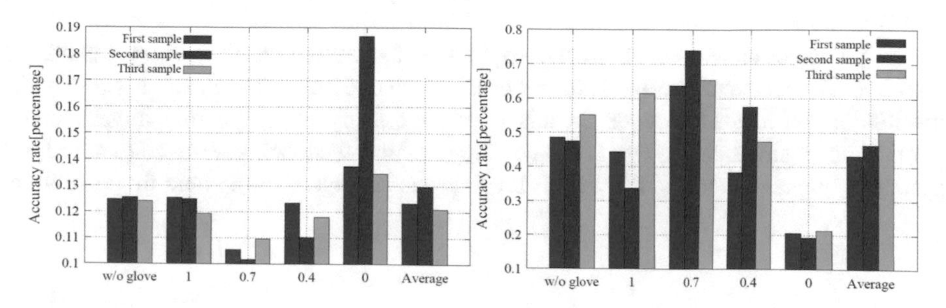

Fig. 5. Experimental results. (a) Average of the S/N ratio. (b) Average of the Sustain.

The experimental results are shown in Fig. 4(a). In that figure, blue represents the first time, red represents the second time, and green represents the third time sample of the correlation coefficient. Figure 4(b) shows the growth ratio from the first sample. From Fig. 4(b), high growth ratio was confirmed from all of the team that haptic assisted by CRL-Glove. In the Fig. 4(a), the team without a glove team shows stable, smooth growth, however, the 0% feedback team shows a decrease in the second sample. From this results, we suggest the master-slave method hasn't a stability on the learning assist, and this conclusion is the same as a conclusion in the paper [1].

Figure 5(a) is the experimental result on the S/N ratio of each team's average, and (b) is each team's average on the sustain. From Fig. 5, high growth ratio can be confirmed from the second sample of 0% feedback team.

However, the correlation ratio of the same term shows low accuracy from Fig. 4(a), we thought they didn't grow in that time.

From Fig. 5(b), a lower growth rate than the average was confirmed from 100% feedback team.

In the CLASS, when the feedback gain was 100%, two operators finger motion is competed by the bilateral control. This force compete has a possibility that effects disturbance for the education. Thus, we thought that experimental results showed not a desirable correlation ratio. This conclusion is also the same as the paper [1].

6 Conclusion and Discussion

This paper proposed a haptic glove: CRL-Glove which controllable each finger joints and apply that glove for the CLASS, and we compared with the two normal education methods previously described. On a future work, by improving movement range along with adding a force observer, this research aims to add a force dimension representation for the operators of the gloves, and then adapt for the CLASS. As a result, our method showed a high growth ratio comparing to the without glove education.

Another objective would be to adjust the force feedback gain in the bilateral system used for the CLASS and to suggest an optimal gain. From the experimental results, we conclude 100% feedback gain is not suitable for the CLASS. This conclusion has corresponded with a suggestion in the paper [1]. Comparing the growth ratio of the sustain and correlation ratio, this paper suggests optimal feedback gain exists between 100% and 70%.

In an experimental result in the paper [1], the master-slave method considered effective. In that experiment, a learning task is only one guitar chord, and it was a static motion of the hand. However, in this paper, we analyzed a dynamic finger motion containing position changing. Thus, we suggest high feedback gain is adaptable for a dynamic finger motion, and low gain feedback is adaptable for a static finger motion.

References

1. Ashimori, K., Igarashi, H., Complemental Learning Assist for Musical Instruments by Haptic Presentation. In: International Conference on Human-Computer Interaction (HCII) 2018 Posters' Extended Abstracts, pp. 11–16 (2018)
2. Iida, W., Ohnishi, K.: Reproducibility and operationality in bilateral teleoperation. In: The 8th IEEE International Workshop on Advanced Motion Control, pp. 217–222 (2004)
3. Yamaoka, S., Nozaki, T., Yashiro, D., Ohnishi, K.: Acceleration control of stacked piezoelectric actuator utilizing disturbance observer and reaction force observer. In: 2012 12th IEEE International Workshop on Advanced Motion Control (AMC), Sarajevo, pp. 1–6 (2012)
4. Ohnishi, K., Shibata, M., Murakami, T.: Motion control for advanced mechatronics. IEEE/ASME Trans. Mechatron. **1**(1), 56–67 (1996)
5. Baldi, T.L., Scheggi, S., Meli, L., Mohammadi, M., Prattichizzo, D.: GESTO: a glove for enhanced sensing and touching based on inertial and magnetic sensors for hand tracking and cutaneous feedback. IEEE Trans. Hum.-Mach. Syst. **47**(6), 1066–1076 (2017)
6. Ma, Z., Ben-Tzvi, P., Danoff, J.: Hand rehabilitation learning system with an exoskeleton robotic glove. IEEE Trans. Neural Syst. Rehabil. Eng. **24**(12), 1323–1332 (2016)
7. Martínez, J., García, A., Oliver, M., Molina, J.P., González, P.: Identifying virtual 3D geometric shapes with a vibrotactile glove. IEEE Comput. Graph. Appl. **36**(1), 42–51 (2016)

Assessment of the Delays in the Delivery of Public Sector Projects in South Africa

Xitshembiso Shivambu and Wellington Didibhuku Thwala(✉)

Department of Construction Management and Quantity Surveying, University of Johannesburg, Johannesburg 2028, South Africa
xitshembiso.shivambu@gmail.com, didibhukut@uj.ac.za

Abstract. South African public sector is experiencing a lot of delays in delivering projects in time, while on the other hand there are a lot of rollovers on the government budget spending, wherein other areas are spending a lot of construction budgeted money on the remedial works of completed houses. These challenges brought to this study the reason to investigate on why there are many delays in the delivery of public construction projects by the main contractors on construction sites. The objectives set to achieve the purpose of this study include to identify the causes of the delays in the construction projects, and to identify the methods of minimizing construction delays in the public construction projects. The findings identified indicate the causes of delays affecting the delivery of public projects, where however the main factors found to be payment delays to the projects, problems with sub-contractors, mistakes during construction, political matters.

Keywords: Delays · Assessment · Projects · Construction · Delivery Public sector

1 Introduction

In the last 15 years, the Gauteng Provincial Department of Local Government and Housing has delivered more than 500,000 units to poor households. However, this delivery has not been sufficient to meet increasing demand for housing in the province. Inevitably, a deficit of 600,000 households continue to live in housing conditions that do not meet acceptable minimum requirements for residential quality (Human Settlements Review 2010). The Department of Housing has three years left to meet its objective of having a nation free of slums by 2014, eradication of informal settlement and it is a deadline it intends to meet. The problem to investigate is to find out why there are many delays in the delivery of public construction projects by the main contractors on construction sites. The aim of this article can be highlighted as follows: to identify the causes of the delays in the construction projects and to identify the methods of minimizing construction delays in the public construction projects.

The delay is a relative term in construction, where construction delay is defined as the time overruns either beyond the completion date specified in a contract, or beyond the date that the parties agreed upon for the delivery of a project (Assaf and Al-Heijji 2006). In the simplified term, Zack (2003) defined delay as an act or event that extends

W. Karwowski and T. Ahram (Eds.): IHSI 2019, AISC 903, pp. 902–908, 2019.
https://doi.org/10.1007/978-3-030-11051-2_138

required time to perform or complete work of the contract manifest itself as additional days of work. Delay time is an added duration to the project estimated time. Assaf and Al-Heijji (2006) further highlighted that delay is a project slipping over its planned schedule and is considered as common problem in construction projects. To the owner, delay means loss of revenue through lack of production facilities and rentable space or a dependence on present facilities.

2 Causes of Delays

Many studies have been done internationally on construction delays especially on the developing and developed countries. Malaysia is one of the developing countries that also suffer from construction delays. Delays give rise to disruption of work and loss of productivity, late completion of project, increased time related costs, and third party claims and abandonment or termination of contract. It is important that general management keep track of project progress to reduce the possibility of delay occurrence or identify it at early stages (Martin 1976). The Government in Malaysia plays a very important role in providing major infrastructures in meeting the socio-economic needs of the nation and uplifting the quality of life and standard of living in the country. Therefore, it is essential to identify the actual causes of delay in order to minimize and avoid the delays and their corresponding expenses (Abd El-Razek et al. 2008). On the study undertaken by Othman et al. (2006) has reviewed the construction time performance of the public projects in Malaysia where it was found to be affected more by variables related to excusable delays than project characteristic variables. They further indicated that In Malaysia that is where most of the delays were caused by excusable factors, the construction time of public sector civil engineering projects can be improved if the occurrence of excusable delays can be minimized. Both excusable and non-excusable delays, according to Chalabi and Camp (1984) affect the project in a negative way.

3 Mitigation of Delays

An analysis is needed to identify the impact of the delay on time and cost followed by taking the appropriate actions to mitigate the delays and minimize the cost required (Clogh 1981). It is important to bring more improvement to the estimated activities duration according to the actual skill levels, unexpected events, efficiency of work time, and mistakes and misunderstandings (Lock 1996). Mitigation efforts are necessary to minimize losses and this can be archived by many procedures such as the protection of uncompleted works, timely and reasonable procurement, and timely changing or cancellation of purchase orders (Bramble and Callahan, 1992). It is very important to predict and economical solutions (Abdul-Rahman and Berwi 2002). Construction project involve more variables and uncertainties than in the product line. This factor increases the probability of delay occurrences in construction project and makes effective manage important to reduce the diversions from the original program. Planning is easiest done in a homogenous task environment under stable condition such as found in production firms than in a construction project and this presents a challenge for managers involved in construction project (Abdul-Rahman et al., 2006). Therefore, to recover the damage caused by the delays, both the delays and the parties responsible for them should be identified. However, delay situations are complex in a nature because multiple delays can occur concurrently and can be cause by more than one party, or by none of the principal parties. In Achieving economic and schedule goals will be possible only by adopting the appropriate control system especially during the construction phase. The use of control system and similar application will cost money, but the potential savings are several times the cost of implementing them through mitigating and even preventing delay during the construction. Advance technique cost more, but offer greater return if properly applied. Time-cost trade-off considerations mean that delays on a large project can easily cause additional costs, therefore if work can be carefully monitored and managed so that it precedes without extra cost the final result would satisfy the client.

4 Methodology

The first step in conducting this research involved the selection of the appropriate method in order to gather information about the causes of delays in the delivery of public construction projects by the Client targeted respondents on construction sites especially public construction projects. A total of one hundred and twenty (120) questionnaires were distributed to targeted respondents from Department of Human Settlement, Consultants and contractors around Gauteng province. Because of the vast geographical area of South Africa such projects were not easily accessible but only 75 questionnaires received back. It was therefore decided to conduct the research in Gauteng. A pilot survey was conducted based on semi-structured and open-ended interviews with targeted respondents from the client', consultant' and contractor's representatives to ascertain how accurate and adequate the data collected would be. The pilot survey revealed the need for the research to incorporate a structured interview questionnaire in order to refine the research instrument to assist with the in-depth collection to assess the causes of delays in the delivery of public construction projects by the main contractors on construction sites, public construction projects. These factors that cause delays were obtained from relevant literature. However in order to ensure that the mission of the study is achieved then data collection needs to be accurate. The research was conducted via telephone, 'face to face' method and mailing through the questionnaires, furthermore having to make follow-ups on the questionnaires to the targeted respondents that received questionnaires.

4.1 Mean Item Score (MIS)

A five point Likert scale was used to determine the environmental performance of the residence. The adopted scale was as follows:

$$1 = \text{Very negative}, 2 = \text{Negative}, 3 = \text{Neutral},$$
$$4 = \text{Positive}, 5 = \text{Very positive}$$

The five-point scale was transformed to mean item score (MIS). The indices were then used to determine the rank of each item. The ranking made it possible to cross compare the relative importance of the items as perceived by the respondents. This method was used to analyse the data collected from the questionnaires survey. The mean item score (MIS) was calculated for each item as follows;

$$\text{MIS} = 1n_1 + 2n_2 + 3n_3 + 4n_4 + 5n_5 \quad (1)$$

ΣN Where;
n_1 Number of respondents for very negative;
n_2 Number of respondents for negative;
n_3 Number of respondents for neutral;
n_4 Number of respondents for positive;
n_5 Number of respondents for very positive;
N Total number of respondents

After mathematical computations, the criteria are then ranked in descending order of their mean item score (from the highest to the lowest).

5 Results and Discussions

Findings from the 75 usable questionnaire revealed that 31% were female, while 69% were male. The average age of the participants of this study is at the mean of 35 year of age, where the youngest respondent is at age 23 years while the oldest is at the age of 62 years of age. The ethnicity that comprises the majority of the respondents was blacks (92%), followed by (6.7%) whites. In order to conclude on the findings of delays both parties involved from different work categories had given their views, where contractors were 44.4%, Project manager/Principal Agent/Engineer were 43.1% and Client representative were respectively 12.5%. The current positions held by the participants from both work categories, where the majority are Project Managers at 29.3%, Site Agents 20%, Managing Directors at 13.3%, site engineers at 10.7%, Construction Project Managers at 6.7%, Architectural members at 5.3%, Construction Managers at 2.7%. Majority of respondents highest education qualification were Post-Matric Certificate are 70.3%, followed by Bachelor Degrees at 16.2% and the minority were Grade 11 or less qualifications at 1.4%. Majority of respondents lengthy of experience is the group that has between 5 to 10 years of experience 38.7%, followed by those that are between 3 to 5 years are at 16%, where the minority are those that are above 15 years at 6.7%. Respondents were further asked to indicate how many projects each person is involve in to get their insight also based on other projects where the majority of respondents are those involve in 3 to 5 projects at the same time at 44%, followed by 1 or 2 projects at 33.3%, when the minority are those that are involve in more than 10 projects at 4%. Respondents were asked to rate the factors that causes delays in their projects based on their experienced, respondents have given their view in this manner as tabulated Table 1.

Table 1. Presents the ranked factors that causes delays in the projects

	Mean	Std. deviation	Rank
Problems with subcontractors	4.04	0.949	1
Payments delays	3.78	1.224	2
Mistakes during the construction stage	3.28	0.843	3
Political matters	3.28	0.923	4
Project variations	3.27	0.932	5
Slow decision making	3.22	0.786	6
Material supply	3.08	0.777	7
Inadequate contractor experience	3.07	0.991	8
Community disruptions	3.05	1.066	9
Equipment availability	2.97	0.707	10
Poor site management	2.96	0.789	11

On the matter of frequency of payment delays from the client on the delays of one month 46.5% of respondents indicated that delays 'often' occur, while only 4.2% says it never happen to their projects. For the Delays of more than two to four months 47.1% regard it to sometime occur on their projects, when only 7.1% responded to never experience this type of delay on their projects. Furthermore on the delays of more than five months 44.3% responded to never experience this delay, when only 1.4% of respondents indicated that they always have this delay to their projects.

Respondents were further requested to identify the effective methods of minimizing delays in construction projects, in which respondents advises includes that; Project Orientated Organizations should understand and practically apply some of project management techniques like Program Evaluation & Review Technique (PERT) & Critical Path Method (CPM) in planning, competent site engineers should be appointed to have a close by look on quality to avoid lot of delays on remedial works, appointment of Competent project Managers to manage projects, accountability amongst team players must be embedded in everyone's mind thus will expose weakest link as soon the boat is not sailing accordingly, community education during public meetings to be conducted some basic technical matters like specifications, materials batches and so on, feasibility study of the project must be properly done, local people must be trained to acquire an artisan skill, cost, quality, time management need to be prioritized.

6 Conclusion

This study investigated the causes of delays in the construction, to find out how the delays can be minimized on sites, with the recommendations of how the both parties can play a role in promoting improvements in the service delivery set by the state government. The findings from the literature survey compared with the primary data revealed that problems with sub-contractors, payment delays to the projects, mistakes during construction, political matters and Project variations are the most influential factors that cause delays in the public construction project. On the very serious note delays continue to be the most critical challenge in the construction industry where most project do not get delivered on the contracted dates, most of the time the project overruns either beyond the completion date specified in a contract, or beyond the date that the parties agreed upon for the delivery of a project. The delays in the public project seem to have become a norm when delays occur on site. Delays always has negative impact to the project because contractors end up over spending, community do not receive their houses as per the agreed period, contractors end up losing chances of winning other projects etc. the networking relationship between all construction parties is very essential and that it always get maintained to allow processes to flow, enhance better communication amongst stakeholders so that every little factor that can have a negative impact to the project or to have potential to cause delays can be dealt with effectively and they be minimized.

References

Abd El-Razek, M.E., Bassion, H.A., Mobarak, A.M.: Causes of delay in building construction projects in Egypt. J. Constr. Eng. Manag. **134**, 11–831 (2008)

Abdullah-Razaki, M., Abdul-Rahman, I., Abdul-Azis, A.A.: Causes of delay in MARA management procurement construction projects. Faculty of Civil and Environmental Engineering, Universiti Tun Hussein Onn Malaysia (2010)

Abdul-Rahman, H., Berawi, M.A.: Developing knowledge management for construction contract management. In: Procurement, 14th International, University of Tokyo (2001)

Alaghbari, W.A.M.: Factors affecting construction speed of industrialized building systems in Malaysia, Master's thesis, University Putra Malaysia, Serdang (2005)

Al-Moumani, H.A.: Construction delay: a quantitative analysis. Int. J. Project Manag. **18**, 51–59 (2000)

Assaf, S.A., Al-Hejji, S.: Causes of delays in large construction projects. Int. J. Project Manag. **24**(4), 349–357 (2006)

Bramble, B.B., Callahan, M.T.: Construction Delays Claims, 2nd edn. Wiley, New York (1992)

Chalabi, A.F., Camp, D.: Causes of delays and overruns of construction projects in developing countries. Proc. CIB **W65**(3), 723–734 (1984)

Clogh, R.R.: Construction Contracting, 4th edn. Wiley, New York (1981)

Department of Human Settlement.: Annual Report of Department of Housing, South Africa, 2011/2012 (2012)

Horner, R.M.W., Talhouni, B.T.: Effects of Accelerated Working, Delays and Disruption on Labor Productivity. CIOB (1995)

Human Settlements Review.: The Use of Alternative Technologies In Low Cost Housing Construction: Why The Slow Pace Of Delivery? vol. 1, no. 1 (2010)

IOL News.: Sexwale Concerned by Rate of Delivery. www.iol.co.za. Accessed on 30 Sept (November 25 2010)

Lock, D.: Project Management, 6th edn. Wiley, New York (1996)

Majid, I.A.: Causes and Effects of Delays in ACEH Construction Industry, Malaysia. Masters of Science in Construction Management: Universiti technologi Malaysia (2006)

Martin, C.C.: Project Management: How to Make It Work, New York (1976)

Nabeemeeah, M.A.: The Causes and Cost Effects of Construction Delays. University of the Witwatersrand, Johannesburg (1996)

Ogunlana, S.O., Promkuntong, K., Jearkjirm, V.: Construction delays in a fast-growing economy: comparing Thailand with other economies. Int. J. Project Manag. **14**(1), 37–45 (1996)

Omran, A., Ailing, O., Pakir, A.H.K., Ramli, M.: Delays factors in construction projects development: the case of Klang Valley, Malaysia. J. Acad. Res. Econ. (2010) **2**(2), 135–158 (2008)

Sambasivan, M., Soon, Y.W.: Causes and effects of delays in Malaysian construction industry. Int. J. Project Manag. **25**, 517–526 (2007)

Zack, J.G.: Schedule Delay Analysis; Is There Agreement? Project Management Institute-College of Performance Management, New Orleans (2003)

Enhancing Cursor Control Using Eye Movements

Muhammad Sohaib Shakir[1(✉)], Amina Akhtar[2], Habiba Kulsum[3], Anam Mukhtar[1], and Iqra Saleem[1]

[1] Department of Information Technology, University of Lahore, Gujrat, Pakistan
hafizsohaib01@gmail.com
[2] Department of Computer Science, University of Gujrat, Gujrat, Pakistan
[3] Department of Computer Science, The Superior College Lahore, Gujrat, Pakistan

Abstract. This paper is about different techniques how we can control our cursor on the computer system without touching the mouse (Hardware). This is possible through the iris of instead of mouse or keyboard. Research that is described in this paper has an aim to reduce the complexities and difficulties that are involved the detection and tracking of the iris for the persons to interact with the computer without touching the mouse or keyboard. The research paper defines operations performed on the images of high resolution of the subject's (Human) eye to compute/calculate the eye movement/tracking and patterns of iris in left & right direction. The detection takes place according to the center of the eye (Origin of the eye). The calculated/computed data from the position of the iris can be used in the technique of application based processing. The Cursor Control Using Iris research has various applications in the field of Human-Computer Interaction (HCI) like mouse cursor control. In this paper, there is a discussion about eye tracking using eyeball mouse.

Keywords: Eye tracking patterns · Eye-blinking computations
Human computer interaction (HCI) · Eye movement with respect to center of eye (Center of eye is as origin from where the tracking starts)

1 Introduction

There has been a glowing attention and interest in developing natural communication between human eye and computer now a day. "Cursor Control using Iris" is a conventional computer screen technology that is also useful for those have only their eyes to have interaction with computer. It is easy for them to have an interaction with computer although they cannot touch the mouse or the keyboard. It is easy, fast and reliable tool for the disabled users to communicate and interact with computer. There is a number of people, who are due to their physical abnormality and disabilities, cannot use the hardware (mouse and the keyboard), and that's why they are fully bounded in their attempt to utilize the computer system as a physically normal user (Human) can

W. Karwowski and T. Ahram (Eds.): IHSI 2019, AISC 903, pp. 909–913, 2019.
https://doi.org/10.1007/978-3-030-11051-2_139

utilize. Because of the disability/immobility of the disables, the movements of head as a means to control the mouse may not be feasible. In such conditions, a unique and somehow better eye movement tracking can be presented; it is an alternate solution for them who are not free to use the computer system with a mouse. There are different new techniques in this era of HCI (Human Computer Interaction) that allow applications to sense the user's iris movement. The core objective of this paper is to express the eye tracking instead of infrared extreme light which is harmful for eyes.

2 The Required Hardware & Software in the Eye Tracking

- Hardware
  - Web Camera to capture movements of eye
  - A Computer System.

- Software
  - C++/CLI
  - Open CV
  - Visual Studio 2012
  - Intel IPP (Integrated Performance Primitives).

Benefits:

- Children can learn easily.
- Cursor of the computer can be controlled easily.
- No external hardware is required.
- Helpful for handicapped or disables.

3 Literature Review

Mouse Control Using Eye Movements can happen by determination of gaze (stare on a particular location) of the user. The gaze of the user is detected by the computer software and then the position of a mouse on the computer screen is controlled.

The system does not work in real time world so well. Because it is less efficient or useful than a professional mouse or keyboard. It has not much facility to check and handle blinks and close eyes. Accurate head movement and eye's iris tracking results are obtained at an imaging processing rate of 3(fps) frames per second. This can be achieved under the different lighting conditions for users, wearing and not wearing glasses. Iris movement tracking is a technique in which an individual's eye movements are measured so that the computer system knows both "**where a person is looking at any given time**" and "**the sequence in which their eyes are shifting from one location to another**". "The Designing and implementation of Human Computer Interface tracking system is based on many eye features".

4 Eye Detection Approaches

Following are the various eye detection approaches that are being discussed in this paper.

5 Regression Approach

Regression approach is that approach in which the image processing tries to reduce/minimize the distance between the predicted (estimated) and actual eye positions. If the distance is minimized then the cursor can easily move in the eye movement's direction. If there is a large distance between eyes and the computer screen's cursor then there are chances of changes in the detection of pupil's diameter. And as a result desired position is not localized.

6 Bach Mode Approach

For human eye's (Iris) movement detection, the Batch mode is used. Iris tracking technique is implemented on the static images for tracking. There are some approaches that are very efficient and there are some approaches that are not much efficient in many cases. When the direction of iris is left, right or center, then this technique simply works. If the location of Human eye's iris is upward or downward, it does not work. In this situation, used gets worry to use the computer system. It is the limitation of Bach Mode Approach. Sometimes, user needs to select an icon vertically from a given collection of icon but he cannot do so because of limitation of the system as the system is designed by following Batch Mode. We want to provide a full pack of easiness to a disable user but in this case, he cannot get complete pack of easiness because of above described limitations.

7 Working

The user has to sit before the screen of a computer system or a laptop computer, an image processing video camera is mounted above the screen to observe the user's eye's iris movements to control cursor on computer screen. The computer system continually processes and analyzes the video image of the eye and detects where the user is looking on the screen. There is no mouse or keyboard that has to be attached to the user's head or body. It is very useful for handicapped persons. No external hardware is attached or required. There should be a person who will see the computer screen and his eye movements will be captured to drag the cursor on the screen of the computer system.

8 Working Diagram

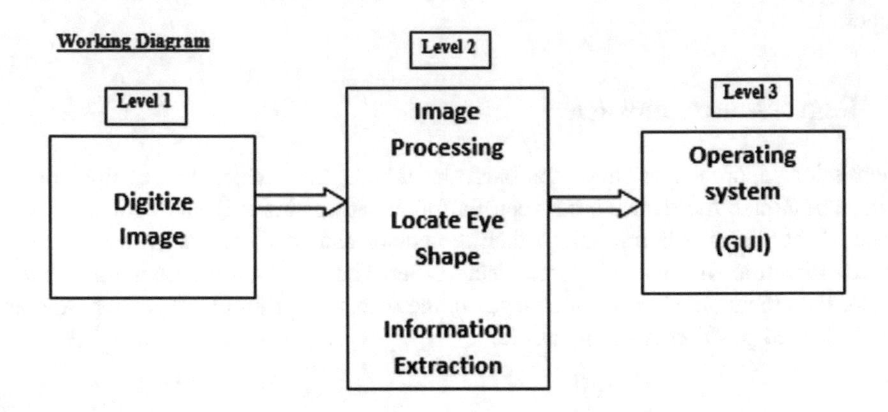

9 Limitation of Eye Blinking

- Due to limitations of video device we get some unwanted noise in the captured image which is mainly caused due to reflection of light from eyeball.
- User has to fully concentrate on cursor and in this way, his/her eyes can get severe effect of that screen radiations of the computer system.

10 Future Work

With the help of a view to providing a complete cursor handling by using eye movement tracking, future work has aim to provide an easy way to perform the scrolling functionalities, selection of the text and drag- drop operations. There are many other operations for example copy/pasting, cut/pasting etc. An eye tracking system should do all the functionalities that are necessary.

In addition, further Optimization (uses fewer resources and gets maximum results) and increment/enhancement of the approaches are important to improve. These techniques in future can also be made i.e. detection of blink through the eye blink such as, Right-click or Left-click. It can be possible in handling difficulties for the handicapped users to reduce their worry to use Computer System.

11 Results

Eye movement tracking detection is an important technique of **Human Computer Interaction** and can be utilized to develop many innovative and modern applications and to interact with the computer system in a more effective way. The user has to only look left or right to move the cursor towards the desired location/direction.

12 Conclusion

The research paper is written to high-light the easiness of disables or handicapped users who are suffering problems to have interaction with the Computer System. Eye is an important part of human body. The handicapped users can see with it but the point is that they can use the Computer System easily if the Computer System has rich machinery to use.

Eyes are very precious part for the disable persons because it is the most frequent and effective part of their body to be used to interact with the Computer System. Some systems have very high radiations. It is difficult to sit for using them. So, the system must have the techniques to detect the eye movement effectively and quickly.

There are many techniques that can be used to build Eye Tracking Detection Systems. The efficient techniques (that can move the cursor rightward, leftward, upward and downward direction) must be used to develop a system. We are trying to discuss as much facilities as we can, for the handicapped users. The other operations that are mentioned above copying, pasting, cutting, selecting and as well saving the text or files must be done in the Computer System. The system must be as effective as it should emit less radiations to be least harmful for the user's eyes. The main objective of writing this paper is to enhance/increase the characteristics of an Eye Tracking system. The system must be user friendly.

References

1. MARGI SADASHIV NARAYAN, 2WARANG PRAVIN RAGHOJI: Viola-jones Algorithm and Houghman Circle Detection Algorithm
2. W. Zhang, Zhang, T.-N., Chang, S.-J.: Eye gaze estimation from the elliptical features of one iris. Opt. Eng. **50**(4), 047003 (2011)
3. Kadyrov, A., Yu, H.: Explore New Eye Tracking and Gaze
4. Locating Methods. 978-1-4799-0652-9/13 $31.00 © 2013 IEEE
5. Jixu Chen Qiang Ji "3D Gaze Estimation with a Single Camera without IR Illumination", 978-1-4244-2175- 6/08/$25.00 ©2008 IEEE [5] Fares Alnajar Theo Geverszz "Calibration-Free Gaze Estimation Using Human Gaze Patterns", ICCV 2013 paper

Causes of Variation Orders in Construction: A Case Study of Polokwane, Limpopo Province

Pfumelani Maluleke, Clinton Aigbavboa, and Wellington Thwala(✉)

Department of Construction Management and Quantity Surveying, University of Johannesburg, Johannesburg 2028, South Africa
didibhukut@uj.ac.za

Abstract. The construction Industry is one of the biggest employment generators in the South Africa. However, the industry suffers delays caused by the occurrence of variation orders which are also evident in other countries worldwide. The current study evaluates the causes of variation orders in Polokwane, Limpopo Province. The information used in this study was obtained from secondary sources through thoroughly conducted literature review and primary sources through questionnaire circulated among construction professionals in Polokwane. Out of 150 circulated questionnaires, only 105 respondents reverted back which denotes 70% of them. Findings showed that design changes, change of plans, change of scope, Inadequate specification, defects in design, defects in Bill of Quantities, change in specification, owner's financial difficulties, errors and omissions in design and Inadequate working drawings were the leading causes of variation orders in Polokwane. Therefore the study adds up to the existing knowledge on the causes variation orders.

Keywords: Change order · Causes · Variation order and construction

1 Introduction

The construction Industry is one of the biggest employment generators in the South Africa. However, the industry suffers delays caused by the occurrence of variation orders which are also evident in other countries world-wide. Ibbs (2012: 67) defines change as any addition, omission or revision to the original contract scope. He further stated that change may influence time and cost adjustment and it happens often in construction projects. Ijaola et al. (2012: 495) defines change in construction as work, procedure, condition or approaches that vary from the original construction plan and specification. As soon as the need for the introduction of a change or variation is acknowledged or discovered, a formal document called 'change order' which describes the scope of change and its effects on time and cost is issued (Desai et al. 2015: 152). In relation to the statement issued by Dickson et al. (2014: 2) states that changes made to the plans are usually implemented through the issuance of a variation or change order which in normal practice is issued by a consultant representing the client or as it would have been brought forth by the contractor. The uniqueness of every project and the uncertainty of the future is what causes changes in construction projects (Ibbs 2012: 67). Due to the increasing nature of change or variation orders in construction projects,

W. Karwowski and T. Ahram (Eds.): IHSI 2019, AISC 903, pp. 914–920, 2019.
https://doi.org/10.1007/978-3-030-11051-2_140

there is a need to investigate the causes of change or variation orders, therefore, the purpose of this study is to investigate the causes of variation orders in Polokwane, Limpopo Province.

2 Causes of Variation Orders

The causative factors of variation orders are those incidents or events that take place, forcing changes to be made in construction projects. According to Desai et al (2015: 156) and Assbeihat et al. (2015: 57), variation orders originate from four groups namely: client, contractor, consultant and other factors. The study by Enshassi et al. (2010: 544) revealed top 10 most important factors causing variations according to their ranks which are as follows: Lack of construction materials and equipment spare parts due to closure and siege, change in design by consultant, lack of consultant's knowledge of available material and equipment, errors and omission in design, conflicts between contract documents, Owner's financial problems, Lack of coordination among project parties, International consultant using inadequate specification to be followed in local conditions (e.g. testing procedure), internal political problems and change in specification by owner. According to Meredith et al. (2010: 167), the 3 basic causes of project changes are: errors in planning of project objectives, increased knowledge of the client or project team regarding the project deliverables and mandate that mainly refers to the government regulations. The study conducted by Desai et al. (2015: 154), listed 10 most common causes of change orders namely: owner's financial problems, change of scope, change in design, change of project schedule, Unavail ability of equipment, defective workmanship, contractor's financial difficulties, weather conditions, Unavailability of skilled manpower and poor design or poor working drawing details. This study has opted to use as bases the study of Ndihokubwayo (2008: 27–28), seeing that it comprised of four detailed origin categories of variation orders.

3 Research Methodology

The information used in this study was obtained from secondary sources through thoroughly conducted literature review and primary sources through questionnaire circulated among construction professionals in Polokwane namely: Architects, Quantity Surveyors, Construction Managers, Construction Project Managers, Civil Engineers, Electrical and Mechanical Engineers. The total number of distributed questionnaires was 150 which was a suitable number considering the City of Polokwane. Out of 150 circulated questionnaires, only 105 respondents reverted back which denotes 70% of them. Mulenga (2014) states that results are suitable for use if the response rate was above 40% as anything below could be biased or have lessor value to the study. In each of the questions, the respondents were requested to indicate the likelihood of each factor causing variation orders in the construction industry of Polokwane. The respondents' judgements on each factor were done with the use of the following 5-point Likert scale in order to determine the **Mean Item Score (MIS)**:

1. = Extremely unlikely
2. = Unlikely
3. = Neutral
4. = Likely
5. = Extremely Likely

The indices obtained were then used to determine the rankings of the causes of variation orders in descending order; and were at large used for the purpose of analysing the findings. Below is the formula used to calculate the MIS for each item:

$$\text{MIS} = \frac{1\text{n}1 + 2\text{n}2 + 3\text{n}3 + 4\text{n}4 + 5\text{n}5}{\sum \text{N}} \quad (1)$$

Where:

n1= Number of respondents for extremely unlikely or never or not at all aware/useful
n2= Number of respondents for unlikely or rarely or slightly aware or useful;
n3= Number of respondents for neutral or sometimes or moderately aware or moderately useful;
n4= Number of respondents for likely or often or strongly aware or extremely useful
n5= Number of respondents for extremely likely or always or fully aware or essentially useful measure.

4 Findings and Discussions

The findings of the study revealed 62.9% of male and 37.1% of female respondents out of a total of 105 respondents. With regards to age, the findings revealed that 3.8% of the respondents were aged between 0 and 25 years, 29.5% of the respondents were aged between 26 and 30 years, 47.6% of the respondents were aged between 31 and 35 years, 8.6% of the respondents were aged between 36 and 40 years, 5.7% of the respondents were aged between 41 and 45 years, 1% of the respondents aged between 46 and 45 years, with 1.9% of them aged between 51 and 55 years, while 1.9% of them were 56 years and older. The findings of the study in relation to occupations/careers revealed 8.6% of the respondents to be Architects, 26.7% of the respondents to be Quantity Surveyors, 17.1% of the respondents to be Project Managers, 14.3% to be Construction Managers, 15.2% to be Construction Project Managers, 16.2% to be Civil Engineers while 1.9% represented other unspecified professionals with the construction industry of Polokwane. The study further revealed that there were no professionals with less than 1 year of experience in the construction industry, and that 20% of the respondents had an experience ranging between 1 and 5 years in the industry, 57.1% of the respondents had an experience ranging between 5 and 10 years, 10.5% of the respondents had an experience ranging between 10 and 15 years, with 5.7% having an experience ranging between 15 and 20 years, while 6.7% were more than 20 years long in the construction industry. The study also showed 0% on respondents without qualifications, with Grade 11 or lower or with Grade 12 (Matric certificate) or with

Doctorates, whilst it also revealed 1.9% of the respondents with Post-Matric Certificate, 25% of the respondents with Post-Matric Diploma, 66.3% of the respondents with Bachelor's degree, while 6.7% of the respondents held Master's degree at their disposal. The findings of the study also revealed that 15.2% of the respondents were working for a Public Sector Clients (Government), 8.6% of the respondents were working for a Private Sector Client, 8.6% of the respondents were working for Quantity Surveying Consultant firms, 6.7% of the respondents were working for Construction Project Management firms, 1.9% of the respondents were working for Consulting Engineering firms (i.e. Mechanical or Electrical), 15.2% of the respondents were working for Consulting Civil Engineering firms, 39% of the respondents were working for Main Contractors, with 2.9% of the respondents working for Sub-contractors while 1.9% of the respondents were working on other unspecified workplaces.

5 Client Related Causes of Variation Orders

Based on the ranking (R) using the calculated standard deviation (SD) and the mean scores ($\overline{X}$) for the recorded top five (5) client related causes of variation orders it was discovered that the leading causes included: change of plans (SD = 0.736; $\overline{X}$ = 4.64; R = 1), Change of scope (SD = 0.719; $\overline{X}$ = 4.59; R = 2), Change in specification (SD = 0.694; $\overline{X}$ = 4.59; R = 3), Financial difficulties (SD = 0.879; $\overline{X}$ = 4.41; R = 4), Change in material or procedures (SD = 0.917; $\overline{X}$ = 4.22; R = 5) . This causes were in agreement with Beshah (2015: 19) where change of scope or plan, change of schedule, financial difficulties, poor project objectives, change in specifications and change of material were identified as client related causes of variation orders. This results were also in agreement with Dickson et al. (2015: 3) where change of scope or plan, owner's financial problems, change of schedule and replacement of material were identified as the client related causes of variation orders. Sunday (2010: 103), also agreed with this results as he listed change or plans or scope, impediment in prompt decision making process, Inadequate project objectives, replacement of materials or procedures as the major client related causes of variation orders. However, this causes were not in agreement with the major causes identified by Assbeihat and Sweis (2015: 60) where owner's instruction to modify designs, owner's additional work instruction and ambiguities and mistakes in specifications and drawings were the most important client related causes of variation orders. The revealed five (5) major client related causes by the findings' analysis were also not in agreement with Pryiantha et al. (2011: 16) where requirement increases, change in mind-choice and change in mind forced were the major client related causes of VOs. Moreover, this results were not in agreement with Meredith et al. (2010: 167) where errors in planning of project objectives, increased knowledge of the client and mandate that mainly refers to government regulations were found to be the major client related causes. Based on the ranking (R) using the calculated standard deviation (SD) and the mean scores ($\overline{X}$) for the recorded top five (5) contractor related causes of variation orders it was discovered that the leading causes included: lack of involvement in design (SD = 1.012; $\overline{X}$ = 4.25; R = 1), lack of required data (SD = 0.873; $\overline{X}$ = 4.23; R = 2), Complex design and technology

(SD = 0.993; $\overline{X}$ = 3. 94; R = 3), Long lead procurement (SD = 0.909; $\overline{X}$ = 3.60; R = 4), Fast track construction (SD = 0.865; $\overline{X}$ = 3.50; R = 5). This causes were in agreement with Beshah (2015: 22–23) where lack of involvement and lack of communication were the major contractor's related causes of VOs. However, this obtained five (5) major causes were not in agreement with Sunday (2010: 103) where differing site conditions, shortage of skilled manpower and contractor's desired profitability were the major contractor related causes of VOs. The causes were also not in agreement with Jadhav and Bhirud (2015: 2231) where the contractor's lack of experience, poor, unskilled manpower and poor planning were identified as contractor related causes of Variation order. Based on the ranking (R) using the calculated standard deviation (SD) and the mean scores ($\overline{X}$) for the recorded top five (5) consultant related causes of variation orders it was discovered that the leading causes included: Design changes (SD = 0.543; $\overline{X}$ = 4.74; R = 1), Inadequate specification (SD = 0.717; $\overline{X}$ = 4.52; R = 2), Defects in design (SD = 0.759; $\overline{X}$ = 4.50; R = 3), Defects in Bill of Quantities (SD = 0.759; $\overline{X}$ = 4.50; R = 3), Errors and omissions in design (SD = 0.671; $\overline{X}$ = 4.43; R = 4) and Inadequate working drawing details (SD = 0.920; $\overline{X}$ = 4.36; R = 5). This causes were in agreement with Muhammad et al. (2015: 94) where conflicting contract documents, change in design and errors and omissions in design were the major consultant related causes. This causes were also in agreement with Sunday (2010: 103) where the major causes listed were Inadequate working drawing details, errors and omissions in design, design discrepancies and conflict between contract documents. The findings were also in agreement with Enshassi et al. (2010: 544) where change in design by consultant, errors and omissions in design and internal political problems were the major consultant related causes. However, this results were not in agreement with Jadhav and Bhirud (2015: 2232) where Unrealistic design periods and design errors, failure by consultant to perform design and supervision effectively, consultant's lack of judgement and experience and obstinate nature of consultant were found to be the major consultant related causes of VOs. Based on the ranking (R) using the calculated standard deviation (SD) and the mean scores ($\overline{X}$) for the recorded top five (5) other related causes of variation orders it was discovered that the leading causes included: Weather conditions (SD = 0.983; $\overline{X}$ = 4.18; R = 1), Unforeseen problems (SD = 0.868; $\overline{X}$ = 3.94; R = 2), Safety considerations (SD = 0.800; $\overline{X}$ = 3.78; R = 3), Change in Government regulations (SD = 0.723; $\overline{X}$ = 3.22; R = 4), and change in economic conditions (SD = 0.812; $\overline{X}$ = 3.11; R = 5). This causes were in agreement with Dickson et al. (2014: 4) where weather conditions, safety considerations, change in government regulations and change in economic conditions were ranked as the major causes of variation orders by other factors. This results were also in agreement with Muhammad et al. (2015: 94) where new government regulations, differing site conditions and weather conditions were the other major factors causing VOs. However, this results were not in agreement with Olsen et al. (2012) where unforeseen conditions and force majeure were the major causes of VOs. Phillips (2013: 271) was also not in agreement with this results as he had external event and risk exposure as his main other factors that could cause VOs.

6 Conclusion and Recommendations

The literature review revealed the leading causes of variation orders which includes: change in scope or plan, financial difficulties of the owner, change in specification, change in material or procedures, weather conditions, unforeseen conditions, change in design by consultant, errors and omissions in design, defects in design, defects in Bill of quantities, inadequate site investigation, inadequate consideration of design, lack of contractors involvement, safety consideration, change in economic conditions, problems on site, impediment in prompt decision making process, inadequate working drawings, change in government regulations, poor procurement process, difficult site conditions, Inadequate scope of work for the contractor, differing site conditions, lack of required data, fast track construction, complex design and technology and lack of communications. The questionnaire survey results revealed the following as the top ten (10) major causes of variation orders in Polokwane, Limpopo Province: design changes, change of plans, change of scope, Inadequate specification, defects in design, defects in Bill of Quantities, change in specification, owner's financial difficulties, errors and omissions in design and Inadequate working drawings. Therefore, It is recommended that the project team should audit/revisit its budgets, designs or documents before issuing them in order to avoid introduction of variation orders which if not conducted, time and cost overruns may be incurred. It is also recommended that, clients should implement the projects that they have well planned for on the bases of finances.

References

Al-dubaisi, A.H.: Change orders in construction projects in Saudi Arabia. M.Sc. (Construction Engineering and Management). King FAHD University of Petroleum and Minerals. Retrieved from http://faculty.kfupmedu.sa/CEM/assaf/students (2000). Accessed 19 June 2017

Assbeihat, J.M., Sweis, G.J.: Factors affecting change orders in public construction projects. Int. J. Appl. Sci. Technology. **5**(6), 56–63 (2015)

Beshah, M.: Assessment of causes and cost impact of change orders on road projects in Ethiopia. M.Sc. (Construction Technology and Management). Addis Ababa University. Retrieved from http://etd.aau.edu.et (2015). Accessed: 18 Mar 2017

Desai, J.N., Pitroda, J., Bhavasar, J.J.: A review on change order and assessing causes affecting change order in construction. Journal of International academic research for Multidisciplinary. **2**(12), 152–162 (2015)

Enshassi, A., Arain, F., Al-Race, S.: Causes of variation orders in construction projects in the Gaza Strip. J. Civ. Eng. Manag. **16**(4), 540–551 (2010)

Ibbs, W.: Construction change: likelihood, severity, and Impact on productivity. J. Leg. Affairs Dispute Resolut. Eng. Constr. **4**(3), 67–73 (2012)

Ijaola, I.A., Iyagba, R.O.: A comparative study of causes of change orders in public construction project in Nigeria and Oman. J. Emerg. Trends Econ. Manag. Sci. **3**(5), 495–501 (2012)

Jadhav, O.U., Bhirud, A.N.: An analysis of causes and effects of change orders on construction projects in Pune. Int. J. Eng. Gen. Sci. **3**(6), 795–799 (2015)

Meredith, J.R., Mantel, S.J.: Project management: a managerial approach, 7th edn. Wiley, Asia (2010)

Muhammad, N.Z., Keyvanfar, A., Majid, M.Z.A., Shafaghaf, A., Magana, A.M., Dankaka, N.S.: Causes of variation order in building and civil engineering projects in Nigeria. J. Technol. (Sci. Eng.) **77**(16), 91–97 (2015)

Mulenga, M.J.: Cost and schedule overruns on construction projects in South Africa. Mtech (Construction Management). University of Johannesburg. Retrieved from https://ujdigispace.uj.ac.za (2014). Accessed 14 July 2017

Olsen, D., Killingsworth, R., Page, B.: Change order causation: who is the guilty party? In: 48th ASC Annual International Conference Proceedings (2012)

Phillips, J.: Project management professional study guide, 4th edn. McGraw-Hill Education, United States of America (2013)

Correction to: A Recursive Co-occurrence Text Mining of the Quran to Build Corpora for Islamic Banking Business Processes

Farhi Marir, Issam Tlemsani, and Munir Majdalwieh

Correction to:
Chapter "A Recursive Co-occurrence Text Mining of the Quran to Build Corpora for Islamic Banking Business Processes" in: W. Karwowski and T. Ahram (Eds.), *Intelligent Human Systems Integration 2019*, AISC 903, https://doi.org/10.1007/978-3-030-11051-2_47

In the original version of the book, the chapter "A Recursive Co-occurrence Text Mining of the Quran to Build Corpora for Islamic Banking Business Processes" was published with the affiliation '2. Issam Tlemsani, Prince Mohammad Bin Fahd University, Khobar, Saudi Arabia' for Issam Tlemsani and the author (Prof. Issam Tlemsani) has provided a belated correction in his affiliation and the same has been updated as '2. Issam Tlemsani, IE Business School, Madrid, Spain, "mail to: i.tlemsani@tcib.org.uk"'. The chapter and book have been updated with the changes.

The updated version of this chapter can be found at
https://doi.org/10.1007/978-3-030-11051-2_47

W. Karwowski and T. Ahram (Eds.): IHSI 2019, AISC 903, p. C1 2020.
https://doi.org/10.1007/978-3-030-11051-2_141

Author Index

W. Karwowski and T. Ahram (Eds.): IHSI 2019, AISC 903, pp. 921–925, 2019.
https://doi.org/10.1007/978-3-030-11051-2